STUDENT'S GUIDE

JAMES C. HILL

California State University, Sacramento

CHEMISTRY THE CENTRAL SCIENCE 11E

BROWN   LeMAY   BURSTEN   MURPHY

PEARSON

Prentice Hall

Upper Saddle River, NJ 07458

Associate Editor: Jennifer Hart

Senior Editor: Andrew Gilfillan

Editor-in-Chief, Science: Nicole Folchetti

Assistant Managing Editor, Science: Gina M. Cheselka

Project Manager, Science: Ashley M. Booth

Supplement Cover Manager: Paul Gourhan

Supplement Cover Designer: Victoria Colotta

Operations Specialist: Amanda A. Smith

Director of Operations: Barbara Kittle

MCAT® is a registered trademark of the American Association of Medical Colleges, which neither sponsors nor endorses this product.

DAT® is a registered trademark of the American Dental Association, which neither sponsors nor endorses this product.

© 2009 Pearson Education, Inc.

Pearson Prentice Hall

Pearson Education, Inc.

Upper Saddle River, NJ 07458

Pearson Prentice Hall™ is a trademark of Pearson Education, Inc.

The author and publisher of this book have used their best efforts in preparing this book. These efforts include the development, research, and testing of the theories and programs to determine their effectiveness. The author and publisher make no warranty of any kind, expressed or implied, with regard to these programs or the documentation contained in this book. The author and publisher shall not be liable in any event for incidental or consequential damages in connection with, or arising out of, the furnishing, performance, or use of these programs.

Printed in the United States of America

10 9 8 7 6 5 4 3 2 1

ISBN-13: 978-0-13-600264-2

ISBN-10: 0-13-600264-1

Pearson Education Ltd., *London*

Pearson Education Australia Pty. Ltd., *Sydney*

Pearson Education Singapore, Pte. Ltd.

Pearson Education North Asia Ltd., *Hong Kong*

Pearson Education Canada, Inc., *Toronto*

Pearson Educación de Mexico, S.A. de C.V.

Pearson Education—Japan, *Tokyo*

Pearson Education Malaysia, Pte. Ltd.

Contents

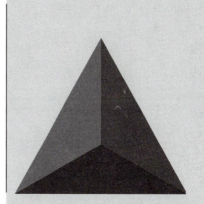

To the Student

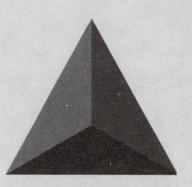

This edition of the *Student's Guide* to the text *Chemistry: The Central Science* by Brown, LeMay, Bursten, and Murphy continues the successful structure of the previous editions.

Many aids are available to help you within these pages:

- Suggestions for material to review
- Learning goals
- Summaries of key ideas and concepts with key points in a "bullet" format
- Sample exercises with problem-solving strategies: Most use *Analyze, Plan,* and *Solve* approach
- End-of-chapter problems, most with detailed solutions
- Integrative questions designed to use several concepts
- Data sufficiency questions: Questions that ask you to qualitatively determine if sufficient information exists to answer the question
- Conceptual and quantitative problems

A new feature appears in this edition: Sectional practice questions based on the types used in the Medical College Admission Test (MCAT) and Dental Admissions Test (DAT). More information is available in the following sections of the *Student's Guide*:

- Introduction to the General Chemistry Questions in the Medical College Admission Test (MCAT) and Dental Admissions Test (DAT)
- Using the *Student's Guide* to Review and Prepare for the General Chemistry Questions in the MCAT or DAT

The following is a suggested method for using this book in conjunction with the Brown, LeMay, Bursten, Murphy text:

1. In the *Student's Guide*, each chapter corresponds to a chapter in the text and is divided into three sections: Overview of the Chapter, Topic Summaries and Exercises, and Self-Study Exercises. By appropriately using these sections, you can learn chemical concepts, theories, facts, and problem-solving techniques.
2. Before attending the first lecture on a particular chapter in the text and before reading the chapter, read the corresponding Overview of the

Chapter in the *Student's Guide*. By doing this, you will gain familiarity with the key topics found in that chapter of the text.

3. After reading the Overview of the Chapter in the *Student's Guide*, you should read the appropriate chapter in the text so that you will be prepared for your instructor's lecture. You will also be equipped to ask appropriate and thoughtful questions.

4. During the lecture, take detailed notes. In most chemistry classes, an instructor chooses to emphasize certain key topics and ideas within a chapter, and the instructor will test your knowledge and your problem-solving skills primarily in those areas. Detailed lecture notes will provide you with a complete record of the material covered in class. Often an instructor will also identify the key points. Be sure to note these for later reference.

5. Once your instructor has begun to discuss a topic, you should study its coverage in the text. At the same time, use the Topic Summaries and Exercises section in the *Student's Guide*. This section contains a summary of the key concepts, theories, and facts associated with each topic listed in the Overview of the Chapter. Further explanations of key material are also included. After each topic summary, there are exercises with detailed solutions; these are similar in style to the sample exercises found in the text. The solutions to the sample exercises often include further explanations of important or difficult material.

6. You can check your understanding of concepts and chemical information in a chapter by answering questions provided in the Self-Study Exercises. Because instructors may use a variety of question formats on tests, the *Student's Guide* provides questions using five common testing formats: matching statements and terms, true-false questions, problems, short-answer questions, and multiple-choice questions.

7. Sectional MCAT and DAT Practice Questions are cumulative for a group of chapters. They can help you check you readiness for a quiz or exam over these chapters. Also, they provide you practice with MCAT passage type questions and MCAT/DAT multiple-choice questions in general chemistry.

You need to be careful when using a calculator to obtain an answer to a problem. The numbers you obtain from a calculator often have six or eight digits, usually more digits than there are in any single number used in a calculation. In Chapter 1 of the text you will learn that you cannot always report all the digits that a calculator shows. Always check the final answer to see if it is reasonable. Is the answer too large or too small? Is the exponent of 10 reasonable? Also, a calculator is only a tool; it cannot provide the logic for solving a problem. Only you can do that.

These suggestions for learning chemistry and doing well in your chemistry course are only that—suggestions. You may already have your own successful strategies for studying. You may want to adapt some of the suggested study techniques so that they complement your own learning methods. The key to being a successful student and learner is to have a consistent study plan and use it!

Acknowledgments

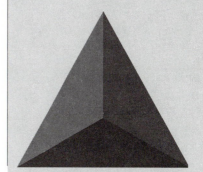

The form and content of the eleventh edition reflects the significant input of students, reviewers, and the editorial staff at Prentice Hall. To those who inspired me, gave me ideas, or noted errors I thank you. I am especially indebted to an outstanding CSU, Sacramento student, Ms. Sarah Bateni, who thoroughly read the chapters and checked most of the problems. Her comments were of significant help in preparing this edition.

I also acknowledge the guidance and support of Jennifer Hart, Associate Editor, who always had a great sense of humor; Andrew Gilfillan, Senior Editor, who provided many thoughtful comments; and Karen Bosch, copyeditor, for her careful proofreading.

Support of one's family is important when writing a book. I thank my wife, Jan, for her love, support, and patience. Without her support, this book would have been far more difficult to prepare.

If you have comments, please send them to me:

Professor James Hill
Department of Chemistry
California State University, Sacramento
6000 J Street
Sacramento, CA 95819
e-mail address: hilljames@csus.edu

Introduction to the General Chemistry Questions in the Medical College Admission Test (MCAT) or Dental Admissions Test (DAT)

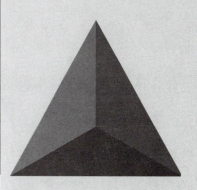

Many students taking general chemistry plan to enter the medical or dental professions. One of the requirements to enroll in medical or dental schools is taking and scoring high on the MCAT or DAT. *The following comments relate to the general chemistry questions on these tests.*

The MCAT has 52 multiple-choice questions in the physical sciences section (general chemistry, basic physics, analytical reasoning, and data interpretation). Analytical reasoning and data interpretation may be included in the questions on general chemistry and basic physics. The questions on general chemistry and basic physics are distributed somewhat equally. You have 70 minutes to answer these questions.

The DAT has a survey of natural sciences section consisting of 100 multiple-choice items, of which 30 are questions based on a year-long course in general chemistry. These questions are very similar to many of the multiple-choice questions found throughout the *Student's Guide*. You have 90 minutes to answer these questions.

The DAT contains a series of multiple-choice questions which are essentially independent of one another. The majority of MCAT questions are based on descriptive passages (essays) containing chemical information, including tables, graphs, and experimental data. Each descriptive passage is followed by five to seven multiple-choice questions. Also there are additional multiple-choice questions which are independent of any descriptive passages.

Topics in general chemistry which are tested on the MCAT include:

- Electronic structures of atoms: Orbitals, Bohr model, spectra, effective nuclear charge, notation systems for electronic structures
- Classification of elements in the periodic table by groups and electronic structure
- Periodic properties: Understanding trends and variations in chemical and physical properties by groups and rows in the periodic table
- Chemical compounds, nomenclature, forms in solution, solubility of compounds and concentrations
- Chemical bonding: Ionic and covalent models, Lewis structures, molecular geometries, electronegativity, polar bonds
- States of matter: Properties, phases, phase equilibrium, solutions and concentrations, colloids
- Gases: Pressure, gas laws, kinetic-molecular theory, non-ideal gas behavior
- Chemical reactions: Balancing, stoichiometry of reactions, oxidation-reduction reactions, oxidation numbers, empirical and molecular formulas, mole concept
- Thermodynamics: Principles, laws, specific heat, energy changes, calorimetry, enthalpy, entropy and free energy
- Chemical kinetics: Reaction rates, rate laws, rate controlling steps, activation energy, catalysts
- Chemical equilibrium: Equilibrium constant, equilibrium concentrations, Le Chatelier's principle, equilibrium and free energy, solubility product-constant, common ion effect, complex ions
- Acid and bases: Definitions, ionization and neutralization reactions, characteristics of strong and weak acids/bases, hydrolysis, pH calculations, titrations, equilibrium calculations including those of buffers
- Electrochemistry: Voltaic and electrolytic cells, cell potentials, and oxidation-reduction reactions.

Topics in general chemistry which are in the MCAT list of topics for physics:

- Atomic and nuclear structure: Nuclear particles, atomic number and weight, isotopes, binding energy, nuclear decay
- Units and dimensions: Metric, conversion of units, dimensional analysis, significant figures, and numerical estimation.

Topics in general chemistry which are tested on the DAT include:

- Stoichiometry: Formulas, balancing equations, moles, density, calculations involving chemical equations
- Atomic and molecular structure: Atomic structure, electronic configurations, bonding, molecular geometry
- Periodic properties: Periodic trends in physical properties and electronic structure, descriptive chemistry, groups of elements
- Gases: Gas laws and kinetic-molecular theory
- Liquids and solids: Phases, phase changes, intermolecular forces, polarity
- Solutions: Properties, forces, concentrations

- Acids and bases: pH, strengths, reaction and calculations
- Chemical equilibrium: Solutions, precipitation, Le Chatelier's principle
- Thermodynamics and Thermochemistry: Laws, energy changes, enthalpy, entropy, and spontaneity
- Chemical kinetics: Reaction rates, rate laws, activation energy, half-lives
- Oxidation-reduction: Reactions, oxidation numbers, balancing, electrochemistry
- Nuclear reactions: Particles, radioactive decay, binding energy, terminology
- Laboratory: Techniques, error analysis, safety and using lab data in analysis.

The multiple-choice questions based on descriptive passages in the MCAT are challenging. They typically integrate information from several different topic areas. Most of the subsequent multiple-choice questions depend on information within the passage. However, some questions may not and will require you to make conclusions based on your knowledge of general chemistry. MCAT questions normally do not require extensive math calculations.

Both tests are administered using a computer-based multiple-choice exam. The tests, however, are quite different in format. Eventually you will need practice answering multiple-choice questions in a computer-based testing environment.

The *Student's Guide* has sets of sectional practice questions which are based on the style of MCAT and DAT questions. As you progress through chapters in *Chemistry: The Central Science,* you will have an opportunity to gain early practice with these types of questions.

The information about the MCAT and DAT is current as of fall 2006. You should go to the ADA or AAMC internet sites for updated information on formats and contents of the tests.

ADA: http://www.ada.org

AAMC: http://www.aamc.org/mcat

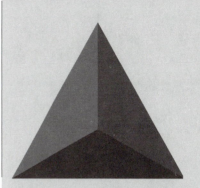

Using the *Student's Guide* to Review and Prepare for the MCAT and DAT

The *Student's Guide* reviews the essential chemical concepts and principles found in the text *Chemistry: The Central Science*. Its primary purpose is to help students learn material in this text. However, you can adapt its contents to help you review and prepare for questions on general chemistry found in the MCAT and DAT.

An outline of an approach you can use to review and prepare for general chemistry questions on the MCAT or DAT follows:

1. Only use chapters which relate to topics covered in the MCAT and DAT.
2. Read the TOPIC SUMMARIES AND EXERCISES in the chapters.
3. Practice the *Multiple-Choice Questions* at the end of each chapter.
4. Take the Sectional MCAT and DAT Practice Questions after reviewing the chapters previous to each set.

The following chapters relate to the topics in general chemistry covered in the MCAT:

Chapter 1: Introduction: Matter and Measurement

Metric units and conversion of units are used in some problems.

Chapter 2: Atoms, Molecules, and Ions

An understanding of atoms, molecules, and ions and how to write or interpret chemical symbols and formulas (molecular and empirical) are needed to answer questions. Naming salts and molecules, or writing chemical formulas from names, is a necessary skill.

Chapter 3: Stoichiometry: Calculations with Chemical Formulas and Equations

All the material in this chapter is appropriate.

Chapter 4: Aqueous Reactions and Solution Stoichiometry

All the material in this chapter is appropriate.

Chapter 5: Thermochemistry

All the material in this chapter is appropriate.

Chapter 6: Electronic Structures of Atoms

All the material in this chapter is appropriate.

Chapter 7: Periodic Properties of the Elements

All the material in this chapter is appropriate.

Chapter 8: Basic Concepts of Chemical Bonding

All the material in this chapter is appropriate.

Chapter 9: Molecular Geometry and Bonding Theories

All the material in this chapter is appropriate except for the two sections on molecular orbitals.

Chapter 10: Gases

All the material in this chapter is appropriate.

Chapter 11: Intermolecular Forces, Liquids and Solids

All the material in this chapter is appropriate.

Chapter 13: Properties of Solutions

All the material in this chapter is appropriate.

Chapter 14: Chemical Kinetics

All the material in this chapter is appropriate.

Chapter 15: Chemical Equilibrium

All the material in this chapter is appropriate.

Chapter 16: Acid–Base Equilibria

All the material in this chapter is appropriate.

Chapter 17: Additional Aspects of Aqueous Equilibria

All the material in this chapter is appropriate. The MCAT sometimes has descriptive passages involving experimental data coupled with the principles of qualitative analysis. The last two sections in this chapter may be useful in preparing for these types of questions.

Chapter 18: Chemical Thermodynamics

All the material in this chapter is appropriate.

Chapter 19: Electrochemistry

All the material in this chapter is appropriate.

Chapter 25: The Chemistry of Life and Biological Chemistry

Questions in general chemistry may use compounds from organic chemistry. It is useful to have an elementary understanding of hydrocarbons including alkanes, alkenes, and alkynes. Your knowledge of organic chemistry is tested in the biological sciences section of the MCAT.

The following chapters relate to the topics in general chemistry covered in the DAT:

All of the chapters listed above in the MCAT. As noted above, Chapter 25 is not specifically covered in the general chemistry portion of either the DAT or MCAT, but the elementary material on hydrocarbons can be useful.

Chapter 21: Nuclear Chemistry

All the material in this chapter is appropriate except for the last section.

After studying the appropriate chapters and practicing taking multiple-choice questions in the *Student's Guide,* you need to personally assess if you have a good grasp of the principles, concepts, and problem-solving skills required to be successful in taking the MCAT or DAT. If you believe you are prepared, then you should take practice tests using a computer since this testing process is not the same as taking a test using paper and pencil. If you do not feel prepared, then you should review the material again and also seek out additional sources of practice questions. *The most important key to success is being well-versed in the concepts, principles, knowledge of basic facts, and problem-solving skills in general chemistry.*

Chapter 1

Introduction: Matter and Measurement

OVERVIEW OF THE CHAPTER

Learning Goals: You should be able to:

1. Distinguish between physical and chemical properties and also between simple physical and chemical changes.
2. Differentiate between the three states of matter.
3. Distinguish between elements, compounds, and mixtures.
4. Give the symbols for the elements discussed in this chapter.

Review: Concept of fraction; exponential notation (see text: Appendix A).

Learning Goal: You should be able to list the basic SI and metric units and the commonly used prefixes in scientific measurements.

Review: Exponential notation (see text Appendix A).

Learning Goals: You should be able to:

1. Determine the number of significant figures in a measured quantity.
2. Express the result of a calculation with the proper number of significant figures.

Learning Goals: You should be able to:

1. Convert temperatures among the Fahrenheit, Celsius, and Kelvin scales.
2. Perform calculations involving density.

Review: Concepts of fraction and ratio.

Learning Goal: You should be able to convert between units by using dimensional analysis.

TOPIC SUMMARIES AND EXERCISES

**MATTER:
ELEMENTS,
COMPOUNDS,
AND MIXTURES**

Matter is any material that occupies space and has mass. Three phases (states) of matter exist: gas, liquid, and solid.

- A sample of matter is either a substance or a mixture.
- **Substances** are either elements or compounds. **Elements** cannot be separated into simpler new substances. **Compounds** consist of two or more elements chemically combined in a definite ratio. A compound can be chemically decomposed into its elements.
- **Mixtures** are physical combinations of two or more substances and are either homogeneous or heterogeneous. A *homogeneous* mixture consists of one phase and a uniform distribution of substances. A *heterogeneous* mixture shows more than one phase and possesses a nonuniform distribution of substances.
- Mixtures can be separated into substances by physical means.

Alterations in matter can involve chemical or physical changes.

- A **chemical change** results in a change in the composition of a substance. A **chemical property** describes the type of chemical change. For example, the property of wood burning is a chemical property.
- A **physical change** does not involve a change in the composition of a substance but rather a change in a **physical property** such as temperature, volume, mass, pressure, or state.

Check your understanding of the new terms you have learned by doing Exercises 1–5.

EXERCISE 1   Identifying characteristics of matter

Match the following characteristics to one or more of the three states of matter: (**a**) has no shape of its own; (**b**) definite shape; (**c**) occupies the total volume of a container; (**d**) partially takes on the shape of a container; (**e**) does not take on the shape of a container; (**f**) readily compressible; (**g**) slightly compressible; (**h**) essentially noncompressible.

SOLUTION: Gas—(**a**), (**c**), (**f**); Liquid—(**a**), (**d**), (**g**); Solid—(**b**), (**e**), (**h**)

EXERCISE 2   Identifying characteristics of matter II

Match the term with the best identifying phrase:

Terms

1. Homogeneous mixture
2. Heterogeneous mixture
3. Mixture
4. Substance
5. Element
6. Compound

Phrases

a. Any kind of matter that is pure and has a fixed composition
b. Cannot be decomposed into simpler substances by chemical changes

c. A solution of uniform composition

d. Can be decomposed into simpler substances by chemical changes

e. Any kind of matter that can be separated into simpler substances by physical means

f. Nonuniform composition

SOLUTION: 1-**c**; 2-**f**; 3-**e**; 4-**a**; 5-**b**; 6-**d**

EXERCISE 3   Writing names or symbols of elements

With the help of the periodic table, write the name or the chemical symbol for each of the following elements: (**a**) F; (**b**) zinc; (**c**) potassium; (**d**) As; (**e**) Al; (**f**) iron; (**g**) helium; (**h**) barium; (**i**) Ne.

SOLUTION: (**a**) fluorine; (**b**) Zn; (**c**) K; (**d**) arsenic; (**e**) aluminum; (**f**) Fe; (**g**) He; (**h**) Ba; (**i**) neon

EXERCISE 4   Identifying changes of matter

Are the following changes physical or chemical: (**a**) the vaporization of solid carbon dioxide; (**b**) the explosion of solid TNT; (**c**) the aging of an egg with a resultant unpleasant smell; (**d**) the formation of a solid when honey is cooled?

SOLUTION: (**a**) A physical change. The form of carbon dioxide is changed from solid to gas. There is no change in its chemical composition. (**b**) A chemical and physical change. The explosion results from a change in the chemical composition of TNT and the formation of a gas. (**c**) A chemical change. A change in the composition of the egg results in the formation of a gas that has an unpleasant smell. (**d**) A physical change. The solid results from the crystallization of dissolved sugars; no change occurs in the chemical form of the sugars.

EXERCISE 5   Recognizing elements, compounds, and mixtures

Classify each of the following as an element, compound, or mixture: (**a**) a 100-percent lead bar; (**b**) wine; (**c**) gasoline; (**d**) carbon dioxide (CO_2).

SOLUTION: (**a**) Lead is an element and cannot be separated by either chemical or physical means into simpler substances. It is listed among the elements in Table 1.2 in the text. (You should know the symbols in Table 1.2.) (**b**) Wine is a mixture of alcohol, other components, and water. The fact that wines contain varying percentages of alcohol attests to their having different compositions. (**c**) We know gasoline must be a mixture because it is available with different compositions and properties (no lead, regular, and different brands with different additives). (**d**) CO_2 is a compound because the ratio of carbon and oxygen atoms is fixed and definite. The name also implies that it is a compound because we do not have such systematic names for mixtures.

PHYSICAL QUANTITIES AND UNITS

A physical property of a sample is measured by comparing it with a standard unit of that property. Measured quantities such as volume, length, mass, and temperature require a number and a reference label, called the unit of measurement. Two systems of unit measurements are shown in Table 1.1 on page 4.

The SI system of units is now the preferred one; however, you will find certain metric units still used. *You must become thoroughly familiar with the units in Table 1.1 before starting the next chapter.*

TABLE 1.1 Metric and SI Units

Physical quantity	Metric unit name	SI unit name[a]
Length	Meter (m)	Meter (m)
Volume	Cubic centimeter (cm^3)[b]	Cubic meter (m^3)
Mass	Gram (g)	Kilogram (kg)
Time	Second (s)	Second (s)
Energy	Calorie (cal)	Joule (J)
Pressure	Atmosphere (atm)	Newton per square meter (N/m^2)

[a] Systeme International d'Unites (SI) or International System of Units.

[b] Chemists commonly use the unit cubic centimeter when dealing with the volume of a solid, but they usually use the unit liter (L) when a substance is a liquid.

A necesary skill requiring proficiency is changing a number with a unit to one with a different unit. We use equivalence relationships between units to do conversions between units. Tables 1.2 and 1.3 give some common equivalences that you will need in this chapter.

Prefixes are used with units to indicate decimal fractions (<1) or multiples (>1) of basic units.

- Example of a decimal fraction: The prefix centi- means $1/100 (= 0.01)$ of a basic unit; thus, $100\ cm = 100 \times 1/100\ m = 1\ m$.
- Example of a multiple: The prefix kilo- means $10^3 (= 1000)$; thus, $1\ km = 1 \times 1000\ m = 1000\ m$.

The commonly used prefixes that you must know are shown in Table 1.4 on page 5. *Memorize them.*

TABLE 1.2 Equivalence Relationships between SI and Metric Units

Physical quantity	Metric unit name	SI unit name	Equivalence
Length	Meter	Meter	Same
Mass	Gram	Kilogram	$1000\ g = 1\ kg$
Time	Second	Second	Same
Energy	Calorie	Joule	$1\ cal = 4.184\ J$
Volume	Cubic centimeter	Cubic meter	$1{,}000{,}000\ cm^3 = 1\ m^3$
Volume	Liter	Cubic meter	$1000\ L = 1\ m^3$
Pressure	Atmosphere	Newton per square meter	$1\ atm = 0.1754\ N/m^2$

TABLE 1.3 Equivalence Relationships between Metric and English Units

Physical quantity	English unit symbol	Metric unit symbol	Equivalence
Mass	lb ($= 16\ oz$)	g	$1\ lb = 453.6\ g$
Length	ft ($= 12\ in.$)	m	$3.272\ ft = 1\ m$
Length	in.	cm	$1\ in. = 2.54\ cm$
Length	mi ($= 5280\ ft$)	m	$1\ mi = 1609\ m$
Volume	qt	L	$1.057\ qt = 1\ L$

TABLE 1.4 Commonly Used Prefixes for Scientific Measurement in Chemistry

Prefix	Fraction or multiple of base unit	Abbreviation
Deci-	$10^{-1}\left(\dfrac{1}{10}\right)$	d
Centi-	$10^{-2}\left(\dfrac{1}{100}\right)$	c
Milli-	$10^{-3}\left(\dfrac{1}{1000}\right)$	m
Micro-	$10^{-6}\left(\dfrac{1}{1{,}000{,}000}\right)$	μ
Nano-	$10^{-9}\left(\dfrac{1}{1{,}000{,}000{,}000}\right)$	n
Pico-	$10^{-12}\left(\dfrac{1}{1{,}000{,}000{,}000{,}000}\right)$	p
Kilo-	$10^{3}\,(1000)$	k
Mega-	$10^{6}\,(1{,}000{,}000)$	M
Giga-	$10^{9}\,(1{,}000{,}000{,}000)$	G

EXERCISE 6   Determining relative magnitudes of quantities

Which quantity of each pair is larger: (**a**) 1 nm or 1 micrometer; (**b**) 1 picogram or 1 cg; (**c**) 1 megagram or 1 milligram?

SOLUTION: Change the pairs so that each quantity is represented by either a fraction or a multiple of the same basic metric unit. Then from their relative magnitudes we can determine which is larger.

(**a**) $1\text{ nm} = 1\text{ nanometer} = 10^{-9}\text{ meter}$
$1\text{ micrometer} = 1\ \mu\text{m} = 10^{-6}\text{ meter}$

One micrometer is larger in value than one nanometer because the fraction $10^{-6}\left(\frac{1}{1{,}000{,}000}\right)$ is larger in magnitude than the fraction $10^{-9}\left(\frac{1}{1{,}000{,}000{,}000}\right)$.

(**b**) $1\text{ picogram} = 1\text{ pg} = 10^{-12}\text{ gram}$

$1\text{ cg} = 1\text{ centigram} = 10^{-2}\text{ gram}$

One centigram is larger in value than one picogram because the fraction $10^{-2}\left(\frac{1}{100}\right)$ is larger in magnitude than the fraction $10^{-12}\left(\frac{1}{1{,}000{,}000{,}000{,}000}\right)$.

(**c**) $1\text{ megagram} = 1\text{ Mg} = 10^{6}\text{ gram}$

$1\text{ mg} = 1\text{ milligram} = 10^{-3}\text{ gram}$

One megagram is larger in value than one milligram because the multiple $10^{6}\,(1{,}000{,}000)$ is larger in magnitude than the fraction $10^{-3}\left(\frac{1}{1{,}000}\right)$.

EXERCISE 7   Recognizing units with measurements

With what types of measurements are the following units associated?

$$g, L, m, km, cm, Mg, pg, cm^3$$

SOLUTION: Mass (g, Mg, pg); volume (L, cm^3); length (cm, m, km). Note that the prefixes such as M- and c- do not change the type of unit. However, the type of unit can be changed if it is raised to some power, as is the case for cm^3. The unit cm^3 means cm $\times$ cm $\times$ cm, which is a unit for volume (V = 1 $\times$ w $\times$ h).

EXERCISE 8   Comparing English to SI System of Units

What is the advantage of the metric system in comparison to the English system?

SOLUTION: In the metric system, all quantities larger or smaller than the basic unit involve multiplication of the basic unit value by some power of 10 (for example, $10^3 = 1000$, $10^{-1} = \frac{1}{10}$, and so on). This is not true of the English system. Smaller or larger quantities of the basic unit in the English system are newly defined units. For example, 4000 qt equals 1000 gal, not 4 "kiloquarts." Many more conversion factors are required in the English unit system than in the metric unit system.

EXERCISE 9   Knowing acceptable SI volume unit

Suggest a reason for the fact that 1 μkL (microkiloliter) is not accepted as an appropriate SI unit for volume.

SOLUTION: The expression 1 μkL involves two prefixes, micro- (μ) and kilo- (k), yielding a compound prefix. This can be confusing, particularly if three or four prefixes are used. Thus, we do not use more than one prefix when expressing numbers. Instead of 1 μkL (microkiloliter), we write 1 mL (milliliter).

UNCERTAINTY IN MEASUREMENTS: SIGNIFICANT FIGURES

Numbers in chemistry are of two types:

- **Exact:** These result from counting objects such as coins or occur as exact numbers in equations or as exact conversion factors.
- **Inexact:** These are obtained from measurements. Uncertainties exist in their values because judgment is required in making measurements.

Measured quantities (inexact numbers) are reported so that the last digit is the first uncertain digit. An uncertain digit is one that requires judgment in determining its value. All certain digits and the first uncertain digit are referred to as **significant figures**. For example:

- 2.86: 2 and 8 are certain and well known. The number 6 is the first that is subject to judgment and is uncertain. The first uncertain digit is assumed to have an uncertainty of $\pm 1 : 2.86 \pm 0.01$. The number 2.86 has three significant figures.
- 0.0020: Zeroes to the left of the first nonzero digit in a number with a decimal point are not significant. The first three zeroes are not significant because they are to the left of the 2 and also define the decimal point. The zero to the right of the 2 is significant. This number has only two significant figures.
- 100: Trailing zeroes that define a decimal point may or may not be significant. Unless stated, assume they are not significant. Therefore, 100 has one significant figure unless otherwise stated; if it is determined from counting objects, it has three significant figures.

Exponential notation is used to remove ambiguity in reporting the number of significant figures a number possesses.

- Only significant digits are shown. The number 0.0020 becomes 2.0×10^{-3}. The zeroes in front of the two in 0.0020 are not significant whereas the trailing zero is significant.

Calculated numbers must show the correct number of significant figures. The rules for doing this are:

1. Addition and Subtraction: The final answer should have the same uncertainty as the quantity in the calculation with the greatest uncertainty. In the following example, the first uncertain digit in each quantity is in bold.

$$
\begin{array}{l}
\quad 32\mathbf{5}.24 \ (\text{uncertainty} = \pm 0.01) \\
+ \ 2\mathbf{1}.4 \ \ (\text{uncertainty} = \pm 0.1) \\
+ \ \underline{1\mathbf{4}5} \quad \ (\text{uncertainty} = \pm 1) \\
\quad 49\mathbf{1}.64 \ (\text{uncertainty in final answer is} \pm 1)
\end{array}
$$

The least precise number is 145 and it controls the number of significant figures in the answer. It has the greatest uncertainty, ±1. Thus 491.64 is rounded to 492 with an uncertainty of ±1.

2. Multiplication and division: When multiplying or dividing numbers, round off the final calculated answer so that it has the same number of significant figures as the least certain number (the one with fewest number of significant figures) in the calculation. A little caution must be used when applying this rule. For example, to divide 101 by 95, you might be tempted to report the final answer to two significant figures because 95 appears to be the least certain number. Yet 95 has almost three significant figures; there is very little difference in error between 1 in 95 and 1 in 101. Thus, in this case it makes more sense to round off the final answer to three significant figures. Use common sense in problems when a number is close in magnitude to 100, 1000, 10,000, and so on.

3. Exact or defined numbers are not used in determining the number of significant figures in a final answer. If you use the equation $A = 4\pi r^2$ to calculate the surface area of a sphere, the number 4 is considered to have an infinite number of significant figures. It therefore not used in determining the uncertainty of the calculated area.

Caution: The final answer of a calculation determined using a calculator often has more digits than any of the numbers in the calculation. You may have to round off the answer to the correct number of significant figures. The rules for rounding off numbers in a calculated answer are:

1. When the number immediately following the last digit to be retained (the first uncertain digit) is less than 5 then the last digit is retained unchanged. If 6.4362 is rounded off to four significant figures it becomes 6.436.

2. When the number immediately following the last digit to be retained is 5* or greater, then increase the last digit by 1. If 6.4366 is to be rounded off to four significant figures it becomes 6.437.

*Note: Your instructor may use an alternative approach: If there are no other numbers or only zeroes beyond the 5, then the last retained digit is increased by 1 if it is odd and left unchanged if it is even. Or if there are numbers other than zero beyond the 5, then the last digit retained is increased by 1. For example, when three significant figures are required, 2.2350 becomes 2.14 (3 is odd and thus it is increased by 1) and 2.1453 becomes 2.15.

Note: Do not round numbers until you have completed your calculation.

EXERCISE 10   Determining uncertain digits

Remembering that measured values are reported to ±1 uncertainty in the last digit, except for those values determined by counting observable objects, or unless otherwise stated, determine the first uncertain digit in each of the following numbers: (a) 10.03 kg; (b) 5 apples; (c) 5.02 ± 0.02 m.

SOLUTION: (a) The 3 in 10.03 kg is uncertain to ±1. (b) This is an exact measured value determined by counting. There is no uncertain digit. (c) The 2 in 5.02 m is uncertain to ±2.

EXERCISE 11   Determining the number of significant figures

The precision of a measurement is indicated by the number of significant figures associated with the reported value. How many significant figures does each number possess: (a) 225; (b) 10,004; (c) 0.0025; (d) 1.0025; (e) 0.002500; (f) 14,100; (g) 14,100.0?

SOLUTION: Try this technique: If a quantity contains a *decimal* point, draw an arrow *starting* at the *left* through all zeroes up to the first nonzero digit; the digits remaining are significant. If the quantity does *not* contain a decimal point, draw an arrow *starting* at the *right* through all zeroes up to the first nonzero digit; the digits remaining are significant.
(a) 225 ⟵ three significant figures (No decimal point—draw arrow to left)
(b) 10,004 ⟵ five significant figures
(c) 0.0025 two significant figures (Contains a decimal point—draw arrow to the right)
(d) ⟶ 1.0025 five significant figures
(e) 0.002500 four significant figures
(f) 14,100 three significant figures; however, because of our lack of knowledge about the significance of the two trailing zeroes, this number also could have four or five significant figures
(g) ⟶ 14,100.0 six significant figures

EXERCISE 12   Writing numbers in scientific notation

Write the numbers in Exercise 11 using scientific notation.

SOLUTION: Move the decimal in the appropriate direction so that it is to the right of the first nonzero digit reported in the number. If the decimal is moved to the left, multiply the resulting quantity by 10 raised to a power that equals the number of digits the decimal is moved past. If the decimal is moved to the right, the power of 10 is again the number of digits the decimal is moved past, but with a negative sign. That is, a number that is greater than 1 will appear as $A.BC \times 10^x$, while one that is less than 1 will appear as $A.BC \times 10^{-x}$. (a) $225 = 2.25 \times 10^2$. The decimal is moved two digits to the left, thus the power of 10 is 2. (b) $10,004 = 1.0004 \times 10^4$. The power of 10 is 4 because the decimal is moved four places to the left. (c) $0.0025 = 0002.5 \times 10^{-3} = 2.5 \times 10^{-3}$. The power of 10 is −3 because the decimal is moved three places to the right. Note that the nonsignificant zeros are omitted. (**d**) 1.0025. We do not write 1.0025×10^0. (**e**) 0.002500 = 0002.500 $\times 10^{-3}$ = 2.500×10^{-3}. The zeros after the 2.5 are written because they define the number of significant figures. (**f**) 14,100 = 1.4100×10^4 = 1.41×10^4 if the zeros in 14,100 are not significant. (**g**) 14,100.0 = 1.41000×10^4.

EXERCISE 13   Rounding answers in calculations

Round the answers in the following problems to the correct number of significant figures:

(**a**) 12.25 + 1.32 + 1.2 = 14.770   (**c**) 12300 + 2.11 = 12302.11
(**b**) 13.7325 − 14.21 = −0.4775

SOLUTION: In each problem, identify the quantity with the greatest uncertainty and use this uncertainty to determine the correct number of significant figures for the answer. (**a**) The 1.2 has the greatest uncertainty, ±0.1. Therefore, the answer must be rounded to one digit to the right of the decimal point: 14.8. (**b**) 14.21 has the greatest uncertainty, ±0.01. Therefore, the answer must be rounded to two digits to the right of the decimal point: −0.48. *Note*: An answer obtained by subtraction may have fewer significant figures than either number used. (**c**) 12300 has an uncertainty of ±100. The trailing zeros are not identified as being significant; therefore, we normally assume that they are not significant. (If the trailing zeros are significant, they should have been identified as 12300., or preferably 1.2300×10^4.) The answer must be rounded at the hundreds place: 12300. Any digit to the right of the hundreds place is assigned a zero value.

EXERCISE 14   Rounding answers in calculations II

Round the final answer in each of the following calculations:

(**a**) (1.256)(2.42) = 3.03952   (**c**) $\dfrac{(1.1)(2.62)(13.5278)}{2.650}$ = 14.712121

(**b**) $\dfrac{16.231}{2.20750}$ = 7.352661

SOLUTION: (**a**) The least precise number in the calculation is 2.42 (three significant figures). The final answer must be rounded to three significant figures: 3.04. (**b**) The least precise number in the calculation is 16.231 (five significant figures). The final answer must be rounded to five significant figures: 7.3527. (**c**) The least precise number in the calculation is 1.1 (two significant figures). The final answer must be rounded to two significant figures: 15.

TEMPERATURE AND DENSITY

Two important concepts discussed in Chapter 1 are temperature and density.

- They are both **intensive properties** as their values are independent of the amount of substance.
- This contrasts with **extensive properties** such as volume and mass, which depend on the amount of substance.

Temperature is a measure of the intensity of heat—the "hotness" or "coldness" of a body.

- Heat is a form of energy. Heat flows from a hot object to a colder one.
- When there is no heat flow between two objects in contact, they have the same temperature.
- Three temperature scales are used: Celsius (°C), Fahrenheit (°F), and Kelvin (K). You need to know how their reference points differ and how to change between them.

	Reference Points		
	Fahrenheit	**Celsius**	**Kelvin**
Freezing point of water	32 °F	0 °C	273.15 K
Boiling point of water	212 °F	100 °C	373.15 K

- *Note that a 1-degree interval is the same on both the Celsius and Kelvin scales, but a 1 °C interval equals a 1.8 °F interval.* The only temperature at which both the Fahrenheit and Celsius scales are equivalent is $-40°(-40 \ °C = -40 \ °F)$. This fact enables us to make conversions between the two scales using the following approach, which is different from the one given in the text.

$$°F \longrightarrow °C \qquad\qquad °C \longrightarrow °F$$

(a) Add 40° to °F = (1)   Add 40° to °C = (1)

(b) $(1) \times \dfrac{1\ °C}{1.8\ °F} = (2)**$ $(1) \times \dfrac{1.8\ °F}{1\ °C} = (2)**^1$

(c) Subtract 40° from (2) = °C   Subtract 40° from (2) = °F

**Notice that only step (b) is different. Step (b) converts Fahrenheit to Celsius or Celsius to Fahrenheit using the relationship that a 1-degree Celsius interval equals a 1.8-degree Fahrenheit interval. Also, do not round off until the calculation is finished.

- The following relationship is used to convert between Celsius and Kelvin temperatures.

$$K = \left(\frac{1\ K}{1\ °C}\right)(°C) + 273.15\ K$$

Density (*d*) measures the amount of a substance (*m*) in a given volume (*V*):

- $d = \dfrac{\text{mass}}{\text{volume}} = \dfrac{m}{V}$
- Density varies with temperature because volume changes with temperature.
- Density can be used to change mass to volume and vice versa for the same substance.
- Chemists commonly use the following units for density: g/mL for liquids, g/cm^3 for solids, and g/L for gases.

EXERCISE 15   Converting temperature to a different scale

The temperature on a spring day is around 22 °C. What is this temperature in degrees Fahrenheit and degrees Kelvin?

SOLUTION: To change 22 °C to °F, first add 40°:

$$22\ °C + 40° = 62\ °C$$

Then multiply by 1.8 °F/1 ° C:

$$(62\ °C)\left(\frac{1.8\ °F}{1\ °C}\right) = 112\ °F$$

Finally, subtract 40° from 112 °F:

$$112° - 40° = 72 \text{ °F}$$

To calculate degrees Kelvin, write the relationship between the two degrees and substitute for °C:

$$K = \left(\frac{1\,K}{1\,°C}\right)(°C) + 273.15\,K = \left(\frac{1\,K}{1\,°C}\right)(22\,°C) + 273.15\,K = 295\,K \text{ (rounded)}$$

Note: The method presented in the text gives the same answer as follows.

$$°F = \left(\frac{1.8\,°F}{1\,°C}\right)(°C) + 32\,°F$$

Substituting 22 °C for °C yields:

$$(°F) = \left(\frac{1.8\,°F}{1\,°C}\right)(22\,°C) + 32\,°F = 39.6\,°F + 32\,°F = 72\,°F \text{ (rounded)}$$

EXERCISE 16   Comparing densities

At 20 °C, carbon tetrachloride and water have densities of 1.60 g/mL and 1.00 g/mL respectively. When water and carbon tetrachloride are poured into the same container, two layers form, one being water and the other carbon tetrachloride. Based on their densities, which one will occupy the lower layer in a container?

SOLUTION: Since carbon tetrachloride and water do not mix together permanently, the heavier substance per unit volume will fall to the bottom of the container. Carbon tetrachloride will be that substance because it has a higher mass per unit volume (density), 1.60 g/mL, than does water, 1.00 g/mL.

EXERCISE 17   Using density in a calculation

Which has the greater mass, 2.0 cm^3 of iron ($d = 7.9$ g/cm^3) or 1.0 cm^3 of gold ($d = 19.32$ g/cm^3)?

SOLUTION: The mass of a substance is related to its density by the equation $d = m/V$. Multiplying both sides of the equation by V yields $m = d \times V$. The mass of 2.0 cm^3 of iron is calculated as follows:

$$m = d \times V = \left(7.9\frac{g}{cm^3}\right)(2.0\,cm^3) = 16\,g$$

The mass of 1.0 cm^3 of gold is calculated similarly:

$$m = d \times V = \left(19.32\frac{g}{cm^3}\right)(1.0\,cm^3) = 19\,g$$

Thus 1.0 cm^3 of gold has a greater mass than 2.0 cm^3 of iron.

DIMENSIONAL ANALYSIS

You need to develop the habit of including units with all measurements in calculations. *Units are handled in calculations as any algebraic symbol:*

- Numbers added or subtracted must have the same units.
- Units are multiplied as algebraic symbols: (2 L)(1 atm) = 2 L-atm
- Units are cancelled in division if they are identical. Otherwise, they are left unchanged: (3.0 m)/(2.0 mL) = 1.5 m/mL.

Dimensional analysis is the algebraic process of changing from one system of units to another. A fraction, called a unit conversion factor, is used to make the conversion. These fractions are obtained from an equivalence between two units. For example, consider the equality 1 in. = 2.54 cm. This equality yields two conversion factors.

$$\frac{1 \text{ in.}}{1 \text{ in.}} = \frac{2.54 \text{ cm}}{1 \text{ in.}} \quad \text{and} \quad \frac{1 \text{ in.}}{2.54 \text{ cm}} = \frac{2.54 \text{ cm}}{2.54 \text{ cm}}$$

$$1 = \frac{2.54 \text{ cm}}{1 \text{ in.}} \quad \text{and} \quad \frac{1 \text{ in.}}{2.54 \text{ cm}} = 1$$

Note that the two conversion factors each equal one and are the inverse of one another. They enable us to convert between units in the equality. For example, to convert from centimeters to inches or vice versa:

$$5.08 \text{ cm} \times \frac{1 \text{ in.}}{2.54 \text{ cm}} = 2.00 \text{ in.}$$

$$4.00 \text{ in.} \times \frac{2.54 \text{ cm}}{1 \text{ in.}} = 10.2 \text{ cm}$$

Note: $\text{given unit} \times \dfrac{\text{new unit}}{\text{given unit}} = \text{new unit}$

EXERCISE 18   Converting units of volume

Convert 10.5 L to milliliters.

SOLUTION: The required operations for converting 10.5 L to milliliters are as follows:

1. State the general relation required to convert units:

   ? mL = (10.5 L)(conversion factor that changes L to mL)

2. Find the conversion factor (or factors) that converts L to mL. In this case it is 1000 mL = 1 L. Using this equivalence relation, determine the appropriate ratio of units that converts L to mL. Because we want to change liters to milliliters, we will have to divide by 1 L:

$$\frac{1000 \text{ mL}}{1 \text{ L}}$$

3. Substitute 1000 mL/1 L for the conversion factor in the equation in step 1 and solve the problem.

$$? \text{ mL} = (10.5 \text{ L})\left(\frac{1000 \text{ mL}}{1 \text{ L}}\right) = 10{,}500 \text{ mL} = 1.05 \times 10^4 \text{ mL}$$

Change the final answer to scientific notation, if necessary. It is sometimes more convenient to change all numbers to scientific notation before doing the mathematics of the problem:

$$? \text{ mL} = (1.05 \times 10^1 \text{ L})\left(\frac{1 \times 10^3 \text{ mL}}{1 \text{ L}}\right) = 1.05 \times 10^4 \text{ mL}$$

4. Check that the units properly cancel to yield the desired unit. In this problem, if we had used the conversion factor 1 L/1000 mL instead of 1000 mL/1 L, the result would have been:

$$? \text{ mL} \neq (1.05 \times 10^1 \text{ L})\left(\frac{1 \text{ L}}{1 \times 10^3 \text{ mL}}\right) \neq 1.05 \times 10^{-2}\frac{\text{L}^2}{\text{mL}}$$

The unit L^2/mL does not equal mL; therefore, we know we have used the wrong conversion factor.

EXERCISE 19   Converting SI units

Convert 6.23 ft^3 to the appropriate SI unit.

SOLUTION: The appropriate SI unit is meter^3. The unit ft^3 is based on an English unit of length, the foot. Use the same method shown in Exercise 18 to convert ft^3 to m^3.

1. $? \text{ m}^3 = (6.23 \text{ ft}^3)$ (conversion factor that changes ft^3 to m^3)
2. From Table 1.3 we find that 3.272 ft = 1 m, but there is no unit conversion given for ft^3 to m^3. What we must recognize is that if

$$1 = \frac{1 \text{ m}}{3.272 \text{ ft}}$$

then we can cube both sides of the expression

$$(1)^3 = \left(\frac{1 \text{ m}}{3.272 \text{ ft}}\right)^3 = \frac{1 \text{ m}^3}{(3.272)^3 \text{ ft}^3} = 1$$

3. Converting units yields:

$$? \text{ m}^3 = (6.23 \text{ ft}^3)\left(\frac{1 \text{ m}^3}{(3.272)^3 \text{ ft}^3}\right) = 0.178 \text{ m}^3$$

EXERCISE 20   Using multiple conversion factors

The Empire State Building in New York City was for many years the tallest building in the world. Its height is 484.7 yd to the top of the lightning rod. Convert this distance to meters.

SOLUTION: The conversion of 484.7 yd to meters is accomplished by the method used in Exercise 18:

1. State the general relation required to convert units:

   $? \text{ m} = (484.7 \text{ yd})$(conversion factor or series of factors that changes yards to meters)

2. Look up the required equivalences in appropriate tables. From the tables in this chapter of the *Student's Guide*, we find that 3.272 ft = 1 m. Because no conversion between yards and feet is given, we will have to go to another source or remember from your past experience that 1 yd = 3 ft. To convert yards to meters we will have to make a series of unit conversions that will look like

$$(\cancel{yd})\left(\frac{ft}{\cancel{yd}}\right)\left(\frac{m}{\cancel{ft}}\right) = m$$

Thus the required conversion factors are

$$\frac{3 \text{ ft}}{1 \text{ yd}} \quad \text{and} \quad \frac{1 \text{ m}}{3.272 \text{ ft.}}$$

3. We could solve the problem in two steps:

$$(484.7 \text{ } \cancel{yd})\left(\frac{3 \text{ ft}}{1 \text{ } \cancel{yd}}\right) = 1454 \text{ ft}$$

$$(1454 \text{ } \cancel{ft})\left(\frac{1 \text{ m}}{3.272 \text{ } \cancel{ft}}\right) = 444.3 \text{ m}$$

However, it is usually simpler to do it all in one step:

$$? \text{ m} = (484.7 \text{ } \cancel{yd})\left(\frac{3 \text{ } \cancel{ft}}{1 \text{ } \cancel{yd}}\right)\left(\frac{1 \text{ m}}{3.272 \text{ } \cancel{ft}}\right) = 444.3 \text{ m}$$

4. Check that the units properly cancel. Inspection of the previous equation shows that they do.

EXERCISE 21   Converting units of measurements having a ratio of units

Florence Griffith Joyner (USA) set a world record in the women's 100 m dash on July 16, 1988, running the distance in 10.49 s. This record has not been broken as of May 30, 2007. Assuming that the distance is exactly 100 m, what is her average speed in miles per hour?

SOLUTION: Joyner's average speed in meters per second is

$$\text{speed} = \frac{\text{distance}}{\text{time}} = \frac{100 \text{ m}}{10.49 \text{ s}} = 9.533 \frac{\text{m}}{\text{s}}$$

The answer has four significant figures because the problem says assume the distance is exact. The units are not requested in the problem; the ratio m/s has to be converted to mi/hr.

1. The conversion of units can be viewed as follows:

$$\frac{\text{m} \rightarrow \text{mi}}{\text{s} \rightarrow \text{min} \rightarrow \text{hr}}$$

Table 1.3 gives us the necessary equivalences between meters and miles and we should know the equivalences between seconds and minutes and minutes and hours:

1 mi = 1609 m
60 s = 1 min
60 min = 1 hr

2. Using these equivalences we can now make the unit conversions by converting meters to miles in the numerator and also seconds to minutes to hours in the denominator:

$$9.533 \frac{\text{m}\left(\dfrac{1 \text{ mi}}{1609 \text{ m}}\right)}{\text{s}\left(\dfrac{1 \text{ min}}{60 \text{ s}}\right)\left(\dfrac{1 \text{ hr}}{60 \text{ min}}\right)} = 21.33 \frac{\text{mi}}{\text{hr}}$$

(3) Alternatively we can do the conversion in a long sequence:

$$\text{speed} = \left(9.533\frac{\text{m}}{\text{s}}\right)\left(\frac{1\text{ mi}}{1609\text{ m}}\right)\left(\frac{60\text{ s}}{1\text{ min}}\right)\left(\frac{60\text{ min}}{1\text{ hr}}\right) = 21.33\frac{\text{mi}}{\text{hr}}$$

SELF-TEST QUESTIONS

Having reviewed key terms in Chapter 1, match key terms with phrases and identify statements as true or false. If a statement is false, indicate why it is incorrect.

Match each phrase with the best term:

1.1 Increases in value with decreasing volume.

1.2 0.01200 contains four of these.

1.3 Heat emitted from burning wood is not an example of this.

1.4 The kg unit in 1.00 kg of silver tells us about this measurement.

1.5 This property does not change with amount of material.

1.6 This property changes with amount of material.

1.7 The temperature of water at 75 °C is an example.

1.8 The freezing of water is an example.

1.9 A chemical reaction is an example.

1.10 The ability of carbon to form carbon dioxide is an example.

1.11 Makes up the composition of a compound.

1.12 HF is an example.

1.13 A temperature scale with the divisions between the freezing point and melting point of water.

Terms:

(**a**) Celsius		(**h**)	Intensive
(**b**) Chemical change		(**i**)	Mass
(**c**) Chemical property		(**j**)	Matter
(**d**) Compound		(**k**)	Physical property
(**e**) Density		(**l**)	Physical change
(**f**) Elements		(**m**)	Significant figures
(**g**) Extensive			

True-False Statements:

1.14 When ice completely melts in a glass of water, there is a change from a heterogeneous *mixture* to a homogeneous one.

1.15 The *SI unit* for mass is the gram.

1.16 A *conversion factor* contains a ratio of units.

1.17 Coke is a *substance*.

1.18 −273.15 °C is equivalent to 0 K.

1.19 A student measures the mass of an object three times: 3.60 g, 3.90 g, and 3.75 g. The object actually has a mass of 3.75 g. The student's measurements show good *precision*.

1.20 In problem 1.19, the average value is 3.75 g. Therefore, the student's measurements gave good *accuracy*.

1.21 A *solution* is a homogeneous mixture of two or more substances.

1.22 The basic unit of mass in the *metric system* is the same as in the SI system of units.

1.23 A *chemical reaction* involves a change in the chemical composition of substances.

1.24 A *change of state* of $N_2(l)$ to $N_2(g)$ is a chemical change.

1.25 10 g of CO_2 *gas* occupies less space than 10 g of CO_2 solid in a one liter flask.

1.26 A *solid* consists of particles closer together than in the gas state.

1.27 A *liquid* is slightly compressible.

1.28 According to the *law of constant composition (definite proportions)*, H_2O and H_2O_2 represent the same substance.

Problems and Short-Answer Questions

1.29 Shown below is a 10 mL graduate cylinder. It is initially filled with 5.0 mL of water. 4.0 g of a solid object with a density of 2.0 g/mL is added. Which figure below best represents the volume of water in the graduate cylinder after the solid is added?

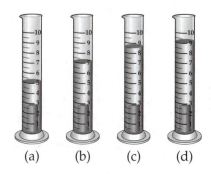

(a)   (b)   (c)   (d)

1.30 Characterize the following dartboard in terms of precision and accuracy of the results.

1.31   Identify the type of matter described by the following figure:

1.32   You calculate your income for last year as $42,125.00. However, when you file your income tax form, you report an income of $42,000.00. You do this because you want to report a rounded-off income figure to two significant figures. Why would an IRS agent be unhappy with your use of significant figures?

1.33   How many significant figures do the following numbers possess?

(a) 20.03 kg   (d) 10 dollar bills
(b) 1.90×10^3 L   (e) 0.00067 cm^3
(c) 120 m

1.34   Convert the following numbers to scientific notation form with the correct number of significant figures.

(a) 0.00067 cm^3 (d) 46900.0 g
(b) 210.0 m   (e) 200 rattlesnakes
(c) 0.040 L

1.35   Complete the following calculations and round off answers to the correct number of significant figures:

(a) $11.020 + 300.0 + 2.0030 =$
(b) $1211 + 2.205 - 1.70 =$
(c) $\dfrac{(1.425 \times 10^2)(2.61 \times 10^3)}{2.89 \times 10^5} =$
(d) $\dfrac{(0.012)(0.100)}{11.0265} =$

1.36   Make the following conversions.

(a) 1 ML to cubic centimeters
(b) $\dfrac{1}{1000}$ g to kg
(c) 8.00 qt to milliliters
(d) 16 lb to grams

1.37   Make the following temperature conversions.

(a) −70 °F in the Antarctic (it gets cold there) to °C
(b) 480 °C at the surface of Venus to °F
(c) 0.95 K, the freezing point of helium gas, to °C

1.38   Calculate the density of bromine, given that a 125.0 mL sample weighs 375.0 g.

1.39   A gold bar has the following dimensions: 2.50 cm × 2.00 cm × 1.50 cm. Assuming that gold can be sold for $400/oz, what is the value of the gold bar? The density of gold is 19.32 g/cm^3, and 1 lb = 453.6 g.

1.40   A column of mercury is contained in a cylindrical tube. This tube has a diameter of 8.0 mm, and the height of the mercury column is 1.20 m. Given that the density of mercury is 13.6 g/cm^3 and that the volume of mercury in the tube can be calculated from the relation $V = \pi r^2 h$ (r radius of the tube and h is the height of mercury column), calculate the mass of mercury present in the cylindrical tube.

1.41   Substances A, B, and C are all liquids and immiscible (do not mix) in each other. It is observed that A lies on top of C and C lies on top of B. Do you have sufficient information to determine the relative densities of A, B, and C? Explain.

1.42   The density of water at 3 °C is greater than its density at 0 °C. What does this information tell you about change in the volume of water from 3 °C to 0 °C? Are water molecules coming closer together or further apart? Explain.

Multiple-Choice Questions

1.43   Which of the following is not correct?

(a) There are 1000 mg in a gram.
(b) There are 100 cm in a meter.
(c) There are 1000 mL in a liter.
(d) There are 100 mm in a centimeter.
(e) There are 1000 μJ in a millijoule.

1.44   Which is the standard unit of volume in the SI system?

(a) meter   (d) milliter
(b) meter^3 (e) gallon
(c) liter

1.45   Which prefix means $\dfrac{1}{1,000,000}$ of a unit?

(a) kilo-   (d) micro-
(b) centi-   (e) nano-
(c) milli-

1.46   2.5 nm equals:

(a) 2.5×10^{-9} m   (d) (a) and (b)
(b) 2.5×10^{-4} mm   (e) (a) and (c)
(c) 2.5×10^{-7} cm

1.47   Which number has exactly four significant figures?

(a) 0.020   (d) 0.020
(b) 2000   (e) 2210
(c) 0.2000

1.48   When 1210.42 is rounded to three significant figures, it becomes:

(a) 1210   (d) 1210.
(b) 121   (e) 210.42
(c) 1000

1.49   Solid carbon dioxide, dry ice, changes directly from a solid to a vapor at 195 K if left in an open container. What is this temperature in degrees Celsius and Fahrenheit?

(a) −78 °C, 468 °F   (d) −108 °C, −78 °F
(b) −108 °C, 468 °F   (e) −78 °C, −108 °F
(c) 468 °C, −108 °F

1.50   Density can be thought of as a conversion factor. For example, the density of aluminum is 2.70 g/cm^3, which can be interpreted to mean 2.70 g of aluminum = 1 cm^3. Using this equivalence, what is the volume occupied by 223.5 g of aluminum?

(a) 603 cm^3 (d) 0.0270 cm^3
(b) 82.8 cm^3 (e) 223.5 cm^3
(c) 0.0121 cm^3

1.51 When a solid substance undergoes a physical change to a liquid, which of the following is always true?

(a) A new substance is formed.
(b) Heat is given off.
(c) A gas is given off.
(d) It vaporizes.
(e) It melts.

1.52 An empty container weighs 15.230 g. When filled with water (density = 1.00 g/mL), it weighs 35.920 g. When filled with an unknown liquid to the same mark as it was filled to with the water, it weighs 36.261 g. What is the density of the unknown liquid?

(a) 1.02 g/mL
(b) 1.02 g/m^3
(c) 1.20 g/mL
(d) 1.20 g/m^3
(e) none of the above

1.53 The maximum speed limit on many interstate highways is 70 mi/hr. How many kilometers can you travel in 4.5 hours at this speed?

(a) 75
(b) 150
(c) 245
(d) 390
(e) 500

1.54 Which is *not* a characteristic property of a compound useful in its identification?

(a) chemical formula
(b) density
(c) melting point temperature
(d) mass
(e) elemental composition

1.55 The copper content of a normal healthy human person is approximately 1.1×10^{-4} percent by mass. How many grams of copper would exist in a person weighing 1.00×10^3 lb (1.0 kg = 2.2 lb)

(a) 0.00050 g
(b) 0.050 g
(c) 0.50 g
(d) 5.0 g
(e) 50.0 g

1.56 A pure solid is heated and it decomposes into two substances, one a liquid and the other a gas. One can conclude with certainty that:

(a) The two products are elements.
(b) One of the two products is an element.
(c) The original solid is not an element.
(d) The liquid is a compound and the gas is an element.
(e) Both products are compounds.

1.57 When 125 mg, 1.2 dg and 1.2223 g are added, how many significant figures does the answer have?

(a) two
(b) three
(c) four
(d) five
(e) six

1.58 An intensive property of matter

(a) depends on the size of the sample.
(b) may depend on a ratio of two extensive properties.
(c) is a property that cannot be easily measured.
(d) does not depend on temperature.
(e) cannot be used to characterize matter.

1.59 A student determines the mass of silver in a sample and does four determinations. The results are 1.75 g, 1.71 g, 1.85 g, and 1.93 g. The true value is 1.81 g. Which statement concerning the results is correct?

(a) High precision and accurate results.
(b) High precision and poor accuracy.
(c) Poor precision and poor accuracy.
(d) Poor precision and accurate results.
(e) Reasonable precision and poor accuracy.

SELF-TEST SOLUTIONS

1.1 (e). **1.2** (m). **1.3** (j). **1.4** (i). **1.5** (h). **1.6** (g). **1.7** (k). **1.8** (l). **1.9** (b). **1.10** (c). **1.11** (f). **1.12** (d). **1.13** (a). **1.14** True.

1.15 False. It is the kilogram. **1.16** True. **1.17** False. It is a mixture of substances. **1.18** True. **1.19** False. If the student's measurements had shown good precision, the masses would have been closer in value, not showing a range of 0.30 g between low and high values, but a much smaller range. **1.20** True. **1.21** True. **1.22** False. In the metric system it is the gram, and in the SI system it is the kilogram. **1.23** True. **1.24** False. A physical change—its composition does not change. **1.25** False. A gas expands to fill the entire vessel holding it. A solid only occupies a limited and lesser amount of space. **1.26** True. **1.27** True. **1.28** False. In the case of water the ratio of hydrogen to oxygen is 2:1, whereas in the case of hydrogen peroxide (H_2O_2) the ratio is 2:2 or 1:1. Thus, they are different substances.

1.29 (b) V = m/d = 4.0 g/2.0 g/mL = 2.0 mL. The object will displace 2.0 mL of water. When added to the original 5.0 mL of water, the total volume becomes 7.0 mL.

1.30 The darts are close to one another; thus the precision is very good. However, the darts are far from the center; thus the throws result in dart positions that are not accurate.

1.31 The figure show two different phases, solid and liquid; thus, it is a heterogeneous mixture.

1.32 The IRS requires that all digits to the left of the decimal be significant when you report your income. You may round off the cents to the nearest dollar. Your actual income of $42,125.00 has five significant figures to the left of the decimal. Your reported income of $42,000.00 implies that the three zeros to the right of the 2 are significant and certain. The IRS agent is unhappy because you underreported your income and underpaid your taxes.

1.33 (a) four; (b) three; (c) two; (d) infinite, because this is an exact number; (e) two, because the zeros immediately to the right of the decimal are not significant.

1.34 (a) 0.00067 cm^3 = 6.7×10^{-4} cm^3;
(b) 210.0 = 2.100×10^2 m;
(c) 0.040 L = 4.0×10^{-2} L;
(d) 46900.0 g = 4.69000×10^4 g;
(e) 200 rattlesnakes = 2.00×10^2 rattlesnakes. Because this is an exact quantity, we need only write 2×10^2 rattlesnakes.

1.35 **(a)** $11.020 + 300.0 + 2.0030 = 313.0230$. The answer can have only one digit to the right of the decimal because 300.0 has the fewest number of digits to the right of the decimal—that is, one. The answer is rounded off to 313.0.

(b) $1211 + 2.205 - 1.70 = 1211.505$. The answer can have no digits to the right of the decimal because 1211 has no digits to the right of the decimal. The answer is rounded off to 1212.

(c) $\dfrac{(1.425 \times 10^2)(2.61 \times 10^3)}{2.89 \times 10^5} = 1.2869377$

The numbers with the fewest significant figures in the calculation are 2.61×10^3 and 2.89×10^5 with three significant figures. The final answer is rounded off to three significant figures: 1.29.

(d) First convert all numbers to exponential notation and then solve:

$$\frac{(0.012)(0.100)}{11.0265} = \frac{(1.2 \times 10^{-2})(1.00 \times 10^{-1})}{1.10265 \times 10^1}$$

$$= 1.08828 \times 10^{-4}$$

The number with the fewest significant figures in the calculation is 1.2×10^{-2}, with two significant figures. The final answer is rounded off to two significant figures: 1.1×10^{-4}.

1.36 **(a)** $(1\,\text{ML})\left(\dfrac{10^6\,\text{L}}{1\,\text{ML}}\right)\left(\dfrac{10^3\,\text{mL}}{1\,\text{L}}\right)\left(\dfrac{1\,\text{cm}^3}{1\,\text{mL}}\right) = 1 \times 10^9\,\text{cm}^3$

(b) $\left(\dfrac{1}{1000}\,\text{g}\right)\left(\dfrac{1\,\text{kg}}{1000\,\text{g}}\right)$

$= (1 \times 10^{-3}\,\text{g})\left(\dfrac{1 \times 10^{-3}\,\text{kg}}{1\,\text{g}}\right) = 1 \times 10^{-6}\,\text{kg}$

(c) $(8.00\,\text{qt})\left(\dfrac{1\,\text{L}}{1.057\,\text{qt}}\right)\left(\dfrac{1000\,\text{mL}}{1\,\text{L}}\right) = 7.57 \times 10^3\,\text{mL}$

(d) $(16\,\text{lb})\left(\dfrac{453.6\,\text{g}}{\text{lb}}\right) = 7.3 \times 10^3\,\text{g}$

1.37 **(a)** $^\circ\text{C} = (-70\,^\circ\text{F} + 40\,^\circ)\left(\dfrac{1\,^\circ\text{C}}{1.8\,^\circ\text{F}}\right) - 40\,^\circ = -57\,^\circ\text{C}$

(b) $^\circ\text{F} = (480\,^\circ\text{C} + 40\,^\circ)\left(\dfrac{1.8\,^\circ\text{F}}{1\,^\circ\text{C}}\right) - 40\,^\circ = 896\,^\circ\text{F}$

(c) $\text{K} = \left(\dfrac{1\,\text{K}}{1\,^\circ\text{C}}\right)(^\circ\text{C}) + 273.15\,\text{K}$ or

$^\circ\text{C} = \left(\dfrac{1\,^\circ\text{C}}{1\,\text{K}}\right)(\text{K} - 273.15\,\text{K})$

$= \left(\dfrac{1\,^\circ\text{C}}{1\,\text{K}}\right)(0.95\,\text{K} - 273.15\,\text{K}) = -272.20\,^\circ\text{C}$

1.38 Density of bromine = mass/volume = $375.0\,g/125.0\,\text{mL} = 3.000\,\text{g/mL}$.

1.39 The mass of the gold bar is calculated from its density using the relation mass = density × volume. The volume of the gold bar equals the product of its length times width times height: $V = (2.50\,\text{cm})(2.00\,\text{cm})$ $(1.50\,\text{cm}) = 7.50\,\text{cm}^3$. Calculating the mass of the gold bar

with this volume: $m = \text{density} \times \text{volume} = (19.32\,\text{g/cm}^3)$ $(7.50\,\text{cm}^3) = 145\,\text{g}$. The value of the gold bar is $\$400/\text{oz} \times 16\,\text{oz/lb} \times 1\,\text{lb}/453.6\,\text{g} \times 145\,\text{g} = \$2046$. (Note: The gold dealer normally does not adhere to significant figure rules!)

1.40 Because the density is expressed in units of cm^3, it is convenient to change all length measurements into that unit. For this problem, the required equivalences are $10\,\text{mm} = 1\,\text{cm}$ and $100\,\text{cm} = 1\,\text{m}$. Calculating the volume of the tube the mercury occupies in the unit of cm^3:

$$V = \pi r^2 h = \pi \left(\frac{8.00\,\text{mm}}{2}\right)^2 \left(\frac{1\,\text{cm}}{10\,\text{mm}}\right)^2$$

$$\times (1.20\,\text{m})\left(\frac{100\,\text{cm}}{1\,\text{m}}\right)$$

$$= 60.3\,\text{cm}^3$$

Mass of mercury = density × volume

$$= \left(13.6\,\frac{\text{g}}{\text{cm}^3}\right)(60.3\,\text{cm}^3)$$

$$= 820\,\text{g}$$

1.41 The information tells you that C is more dense than A as A lies on top of C. Furthermore you know that B is more dense than C as C lies on top of B. If density(A) < density(C) and density(C) < density(B), it must also be true that density(A) < density(B). Thus, the problem provides sufficient information to conclude that the trend in relative densities is density(A) < density(B) < density(C).

1.42 Density = mass/volume   or   volume = mass/density. If the density of water decreases, then the volume increases because mass is divided by a smaller number as the temperature decreases. An increase in volume suggests that the molecules must be further apart from one another as the temperature decreases. If they were more closely associated there would be less space between molecules and the volume would decrease.

1.43 **(d)** $100\,\text{mm} = (100\,\text{mm})(1\,\text{m}/1000\,\text{mm})$ $= 1/10\,\text{m};\ 1\,\text{cm} = 1/100\,\text{m}$. Thus the two quantities are not equal.

1.44 **(b)**

1.45 **(d)**

1.46 **(e)**

$$(2.5\,\text{nm})\left(\frac{10^{-9}\,\text{m}}{1\,\text{nm}}\right) = 2.5 \times 10^{-9}\,\text{m}$$

$$(2.5\,\text{nm})\left(\frac{10^{-9}\,\text{m}}{1\,\text{nm}}\right)\left(\frac{100\,\text{cm}}{1\,\text{m}}\right) = 250 \times 10^{-9}\,\text{cm}$$

$$= 2.5 \times 10^{-7}\,\text{cm}$$

1.47 **(c)** The zeroes after the two in 0.2000 are significant because they do not define the decimal point: 2.000×10^{-1}.

1.48 **(a)** The number of significant figures is counted from the left and trailing zeroes in a number without a decimal point are considered not significant.

1.49 **(e)**

1.50 **(b)** $(223.5 \text{ g})(1 \text{ cm}^3/2.70 \text{ g}) = 82.8 \text{ cm}^3$

1.51 **(e)**

1.52 **(a)** Mass of water = 35.920 g – 15.230 g = 20.690 g; mass of liquid = 36.261 g – 15.230 g = 21.031 g; volume occupied by liquid = (mass water) (1.00 mL/1 g) = 20.7 mL; density of liquid = m/V = 21.031 g/20.7 mL = 1.02 g/mL.

1.53 **(d)** $\left(70 \, \dfrac{\text{mi}}{\text{hr}}\right)(3.5 \text{ hr})\left(1.609 \, \dfrac{\text{km}}{\text{mi}}\right)$
$= 390 \text{ km (rounded to two significant figures)}$

1.54 **(d)** A compound has a specific formula and elemental composition, density, and melting point. Its mass depends on the amount present and is not useful for identification.

1.55 **(c)**
$(1.00 \times 10^3 \text{ lb})\left(\dfrac{1000 \text{ g}}{2.2 \text{ lb}}\right)(1.1 \times 10^{-4}/100) = 0.50 \text{ g}$

1.56 **(c)** The original solid is pure and it must be a compound because it can be chemically decomposed into simpler substances. You are not given specific information about the composition of the products and thus you cannot make any definite conclusions about them.

1.57 **(b)** First convert all numbers to the same units: 125 mg is 0.125 g; 1.2 dg is 0.12 g; and 1.2223 g is unchanged. The sum of these numbers is 1.4673 g. The final answer must be rounded to two digits to the right of the decimal: 1.47 g. Thus, the answer has three significant figures.

1.58 **(b)** Density is an example of an intensive property that depends on the ratio of two extensive properties, mass and volume. An intensive property does not depend on the amount of substance present.

1.59 **(d)** The average of the measurements is 1.81 g, which is the same as the true value. Therefore, the accuracy is high. The range of measured values is 1.75 g – 1.93 g, or a difference of 0.18 g, or 10% of the average value, which is poor precision. Precision is a measure of how close the four masses are to each other.

Chapter

2

Atoms, Molecules, and Ions

OVERVIEW OF THE CHAPTER

2.1, 2.2, 2.3 ATOMS

Learning Goals: You should be able to:

1. Describe the composition of an atom in terms of protons, neutrons, and electrons.
2. Give the approximate size, relative mass, and charge of an atom, proton, neutron, and electron.
3. Write the chemical symbol for an element, having been given its mass number and atomic number, and perform the reverse operation.
4. Describe the properties of the electron as seen in cathode rays. Describe the means by which J.J. Thomson determined the ratio e/m for the electron.
5. Describe Millikan's oil-drop experiment and indicate what property of the electron he was able to measure.
6. Cite the evidence from studies of radioactivity for the existence of subatomic particles.
7. Describe the experimental evidence for the nuclear nature of the atom.

2.4, 2.5 PERIODIC TABLE: ARRANGEMENT OF ATOMS AND ATOMIC WEIGHTS

Learning Goals: You should be able to:

1. Use the unit of atomic mass unit (amu) in calculation of masses of atoms.
2. Define the term atomic weight and calculate the atomic weight of an element given its natural distribution of isotopes and isotopic masses.
3. Use the periodic table to determine the atomic number, atomic symbol, and atomic weight of an element.
4. Define the terms group and period and recognize the common groups of elements.
5. Use the periodic table to predict whether an element is metallic, nonmetallic, or metalloid.

2.6, 2.7 MOLECULES, MOLECULAR COMPOUNDS, IONIC COMPOUNDS, AND IONS

Learning Goals: You should be able to:

1. Define the term molecule and recognize which elements typically combine to form molecules.
2. Distinguish between empirical and molecular formulas.
3. Draw the structural and ball-and-stick formulas of a substance given its chemical formula and the linkage between atoms.

4. Use the periodic table to predict the charges of monatomic ions of non-transition elements.
5. Write the symbol and charge for an atom or ion having been given the number of protons, neutrons, and electrons and perform the reverse operation.
6. Determine whether a substance is likely to be ionic or molecular.
7. Write the simplest formula of an ionic compound having been given the charges of ions from which it is made.

Learning Goals: You should be able to:

1. Write the name of a simple inorganic compound having been given its chemical formula and perform the reverse reaction.
2. Write and name common polyatomic ions.
3. Write and name acids based on anions whose names end in -ide, -ate, and -ite.
4. Write the name of simple binary molecular compounds and perform the reverse operation.
5. Define the terms hydrocarbon, alkane, and alcohol and be able to name simple alkanes and alcohols, having been given the chemical formula, and perform the reverse operation.

2.8, 2.9 NOMENCLATURE: SIMPLE INORGANIC, MOLECULAR, AND ORGANIC COMPOUNDS

TOPIC SUMMARIES AND EXERCISES

This chapter focuses on the structure of atoms, elements in the periodic table, how atoms combine to form substances, and naming inorganic compounds. The work of many scientists, such as John Dalton, J.J. Thompson, and R. Millikan, provided experimental information for development of a modern understanding of the structure of atoms. You should know the contributions of the key historical figures discussed in this chapter: See Exercise 2.

ATOMS

Atoms are the smallest particles of an element that have all the same chemical properties of that element. Some general characteristics about the structure of atoms are:

- The mass of a single atom is extremely small, less than 10^{-21} g.
- The center of an atom contains a small **nucleus**. A nucleus contains **protons** and **neutrons**, which make up most of the mass of an atom.
- The volume of an atom mostly consists of **electrons**.
- Protons (positively charged) and neutrons (no charge) have essentially the same mass. Electrons (negatively charged) are significantly less massive.

All atoms of the same element have the same number of protons.

- Atoms that have the same number of protons but differ in their number of neutrons are called **isotopes**.
- The general symbol for an isotope is

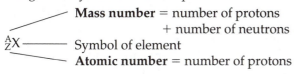

Mass number = number of protons
+ number of neutrons
$^A_Z X$ —— Symbol of element
Atomic number = number of protons

Note: The atomic number Z defines an element; it also tells you how many electrons a neutral atom of that element possesses. Thus, the number of neutrons equals A - Z.

EXERCISE 1   Writing isotopic symbols

Write the nuclear-isotope symbols for the four isotopes of sulfur with 16, 17, 18, and 20 neutrons, respectively.

SOLUTION: The atomic number of sulfur is 16, and all of its isotopes will have 16 protons. The mass number of each isotope is the sum of its number of neutrons plus its number of protons: $16 + 16 = 32$; $16 + 17 = 33$; $16 + 18 = 34$; and $16 + 20 = 36$. The nuclear-isotope symbols are $^{32}_{16}S$, $^{33}_{16}S$, $^{34}_{16}S$, and $^{36}_{16}S$. Note that the subscript before each elemental symbol (that is, the atomic number) is the same for all isotopes because the number of protons is invariant for sulfur.

EXERCISE 2   Developments in the history of atomic structure

The composition of the atom was determined from the experiments of many individuals. For each experiment, state the principal person associated with the experiment, the particle studied, and the property of the particle determined: (**a**) α scattering; (**b**) cathode ray; (**c**) oil drop.

SOLUTION: (**a**) E. Rutherford determined the relative volume and mass of the nucleus in the atom by observing the scattering of α particles. (**b**) J. J. Thomson determined the e/m ratio of the electron by using a cathode-ray tube. (**c**) R. Millikan determined the charge of an electron by observing the motion of charged oil drops in an electric field.

EXERCISE 3   Using isotopic symbols to determine nuclear composition

Find the number of protons, electrons, and neutrons in the following isotopes:

$$(\text{a}) \; ^{40}_{20}Ca; \quad (\text{b}) \; ^{238}_{92}U.$$

SOLUTION: (**a**) The number of protons is the subscript number shown in the isotopic symbol: 20. The number of electrons in a neutral atom is the same as the number of protons: 20. The number of neutrons equals the mass number minus the number of protons. This is the superscript number (mass number) minus the subscript number: $40 - 20 = 20$. (**b**) Preceding as in (**a**), the number of protons is 92, the number of electrons is 92, and the number of neutrons is $238 - 92 = 146$.

EXERCISE 4   Changing the number of electrons in an isotope

If an electron is added or removed from ^{35}Cl, what changes occur to the isotope?

SOLUTION: The isotope is still ^{35}Cl because the number of protons and the number of neutrons do not change; however, the number of electrons does. If electrons, which are negatively charged, are added to an isotope, the isotope gains negative charge. Therefore if one electron is added to ^{35}Cl, it becomes $^{35}Cl^-$. Similarly, if an electron is removed, the isotope loses negative charge and becomes positively charged because there are more protons than electrons: $^{35}Cl^+$.

PERIODIC TABLE: ARRANGEMENT OF ATOMS AND ATOMIC WEIGHTS

The masses of isotopes are measured relative to the atomic mass of $^{12}_{6}C$, which is defined to have a mass of exactly 12 **atomic mass units** (amu). A modern mass spectrometer is used to determine accurate relative masses.

- Except for a few elements, most elements have a distribution of naturally occurring isotopes. The concept of atomic weight is based on determining the average of the masses of naturally occurring isotopes weighted according to their abundances:

  atomic weight = % abundance × mass of isotope 1

  $\qquad$ + % abundance × mass of isotope 2 + $\cdots$

  $\qquad$ + % abundance × mass of isotope n.

- The relationship between amu and grams is 1.66054×10^{-24} g = 1 amu.

The periodic table lists elements in order of increasing atomic number. It is very useful in helping us to organize trends in the chemical and physical properties of elements. The manner in which elements are grouped in vertical columns (families) and horizontal rows (periods) helps you to remember that

- *Metals* are on the left side and middle (except hydrogen).
- *Nonmetals* are on the right side.
- *Metalloids* have properties of both metals and nonmetals. They are identified by the dark diagonal line toward the right side of the periodic table.
- Elements in vertical columns, called *families* or *groups*, exhibit similar chemical and physical properties, whereas elements in a horizontal row (*period*) exhibit different properties. The names of groups in the periodic table and the general location of metals, metalloids, and nonmetals are shown in Figure 2.1.

EXERCISE 5   Understanding the concept of average atomic weight

(a) What is the sum of percentages of naturally occurring isotopes for an element?
(b) The following problem shows the concept of atomic weights applied to a different situation. What is the average weight of a class of students given the following information about the class: (1) The average woman weighs 122 pounds; (2) the average man weighs 165 pounds; and (3) the percentage of men in the class is 45.0%?

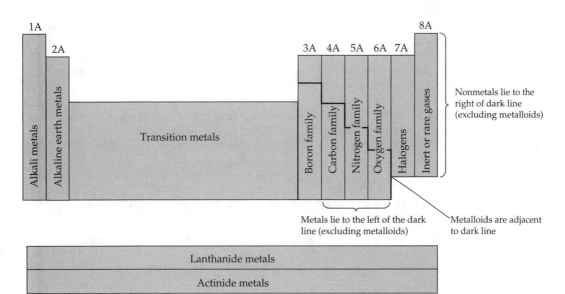

▲ **FIGURE 2.1** Families in the periodic table. Note that hydrogen, the first element in family 1A, is not an alkali metal.

SOLUTION: (a) The sum of percentages must add to 100%; however, in experimental situations it rarely adds exactly to this value because of experimental error. (b) The average atomic weight of the class is calculated from the relation:

$$\text{average weight} = \% \text{ men} \times \text{average mass of men}$$
$$+ \% \text{ women} \times \text{averagemass of women}$$

We are given the percentage of men but not that of women. We can calculate the percentage of women from relation:

$$100.0\% = \% \text{ men} + \% \text{ women} = 45.0\% + \% \text{ women}$$

$$\% \text{ women} = 100\% - 45.0\% = 55.0\%$$

Therefore, the average weight of the class is:

$$\text{average weight} = (0.450)(165 \text{ pounds}) + (0.550)(122 \text{ pounds}) = 141 \text{ pounds}$$

EXERCISE 6   Calculating the atomic weight of an element

What is the atomic weight of antimony if it has only two naturally occurring isotopes, Sb-121 with an isotopic mass of 120.904 amu and an abundance of 57.21% and Sb-123 with an isotopic mass of 122.904 amu and an abundance of 42.79%?

SOLUTION: We can use the concept of average atomic mass discussed in the previous exercise to solve for the atomic weight of antimony:

$$\text{Atomic weight of Sb} = 120.904 \text{ amu}(0.5721) + 122.904 \text{ amu}(0.4279)$$

$$= 69.17 \text{ amu} + 52.59 \text{ amu}$$

$$= 121.76 \text{ amu}$$

EXERCISE 7   Using a periodic table

The position of an element in the periodic table is a function of its atomic composition. Elements are arranged in order of increasing atomic number. The atomic number of an atom equals its number of protons. By referring to a periodic table, determine which *neutral* atom: (a) contains 50 protons; (b) contains 17 electrons; (c) has an atomic number of 56.

SOLUTION: (a) Tin (Sn). Its atomic number, 50, corresponds to 50 protons. (b) Chlorine (Cl). Its atomic number corresponds to 17 protons. This is also the number of electrons in the neutral atom. (c) Barium (Ba). Its atomic number is 56.

EXERCISE 8   Identifying the family of an element in the periodic table

What is the name of the family in which the following elements occur: I, Na, S, and Ca?

SOLUTION: By referring to Figure 2.1 and a periodic table we find the requested information: I, halogen; Na, alkali metal; S, oxygen family; and Ca, alkaline-earth metal.

Molecular compounds consist of small particles called molecules.

- A **molecule** is a small particle consisting of two or more atoms combined together in a discrete unit. Molecules have their own chemical and physical properties, which differ from the properties of the elements forming them.
- Molecules typically consist of nonmetallic elements.
- You should know the seven elements that form homonuclear diatomic molecules: H_2, N_2, O_2, F_2, Cl_2, Br_2, and I_2.
- Subscripts in a **molecular formula** tell you how many atoms are actually present

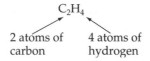

$$C_2H_4$$

2 atoms of carbon   4 atoms of hydrogen

- An **empirical formula** shows only the simplest whole number ratio of atoms. For example, C_2H_4 is a molecular formula; its empirical formula, CH_2, is obtained by dividing the subscripts in the molecular formula by 2: $CH_2 = C_{2/2}H_{4/2}$
- A **structural formula** shows "how atoms are joined together"—that is, the relative orientation and position of bonded atoms. For example, the structural formula of C_2H_6 shows the following arrangement of atoms:

An **ionic compound** consists of positively charged ions (cations) and negatively charged ions (anions).

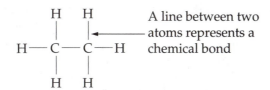

A line between two atoms represents a chemical bond

- Ionic compounds typically contain a metal and a nonmetal.
- **Cations** are positively charged ions of metallic elements formed by the loss of electrons; e.g., Na^+ is formed by Na losing one electron.
- **Anions** are negatively charged ions of nonmetallic elements formed by the addition of electrons; e.g., S^{2-} is formed by S gaining two electrons.
- Ionic compounds do not contain unique identifiable pairs of ions. Therefore, empirical formulas are used to describe their composition.
- Ionic compounds may contain polyatomic ions such as NO_3^- as in $NaNO_3$. Polyatomic ions are like molecules except that they carry a charge.
- You should know the most common charge carried by elements in the following families:

Alkali metals	Alkaline-earth metals	Transition metals	Halogens	Oxygen family
1+	2+	2+, 3+	1−	2−

EXERCISE 9   Interpreting chemical formulas

Interpret the following chemical formulas: (a) N_2; (b) NH_3; (c) NH_4^+; (d) H_2SO_4.

SOLUTION: (a) N_2 is a homonuclear (having atoms of the same type) molecule composed of two nitrogen atoms; this is the elemental form of nitrogen. (b) NH_3 is a heteronuclear (having more than one type of atom) molecule composed of one nitrogen atom and three hydrogen atoms. (c) NH_4^+ is an ion that has a 1+ charge and is composed of one nitrogen atom and four hydrogen atoms. (d) H_2SO_4 is a heteronuclear molecule composed of two hydrogen atoms, one sulfur atom, and four oxygen atoms.

EXERCISE 10   Writing empirical and chemical formulas

(a) Do C_2H_4 and 2 CH_2 represent the same substance? (b) What are the empirical formulas for B_2H_6, H_2O, and $C_2H_2O_4$?

SOLUTION: (a) The formulas do not represent the same substance. C_2H_4 represents the molecular formula for the substance acetylene, a gas used in welding. 2 CH_2 represents two empirical formulas of acetylene. (b) To determine an empirical formula, divide each subscript of a molecular formula by the largest whole number that goes into each subscript (the greatest common denominator)

$$B_{2/2}H_{6/2} = BH_3$$

There is no way to further reduce the subscripts in H_2O—this is the molecular and empirical formula for water.

$$C_{2/2}H_{2/2}O_{4/2} = CHO_2$$

EXERCISE 11   Writing chemical formulas of ionic substances

Write the formula for: (a) a neutral polyatomic compound consisting of one barium ion with a 2+ charge (written Ba^{2+}) and chlorine ions with a 1− charge (written Cl^-). (b) A polyatomic ion that has a 1− charge and consists of one boron ion with a 3+ charge (written B^{3+}) and fluorine ions with a 1− charge (written F^-).

SOLUTION: (a) The sum of positive and negative charges for the ions must equal zero in a neutral compound. Two Cl^- ions are required to balance the 2+ charge of Ba^{2+}: $BaCl_2$. (b) The sum of positive and negative charges for the ions must equal the 1− charge of the compound. Four F^- ions are required so that, when they are combined with B^{3+}, the sum of the charges is 1−: $BF_4^-[+3 + 4(-1) = -1]$.

EXERCISE 12   Determining the number of electrons possessed by a molecule or ion

Determine the total number of electrons in each of the following: (a) NH_3; (b) NH_4^+; (c) NH_2^-.

SOLUTION: Determine the number of electrons for each atom in each chemical formula, sum these numbers, and then adjust the number for the charge shown. For each negative charge, one electron is added; for each positive charge, one electron is subtracted. (a) The number of electrons for each atom is the same as the number of protons. Each hydrogen atom has one proton and each nitrogen atom has 7 protons. Thus, for NH_3 the number of electrons is $(1 \times 7) + (3 \times 1) = 10$ electrons. The charge of the compound is zero, thus no adjustment for charge is necessary. (b) The number of electrons for NH_4^+ with no charge is $(1 \times 7) + (4 \times 1) = 11$ electrons.

However, the species has a positive charge, thus one electron must be removed so that there are more protons than electrons: $11 - 1 = 10$ electrons for NH_4^+. (c) The number of electrons for NH_2^- with no charge is $(1 \times 7) = (2 \times 1) = 9$ electrons. The species has a charge of $1-$, thus one electron must be added so that there are more electrons than protons: $9 + 1 = 10$ electrons.

EXERCISE 13   Writing binary chemical formulas

Write the expected formula when the following elements combine to form compounds: (a) Al and O; (b) B and Cl; (c) Ca and F; (d) Na and I; (e) S and O; and (f) H and Ca.

SOLUTION: (a) Al forms the 3+ state in salts. Oxygen forms the oxide ion, O^{2-}: Al_2O_3. (b) Boron is a member of the family containing aluminum; thus, you should expect it to behave in a similar manner to aluminum. BCl_3. (c) Calcium forms Ca^{2+} and flourine forms F^-: CaF_2. (d) Na forms Na^+ and iodine in the presence of metals forms I^-: NaI. (e) Sulfur and oxygen, both nonmetals, form nonmetal oxides: For example, $SO_2(g)$ and $SO_3(g)$. (f) Ca is an active metal. In the presence of an active metal hydrogen forms the hydride ion, H^-: CaH_2.

Chemists not only write chemical formulas for compounds but also give them reasonably systematic names. Rules for naming simple inorganic and binary molecular compounds are found in Section 2.8 in the text. *These rules should be committed to memory; and you should then practice them by naming anions, cations, ionic compounds, oxyanions, acids, and simple binary molecular compounds.* Attempt Exercises 13–19 *after* you have reviewed nomenclature rules. The solutions to many of the exercises provide explanations which reinforce what you have learned.

In Section 2.9 the topic of organic chemistry is introduced with a focus on hydrocarbons and the alcohol functional group. *Hydrocarbons* are compounds containing hydrogen and carbon atoms.

NOMENCLATURE: SIMPLE INORGANIC, MOLECULAR, AND ORGANIC COMPOUNDS

* One class of hydrocarbons is called an *alkane*, in which each carbon atom is bonded to four other atoms; all bonds are single. You need to learn the prefixes for alkanes in Table 2.6 in the text.
* A *functional group* is a nonhydrogen atom or group of atoms that replaces a hydrogen atom in a hydrocarbon. *Alcohols* contain the – OH functional group and their names end in *-ol*.

EXERCISE 14   Correcting chemical formulas when they are written incorrectly

The following chemical formulas are written incorrectly. Correct them so that they are in the proper form: (a) ClNa; (b) CaClCl; (c) $NH_4NH_4SO_4$; (d) H^2S; (e) FXeF.

SOLUTION: (a) NaCl. The cation (Na^+) is always placed before the anion (Cl^-). (b) $CaCl_2$. A subscript is used to indicate the number of Cl^- ions present. (c) $(NH_4)_2SO_4$. If there is more than one type of polyatomic ion present, its formula is enclosed in parentheses, and subscripts are used as necessary. (d) H_2S. The 2 must be a subscript when it indicates the number of the same type of atoms or ions present. (e) XeF_2.

EXERCISE 15   Naming cations

Name the following cations: (a) H^+; (b) Na^+; (c) Be^{2+}; (d) Al^{3+}; (e) Co^{2+}; (f) Co^{3+}.

SOLUTION: Monatomic cations have the same names as the elements from which they are formed. Some monatomic cations commonly exist in two common ion states; for example, Fe^{2+} and Fe^{3+}. When naming such ions, indicate the charge of the ion in Roman numerals enclosed within parentheses after the name of the element. Sometimes an older method is used, in which the suffix -*ic* indicates the higher charge and -*ous* the lower one. This method is further complicated by the fact that the Latin name of the element is often used. Thus Fe^{2+} is iron(II) or ferrous, and Fe^{3+} is iron(III) or ferric. When in doubt, use the newer method. (**a**) Hydrogen. (**b**) Sodium. (**c**) Beryllium. (**d**) Aluminum. (**e**) Cobalt(II) or cobaltous. (**f**) Cobalt(III) or cobaltic.

EXERCISE 16   Naming anions

Name the following anions: (**a**) F^-; (**b**) S^{2-}; (**c**) OH^-; (**d**) CN^-; (**e**) PO_4^{3-}; (**f**) PO_3^{3-}; (**g**) IO^-; (**h**) IO_2^-; (**i**) IO_3^-; (**j**) IO_4^-; (**k**) HS^-; (**l**) $H_2PO_4^-$.

SOLUTION: (**a**) Fluoride. Monatomic anions have -*ide* added to the stem of the element's name. (**b**) Sulfide. (**c**) Hydroxide. Treated as a monatomic anion. (**d**) Cyanide. Treated as a monatomic anion. (**e**) Phosphate. When two oxyanions exist, as in the cases of PO_4^{3-} and PO_3^{3-}, the one with more oxygen atoms ends in -*ate*. (**f**) Phosphite. The one with fewer oxygen atoms ends in -*ite*. (**g**) Hypoiodite. The prefix *hypo-* indicates one less oxygen atom than in IO_2^-. (**h**) Iodite. (**i**) Iodate. (**j**) Periodate. The prefix *per-* indicates one more oxygen atom than in IO_3^-. (**k**) Hydrogen sulfide. *Hydrogen* is added to the name of the anion to indicate that there is one hydrogen atom present. (**l**) Dihydrogen phosphate. *Dihydrogen* is added to phosphate to indicate that there are two hydrogen atoms present.

EXERCISE 17   Naming ionic and molecular compounds

Name the following compounds: (**a**) ZnS; (**b**) $(NH_4)_2SO_4$; (**c**) $FeCl_2$; (**d**) $KClO_4$; (**e**) $SnCl_2$; (**f**) PCl_3; (**g**) SF_6; (**h**) CO.

SOLUTION: Remember: When naming ionic compounds cations are named before anions; in molecular compounds the atoms are named in the order found in the molecular formula. For binary molecular compounds, the first atom is given its elemental name, and the second is named as if it were a negative ion. (**a**) Zinc sulfide. Zinc exists only in the 2+ state. (**b**) Ammonium sulfate. The use of *di-* before *ammonium* to indicate two NH_4^+ ions is not necessary because the −2 charge of SO_4^{2-} automatically requires two NH_4^+ ions. (**c**) Iron(II) chloride or, using the older method, ferrous chloride. (**d**) Potassium perchlorate. (**e**) Tin(II) chloride or stannous chloride. (**f**) Phosphorus trichloride. A compound formed from two nonmetallic elements is named as if it were an ionic compound, with the appropriate prefixes added to indicate how many atoms are present. (**g**) Sulfur hexafluoride. The prefix *hexa-* indicates six fluorine atoms. (**h**) Carbon monoxide.

EXERCISE 18   Naming acids

Name the following acids in water: (**a**) HCN; (**b**) H_2S; (**c**) H_2CO_3; (**d**) H_3PO_4.

SOLUTION: (**a**) Hydrocyanic acid. For hydrogen acids formed from monatomic anions, *hydro-* is added to the name of the anion—in this case, *cyanide*—and the -*ide* ending of the anion is changed to an -*ic* ending. (**b**) Hydrosulfuric acid. Again, the prefix *hydro-* indicates a hydrogen acid, and -*ic* is added to the end of sulfur. (**c**) Carbonic acid. When an acid is derived from a polyatomic anion whose name ends in -*ate*—*carbonate* (CO_3^{2-}) in this case—the ending -*ic* is added to the name of the central atom of the polyatomic anion, and no prefix is used. (**d**) Phosphoric acid. Its name is derived from the polyatomic anion *phosphate* (PO_4^{3-}). The -*ate* ending is dropped and is changed to -*oric*. The -*ic* ending for acids is associated with acids formed from anions ending in -*ate*.

EXERCISE 19 Naming molecular compounds

Name the following substances: (a) SF_6; (b) NO; (c) N_2O; (d) P_2O_5.

SOLUTION: We will use the prefixes in Table 2.6 of the text to name the substances. (a) sulfur hexaflouride; (b) nitrogen monoxide; (c) dinitrogen monoxide; (d) diphosphorus pentaoxide.

EXERCISE 20 Writing names and chemical formulas of alkanes

Consider the alkane called butane. (a) Assuming the carbon atoms are in a straight chain, write a structural formula for butane. (b) What is its molecular formula? (c) If one of the hydrogen atoms attached to a carbon atom at the end of the chain is replaced with – OH, what is its name?

SOLUTION: (a) The ending *-ane* in propane tells you that it is an alkane. Referring to Table 2.6 in the text we read that the prefix *but-* means four carbon atoms. We can write

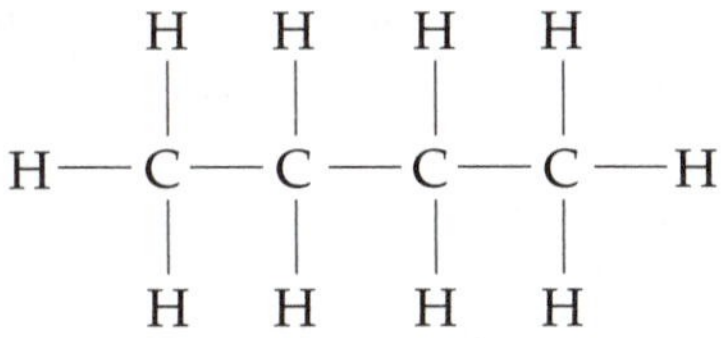

(b) With the structural formula written we can determine the molecular formula by counting the number of carbon and hydrogen atoms: C_4H_{10}. (c) If one of the hydrogen atoms at the end of the carbon chain is replaced by – OH, an alcohol functional group, the following alcohol is produced: Butanol.

SELF-TEST QUESTIONS

Key Terms

Having reviewed key terms in Chapter 2, match key terms with phrases and identify statements as true or false. If a statement is false, indicate why it is incorrect.

Match each phrase with the best term:

2.1 A general name for a positive ion.
2.2 A sulfide ion is an example of this species.
2.3 The least massive particle in an atom as discussed in the chapter.
2.4 A particle possessing no charge in the nucleus.
2.5 Helium atom with a +2 charge.
2.6 $^{85}_{37}Rb$ and $^{87}_{37}Rb$ are examples.
2.7 The number 85 in $^{85}_{37}Rb$.
2.8 The number 37 in $^{85}_{37}Rb$.
2.9 HCl is an example of this species.
2.10 Occupies a small volume of an atom.
2.11 NO_3^- is an anion and also called this type of particle.
2.12 Shows smallest whole-number ratio of atoms.
2.13 Shows arrangement of atoms and bonds.
2.14 Name of any type of charged atom.
2.15 H_2O_2 is an example of this type of chemical formula.
2.16 Contains a beam of electrons.
2.17 Contains two atoms bonded together which act as a unit.
2.18 Emits gamma rays.
2.19 Particles of an element having the same atomic number.

Terms:

(a) alpha particle
(b) anion
(c) atomic number
(d) atoms
(e) cation
(f) cathode ray
(g) diatomic molecule
(h) electron
(i) empirical formula
(j) ion
(k) isotopes
(l) mass number
(m) molecule
(n) molecular formula
(o) neutron
(p) nucleus
(q) polyatomic
(r) radioactivity
(s) structural formula

True-False Statements

2.20   If a substance spontaneously emits alpha particles, we say it is *radioactive*.

2.21   The *atomic number* of a specific element is always constant.

2.22   The phrase "cobalt-60" means an isotope of cobalt that has a *mass number* of 60.

2.23   One *Angstrom* is equivalent to 10^{-8} m.

2.24   Elements in the *periodic table* are sequentially arranged in order of increasing mass number.

2.25   One *electronic charge* is equivalent to one coulomb.

2.26   He, Ne, and Ar are all members of a *family* of elements.

2.27   An example of the *law of multiple proportions* is the existence of NO_2 and NO.

2.28   Hg is a *nonmetallic* element.

2.29   Sr is a *metallic* element.

2.30   Antimony is a *metalloid*.

2.31   To specify a *nuclide*, only its atomic number is needed.

2.32   A *structural formula* of a substance shows the simplest ratio of elements in the chemical formula.

2.33   An important aspect of a *ball-and-stick model* is the showing of approximate bond angles.

2.34   NO is an *ionic* substance.

2.35   KI contains a *cation*.

2.36   The charge of the *anion* in $CaCl_2$ is $1-$.

2.37   An example of a *polyatomic ion* is the bromide ion in $AlBr_3$.

2.38   Methane, CH_4, is a *hydrocarbon*.

2.39   An *alkane* contains a carbon–carbon double bond.

2.40   Ethanol is an *alcohol* with two carbon atoms and an –OH group.

Problems and Short-Answer Questions

2.41   Which element in the following periodic table is most likely to form an anion?

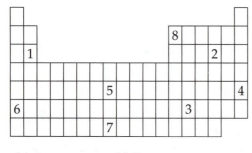

(a) 1	(c) 3
(b) 2	(d) 4

2.42   Which element in the previous periodic table has the largest atomic weight?

(a) 5	(c) 7
(b) 6	(d) 8

2.43   John Dalton's model for the atom includes the postulate that all atoms of a given element are identical. Contrast this postulate to what we know about atoms today. Give at least one example to support your statements.

2.44   What are the three primary particles that comprise an atom of an element? How do they differ in charge and mass? What is the approximate range of diameters for atoms?

2.45   What does the symbol $^{201}_{80}X$ tell you about the isotope? From this information can you identify the element X?

2.46   Various columns in the periodic table are given names. Which groups are given the following names: Alkali metals, halogen, noble gases, and coinage metals?

2.47   What is a molecule? Can an element form a molecule or can molecules only form from combinations of different elements? What are seven important diatomic elements?

2.48   In the ensuing chapters you will often encounter ions. Does an atom become an ion by a change in the nuclear or electronic structure? Which is more likely to become a cation, potassium or sulfur? Which is more likely to become an anion, calcium or bromine?

2.49   What are the most likely charges of the elements Ca, P, and I? How does the periodic table help you answer this question?

2.50   If you are given a chemical formula such as FeO or ICl, is it possible to predict whether the substance is ionic or molecular? Explain.

2.51   Critique the following statements: The name of $FeCl_3$ is iron chloride. The name of NO_2 is nitrogen dioxygen.

2.52   Name the following anions: H^-; Cl^-; CN^-; $C_2H_3O_2^-$; and CrO_4^{2-}.

2.53   What is an acid? What clue in a chemical formula may help you identify a substance as an acid? When naming the binary acid HCl we use the prefix *hydro-* and add the *-ic* ending to the nonmetal to form hydrochloric acid. Why can't we simply name it chloric acid?

2.54   The name of a particular alcohol is hexanol. Is there sufficient information to determine both the number of carbon atoms and to which carbon atom the alcohol functional group is attached? If not, what additional information is needed?

2.55   The prefix cyclo- in the name of an alkane tells us that the carbon atoms form a ring. Is there sufficient information given to conclude that the alkane cyclopentane has ten hydrogen atoms? If not, what additional information is needed?

2.56   The chlorine (Cl) atom occurs in combination with many other elements.

(a) A $^{37}_{17}Cl^-$ nuclide contains how many protons, electrons, and neutrons?

(b) What is chlorine's elemental form?

(c) In the elemental state, chlorine is a green gas at room temperature. Is this color a physical or chemical property?

(d) Write the formula of a compound containing Cl^- with Cr^{3+}.

(e) 1 g of KCl is placed in 100 mL of water, and it dissolves with stirring. Identify the combination as one or more of the following: element, compound, heterogeneous mixture, homogeneous mixture, or solution.

2.57 Identify the following elements if an atom of each has the following:

(a) A 1+ charge and 11 protons;

(b) a 2+ charge and 36 electrons;

(c) a 2$^-$ charge and an atomic number of 34.

2.58 Bromine has two naturally occurring isotopes: ^{79}Br (78.918 amu) and ^{81}Br (80.916 amu). The atomic weight of Br is 79.904 amu. What are the fractional abundances of ^{79}Br and ^{81}Br?

2.59 Complete the following table:

Symbol	Number of protons	Number of neutrons	Number of electrons	Charge
$^{90}_{38}Sr^{2+}$				
	92	143		0
		10	10	1$-$
	46	60		2+

2.60 Predict the chemical formulas of the compounds most likely to form when the following combine and indicate if the predicted substances are ionic or molecular: (a) Al and nitrate ion; (b) Mg and phosphate ion; (c) ammonium ion and Br.

2.61 Name the following compounds:

(a) $CaSO_4$

(b) PF_5

(c) KBr

(d) $KHSO_4$

(e) Na_2S

(f) H_2SO_4

(g) CO_2

(h) $HClO_4$

(i) $NaClO_3$

(j) $Cu(CN)_2$

2.62 Using the periodic table, determine which ions are not likely:

(a) Cl^+

(b) Cs^+

(c) S^{2-}

(d) Rb^-

2.63 Write the formula for each of the following:

(a) tin(IV) chloride;   (d) dinitrogen trioxide;

(b) chromium(III) hydroxide;   (e) cobalt(III) oxide;

(c) cesium cyanide;   (f) calcium phosphate;

(g) osmium tetraoxide;   (i) hypobromous acid;

(h) mercury(II) bromide;   (j) hydroselenic acid.

2.64 Write the chemical formula for or give the name of each of the following oxyanions and oxyacids:

(a) $HClO$

(b) ClO^-

(c) perchloric acid

(d) perchlorate ion

(e) $HMnO_4$

(f) MnO_4^-

(g) H_2SO_3

(h) SO_3^{2-}

2.65 Write the chemical formulas for:

(a) ethane

(b) propane

(c) butanol

(d) methanol

Multiple-Choice Questions

2.66 Which experimental observation evidence indicates that cathode rays consist of electrons?

(a) Their path is curved in a magnetic field;

(b) Gravity significantly affects their path over a short distance;

(c) They have no effect on an electroscope;

(d) They are assigned a negative charge;

(e) All of the above are indications.

2.67 Which particle helped to explain discrepancies in atomic weights observed by early nineteenth century chemists?

(a) neutron

(b) electron

(c) proton

(d) atom

(e) molecule

2.68 The development of modern atomic theory took a leap forward with the proof that an atom is mostly empty space. Which of the following experiments was responsible for this proof?

(a) Millikan's oil drop

(b) cathode-ray deflection in a magnetic field

(c) the bombardment of gold foil with alpha particles

(d) separation of gases by gas chromatography

(e) the Wilson cloud chamber.

2.69 Which of the following cations are *correctly* named: (1) Sn^{2+}—tin(II) ion; (2) NH_4^+—ammonium ion; (3) Ca^{2+}—calcium ion?

(a) 1 and 2

(b) 2 and 3

(c) 1 and 3

(d) all of them

(e) none of them

2.70 Which of the following anions are *incorrectly* named: (1) IO^-—hyperiodate ion; (2) SO_3^{2-}—sulfite ion; (3) ClO_3^-—chlorate ion; (4) Cl^-—chloride ion; (5) NO_2^-—nitrate ion?

(a) 1 and 2

(b) 1 and 3

(c) 1 and 5

(d) all of them

(e) none of them

2.71 Which of the following compounds are *incorrectly* named: (1) $FeCl_3$—ferric chloride; (2) $HgCl_2$—mercurous chloride; (3) SnS_2—tin(IV) sulfide; (4) $KClO_2$—potassium chlorate?

(**a**) 1 and 2

(**b**) 2 and 4

(**c**) 3 and 4

(**d**) all of them

(**e**) none of them

2.72 Which of the following acids in water are *correctly* named: (1) H_2S—hydrosulfuric; (2) $HClO_4$—perchloric; (3) H_3PO_3—phosphorous; (4) HNO_2—nitrous; (5) H_2CO_3—carbonic?

(**a**) 1 and 2

(**b**) 2, 3, and 4

(**c**) 3, 4, and 5

(**d**) all of them

(**e**) none of them

2.73 Which of the following is the formula of phosphoric acid?

(**a**) H_2CO_3

(**b**) H_3P

(**c**) H_3PO_4

(**d**) H_3PO_2

(**e**) H_2PO_3

2.74 Which of the following is the most likely formula of the compound formed between Sr and S?

(**a**) SrS

(**b**) Sr_2S

(**c**) SrS_2

(**d**) Sr_2S_3

(**e**) Sr_3S_2

2.75 Which pair represents the correct chemical formulas for iron(II) oxide and iron(III) oxide?

(**a**) FeO and Fe_3O_2

(**b**) Fe_2O and Fe_2O_3

(**c**) FeO and Fe_2O_3

(**d**) FeO_2 and FeO

(**e**) FeO_2 and FeO_2

2.76 The oxyanions of the 3rd period have at most four oxygen atoms. Which of the following represents the correct chemical formulas for the perchlorate, sulfate, phosphate, and silicate ions? The silicate ion is not given in Chapter 2, but you can predict it by analogy.

(**a**) ClO^-, SO_4^{2-}, PO_4^{3-}, SiO_3^{3-}

(**b**) ClO_4^-, SO_4^{2-}, PO_4^{3-}, SiO_3^{3-}

(**c**) ClO_3^-, SO_4^{2-}, PO_4^{3-}, SiO_4^{3-}

(**d**) ClO_4^-, SO_4^{2-}, PO_4^{3-}, SiO_3^{4-}

(**e**) ClO_4^-, SO_4^{2-}, PO_4^{3-}, SiO_4^{4-}

2.77 What do these elements have in common: Na, K, and Cs?

1. metallic

2. nonmetallic

3. form cations

4. form anions

5. alkali metals

6. alkaline-earth metals

7. halogens

(**a**) 1 and 3

(**b**) 2 and 4

(**c**) 1, 4, and 7

(**d**) 1, 3, and 5

(**e**) 2, 4, and 6

2.78 What is the charge of the metal ion in $Co_3(PO_4)_2$?

(**a**) 1+

(**b**) 2+

(**c**) 3+

(**d**) 2−

(**e**) 3−

2.79 The use of mercury salts has been minimized in general chemistry laboratories because of their potential toxicity and damage to the environment if not disposed of properly. Which of the following represents the correct chemical formulas for mercury(I) chloride and mercury(II) sulfide?

(**a**) HgCl and HgS_2

(**b**) $HgCl_2$ and HgS_2

(**c**) Hg_2Cl and HgS_2

(**d**) Hg_2Cl_2 and HgS

(**e**) Hg_2Cl_2 and Hg_2S

2.80 Which is not an alkane?

(**a**) C_3H_8

(**b**) CH_4

(**c**) C_4H_{10}

(**d**) C_5H_8

(**e**) C_8H_{18}

SELF-TEST SOLUTIONS

2.1 (e). **2.2** (b).**2.3** (h). **2.4** (o). **2.5** (a). **2.6** (k). **2.7** (l).**2.8** (c). **2.9** (g). **2.10** (p). **2.11** (q). **2.12** (i). **2.13** (s). **2.14** (j). **2.15** (n). **2.16** (f). **2.17** (m). **2.18** (r). **2.19** (d). **2.20** True. **2.21** True. **2.22** True. **2.23** False. It equals 10^{-10} m. **2.24** False. In order of increasing atomic number. **2.25** False. 1.602×10^{-19} C = 1 electronic charge. **2.26** True. **2.27** True. **2.28** False. A metal. **2.29** True. **2.30** True. **2.31** False. Both its atomic number and mass number must be specified. **2.32** False. A structural formula is a picture that shows how atoms are attached within a molecule and it normally shows the actual number of elements, a molecular formula. **2.33** True. **2.34** False. It is a covalent substance. An ionic substance contains a metal and a nonmetal. **2.35** True. A cation is an ion with a positive charge, K^+. **2.36** True. **2.37** False. A polyatomic ion consists of atoms joined together as in a molecule but it contains a charge. The bromide ion consists of only one element; a polyatomic ion will have two or more different elements. **2.38** True. **2.39** False. An alkane contains all single carbon-carbon bonds. **2.40** True. **2.41** (b) Element "2," sulfur, forms the sulfide ion. Element "4" is a noble gas and does not readily form ions. **2.42** (c) Element "7" is Hs, hassium.

2.43 John Dalton believed that atoms were the basic building blocks of all matter and that all atoms of a particular element were identical. This concept survived until the discovery of isotopes. Isotopes are atoms of the same element that differ in the number of neutrons. For example, the element uranium has three naturally occurring isotopes: $^{233}_{92}U$, $^{235}_{92}U$, $^{238}_{92}U$.

2.44 The three primary particles are the proton, electron, and neutron. A proton has a positive charge and a mass of 1.0073 amu; an electron has a negative charge and a mass of 5.486×10^{-4} amu; and a neutron has no charge and a mass of 1.0087 amu. Most atoms have diameters between 1×10^{-10} m and 5×10^{-10} m.

2.45 The subscript tells you the atomic number of the element. In this case, the atomic number 80 uniquely identifies the element mercury. The superscript tells you the sum of protons and neutrons, 201. The difference, 201–80, indicates that the atom has 121 neutrons.

2.46 Refer to Table 2.3 in the text for a complete collection of names. The alkali metals form group 1A, the halogens form group 7A, the noble gases form group 8A, and the coinage metals—copper, silver, and gold—belong to group 1B.

2.47 A molecule is a substance that contains two or more atoms tightly bound together and these combined atoms act as a unit with its own unique properties. In the text you encounter six important elements that form diatomic molecules: H_2; O_2; N_2; F_2; Cl_2; Br_2; and I_2. A molecule can be composed of identical elements if they are combined and form a discrete unit.

2.48 An atom becomes an ion by the addition of electrons to form an anion or the removal of electrons to become a cation. A change in the number of protons in the nucleus would alter the identity of the element to a different one. Cations are generally formed by metals; thus, potassium is more likely to form a cation. Anions are generally formed by nonmetals; thus, bromine is more likely to become an anion.

2.49 Ca^{2+}, P^{3-}, and I^-. For an active metal, groups 1(1A) and 2(2A), the most likely positive charge is the number of the group in which it resides. For a nonmetal the most likely negative charge is the difference between its group number and the number of the last family of elements in the periodic table.

2.50 In Chapter 2 you are given some general statements about ionic and molecular compounds that help you predict the type. Ionic compounds are generally formed from combinations of metals and nonmetals, whereas molecular compounds are generally composed of nonmetals only. If you do this analysis, then FeO should be ionic and ICl should be molecular. Some substances do not follow these observations. For example, $BeCl_2$ has a significant amount of molecular character.

2.51 Iron is a transition element and exhibits more than one charge state. For elements of this type we indicate the charge as a Roman numeral, in parentheses, after the name of the element. Thus, $FeCl_3$ is named as iron(III) chloride. NO_2 is a molecular compound and for molecular compounds we name the first element as it is named in the periodic table. The second element is named as if it were a negative ion. For both elements the number of atoms is indicated using the appropriate prefix. NO_2 is named as nitrogen dioxide.

2.52 Hydride ion; chloride ion; cyanide ion; acetate ion; and chromate ion.

2.53 An acid is a hydrogen-containing compound that can form hydrogen ions (H^+) when dissolved in water. Generally, a compound is identified as an acid if it begins with a hydrogen element such as HCl, HNO_3, or $HC_2H_3O_2$. The name chloric acid refers to an oxyacid with the formula $HClO_3$. The prefix *hydro-* is used with binary acids to distinguish them from oxyacids.

2.54 There is sufficient information to determine hexanol has six carbon atoms. The prefix *hex-* means six carbon atoms in a hydrocarbon. However, there is

insufficient information to determine the location of the –OH group. It could be at the end of the carbon chain but it might also be attached to an interior carbon atom with the –OH group. Additional information is needed in the name to identify the location of the carbon atom. In Chapter 2.25 additional nomenclature rules are provided to locate the carbon atom attached to the –OH group.

2.55 There is sufficient information. The prefix *pent-* in pentane tells us that it contains five carbon atoms. If we form a structural formula with five carbon atoms in a cyclic chain and then add hydrogen atoms so each carbon atom has four bonds, we obtain

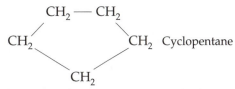

From the structural picture we count ten hydrogen atoms.

2.56 (a) $^{37}_{17}Cl^-$ has 18 electrons (17 electrons for a neutral $^{37}_{17}Cl$ and one for the negative charge), 17 protons, and 20 neutrons.

(b) Cl_2 (similar to F_2, Br_2, and I_2).

(c) Physical. Its color does not determine its chemical behavior.

(d) $CrCl_3$. Three Cl^- ions are needed to balance the 3+ charge of Cr^{3+}.

(e) It is a homogeneous mixture, that is, a solution.

2.57 (a) The atomic number of the unknown element equals its number of protons: 11. The charge results from a loss of electrons, not the addition of protons. Na has an atomic number of 11.

(b) The 2+ charge results from the loss of two electrons. Thus in the neutral atom there are $36 + 2 = 38$ electrons and 38 protons. It has the atomic number 38, which is the atomic number of Sr.

(c) Se has an atomic number of 34.

2.58 Let x and y equal the fractional abundances of ^{79}Br and ^{81}Br, respectively. Then $x + y = 1$, as the sum of the fractional abundances must equal unity.

The fractional abundances, x and y, are solved for as follows:

$$AW(Br) = x(\text{amu of}^{79}Br) + y(\text{amu of}^{81}Br)$$

$$79.904 \text{ amu} = x(78.918 \text{ amu}) + y(80.916 \text{ amu})$$

From the relationship $x + y = 1$ we write: $y = 1 - x$

$$79.904 \text{ amu} = x(78.918 \text{ amu}) + (1 - x)(80.916 \text{ amu})$$

Solving for x gives: 0.50651; and thus
$y = 1 - 0.50651 = 0.49349$

Thus elemental bromine consists of 50.651% ^{79}Br and 49.349% ^{81}Br.

2.59 This table is easily completed if you remember that the superscript number with a nuclide symbol equals the number of protons + the number of neutrons and that the subscript number equals the number of protons. The difference between the former and latter numbers equals the number of neutrons. Also, the number of electrons equals the number of protons adjusted for the charge of the nuclide.

Symbol	Number of protons	Number of neutrons	Number of electrons	Charge
$^{90}_{38}Sr^{2+}$	38	$90 - 38$ $= 52$	$38 - 2$ $= 36$	$2+$
$^{235}_{92}U$	92	143	92	0
$^{19}_{9}F^{-}$	9	10	$9 + 1$ $= 10$	$1-$
$^{106}_{46}Pd^{2+}$	46	60	$46 - 2$ $= 44$	$2+$

2.60 (a) $Al(NO_3)_3$.

(b) $Mg_3(PO_4)_2$.

(c) NH_4Br. They are all ionic substances. A molecular compound has all nonmetallic elements.

2.61 (a) calcium sulfate;

(b) phosphorus pentafluoride;

(c) potassium bromide;

(d) potassium hydrogen sulfate;

(e) sodium sulfide;

(f) sulfuric acid;

(g) carbon dioxide;

(h) perchloric acid;

(i) sodium chlorate;

(j) copper(II) cyanide.

2.62 (a) Chlorine is a nonmetal and thus is expected to form an ion with a *negative* charge;

(b) Cs^+ is correct because metals tend to form cations;

(c) S^{2-} is correct because nonmetals tend to form anions;

(d) Rb is a metal and thus is expected to form a cation, not an anion.

2.63 (a) $SnCl_4$;

(b) $Cr(OH)_3$;

(c) $CsCN$;

(d) N_2O_3;

(e) Co_2O_3;

(f) $Ca_3(PO_4)_2$;

(g) OsO_4;

(h) $HgBr_2$;

(i) $HBrO$;

(j) H_2Se.

2.64 (a) hypochlorous acid;

(b) hypochlorite ion;

(c) $HClO_4$;

(d) ClO_4^-;

(e) permanganic acid;

(f) permanganate ion;

(g) sulfurous acid;

(h) sulfite ion.

2.65 (a) C_2H_6

(b) C_3H_8

(c) C_4H_8OH

(d) CH_3OH

2.66 (a)

2.67 (a) Differing mixtures of isotopic nuclides (which differ in the number of neutrons) in samples of elements led to discrepancies in measurements of atomic weights.

2.68 (c). **2.69** (d). **2.70** (c). **2.71** (b). **2.72** (d). **2.73** (c). **2.74** (a).

2.75 (a) The oxide ion is O^{2-}. With iron(II), only one oxygen is needed to have the sum of charges equal zero. With iron(III), two iron(III) ions and three oxide ions are needed to have the sum of charges equal zero: $2(3+) + 3(2-) = 0$.

2.76 (e) All of the polyatomic ions in the question end in -ate. Thus, they possess the maximum number of oxygen atoms, four. The trend in their charges shows the effect of the first element being in different families.

2.77 (d)

2.78 (b) To determine the charge of the cobalt ion, you must first determine the charge of the polyatomic ion, the phosphate ion; it is $3-$. Then apply the rule that the sum of charges for a substance must be zero. With three cobalt ions and two phosphate ions, the only way for this rule to apply is for the cobalt ion to be $2+: 3(2+) = 0$.

2.79 (d) The mercury(I) ion is unusual in that it occurs in nature in a diatomic form.

2.80 (d) The general formula of a simple alkane is C_nH_{2n+2}. Only (d) does not obey this rule.

Stoichiometry: Calculations with Chemical Formulas and Equations

Chapter

3

OVERVIEW OF THE CHAPTER

Review: Elements and compounds (1.1,1.2); formulas (2.6,2.7); nomenclature (2.8).

Learning Goals: You should be able to:

1. Balance chemical equations.
2. Predict the products of a chemical reaction, having seen a suitable analogy.
3. Predict the products of the combustion reactions of hydrocarbons and simple compounds containing C, H, and O atoms.

3.1, 3.2 CHEMICAL EQUATIONS: BALANCING AND PREDICTING PRODUCTS OF REACTIONS

Learning Goals: You should be able to:

1. Calculate the formula weight of a substance given its chemical formula.
2. Calculate the molecular weight of a molecular substance given its chemical formula.
3. Recognize when to use formula weights and molecular weights in calculations.
4. Calculate the molar mass of a substance from its chemical formula.
5. Interconvert the number of moles of a substance and its mass.
6. Use Avogadro's number and molar mass to calculate the number of particles making up a substance and *vice versa*.

3.3, 3.4 FORMULA WEIGHT, MOLECULAR WEIGHT, AND THE MOLE

Review: Empirical and molecular formulas (2.6).

Learning Goals: You should be able to:

1. Calculate the empirical formula of a compound, having been given appropriate analytical data such as elemental percentages or the quantity of CO_2 and H_2O produced by combustion.
2. Calculate the molecular formula, having been given the empirical formula and molecular weight.

3.5 DETERMINING EMPIRICAL AND MOLECULAR FORMULAS

3.6, 3.7 CHEMICAL EQUATIONS: MASS AND MOLE RELATIONSHIPS

Review: Dimensional analysis (1.6); rounding numbers (1.5).

Learning Goals: You should be able to:

1. Calculate the mass of a particular substance produced or used in a chemical reaction (mass–mass problem).
2. Determine the limiting reagent in a reaction.
3. Calculate the theoretical and actual yields of chemical reactions given the appropriate data.

TOPIC SUMMARIES AND EXERCISSE

CHEMICAL EQUATIONS: BALANCING AND PREDICTING PRODUCTS OF REACTIONS

Chemical equation such as $2\,C + O_2 \rightarrow 2\,CO$:

- Describe chemical processes involving **reactants** (left side of arrow) to form **products** (right side of arrow).
- Should be balanced.
- Provide a means for calculating mass relationships among products and reactants.

When you balance chemical equations, you must keep the following requirements in mind:

- Formulas of substances must be correctly written.
- The number of atoms of each type of element must be the same on both sides of the arrow.
- Only coefficients in front of substances may be adjusted to change the number of atoms on the reactant or product side. Subscripts in chemical formulas must not be changed.
- The sum of charges of ions on the left side of the arrow must be the same on the right side.

See Exercise 1 for an approach to balancing chemical reactions.

An important skill to develop is the ability to predict the products of simple chemical reactions having been given only the reactants. In this chapter we look at several classes of chemical reactions to help us develop this skill: combustion, combination, and decomposition.

Combustion reactions produce a flame and usually involve oxygen as a reactant. For example:

$$C_2H_4(g) + 3\,O_2(g) \longrightarrow 2\,CO_2(g) + 2\,H_2O(l)$$

Means a substance is a gas   Means a substance is a liquid

- The text highlights the combustion of **hydrocarbons**, carbon-, and hydrogen-containing compounds. When hydrocarbons are combusted in the presence of oxygen, carbon is converted to CO_2 and hydrogen is converted to H_2O. If oxygen is also present in a compound containing C and H, for example CH_3OH, the oxygen is used along with O_2 in balancing the equation.

Combination reactions involve forming one product from two or more reactants. For example:

$$CaO(s) + CO_2(g) \longrightarrow CaCO_3(s)$$

Decomposition reactions are the reverse of combination reactions: One reactant breaks down (decomposes) into two or more substances. For example:

$$2 H_2O(l) \longrightarrow O_2(g) + 2 H_2(g)$$

E X E R C I S E 1   Balancing chemical equations

Balance the following reactions:

(a) $P_4(s) + O_2(g) \longrightarrow P_4O_{10}(s)$
(b) $SF_4(g) + H_2O(l) \longrightarrow SO_2(g) + HF(g)$

SOLUTION: *Analyze*: We are given chemical reactions which are not balanced and we are asked to determine the balancing coefficients for each substance.

Plan: A chemical reaction is balanced when the number of atoms of each type are the same on both sides of the arrow. If charged species are present, then the sum of charges on the left must also equal the sum of charges on the right. The first step is to count the number of atoms to determine if the given reaction is already balanced. If it is not balanced then we have to place coefficients in *front* of substances so that the number of atoms of each type are the same for both reactants and products.

Solve: (a) By inspection we see that the number of atoms are not balanced in the chemical equation.

$$P_4(s) + O_2(g) \longrightarrow P_4O_{10}(s)$$

An inventory of atoms shows:

	No. of P Atoms	No. of O Atoms
Reactants	4	2
Products	4	10

Although the number of phosphorus atoms is balanced, the number of oxygen atoms is not. 10 oxygen atoms, that is, 5 O_2, are needed on the reactant side to equal the 10 oxygen atoms on the product side. The balanced chemical equation is

$$P_4(s) + 5 O_2(g) \longrightarrow P_4O_{10}(s)$$

(b) An inventory of atoms in the chemical equation

$$SF_4(g) + H_2O(l) \longrightarrow SO_2(g) + HF(g)$$

shows:

	No. of S Atoms	No. of F Atoms	No. of H Atoms	No. of O Atoms
Reactants:	1	4	2	1
Products:	1	1	1	2

Starting with the atom that is present in the greatest number on the reactant side, fluorine, we can balance the fluorine atoms on the product side with 4 HF. The inventory is now

	No. of S Atoms	No. of F Atoms	No. of H Atoms	No. of O Atoms
Reactants	1	4	2	1
Products:	1	4	4	2

The hydrogen atoms can be balanced with 2 H_2O on the reactant side. The balancing of the hydrogen atoms also causes the oxygen atoms to be balanced. The balanced equation is

$$SF_4(g) + 2\,H_2O(l) \longrightarrow SO_2(g) + 4\,HF(g)$$

Check: If we have done our tables correctly the number of atoms of each element should be the same on both sides of the arrow. However, it is advisable to look at each balanced chemical equation and recheck that this is true. For example, in (a) we can count that there are 4 P atoms on each side of the arrow and 10 oxygen atoms on each side of the arrow.

EXERCISE 2 Writing and balancing a combustion reaction

Write the balanced chemical equation for the combustion of butane, C_4H_{10} in air.

SOLUTION: *Analyze:* We are given butane, C_4H_{10}, and are asked to write the balanced chemical equation for its combustion reaction.

Plan: The first step is to write the skeletal equation that describes the combustion reaction for butane, a hydrocarbon. We can do this by using oxygen gas, a component of air, as a reactant. The typical products formed when a hydrocarbon is combusted are carbon dioxide and water in the gaseous state. After writing the skeletal equation we can then balance it.

Solve: The skeletal chemical equation is

$$C_4H_{10}(g) + O_2(g) \longrightarrow CO_2(g) + H_2O(l)$$

When balancing combustion reactions of simple hydrocarbons, first balance the carbon and hydrogen atoms without considering oxygen. Then balance the oxygen atoms using $O_2(g)$. There are 4 carbon atoms, therefore, place a 4 in front of CO_2; there are 10 hydrogen atoms in butane, therefore, place a 5 in front of H_2O:

$$C_4H_{10}(g) + O_2(g) \longrightarrow 4\,CO_2(g) + 5\,H_2O(l)$$

The products contain a total of 13 oxygen atoms, 8 from 4 carbon dioxide molecules and 5 from 5 water molecules. Place a 13/2 in front of $O_2(g)$ to balance the 13 oxygen atoms:

$$C_4H_{10}(g) + \frac{13}{2}\,O_2(g) \longrightarrow 4\,CO_2(g) + 5\,H_2O(l)$$

It is more convenient to deal with whole numbers; it is usually customary (but not always required) to clear the fractions by multiplying the entire equation by a number that cancels the denominators. In this case, multiply the entire equation by two to eliminate the denominator in 13/2:

$$2\,C_4H_{10}(g) + 13\,O_2(g) \longrightarrow 8\,CO_2(g) + 10\,H_2O(l)$$

Check: A check of the number of atoms shows that there are 8 carbon atoms, 20 hydrogen atoms, and 36 oxygen atoms on both sides of the arrow.

EXERCISE 3 Completing and balancing chemical reactions

Complete and balance the following reactions, and indicate the phases of each substance: (a) $SbBr_3 + H_2S \rightarrow$ (b) $LiH + H_2O \rightarrow$ (c) $CO_2 + Na \rightarrow$ (d) $FeS + O_2 \rightarrow$ Use the following reactions as examples:

(1) $3\,CO_2(g) + 4\,K(s) \rightarrow 2\,K_2CO_3(s) + C(s)$
(2) $CaH_2(s) + 2\,H_2O(l) \rightarrow Ca(OH)_2(aq) + 2\,H_2(g)$

(3) $2\,SbCl_3(s) + 3\,H_2S(g) \rightarrow Sb_2S_3(s) + 6\,HCl(g)$

(4) $2\,ZnS(s) + 3\,O_2(g) \rightarrow 2\,ZnO(s) + 2\,SO_2(g)$

SOLUTION: *Analyze:* We are given four incomplete chemical reactions and are asked to complete and balance their chemical equations.

Plan: The reactions can be completed by finding an analogous reaction among those given. Analogous compounds or elements that are from the same family or that are otherwise similar in nature often exist in the same phase.

Solve: We can complete the chemical equation given in (**a**) by observing that H_2S is a reactant and $H_2S(g)$ is also a reactant in the sample chemical equation (3). Furthermore the other reactant in the sample chemical equation (3) is an antimony halide as in (**a**). Repeating this process for each incomplete chemical equation gives the following results:

Balanced equation	Analogous reaction
(a) $2\,SbBr_3(s) + 3\,H_2S(g) \rightarrow Sb_2S_3(s) + 6\,HBr(g)$	(3)
(b) $LiH(s) + H_2O(l) \rightarrow LiOH(aq) + H_2(g)$	(2)
(c) $3\,CO_2(g) + 4\,Na(s) \rightarrow 2\,Na_2CO_3(s) + C(s)$	(1)
(d) $2\,FeS(s) + 3\,O_2(g) \rightarrow 2\,FeO(s) + 2\,SO_2(g)$	(4)

Comment: It takes reading and practice to develop our knowledge of chemical reactions. As this knowledge base grows we will be able to predict the products of many chemical reactions.

Substances come in a variety of forms. One form is molecular: A molecule is the smallest particle of a pure substance that has the composition and properties of the pure substance and also has an independent existence. Another is ionic: An ionic substance consists of particles with charges and there is no discrete smaller unit with an independent existence. In this chapter we learn how to calculate the masses of molecular and ionic substances.

FORMULA WEIGHT, MOLECULAR WEIGHT, AND THE MOLE

- The term **formula weight** refers to the sum of atomic weights of the atoms in a substance. It can be used with both molecular and ionic substances.
- The term **molecular weight** refers to the sum of atomic weights of the atoms in a molecular substance. For example, the molecular weight of a molecule of water, H_2O, is:

$$2(\text{AW of H}) + 1(\text{AW of O}) = (2 \text{ atoms H})\left(\frac{1.01 \text{ amu}}{1 \text{ atom H}}\right)$$

$$+ (1 \text{ atom O})\left(\frac{16.00 \text{ amu}}{1 \text{ atom O}}\right)$$

$$= 18.02 \text{ amu}$$

The percent composition of a substance refers to the percent by mass contributed by each element in the substance.

- The sum of percent by mass of all elements in a substance equals 100%.
- % by mass of an element $= \dfrac{\text{mass of an element in substance}}{\text{formula weight of substance}} \times 100$

Chemists do not ordinarily work with single molecules or atoms, but rather with trillions upon trillions of them. To facilitate the counting and weighing of such samples, a quantity called the mole has been defined.

- A **mole** of any type of particle equals the number of ^{12}C atoms in exactly 12 g of ^{12}C. Thus a mole represents a certain number of objects, just like a dozen represents 12.
- In 12 g of ^{12}C, there are 6.022×10^{23} atoms; this number is given the name **Avogadro's number** (symbol is N, and unit is g/mol).
- Thus a mole of water contains 6.022×10^{23} molecules of water, and a mole of NaCl contains 6.022×10^{23} sodium ions and 6.022×10^{23} chloride ions. Note that the total number of ions in one mole of NaCl is $2(6.022 \times 10^{23})$ ions or 1.204×10^{24} ions.
- The term **molar mass** is used to describe the mass in grams of one mole of a substance.

We can use mass–quantity relationships as conversion factors. Examples of equivalences that can be used are

$$
\begin{array}{ll}
1\ ^{12}C \text{ atom} = 12 \text{ amu} & 1 \text{ mol } ^{12}C = 12 \text{ g} \\
1\ Cl_2 \text{ molecule} = 70.90 \text{ amu} & 1 \text{ mol } Cl_2 = 70.90 \text{ g} \\
1\ BaCl_2 \text{ formula} = 208 \text{ amu} & 1 \text{ mol } BaCl_2 = 208 \text{ g}
\end{array}
$$

EXERCISE 4   Calculating molecular and formula weights

Calculate the molecular or formula weights for: (a) NO_3^-; (b) $C_{21}H_{30}O_2$.

SOLUTION: *Analyze:* We are given the chemical formulas for an ion, (a), and a molecular substance, (b), and are asked to calculate their formula weights.

Plan: First look up the atomic weights of all elements in the substances. Then calculate the formula or molecular weight of each substance by multiplying each atomic weight by the number of atoms in the chemical formula and summing these numbers. This procedure gives the formula weight for (a) since it is an ion and the molecular weight for (b) since it is a molecular substance.

Solve:

(a)

$$\text{N: (1 atom N)}\left(\frac{14.01 \text{ amu}}{1 \text{ atom N}}\right) = 14.01 \text{ amu}$$

$$\text{O: (3 atoms O)}\left(\frac{16.00 \text{ amu}}{1 \text{ atom O}}\right) = \underline{48.00 \text{ amu}}$$

$$\text{Formula weight of } NO_3^- = 62.01 \text{ amu}$$

(b)

$$\text{C: (21 atoms C)}\left(\frac{12.01 \text{ amu}}{1 \text{ atom C}}\right) = 252.21 \text{ amu}$$

$$\text{H: (30 atoms H)}\left(\frac{1.01 \text{ amu}}{1 \text{ atom H}}\right) = 30.30 \text{ amu}$$

$$\text{O: (2 atoms O)}\left(\frac{16.00 \text{ amu}}{1 \text{ atom O}}\right) = \underline{32.00 \text{ amu}}$$

$$\text{Molecular weight of } C_{21}H_{30}O_2 = 314.51 \text{ amu}$$

Check: We can estimate the formula weight for (a) by adding 14 amu and 3×16 amu and finding it is 62 amu, which is close to the calculated value in the solution. Similarly we can estimate the molecular formula for (b) by adding 20×12 amu and 30×1 amu and 2×20 amu and finding it is 310 amu, which is close to the calculated molecular weight.

EXERCISE 5   Calculating molecular weight, number of molecules and moles, and percentage of an element in a molecule

Answer the following with respect to ethanol, C_2H_6O: (**a**) What is its molecular weight? (**b**) What is the mass of 1 mol of ethanol molecules? (**c**) Calculate the number of moles of ethanol in 1.00 g. (**d**) Calculate the number of molecules in 1.00 g of ethanol. (**e**) Calculate the percentage of carbon in one molecule of ethanol.

SOLUTION: (**a**) *Analyze:* We are asked to calculate for ethanol, a molecular substance, its molecular weight.

Plan: We can use the same approach as in Exercise 4.

Solve: The molecular weight of ethanol is calculated as follows:

$$C: (2 \text{ atoms C})\left(\frac{12.01 \text{ amu}}{1 \text{ atom C}}\right) = 24.02 \text{ amu}$$

$$H: (6 \text{ atoms H})\left(\frac{1.01 \text{ amu}}{1 \text{ atom H}}\right) = 6.06 \text{ amu}$$

$$O: (1 \text{ atom O})\left(\frac{16.00 \text{ amu}}{1 \text{ atom O}}\right) = 16.00 \text{ amu}$$

$$\text{Molecular weight of } C_2H_6O = 46.08 \text{ amu}$$

Check: We can estimate the molecular weight by adding 2×12 amu, 6×1 amu, and 1×16 amu and finding it is 46 amu, which is close to the calculated molecular weight.

(**b**) *Analyze:* We are asked to calculate the mass of one mole of ethanol.

Plan: Use the definition of molar mass: The mass of one mole of a substance is its molecular weight expressed in grams.

Solve: The mass of 1 mol of ethanol is the weight of one molecule expressed in grams, 46.08 g.

(**c**) *Analyze*: You are given a mass of ethanol; you are asked to solve for the number of moles.

Plan: You need a conversion factor that will change 1.00 g of ethanol to moles. The molar mass relationship yields the necessary conversion factor:

$$46.08 \text{ g ethanol} = 1 \text{ mol ethanol}$$

Since you want to carry out the conversion *grams* → *moles* (the symbol → indicates a unit conversion), the conversion factor you need to use is 1 mol ethanol/46.08 g ethanol.

Solve: Moles ethanol $= (1.00 \text{ g ethanol})\left(\dfrac{1 \text{ mol ethanol}}{46.08 \text{ g ethanol}}\right) = 0.0217 \text{ mol ethanol}$

Check: The answer is significantly less than one mole, which is rational because the given mass of ethanol is less than the mass of one mole.

(**d**) *Analyze*: You are asked for the number of molecules in 1.00 g of ethanol. This will require the number of moles of ethanol which is calculated in (**b**).

Plan: Most problems that ask for the number of particles, such as molecules or atoms, will require you at some point in the calculation to use Avogadro's number to convert the number of moles of the substance to number of particles. The two relationships needed for the conversion of *grams* → *moles* → *molecules* are:

$$46.08 \text{ g ethanol} = 1 \text{ mol ethanol} \qquad [\textit{Molar Mass}]$$

$$1 \text{ mol ethanol} = 6.022 \times 10^{23} \text{ molecules ethanol} \quad [\text{Avogadro's number}]$$

Solve: $\text{Molecules ethanol} = (1.00 \text{ g ethanol})\left(\dfrac{1 \text{ mol ethanol}}{46.08 \text{ g ethanol}}\right)$

$$\times \left(\dfrac{6.022 \times 10^{23} \text{ molecules ethanol}}{1 \text{ mol ethanol}}\right)$$

$$= 1.31 \times 10^{22} \text{ molecules ethanol}$$

Check: The answer is less than 6.02×10^{23}, which is rational given that the number of moles is less than one.

(**e**) *Analyze*: You are asked for the percentage of carbon in one molecule of ethanol. This will require the use of the molecular weight of ethanol and the number of atoms of carbon in a molecule of ethanol.

Plan: The percentage of any element in a molecular compound is the mass of that element in one molecule divided by the molecular weight of the molecule and multiplied by 100:

$$\% \text{ C} = \dfrac{\left(\begin{array}{c}\text{number C atoms}\\ \text{per molecule}\end{array}\right)(\text{AW of C})}{(\text{mass of one C}_2\text{H}_6\text{O molecule})} \times 100$$

Solve: $\% \text{ C} = \dfrac{\left(\dfrac{2 \text{ atoms C}}{1 \text{ molecule C}_2\text{H}_6\text{O}}\right)\left(\dfrac{12.01 \text{ amu}}{1 \text{ atom C}}\right)}{\dfrac{46.08 \text{ amu}}{1 \text{ molecule C}_2\text{H}_6\text{O}}} \times 100 = 52.13\%$

Alternatively, the percentage of carbon in C_2H_6O can be solved for as follows:

$$\left(\dfrac{2 \text{ mol C}}{1 \text{ mol C}_2\text{H}_6\text{O}}\right)\left(\dfrac{12.01 \text{ g C}}{1 \text{ mol C}}\right)\left(\dfrac{1 \text{ mol C}_2\text{H}_6\text{O}}{46.08 \text{ g C}_2\text{H}_6\text{O}}\right) = \dfrac{0.5213 \text{ g C}}{1 \text{ g C}_2\text{H}_6\text{O}} = \dfrac{52.13 \text{ g C}}{100 \text{ g C}_2\text{H}_6\text{O}}$$

The last ratio is equivalent to saying that the percentage of carbon in C_2H_6O is 52.13%.

Check: Carbon is present in the greatest total mass in the molecule and thus it should represent a significant percentage. Also the percentage is less than 100. Except for an element by itself, the percentage of an element in a compound must be less than 100%.

Exercise 6   Calculating the mass of a substance given the mass of an element in its chemical formula

A sample of $Na_2B_4O_7$ contains 0.3478 g of sodium. What is the mass of this sample?

SOLUTION: *Analyze*: We are asked for the mass of a salt, $Na_2B_4O_7$, and we are given the mass of sodium in the sample.

Plan: We can calculate the mass of $Na_2B_4O_7$ by recognizing that the mass of the sample is related to the mass of the sodium in the following manner:

$$\text{Mass of sample} = (\text{mass of Na}) \times \left(\begin{array}{c}\text{conversion factors that}\\ \text{change g Na to g Na}_2\text{B}_4\text{O}_7\end{array}\right)$$

To convert from grams Na to grams $Na_2B_4O_7$, we will need to go through the following series of conversions:

Grams Na $\longrightarrow$ *moles* Na $\longrightarrow$ *moles* $Na_2B_4O_7$ $\longrightarrow$ *grams* $Na_2B_4O_7$

The required equivalences are

1 mol $Na_2B_4O_7$ = 2 mol Na   [from formula of $Na_2B_4O_7$]*

1 mol $Na_2B_4O_7$ = 201.24 g $Na_2B_4O_7$ [from molar mass of $Na_2B_4O_7$]

1 mol Na = 23.00 g Na   [from molar mass of Na]

Solve: Mass of sample = $(0.3478 \text{ g Na}) \left(\dfrac{1 \text{ mol Na}}{23.00 \text{ g Na}} \right)$

$$\times \left(\frac{1 \text{ mol Na}_2\text{B}_4\text{O}_7}{2 \text{ mol Na}} \right) \left(\frac{201.24 \text{ g Na}_2\text{B}_4\text{O}_7}{1 \text{ mol Na}_2\text{B}_4\text{O}_7} \right)$$

$$= 1.522 \text{ g Na}_2\text{B}_4\text{O}_7$$

Check: The answer is larger than the mass of sodium, and this is rational given that the mass of any element in a compound is less than the mass of the compound.

In Chapter 2 we learned that an empirical formula shows the simplest whole-number ratio of atoms. A molecular formula shows the actual number of atoms. For example, CH_2 is the empirical formula for C_2H_4. An empirical formula is typically determined from percent composition data. *The subscripts in an empirical formula are calculated as follows:*

DETERMINING EMPIRICAL AND MOLECULAR FORMULAS

- Convert the mass percent of each element to grams using an arbitrarily chosen sample size, such as 100 g.
- Determine the number of moles of each element.
- Determine the simplest whole-number ratio of atoms in the compound by dividing the number of moles of each element by the number of moles of the element having the smallest number of moles.
- If the ratios are not whole numbers, multiply the ratios by an integer that clears the denominators of the fractions. For example, the numbers 0.50 and 1.75 are expressed as fractions: $\frac{1}{2}$ and $\frac{7}{4}\left(1\frac{3}{4}\right)$. If they are multiplied by four, the ratios are converted to whole numbers: $4 \times \frac{1}{2} = 2$ and $4 \times \frac{7}{4} = 7$. If a ratio is very near a whole number or fraction, such as 1.05 or 1.55, assume that they can be expressed as 1.00 and 1.50 because of experimental error.

To determine the molecular formula of a compound, we need its molecular weight:

- Calculate the number of empirical formula units making up the molecular formula by dividing the mass of one mole of the substance by the mass of one empirical formula.
- To determine the molecular formula, multiply the subscripts of the empirical formula by the number of empirical formula units making up the molecular formula.

EXERCISE 7   Determining the empirical and molecular formulas of a compound

A compound contains only the elements Al and O. Its elemental composition is determined to be 53.0% aluminum and 47.0% oxygen. The mass of one mole of the compound is 102 g. What is the empirical formula of the compound? What is the molecular formula?

---

*The conversion factor 1 mol $Na_2B_4O_7$/2 mol Na is included in the problem to reflect the fact that there are two sodium atoms per $Na_2B_4O_7$ formula unit.

SOLUTION: *Analyze:* We are asked to calculate the empirical and molecular formula of a compound containing Al and O. The percent composition data provides information for determining the empirical formula, and molar mass permits calculation of the molar formula.

Plan: Follow the outline for determining empirical formulas from percent composition data. First convert the mass percent of each element to grams in an arbitrarily chosen sample size, for example, 100 g. Then determine the number of moles of each element in the 100 gram sample and the ratio of moles. This gives the empirical formula. To determine the molecular formula, divide the molar mass by the empirical mass and multiply the subscripts of the empirical formula by this number.

Solve:

$$\text{Grams Al} = (100\text{-g sample})\left(\frac{53.0 \text{ g Al}}{100 \text{ g}}\right) = 53.0 \text{ g Al}$$

$$\text{Grams O} = (100\text{-g sample})\left(\frac{47.0 \text{ g O}}{100 \text{ g}}\right) = 47.0 \text{ g O}$$

Next determine the number of moles of each element in 100 g of the sample:

$$\text{Moles Al} = (53.0 \text{ g Al})\left(\frac{1 \text{ mol Al}}{27.0 \text{ g Al}}\right) = 1.96 \text{ mol Al}$$

$$\text{Moles O} = (47.0 \text{ g O})\left(\frac{1 \text{ mol O}}{16.0 \text{ g O}}\right) = 2.94 \text{ mol O}$$

To determine the empirical formula of the compound, calculate the simplest whole-number ratio of atoms in the compound. This is done by dividing the number of moles of each element by the number of moles of the element having the *smallest* number of moles. In this case Al has the fewer number of moles; thus you divide by 1.96:

$$Al_{\frac{1.96}{1.96}}O_{\frac{2.94}{1.96}} = Al_{1.00}O_{1.50}$$

The subscripts are not all integers. You must multiply them by an integer that will convert 1.50 into an integer. Inspection should convince you that if you multiply by 2 you will convert 1.50 to 3.00 and 1.00 to 2.00. The empirical formula is therefore Al_2O_3.

To determine the molecular formula, divide the mass of one mole by the mass of one empirical formula unit. The mass of one empirical formula unit for Al_2O_3 is 102 g, the same as the mass of one mole. Thus, the empirical and molecular formulas are identical, Al_2O_3.

Check: The formula Al_2O_3 is reasonable because aluminum has a charge of 3+ as an ion and oxygen has a charge of 2− as an ion; the sum of charges adds to zero for the formula and is consistent with observation about charges.

CHEMICAL EQUATIONS: MASS AND MOLE RELATIONSHIPS

A balanced chemical equation gives us information about the relative number of moles of reactants and products. The masses of substances in the reaction are determined from the mode concept. The branch of chemistry that deals with these quantitative relationships is termed **stoichiometry**.

To understand the above statements, examine the following balanced chemical reaction and see what mass–mole relationships can be derived from it:

$$CH_4(g) + 2\,O_2(g) \longrightarrow CO_2(g) + 2\,H_2O(l)$$

- On the atomic-molecular level, the equation states:

1 molecule CH_4 + 2 molecules $O_2 \longrightarrow$
$$1 \text{ molecule } CO_2 + 2 \text{ molecules } H_2O$$

- Or, since an Avogadro's number of molecules is equivalent to a mole of molecules:

$$1 \text{ mol } CH_4 + 2 \text{ mol } O_2 \longrightarrow 1 \text{ mol } CO_2 + 2 \text{ mol } H_2O$$

- The numerical coefficients in front of the reactants and products show the ratio of moles in which the chemical substances react. For example, because 2 mol of O_2 is required to react with 1 mol of CH_4, then we know that 4 mol O_2, are required to react with 2 mol of CH_4—that is, O_2 always reacts with CH_4 in a 2:1 mole ratio. You can represent this stoichiometry by the statement

$$1 \text{ mol } CH_4 \simeq 2 \text{ mol } O$$

where the symbol $\simeq$ means a stoichiometrically equivalent quantity *in terms of the given reaction*. Similarly, you can represent the stoichiometric ratio for the formation or products as

$$1 \text{ mol } CO_2 \simeq 2 \text{ mol } H_2O$$

That is, 1 mole of CO_2 forms for every 2 moles of water that forms.

- Other stoichiometric ratios can be derived from the balanced chemical reaction. For example:

$$1 \text{ mol } CH_4 \simeq 1 \text{ mol } CO_2$$
$$2 \text{ mol } O_2 \simeq 1 \text{ mol } CO_2$$
$$1 \text{ mol } CH_4 \simeq 2 \text{ mol } H_2O$$
$$2 \text{ mol } O_2 \simeq 2 \text{ mol } H_2O$$

- All of the above stoichiometrically equivalent statements can be converted to mass equivalences by converting a mole of a substance to its molar mass.

Various kinds of chemical problems involving stoichiometry are encountered in chemical practice.

- One important type involves a **limiting reactant**. In many reactions, one or more substances are in excess and therefore some will be left over when the reaction is completed. The substance that is completely consumed determines the amount of product formed and is called the limiting reactant. Exercise 11 explores how to solve limiting reactant problems.
- Most chemical reactions are not 100% efficient; they do not produce as much product as expected from the stoichiometry. The extent of a reaction is given by **percent yield**:

$$\text{Percent yield} = \frac{\text{actual yield}}{\text{theoretical yield}} \times 100$$

See Exercise 11.

EXERCISE 8   Using stoichiometry to determine the amount of a product formed in a chemical reaction I

How many moles of Al_2O_3 are produced when 0.50 mol Al reacts with an excess of PbO_2? The balanced chemical equation is

$$4\,Al(s) + 3\,PbO_2(s) \longrightarrow 2\,Al_2O_3(s) + 3\,Pb(s)$$

SOLUTION: *Analyze*: We are given the number of moles of Al and asked how many moles of Al_2O_3 are produced. This will require a stoichiometric conversion using the chemical reaction.

Plan: Determine the stoichiometric equivalences that allow us to make the transformation:

$$\text{Moles Al} \longrightarrow \text{moles } Al_2O_3$$

The stoichiometric equivalence derived from the balanced chemical equation that relates moles of Al to moles of Al_2O_3 is:

$$4\,\text{mol Al} \simeq 2\,\text{mol } Al_2O_3$$

The problem states that there is sufficient PbO_2 to react with 0.50 mol Al. Thus the amount of PbO_2 present does not have to be considered.

Solve: $\text{Moles } Al_2O_3 = (0.50\text{ mol Al})\left(\dfrac{2\text{ mol } Al_2O_3}{4\text{ mol Al}}\right) = 0.25\text{ mol } Al_2O_3$

Thus when 0.50 mol Al reacts with an excess of PbO_2, 0.25 mol Al_2O_3 is produced.

Check: Note that the calculation uses units with the name of the substance. This ensures that the calculated value is associated with the correct substance. If you, for example, inadvertently invert the stoichiometric ratio, you will find that the units with substances do not properly cancel. This is one of the best ways of checking when doing stoichiometric problems.

EXERCISE 9   Using stoichiometry to determine the amount of a reactant needed in a chemical reaction

Determine how many grams of HI are required to form 1.20 moles of H_2 when HI reacts according to the balanced chemical equation

$$2\,HI(g) \longrightarrow H_2(g) + I_2(g)$$

SOLUTION: *Analyze*: We are given the number of moles of a product, H_2, that forms and asked to determine how many grams of HI are required. Again, as in the previous problem, we will need stoichiometric ratios.

Plan: This problem requires you to work from products to reactants as follows:

$$\text{Moles } H_2 \longrightarrow \text{moles HI} \longrightarrow \text{grams HI}$$

From this sequence of proposed conversions, we see that you need both mole and mass stoichiometric equivalences. These are

$$2\,\text{mol HI} \simeq 1\,\text{mol } H_2 \text{ (from balanced equation)}$$

$$127.91\text{ g HI} = 1\text{ mol HI (from molar mass of HI)}$$

Solve: The problem can be solved in two steps:

$$\text{Moles } H_2 \longrightarrow \text{moles HI:}$$

$$\text{moles HI} = (1.20\text{ mol } H_2)\left(\dfrac{2\text{ mol HI}}{1\text{ mol } H_2}\right) = 2.40\text{ mol HI}$$

and

$$Moles\ HI \longrightarrow grams\ HI:$$

$$grams\ HI = (2.40\ mol\ HI)\left(\frac{127.91\ g\ HI}{1\ mol\ HI}\right) = 307\ g\ HI$$

Alternatively, the two steps can be combined into one as follows:

$$Grams\ HI = (1.20\ mol\ H_2)\left(\frac{2\ mol\ HI}{1\ mol\ H_2}\right)\left(\frac{127.91\ g\ HI}{1\ mol\ HI}\right) = 307\ g\ HI$$

In the *Student's Guide* the latter approach will be the one usually followed when solving stoichiometry problems.

Check: The units associated with the final answer have the correct substance indicated. An estimate of 2.5×128 in the last step gives about 300, which is the magnitude of the final answer.

EXERCISE 10   Using stoichiometry to determine the amount of a product formed in a chemical reaction II

Determine how many grams of PCl_3 are produced when 2.80 g of Cl_2 reacts with a sufficient quantity of P_4 according to the chemical reaction

$$P_4(s) + 6\ Cl_2(g) \longrightarrow 4\ PCl_3(l)$$

SOLUTION: The *Analysis* and *Plan* for this problem follow the same format as in the previous problems. We are given the grams of chlorine gas and asked to calculate the grams of product formed. We will need stoichiometric ratios (equivalences) as follows:

$$Grams\ Cl_2 \longrightarrow moles\ Cl_2 \longrightarrow moles\ PCl_3 \longrightarrow grams\ PCl_3$$

The necessary stoichiometric equivalence is

$$6\ mol\ Cl_2 \simeq 4\ mol\ PCl_3$$

Also, we will need the following molar mass equivalences:

$$70.90\ g\ Cl_2 = 1\ mol\ Cl_2$$

$$137.32\ g\ PCl_3 = 1\ mol\ PCl_3$$

Solve: The grams of PCl_3 produced are calculated using the sequence of conversions previously given:

$$Grams\ PCl_3 = (2.80\ g\ Cl_2)\left(\frac{1\ mol\ Cl_2}{70.90\ g\ Cl_2}\right)$$

$$\times \left(\frac{4\ mol\ PCl_3}{6\ mol\ Cl_2}\right)\left(\frac{137.32\ g\ PCl_3}{1\ mol\ PCl_3}\right) = 3.62\ g\ PCl_3$$

Check: The final unit has the correct substance, PCl_3, identified. This is the unknown in the question.

EXERCISE 11   Determining the limiting reactant in a chemical reaction and the percent yield

(a) What is the limiting reactant when 10.0 g of C_2H_6 reacts with 50.0 g of O_2 according to the chemical equation

$$2\ C_2H_6(g) + 7\ O_2(g) \longrightarrow 4\ CO_2(g) + 6\ H_2O(l)$$

(b) Calculate the percent yield of the reaction

$$2 \, Al(OH)_3(s) + 3 \, H_2SO_4(aq) \longrightarrow Al_2(SO_4)_3(s) + 6 \, H_2O(l)$$

given that 205 g of $Al(OH)_3$ reacts with 751 g of H_2SO_4 to yield 252 g of $Al_2(SO_4)_3$.

SOLUTION: (a) *Analyze*: We are given masses of the two reactants and asked to determine the limiting reactant. This means we have to identify the limiting reactant and provide an explanation.

Plan: How do we know when a problem involves a limiting reactant? If the quantities of two or more reactants are given, then we should assume it is a limiting reactant problem until we show otherwise. If the quantity of only one reactant is given, then we can assume all other reactants are in excess. There are two general methods for doing limiting reactant problems:

1. Choose one of the reactants and calculate the stoichiometric quantities required for the other reactants to react with it. We then compare the calculated quantities to the given quantities to determine the limiting reactant. If a given quantity is smaller than the calculated quantity, then that reactant is the limiting reactant. If all calculated quantities are equal to or smaller than the given quantities, then the reference reactant used to calculate all other quantities is the limiting reactant.
2. Choose a product in the reaction. If there is a question about theoretical yield, choose the product needed to calculate the theoretical yield. Do not change the chosen product in the following calculations—it must be fixed for proper interpretation of the results. Use each reactant and separately calculate the theoretical amount of the chosen product. *The reactant yielding the smallest quantity of the chosen product is the limiting reactant.*

Procedure (1) above will be used to solve this problem. Calculate the exact mass of O_2 required to react with 10.0 g of C_2H_6. If the available mass of oxygen is greater than the calculated mass of oxygen, then C_2H_6 is the limiting reactant. Conversely, oxygen is the limiting reactant if its available mass is less than the calculated mass. The sequence of conversions required to make this determination is

$$Grams \; C_2H_6 \rightarrow moles \; C_2H_6 \rightarrow moles \; O_2 \rightarrow grams \; O_2$$

The required stoichiometric and molar mass equivalences needed for these conversions are

$$2 \, mol \, C_2H_6 \approx 7 \, mol \, O_2$$

$$30.08 \, g \, C_2H_6 = 1 \, mol \, C_2H_6$$

$$32.00 \, g \, O_2 = 1 \, mol \, O_2$$

Solve: Following the sequence of steps shown above,

$$Grams \, O_2 = (10.0 \, g \, C_2H_6)\left(\frac{1 \, mol \, C_2H_6}{30.08 \, g \, C_2H_6}\right)$$

$$\times \left(\frac{7 \, mol \, O_2}{2 \, mol \, C_2H_6}\right)\left(\frac{32.00 \, g \, O_2}{1 \, mol \, O_2}\right) = 37.2 \, g \, O_2$$

The available mass of O_2, 50.0 g, is greater than the calculated mass (37.2 g) of O_2 required to react completely with 10.0 g of C_2H_6. Therefore, O_2 is in excess, and C_2H_6 is the limiting reagent.

Check: The magnitude of the answer is in agreement with a rough estimation for the calculation: $10\left(\frac{1}{30}\right)\left(\frac{7}{2}\right)\left(\frac{30}{1}\right) = 35$. The units are also correct.

(b) *Analyze* and *Plan*: This problem asks for the percent yield of one of the products; thus, we will need to calculate the theoretical amount of aluminum sulfate. It also may be a limiting reactant problem because we are given the masses of the reactants. To calculate the theoretical amount of aluminum sulfate and also to determine the limiting reactant, we will use procedure (2) because it requires us to calculate the amount of a product. Each reactant is assumed to be the limiting reactant and the theoretical amount of aluminum sulfate is calculated. The limiting reactant will produce the *fewer* number of moles of aluminum sulfate.

Solve: First, assume aluminum hydroxide is the limiting reactant and convert the grams of aluminum hydroxide to grams of aluminum sulfate:

$$\text{g } Al_2(SO_4)_3 = (205 \text{ g } Al(OH)_3)\left(\frac{1 \text{ mol } Al(OH)_3}{78.0 \text{ g}}\right)$$

$$\times \left(\frac{1 \text{ mol } Al_2(SO_4)_3}{2 \text{ mol } Al(OH)_3}\right)\left(\frac{342.1 \text{ g } Al_2(SO_4)_3}{1 \text{ mol } Al_2(SO_4)_3}\right) = 450 \text{ g } Al_2(SO_4)_3$$

Then assume sulfuric acid is the limiting reactant and determine the grams of aluminum sulfate produced:

$$\text{g } Al_2(SO_4)_3 = (751 \text{ g } H_2SO_4)\left(\frac{1 \text{ mol } H_2SO_4}{98.1 \text{ g}}\right)$$

$$\times \left(\frac{1 \text{ mol } Al_2(SO_4)_3}{3 \text{ mol } H_2SO_4}\right)\left(\frac{342.1 \text{ g } Al_2(SO_4)_3}{1 \text{ mol } Al_2(SO_4)_3}\right) = 873 \text{ g } Al_2(SO_4)_3$$

$Al(OH)_3$ is the limiting reagent because it yields the smaller amount of $Al_2(SO_4)_3$ and thus the theoretical amount of $Al_2(SO_4)_3$ is 450 g. The percent yield is

$$\text{Percent yield} = \frac{252 \text{ g}}{450 \text{ g}} \times 100 = 56.0\%$$

Check: The magnitudes of the two answers agree with a rough calculation for each step, $200\left(\frac{1}{80}\right)\left(\frac{1}{2}\right)\left(\frac{350}{1}\right) \simeq 400$ and $750\left(\frac{1}{100}\right)\left(\frac{1}{3}\right)\left(\frac{350}{1}\right) \simeq 800$. The units are also correct.

Note: In Exercises 8 through 11 every sequence of conversions involved moles. The unit mole is the central focus point for most stoichiometric conversions. The reason is that if you know the number of moles of a substance, you can transform it to either grams or number of particles.

Number of particles	$\rightleftharpoons$	Moles	$\rightleftharpoons$	Grams
	Avogadro's number		Molar mass	

Avogadro's number is used to convert between number of particles and number of moles. The mass of one mole of a substance (molar mass) is used to convert between number of moles and grams. Keep the above picture in mind when you do stoichiometric problems.

SELF-TEST QUESTIONS

Key Terms

Having reviewed key terms in Chapter 3, match key terms with phrases and identify statements as true or false. If a statement is false, indicate why it is incorrect.

Match each phrase with the best term:

3.1   Equal numbers of atoms of each element occur on both sides of a reaction arrow.

3.2   Oxygen is typically a reactant in this type of reaction.

3.3 6.022×10^{23} is given this historical name.

3.4   The name given to the mass in amu of a compound containing all nonmetals.

3.5   The name given to the mass in amu of any compound.

3.6   In the following reaction, when 6 mol of Cu and 18 mol of HNO_3 are initially present, we give this term to Cu.

$$3\,Cu + 8\,HNO_3 \longrightarrow 3\,Cu(NO_3)_2 + 2\,NO + 4\,H_2O$$

3.7   The total mass of materials in a chemical reaction does not change.

3.8   The name given to the maximum amount of a product obtainable in a chemical reaction.

Terms:

    (**a**) Avogadro's number
    (**b**) balanced chemical reaction
    (**c**) combustion reaction
    (**d**) formula weight
    (**e**) law of conservation of mass
    (**f**) limiting reactant
    (**g**) molecular weight
    (**h**) theoretical yield

True-False Statements:

3.9   The area of study of *stoichiometry* involves measuring concentrations and percent yield, for example.

3.10   *Reactants* can be gases and liquids.

3.11   *Products* can be liquids and solids.

3.12   To determine if a reaction is a *combination reaction*, you would look to see if two or more products are formed.

3.13   To determine if a reaction is a *decomposition reaction*, you would look to see if two or more reactants exist.

3.14   A *mole* is a direct measure of mass.

3.15   The mass of one mole of a substance is known as *molar mass*.

3.16   Percent yield is defined as

$$\frac{\text{amount of product formed}}{\text{initial amount of reactant}} \times 100$$

Problems and Short-Answer Questions

3.17   In the following boxes unshaded spheres represent oxygen atoms, spheres with a line represent nitrogen atoms and fully shaded spheres represent bromine. Which box contains spheres that represent reactants and which box represents the product for the following reaction: $2\,NO(g) + Br_2(g) \longrightarrow 2\,NOBr(g)$?

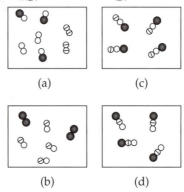

3.18   Unshaded spheres represent atom X and shaded spheres represent atom Y in the following diagram. The diagram shows reactants converted to product in the correct stoichiometric ratio. How many moles of product form if 2.0 mol X_2 react with 1.0 mol Y_2? If there is a limiting reactant, identify it.

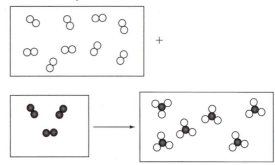

3.19   Balance the following equations:
    (**a**) $AgNO_3(aq) + CaCl_2(aq) \longrightarrow AgCl(s)$
                               $+ Ca(NO_3)_2(aq)$
    (**b**) $VO(s) + Fe_2O_3(s) \longrightarrow FeO(s) + V_2O_5(s)$
    (**c**) $Na(s) + H_2O(l) \longrightarrow NaOH(aq) + H_2(g)$
    (**d**) $NH_4NO_3(s) \longrightarrow N_2O(g) + H_2O(l)$
    (**e**) $MnO_2(s) + HCl(ag) \longrightarrow Cl_2(g) + MnCl_2(aq)$
                               $+ H_2O(l)$

3.20   Write balanced equations for the following reactions:
    (**a**) aqueous silver nitrate reacts with aqueous copper(II) chloride to form insoluble silver chloride and aqueous copper(II) nitrate;
    (**b**) metallic aluminum reacts with oxygen gas to form solid aluminum oxide;
    (**c**) aqueous barium chloride reacts with aqueous potassium sulfate to form solid barium sulfate and aqueous potassium chloride;
    (**d**) solid magnesium chloride reacts with aqueous sodium hydroxide to yield insoluble magnesium hydroxide and aqueous sodium chloride;

(e) solid potassium chlorate decomposes to solid potassium chloride and oxygen gas.

3.21 A sample of a B,H-containing compound is analyzed and it contains 78.2% boron. Is this sufficient information to determine the empirical and molecular formulas of the compound? Explain without doing calculations. In your explanation identify any information that is needed and why.

3.22 Freon-12 has the formula CCl_2F_2. If you are told that a sample has 3.00×10^{20} molecules of freon-12, is this sufficient information to determine the mass of the sample? Explain without doing calculations. In your explanation identify any information that is needed and why.

3.23 You are told that the percent yield for the following reaction is 79.1%: $C_6H_{12}O_6(s) \longrightarrow 2\,C_6H_6O(l) + CO_2(g)$. Is this sufficient information to determine the amount of $C_6H_{12}O_6(s)$ that reacted? Explain without doing calculations. In your explanation identify any information that is needed and why.

3.24 Ethylene, C_2H_4, is used to make the plastic polyethylene.

    (a) What are its molecular weight and formula weight?

    (b) How many moles of C_2H_4 are there in 3.20×10^{-2} g?

    (c) How many molecules of C_2H_4 are there in 3.20×10^{-2} g?

3.25 If a certain quantity of $NO(g)$ has a mass of 5.00 g, what is the quantity of H_2O containing the same number of molecules?

3.26 What is the density of $F_2(g)$ in grams per cubic centimeter if at 0 °C one mole of it occupies 22.4 L?

3.27 Answer the following questions using the reaction

$$2\,KClO_3(s) \longrightarrow 2\,KCl(s) + 3\,O_2(g)$$

    (a) How many moles of $KClO_3$ are required to produce 9 moles of O_2?

    (b) How many moles of KCl are produced if 4.0 moles of O_2 form?

    (c) How many moles of KCl form from the decomposition of 3.0 moles of $KClO_3$?

3.28 The main constituent of gallstones is cholesterol. Cholesterol may have a role in heart attacks and blood clot formation. Its elemental percentage composition is 83.87% C, 11.99% H, and 4.14% O. It has a molecular weight of 386.64 amu. Calculate its empirical and molecular formulas.

3.29 What is the empirical formula of compound such that 200.0 g of it contains 87.2 g of P and 112.8 g of O?

3.30 A stimulant of the nervous system found in coffee, tea, and cola is caffeine. Caffeine contains the following weight percentage of elements: 49.5% C; 28.9% N; 16.5% O; 5.2% H. What is the empirical formula of caffeine?

3.31 $H_2C_2O_4$ and $KMnO_4$ react according to the equation

$$5\,H_2C_2O_4 + 2\,KMnO_4 \longrightarrow 4\,H_2O + 10\,CO_2 \\ + 2\,MnO_2 + 2\,KOH$$

How many grams of CO_2 are formed when 10.05 g of $H_2C_2O_4$ and 26.72 g of $KMnO_4$ are mixed together?

3.32 Silicon carbide, SiC, is an important industrial abrasive. It is formed by the reaction of SiO_2 and carbon at high temperatures:

$$SiO_2 + 3\,C \longrightarrow SiC + 2\,CO$$

    (a) Calculate the number of moles of silicon carbide formed when 5.00 g of carbon reacts with an excess of SiO_2.

    (b) What is the minimum amount of carbon required to react with 25.0 g of SiO_2?

3.33 The reaction for the production of iron from the reduction of the ore hematite, Fe_2O_3, is as follows:

$$Fe_2O_3 + 3\,CO \longrightarrow 2\,Fe + 3\,CO_2$$

    (a) If the reaction yields 4.52 g of CO_2, how many grams of Fe are also formed?

    (b) How many grams of Fe are formed from 7.25 g of Fe_2O_3 and 6.00 g of CO?

3.34 Answer the following questions using the reaction

$$2\,Na(s) + 2\,H_2O(l) \longrightarrow 2\,NaOH(aq) + H_2(g)$$

    (a) If 1.0 mol Na and 1.0 mol H_2O react and the yield is 85.0%, is there a limiting reactant? Does the reaction go to completion?

    (b) If 4.0 mol Na reacts with 6 mol H_2O, what is the limiting reactant and how many moles of the other reactant remain?

    (c) If in problem (b) you had been given 5.0 g Na and 10.0 g H_2O and you were asked to solve for the number of grams of excess reactant, how would the problem have been solved? Explain in words without doing a numerical calculation.

Integrative Exercises

3.35 A 3.15 g sample of KCl contains chlorine-35 and chlorine-37. The fractional abundance of chlorine-35 is 0.75771. This sample is reacted with excess $AgNO_3$ in water to form AgCl in a yield of 85.5%. (a) What is the mass of AgCl? (b) What is the mass of the chlorine-35 atoms in the AgCl?

3.36 A student prepared a sample of an unknown substance containing only nitrogen and oxygen. After studying the composition of the substance the student named it nitrogen (IV) oxide. Other students also studied the substance and found that it contained 30.45% N and 69.55% O and had a molar mass of 91.98 g. Did the original student properly name the compound? Explain.

3.37 Urea is an important compound because it is used as a fertilizer. It is commercially produced from two gases, each containing two different elements; water is the other product formed. The carbon-containing gas is colorless, a major component of exhaled air, and extinguishes a flame. The percentage composition of urea is 20.0% C, 46.6% N, 26.6% O, and 6.7% H, and it has a molar mass of 60.1 g. Given this information and your knowledge of gases obtained from the text, determine the two gases used in the production of urea and write the balanced chemical equation.

3.38 How do the formula or molecular weights of the following differ: $C^{16}O_2$, $C^{18}O_2$, and $K_2^{18}O$? Explain.

Multiple-Choice Questions

3.39 Which of the following equations does not obey the law of conservation of mass?
1. $C_6H_{12}O_6 + 6\,O_2 \rightarrow 6\,CO_2 + 6\,H_2O$
2. $C_2H_6 + 7\,O_2 \rightarrow 4\,CO_2 + 6\,H_2O$
3. $BCl_3 + H_2O \rightarrow H_3BO_3 + 3\,HCl$

(a) (1)   (d) (1) and (2)
(b) (2)   (e) (2) and (3)
(c) (3)

3.40 0.400 moles of a substance weighs 17.6 g. What is its molar mass?

(a) 53.2 g/mol   (d) 7.04 g/mol
(b) 44.0 g/mol   (e) 3.52 g/mol
(c) 17.6 g/mol

3.41 What is the weight percentage of Al in Al_2O_3?

(a) 26.46%   (d) 47.08%
(b) 20.93%   (e) 53.21%
(c) 52.92%

3.42 In photosynthesis, $CO_2(g)$ and $H_2O(l)$ are converted into glucose, $C_6H_{12}O_6$ (a sugar), and O_2. If 0.256 mol of $C_6H_{12}O_6$ is formed by the reaction of CO_2 with water, how many grams of CO_2 would be needed?

(a) 67.6 g   (d) 76.6 g
(b) 11.3 g   (e) 0.256 g
(c) 0.0349 g

3.43 Which of the following pairs of substances, with their indicated quantities, contain(s) the same number of atoms: (1) 0.50 mol HCl, 0.50 mol He; (2) 0.20 mol H_3PO_4, 0.80 mol N_2; (3) 0.45 mol HNO_3, 0.45 mol HNO_2?

(a) (1)   (d) all of them
(b) (2)   (e) none of them
(c) (3)

3.44 What is the mass in grams of 4.25×10^{20} molecules of H_2O?

(a) 0.127 g   (d) 142 g
(b) 0.0127 g   (e) 4.25×10^{20} g
(c) 1420 g

3.45 Which of the following is the limiting reagent if three moles of H_2O and eight moles of NO_2 are available in the reaction:

$$3\,NO_2(g) + H_2O(l) \longrightarrow 2\,HNO_3(l) + NO$$

(a) NO_2 (c) HNO_3
(b) H_2O (d) NO

3.46 What is the percentage yield of C_2H_2 when 50.0 g of $CaC_2(s)$(molar mass = 64.01 g) reacts with an excess of water to yield 13.5 g of $C_2H_2O(l)$(molar mass = 26.04 g) according to the following reaction:

$$CaC_2(s) + 2\,H_2(l) \longrightarrow Ca(OH)_2(s) + C_2H_2(g)$$

(a) 27.0%   (c) 66.5%
(b) 51.8%   (d) 82.5%

3.47 What is the empirical formula of benzoic acid if it has a mass percentage composition of 69% carbon, 5% hydrogen, and 26% oxygen?

(a) $C_7H_6O_2$ (c) C_5H_6O
(b) $C_6H_7O_2$ (d) C_3H_4O

3.48 What is the coefficient in front of CO_2 when the following chemical equation is balanced?

$$C_3H_7OH(l) + O_2(g) \longrightarrow CO_2(g) + H_2O(l)$$

(a) 2   (c) 6   (e) 9
(b) 4   (d) 7

3.49 Which characteristic of a chemical formula typically informs you that it is *not* a molecular formula?

(a) All elements have no subscripts (that is, a subscript of one).
(b) All elements have subscripts greater than one.
(c) All elements are the same type.
(d) One of the elements is a metal.
(e) One of the elements is a nonmetal.

SELF-TEST SOLUTIONS

3.1 (b) **3.2** (c) **3.3** (a) **3.4** (g) **3.5** (d) **3.6** (f) **3.7** (e) **3.8** (h) **3.9** True. **3.10** True. Also, solids. **3.11** True. Also gases. **3.12** False. A single product is formed from two reacting substances. **3.13** False. A single reactant decomposes. **3.14** False. It is a counting number—like a dozen. It is not a measure of mass, although if you know the number of moles of a chemical substance and its formula, you can calculate its mass. **3.15** True. **3.16** False. It is

$$\frac{\text{amount of product formed}}{\text{theoretical amount of product}} \times 100$$

3.17 Box **b** represents the reactants and box **c** represents the product. Note in box **d** that molecules are of the type BrNO, which is not the correct arrangement of atoms shown in the chemical formula of the product, NOBr.

3.18 The first box has nine molecules of X_2 and the second box has three molecules of Y_2. These react to form six molecules of YX_3 in the third box. This can be written as $9X_2 + 3Y_2 \rightarrow 6YX_3$. Dividing by three gives the balanced chemical equation: $3X_2 + Y_2 \rightarrow 2YX_3$. The limiting reactant is X_2 because only 2.0 moles of X_2 are available to react with 1.0 mole of Y_2. The stoichiometry of the chemical reaction requires 3 mol of X_2 to react with 1 mol of Y_2. The number of moles of YX_3 formed is $(2.0\ \text{mol}\ X_2)(2\ \text{mol}\ YX_3/3\ \text{mol}\ X_2) = 1.3\ \text{mol}\ YX_3$.

3.19 (a) $2\,AgNO_3(aq) + CaCl_2(aq) \rightarrow 2\,AgCl(s)$
$+ Ca(NO_3)_2(aq)$

(b) $2\,VO(s) + 3\,Fe_2O_3(s) \rightarrow 6\,FeO(s) + V_2O_5(s)$
(c) $2\,Na(s) + 2\,H_2O(l) \rightarrow 2\,NaOH(aq) + H_2(g)$
(d) $NH_4NO_3(s) \rightarrow N_2O(g) + 2\,H_2O(l)$

(e) $MnO_2(s) + 4\,HCl(aq) \rightarrow Cl_2(g) + MnCl_2(aq)$
$+ 2\,H_2O(l)$

3.20 (a) $2\,AgNO_3(aq) + CuCl_2(aq) \rightarrow 2\,AgCl(s)$
$+ Cu(NO_3)_2(aq)$

(b) $4\,Al(s) + 3\,O_2(g) \rightarrow 2\,Al_2O_3(s)$

(c) $BaCl_2(aq) + K_2SO_4(aq) \rightarrow BaSO_4(s)$
$+ 2\,KCl(aq)$

(d) $MgCl_2(s) + 2\,NaOH(aq) \rightarrow Mg(OH)_2(s)$
$+ 2\,NaCl\,(aq)$

(e) $2\,KClO_3(s) \rightarrow 2\,KCl(s) + 3\,O_2(g)$

3.21 The empirical formula can be determined by (1) calculating the percent oxygen in the compound by subtracting the 78.2% of boron from 100%, (2) determining the mass of each element in an arbitrary sample size from the percent composition data, (3) calculating the relative number of moles of each element in the sample and (4) finally the empirical formula. However, the molecular formula cannot be determined without knowing the molar mass of the compound.

3.22 There is sufficient information. Avogadro's number is used to calculate the number of moles of freon-12 $[(\text{no. molecules})(\frac{1\,\text{mol}}{6.02 \times 10^{23}\,\text{molecules}})]$. The molar mass is calculated from the chemical formula and the mass of the sample is therefore (no. moles)(molar mass).

3.23 There is insufficient information. Percent yield

$= 79.1\% = \frac{\text{amount of product formed}}{\text{theoretical amount of product that should form}} \times 100.$

There are two unknowns in the relationship and neither is provided. If you are given the actual amount of product that forms, you can then calculate the theoretical amount of product that should form. You can then do a stoichiometric conversion to calculate the original quantity of the reactant using the theoretical amount of product:

$$g\;product \longrightarrow mol\;product \longrightarrow mol\;reactant$$
$$\longrightarrow g\;reactant.$$

3.24 (a) Molecular and formula weights are the same; 28.04 amu.

(b) Moles $C_2H_4 = (3.20 \times 10^{-2}\,g\,C_2H_4)$
$\times \left(\dfrac{1\,\text{mol}\,C_2H_4}{28.04\,g\,C_2H_4}\right)$
$= 1.14 \times 10^{-3}\,\text{mol}\,C_2H_4$

(c) Molecules $C_2H_4 = (1.14 \times 10^{-3}\,\text{mol}\,C_2H_4)$
$\times \left(\dfrac{6.022 \times 10^{23}\,\text{molecules}\,C_2H_4}{1\,\text{mol}\,C_2H_4}\right)$
$= 6.87 \times 10^{20}\,\text{molecules}\,C_2H_4$

3.25 Since equal numbers of moles contain equal numbers of molecules, we want a quantity of H_2O such that it is equivalent in number of moles to the number of moles of NO in the given sample weight. The conversions required are $grams\,NO \rightarrow moles\,NO \rightarrow moles\,H_2O \rightarrow grams\,H_2O$. The required equivalences are 1 mol H_2O = 1 mole NO;

30.01 g NO = 1 mol NO; 18.02 g H_2O = 1 mol H_2O. Hence,

Grams $H_2O = (5.00\,g\,NO)\left(\dfrac{1\,\text{mol NO}}{30.01\,g\,NO}\right)\left(\dfrac{1\,\text{mol}\,H_2O}{1\,\text{mol NO}}\right)$
$\times \left(\dfrac{18.02\,g\,H_2O}{1\,\text{mol}\,H_2O}\right) = 3.00\,g\,H_2O$

3.26 The definition of density is: density = mass/volume. The mass of 1 mol of F_2 occupying 22.4 L is the molecular weight of F_2 in grams, which equals 38.0 g. The density of $F_2(g)$ is

$$\text{Density} = \frac{38.0\,g}{(22.4\,L)\left(\dfrac{1000\,\text{mL}}{1\,L}\right)\left(\dfrac{1\,\text{cm}^3}{1\,\text{mL}}\right)}$$
$$= 1.70 \times 10^{-3}\,\frac{g}{\text{cm}^3}$$

3.27 (a) mol $KClO_3 = (9.0\,\text{mol}\,O_2)(\frac{2\,\text{mol}\,KClO_3}{3\,\text{mol}\,O_2})$
$= 6.0\,\text{mol}\,KClO_3$

(b) mol $KCl = (4.0\,\text{mol}\,O_2)(\frac{2\,\text{mol}\,KCl}{3\,\text{mol}\,O_2})$
$= 2.7\,\text{mol}\,KCl$

(c) mol $KCl = (3.0\,\text{mol}\,KClO_3)(\frac{2\,\text{mol}\,KCl}{2\,\text{mol}\,KClO_3})$
$= 3.0\,\text{mol}\,KCl$

3.28 The number of moles of each element in an arbitrary sample size of 100 g is as follows:

Moles $C = (83.37\,g\,C)\left(\dfrac{1\,\text{mol}\,C}{12.01\,g\,C}\right) = 6.983\,\text{mol}\,C$

Moles $H = (11.99\,g\,H)\left(\dfrac{1\,\text{mol}\,H}{1.008\,g\,H}\right) = 11.89\,\text{mol}\,H$

Moles $O = (4.14\,g\,O)\left(\dfrac{1\,\text{mol}\,O}{16.00\,g\,O}\right) = 0.260\,\text{mol}\,O$

The mole ratio of O:H:C is 0.260/0.260:11.89/0.260:6.983/0.260, or 1:45.7:26.9. The empirical formula is $C_{27}H_{46}O$. The empirical formula weight is 386.64 amu. This is the same as the actual molecular weight: thus the molecular formula is $C_{27}H_{46}O$.

3.29 From the masses of P and O in the sample, we can calculate the number of moles of each in the sample:

Moles $P = (87.2\,g\,P)\left(\dfrac{1\,\text{mol}\,P}{30.97\,g\,P}\right) = 2.82\,\text{mol}\,P$

Moles $O = (112.8\,g\,O)\left(\dfrac{1\,\text{mol}\,O}{16.00\,g\,O}\right) = 7.05\,\text{mol}\,O$

The mole ratio of P to O is 2.82/2.82:7.05/2.82, or 1:2.5, or $1:2\frac{1}{2}$ or 2:5. The empirical formula is P_2O_5.

3.30 The moles of each element in an arbitrary sample size of 100 g are as follows:

Moles $C = (49.5\,g\,C)\left(\dfrac{1\,\text{mol}\,C}{12.01\,g\,C}\right) = 4.12\,\text{mol}\,C$

Moles $N = (28.9\,g\,N)\left(\dfrac{1\,\text{mol}\,N}{14.01\,g\,N}\right) = 2.06\,\text{mol}\,N$

$$\text{Moles O} = (16.5 \text{ g O})\left(\frac{1 \text{ mol O}}{16.0 \text{ g O}}\right) = 1.03 \text{ mol O}$$

$$\text{Moles H} = (5.2 \text{ g H})\left(\frac{1 \text{ mol H}}{1.01 \text{ g H}}\right) = 5.15 \text{ mol H}$$

The ratio of moles of O:N:C:H is 1.03/1.03:2.03/1.03:4.12/1.03: 5.15/1.03, or 1:2:4:5. The empirical formula is $C_4H_5N_2O$.

3.31 Reaction equation: $5 \text{ H}_2\text{C}_2\text{O}_4 + 2 \text{ KMnO}_4 \rightarrow 4 \text{ H}_2\text{O} + 10 \text{ CO}_2 + 2 \text{ MnO}_2 + 2 \text{ KOH}$. Let us determine if $KMnO_4$ is in excess. Conversions required: *Grams* $H_2C_2O_4 \rightarrow$ *moles* $H_2C_2O_4 \rightarrow$ *moles* $KMnO_4 \rightarrow$ *grams* $KMnO_4$. Equivalences needed for this calculation: 5 mol $H_2C_2O_4 \simeq 2$ mol $KMnO_4$; 90.4 g $H_2C_2O_4 = 1$ mol $H_2C_2O_4$; 158.0 g $KMnO_4 = 1$ mol $KMnO_4$. Calculating the amount of $KMnO_4$ required to react with 10.05 g $H_2C_2O_4$:

$$\text{Grams KMnO}_4 = (10.05 \text{ g H}_2\text{C}_2\text{O}_4)\left(\frac{1 \text{ mol H}_2\text{C}_2\text{O}_4}{90.04 \text{ g H}_2\text{C}_2\text{O}_4}\right)$$

$$\times \left(\frac{2 \text{ mol KMnO}_4}{5 \text{ mol H}_2\text{C}_2\text{O}_4}\right)\left(\frac{158.0 \text{ g KMnO}_4}{1 \text{ mol KMnO}_4}\right)$$

$$= 7.06 \text{ g KMnO}_4$$

$KMnO_4$ is in excess because 26.72 g is available but only 7.06 g is required. To calculate the mass of CO_2 formed, we must make the following conversions: *grams* $H_2C_2O_4 \rightarrow$ *moles* $H_2C_2O_4 \rightarrow$ *moles* $CO_2 \rightarrow$ *grams* CO_2. Additional equivalences needed are 5 mol $H_2C_2O_4 \simeq 10$ mol CO_2; 44.01 g $CO_2 = 1$ mol CO_2. Hence,

$$\text{Grams CO}_2 = (10.05 \text{ g C}_2\text{H}_2\text{O}_4)\left(\frac{1 \text{ mol H}_2\text{C}_2\text{O}_4}{90.05 \text{ g H}_2\text{C}_2\text{O}_4}\right)$$

$$\times \left(\frac{10 \text{ mol CO}_2}{5 \text{ mol H}_2\text{C}_2\text{O}_4}\right)\left(\frac{44.1 \text{ g CO}_2}{1 \text{ mol CO}_2}\right)$$

$$= 9.825 \text{ g CO}_2$$

3.32 Reaction equation: $SiO_2 + 3 \text{ C} \rightarrow SiC + 2 \text{ CO}$.

(a) Conversions: *grams* $C \rightarrow$ *moles* $C \rightarrow$ *moles* SiC. Equivalences needed: 12.01 g C = 1 mol C; 3 mol C $\simeq$ 1 mol SiC.

$$\text{Moles SiC} = (5.00 \text{ g C})\left(\frac{1 \text{ mol C}}{12.01 \text{ g C}}\right)\left(\frac{1 \text{ mol SiC}}{3 \text{ mol C}}\right)$$

$$= 0.139 \text{ mol SiC}$$

(b) Conversions: *Grams* $SiO_2 \rightarrow$ *moles* $SiO_2 \rightarrow$ *moles* $C \rightarrow$ *grams* C. Additional equivalences needed: 1 mol SiC $\simeq$ 3 mol C; 60.09 g $SiO_2 = 1$ mol SiO_2.

$$\text{Grams C} = (25.00 \text{ g SiO}_2)\left(\frac{1 \text{ mol SiO}_2}{60.09 \text{ g SiO}_2}\right)$$

$$\times \left(\frac{3 \text{ mol C}}{1 \text{ mol SiO}_2}\right)\left(\frac{12.01 \text{ g C}}{1 \text{ mol C}}\right)$$

$$= 14.99 \text{ g C}$$

3.33 Reaction equation: $Fe_2O_3 + 3 \text{ CO} \rightarrow 2 \text{ Fe} + 3 \text{ CO}_2$.

(a) Conversions: *Grams* $CO_2 \rightarrow$ *moles* $CO_2 \rightarrow$ *moles* $Fe \rightarrow$ *grams* Fe. Equivalences needed: 44.01 g $CO_2 = 1$ mol CO_2; 2 mol Fe $\simeq$ 3 mol CO_2; 55.85 g Fe = 1 mol Fe.

$$\text{Grams Fe} = (4.52 \text{ g CO}_2)\left(\frac{1 \text{ mol CO}_2}{44.01 \text{ g CO}_2}\right)$$

$$\times \left(\frac{2 \text{ mol Fe}}{3 \text{ mol CO}_2}\right)\left(\frac{55.85 \text{ g Fe}}{1 \text{ mol Fe}}\right)$$

$$= 3.82 \text{ g Fe}$$

(b) This is another limiting reactant problem. Let us calculate the amount of CO that reacts with 7.25 g of Fe_2O_3. The required conversions are: *Grams* $Fe_2O_3 \rightarrow$ *moles* $Fe_2O_3 \rightarrow$ *moles* $CO \rightarrow$ *grams* CO. Equivalences needed: 159.7 g $Fe_2O_3 = 1$ mol Fe_2O_3; 1 mol $Fe_2O_3 \simeq 3$ mol CO; 28.01 g CO = 1 mol CO.

$$\text{Grams CO} = (7.25 \text{ g Fe}_2\text{O}_3)\left(\frac{1 \text{ mol Fe}_2\text{O}_3}{159.7 \text{ g Fe}_2\text{O}_3}\right)$$

$$\times \left(\frac{3 \text{ mol CO}}{1 \text{ mol Fe}_2\text{O}_3}\right)\left(\frac{28.01 \text{ g CO}}{1 \text{ mol CO}}\right)$$

$$= 3.81 \text{ g CO}$$

Since 6.00 g of CO is available and only 3.82 g is required, CO is in excess. To calculate the amount of Fe formed, we need to make the following conversions. *Grams* $Fe_2O_3 \rightarrow$ *moles* $Fe_2O_3 \rightarrow$ *moles* $Fe \rightarrow$ *grams* Fe. Additional equivalences needed: 1 mol $Fe_2O_3 \simeq 2$ mol Fe; 55.85 g Fe = 1 mol Fe.

$$\text{Grams Fe} = (7.25 \text{ g Fe}_2\text{O}_3)\left(\frac{1 \text{ mol Fe}_2\text{O}_3}{159.7 \text{ g Fe}_2\text{O}_3}\right)$$

$$\times \left(\frac{2 \text{ mol Fe}}{1 \text{ mol Fe}_2\text{O}_3}\right)\left(\frac{55.85 \text{ g Fe}}{1 \text{ mol Fe}}\right)$$

$$= 5.07 \text{ g Fe}$$

3.34 (a) The reactants combine in a one-to-one mole ratio, and the given ratio of moles of the reactants is 1.0, thus there is no limiting reactant. The percent yield tells you that only 85% of the theoretical maximum amount of the products formed; the reaction did not go to completion.

(b) The ratio of the given number of moles is $\frac{6.0 \text{ mol H}_2\text{O}}{4.0 \text{ mol Na}}$ or 1.5. This is larger than the stoichiometric ratio of 1.0. The larger ratio occurs because the numerator is relatively larger than the denominator term. Thus, water must be present in a larger quantity than needed. Na is the limiting reactant. The number of moles of water that reacts is also 4.0 mol because the stoichiometric equivalence is $\frac{2.0 \text{ mol H}_2\text{O}}{2.0 \text{ mol Na}}$. Therefore, the number of moles of water remaining unreacted is 6.0 mol H_2O − 4.0 mol H_2O = 2.0 mol H_2O.

(c) The only difference would be an extra step in converting the number of grams of each reactant to number of moles. You would do the same analysis using the number of moles of reactants to determine the limiting reactant and the number of moles of excess reactant. Finally, you would convert the number of moles of excess reactant to grams of excess reactant.

3.35 **(a)** The number of moles of KCl is (3.15 g) $(1 \text{ mol}/74.55 \text{ g}) = 0.0423 \text{ mol KCl}$. The number of moles of Cl in the sample is the same because there is a 1:1 mole ratio of K:Cl. The number of grams of Cl in the sample is $(0.0423 \text{ mol})(35.45 \text{ g}/1 \text{ mol}) = 1.50 \text{ g Cl}$. The theoretical quantity of AgCl is the amount produced assuming the reaction yields 100%. Therefore, the number of moles of Cl in the theoretical sample is 0.0423 mol Cl assuming all of the chlorine atoms in KCl are combined with silver atoms. However, since the yield is less than 100%, not all chlorine atoms are combined with silver atoms. The number of moles of Cl combined with AgCl is equal to: (percent yield in fractional form)(theoretical number of moles of Cl in AgCl) $= (0.855)(0.0423 \text{ mol Cl}) = 0.0362 \text{ mol Cl}$. This number also represents the number of moles of AgCl formed since the Ag:Cl mole ratio is 1:1. The number of grams of AgCl formed is thus: (0.0362 molAgCl) $(143.32 \text{ g}/1 \text{ mol}) = 5.19 \text{ g AgCl}$.

(b) The number of grams of chlorine-35 in the sample of AgCl is: (fractional abundance of Cl-35) $\times$ (mass Cl in *AgCl sample*) $= (0.75771)(0.0362 \text{ mol Cl})(35.45 \text{ g}/1 \text{ mol})$ $= 0.972 \text{ g } Cl\text{-}35$.

3.36 To determine if the correct name is used, you need to determine the molecular formula of the unknown substance. First calculate the empirical formula by assuming a 100.0 g sample:

$$\text{mol N} = (100.0 \text{ g})(0.3045)(1 \text{ mol N}/14.0067 \text{ g})$$
$$= 2.165 \text{ mol N}$$

$$\text{mol O} = (100.0 \text{ g})(0.6955)(1 \text{ mol O}/15.994 \text{ g})$$
$$= 4.349 \text{ mol O}$$

Thus, the empirical formula is $N_{2.165}O_{4.349}$ or NO_2. The mass of this empirical formula is 45.99 g. The number of empirical units in one molecular formula is $(91.98 \text{ g})/(45.99 \text{ g}) = 2.000$ or simply 2. Therefore, the molecular formula is N_2O_4. The proper name of this substance is dinitrogen tetraoxide. The name nitrogen(IV) oxide, or NO_2, only gives the name of the empirical formula.

3.37 First determine the empirical formula of urea from the composition data. Assume a 100 g sample to determine the relative number of moles of each element. C: $(20.0 \text{ g})(1 \text{ mol}/12.01 \text{ g}) = 1.67 \text{ mol C}$. N: $(46.6 \text{ g})(1 \text{ mol}/14.01 \text{ g}) = 3.33 \text{ mol N}$. O: $(26.6 \text{ g})(1 \text{ mol}/16.0 \text{ g}) = 1.66 \text{ mol O}$. H: $(6.7 \text{ g})(1 \text{ mol}/1.01 \text{ g}) = 6.6 \text{ mol H}$. Divide by the smallest number of moles to determine the ratio of moles of atoms in the empirical formula: 1.66 mol is the smallest. CN_2OH_4 is the result. The molar mass of this empirical formula is 60.1 g, which is the same as the molar mass. Therefore, this formula also represents the molecular formula. The reactants used in the chemical reaction to form urea are two gases in which each contain two different elements; thus the possible combinations of elements are CN, CH, CO, NO, HO, and HN. From your reading of the text, three carbon-containing gases that you may have encountered are CH_4 (methane), CO (carbon monoxide), and CO_2 (carbon dioxide). Methane is a combustible gas and would not extinguish a flame. Of the two carbon oxides, carbon dioxide is the logical choice since it is a gas produced in the

lungs, is exhaled during breathing, and can extinguish flames. If carbon dioxide is one reactant, then the other reactant must contain the elements N and H so that urea can be produced. The most likely gas is ammonia, NH_3. Other combinations such as N_2H_4 are not commonly used gases. Thus, the balanced chemical reaction must be:

$$2 \text{ NH}_3 + \text{CO}_2 \longrightarrow \text{CO(NH}_2)_2 + \text{H}_2\text{O}$$

3.38 The formula weight (and the molecular weight) of $C^{18}O_2$ is larger than that of $C^{16}O_2$ because the oxygen-18 isotope has a greater isotopic mass than that of the oxygen-16 isotope. $K_2{}^{18}O$ has a greater formula weight compared to the other two substances because of the larger mass of the potassium atom coupled with two atoms of it present in the substance. The term molecular weight is not used with $K_2{}^{18}O$ because it is ionic and no molecular unit exists.

3.39 **(e)** $2 \text{ C}_2\text{H}_6 + 7 \text{ O}_2 \longrightarrow 4 \text{ CO}_2 + 6 \text{ H}_2\text{O}$
$$\text{BCl}_3 + 3 \text{ H}_2\text{O} \longrightarrow \text{H}_3\text{BO}_3 + 3 \text{ HCl}$$

3.40 **(b)** Molar mass $= 17.6 \text{ g}/0.400 \text{ mol} = 44.0 \text{ g/mol}$

3.41 **(c)**

$$\% \text{ Al} = \left(\frac{(2 \text{ mol Al})\left(\dfrac{26.98 \text{ g Al}}{1 \text{ mol Al}} \right)}{101.96 \text{ g Al}_2\text{O}_3} \right) \times 100$$
$$= 52.92\%$$

3.42 **(a)** Chemical equation:

$$6 \text{ CO}_2 + 6 \text{ H}_2\text{O} \longrightarrow \text{C}_6\text{H}_{12}\text{O}_6 + 6 \text{ O}_2.$$

$$\text{Grams CO}_2 = (0.256 \text{ mol C}_6\text{H}_{12}\text{O}_2)$$
$$\times \frac{(6 \text{ mol CO}_2)}{(1 \text{ mol C}_6\text{H}_{12}\text{O}_2)} \times \frac{(44.01 \text{ g CO}_2)}{(1 \text{ mol CO}_2)}$$
$$= 67.6 \text{ g CO}_2$$

3.43 **(b)**

$$\left(\frac{8 \text{ mol of atoms in H}_3\text{PO}_4}{1 \text{ mol H}_3\text{PO}_4} \right) \times (0.20 \text{ mol H}_3\text{PO}_4)$$
$$= 1.6 \text{ mol of atoms}$$
$$\left(\frac{2 \text{ mol of atoms in N}_2}{1 \text{ mol N}_2} \right)(0.80 \text{ mol N}_2)$$
$$= 1.6 \text{ mol of atoms}$$

3.44 **(b)**

$$\text{Grams H}_2\text{O} = (4.25 \times 10^{20} \text{ molecules})$$
$$\times \left(\frac{1 \text{ mol}}{6.022 \times 10^{23} \text{ molecules}} \right)\left(\frac{18.02 \text{ g H}_2\text{O}}{1 \text{ mol H}_2\text{O}} \right)$$
$$= 0.0127 \text{ g H}_2\text{O}$$

3.45 **(a)** NO_2 and H_2O react in a 3:1 ratio of moles. The given moles are in the ratio 8:3, or $2\frac{2}{3}:1$. The last ratio shows that there is an insufficient quantity of NO_2 to react completely with water.

3.46 **(c)** $50.0 \text{ g} \times (1 \text{ mol}/64.1 \text{ g}) \times (26.04 \text{ g C}_2\text{H}_2/1 \text{ mol})$ $= 20.3 \text{ g}$. $(13.5 \text{ g}/20.3 \text{ g}) \times 100 = 66.5\%$

3.47 **(a)**

3.48 **(c)**

3.49 **(d)**

Sectional MCAT and DAT Practice Questions I

Hydrocarbons are substances containing only carbon and hydrogen atoms. Alkanes are hydrocarbons containing a chain of carbon atoms linked together by single C-C bonds. Alkanes are found in the solar system; for example they are found in the tails of comets and form a key component of the atmospheres of Jupiter and Saturn.

Natural gas and oil deposits on Earth are the primary commercial sources of alkanes. Natural gas consists mostly of methane and ethane with small quantities of propane and butane. Natural gas was formed from high temperatures and pressures acting on dead marine animals and plants which sank to the bottoms of ancient oceans and then were covered with sediment. Forming methane in this situation is simplistically represented by the following reaction.

$$C_6 H_{12} O_6 \xrightarrow{\text{Heat and Pressure}} CH_4 + CO_2$$

Equation 1

TABLE 1 Formula, density, melting point of four alkanes

Name	Formula	Density	Melting point (K)
Methane	CH_4	0.717 kg/m³, gas	90.6
Ethane	C_2H_6	1.212 kg/m³, gas	90.3
Propane	C_3H_8	1.83 kg/m³, gas	85.5
Butane	C_4H_{10}	2.52 g/L, gas (15 °C, 1 atm)	134.9

Melting point is the temperature at which a crystalline solid changes its physical state from solid to a liquid. Boiling point is the temperature at which a liquid changes its physical state to a gas at a given pressure.

The molar masses of carbon and hydrogen are 12.01 g/mol and 1.01 g/mol, respectively. Avogadro's number is 6.02×10^{23}. At zero Kelvin the equivalent centigrade temperature is $-273.15\,°C$. A liter is equivalent to 10^{-3} m³.

1. What is the molecular formula of a ten-carbon alkane?
   (a) $C_{10}H_{10}$ (b) $C_{10}H_{12}$ (c) $C_{10}H_{22}$ (d) $C_{10}H_{30}$
2. The melting point of propane in degrees centigrade is:
   (a) 187.7   (b) -85.5 (c) -187.7 (d) -358.7
3. The density of butane in units of kg/m³ is:
   (a) 2.52×10^6 (b) 2.52×10^3 (c) 2.52   (d) 2.52×10^{-3}

4. The number of significant figures in the density of methane is:
    (a) infinite   **(b)** 2   (c) 3   **(d)** 4

5. What is the empirical formula for butane?
    (a) CH   **(b)** CH_2 (c) C_2H_5 **(d)** C_4H_{10}

6. How many total atoms of hydrogen in methane molecules are shown when Equation 1 is balanced?
    (a) 16   **(b)** 12   (c) 8   **(d)** 4

7. How many molecules of ethane are contained in 60.12 grams?
    (a) 3.01×10^{23} **(b)** 6.02×10^{23}
    (c) 1.20×10^{24} **(d)** 2.4×10^{24}

Questions 8 through 12 are **not** based on a descriptive passage.

---

8. Which is an ionic compound?
    (a) CO_2 **(b)** PCl_3 (c) NO_2 **(d)** $BaCl_2$

9 How many electrons does $^{24}_{12}Mg^{2+}$ possess?
    (a) 10   **(b)** 12   (c) 14   **(d)** 24

10 Which pair represents the chemical formulas for titanium (II) oxide and titanium (IV) oxide?
    (a) TiO and TiO_2 **(b)** Ti_2O and Ti_4O
    (c) TiO_2 and TiO_4 **(d)** Ti_2O_2 and TiO_4

11 Which is correctly named?
    (a) H_3PO_4—phosphorous acid
    (b) PCl_3—phosphorus chloride
    (c) NO_2—nitrogen dioxide
    (d) SO_3—sulfur trioxygen

12. Which prefix correctly matches 10^{-3}?
    (a) mega-   **(b)** deci-   (c) micro-   **(d)** milli-

ANSWERS

1. (c) The formulas of the given alkanes in Table 1 show a general trend: C_nH_{2n+2} where n is the number of carbon atoms. When n is ten, the formula is $C_{10}H_{22}$.

2. (c) The descriptive passage tells us that 0 K = $-273.15\,°C$. Thus, K = °C + 273.15, or 85.5 K − 273.15 = $-187.7\,°C$.

3. (c) $2.52\dfrac{g}{L}\left(\dfrac{1\,kg}{10^3\,g}\right)\left(\dfrac{1\,L}{10^{-3}\,m^3}\right) = 2.52\dfrac{kg}{m^3}$

4. (c) 0.717 has three significant figures; the zero before the decimal is not significant.

5. (c) Diving the subscripts in C_4H_{10} by two gives lowest whole-number ratio of atoms, C_2H_5.

6. (b) When the equation is balanced, the balancing coefficient for methane is three. Thus three CH_4 molecules contain a total of
3 molecules × 4 atoms H/molecule of hydrogen.

7. (c) $60.12\,g\left(\dfrac{1\,mol}{30.08\,g}\right)\left(\dfrac{6.02 \times 10^{23}\,molecules}{mole}\right) = 1.20 \times 10^{24}\,molecules$

8. (d) An ionic compound consists of a metal cation and a nonmetal anion.

9. (a) The atomic number is 12, which corresponds to the number of electrons in the neutral atom. It is a 2+ ion; thus, it has lost two electrons and has a total of 10 electrons.

10. (a) The oxide ion has a 2− charge or oxidation number. Thus, with titanium atoms of 2+ and 4+ oxidation numbers, the formulas are correctly shown in **(a)**.

11. (c) The others are correctly named as: **(a)** phosphoric acid; **(b)** phosphorus trichloride; and **(d)** sulfur trioxide.

12. (d)

Aqueous Reactions and Solution Stoichiometry

OVERVIEW OF THE CHAPTER

Review: Solutions (1.2), Nomenclature of acids, bases, and salts (2.8).

Learning Goals: You should be able to:

1. Predict whether a substance is a nonelectrolyte, strong electrolyte, or weak electrolyte from its chemical behavior.
2. Predict the ions formed by electrolytes when they dissociate or ionize.
3. Identify substances as acids, bases, or salts.

Learning Goals: You should be able to:

1. Use solubility rules to predict whether a precipitate forms when electrolyte solutions are mixed.
2. Predict the products of metathesis reactions (including both neutralization and precipitation reactions) and write balanced chemical equations for them.
3. Identify the spectator ions and write the net ionic equations for solution reactions, starting with their molecular equations.

Learning Goals: You should be able to:

1. Determine whether a chemical reaction involves oxidation and reduction.
2. Assign oxidation numbers to atoms in molecules and ions.
3. Use the activity series to predict whether a reaction will occur when a metal is added to an aqueous solution of either a metal salt or an acid, and write the balanced molecular and net ionic equations for the reaction.

Learning Goals: You should be able to:

1. Calculate molarity, solution volume, or number of moles of solute given any two of these quantities.
2. Calculate the volume of a more concentrated solution that must be diluted to obtain a given quantity of a more dilute solution.

4.6 SOLUTION STOICHIOMETRY

Review: Stoichiometry of chemical reactions (3.6), limiting reactants (3.7).

Learning Goals: You should be able to:

1. Calculate the volume of a solution required to react with a volume of a different solution using molarity and the stoichiometry of the reaction.
2. Calculate the amount of a substance required to react with a given volume of a solution using molarity and the stoichiometry of the reaction.
3. Calculate the concentration or mass of solute in a sample from titration data.

TOPIC SUMMARIES AND EXERCISES

AQUEOUS SOLUTIONS: ELECTROLYTES, ACIDS, AND BASES

In Chapter 2 of the text we learned that a solution is a homogeneous mixture. Key definitions include:

- A solution contains a solvent and solute(s).
- The **solvent** is the substance in a mixture that maintains its physical state and usually is in the greatest amount.
- Whatever else is dissolved in the solution is called the **solute**. A solute may or may not maintain its physical state.

An **electrolyte** is a substance that causes a solution to be a better electrical conductor than the pure solvent. This occurs when the solute forms ions in the solution. The different types of electrolytes are summarized below:

1. **Strong electrolytes:** Effectively ionize or dissociate 100% in a solvent. Examples are:

   - Most ionic compounds (salts)
   - Strong acids and bases

2. **Weak electrolytes:** Incompletely ionize or dissociate ($<$100%) in a solvent. Examples are:

   - Weak acids
   - Weak bases

3. **Nonelectrolytes:** These do not ionize or dissociate in a solvent.

   - If a substance is not one of the above examples, we can assume it is a nonelectrolyte in water.
   - Sugar and alcohol (ethanol) are two examples.

Do not confuse solubility with whether a substance is a weak or strong electrolyte.

- Solubility refers to the quantity of a solute dissolved in a solvent.
- The form of a solute in solution determines whether it is an electrolyte. $BaSO_4$ is only very slightly soluble in water. However, essentially all of

the dissolved $BaSO_4$ exists in solution as $Ba^{2+}(aq)$ and $SO_4^{2-}(aq)$; therefore it is a strong electrolyte. Unless otherwise noted we will assume insoluble (slightly soluble) salts are strong electrolytes in water.

Acids, bases, and salts are commonly encountered in your general chemistry course. Therefore, it is important that you become thoroughly familiar with common acids and bases and know their properties.

- **Acids** are substances that contain one or more hydrogen atoms that ionize to form $H^+(aq)$ in water. The chemical formulas of acids usually have the element hydrogen listed as the first element. Acids that lose only one hydrogen atom are referred to as *monoprotic* and those that lose two hydrogen atoms are *diprotic*. Two types of acids are discussed:

  1. **Strong acids:** These are strong electrolytes. Memorize the acids listed in Table 4.2 in the text. Note that hydrochloric acid (HCl) is an example of a monoprotic acid, and sulfuric acid (H_2SO_4) is an example of a diprotic.
  2. **Weak acids:** These are weak electrolytes. Examples of these are acetic acid ($HC_2H_3O_2$), hydrocyanic acid (HCN), hydrofluoric acid (HF), and hydrosulfuric acid (H_2S).

- **Bases** are substances that accept H^+ ions in chemical reactions.

  1. **Strong bases:** Metallic hydroxides that dissociate 100% in water such as NaOH.
  2. **Weak bases:** The most common one you will encounter is ammonia (NH_3) and occasionally amines (RNH_2, R_2NH, and R_3N where R represents a carbon-based group—you will learn more about these in a future chapter).

- **Neutralization** is a reaction in which an acid reacts with a base to form water and a **salt** (contains cation of base and anion of acid). For example:

$$\begin{array}{ccc} \text{Acid} & \text{Base} & \text{Salt} \\ HBr(aq) + KOH(aq) & \longrightarrow & H_2O(l) + KBr(aq) \end{array}$$

Means a substance is dissolved in water	Means water is liquid and in this example it is also the solvent

When writing dissociation or ionization reactions of strong and weak electrolytes, we use single and double arrows:

- Single arrow: Strong electrolytes or strong acids and bases

$$HCl(g) \xrightarrow{H_2O} H^+(aq) + Cl^-(aq)$$

- Double arrows: Weak electrolytes or weak acids and bases

$$HC_2H_3O_2(aq) \rightleftharpoons H^+(aq) + C_2H_3O_2^-(aq)$$

EXERCISE 1   Identifying types of electrolytes

Classify each of the following as a strong electrolyte, weak electrolyte, or non-electrolyte in water: HBr; H_2S; NH_3; $Ba(OH)_2$; KCl; C_6H_6; I_2.

SOLUTION: *Analyze*: We are given the formulas of seven substances and asked to predict what type of electrolyte each is in water.

Plan: We can use Tables 4.2 and 4.3 in the text to help us.

Solve: $Ba(OH)_2$ and KCl are ionic (salts) and therefore they are strong electrolytes. You can use Table 4.2 to determine if a H_nX compound is an acid. HBr is a strong acid and thus a strong electrolyte. H_2S is a weak acid and thus a weak electrolyte (it is not listed as a strong acid, thus we can reasonably conclude it is a weak acid). NH_3 is listed as a weak base and thus is a weak electrolyte. C_6H_6 and I_2 are not found in the tables; however both are molecular substances because they do not contain metals. They are not listed as acids or bases (the H of C_6H_6 comes after the carbon and this suggests that it is not acidic; also iodine is a diatomic element) and therefore this suggests they are nonelectrolytes.

Comment: It is important to learn the information contained in Tables 4.2 and 4.3. This may require using flash cards or other learning techniques. We will be using this information frequently in answering questions.

EXERCISE 2   Identifying acids and bases and describing their reactions in water

(**a**) Identify each of the following as an acid or base and write its reaction with water: HF(*g*); H_2SO_4(*l*); NaOH(*s*); $Ba(OH)_2$(*s*). (**b**) Write the neutralization reaction between $Ba(OH)_2$(*s*) and HNO_3(*aq*).

SOLUTION: *Analyze*: (**a**) We are asked to determine whether the substances are acids and bases and then write an equation that shows how they react with water.

(**b**) We are given two substances and asked to write a neutralization reaction.

Plan: (**a**) The acids have a hydrogen atom listed as the first element in the chemical formulas and the bases contain one or more hydroxide units. Strong acids and bases completely ionize and weak acids and bases only partially ionize; single and double arrows are used to show the differences. (**b**) A neutralization reaction between an acid and base yields a salt and water. We have to use the information in the chapter to determine which is an acid and base; the approach for this task is similar to that in (**a**).

Solve: (**a**) Using the approach in the *Plan*, we obtain:

Comment: As in the previous problem we need specific information to answer this type of question. Drill and practice in learning the names and properties of acids

Compound	Type	Reaction with Water
HF	Acid	$HF(g) \xrightleftharpoons{H_2O} H^+(aq) + F^-(aq)$
H_2SO_4	Acid (diprotic)	$H_2SO_4(l) \xrightarrow{H_2O} H^+(aq) + HSO_4^-(aq)$
		$HSO_4^-(aq) \rightleftharpoons H^+(aq) + SO_4^{2-}(aq)$
NaOH	Base	$NaOH(s) \xrightarrow{H_2O} Na^+(aq) + OH^-(aq)$
$Ba(OH)_2$	Base	$Ba(OH)_2(s) \xrightarrow{H_2O} Ba^{2+}(aq) + 2\,OH^-(aq)$

and bases are important to success in general chemistry.

(b) Using the approach in the *Plan*, we obtain:

$$2\,HNO_3(aq) \;+\; Ba(OH)_2(s) \;\longrightarrow\; 2\,H_2O(l) \;+\; Ba(NO_3)_2(aq)$$

Acid   Base   Water   Salt

Note that the salt formed consists of the cation of the base (Ba^{2+}) and the anion of the acid (NO_3^-). Also notice that because Ba has a 2+ charge two NO_3^- ions are needed to balance the 2+ charge when writing the formula of $Ba(NO_3)_2$.

Section 4.2 of this chapter introduces reactions in water that result in the formation of a solid substance, a precipitate. A precipitate is insoluble in the solvent. One of the goals in this section is to learn which ionic substances are soluble and insoluble in water. A table of solubilities is given in Table 4.1 in the text. Note that all ionic compounds of the alkali metals (group 1A), any ionic compound containing the nitrate ion (NO_3^-) or acetate ion ($C_2H_3O_2^-$), and any ionic compound containing the ammonium ion (NH_4^+) are soluble in water. The other rules are easier to remember if we know these.

Another goal in this section of the text is to predict whether a precipitate forms when two aqueous solutions containing ionic substances are mixed together. How do we do this?

1. If we are given only the reactants, we can carry out an exchange reaction (*metathesis* reaction) as follows:

$$AX + BY \longrightarrow AY + BX$$

2. Identify the phase of each reactant. Soluble substances are assigned an aqueous phase, (*aq*). Insoluble substances in water are assigned a solid phase, (*s*). Use solubility rules to help you identify soluble and insoluble substances.
3. Balance the chemical equation.

Another way of writing reactions involving ions in solution is writing net ionic equations. **Net ionic equations** show only ions, solids, gases, and weak or nonelectrolytes involved in chemical reactions.

• A chemical equation showing the *complete* chemical formulas of reactants and products, such as $HNO_3(aq) + NH_3(aq) \longrightarrow NH_4NO_3(aq)$, is called a **molecular equation**.
• To form a net ionic equation from a molecular equation, rewrite the molecular equation so that all *soluble strong* electrolytes are written in their ion form in solution. Nonelectrolytes, weak electrolytes, gases, and insoluble salts are not changed. Eliminate all ions common to both reactants and products; these are termed **spectator ions**. The prior example of a molecular equation becomes

$$H^+(aq) + \cancel{NO_3^-}(aq) + NH_3(aq) \longrightarrow NH_4^+(aq) + \cancel{NO_3^-}(aq)$$

HNO$_3$ is a
strong electrolyte

NH$_3$ is a
weak electrolyte
and is not altered

NH$_4$NO$_3$ is a
strong electrolyte

• The ionic reaction that remains after removing spectator ions is the net ionic equation:

$$H^+(aq) + NH_3(aq) \longrightarrow NH_4^+(aq)$$

PRECIPITATION REACTIONS: IONIC EQUATIONS

EXERCISE 3   Using solubility rules

Without referring to Table 4.1 in the text, identify the following salts as soluble or insoluble in water: (a) $Fe(NO_3)_3$; (b) $PbCl_2$; (c) $CaBr_2$; (d) $BaSO_4$; (e) Na_2SO_4; (f) K_2CO_3.

SOLUTION: *Analyze*: We are asked to identify the solubility of several salts in water based on having learned the rules of solubility.

Plan: We can apply the rules of solubility and our knowledge of the properties of salts to answering the question. We should look carefully at the cation and anion of each salt to help us determine if it is soluble or insoluble.

Solve: (a) Soluble. All nitrate salts are soluble; (b) Insoluble. This is one of the three exceptions to the rule that chloride salts are soluble. $AgCl$, Hg_2Cl, and $PbCl_2$ are insoluble. (c) Soluble. All bromide salts are soluble except $AgBr$, Hg_2Br_2, $PbBr_2$, and $HgBr_2$; (d) Insoluble. $BaSO_4$ is one of the exceptions to the rule that sulfate salts are soluble; (e) Soluble. All alkali metal salts are soluble; (f) Soluble. All alkali metal salts are soluble.

EXERCISE 4   Writing net ionic equations I

Write a net ionic equation for the following reaction:

$$NaCl(aq) + AgNO_3(aq) \longrightarrow AgCl(s) + NaNO_3(aq)$$

SOLUTION: *Analyze*: We are asked to write a net ionic equation having been given a molecular equation.

Plan: Rewrite the molecular equation so that *soluble, strong* electrolytes are in their ion form in water. Eliminate any spectator ions and write the net ionic equation:

Solve:

$$\cancel{Na^+}(aq) + Cl^-(aq) + Ag^+(aq) + \cancel{NO_3^-}(aq) \longrightarrow AgCl(s) + \cancel{Na^+}(aq) + \cancel{NO_3^-}(aq)$$

$AgCl$ is an insoluble salt and remains in a combined form. All other species are strong electrolytes. Na^+ and NO_3^- ions are spectator ions. The net ionic equation is

$$Ag^+(aq) + Cl^-(aq) \longrightarrow AgCl(s)$$

Check: The number of each type of atom is the same on both sides of the arrow and the charges balance (zero on both sides of the arrow). No spectator ion is shown. Chemical formulas are correct.

EXERCISE 5   Writing net ionic equations II

Complete the following ionic reactions, and write net ionic equations for them.

(a) $Pb(NO_3)_2(aq) + KBr(aq) \longrightarrow$
(b) $NiCl_2(aq) + Na_3PO_4(aq) \longrightarrow$

SOLUTION: *Analyze*: We are given reactants in each problem and asked to complete the reactions and also write net ionic equations.

Plan: First determine the nature of the reactants. Are they salts? If they are both salts, then we can predict the products by doing a double displacement of the ions. After doing the double displacement reaction and correctly writing the chemical formulas of the products (use the charges of the ions to help you determine the subscripts for each ion in a chemical formula), we should identify the phases of the products. Use Table 4.3 in the text to help you identify insoluble salts. To write a net ionic equation, separate any soluble, strong electrolyte in the molecular equation into its component ions and then eliminate spectator ions.

Solve: **(a)** $Pb(NO_3)_2(aq) + 2KBr(aq) \longrightarrow PbBr_2(s) + 2KNO_3(aq)$

$PbBr_2$ is an insoluble salt; KNO_3 is a strong electrolyte and a soluble salt.

$$Pb^{2+}(aq) + 2NO_3^-(aq) + 2K^+(aq) + 2\,Br^-(aq) \longrightarrow PbBr_2(s) + 2K^+(aq) + 2NO_3^-(aq)$$

Net ionic equation:

$$Pb^{2+}(aq) + 2\,Br^-(aq) \longrightarrow PbBr_2(s)$$

(b) $3NiCl_2(aq) + 2Na_3PO_4(aq) \longrightarrow Ni_3(PO_4)_2(s) + 6NaCl$

$Ni_3(PO_4)_2$ is an insoluble salt; NaCl is a strong electrolyte and a soluble salt.
$$3\,Ni^{2+}(aq) + 6Cl^-(aq) + 6\,Na^+(aq) + 2\,PO_4^{3-}(aq) \longrightarrow Ni_3(PO_4)_2(s) + 6Na^+(aq) + 6Cl^-(aq)$$

Net ionic equation:

$$3\,Ni^{2+}(aq) + 2\,PO_4^{3-}(aq) \longrightarrow Ni_3(PO_4)_2(s)$$

Check: The number of each type of atom is the same on both sides of the arrow and the charges balance (zero on both sides of the arrow) for each net ionic equation. All ions reacting are in a different form in the product. No spectator ion is shown. Also check that the chemical formulas are correct. This means you must determine the correct charges of ions and ensure that the sum of charges in a compound is zero. For example, $PbBr_2$ consists of Pb^{2+} and Br^-; two Br^- ions are needed to balance the positive two charge of Pb^{2+}. This shows that the formula is correct based on charge balance. This check should be done for every substance formed in a net ionic equation.

OXIDATION AND REDUCTION: OXIDATION NUMBERS AND ACTIVITY SERIES

Section 4.4 introduces an important type of chemical reaction: oxidation-reduction. Oxidation-reduction reactions involve the transfer of electrons between substances.

- **Oxidation** occurs when an atom loses electrons. When an atom loses electrons it gains positive charge. Thus, when a Ca atom loses two electrons, it forms Ca^{2+}. A substance that has lost electrons is said to be *oxidized*.
- **Reduction** is the gain of electrons by an atom. When an atom gains electrons, it loses positive charge (i.e., it becomes more negatively charged). Thus, when an oxygen atom gains two electrons, it forms O^{2-}. A substance that has gained electrons is said to be *reduced*.
- Oxidation and reduction occur together in a chemical reaction.

The concept of oxidation number helps us determine whether a reaction involves oxidation and reduction. Oxidation numbers help us keep track of the change of electrons in reactions. The terms oxidation and reduction can also be stated as follows:

- Oxidation occurs when the oxidation number of an atom increases.
- Reduction occurs when the oxidation number of an atom decreases.

Oxidation number reflects the charge assigned to an atom in a particular bonding situation. It is also a tool that helps us keep track of electrons in a chemical reaction.* The terms *oxidation number* and *oxidation state* are closely related. An atom in an oxidation state of +2 is said to have a +2 oxidation number. An atom may have several possible oxidation states, positive and negative, depending on the identity of the atoms bonded to it.

A few general rules have been developed to aid you in assigning oxidation numbers to atoms in compounds. *You must memorize these now.*

1. The oxidation number of an element in its elementary or uncombined state is zero. For example, in Cl_2 each chlorine atom has an oxidation number of zero.

2. In an ionic compound, the oxidation number of a monatomic ion is the same as its charge. For example, in KCl the potassium has an oxidation number of +1 and the chloride ion has one of −1.

3. Certain elements almost always have the same oxidation number in their compounds. These elements are:
   (a) Group 1A elements (Li, Na, K, Rb, Cs), with an oxidation number of +1;
   (b) Group 2A elements (Be, Mg, Ca, Sr, Ba), +2;
   (c) Group 3A elements (B, Al), +3;
   (d) Fluorine, chlorine, bromine, iodine, −1 in binary compounds with metals (other states are possible when combined with non-metals);
   (e) Hydrogen: +1 (except for metallic hydrides, where its oxidation state is −1, as in CaH_2 and NaH);
   (f) Oxygen: −2 (except in peroxide compounds, where the oxidation state is −1 (e.g., H_2O_2); in superoxide ions $[O_2^-]$, where it is $-\frac{1}{2}$; and when bonded to F $[+2$ in $OF_2]$).

4. In an AB_y compound, the more electronegative element is assigned the negative oxidation number; the less electronegative one is assigned a positive oxidation number (for example, in SF_4, F is assigned −1 and thus S is +4).

5. In a neutral compound, the sum of oxidation numbers of all atoms is zero; in a compound with a charge, the sum is equal to the charge of that compound.

The last part of Section 4.4 examines the chemical reactions of metals with acids and salts. Metals (M) participate in oxidation–reduction reactions involving acids (HX) or salts (BX). These typically are displacement reactions.

For example, the reaction of a metal with an acid or salt is exemplified by the following general reaction:

- $M + HX \rightarrow MX + X$ or $M + BX \rightarrow MX + B$
- Note that M displaces X from HX (acid) or B from BX (salt)
- Many metals react with strong acids such as HCl, HNO_3, and H_2SO_4 to form a salt and $H_2(g)$ in a displacement reaction. For example, consider the following example with the oxidation numbers shown for each element:

$$\overset{0}{Zn}(s) + \overset{+1\ \overset{+6}{\ }\ -2}{H_2SO_4}(aq) \longrightarrow \overset{+2\ \overset{+6}{\ }\ -2}{ZnSO_4}(aq) + \overset{0}{H_2}(g)$$

---

*Caution: Oxidation number may be related to the charge of an atom or it may have no relationship. For example, sulfur in $Na_2S_4O_6$ has an oxidation number of $+2\frac{1}{2}$, but an atom cannot have $\frac{1}{2}$ an electron, only an integral number.

Zn is oxidized because its oxidation number increases from 0 to +2. Hydrogen is reduced because its oxidation number decreases from +1 to 0.

Metals may also displace another metal from a salt.

- For example, $Zn(s) + CuSO_4(aq) \rightarrow Cu(s) + ZnSO_4(aq)$
- Displacement reactions of this type occur if

$$A + BX \longrightarrow AX + B$$

must be more easily oxidized than B

- An **activity series** arranges metals in order of decreasing ease of oxidation. Refer to Table 4.5 in the text. A metallic element is able to displace ions of elements below it from their compounds.
- Note in Table 4.5 that Li is the most active element, whereas Au is the least active. Li can displace all elements below it from their compounds. Au is essentially nonreactive.
- H_2 is low in the activity series. Metals above it can displace H^+ from acids to form $H_2(g)$. Only Li, K, Ba, Ca, and Na will react with cold water: Others require steam.

EXERCISE 6   Determining oxidation numbers

State the most common oxidation number(s) for each of the following elements and give an example of a compound in which an atom of the element has that oxidation number: (**a**) Li; (**b**) Sr; (**c**) Al; (**d**) Ag; (**e**) N; (**f**) O; (**g**) F; (**h**) Zn; (**i**) S; (**j**) H.

SOLUTION: *Analyze*: We are to determine the most common oxidation number(s) for 10 elements.

Plan: We may have to read the text, use the index in the text or apply principles of the periodic table to answer this question. The text discusses common ion charges as a function of the column number of the periodic table. You should learn these principles and apply them to this question.

Solve: (**a**) +1: LiF. All atoms in group 1A have a +1 oxidation number. (**b**) +2: SrO. All atoms in group 2A have a +2 oxidation number. (**c**) +3: $AlCl_3$. (**d**) +1: AgCl or Ag_2O. (**e**) −3: NH_3 + 5: NO_3^-. (**f**) −2: CaO. (**g**) −1: HF. (**h**) +2: ZnS. (**i**) +6: SF_6. −2: H_2S. (**j**) +1: HCl. −1: NaH (only with metallic elements).

Comment: We need to know the most common oxidation state(s) for the majority of representative elements in the first four periods. Assuming we have used the correct oxidation states we should check that chemical formulas are correctly written. This means that the sum of oxidation states in a compound is zero. For example, H_2S consists of hydrogen with a +1 oxidation state and sulfur with a −2 oxidation state; two hydrogen atoms are needed to balance the negative two oxidation state of sulfur. This shows that the formula is correct based on the sum of oxidation states. This should be done for every substance.

EXERCISE 7   Determining oxidation numbers

Determine the oxidation state of nitrogen in each of the following: (a) NH_3; (b) N_2O_4; and (c) $NaNO_3$.

SOLUTION: *Analyze*: We are given three substances and asked to determine the oxidation number of nitrogen in each.

Plan: We can apply the rules of oxidation numbers to the other element in each compound. If we know its oxidation number then we can calculate the oxidation number of nitrogen by applying the principle that the sum of oxidation numbers for a compound is zero.

Solve: (**a**) The oxidation number of hydrogen is +1 when bonded to a nonmetal such as nitrogen. To calculate the oxidation number of nitrogen we use the rule that the sum of oxidation numbers is zero for this molecular substance: $0 = ? + 3(+1)$. Solving for the unknown oxidation number of nitrogen gives -3. (**b**) Oxygen typically has an oxidation number of -2 when bonded to nonmetals or metals. Again, since the sum of oxidation numbers is zero: $0 = 2(?) + 2(-2)$. Solving for the unknown oxidation number of nitrogen gives $+2$. (**c**) Sodium is a metal and in compounds it carries a 1+ ion charge; thus, its oxidation number is +1. This means that the NO_3 group must carry a -1 charge (it is the nitrate ion). Solving, we have $-1 = ? + 3(-2)$ and the unknown oxidation number of nitrogen is $+5$. Note that the -1 is used to the left of the equal sign because the nitrate ion carries a 1− charge.

Comment: Note that you must know the rules for assigning oxidation numbers of elements to do this type of problem.

EXERCISE 8   Writing single displacement reactions of active metals

Write a balanced single displacement equation illustrating each of the following: (**a**) a metal displacing H^+ from H_2O; (**b**) a metal displacing H^+ from an acid; and (**c**) a metal replacing copper in a copper (II) nitrate solution.

SOLUTION: *Analyze*: We are asked to write balanced single displacement reactions for three metals under different conditions. Also, we are to choose a metal that will react.

Plan: To answer these types of questions we will need to find analogous reactions in the text or use the activity series to help us find appropriate metals and to predict products of the reaction. Table 4.5 in the text provides information about the activities of metals. A metal that will react with the other species given in the problem has to have a higher activity.

Solve: (**a**) Choose a metal in the activity series that is above hydrogen. Li, K, Ba, Sr, Ca, and Na will liberate H_2 in cold water (Mg, Al, Mn, Zn, and Fe require steam). For example,

$$Mg(s) + 2\,H_2O(g) \longrightarrow Mg(OH)_2(s) + H_2(g)$$

(**b**) Also choose a metal that is higher in the activity series than hydrogen. For example, we can use magnesium

$$Mg(s) + H_2SO_4(aq) \longrightarrow MgSO_4(aq) + H_2(g)$$

(**c**) Choose an element higher in the activity series than copper. For example,

$$Zn(s) + CuSO_4(aq) \longrightarrow ZnSO_4(aq) + Cu(s)$$

Comment: Answering this type of question requires interpreting information in a table.

EXERCISE 9   Using the activity series of metals to predict if a reaction occurs

Use the activity series to predict which of the following reactions will occur.

(**a**) $Hg(l) + MnSO_4(aq) \longrightarrow HgSO_4(s) + Mn(s)$
(**b**) $2\,Ag(s) + H_2SO_4(aq) \longrightarrow Ag_2SO_4(aq) + H_2(g)$

(c) $Ca(s) + 2 H_2O(l) \longrightarrow Ca(OH)_2(aq) + H_2(g)$

SOLUTION: *Analyze*: We are given molecular equations and asked to use the activity series to predict if the reactions occur.

Plan: All of the reactions are examples of displacement reactions:

$$M + M'X \longrightarrow MX + M'.$$

For these reactions to occur, M must be higher in the activity series than M'. Refer to Table 4.5 in the text for required information.

Solve: **(a)** Hg lies below Mn in the activity series; thus, the reaction does not occur. **(b)** Ag lies below hydrogen in the activity series; thus, the reaction does not occur. **(c)** Ca lies above hydrogen in the activity; thus, the reaction occurs.

Check: The *Plan* identifies all reactions as single displacement reactions and that they will occur if the metal reacting is more active than the first element in the compound reacting. Review Table 4.5 in the text. A review supports the conclusions in the problem.

The term **concentration** refers to the amount of a solute dissolved in a given quantity of solvent or solution. A quantitative way of expressing the concentration of a solute in solution is by using **molarity**.

- **Molarity** (M) is used to measure the amount of solute in a solution.

- $M = \dfrac{\text{number of moles of solute}}{\text{volume of the solution in } \textit{liters}}$

- The previous equation can be algebraically changed to a form that is useful when calculating the number of moles of a solute in a solution.

$$n_{\text{moles of solute}} = M_{\text{solution}} \times V_{\text{liters}}$$

- Molarity, like density, can be used as a conversion factor to change between volume and number of moles.

Solutions of known concentration may be diluted with the solvent to produce a more diluted (less concentrated) solution.

- When dilution of a solution occurs by adding solvent we can use the molarity concept to solve for the molarity of the diluted solution. Note that the *number of moles of solute does not change when only solvent is added to a solution*. Therefore, the number of moles of solute before dilution equals the number of moles of solute after dilution. We can write the following relationships

$$n_i = n_f$$

and by substitution of MV for n we have

$$M_i V_i = M_f V_f$$

The subscript *i* refers to the undiluted solution and the subscript *f* refers to the final diluted solution. This relationship is often used in solution-type problems. *Be sure you understand its use*. See Exercises 13 and 14.

EXERCISE 10   Calculating molarity

What is the molarity of an ethanol (C_2H_6O) solution containing 10.0 g of ethanol in water with a total volume of 100 mL?

SOLUTION: *Analyze*: We are given the volume of an ethanol solution and the mass of ethanol in it and asked to calculate the molarity of the solution.

CONCENTRATIONS OF SOLUTIONS

We will need the definition of molarity to solve for the requested infor–mation.

Plan: The definition of molarity is $M = \dfrac{\text{moles of solute}}{\text{volume of solution in liters}}$

We need the number of moles of solute, ethanol, and the volume of the solution in liters to solve for M. We will need the equivalence relations between moles and grams of ethanol and between milliliters and liters:

$$46.07 \text{ g } C_2H_6O = 1 \text{ mol } C_2H_6O \quad [\text{Molar Mass}]$$
$$1000\text{mL} = 1\text{L}$$

Solve:

$$M_{C_2H_6O} = \frac{(10.0 \text{ g } C_2H_6O)\left(\dfrac{1 \text{ mol } C_2H_6O}{46.07 \text{ g } C_2H_6O}\right)}{(100 \text{ mL})\left(\dfrac{1 \text{ L}}{1000 \text{ mL}}\right)}$$

$$= \frac{0.217 \text{ mol } C_2H_6O}{0.100 \text{ L}}$$

$$= 2.17 M$$

Check: The answer is a reasonable concentration in terms of its magnitude. If it had been a very large or very small number it is advisable to redo the problem. If we estimate the answer based on the setup of the problem,

$(10)\left(\dfrac{1}{50}\right)/(100)\left(\dfrac{1}{1000}\right) = 2$, we obtain an answer close to the calculated value.

The units cancel correctly and give the correct final unit.

EXERCISE 11   Using molarity to calculate grams of a solute

How many grams of HCl are contained in exactly 500 mL of a 0.250 M HCl solution?

SOLUTION: *Analyze*: We are given the molarity of a HCl solution and its volume and asked to calculate the number of grams of HCl in the solution. We will need to use the definition of molarity to solve for the requested information.

Plan: HCl is the solute and the number of grams can be calculated if the number of moles of it is known. We can use the definition of molarity to calculate the number of moles of HCl from the volume of the solution. We can use the follow-ing sequence of conversions to obtain the desired unit of grams:

$$\textit{Volume} \text{ HCl} \longrightarrow \textit{moles} \text{ HCl} \longrightarrow \textit{grams} \text{ HCl}$$

These conversions require the following relationships:

$$0.250 \text{ mol HCl} \approx 1 \text{ L HCl solution} \quad [\text{from molarity definition}]$$
$$1000 \text{ mL} = 1 \text{ L}$$
$$36.45 \text{ g HCl} = 1 \text{ mol HCl} \quad [\text{Molar Mass}]$$

Solve: Following the sequence of conversion steps above we convert 500 mL of the solution to grams of HCl.

$$\text{Grams HCl} = (500 \text{ mL})\left(\frac{1 \text{ L}}{1000 \text{ mL}}\right)\left(\frac{0.250 \text{ mol HCl}}{1 \text{ L}}\right)\left(\frac{36.45 \text{ g HCl}}{1 \text{ mol HCl}}\right)$$

$$= 4.56 \text{ g HCl}$$

Check: The answer of about 4 g is not unreasonable. An estimate of the calculation gives $(500)\left(\dfrac{1}{1000}\right)(0.25)(40) = 5$ which is close to the calculated value. The units cancel correctly and give the correct final unit.

EXERCISE 12   Calculating the molarity of ions in a salt solution

What is the molarity of Na^+ ions in a 0.02 M Na_3PO_4 solution?

SOLUTION: *Analyze*: We are given the concentration of a salt solution, sodium phosphate, and asked to determine the molarity of sodium ions in it.

Plan: We need to determine whether sodium phosphate is a weak or strong electrolyte. It is a salt and therefore a strong electrolyte in water. This means it dissociates completely into sodium and phosphate ions. We will write the dissociation reaction so that we can determine how many moles of sodium ion form when one mole of the salt dissociates. This will permit us to determine the molarity of the sodium ions.

Solve: Na_3PO_4 dissociates completely in H_2O according to the reaction:

$$Na_3PO_4 \longrightarrow 3\,Na^+ + PO_4^{3-}$$

For every 1 mol of Na_3PO_4 that ionizes, 3 mol of Na^+ ions form. Thus, a 0.020 M Na_3PO_4 solution is $3(0.02\ M) = 0.06\ M$ in Na^+ ions.

Comment: This type of problem requires us to know which substances are salts (contain metal and nonmetal) and how salts dissociate in water. This also requires that we know or can determine the charges of metal and nonmetal ions in salts. A review of the dissociation reaction shows that the number and types of atoms as well as charge are conserved.

EXERCISE 13   Diluting a solution and calculating the volume required to achieve a given molarity

1.00 L of 6.00 M HCl is used to prepare exactly 100 mL of 1.00 M HCl. How many milliliters of the 6.00 M HCl must be diluted with water in order to prepare the 1.00 M HCl solution?

SOLUTION: *Analyze*: This is a typical dilution problem that is often encountered in a laboratory. We are given a 6.00 M HCl solution and we are asked what quantity (an aliquot or a portion) of this solution must be diluted with water to prepare 100 mL of a 1.00 M HCl solution.

Plan: The key to this type of problem is to understand that the number of moles of HCl before dilution must equal the same number after dilution because only water (solvent) is added. The relation between the molarity of a solute before dilution and after dilution is

$$\text{Moles solute}_i = \text{moles solute}_f$$

or

$$M_iV_i = M_fV_f$$

The subscript *i* means the initial conditions before dilution, and *f* means the final conditions after dilution. The volumes may be expressed in units of milliliters or liters providing both volumes have the same unit. The initial volume of 6.00 M HCl required to form a 1.00 M solution by dilution is

Solve:

$$V_i = \frac{M_f V_f}{M_i} = \frac{(1.00\ M)(100\ \text{mL})}{6.00\ M} = 16.7\ \text{mL}$$

Thus, 16.7 mL of 6.00 M HCl diluted to 100 mL with H_2O produces a 1.00 M HCl solution.

Check: The answer of 16.7 mL is less than the final volume of the solution, 100 mL. An estimate of the calculation, $(1)(100)/5 = 20$, is approximately the calculated value. The units cancel properly and the final unit is correct.

EXERCISE 14   Calculating the molarity of a solution after it is diluted

What is the molarity of a solution of NaOH formed by diluting 125 mL of a 3.0 M NaOH solution to exactly 500 mL?

SOLUTION: *Analyze*: We are given 125 mL of a 3.0 M NaOH solution and asked to calculate the molarity of the solution after it is diluted to 500 mL. Recognize that the number of moles of NaOH does not change when only solvent is added.

Plan: When the number of moles of solute does not change upon dilution we can use the relation:

$$M_i V_i = M_f V_f$$

to solve for the molarity of the diluted solution

$$M_f = M_i\left(\frac{V_i}{V_f}\right)$$

Solve: $M_f = (3.0\ M)\left(\dfrac{125\ \text{mL}}{500\ \text{mL}}\right) = 0.75\ M$

Note: *The molarity of a diluted solution must always be less than that of the initial solution because the ratio V_i/V_f is always smaller than one.*

Check: The calculated concentration is less than the initial concentration, which agrees with the concept that adding a solvent reduces the concentration of a solution. An estimate of the calculation, $(3)(1/5) = 0.6$, is approximately the calculated value. The units cancel properly and the final unit is correct.

SOLUTION STOICHIOMETRY

In Section 3.6 we learned how to use the stoichiometry of a chemical reaction to convert the amount of a substance in a chemical reaction to the amount of a different one. This requires the use of the balancing coefficients in the chemical reaction and molar mass. In Section 4.6 we learn how to use molarity and volume in solution stoichiometry and chemical analysis. A key relationship when doing solution stoichiometry problems is

$$n_{\text{moles of solute}} = M_{\text{solution}} \times V_{\text{liters}}$$

A laboratory procedure for determining the concentration of a solute in a solution using a solution of known concentration is called a **titration**:

- **Titration** is a laboratory procedure used in chemical analyses and involves measuring precisely the volume of a solution whose molarity is known (called a *standard* solution) required to react with a volume of a solution whose concentration is unknown. The stoichiometry of the reaction provides information about the ratio of moles between the standard and the substance whose concentration is unknown.

- The **equivalence point** of a titration is when chemically equivalent, or stoichiometric, amounts of reactants have reacted. The **end point** is the experimentally determined condition when an indicator or other signalling device indicates that the end of the reaction has occurred.
- The stoichiometry of a reaction is important in determining the concentration of the unknown solution. For example, the reaction between sodium hydroxide and hydrochloric acid is:

$$\text{NaOH}(aq) + \text{HCl}(aq) \longrightarrow \text{H}_2\text{O}(i) + \text{NaCl}(aq)$$

Note that they react in a 1:1 mole ratio. However, in the titration of sodium hydroxide with sulfuric acid

$$2\,\text{NaOH}(aq) + \text{H}_2\text{SO}_4(aq) \longrightarrow 2\,\text{H}_2\text{O}(l) + \text{Na}_2\text{SO}_4(aq)$$

note that the base and acid react in a 2:1 mole ratio.
- When doing problems involving titrations you need to write the chemical reaction to determine the stoichiometry between the standard and the unknown substance.

EXERCISE 15   Determining the volume of a solution required to react with another solution

What volume of 0.250 *M* HCl is required to react completely with 25.00 mL of 0.500 *M* NaOH?

SOLUTION: *Analyze*: We are asked to determine the volume of a 0.250 *M* HCl solution necessary to neutralize a specified volume of 0.500 *M* NaOH, 25.00 mL. This is a stoichiometry problem involving concentrations.

Plan: Since it is a stoichiometry problem we should write the neutralization reaction between the acid HCl and the base NaOH. In a neutralization reaction water and a salt form.

$$\text{HCl}(aq) + \text{NaOH}(aq) \longrightarrow \text{NaCl}(aq) + \text{H}_2\text{O}(l)$$

To solve this problem we can carry out the following conversions:

$$\textit{Volume NaOH} \longrightarrow \textit{moles NaOH} \longrightarrow \textit{moles HCl} \longrightarrow \textit{Volume HCl}$$

We know the molarity of the solutions and from the stoichiometry of the chemical reaction that

$$1 \text{ mol HCl} \doteqdot 1 \text{ mol NaOH}$$

Solve: We first use the molarity concept to solve for the number of moles of sodium hydroxide used in the chemical reaction.

$$M_{\text{NaOH}}V_{\text{NaOH solution}} = \text{moles of NaOH}$$

$$(0.500 \text{ M})(0.02500 \text{ L}) = 0.0125 \text{ mol NaOH}$$

Next, convert the number of moles of NaOH to the number of moles of HCl that reacts with the NaOH using the balancing coefficients in the chemical reaction.

$$\text{mol HCl} = (0.0125 \text{ mol NaOH})\left[\frac{1 \text{ mol HCl}}{1 \text{ mol NaOH}}\right] = 0.0125 \text{ mol HCl}$$

Finally, convert the number of moles of HCl to the volume of the initial HCl solution required in the reaction by once again using molarity.

$$M_{\text{HCl}}V_{\text{HCl solution}} = \text{moles of HCl}$$

$$(0.250 \text{ M})V = 0.0125 \text{ mol HCl}$$

$$V = \frac{0.0125 \text{ mol HCl}}{0.250 \text{ M}} = 0.0500 \text{ L} = 50.0 \text{ mL HCl}$$

Alternatively, we can solve for the volume in one step by combining the individual steps as we learned in Section 3.6. This is not shown here but you should try this approach for additional practice.

Check: The chemical equation is balanced and it appropriately represents a neutralization reaction (formation of water and salt from an acid and base reacting). An estimate of the number of moles of HCl is $(0.01)(1) = 0.01$ and it agrees closely with the calculated value of 0.0125; and for the volume, $(0.01/.3) = 0.03$, it agrees closely with the calculated value of 0.05. The units cancel properly and the final unit is correct.

SELF-TEST QUESTIONS

Key Terms

Having reviewed key terms in Chapter 4, match key terms with phrases and identify statements as true or false. If a statement is false, indicate why it is incorrect.

Match each phrase with the best term:

4.1   Metals easily oxidized are high in this list.

4.2   A process in which an element loses electrons.

4.3   A process in which an element gains electrons.

4.4   A substance that produces hydroxide ions in water.

4.5   A substance that decreases the concentration of hydroxide ions in water.

4.6   This concentration term decreases in magnitude with increasing volume of the solution.

4.7   HF is an example of this type of electrolyte in water.

4.8   HI is an example of this type of electrolyte in water.

4.9   These substances cause an increase in the conductivity of a solvent.

4.10   The point in a titration when one volume of 0.10 *M* sulfuric acid reacts exactly with two volumes of 0.10 *M* sodium hydroxide.

4.11   A solution whose concentration is exactly known.

4.12   A method for determining concentrations that uses volume.

4.13   A dye that changes color as a function of pH.

4.14   Ions in a chemical reaction that do not undergo change.

4.15   A reaction in which weak electrolytes are shown in a molecular form and soluble strong electrolytes are not.

4.16   Strong electrolytes, weak electrolytes, and gases are left in an unionized form.

4.17   A reaction that produces salt and water.

Terms:

(**a**) acid
(**b**) activity series
(**c**) base
(**d**) concentration
(**e**) electrolytes
(**f**) equivalence point
(**g**) indicator
(**h**) molecular equations
(**i**) net ionic equation

(**j**) neutralization
(**k**) oxidation
(**l**) reduction
(**m**)spectator ions
(**n**) standardized solution
(**o**) strong electrolyte
(**p**) titration
(**q**) weak electrolytes

True-False Statements:

4.18   The *molarity* of a solution containing 7.46 g of KCl in 0.500 liters of a solution is 0.200 *M*.

4.19   An example of a *metathesis reaction* is

$$HNO_3(aq) + KOH(aq) \longrightarrow H_2O(l) + KNO_3(aq)$$

4.20   In the following *precipitation reaction*

$$CaI_2 + 2\,AgNO_3 \longrightarrow Ca(NO_3)_2 + 2\,AgI$$

the insoluble material (the precipitate) is $Ca(NO_3)_2$.

4.21   The term *aqueous solution* means water is the solvent.

4.22   3 g of sodium metal and a large amount of liquid ammonia form a solution. Thus, sodium is the *solvent*.

4.23   A solution can have only one *solute*.

4.24   When a solution is *diluted* with a solvent, the number of moles of solute decrease.

4.25   Ethanol is a *nonelectrolyte* in water.

4.26   In a state of *chemical equilibrium*, the concentrations of reactants and products remain constant.

4.27   A *base* in water always forms hydroxide ion.

4.28   Sulfuric acid is a *strong acid*.

4.29   Hydroiodic acid is a *weak acid*.

4.30   Barium hydroxide is a *weak base*.

4.31   When a strong acid reacts with a strong base, a *salt* is produced.

4.32 The reaction of lead nitrate with sodium chloride in water produces a white *precipitate*.

4.33 The *solubility* of nickel(II) phosphate in water is high.

4.34 The *oxidation number* of iodine in IF_3 is -3.

Problems and Short-Answer Questions

4.35 The following two pictures represent HX and HA. Which one is the stronger acid? Explain your reasoning.

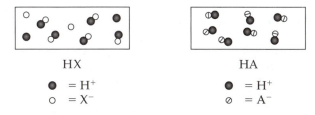

HX   HA

● = H^+ ● = H^+
○ = X^- ⊘ = A^-

4.36 An aqueous solution containing the cation X^+ (shaded spheres) is added to an aqueous solution containing anion Y^{n-} (unshaded spheres). The following diagram represents the final solution. Given this information write a balanced chemical equation. Explain your reasoning.

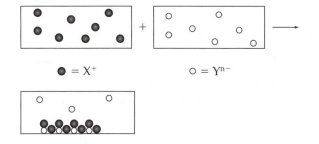

● = X^+ ○ = Y^{n-}

4.37 Are the following compounds ionic or molecular? Justify your answer.

(a) $CaCl_2$ (c) I_2
(b) $Fe(NO_3)_3$ (d) $HC_2H_3O_2$

4.38 For the compounds in **4.37** describe their electrolytic behavior in water. Describe the primary species that exist in solution.

4.39 Rank the following 0.1 M solutions in order of increasing conductivity in water: $C_6H_{12}O_6$ (glucose), KI, HI, and $HC_2H_3O_2$.

(a) When does an oxidation number actually reflect the charged state of an element in a substance?
(b) When is an oxidation number always zero?
(c) What is the difference between an oxide ion and a peroxide ion in terms of oxidation numbers?
(d) When assigning oxidation numbers why must you have knowledge of the charge of the substance?

4.41 Which element has the highest oxidation number in the following group of substances? $KMnO_4$, H_2SO_4, $NaHCO_3$, $Ca(ClO_3)_2$.

4.42 In the activity series (Table 4.5 in the text) copper is below hydrogen gas.

(a) What does this tell you about the reactivity of copper compared to hydrogen gas?
(b) Will hydrogen gas react with copper metal or copper(II) ions?

4.43 10.00 mL of a base containing one hydroxide unit is titrated with 25.00 mL of a 0.30 M HCl solution. Is this sufficient information to determine the molar mass of the base? Explain without doing calculations. In your explanation identify any information that is needed and why.

4.44 A solution contains either Ba^{2+}, Pb^{2+}, or Ag^+. The metal ion reacts with a solution of sodium sulfate to form a white precipitate but it does not react with a solution of sodium chloride. Is this sufficient information to determine the identity of the metal ion? Explain. In your explanation identify any information that is needed and why.

4.45 Critique the following statement: A 1.2 M aqueous solution of $AgNO_3$ contains 1.2 moles of $AgNO_3$ in 1.0 L of water.

4.46 A student wants to prepare 100.0 mL of a 0.100 M NaCl solution and has available the following: (1) A bottle of reagent grade NaCl; (2) 2.00 L of 0.0500 M NaCl; and (3) 2.00 L of 1.00 M NaCl.

(a) Can the student readily prepare the desired solution from each sample of NaCl by itself and with water? Justify your response.
(b) In words outline how the student could prepare the desired solutions based on your responses to (a).

4.47 Determine the following for a sucrose ($C_{12}H_{22}O_{11}$, molar mass = 342 g/mol) solution:

(a) The number of moles of sucrose in exactly 200 mL of a 0.100 M solution.
(b) The volume of a 1.25 M solution containing 0.100 moles of sucrose.
(c) The molarity of a solution containing 10 g of sucrose in exactly 500 mL.

4.48 Answer the following questions about a 1.50 M HCl solution:

(a) What is the new molarity if exactly 200 mL of the solution is diluted to 1.000 L?
(b) How many milliliters of the solution must be diluted to exactly 500 mL to form a 0.100 M HCl solution?
(c) How many moles of HCl are present in a diluted solution if exactly 100 mL of the original solution are diluted to 2.000 L?

4.49 Identify the acid and base in each of the following reactions:

(a) $NH_4^+(aq) + H_2O(l) \rightleftharpoons NH_3(aq) + H^+(aq)$

(b) $HC_2H_3O_2(aq) + H_2O(l) \rightleftharpoons$
$$H^+(aq) + C_2H_3O_2^-(aq)$$
(c) $HF(aq) + NaOH(aq) \rightarrow NaF(aq) + H_2O(l)$

4.50 Complete and balance the following neutralization reactions. Underline the acid reactant with one line and the base reactant with two lines.

(a) $Ba(OH)_2(aq) + HBr(aq) \longrightarrow$
(b) $NH_3(aq) + H_2SO_4(aq) \longrightarrow$

4.51 Write balanced net ionic equations for each of the following reactions:

(a) $NaC_2H_3O_2(aq) + HCl(aq) \longrightarrow$
(b) $KCl(aq) + Hg_2(NO_3)_2(aq) \longrightarrow$
(c) $Ni(NO_3)_2(aq) + CuSO_4(aq) \longrightarrow$
(d) $H_2SO_4(aq) + KOH(aq) \longrightarrow$
(e) $HC_2H_3O_2(aq) + NaOH(aq) \longrightarrow$

4.52 Using solubility rules or reasonable extensions of them, predict whether each of the following compounds is soluble in water:

(a) $PbBr_2$
(b) $CsCl$
(c) $Cu(C_2H_3O_2)_2$
(d) $Mn(OH)_2$
(e) $Ca_3(PO_4)_2$
(f) ZnS
(g) Hg_2Cl_2
(h) Ag_2SO_4
(i) K_2SO_4

4.53 Identify the following substances as strong or weak acids or bases in water:

(a) HCl
(b) H_2S
(c) NH_3
(d) KOH
(e) $HClO_4$
(f) HF
(g) H_3PO_4

4.54 What is the molarity of a nitric acid solution if 36.00 mL of it reacts completely with 2.00 g of NaOH?

4.55 What is the molarity of a solution of H_3PO_4 if 50.00 mL of it is titrated with 25.86 mL of 0.1201 M NaOH? Assume that all three hydrogens in H_3PO_4 react with NaOH.

4.56 Using the activity series (Table 4.5 in the text), write balanced chemical equations for the following reactions: (a) Mg is added to a solution of silver nitrate; (b) iron metal is added to a solution of copper (II) sulfate; (c) silver is added to a solution of zinc nitrate; (d) hydrogen gas is passed over a sample of zinc oxide; (e) hydrogen gas is passed over mercuric oxide.

Integrative Exercises

4.57 10.30 mL of 0.1500 M HCl is reacted with 11.25 mL of 0.1355 M NaOH.

(a) What is the limiting reactant?
(b) How many grams of NaCl are formed?

4.58 A 20.05 mL sample of vinegar (an aqueous solution of acetic acid, $HC_2H_3O_2$) has a density of 1.061 g/mL. The vinegar is titrated completely with 40.10 mL of 0.4100 M KOH. What is the percentage by mass of acetic acid in the vinegar?

4.59 An excess of silver(I) nitrate reacts with 100.0 mL of a barium bromide solution to give 0.300 g of a solid compound.

(a) Write the net ionic equation for the chemical reaction.
(b) What is the molarity of the barium bromide solution?

4.60 How many K^+ ions are there in 200.0 mL of a 0.20 M K_2SO_4 solution?

Multiple-Choice Questions

4.61 Which of the following represents the net ionic equation for $NH_3(aq) + HBr(aq) \longrightarrow$?

(a) $NH_3(aq) + HBr(aq) \longrightarrow NH_4^+(aq) + Br^-(aq)$
(b) $NH_3(aq) + HBr(aq) \longrightarrow NH_4Br(aq)$
(c) $NH_3(aq) + H^+(aq) + Br^-(aq) \longrightarrow NH_4Br(aq)$
(d) $NH_3(aq) + H^+(aq) \longrightarrow NH_4^+(aq)$

4.62 Which is a spectator ion in the following reaction in water? $ZnCl_2 + NaOH \longrightarrow$

(a) $Zn^{2+}(aq)$
(b) $Cl^-(aq)$
(c) $OH^-(aq)$
(d) There is no spectator ion

4.63 Which of the following aqueous solutions has the greatest total concentration of ions present?

(a) 0.01 M HCl
(b) 0.05 M $CaBr_2$
(c) 0.02 M $Ca(NO_3)_2$
(d) 0.02 M $Al(NO_3)_3$

4.64 Which of the following metal ions form sulfate compounds that are insoluble in water: (1) K^+, (2) Ba^{2+}, (3) Ag^+, (4) Mg^{2+}, (5) Pb^{2+}?

(a) (1) and (2)
(b) (3) and (4)
(c) (1) and (3)
(d) (2) and (5)
(e) (3), (4), and (5)

4.65 Which of the following metal ions form carbonate salts that are insoluble in water: (1) Na^+, (2) Ca^{2+}, (3) NH_4^+, (4) Pb^{2+}, (5) K^+?

(a) (1) and (2)
(b) (3) and (4)
(c) (2) and (4)
(d) (3) and (5)
(e) all of them

4.66 Which atom undergoes reduction in the following reaction? $2 Fe_2O_3(s) + 3 CO(g) \longrightarrow 2 Fe(s) + 3 CO_2(g)$

(a) Fe
(b) O
(c) C
(d) It is not an oxidation–reduction reaction

4.67 What are the products of the neutralization reaction between hydrogen iodide and calcium hydroxide in water?

(a) calcium and water
(b) CaI_2
(c) CaI_2 and water
(d) water

4.68 What is the molarity of an aqueous HBr solution if 35.00 mL is neutralized with 70.00 mL of a 0.500 M NaOH solution?

(a) 0.250 M
(b) 0.500 M
(c) 0.750 M
(d) 1.00 M

4.69 How many grams of KCl (molar mass = 74.6 g/mol) are contained in 500 mL of a 0.250 *M* KCl aqueous solution?

(a) 0.125   (c) 125
(b) 9.33   (d) 9330

4.70 What is the molarity of a solution consisting of 1.25 g of NaOH in enough water to form 250 mL of solution?

(a) 1.25 *M*   (d) 1.25×10^{-4} *M*
(b) 0.800 *M*   (e) 0.125 *M*
(c) 8.00 *M*

4.71 A 300-mL solution of 3.0 *M* HCl is diluted to 2.0 L. What is the final molarity?

(a) 0.45 *M*   (d) 0.05 *M*
(b) 450 *M*   (e) 0.50 *M*
(c) 20 *M*

4.72 A 3.000-g sample of a soluble chloride is titrated with 52.60 mL of 0.2000 *M* $AgNO_3$. What is the percentage of chloride in the sample?

(a) 37.29%   (d) 37.29%
(b) 24.86%   (e) 18.65%
(c) 12.43%

4.73 Which substance contains Mn in a +7 oxidation state? (1) MnO (2) $KMnO_4$ (3) MnO_2 (4) Mn_2O_7

(a) (1)   (c) (3) and (4)
(b) (2)   (d) (2) and (4)

SELF-TEST SOLUTIONS

4.1 (b) **4.2** (k) **4.3** (l) **4.4** (c) **4.5** (a) **4.6** (d) **4.7** (q) **4.8** (o) **4.9** (e) **4.10** (f) **4.11** (n) **4.12** (p) **4.13** (g) **4.14** (m) **4.15** (i) **4.16** (h) **4.17** (j) **4.18** True. **4.19** True. **4.20** False. All nitrate salts are soluble in water; silver iodide is the insoluble substance. **4.21** True. **4.22** False. Ammonia is the solvent because it is in greatest quantity and has the phase of the solution. **4.23** False. One or more solutes. **4.24** False. The number of moles of solvent increase upon dilution, but the number of moles of solute does not change. **4.25** True. **4.26** True. **4.27** True. **4.28** True. **4.29** False. It is a strong acid. **4.30** False. It is a strong base. Metallic hydroxides generally are strong bases. **4.31** True. **4.32** True. **4.33** False. According to solubility rules, most metallic phosphates are insoluble, and nickel is not one of the exceptions.

4.34 False. Fluorine is always assigned a −1 oxidation number; thus, the oxidation number of iodine is +3.

4.35 HX is the stronger acid, although it is a weak acid. In the HX solution we see undissociated HX particles and separate ions, however it is only partial dissociation. In the HA solution we see that all the particles are undissociated, thus HA is not acting as an acid.

4.36 We see in the first solution eight X^+ ions and in the second solution seven Y^{n-} ions. In the final solution we see most of the ions have come together as particles in a precipitate and some excess Y^{n-} ions remain. In the final solution eight X^+ ions have combined with four Y^{n-} ions.

This means that the charge of Y^{n-} is Y^{2-}; that is, eight positive charges must be balanced by eight negative charges in the balanced chemical equation. The balanced chemical equation is: $8 X^-(aq) + 4 Y^{2-}(aq) \rightarrow 4 X_2Y(s)$

4.37 An ionic compound will consist of a metal cation and a nonmetal anion. If a substance contains a metal, it is very likely ionic. Molecular compounds generally contain all nonmetals:

(a) Ionic.

(b) Ionic. The nitrate ion is a polyatomic ion containing nonmetals.

(c) Molecular.

(d) Molecular.

4.38 (a) A strong electrolyte; it dissociates completely to form hydrated Ca^{2-} and Cl^- ions. Ionic substances are strong electrolytes in water except for a few cases.

(b) A strong electrolyte; it dissociates completely into hydrated Fe^{3+} and NO_3^- ions.

(c) Molecular iodine is not very soluble in water unless iodide ion is also present. The form that exists in water is $I_2(aq)$.

(d) Acetic acid is a weak acid and only slightly ionizes. The primary species in water is $HC_2H_3O_2(aq)$.

4.39 glucose[non-electrolyte]<$HC_2H_3O_2$ [weak acid] <KI[ionic] ≈ HI[strong acid]

4.40 (a) When an element is an ion in an ionic substance the oxidation number equals the charge of the ion.

(b) When an element is in its natural elementary state (e.g., O_2) it is assigned an oxidation number of zero.

(c) The oxide ion is O^{2-} and the superoxide ion is O_2^{2-}. The oxidation number for the oxygen in the oxide ion is −2 and in the superoxide ion it is −1.

(d) When assigning oxidation numbers in a substance you must account for the requirement that the sum of oxidation numbers must equal the charge of the substance.

4.41 Manganese in $KMnO_4$ has the highest oxidation number, $+7 [+1, +7, 4(-2)]$. Sulfur in sulfuric acid has a +6 oxidation number, carbon has a +4 oxidation number in $NaHCO_3$, and chlorine has a +5 oxidation number in ClO_3^- in $Ca(ClO_3)_2$.

(a) Elements above hydrogen gas are more reactive than hydrogen and elements below it are less reactive with respect to oxidation. Thus, copper is less reactive as an element compared to hydrogen gas.

(b) Hydrogen gas will react with copper(II) ions to form hydrogen ions and metallic copper.

4.43 No. The reaction between the base and acid is 1:1 because the base has only one hydroxide unit and HCl is a monoprotic acid. Thus, you can determine the molarity of the base at the equivalence point using the relation: moles of acid or base = $M_A V_A = M_B V_B$. The molarity of

the base equals $\frac{\text{moles of base}}{\text{volume of base in liters}}$; this permits you to calculate the number of moles of base.

The definition of moles is $\frac{\text{grams}}{\text{molar mass}}$; you cannot calculate the molar mass because you are not given the grams of base.

4.44 Yes, the metal ion is barium. Only barium ion reacts with sulfate ion to form a precipitate and the other two ions react with chloride ion to form white precipitates but barium does not.

4.45 The definition of molarity is the number of moles of solute divided by the volume of the *solution* in liters. In the statement the reference is to the volume of water, which is the solvent not the solution.

4.46 (a) Samples (1) and (3) can be used to prepare the desired solution by adding a solid to a volumetric flask and adding water to the mark or by diluting the more concentrated solution. Sample (2) cannot be used because its concentration is less than that of the desired solution. Removing solvent by evaporation will concentrate the solution but it would not be considered "readily" prepared and its accuracy for dilution to form the final solution is questionable.

(b) For sample (1) the amount of NaCl needed to prepare 100.0 mL of a 0.100 M NaCl solution is calculated using the definition of molarity. The required amount of solid is carefully weighed and added to a 100.0 mL volumetric flask. Water is carefully added and the solution is swirled carefully. Water continues to be added until the 100 mL mark is reached. Sample (3) is used to prepare the desired solution by taking an aliquot of it and diluting the aliquot in a 100 mL volumetric flask to the mark. The volume of the aliquot is determined by the relationship: $(MV)_{\text{concentrated solution}} = (MV)_{\text{diluted solution}}$. The volume of the concentrated solution, the aliquot, is calculated from the molarity of the concentrated solution (1.00 M NaCl), the required molarity of the diluted NaCl (0.100 M), and the volume of the diluted solution (100.0 mL).

4.47 (a) mol sucrose = M × V(liters)

$$= (0.100 \text{ M})(0.200 \text{ L})$$

$$= 2.00 \times 10^{-2} \text{ mol.}$$

(b) V = mol/M = 0.100 mol/1.25 M = 0.0800 L.

(c) M = (g/molar mass)/V

$$= (10 \text{ g}/342 \text{ g/mol})/0.500 \text{ L}$$

$$= 0.058 \text{ molar.}$$

4.48 (a) $M_f = M_i(V_i/V_f) = 1.50 \text{ M} (200 \text{ mL}/1000 \text{ mL})$

$$= 0.300 \text{ M.}$$

(b) $V_i = V_f(M_f/M_i) = 500 \text{ mL} (0.100M/1.50M)$

$$= 33.3 \text{ mL.}$$

(c) The number of moles does not change upon dilution, only the concentration.
Moles = $M \times V$ = $(1.50 \text{ M})(0.100 \text{ L})$ = 0.150 mol HCl.

4.49 (a) acid: NH_4^+; base: H_2O.

(b) acid: $HC_2H_3O_2$; base: H_2O.

(c) acid: HF; base: NaOH.

4.50 (a) $Ba(OH)_2(aq) + 2 HBr(aq) \rightarrow$
$$BaBr_2(aq) + 2 H_2O(l);$$

(b) $2 NH_3(aq) + H_2SO_4(aq) \rightarrow (NH_4)_2SO_4(aq).$

4.51 (a) $H^+(aq) + C_2H_3O_2^-(aq) \rightarrow HC_2H_3O_2(aq);$

(b) $Hg_2^{2+}(aq) + 2 Cl^-(aq) \rightarrow Hg_2Cl_2(s);$

(c) No reaction—predicted products $NiSO_4$ and $Cu(NO_3)_2$ are both soluble and strong electrolytes in water;

(d) $H^+(aq) + HSO_4^-(aq) + 2 OH^-(aq)$
$$\rightarrow 2 H_2O(l) + SO_4^{2-}(aq);$$

(e) $HC_2H_3O_2(aq) + OH^-(aq) \rightarrow$
$$H_2O(l) + C_2H_3O_2^-(aq).$$

4.52 Table 4.1 is used to reach the following conclusions:

(a) insoluble, an exception to rule that all bromide salts are soluble;

(b) soluble;

(c) soluble;

(d) insoluble;

(e) insoluble;

(f) insoluble;

(g) insoluble;

(h) soluble;

(i) soluble.

4.53 (a) strong acid;   (e) strong acid;

(b) weak acid;   (f) weak acid;

(c) weak base;   (g) weak acid.

(d) strong base;

4.54 Reaction: $HNO_3 + NaOH \rightarrow NaNO_3 + H_2O$.
Conversions required: *grams* NaOH → *moles* NaOH → *moles* HNO_3 → *molarity* HNO_3. Equivalences: 40.00 g $NaOH$ = 1 *mol* NaOH; 1 mol NaOH $\cong$ 1 mol HNO_3 (from neutralization reaction).

$$M \text{ HNO}_3 = \frac{\text{moles HNO}_3}{\text{volume in liters}}$$

$$= \frac{(2.00 \text{ g NaOH})\left(\dfrac{1 \text{ mol NaOH}}{40.00 \text{ g NaOH}}\right)\left(\dfrac{1 \text{ mol HNO}_3}{1 \text{ mol NaOH}}\right)}{(36.00 \text{ mL})\left(\dfrac{1 \text{ L}}{1000 \text{ mL}}\right)}$$

$$= \frac{1.39 \text{ mol HNO}_3}{1 \text{ L}} = 1.39 \text{ molar}$$

4.55 Reaction equation:
$H_3PO_4 + 3 NaOH \rightarrow Na_3PO_4 + 3 H_2O$. Conversions required: *milliliters* NaOH → *moles* NaOH → *moles* H_3PO_4. Equivalences: 0.2101 mol NaOH = 1 L of NaOH solution; 1 mol H_3PO_4 $\cong$ 3 mol NaOH.

$$\text{Moles H}_3\text{PO}_4 = (25.86 \text{ mL NaOH})\left(\frac{1 \text{ L}}{1000 \text{ mL}}\right)$$

$$\times \left(\frac{0.1201 \text{ mol NaOH}}{1 \text{ L}}\right)\left(\frac{1 \text{ mol H}_3\text{PO}_4}{3 \text{ mol NaOH}}\right)$$

$$= 1.035 \times 10^{-3} \text{ mol}$$

$$M\ H_3PO_4 = \frac{\text{moles } H_3PO_4}{\text{volume in liters}}$$

$$= \frac{1.035 \times 10^{-3}\ \text{mol}}{0.0500\ \text{L}}$$

$$= 0.02070\ M$$

4.56 **(a)** $Mg(s) + 2\ AgNO_3(aq) \rightarrow$
$$Mg(NO_3)_2(aq) + 2\ Ag(s);$$

(b) $Fe(s) + CuSO_4(aq) \rightarrow FeSO_4(aq) + Cu(s);$

(c) No reaction. Ag lies below zinc in the activity series.

(d) No reaction. H_2 gas lies below Zn in the activity series.

(e) $H_2(g) + HgO(s) \rightarrow H_2O(l) + Hg(l)$.

4.57 **(a)** The neutralization reaction is $HCl(aq) + NaOH(aq) \longrightarrow H_2O(aq) + NaCl(aq)$. The number of moles of each reactant is:

$$\text{mol HCl} = M \times V = (0.01030\ \text{L})(0.1500\ M)$$
$$= 0.001545\ \text{mol HCl}$$

$$\text{mol NaOH} = M \times V = (0.01125\ \text{L})(0.1355\ M)$$
$$= 0.001524\ \text{mol NaOH}$$

The limiting reactant is NaOH because HCl and NaOH react in a 1:1 mole ratio. Since the number of moles of NaOH is smaller, only 0.001524 mol HCl are needed; HCl is in excess.

(b) The number of moles of NaCl formed is the same number as that for NaOH, the limiting reactant, since there is a 1:1 mole ratio between NaOH and NaCl. Therefore the number of grams of NaCl is: $(0.001524\ \text{mol NaCl})(58.44\ \text{g}/1\ \text{mol}) = 0.08906\ \text{g NaCl}$.

4.58 To determine the percentage by mass of acetic acid, $HC_2H_3O_2$, in the vinegar sample you need to calculate the mass of acetic acid. The reaction between acetic acid and KOH is

$$HC_2H_3O_2(aq) + KOH(aq) \longrightarrow$$
$$H_2O(aq) + KC_2H_3O_2(aq)$$

The number of moles of KOH can be calculated from the relation: $M \times V = \text{moles}$

$$\text{mol KOH} = (0.04010\ L)(0.4100\ M) = 0.01644\ \text{mol KOH}$$

The number of moles of acetic acid is the same as for KOH because they react in a 1:1 mole ratio. Therefore, the number of grams of acetic acid is:

$$(0.01644\ \text{mol } HC_2H_3O_2)(60.04g/1\ \text{mol})$$
$$= 0.9871\ g\ HC_2H_3O_2.$$

The percentage by mass of acetic acid is: $[(\text{mass } HC_2H_3O_2/\text{mass vinegar solution})] \times 100$. The mass of vinegar solution is calculated using density: $(20.05\ \text{mL})(1.061\ \text{g}/1\ \text{mL}) = 21.27\ \text{g vinegar}$. Thus, the percentage by mass of acetic acid is:

$$[0.9871\ g\ HC_2H_3O_2/21.27\ \text{g vinegar}] \times 100 = 4.641\%$$

4.59 **(a)** $BaBr_2(aq) + 2\ AgNO_3(aq) \rightarrow Ba(NO_3)_2(aq)$
$$+ 2\ AgBr(s)$$

Net ionic equation: $Ag^+(aq) + Br^-(aq) \rightarrow AgBr(s)$

(b) moles $BaBr_2 =$

$$(0.300\ g\ AgBr)\left(\frac{1\ \text{mol AgBr}}{187.77\ g\ AgBr}\right)\left(\frac{1\ \text{mol } BaBr_2}{2\ \text{mol AgBr}}\right)$$

$$= 7.99 \times 10^{-4}\ \text{mol } BaBr_2$$

$$\text{Molarity} = \frac{7.99 \times 10^{-4}\ \text{mol } BaBr_2}{0.1000\ \text{L}} = 7.99 \times 10^{-3}\ \text{molar}$$

4.60 Potassium sulfate is a strong electrolyte:

$$K_2SO_4(aq) \longrightarrow 2\ K^+(aq) + SO_4{}^{2-}(aq).$$

$$\text{moles } K^+ = \left(0.20\frac{\text{mol}}{\text{L}}\right)$$

$$\times (0.200\ \text{L})\left(\frac{2\ \text{mol } K^+\ \text{ions}}{1\ \text{mol } K_2SO_4}\right)\left(\frac{6.02 \times 10^{23}\ \text{ions}}{1\ \text{mol ions}}\right)$$

$$= 4.8 \times 10^{22}\ K^+\ \text{ions}.$$

4.61 **(d)** HBr and NH_4Br are both strong electrolytes.
4.62 **(b)**
4.63 **(b)** $3 \times 0.05\ M = 0.15\ M$ in total ions. Note that although **(d)** has four ions per formula, the total ion concentration is $0.08\ M$. **4.64** **(d)** **4.65** **(c)** **4.66** **(a)** **4.67** **(c)** **4.68** **(d)** **4.69** **(b)**
4.70 **(e)**

$$M\ NaOH = \frac{\text{mol NaOH}}{\text{volume in liters}}$$

$$= \frac{(1.25\ g\ NaOH)\left(\dfrac{1\ \text{mol NaOH}}{40.01\ g\ NaOH}\right)}{(250\ \text{mL})\left(\dfrac{1\ L}{1000\ \text{mL}}\right)}$$

$$= 0.125\ M$$

4.71 **(a)**

$$M_iV_i = M_fV_f$$

$$M_f = M_i\frac{V_i}{V_f}$$

$$= (3.0\ M)\left[\frac{300\ \text{mL}}{(2.0\ L)\left(\dfrac{1000\ \text{mL}}{L}\right)}\right]$$

$$= 0.45\ M$$

4.72 **(c)** The reaction of interest is $Cl^-(aq) + AgNO_3(aq) \rightarrow AgCl(s) + NO_3{}^-(aq)$. We will need to make the following conversion: *Milliliters* $AgNO_3 \rightarrow$ *moles* $AgNO_3 \rightarrow$ *moles* Cl^- $\rightarrow$ *grams* Cl^-. Hence,

Grams $Cl^- = (52.60 \text{ mL AgNO}_3)\left(\dfrac{1 \text{ L}}{1000 \text{ mL}}\right)$

$\times \left(\dfrac{0.2000 \text{ mol AgNO}_3}{1 \text{ L AgNO}_3}\right)\left(\dfrac{1 \text{ mol Cl}^-}{1 \text{ mol AgNO}_3}\right)$

$\times \left(\dfrac{35.45 \text{ g Cl}^-}{1 \text{ mol Cl}^-}\right) = 0.3729 \text{ g Cl}^-$

% Cl^- in sample $= \left(\dfrac{\text{mass of Cl}^- \text{ in sample}}{\text{sample mass}}\right) \times 100$

% Cl^- in sample $= \left(\dfrac{0.3729 \text{ g Cl}^-}{3.000 \text{ g sample}}\right) \times 100$

$= 12.43\%$

4.73   **(d)**

Chapter 5

Thermochemistry

OVERVIEW OF THE CHAPTER

5.1, 5.2 THERMODYNAMICS: THE FIRST LAW AND INTERNAL ENERGY CHANGES

Review: Dimensional Analysis (1.6).

Learning Goals: You should be able to:

1. Give examples of different forms of energy.
2. List the important units in which energy is expressed and convert from one to another.
3. Define the first law of thermodynamics both verbally and by means of an equation.
4. Describe how the change in internal energy of a system is related to the exchanges of heat and work between the system and its surroundings.
5. Define the term *state function* and describe its importance in thermochemistry.

5.3, 5.4 ENTHALPY: ENTHALPIES OF REACTION

Review: Meaning of chemical equations (3.1); stoichiometric calculations (3.6).

Learning Goals: You should be able to:

1. Define enthalpy, and relate the enthalpy change in a process occurring at constant pressure to the heat added to or lost by the system during the process.
2. Sketch an energy diagram such as that shown in Figure 5.14 of the text, given the enthalpy changes in the processes involved, and associate the sign of ΔH with whether the process is exothermic or endothermic.
3. Calculate the quantity of heat involved in a reaction at constant pressure given the quantity of reactants and the enthalpy change for the reaction on a mole basis.

5.5, 5.8 CALORIMETRY: FUEL VALUES

Learning Goals: You should be able to:

1. Define the terms *heat capacity* and *specific heat*.
2. Calculate any one of the following quantities given the other three: heat, quantity of material, temperature change, and specific heat.

3. Calculate the heat capacity of a calorimeter, given the temperature change and quantity of heat involved; also calculate the heat evolved or absorbed in a process from a knowledge of the heat capacity of the system and its temperature change.
4. Define the term *fuel value*; calculate the fuel value of a substance given its heat of combustion or estimate the fuel value of a material given its composition.
5. List the major sources of energy on which humankind must depend, and discuss the likely availability of these for the foreseeable future.

5.6 HESS'S LAW

Learning Goal: You should be able to:

1. State Hess's law, and apply it to calculate the enthalpy change in a process, given the enthalpy changes in other processes that could be combined to yield the reaction of interest.

5.7 ENTHALPIES OF FORMATION: CALCULATING HEATS OF REACTIONS

Learning Goals: You should be able to:

1. Define and illustrate what is meant by the term *standard state*, and identify the standard states for the elements carbon, hydrogen, and oxygen.
2. Define the term *standard heat of formation*, and identify the type of chemical reaction with which it is associated.
3. Calculate the enthalpy change in a reaction occurring at constant pressure, given the standard enthalpies of formation of each reactant and product.

TOPIC SUMMARIES AND EXERCISES

THERMODYNAMICS: THE FIRST LAW AND INTERNAL ENERGY CHANGES

In Chapter 3 of the text, we learned about the law of conservation of mass and how applying it to chemical reactions enables us to solve stoichiometric problems. In this chapter, we will learn about energy, how it is conserved, and its measurement.

What is energy? Energy can be conceived as the ability to do work or to transfer heat.

- **Mechanical work** $(w) = $ force(f) applied to an object times the distance (d) the object is moved by the force: $w = f \times d$. Energy enables you to supply the required force.
- Chemical compounds possess energy in two principal forms: Potential and kinetic energy. **Potential energy** $(= mgh)$ is energy stored by an object as a result of its position and mass. **Kinetic energy** $(= 1/2mv^2)$ is associated with motion. Except in the case of a solid at 0 K, the particles that make up a substance (gas, liquid, or solid) undergo motion.
- Kinetic and potential energy are interconvertible; however, their total energy is constant during all changes as required by the first law of thermodynamics.
- Energy is conserved in physical and chemical processes—that is, in any change that occurs in nature the total energy of the universe remains constant. This is a statement of the law of conservation of energy, which is also called the **first law of thermodynamics**.

- The term **thermodynamics** refers to the study of the forms of energy and the changes energy undergoes.

  The universe is divided for purposes of studying energy changes into a system and its surroundings.

- The **system** is arbitrarily defined, but it is usually that part of the universe that is being studied. Everything else is termed the **surroundings**.
- An open system permits an exchange of matter and energy between the system and the surroundings. In a closed system energy but not matter is exchanged with the surroundings. In an isolated system there is no exchange of matter and energy. An insulated thermos approximates a closed system.
- A system can interact with its surroundings in two ways: (1) By doing work (w) on the surroundings or by having work done on it by the surroundings; and (2) by exchanging heat (q) with the surroundings.

  The sum of all the kinetic and potential energies of a system is called **internal energy**. A change in the internal energy of a system is ΔE (Δ means a change in).

- $\Delta E = E_{final} - E_{initial}$

  internal energy   internal energy
  at end of change   at start of change
- ΔE = positive number—system has gained energy.
- ΔE = negative number—system has lost energy.

  Energy is exchanged between the system and surroundings as heat (q) and work (w):

  $$\Delta E = q + w \text{ (a statement of the first law of thermodynamics)}$$

	+	−
q (heat)	added to system	given off by system
w (work)	done on system	done by system

The relationships between the signs of q and w and the type of energy exchange are summarized in the following table:

The internal energy of a substance is a state function.

- A **state function** depends only on its present condition or state (specified by temperature, pressure, moles, etc.) and not on the manner in which the state of the system was reached.
- Δ(state function) = final state value − initial state value.

You should know the following units of energy:

- **Calorie:** The amount of energy required to raise the temperature of 1 g of water by 1 °C from 14.5 ° to 15.5 °C.
- **Joule:** 1 cal = 4.184 J. Joule is the SI unit for energy.

EXERCISE 1   Identifying energy as potential or kinetic

Classify the following as possessing kinetic energy, potential energy, or both: (a) two stationary charged particles that are adjacent to one another; (b) an arrow moving through the air.

SOLUTION: (a) The particles have only potential energy because they are stationary. (b) The arrow has kinetic energy because it is not stationary and potential energy because of its position with respect to the surface of the Earth.

EXERCISE 2   Calculating velocity knowing kinetic energy

Calculate the velocity (v) of an electron whose mass is 9.107×10^{-28} g and whose kinetic energy (E_k) is 1.585×10^{-17} J. (1 J $= 1$ kg-m^2/sec^2)

SOLUTION: *Analyze*: We are asked to calculate the velocity of an electron given both its mass and kinetic energy. We will need a relationship between these three terms.

Plan: The kinetic energy of a moving particle is $E_k = \frac{1}{2}mv^2$. This expression relates the given quantities, mass and energy, to velocity. When using equations you must consider the units. The unit for energy is joule, for mass it is kilogram, and for velocity it is meter/second.

Solve: We will have to convert mass from grams to kilograms and also joule to $\frac{\text{kg-m}^2}{\text{s}^2}$ so that the units of velocity can be calculated in m/s:

$$E_k = \tfrac{1}{2}mv^2$$

$$1.585 \times 10^{-17}\,\text{J} = \tfrac{1}{2}(9.107 \times 10^{-31}\,\text{kg})v^2$$

Solving for v yields

$$v = \sqrt{\frac{2E_k}{m}} = \sqrt{\frac{(2)(1.585 \times 10^{-17}\,\text{kg-m}^2/\text{sec}^2)}{9.107 \times 10^{-31}\,\text{kg}}} = 5.899 \times 10^6 \text{ m/sec}$$

Check: The units cancel correctly to give the desired unit of m/s. An estimate of the final answer using the last equation gives $\sqrt{\dfrac{10^{-17}}{10^{-30}}} = \sqrt{10^{13}} \simeq 10^6$ which is of the same magnitude as the exact calculation.

EXERCISE 3   Calculating internal energy

Calculate the internal energy change in a system in which the following changes occur: (a) 100 g of MgO(s) is heated from 50 °C to 100 °C, a process that requires 870 J; no work is done; and (b) a gas expands slowly while heating, and does 200 J of work and gains 350 J of heat.

SOLUTION: *Analyze*: We are asked to calculate the internal energy change when two different changes occur, one involving a process that requires energy and no work is done [problem (a)] and the other involves a gas expanding, gaining energy and doing work [problem (b)]. We will need a relationship between internal energy and work and energy flow.

Plan: Both problems require us to substitute the appropriate data with correct signs into the equation $\Delta E = q + w$. We will have to analyze each problem to determine the signs of q and w.

Solve: (**a**) Raising the temperature of MgO requires heat; thus the sign of q is positive. w is zero as stated in the question.

$$\Delta E = q + w = 870\,J + 0 = 870\,J$$

(**b**) Since the system gains heat, q is positive. Because the gas does work on the surroundings, w is negative.

$$\Delta E = q + w = 350\,J + (-200\,J) = 150\,J$$

Check: The numbers added have the same units and the final answer has the correct unit. A check of the numbers given in the question and used in the problem agree.

Chemical reactions in the laboratory or in the environment commonly occur at constant external pressure, most often under the constant pressure of the atmosphere. **Enthalpy,** is the sum of the internal energy of a system and the product of the pressure and volume of a system. In Chapter 4 we focus on enthalpy changes at constant pressure. An example of this condition is when we carry out reactions in an open beaker and the atmospheric pressure remains constant during the change.

ENTHALPY: ENTHALPIES OF REACTION

- $\Delta H = \Delta E + P\Delta V = q_p$ (the subscript p means constant pressure)
- $\Delta H = H_{final} - H_{initial}$ (state function)
- ΔH = positive number—a process gains heat: **endothermic**;
- ΔH = negative number—a process gives off heat: **exothermic**.
- For a chemical reaction the following relation exists:

$$\Delta H_{rxn} = H(\text{products}) - H(\text{reactants})$$

See Figure 5.1 for a pictorial representation.

The physical states of reactants and products must be specified when you give an enthalpy change for a particular reaction.

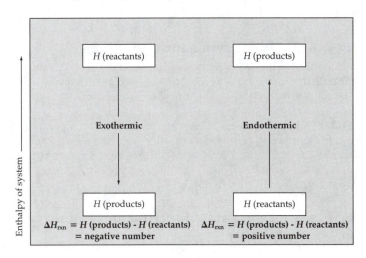

▲ **FIGURE 5.1** Energy diagram for endothermic and exothermic reactions.

- The enthalpy content of a substance varies as its physical state changes.
- A thermochemical equation is used to describe the physical states of reactants and products and the associated enthalpy change. The subscript with ΔH specifies the temperature in Kelvin at which the reaction is carried out. The enthalpy change usually is written to the right. For example,

$$CaO(s) + CO_2(g) \longrightarrow CaCO_3(s) \qquad \Delta H_{298} = -178 \text{ kJ}$$

What are the relationships between (1) the sign of ΔH and the direction of a reaction and between (2) the value of ΔH and the stoichiometry of a reaction?

- The enthalpy change for a reaction is equal in magnitude but opposite in sign to ΔH for the reverse reaction. For example,

$$CaO(s) + CO_2(g) \longrightarrow CaCO_3(s) \qquad \Delta H_{298} = -178 \text{ kJ}$$
$$CaCO_3(s) \longrightarrow CaO(s) + CO_2(g) \qquad \Delta H_{298} = 178 \text{ kJ}$$

- Enthalpy is an extensive property of a system, just as is mass. Therefore, if you change the stoichiometry of a reaction, you also change the enthalpy. For example, if the number of moles of $CaCO_3(s)$ decomposing is changed to two moles, then the enthalpy change must also be multiplied by two:

$$2 \times [CaCO_3(s) \longrightarrow CaO(s) + CO_2(g)] \qquad \Delta H_{298} = 2 \times 178 \text{ kJ}$$
$$2\,CaCO_3(s) \longrightarrow 2\,CaO(s) + 2\,CO_2(g) \qquad \Delta H_{298} = 356 \text{ kJ}$$

EXERCISE 4   Identifying system and surroundings and sign of energy change

You observe an ice cube melting in a glass of water. What are the system and the surroundings for this observed change? Is the melting of ice an exothermic or endothermic process?

SOLUTION: We can label what we observe, the glass containing the water and ice, as the system and everything else as the surroundings. We know that the application of heat to ice causes it to melt. Because the melting of ice requires heat, the change is said to be endothermic.

EXERCISE 5   Identifying enthalpy changes

For which one of the following processes does the heat evolved or absorbed equal the enthalpy change: The combustion of gasoline in a closed container or the boiling of water at 100 °C and at atmospheric pressure?

SOLUTION: Only for a process at constant pressure and temperature does the enthalpy change equal q_p. The boiling of water occurs at a constant atmospheric pressure and temperature; thus the heat change equals q_p, enthalpy. (As the gasoline is combusted in the closed container, the pressure changes because gases are evolved during combustion.)

EXERCISE 6   Using state function and extensive properties of enthalpy

The reaction

$$2\,H_2(g) + O_2(g) \longrightarrow 2\,H_2O(l)$$

occurs at 25 °C with a ΔH value of -571.66 kJ. (**a**) Is this an exothermic or endothermic reaction? (**b**) What is ΔH for the reaction

$$2\,H_2O(l) \longrightarrow 2\,H_2(g) + O_2(g)$$

(**c**) What is ΔH for the reaction when 0.500 mol of $H_2O(l)$ is formed?

SOLUTION: *Analyze*: We are given a chemical reaction and the magnitude and sign of ΔH. We are asked to determine whether the reaction is exothermic or endothermic, the value of ΔH when the reaction is reversed, and the value of ΔH when a half mole of water forms instead of two moles as shown in the chemical reaction.

Plan: We need to remember that the sign of an exothermic reaction is negative and an exothermic reaction is positive; that enthalpy is a state function and therefore its value depends only on the differences between the enthalpies of the final and initial states; and finally, enthalpy is an intensive property which means its value depends on the quantity of material forming or reacting.

Solve: (**a**) This is an exothermic reaction because the sign of ΔH is negative. (**b**) 571.66 kJ. This reaction is the reverse of the first one, and so the enthalpy change is equal in magnitude but opposite in sign to that of the original reaction. (**c**) The heat evolved in the original reaction is for the formation of 2 mol of $H_2O(l)$; the heat evolved in (**c**) is for the formation of 0.50 mol of $H_2O(l)$:

$$\Delta H = \left(\frac{-571.66\text{ kJ}}{2\text{ mol }H_2O}\right)(0.500\text{ mol }H_2O) = -143\text{ kJ}$$

Check: Checking (**a**) and (**b**) involves a review of the questions and ensuring that the numbers and information agree in the answer, which they do. An estimate of the final answer in (**c**) gives $\frac{-600}{2} \times 0.5 = -150$, which is of the same magnitude as the exact calculation. The negative sign also agrees with the conclusion in (**a**).

Calorimetry is an experimental technique for measuring heat flow in a thermochemical change. Temperature changes are accurately measured and converted to heat energy by using heat capacities.

CALORIMETRY: FUEL VALUES

- **Heat capacity** (C) represents the amount of heat required to raise the temperature of a given amount of a substance by 1 K.

- $C = \dfrac{q}{\Delta T} \quad \begin{array}{l} \leftarrow \text{ heat flow} \\ \leftarrow \text{ temperature change} = T_{final} - T_{initial} \end{array}$

- Heat flow $= q = C \times \Delta T \longleftarrow$ *Key concept in calorimetry*

The text discusses two types of heat capacities: Molar heat capacity and specific heat.

- **Molar heat capacity**, C_m, is the heat capacity of one mole of a substance. It has the units of joules per mole per degree (J/mol-K).

$$C_m = \frac{q}{\Delta T} \quad \leftarrow J/mol$$

- **Specific heat**, C_s, is the heat capacity for one gram of a substance. It has the units of joules per gram per degree (J/g-K).

$$C_s = \frac{q}{\Delta T} \quad \leftarrow J/g$$

A *key equation in calorimetry* is:

$$\text{Heat flow } (q) \text{ for a given mass} = m \times C_s \times \Delta T$$

One type of calorimeter for measuring heat flow is a **bomb calorimeter**.

- Heat flow (q) is measured under *constant volume* conditions. Therefore,

$$q_v = \Delta E.$$

- Heat flow from the reaction vessel, called the bomb, raises the temperature of the bomb and the surrounding water bath. Therefore, the heat evolved is related to the temperature increase as follows:

$$q_{\text{evolved}} = -C_{cal} \times \Delta T$$

  where C_{cal} is the heat capacity of the calorimeter.
- C_{cal} is determined by measuring ΔT for combustion of a given mass of a standard (i.e., a reaction whose q_v is known).

Reactions at constant pressure can be studied in a general chemistry laboratory using an insulated styrofoam cup as an inexpensive calorimeter.

- At *constant pressure* conditions, enthalpy is measured.
- If heat is evolved by a chemical reaction in a solution, the temperature of the solution increases. Conversely, if heat is absorbed by a chemical reaction, the temperature of the solution decreases.
- The heat change in the solution is followed by measuring the temperature change of the solution. Since the mass of the solution is known and its heat capacity is either measured using a standard reaction or is assumed to be approximately that of the solvent, usually water, the heat change of the solution is determined by the relation:

$$q_{\text{solution}} = (\text{mass of the solution})(\text{specific heat of the solution})\Delta T$$

- Therefore, the heat change of the chemical reaction is:

$$q_{rxn} = -q_{\text{solution}} = -\text{m}_{\text{solution}} C_s \Delta T$$

The foods we eat and the fuels we use to run our machines are potential energy sources. When they are combusted, the heat released can be used by our bodies or by machines to do work. The heat evolved during the combustion of a gram of food or fuel is known as the **fuel value** of the substance.

- Fuel values are always exothermic quantities. They are reported as positive quantities in tables and thus must have a negative sign included when numbers are used in calculations.

EXERCISE 7   Using specific heat to calculate heat change

The specific heat of $NH_3(l)$ is 4.381 J/g-K. Calculate the heat required to raise the temperature of 1.50 g of $NH_3(l)$ from 213.0 K to 218.0 K.

SOLUTION: *Analyze*: We are asked to calculate the heat required to raise the temperature of 1.50 g of ammonia from 213.0 K to 218.0 K and given the specific heat of ammonia. We will need a relationship between specific heat, mass, and temperature.

Plan: The heat (q) required to raise the temperature of any substance can be calculated using the relationship:

$$q = mC_s\Delta T$$
$$q = mC_s \text{ (final temperature-initial temperature)}$$

Solve: We can solve for q by appropriate substitution into the previous equation:

$$q = (1.50 \text{ g})\left(4.381\frac{J}{g\text{-}K}\right)[218.0 \text{ K} - (213.0 \text{ K})]$$

$$q = (1.50 \text{ g})\left(4.381\frac{J}{g\text{-}K}\right)(5.0 \text{ K}) = 33 \text{ J}$$

Check: The positive sign agrees with information in the problem that heat is added. The value is not abnormally large or small based on the given data. An estimate of the final answer gives $(2)(4)(5) = 40$, which is of the same magnitude as the exact calculation.

EXERCISE 8   Calculating heat of combustion

0.800 g of sulfur is combusted to form SO_2. All of the heat evolved is used to raise the temperature of 100.0 g of water by 17.8 K. **(a)** Write the combustion reaction for sulfur. **(b)** Calculate the heat of combustion of 1 mol of sulfur.

SOLUTION: *Analyze*: The first part of the problem asks us to write the combustion reaction for sulfur. We need to know what a combustion reaction is. The second part asks us to calculate the heat of combustion of one mole of sulfur given that 0.800 g of sulfur is combusted and the heat evolved raises the temperature of 100.0 g of water by 17.8 K.

Plan: **(a)** Combustion is the reaction of a substance with oxygen with the release of heat. We will need to write a reaction of sulfur with oxygen to form sulfur dioxide. **(b)** The heat evolved by the combustion reaction is transferred completely to water and its temperature rises. Therefore,

$$q_{combustion} + q_{water} = 0$$

$$q_{combustion} = -q_{water} = -m_{water}C_{s,water}\Delta T$$

The specific heat of water is obtained from tables or the text; it is 4.184 J/g-K. The heat of combustion of sulfur for 0.800 g S is calculated. This is then converted to one mole of sulfur by recognizing that one mole of sulfur has a molar mass of 32.0 g.

Solve: **(a)** The combustion of sulfur involves its reaction with O_2 to form an oxide of sulfur: $S(s) + O_2(g) \rightarrow SO_2(g)$. **(b)** The heat evolved when sulfur is combusted raises the temperature of water. Therefore,

$$q_{combustion} = -q_{water}$$

$$q_{water} = m_{water} \, C_{s,water}\Delta T$$

$$q_{water} = (100.0 \text{ g})\left(4.184\frac{J}{g-K}\right)(17.8 \text{ K}) = 7450 \text{ J} = 7.45 \text{ kJ}$$

$$q_{combustion} = -7.45 \text{ kJ}$$

This is the heat of combustion when 0.80 g of sulfur reacts with oxygen. To determine the molar heat of combustion, we must make the following conversion using the molar mass of sulfur:

$$\Delta H_{combustion} = \left(\frac{-7.45 \text{ kJ}}{0.800 \text{ g S}}\right)\left(\frac{32.0 \text{ g S}}{1 \text{ mol S}}\right) = -298\frac{\text{kJ}}{\text{mol}}$$

Check: The negative sign agrees with information in the problem that heat is evolved in a combustion reaction. The positive sign of q_{water} agrees with the observation that water absorbs the energy from the combustion reaction. An estimate of calculations, $(100)(4)(18) = 7200$ and $\left(\frac{-7}{1}\right)(32) = -224$, gives magnitudes of numbers in agreement with those in the exact calculations.

EXERCISE 9　Calculating temperature change in a bomb calorimeter

A 1.05 g sample of benzoic acid is combusted completely in a bomb calorimeter. Calculate the temperature change given that the heat capacity of the calorimeter is 1.80 kJ/K and the heat of combustion of benzoic acid is −26.4 kJ/g.

SOLUTION: *Analyze*: 1.05 g of benzoic acid is combusted in a bomb calorimeter (a constant volume device) and we are asked to calculate the temperature change given the heat of combustion of benzoic acid (based on one mole) and the heat capacity of the bomb calorimeter.

Plan: We can first calculate the heat evolved when 1.05 g of benzoic acid is combusted from its known heat of combustion. Then we can calculate the temperature change of the calorimeter by the relationship: $q_{\text{benzoic acid}} + q_{\text{calorimeter}} = 0$. The heat absorbed by the calorimeter from the heat of combustion is related to the temperature change by the relationship: $q_{\text{benzoic acid}} = -C_{cal}\Delta T$. The problem gives the value of the heat capacity of the calorimeter.

Solve: First solve for the heat of combustion of benzoic acid:

$$q_{\text{benzoic acid}} = \Delta H_{combustion} \times m_{\text{benzoic acid}} = (-26.4 \text{ kJ/g})(1.05 \text{ g}) = -27.7 \text{ kJ}$$

The heat evolved from the combustion of benzoic acid raises the temperature of the calorimeter.

$$q_{\text{benzoic acid}} = -C_{cal} \times \Delta T$$
$$-27.7 \text{ kJ} = -(1.80 \text{ kJ/K}) \times \Delta T$$

Solving for ΔT,

$$\Delta T = \frac{-27.7 \text{ kJ}}{-1.80 \text{ kJ/K}} = 15.4 \text{ K}$$

Check: ΔT is positive, which agrees with the information that heat is evolved in a combustion reaction. An estimate of calculations, $(-26)(1) = -26$ and $\frac{-28}{-2} = 14$ gives magnitudes of numbers in agreement with those in the exact calculations.

EXERCISE 10　Calculating fuel values of food

Fudge is about 2 percent protein, 11 percent fat, and 81 percent carbohydrate. The average fuel values of these substances are as follows: protein, 17 kJ/g; fat, 38 kJ/g; carbohydrate, 17 kJ/g. What is the fuel value of 10 g of fudge?

SOLUTION: *Analyze*: We are asked to calculate the fuel value of 10 g of fudge given the mass percent of its components and their respective fuel values.

Plan: The fuel value of fudge is a weighted average of each component's fuel value.

Fuel value = (mass of fudge)[(protein fraction × its average fuel value)
+ (fat fraction × its average fuel value)
+ (carbohydrate fraction × its average fuel value)]

Solve: We can solve for the fuel value of fudge by substituting the appropriate quantities into the previous equation:

$$\text{Fuel value of fudge} = (10\ \text{g})\left[(0.02)\left(17\frac{\text{kJ}}{\text{g}}\right) + (0.11)\left(38\frac{\text{kJ}}{\text{g}}\right) + (0.81)\left(17\frac{\text{kJ}}{\text{g}}\right)\right]$$

$$\text{Fuel value of fudge} = (10\ \text{g})\left(18\frac{\text{kJ}}{\text{g}}\right) = 180\ \text{kJ}$$

The food value of 10 g of fudge in calories is therefore as follows:

$$(180\ \text{kJ})\left(\frac{10^3\ \text{J}}{1\ \text{kJ}}\right)\left(\frac{1\ \text{cal}}{4.184\ \text{J}}\right) = 4.3 \times 10^4\ \text{cal}$$

Or since 1 kcal = 1 Calorie, the food value is 43 Calories. (The Calorie, capitalized, is the unit used in nutrition.)

Check: The sign of the final answer agrees with the idea that food values are reported as positive numbers. An estimate of calculations, $10[(0.01)(15) + (0.1)(40) + 1(20)] = 204$ and $(200)(1000)\left(\frac{1}{5}\right) = 40,000$ gives magnitudes of numbers in agreement with those in the exact calculations.

HESS'S LAW

Hess's law is a statement of an important principle of thermochemistry: *The enthalpy change for a chemical reaction is the same whether the reaction takes place in one step or in several steps.* ΔH_{rxn} equals the sum of enthalpy changes for each step. Use the following procedure to calculate an enthalpy change for a reaction, having been given known thermochemical equations:

1. Write the thermochemical equation for the desired chemical process.
2. The goal is to write several thermochemical equations which when added together give the desired thermochemical equation. Thus, you have to write each thermochemical equation so a desired reactant is on the left side of an arrow and a desired product is on the right side of an arrow.
3. If a desired reactant or product is on the wrong side of a given thermochemical equation, reverse the thermochemical equation. You also must change the sign of ΔH.
4. Multiply the known thermochemical equations by the appropriate coefficients so that the reactants and products have the same coefficients as they have in the desired thermochemical equation.
5. Algebraically add the thermochemical equations after step 4. If an unwanted species does not disappear, you must now include another thermochemical equation such that when it is added to the others the unwanted species cancels. Be sure that the reactants and products remain balanced.
6. Add ΔH's for each thermochemical equation to obtain ΔH for the desired reaction.

EXERCISE 11 Using Hess's law to determine enthalpy change I

Calculate the enthalpy change for the reaction

$$C(s) + \tfrac{1}{2} O_2(g) \longrightarrow CO(g)$$

given the following information:

(a) $C(s) + O_2(g) \longrightarrow CO_2(g)$ $\quad\quad\quad \Delta H = -393.5 \text{ kJ}$

(b) $CO(g) + \tfrac{1}{2} O_2(g) \longrightarrow CO_2(g)$ $\quad\quad \Delta H = -283.0 \text{ kJ}$

SOLUTION: *Analyze*: We are asked to calculate the enthalpy change for the combustion of carbon given two other reactions and their enthalpy changes. This is a typical type of Hess's Law problem.

Plan: We can apply Hess's Law to this problem if we can determine how to combine the two reactions (a) and (b) to produce the desired reaction. Since we want $CO(g)$ as a product, we can see by inspection of reaction (b) that it will have to be reversed so that $CO(g)$ is changed from a reactant to a product. If you add equation (a) and the reverse of equation (b), the desired equation is obtained:

Solve:

$$
\begin{array}{ll}
C(s) + O_2(g) \longrightarrow CO_2(g) & \Delta H = -393.5 \text{ kJ} \\
\underline{CO_2(g) \longrightarrow CO(g) + \tfrac{1}{2}O_2(g)} & \underline{\Delta H = 283.0 \text{ kJ}} \\
C(s) + \tfrac{1}{2}O_2(g) \longrightarrow CO(g) & \Delta H = -110.5 \text{ kJ}
\end{array}
$$

Note that the sign of ΔH for reaction (b) must be changed because the reaction has been reversed.

Check: A check of the reactants and products in each step and the associated signs of ΔH should be done to ensure no mistake has been made. This check shows the two steps and their signs and values of ΔH are in agreement. An estimate of the sum is $-400 + 300 = -100$, which is of the same magnitude as in the exact calculation.

EXERCISE 12 Using Hess's law to determine enthalpy change II

Calculate ΔH for the reaction

$$C_2H_4(g) + H_2(g) \longrightarrow C_2H_6(l)$$

given the following data:

(a) $C_2H_4(g) + 3 O_2(g) \longrightarrow 2 CO_2(g) + 2 H_2O(l)$ $\quad \Delta H = -845.2 \text{ kJ}$

(b) $2 H_2(g) + O_2(g) \longrightarrow 2 H_2O(l)$ $\quad\quad\quad\quad\quad\quad \Delta H = -571.7 \text{ kJ}$

(c) $2 C_2H_6(l) + 7 O_2(g) \longrightarrow 4 CO_2(g) + 6 H_2O(l)$ $\quad \Delta H = -1987.4 \text{ kJ}$

SOLUTION: *Analyze*: This problem is similar to the previous one. We are asked to calculate the enthalpy of reaction given three chemical reactions and their enthalpies of reactions.

Plan: We must find a way to write the three reactions so that when they are added the required chemical reaction results. By inspection we see that equation (a) is not changed as it has C_2H_4 as a reactant. Equation (c) must be reversed and divided by 2 so that 1 mol of $C_2H_6(l)$ is produced, since it is the product of the required reaction. Chemical equation (**b**) must be divided by 2 so that only 1 mol of $H_2(g)$ appears as a reactant as in the required reaction.

Solve: Following the plan gives us

$$
\begin{aligned}
C_2H_4(g) + 3\,O_2(g) &\longrightarrow 2\,CO_2(g) + 2\,H_2O(l) & \Delta H &= -845.2 \text{ kJ} \\
H_2(g) + \tfrac{1}{2}O_2(g) &\longrightarrow H_2O(l) & \Delta H &= -285.9 \text{ kJ} \\
2\,CO_2(g) + 3\,H_2O(l) &\longrightarrow \tfrac{7}{2}O_2(g) + C_2H_6(l) & \Delta H &= +993.7 \text{ kJ} \\
\hline
C_2H_4(g) + H_2(g) &\longrightarrow C_2H_6(l) & \Delta H &= -137.4 \text{ kJ}
\end{aligned}
$$

Note: The ΔH value associated with equation (**b**) has been divided by 2 because equation (**b**) has been divided by 2. Similarly, the ΔH value associated with equation (**c**) has been divided by 2; its sign has also changed because equation (**c**) has been reversed.

Check: The same check as done in Exercise 11 is used for this problem. An estimate of the sum is $-800 - 300 + 1000 = -100$, which is of the same magnitude as in the exact calculation.

EXERCISE 13   Relating Hess's law to state function properties

Explain how Hess's law of heat summation depends on the fact that enthalpy is a state function.

SOLUTION: The enthalpy change for a reaction depends only on the amount of matter that undergoes a change and the initial states of the reactants and the final states of the products. Thus $\Delta H = H_{\text{final}} - H_{\text{initial}}$, and the value does not depend on any intermediate steps that might be involved. Therefore, if a particular reaction is carried out in a series of steps, the sum of ΔH's for the series of reactions must be the same as the enthalpy change for the given one-step reaction.

ENTHALPIES OF FORMATION: CALCULATING HEATS OF REACTIONS

We can use standard enthalpies of formation to calculate enthalpy changes of reactions. Before doing such calculations, we need to know the following about enthalpies of formation:

- The enthalpy change associated with the formation of one mole of a compound from its constituent elements is termed **enthalpy of formation** (ΔH_f). For example, at 100 °C and standard atmospheric pressure, the thermochemical equation for the formation of $H_2O(g)$ is

$$
H_2(g) + \tfrac{1}{2}O_2(g) \longrightarrow H_2O(g) \quad \Delta H_f = -242.3 \text{ kJ/mol}
$$

- The value of ΔH_f for a compound depends on the physical state of the compound, pressure, and temperature.

To enable chemists to make appropriate comparisons of ΔH_{rxn} values, a standard set of conditions is defined.

- The **standard state** for a substance is its most stable form at the given temperature and pressure. The pressure usually is standard atmospheric pressure and the temperature is 25 °C; however, other pressures and temperatures can be specified.
- The **standard enthalpy of formation** of a compound, ΔH°_f, is the heat of formation of *one mole* of the compound in its *standard state* formed from its constituent elements in their *standard states*.

- The standard enthalpy of formation of any element in its most stable form is by convention taken to be zero: $(\Delta H_f^\circ)_{298}$ (O_2) = 0 kJ/mol, but $(\Delta H_f^\circ)_{298}$ (O_3) = 142.7 kJ/mol because O_3 is not the most stable form of elemental oxygen at 25 °C and 1 atm.

A simple method for determining enthalpies of reaction from standard enthalpies of formation is

- ΔH_{rxn}° = (sum of enthalpies of formation of products)
    $-$(sum of the enthalpies of formation of reactants)

which is sometimes written as

$$\Delta H_{rxn}^\circ = \Sigma n \Delta H_f^\circ \text{ (products)} - \Sigma m \Delta H_f^\circ \text{ (reactants)}$$

where Σ (sigma) means "the sum of" and n and m are the stoichiometric coefficients of the chemical equation.
- Keep in mind that $\Delta H_f^\circ = 0$ for elements in their standard states.

EXERCISE 14   Writing reactions associated with standard enthalpies of formation

The standard state of an element or compound is its most stable form at 25 °C and standard atmospheric pressure. Write the reaction associated with the standard enthalpy of formation for each of the following: (a) AgCl(s); (b) $NH_3(g)$; (c) $Na_2CO_3(s)$.

SOLUTION: *Analyze*: We are asked to write the standard enthalpy of formation reaction for three compounds. We need to know what the term standard enthalpy of formation means.

Plan: The standard enthalpy of formation (ΔH_f°) is the enthalpy change for the formation of 1 mol of a compound from its elements in their standard states at 25 °C and standard atmospheric pressure. Develop balanced chemical reactions that have as reactants the elements comprising each compound. Each element should be in its most stable form at 25 °C and at 1 atmosphere pressure.

Solve:

(a) $Ag(s) + \frac{1}{2} Cl_2(g) \longrightarrow AgCl(s)$

(b) $\frac{1}{2} N_2(g) + \frac{3}{2} H_2(g) \longrightarrow NH_3(g)$

(c) $2 Na(s) + C(graphite) + \frac{3}{2} O_2(g) \longrightarrow Na_2CO_3(s)$

Check: The three heats of formation reactions all have elements as reactants and a single compound as a product. Also each product is only one mole. A check of charge and mass balance shows that the reactions are balanced.

EXERCISE 15   Calculating the standard enthalpy change for a reaction

Calculate the standard enthalpy change for the reaction

$$CH_2N_2(s) + O_2(g) \longrightarrow CO(g) + H_2O(l) + N_2(g)$$

at 25 °C and standard atmospheric pressure, given the following standard heats of formation: $CO(g)$, -110.5 kJ/mol; $CH_2N_2(s)$, $+62.4$ kJ/mol; $H_2O(l)$, -285.8 kJ/mol.

(Remember that the standard heat of formation for any element in its standard state is zero.)

SOLUTION: *Analyze*: We are asked to calculate the enthalpy change for a specified reaction and are given standard heats of formation for all substances. There is a note that you have to remember that the standard enthalpy of formation for any element in its standard state is zero.

Plan: The standard enthalpy for a reaction at 25 °C and 1 atm is calculated using the relation:

$$\Delta H^{\circ}_{rxn} = \Sigma n \Delta H^{\circ}_{f} \text{ (products)} - \Sigma m \Delta H^{\circ}_{f} \text{ (reactants)}$$

$$\Delta H^{\circ}_{rxn} = [\Delta H^{\circ}_{f}(CO) + \Delta H^{\circ}_{f}(H_2O) + \Delta H^{\circ}_{f}(N_2)] - [\Delta H^{\circ}_{f}(CH_2N_2) + \Delta H^{\circ}_{f}(O_2)]$$

Solve: Substituting for the values of the standard heats of formation in the previous equation gives:

$$\Delta H^{\circ}_{rxn} = \left[(1 \text{ mol})\left(-110.5\frac{kJ}{mol}\right) + (1 \text{ mol})\left(-285.8\frac{kJ}{mol}\right) \right.$$

$$\left. + (1 \text{ mol})\left(0\frac{kJ}{mol}\right) \right] - \left[(1 \text{ mol})\left(+62.4\frac{kJ}{mol}\right) + (1 \text{ mol})\left(0\frac{kJ}{mol}\right) \right]$$

$$\Delta H^{\circ}_{rxn} = -458.7 \text{ kJ}$$

Check: First check that the calculation of heat of reaction indeed involves values associated with products less values associated with reactants. Then check that the number of moles of each species in the chemical reaction is used in the calculation. Finally, do an estimate of the answer:

$$[(1)(-100) + (1)(-300) + 0 - 50 + 0] = -450.$$

This is in close agreement with the exact calculation.

SELF-TEST QUESTIONS

Key Terms

Having reviewed key terms in Chapter 5, match key terms with phrases and identify statements as true or false. If a statement is false, indicate why it is incorrect.

Match each phrase with the best term:

5.1   A baseball possesses this type of energy before it is thrown.

5.2   Molecules in motion have this type of energy.

5.3   A positive sign is associated with this type of energy change.

5.4   The name given to the region observed in an experiment.

5.5   Without this we could not do work.

5.6   The SI unit for energy.

5.7   Heat flow measured in a constant pressure system.

5.8   The heat flow that occurs when a substance is formed from its elements in their standard states at one atmosphere pressure.

5.9   Energy is conserved in all processes.

5.10   The energy change measured in a bomb calorimeter.

5.11   Its value is independent of the pathway.

5.12   Its unit is J/g- ° C.

5.13   The study of heat flow and processes involving energy changes.

5.14   Properties observable with our senses.

5.15   A method for measuring heat flow.

5.16   If it flows out of a system the temperature of a system decreases.

5.17   The universe minus a system.

Key Terms:

(a) calorimetery
(b) endothermic
(c) energy
(d) enthalpy
(e) first law of thermodynamics
(f) heat
(g) internal energy
(h) joule
(i) kinetic energy
(j) macroscopic
(k) potential energy
(l) specific heat
(m) standard enthalpy of formation
(n) state function
(o) surroundings
(p) system
(q) thermodynamics

True-False Statements:

5.18 *Syngas* is primarily a mixture of CH_4, H_2, and CO.

5.19 The validity of *Hess's law* arises from the fact that enthalpy is a state function.

5.20 The *specific heat capacity* of water is 4.18 J/g- °C. To convert the specific heat capacity of water to molar heat capacity, you would divide the specific heat capacity by the molar mass of water.

5.21 Proteins and carbohydrates have similar average *fuel values*, about 17 kJ/g.

5.22 *Thermochemistry* is the study of chemical reactions and their energy changes.

5.23 The greater the *force* applied to an object when it is moved, the greater the work.

5.24 In one minute, a 100-watt light bulb requires a *power* input of 100 J to operate.

5.25 One *calorie* is the same as the Calorie.

5.26 In an insulated container, an *exothermic* reaction would increase the temperature of the system.

5.27 The value of a *state function* depends on the particular history of the sample.

5.28 *Enthalpies of formation* for compounds are generally exothermic quantities.

5.29 An example of a constant pressure *calorimeter* is the "coffee cup" calorimeter described in the text.

5.30 *Heat capacity* is a positive number.

5.31 The *molar heat capacity* of water is calculated by taking its specific heat capacity and dividing by 18 g, the molar mass of water.

5.32 *Specific heat* of a substance is based on the heat capacity of 1 g of the material.

5.33 *Fossil fuels* include gasoline.

5.34 A major component of *natural gas* is methane.

5.35 *Coal* contains hydrocarbons of low molecular weight.

Problems and Short-Answer Questions

5.36 The following figure represents a phase change. What are the initial and final phases? What is the sign of ΔH? Explain.

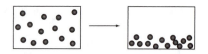

5.37 Using the following figure answer the following questions: (a) Which energy change, A, B, or C, corresponds to the enthalpy of formation of carbon dioxide gas? (b) What is the mathematical relationship between energy changes A, B, and C? (c) Write the net chemical reaction for step C.

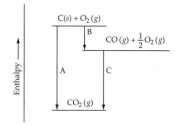

5.38 How would an increase in the internal energy of a system affect the following: temperature, phase change, and chemical reaction? For each explain the type and direction of change that can occur.

5.39   7.5 g of a metal at 50.0 °C is placed in water at 25.0 °C and the temperature of the water rises by 4.5 °C. The specific heat of water is $4.184 \frac{J}{K \cdot g}$. Without doing a calculation, is there sufficient information to determine the specific heat capacity of the metal? If there is insufficient information, what other data is needed and why?

5.40 The combustion of acetylene gas is a source of intense heat for welding. The reaction is:

$$2\,C_2H_5(g) + 5\,O_2(g) \longrightarrow 4\,CO_2(g) + 2\,H_2O(g)$$

The heat of combustion for acetylene is −1300 kJ/mol. Without doing a calculation, is there sufficient information to determine the heat liberated when 0.780 g of acetylene is burned? If there is insufficient information, what other data is needed and why?

5.41 Determine the enthalpy of reaction for

$$CH_3COCH_3(l) + 2\,O_2(g) \longrightarrow CH_3COOH(l) + CO_2(g) + H_2O(g)$$

using Hess's law of constant heat summation and given:

$$CH_3COCH_3(l) + 4\,O_2(g) \longrightarrow 3\,CO_2(g) + 3\,H_2O(l)$$
$$\Delta H = -1787.0 \text{ kJ}$$

$$H_2O(l) \longrightarrow H_2O(g) \qquad \Delta H = -44.1 \text{ kJ}$$

$$CH_3COOH(l) + 2\,O_2(g) \longrightarrow 2\,CO_2(g) + 2\,H_2O(l)$$
$$\Delta H = -835 \text{ kJ}$$

5.42 Calculate the standard enthalpy of formation of $PCl_5(g)$, given that $\Delta H° = -136.0$ kJ for the reaction

$$PCl_5(g) + H_2O(g) \longrightarrow POCl_3(g) + 2\,HCl(g)$$

and the following standard enthalpies of formation: $H_2O(g)$, -241.8 kJ/mol; $POCl_3(g)$, -592 kJ/mol; $HCl(g)$, -92.5 kJ/mol.

5.43 The combustion of 4.086 g of In increased the temperature of a calorimeter by 1.626 °C. The energy equivalent (the amount of energy required to raise the temperature by 1 °C) of the calorimeter is 10.095 kJ. What is the heat evolved per mole of In?

5.44 Calculate the heat evolved when 0.500 g of $Cl_2(g)$ reacts with an excess of $HBr(g)$ to form $HCl(g)$ and $Br_2(l)$, given the following standard enthalpies of formation (kJ/mol): $HCl(g)$, -92.30; $HBr(g)$, -36.20.

5.45 The fuel value for bituminous coal is 32 kJ/g. The fission fuel value of ^{235}U is 3.17×10^{-11} kJ/atom. How many grams of coal must be combusted to equal the energy released by the fission of 1 g of ^{235}U in a nuclear power plant?

5.46 Benzene, C_6H_6, is a chemical that has been used extensively in chemical laboratories. However, recent toxicological evidence suggests that it is capable of causing cancer, and thus chemists are limiting its use. Calculate the enthalpy of combustion of $C_6H_6(l)$, given the following standard enthalpies of formation: $C_6H_6(l)$, 49.0 kJ/mol; $CO_2(g)$, -395.5 kJ/mol; $H_2O(l)$, -285.9 kJ/mol.

5.47 In some processes, the system is totally isolated from the surroundings. If such a process is exothermic, what do you think happens to the heat energy? What device commonly found in a house enables one to approximate these conditions?

5.48 In some processes, the temperature of the system does not change. In an exothermic process that occurs under such conditions, what happens to the energy released?

5.49 A small "coffee cup" calorimeter contains 110 g of H_2O at 22.0 °C. A 100-g sample of lead is heated to 90.0 °C and then placed in the water. The contents of the calorimeter come to a temperature of 23.9 °C. What is the specific heat of lead?

Integrative Exercises

5.50 Hydrazine, N_2H_4, is used as a fuel in rockets. **(a)** Hydrazine can thermally decompose in the absence of air as follows:

$$3\,N_2H_4(l) \longrightarrow 4\,NH_3(g) + N_2(g)$$

If ΔH_f° for ammonia gas is -46.3 kJ/mol and for liquid hydrazine it is 50.6 kJ/mol, what is ΔH° for the reaction at 25 °C? **(b)** If 10.0 g of hydrazine reacts, what is the heat evolved? **(c)** Does the combustion of hydrazine in oxygen at 25 °C generate more or less heat than the reaction in **(a)**? The combustion reaction forms nitrogen gas and liquid water. The heat of formation of $H_2O(l)$ is -285.8 kJ/mol.

5.51 50.00 mL of 1.00 M HCl is added to 50.0 mL of 0.95 M KOH in a "coffee-cup" calorimeter and stirred. A change in temperature is observed from 26.0 °C to 32.6 °C. What is the heat of reaction per mole of HCl? Assume that the specific heat of the solution is the same as that of water, 4.184 J/g- °C, the density of all solutions is 1.00 g/mL, and that no heat is lost to the "coffee-cup" and surroundings.

Multiple-Choice Questions

5.52 In Chapter 5.11 you will learn more about phase changes and factors affecting forces between particles of a substance. When a substance undergoes a phase transition from solid to liquid with application of heat the temperature remains constant until all the solid becomes liquid. With further heating the temperature rises. **(a)** Why is the temperature constant during the phase transition? **(b)** 25.0 g of ice at 0 °C is added to 50.0 g of water at 50 °C. It takes $0.33\ \frac{kJ}{g}$ to change ice to water at the melting point. What is the final temperature of the mixture after all of the ice has melted? Assume that no heat is lost to the surroundings.

5.53 Given the following information:

$$Ag_2O(s) \longrightarrow 2\,Ag(s) + 1/2\,O_2(g) \qquad \Delta H^\circ = 31.05\ kJ$$

which of the following statements concerning the reaction are true: (1) heat is released; (2) heat is absorbed; (3) reaction is exothermic; (4) reaction is endothermic; (5) products have higher enthalpy content than reactants; (6) reactants have higher enthalpy content than products?

(a) 1, 3, and 5	**(d)** 1 and 3
(b) 2, 4, and 6	**(e)** 2, 4, and 5
(c) 1, 3, and 6	

5.54 Given the chemical equation

$$S(s) + O_2(g) \longrightarrow SO_2(g) \qquad \Delta H = -297.1\ kJ/mol$$

how much heat is associated with the formation of 7.99 g of $SO_2(g)$?

(a) -586 kJ	**(d)** -37.1 kJ
(b) 37.1 kJ	**(e)** -293 kJ
(c) 293 kJ	

5.55 How much heat is required to convert 100 g of water at 40 °C to water vapor at 100 °C? The heat capacity of water is 4.184 J/g- °C, and the heat required to vaporize water is 2.26 kJ/g.

(a) 227 kJ	**(d)** 25.1 kJ
(b) 418 kJ	**(e)** 251 kJ
(c) 226 kJ	

5.56 In a hydrogen bomb, the requisite high temperature for initiating the nuclear fusion comes from the following exothermic reaction, where ^1_0n represents a neutron:

$$^2_1H + ^3_1H \longrightarrow ^4_2He + ^1_0n$$

During the above reaction, mass is converted to energy. For each 1 g of reactants (^2_1H and ^3_1H), 0.0188 g of mass is converted to energy. Using Einstein's equation $E = mc^2$, calculate how much energy is released when 0.0188 g of mass is converted to energy. (The symbol c represents the speed of light; it has a value of 3.00×10^8 m/sec.)

(a) 5.67×10^{11} J (d) 1.70×10^{15} kJ
(b) 5.67×10^{12} J (e) 1.70×10^{12} kJ
(c) 1.70×10^{12} J

5.57 If it takes 46 cal to vaporize 1 g of a substance at 106 °C, then which of the following is true: (1) 46 cal will be required to vaporize 1 g of the substance at 53 °C; (2) 46 cal will be required to change 1 g of the vapor to liquid at 106 °C; (3) the temperature of the vapor must be higher than that of the liquid; (4) 46 cal will vaporize 1 g of the substance at 100 °C; (5) 46 cal of heat will be given off when 1 g of the vapor condenses at 106 °C?

(a) 1 (d) 4
(b) 2 (e) 5
(c) 3

5.58 If 1.00 g of kerosene liberates 46.0 kJ of heat when it is burned, to what temperature can 0.250 g of kerosene heat 75.00 cm^3 of water at 25.0 °C?

(a) 62.0 °C (d) 61.6 °C
(b) 6.10 °C (e) 63.0 °C
(c) 62.6 °C

5.59 Given the following data:

$$2\,SO_2(g) + O_2(g) \longrightarrow 2\,SO_3(g) \quad \Delta H = -196.7\ kJ/mol$$

$$SO_3(g) + H_2O(l) \longrightarrow H_2SO_4(l) \quad \Delta H = -130.1\ kJ/mol$$

what is the enthalpy of reaction for

$$2\,SO_2(g) + O_2(g) + 2\,H_2O(l) \longrightarrow 2\,H_2SO_4(l)$$

(a) 66.6 kJ (d) 456.9 kJ
(b) 326.7 kJ (e) −456.9 kJ
(c) −326.7 kJ

5.60 Given the following:

$$C_2H_5OH(l) + 3\,O_2(g) \longrightarrow 2\,CO_2(g) + 3\,H_2O(l)$$
$$\Delta H = -1366.9\ kJ$$

$$CH_3CO_2H(l) + 2\,O_2(g) \longrightarrow 2\,CO_2(g) + 2\,H_2O(l)$$
$$\Delta H = -869.9\ kJ$$

what is the enthalpy of reaction for

$$C_2H_5OH(l) + O_2(g) \longrightarrow CH_3CO_2H(l) + H_2O(l)$$

(a) 497.0 kJ (c) 2237 kJ
(b) −497.0 kJ (d) −2237 kJ
(e) insufficient data provided to answer question

5.61 Which of the following is in a standard state at 100 °C and 1 atm pressure?

(a) $H_2O(g)$ (d) $H_2O(s)$
(b) $CO_2(s)$ (e) $Fe(s)$
(c) $Cl_2(l)$

5.62 Which statement concerning the standard states and standard enthalpies is correct?

(a) The standard state of a substance is specified only for 25 °C.
(b) The enthalpy of formation for an element in its standard state is 0 kJ/mol.
(c) The standard enthalpy of formation for a substance is endothermic.
(d) The symbol for a standard enthalpy is ΔH.
(e) The standard enthalpy of reaction is given by the relationship:

$$\Delta H^{\circ}_{rxn} = \sum \Delta H^{\circ}_{reactants} - \sum \Delta H^{\circ}_{products}$$

5.63 5.00 kg of a hot metal at 200.0 °C is added to 25.0 kg of water at 30.0 °C. What is the final temperature of the metal? The specific heat of the metal is 0.800 J/g- °C and for water it is 4.184 J/g- °C.

(a) 25.3 °C (d) 41.3 °C
(b) 34.3 °C (e) 48.3 °C
(c) 36.3 °C

5.64 The molar heat of combustion of methanol, CH_3OH (molar mass = 32.0 g/mol) is −757 kJ/mol. If 2.00 g of methanol is placed in a bomb calorimeter with a heat capacity of 6.50 kJ/°C, what is the temperature rise when the sample is combusted?

(a) 1.34 °C (d) 12.78 °C
(b) 7.51 °C (e) 15.67 °C
(c) 9.56 °C

5.65 10.00 g of Mg(s) was completely burned in chlorine gas to give pure $MgCl_2(s)$ and 131.9 kJ of heat was evolved. What is the molar enthalpy of formation of $MgCl_2(s)$?

(a) −320.6 kJ (c) −641.2 kJ
(b) 320.6 kJ (d) 641.2 kJ
(e) Additional information is required.

5.66 0.50 mol of ammonium nitrate is added to 50.0 mL of water in a thermally insulated reaction vessel. Both are initially at 20.0 °C. After stirring, the temperature of the solution is found to be less than 10° C. Which statement best explains the temperature change?

(a) Heat is evolved from the system to the surroundings.
(b) Heat is absorbed from the surroundings by the system.
(c) Heat is absorbed by ammonium nitrate when it dissolves and becomes hydrated.
(d) Heat is evolved by ammonium nitrate when it dissolves and becomes hydrated.
(e) None of the above.

5.67 What is the standard enthalpy of formation of NO(g) given that the enthalpy of formation of $NO_2(g)$ is 32 kJ/mol and the enthalpy of reaction for

$$2\,NO(g) + O_2(g) \longrightarrow 2\,NO_2(g)$$

is −116 kJ at standard state conditions?

(a) −45 kJ

(b) −10 kJ

(c) 26 kJ

(d) 52 kJ

(e) 90 kJ

SELF-TEST SOLUTIONS

5.1 (k). 5.2 (i). 5.3 (b). 5.4 (p). 5.5 (c). 5.6 (h). 5.7 (d). 5.8 (m). 5.9 (e). 5.10 (g). 5.11 (n). 5.12 (l). 5.13 (q). 5.14 (j). 5.15 (a). 5.16 (f). 5.17 (o). 5.18 True. 5.19 True. 5.20 True. 5.21 False. The specific heat capacity is multiplied by 18 g/1 mol. Note that grams cancel and mole is now a unit in the denominator. 5.22 True. 5.23 True. 5.24 True. 5.25 False. A watt is 1 J/s. Thus, in one minute, a 100-watt lightbulb requires 100 J/s × 60 s or 6,000 J. 5.26 False. One Calorie (a big calorie) is 1000 calories. 5.27 True. 5.28 False. It does not depend on how the state of the system was achieved. 5.29 True. 5.30 True. 5.31 True. It is a positive number because heat is required to raise the temperature of a substance. 5.32 False. The unit of molar heat capacity is J/mol-K. Thus, to convert J/g-K to J/mol-K, you have to multiply by 18 g/1 mol. 5.33 True. 5.34 True. 5.35 True. 5.36 False. Coal contains hydrocarbons of high molecular weight, otherwise, coal would be a liquid or gas. 5.37 The initial phase is a gas and the final phase is a liquid. The sign of ΔH is negative because heat must be extracted from the gas phase to reduce the kinetic energies of the particles to permit them to coalesce and condense to a liquid state 5.38 (a) A: Carbon dioxide gas is formed from its constituent elements. (b) A = B + C. This is an application of Hess's law. (c) $CO(g) + \frac{1}{2} O_2(g) \longrightarrow CO_2(g)$ 5.39 If the internal energy of a system increases, then the temperature increases. For a phase change, melting or vaporization can occur. When internal energy increases, the particles of a system move at a faster rate (increase in kinetic energy). This increased kinetic energy results in sufficient energy to overcome the forces between particles and melting or vaporization can happen. A chemical reaction can occur when a system gains internal energy. If this reaction occurs it is by absorption of energy and thus is endothermic. 5.40 There is insufficient information. The heat absorbed by water when the metal is added is calculated by the relation:
$q = m_{water} C_{s,metal} \Delta T$. This requires the mass of water, which is not given. If q_{water} can be calculated, then the specific heat capacity of the metal can be calculated by the relation: $q_{metal} = -q_{water}$ where $q_{metal} = m_{metal}C_{s,metal} \Delta T$.

5.41 There is sufficient information. The mass of acetylene can be converted to moles of acetylene. Then the heat liberated is simply the heat of combustion (in kJ/mol) times the number of moles of acetylene. Note that the number of moles cancels in the calculation.

5.42 The appropriate arrangement of the three reactions to yield the desired reaction is

$$CH_3COCH_3(l) + 4 O_2(g) \longrightarrow 3 CO_2(g) + 3 H_2O(l)$$
$$\Delta H = -1787 \text{ kJ}$$

$$H_2O(l) \longrightarrow H_2O(g) \qquad \Delta H = -44.1 \text{ kJ}$$

$$2 CO_2(g) + 2 H_2O(l) \longrightarrow CH_3COOH(l) + 2 O_2(g)$$
$$\Delta H = +835 \text{ kJ}$$

$$CH_3COCH_3(l) + 2 O_2(g) \longrightarrow CO_2(g) + H_2O(g)$$
$$+ CH_3COOH(l) \quad \Delta H = -966 \text{ kJ}$$

The enthalpy of reaction is −966 kJ.

5.43 The enthalpy of reaction for $PCl_5(g) + H_2O(g) \longrightarrow POCl_3(g) + 2 HCl(g)$ is related to the standard heats of formation of the products and reactants by the relation: $\Delta H^\circ_{rxn} = [\Delta H^\circ_f(POCl_3) +$

$2 \Delta H^\circ_f(HCl)] - [\Delta H^\circ_f(PCl_5) - \Delta H^\circ_f(H_2O)]$. Substituting the given data into this equation yields: −136.0 kJ
$= -592 \text{ kJ} + 2 (-92.5 \text{ kJ}) - \Delta H^\circ_f(PCl_5) - (-241.8 \text{ kJ})$.
Solving for $\Delta H^\circ_f(PCl_5)$ gives $\Delta H^\circ_f = -399$ kJ.

5.44 The heat produced by the combustion of 4.086 g of In is: heat = (energy equivalent)(temperature rise)
$= (10.095 \text{ kJ/°C})(1.626 °C) = 16.41 \text{ kJ}$. The heat evolved per mol of In is the heat produced per gram of indium times the mass of a mol of indium:

$$\text{Heat per mole} = -\left(\frac{16.41 \text{ kJ}}{4.086 \text{ g}}\right)\left(\frac{114.82 \text{ g}}{1 \text{ mol}}\right)$$
$$= -461.1 \text{ kJ/mol}$$

5.45 The chemical reaction is described by $Cl_2(g) + 2 HBr(g) \longrightarrow 2HCl(g) + Br_2(l)$. The heat evolved per mole of Cl_2 is: $\Delta H^\circ_{rxn} = \Delta H^\circ_f(Br_2) + 2 \Delta H^\circ_f(HCl)$
$- \Delta H^\circ_f(Cl_2) - 2 \Delta H^\circ_f(HBr) = 0 \text{ kJ} + 2(-92.30 \text{ kJ}) -$

$0 \text{ kJ} - 2(-36.20 \text{ kJ}) = -112.2 \text{ kJ}$. The heat evolved when 0.500 g of Cl_2 reacts is

$$\Delta H^\circ = (\Delta H^\circ_{rxn})(\text{moles } Cl_2)$$

$$\Delta H^\circ = \left(\frac{-112.2 \text{ kJ}}{1 \text{ mol}}\right)\left(\frac{0.500 \text{ g}}{70.90 \text{ g/mol}}\right) = 0.791 \text{ kJ}$$

5.46 We first need to calculate the heat evolved when 1 g of ^{235}U undergoes fission. The fission fuel value is given as 3.17×10^{-11} kJ/atom. The necessary equivalences that convert this number to kJ/g are: 6.022×10^{23} atoms = 1 mol ^{235}U; 1 mol ^{235}U = 235 g. This fission fuel value for ^{235}U, expressed in kJ/g, is

$$\left(3.17 \times 10^{-11} \frac{\text{kJ}}{\text{atom}}\right)\left(6.022 \times 10^{23} \frac{\text{atoms}}{\text{mol}}\right)\left(\frac{1 \text{ mol}}{235 \text{ g}}\right)$$
$$= 8.12 \times 10^{10} \text{ kJ/g}$$

The fission fuel value for 1 g is 8.12×10^{10} kJ. The mass of coal required to produce the same energy is calculated as follows: 8.12×10^{10} kJ = (fuel value of coal)(mass of coal) = $(32 \, kJ/g) \times$ (mass of coal). Solving for the mass of coal gives a mass of 2.5×10^9 g (which equals 5.5×10^6 lb). The high fuel value per gram for ^{235}U versus the low fuel value for coal is one reason for the great interest in using ^{235}U rather than coal to generate electricity. **5.47** The thermochemical equation for the combustion of benzene is $2 \, C_6H_6(l) + 15 \, O_2(g) \longrightarrow 12 \, CO_2(g) + 6 \, H_2O(l)$. We can calculate the heat of combustion of C_6H_2 using the relation

$$\Delta H^\circ_{rxn} = [12 \, \Delta H^\circ_f(CO_2) + 6 \, \Delta H^\circ_f(H_2O)]$$

$$-[2 \, \Delta H^\circ_f(C_6H_6) + 15 \, \Delta H^\circ_f(O_2)]$$

$$\Delta H^\circ_{rxn} = \left[(12 \text{ mol})\left(-395.5\frac{kJ}{mol}\right) + (6 \text{ mol})\left(-285.9\frac{kJ}{mol}\right)\right]$$

$$-\left[(2 \text{ mol})\left(49.0\frac{kJ}{mol}\right) + (15 \text{ mol})\left(0\frac{kJ}{mol}\right)\right]$$

$$\Delta H^\circ_{rxn} = (-4746 \text{ kJ} - 1715 \text{ kJ}) - (98.0 \text{ kJ} + 0)$$

$$= -6559 \text{ kJ}$$

This is the heat of combustion for 2 mol of C_6H_6. For 1 mol of C_6H_6, the enthalpy is $-6559 \text{ kJ}/2 = -3280$ kJ. **5.48** Since the heat energy released is contained in the system, the heat must be absorbed by the substances present in the system in such a way that their kinetic energy is increased. This increased kinetic motion is reflected by a temperature rise within the system. A closed thermos jug is a device that approximates this type of system. **5.49** For a process of this sort to occur all the heat released during the exothermic change must be given off to the surroundings. Thus the temperature of the system remains constant, but the temperature of the surroundings increases.

5.50   Heat lost $= -$heat gained *or*

$$q_{lead} = -q_{H_2O} \text{ where } q = mC\Delta T$$

$$(100g)(C)(23.9\,°C - 90.0\,°C)$$

$$= -(110 \text{ g})(4.184 \text{ J/g-°C})(23.9\,°C - 22.0\,°C)$$

Solving for C gives $C = 0.134$ J/g-°C.

5.51 (a)$\Delta H^\circ_{rxn} = 4 \, \Delta H^\circ_f(NH_3) + \Delta H^\circ_f(N_2) - 3\Delta H^\circ_f(N_2H_4)$

$$\Delta H^\circ_{rxn} = 4 \text{ mol}(-46.3 \text{ kJ/mol}) + (0)$$
$$- 3 \text{ mol} (50.6 \text{ kJ/mol}) = -337 \text{ kJ}$$

(b) The heat evolved in **(a)** is for the reaction of three moles of hydrazine; for one mole it is -112 kJ. The number of moles of hydrazine is: $(10.0 \text{ g})(1 \text{ mol}/32.06 \text{ g}) = 0.312$ mol. Thus, the heat evolved for 10.0 g of hydrazine is: $(-112 \text{ kJ/mol})(0.312 \text{ mol}) = -34.9$ kJ. **(c)** First write the combustion reaction:

$$N_2H_4(l) + O_2(g) \longrightarrow N_2(g) + 2 \, H_2O(l)$$

$$\Delta H^\circ_{rxn} = \Delta H^\circ_f(N_2) - 2\Delta H^\circ_f(H_2O) - \Delta H^\circ_f(N_2H_2)$$
$$- \Delta H^\circ_f(O_2)$$

$$\Delta H^\circ_{rxn} = (0) + 2 \text{ mol}(-285.8 \text{ kJ/mol})$$
$$- 1 \text{ mol}(50.6 \text{ kJ/mol}) - (0) = -622.2 \text{ kJ}$$

The combustion of hydrazine in oxygen evolves significantly more heat than the decomposition of one mole of hydrazine as calculated in **(b)**.

5.52   The reaction is $HCl(aq) + KOH(aq) \longrightarrow KCl(aq) + H_2O(l)$. Note that HCl and KOH react in a 1:1 mole ratio. The heat evolved by the reaction is absorbed by the solution, thus:

$$-q_{rxn} = q_{cal} \text{ } or \text{ } q_{rxn} = -q_{cal}$$

$$= -m_{solution}C_{s,solution}\Delta T$$

The mass of the solution is determined from the volume of the solution and its density. Since the density is 1.00 g/mL, the mass is the total volume (50 mL + 50 mL) expressed in grams: 100 g. The specific heat, s is 4.184 J/g-°C. The heat of reaction is thus:

$$q_{rxn} = -(100 \text{ g})(4.184 \text{ J/g-°C})(32.6\,°C - 26.0\,°C)$$

$$= -2.8 \times 10^3 \text{ J}$$

But this is not the heat per mole of HCl. The next question to answer is how many moles of HCl reacted? The number of moles of HCl $= M \times V = (0.05000 \, L)(1.00 \, M)$ = 0.0500 moles HCl. The number of moles of KOH that reacted is: $(0.5000 \, L)(0.95 \, M)$ = 0.048 moles KOH. This is a limiting reactant situation. Only 0.048 moles of KOH are available to react with 0.0500 moles of HCl. Thus, only 0.048 moles HCl can react. The energy evolved per mole of HCl is:

$$(-2.8 \times 10^3 \text{ J}/0.048 \text{ mol HCl}) = -58 \times 10^3 \text{ J/mol} =$$
$$-58 \text{ kJ/mol.}$$

5.53   **(a)** When a substance is transformed from a solid to a liquid, forces between the solid particles must first be overcome before the liquid increases in temperature. Thus, at the melting point heat is used to transform the solid to liquid and the temperature remains constant. Once the solid is entirely melted, added heat increases the average kinetic energy of the particles in the liquid and the temperature rises. **(b)** The energy needed to melt the ice and raise its temperature (as a liquid) is obtained from the cooling of the water originally at 50.0 °C. Since there is no heat exchanged between the system and surroundings, we can write

$$q_{ice \, melting} + q_{raise \, temperature \, of \, water \, formed \, from \, ice}$$
$$+ q_{cooling \, of \, water} = 0$$

The first term is (mass ice)(heat to melt one gram of ice); the second term is (mass ice)(specific heat of water)(final temperature-initial temperature); and the final term is (mass water)(specific heat of water)(final temperature-initial temperature). Let y be the final temperature of the mixture.

$$(25.0 \text{ g ice})\left(\frac{0.33 \text{ kJ}}{g}\right) + (25.0 \text{ g water})\left(4.184\frac{J}{g\text{-}K}\right)(y - 298K)$$

$$+ (50.0 \text{ g water})\left(4.184\frac{J}{g\text{-}K}\right)(y - 348 \text{ K}) = 0.$$

Solving for y gives $y = 305 \text{ K}$ or 7 °C

5.54 (e)

5.55 (d) $7.99 \text{ g SO}_2\left(\frac{1 \text{ mol SO}_2}{64.06 \text{ g SO}_2}\right)\left(\frac{-291.1 \text{ kJ}}{1 \text{ mol SO}_2}\right) = -37.1 \text{ kJ}$

5.56 (e) Total heat = heat required to raise temperature of water + heat required to vaporize water

$$= (100 \text{ g})\left(4.184\frac{J}{g\text{-}°C}\right)\left(\frac{1 \text{ kJ}}{1000 \text{ J}}\right)(100 °C - 40 °C)$$

$$+ (100 \text{ g})\left(2.26\frac{kJ}{g}\right) = 251 \text{ kJ}$$

5.57 (c) $E = mc^2$

$$= (0.0188 \text{ g})\left(\frac{1 \text{ kg}}{1000 \text{ g}}\right)\left(3.00 \times 10^8\frac{m}{sec}\right)^2$$

$$= (1.88 \times 10^{-5} \text{ kg})\left(9.00 \times 10^{16}\frac{m^2}{sec^2}\right)$$

$$= 1.70 \times 10^{12}\frac{kg - m^2}{sec^2} = 1.70 \times 10^{12} \text{ J}$$

5.58 (e)
5.59 (d) Mass of 75.00 cm^3 of water = 75.00 g Heat liberated by kerosene

$$= (0.250 \text{ g})\left(46.0\frac{kJ}{g}\right)\left(\frac{1000 \text{ J}}{1 \text{ kJ}}\right) = 11.500 \text{ J}$$

Heat flow = (mass of water)(C of water)(ΔT)

$$11.500 \text{ J} = (75.0 \text{ g})\left(4.184\frac{J}{g\text{-}°C}\right)(T - 25.0 °C)$$

$$T - 25.0 °C = 36.6 °C \qquad T = 61.6 °C$$

5.60 (e)

$2 \text{ SO}_2(g) + \text{O}_2(g) \longrightarrow 2 \text{ SO}_3(g) \qquad \Delta H = -196.7 \text{ kJ}$

$2 \text{ SO}_3(g) + 2 \text{ H}_2\text{O}(l) \longrightarrow 2 \text{ H}_2\text{SO}_4(l) \quad \Delta H = -260.2 \text{ kJ}$

$2 \text{ SO}_2(g) + \text{O}_2(g) + 2 \text{ H}_2\text{O}(l) \longrightarrow 2 \text{ H}_2\text{SO}_4(l)$
$$\Delta H = -456.9 \text{ kJ}$$

5.61 (b)

$\text{C}_2\text{H}_5\text{OH}(l) + 3 \text{ O}_2(g) \longrightarrow 2 \text{ CO}_2(g) + 3 \text{ H}_2\text{O}(l)$
$$\Delta H = -1366.9 \text{ kJ}$$
$2 \text{ CO}_2(g) + 2 \text{ H}_2\text{O}(l) \longrightarrow \text{CH}_3\text{CO}_2\text{H}(l) + 2 \text{ O}_2(g)$
$$\Delta H = -866.9 \text{ kJ}$$

$\text{C}_2\text{H}_5\text{OH}(l) + \text{O}_2(g) \longrightarrow \text{CH}_3\text{CO}_2\text{H}(l) + \text{H}_2\text{O}(l)$
$$\Delta H = -497.0 \text{ kJ}$$

5.62 (a) Note that the temperature is 100 °C.

5.63 (b)

5.64 (c) In a thermally insulated system $q_{metal} + q_{water} = 0$. Therefore, $-q_{metal} = q_{water} - (5.00 \times 10^3 \text{ g metal})(0.800 \text{ J/g-°C})(T_f - 200.0 °C) = (2.50 \times 10^4 \text{ g})(4.184 \text{ J/g-°C})(T_f - 30.0 °C)$. Solving for $T_f = 36.3 °C$

5.65 (b)
mol methanol = $2.00 \text{ g}(1 \text{ mol}/31.0 \text{ g}) = 0.0645 \text{ mol}$ heat evolved by combustion of methanol = $(0.0625 \text{ mol} methanol)(-757 \text{ kJ}/\text{mol}) = -48.8 \text{ kJ}$

$$-q_{combustion} = q_{cal}$$

$$-(-48.8 \text{ kJ}) = (6.50 \text{ kJ}/°C)(\Delta T)$$

$$\Delta T = 7.51 °C$$

5.66 (a) First write the heat of formation reaction for magnesium chloride:

$$\text{Mg}(s) + \text{Cl}_2(g) \longrightarrow \text{MgCl}_2(s)$$

The stoichiometry of the reaction enables you to calculate the number of moles of magnesium chloride formed from the initial mass of magnesium metal.

$$\text{mol Mg} = (10.00 \text{ g})\left(\frac{1 \text{ mol Mg}}{24.31 \text{ g}}\right) = 0.4114 \text{ mol Mg}$$

$$\text{mol MgCl}_2 = (0.4114 \text{ mol Mg})\left(\frac{1 \text{ mol MgCl}_2}{1 \text{ mol Mg}}\right)$$

$$= 0.4114 \text{ mol MgCl}_2$$

$$\Delta H_f° = \frac{-131.9 \text{ kJ}}{0.4114 \text{ mol MgCl}_2} = -320.6 \text{ kJ}/\text{mol}$$

5.67 (c) For the temperature of the solution to decrease, the solvent water must lose energy. The process of ammonium and nitrate ions in solid ammonium nitrate dissolving and becoming hydrated must therefore absorb energy from water.

5.68 (e)

$$\Delta H_{reaction}° = 2\Delta H_f°(\text{NO}_2) - 2\Delta H_f°(\text{NO}) - \Delta H_f°(\text{O}_2)$$

$$-116 \text{ kJ} = 2 \text{ mol} (32 \text{ kJ}/\text{mol}) - 2 \text{ mol}(\Delta H_f°(\text{NO})) - 0$$

$$\Delta H_f°(\text{NO}) = 90 \text{ kJ}$$

Chapter

6

Electronic Structures of Atoms

OVERVIEW OF THE CHAPTER

6.1 ELECTROMAGNETIC RADIATION

Learning Goals: You should be able to:

1. Describe the wave properties and characteristic speed of propagation of radiant energy (electromagnetic radiation).
2. Use the relationship $\lambda v = c$, which relates the wavelength (λ) and the frequency (v) of radiant energy to its speed (c).

6.2 QUANTIZATION OF ENERGY

Learning Goals: You should be able to:

1. Explain the essential feature of Planck's quantum theory, namely, that the smallest increment, or quantum, of radiant energy of frequency, v, that can be emitted or absorbed is hv, where h is Planck's constant.
2. Explain how Einstein accounted for the photoelectric effect by considering the radiant energy to be a stream of particle-like photons striking a metal surface. In other words, you should be able to explain all the observations about the photoelectric effect using Einstein's model.

6.3 LINE SPECTRA AND THE BOHR MODEL

Learning Goals: You should be able to:

1. Explain the origin of the expression *line spectra*.
2. List the assumptions made by Bohr in his model of the hydrogen atom.
3. Explain the concept of an allowed energy state and how this concept is related to the quantum theory.
4. Calculate the energy differences between any two allowed energy states of the electron in hydrogen.
5. Explain the concept of ionization energy.

6.4, 6.5, 6.6 PRINCIPLES OF MODERN QUANTUM THEORY

Learning Goals: You should be able to:

1. Calculate the characteristic wavelength of a particle from a knowledge of its mass and velocity.
2. Describe the uncertainty principle and explain the limitation it places on our ability to define simultaneously the location and momentum of a subatomic particle, particularly an electron.

3. Explain the concepts of orbital, electron density, and probability as used in the quantum-mechanical model of the atom: Explain the physical significance of Ψ^2.
4. Describe the quantum numbers, *n*, *l*, and m_l used to define an orbital in an atom and list the limitations placed on the values each may have.
5. Describe the shapes of the *s*, *p*, and *d* orbitals.

Learning Goals: You should be able to:

1. Explain why electrons with the same value of principal quantum number *(n)* but different values of the azimuthal quantum number *(l)* possess different energies.

Learning Goals: You should be able to:

1. Explain the concepts of electron spin and the electron spin quantum number.
2. State the Pauli exclusion principle and Hund's rule, and illustrate how they are used in writing the electronic structures of the elements.
3. Write the electron configuration for any element.
4. Write the orbital diagram representation for electron configurations of atoms.

Learning Goals: You should be able to:

1. Describe what we mean by the *s*, *p*, *d*, and *f* blocks of elements.
2. Write the electron configuration and valence electron configuration for any element once you know its place in the periodic table.

TOPIC SUMMARIES AND EXERCISES

When you sense the warmth of a fire in a fireplace you are feeling what scientists call **radiation**, or **electromagnetic radiation**. The fire gives off light (visible radiation) and heat (thermal radiation). Both types of electromagnetic radiation exist in the form of **electromagnetic waves**.

*   Note in Figure 6.3 in the text the following characteristics of electromagnetic waves: wavelength (λ) and amplitude.
*   Electromagnetic waves move through a vacuum at 3×10^8 m/s, the "speed of light."
*   Electromagnetic waves differ from one another by their **frequency** (*v*). This is the number of cycles per second (1 hertz = 1 cycle/s).

Chemists often characterize electromagnetic radiation by its **wavelength**, λ.

* λ measures the distance between two adjacent maxima (peaks) in a periodic wave.
*   A key relationship between wavelength, λ, and frequency is:

$$v = \frac{c}{\lambda}$$

where *c* is the speed of light, 3.00×10^8 m/s.

6.7 ENERGIES OF ORBITALS IN MANY-ELECTRON ATOMS

6.8 ELECTRONIC STRUCTURE OF MANY-ELECTRON ATOMS

6.9 THE PERIODIC TABLE: ELECTRON CONFIGURATIONS AND VALENCE ELECTRONS

ELECTROMAG-NETIC RADIATION

• Examples of wavelength units are in the following table

Unit	Symbol	Length
Nanometer	nm	1×10^{-9} m
Angstrom	Å	1×10^{-10} m
Picometer	pm	1×10^{-12} m

EXERCISE 1   Characterizing the properties of a wave

Figure 6.1 shows two periodic waves, 1 and 2. (a) What do the distances a and b correspond to? (b) Which electromagnetic radiation, 1 or 2, has the greater frequency?

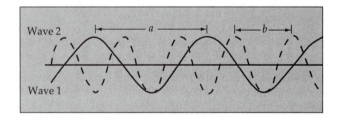

▲ **FIGURE 6.1** Two periodic waves

SOLUTION: (a) The distance a corresponds to the wavelength for wave 1, and similarly, the distance b corresponds to the wavelength for 2. (b) Frequency is defined as the number of times per second that maxima of a periodic wave pass a given point. We can see by inspection of Figure 6.1 that the adjacent maxima of wave 2 are closer together than those of wave 1. Because of this closer proximity of adjacent maxima, wave 2 will have more maxima passing a given point in one second than will wave 1. Thus wave 2 has the higher frequency.

EXERCISE 2   Converting units of wavelength

Convert 2×10^{-10} m to nanometer.

SOLUTION: *Analyze*: We are asked to convert 2×10^{-10} m to nanometer.

Plan: We will do the following conversions: *meter → nanometers*. The conversion is done with the appropriate numerical value of the prefix.

Solve: The unit conversion is made as follows: A nanometer is equal to 1×10^{-9} meter; therefore, the conversion from meter to nanometer is

$$\text{Nanometers} = (2 \times 10^{-10} \text{ m})\left(\frac{1 \text{ nm}}{1 \times 10^{-9} \text{ m}}\right) = 2 \times 10^{-1} \text{ nm}$$

Check: A check of the units show that they cancel to give the desired unit.

EXERCISE 3   Determining wavelength from frequency of electromagnetic radiation

Calculate the wavelength of electromagnetic radiation that has a frequency of 9.22×10^{17}/s.

SOLUTION: *Analyze*: We are asked to calculate a wavelength of electromagnetic energy given its frequency.

Plan: The relationship between frequency and wavelength of electromagnetic radiation is:

$$v = \frac{c}{\lambda}$$

where v has units of s^{-1}. We want to calculate the wavelength, λ. It is convenient to change equation so the term λ is on the left-hand side of the equation.

Solve: Rearranging the equation and substituting the known values of c and v gives:

$$\lambda = \frac{c}{v} = \frac{3.00 \times 10^8 \text{ m/s}}{9.22 \times 10^{17} / \text{s}} = 3.25 \times 10^{-10} \text{ m}$$

Check: An estimate of wavelength gives $\frac{10^8}{10^{18}} = 10^{-10}$, which has the same magnitude as the calculated value. A check of the units shows they properly cancel to give the desired unit.

During the nineteenth century, theories of classical physics were unable to explain certain physical phenomena. For example, the color of an extremely hot iron bar was predicted to be blue; yet it is red. Max Planck proposed that light, which had been previously thought to consist of a continuous collection of electromagnetic waves, consisted, instead, of bundles of energy.

QUANTIZATION OF ENERGY

- Each bundle of energy has a discrete (individually distinct or separate) value.
- The smallest allowed increment of energy gained or lost is called a **quantum**.
- *Planck's relationship* between the smallest amount of energy gained or lost and the frequency of the associated radiation is

$$E = hv$$

$$h = 6.63 \times 10^{-34} \text{ J-s (Planck's constant)}$$

Acceptance of Planck's quantum theory was slow until Albert Einstein used it successfully in 1905 to explain the **photoelectric effect**. This effect occurs when a light shines on the surface of a clean metal and electrons are ejected from the surface.

- Einstein proposed that a beam of light consists of a collection of small particles of energy, called **photons**.
- A photon can interact with an electron on a metal surface and transfer its energy to the electron.
- An electron is ejected if it gains sufficient energy to overcome the forces binding the electron to the surface. Energy beyond the binding energy appears as the kinetic energy of the emitted electron.

EXERCISE 4 Determining the energy of a photon I

(a) Does the energy of a photon increase or decrease with increasing wavelength?
(b) In Exercise 1, which electromagnetic wave, 1 or 2, has the greater energy?

SOLUTION: *Analyze*: Question (a) asks us to describe how the energy of a photon depends on wavelength. We will have to look up or find a relationship between the two. Question (**b**) asks us to apply the information we learned in question (a) to two waves described in Exercise 1.

Plan: (**a**) The relation between energy and wavelength is derived from the following two equations:

$$E = hv \quad \text{and} \quad v = \frac{c}{\lambda}$$

Substituting c/λ for v in Planck's equation yields

$$E = hv = h\frac{c}{\lambda}$$

(**b**) In Exercise 1 we determined which wave had the greater frequency. We can use the relationship above to determine which has the greater energy.

Solve: (**a**) By inspection of the derived relationship we see that the energy of a photon is *inversely* proportional to its wavelength; thus, its energy decreases with increasing wavelength. (**b**) In Exercise 2 we found that wave 2 has the greater frequency; since energy is *directly* proportional to frequency, wave 2 has the greater energy.

Comment: The concept that energy of a wave is inversely proportional to its wavelength is often encountered when discussions of energy and spectra occur.

EXERCISE 5   Determining the energy of a photon II

Calculate the energy of an X-ray photon with a wavelength of 3.00×10^{-10} m.

SOLUTION: *Analyze*: We are asked to calculate the energy of an X-ray photon with a specified wavelength. The information in the previous Exercise is useful for doing this problem.

Plan: Use the relationship $E = \dfrac{hc}{\lambda}$ to solve for energy.

Solve: Substitute the given or known values of h, c, and λ into this relationship:

$$E = \frac{hc}{\lambda} = \frac{(6.63 \times 10^{-34} \text{ J-sec})(3.00 \times 10^{8} \text{ m/sec})}{3 \times 10^{-10} \text{ m}} = 6.63 \times 10^{-16} \text{ J}$$

Check: An estimate of the calculated energy gives $\dfrac{10^{-33}(10^{8})}{10^{-10}} = 10^{-15}$ which has the same order of magnitude as the solution. A check of the units show that they properly cancel to give J, the desired unit.

LINE SPECTRA AND THE BOHR MODEL

In the 1800s, Josef Fräunhofer and others had shown that light emitted from the sun, or from hydrogen atoms heated in a partial vacuum, was not a continuum.

- A continuum is a collection of electromagnetic waves of all possible wavelengths in a given energy region.
- A prism separates light emitted from the sun, or from heated atoms, into discrete and noncontinuous wavelength components.
- A collection of discrete lines or wavelengths on a photographic plate is called a **line spectrum**. See Figure 6.13 in the text.
- Each gaseous monatomic element has its own characteristic line spectra.

In 1914, Niels Bohr proposed a model for the hydrogen atom that enabled scientists to explain hydrogen line spectra. Key ideas in **Bohr's model** of the hydrogen atom are:

- An electron moves in a circular path, called an **orbit**.
- Only orbits restricted to certain energies and radii are permitted.
- Each allowed orbit is assigned an integer n, known as the **principal quantum number**:

$$n = \underbrace{1, 2, 3, 4, \ldots, \infty}_{\text{range of values of } n}$$

- The energy of an orbit in a hydrogen atom is quantized:

$$E_n = -R_H\left(\frac{1}{n^2}\right) \qquad n = \text{principal quantum number of an orbit}$$

R_H = Rydberg constant = $2.18 \times 10^{-18}\,\text{J}$

- Note that as the principal quantum number increases the energy of an orbit increases (becomes a smaller negative number).
- The orbit of lowest energy, $n = 1$, is called the **ground state**. All others are termed **excited states**.

Emission or absorption of radiant energy from a hydrogen atom occurs when an electron moves from one orbit to another. The energy change is

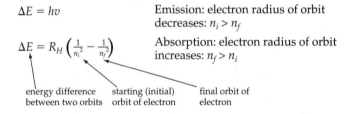

$\Delta E = h\nu$ Emission: electron radius of orbit decreases: $n_i > n_f$

$\Delta E = R_H\left(\frac{1}{n_i^2} - \frac{1}{n_f^2}\right)$ Absorption: electron radius of orbit increases: $n_f > n_i$

energy difference between two orbits   starting (initial) orbit of electron   final orbit of electron

EXERCISE 6   Identifying the postulates of Bohr's model of a hydrogen atom

What are the postulates of Niels Bohr's model for the hydrogen atom?

SOLUTION: The postulates are: (**a**) An electron moves in a circular path about the nucleus and does not collapse into the nucleus. This movement about the nucleus is characterized by the electron's orbit. (**b**) The energy of an electron can have only certain allowed values; its energy is quantized. (**c**) An electron can move from one circular path to another circular path only when it absorbs or emits a photon corresponding to the exact energy difference between the two quantized energy states.

EXERCISE 7   Meaning of sign of E_n and ionization

What is the meaning of the negative sign in the equation $E_n = -R_H(1/n^2)$? What is an excited state? When an electron is ionized, what is its final principal quantum number?

SOLUTION: The negative sign means that energy is required to remove the electron from the nucleus—that is, when the electron and nucleus are combined, energy is released. An alternative explanation is that the electron and nucleus in the atom are more stable than the separated electron and nucleus. Any orbit of the hydrogen atom with $n \geq 2$ is considered a higher energy orbit and is said to be an "excited state." Ionization corresponds to a transition to a final state of $n = \infty$.

EXERCISE 8   Determining energy change when an electron changes its orbit

An electron moves from an $n = 1$ Bohr hydrogen orbit to an $n = 2$ Bohr hydrogen orbit. (a) What is the energy associated with this transition? (b) Is this electronic transition accompanied by the emission or absorption of energy?

SOLUTION: *Analyze*: In the first part of the question we are asked to determine the energy associated with an electron moving from the $n = 1$ Bohr orbit to the $n = 2$ Bohr orbit. We will need a relationship between energy and the principal quantum number of a Bohr orbit. The second part asks us to determine whether energy is absorbed or emitted when an electron changes from a lower numbered Bohr orbit to a higher one.

Plan: (a) The energy change associated with an electron moving from one Bohr orbit to another is given by the relation:

$$\Delta E = R_H \left(\frac{1}{n_i^2} - \frac{1}{n_f^2} \right)$$

where n_i is the principal quantum number of the initial orbit, n_f is the principal quantum number of the final orbit, and R_H is the Rydberg constant, 2.18×10^{-18} J.

(b) The sign of ΔE tells us whether energy is absorbed (positive sign) or emitted (negative sign).

Solve: (a) Solve for ΔE by substituting into the previous equation the values $n_i = 1$ (the initial orbit) and $n_f = 2$ (the final orbit):

$$\Delta E = R_H \left(\frac{1}{n_i^2} - \frac{1}{n_f^2} \right) = (2.18 \times 10^{-18} \text{ J}) \left(\frac{1}{1^2} - \frac{1}{2^2} \right)$$

$$= (2.18 \times 10^{-18} \text{ J}) \left(1 - \frac{1}{4} \right) = (2.18 \times 10^{-18} \text{ J}) \left(\frac{3}{4} \right)$$

$$= 1.63 \times 10^{-18} \text{ J}$$

(b) The sign of the value for ΔE is positive; thus, the transition requires energy. For this transition to occur, a photon with an energy of at least 1.63×10^{-18} J would have to be absorbed by the electron.

Check: An estimate of energy gives $2x \ 10^{-18}\left(1 - \frac{1}{4}\right) = 1.5 \times 10^{-18}$ which is close to the calculated value. The appropriate unit, **J**, is also shown.

PRINCIPLES OF MODERN QUANTUM THEORY

As discussed in the previous topic summary sections, electromagnetic radiation has both wave and particle aspects. In 1924, Louis de Broglie postulated that all material particles have wave properties.

• The result of his conjecture was the **de Broglie relation:**

$$\lambda = \frac{h}{p} = \frac{h}{mv}$$

where λ is the wavelength of the matter wave, h is Planck's constant, p is the momentum of the moving particle (which equals $m \times v$), m is the mass of the particle, and v is the velocity of the particle.

* Note that for large particles, such as a ball, m is very large; therefore λ is extremely small and not measurable. For small particles, such as an electron, λ is larger and measurable.

Application of wave-equation principles, the de Broglie postulate, and the uncertainty principle to the properties of an electron moving about a hydrogen nucleus eventually led to the derivation of mathematical functions (given the symbol ψ) that describe the wave behavior of electrons.

* See Figure 6.16 in the text. Large values of ψ^2 represent a high probability of finding an electron at a particular point in space. This density of dots is referred to as an electron-density distribution.
* An **orbital** is a one-electron wave function and has an associated allowed energy state. An orbital also is represented as an electron-density distribution in space.

Each hydrogen-like orbital is described by three characteristic **quantum numbers**: n, l, and m_l.

* All three are required to completely characterize a hydrogen orbital. Their names and their relationships to one another are summarized in Table 6.1.
* A collection of orbitals having the same value of n is called an **electron shell**.
* A **subshell** consists of all orbitals with the same n and l values.
* For a hydrogen atom, all orbitals with the same n values have the same energy. This is not true for multi-electron atoms.

A shorthand notation has been devised to describe each hydrogen-like orbital in terms of its n and l quantum numbers.

* In this notation system, the principle quantum number, n, appears in front of the l value for the orbital, where the value of l is designated by a letter:

TABLE 6.1 Relationship among Quantum Numbers n, l, and m_l

Quantum	Name number	Dependence on other quantum numbers	Range of values
n	Principal	None	$1, 2, 3, \ldots, \infty$
l	Angular momentum	Integral values that depend on the value of n. For each n quantum number, there are n number of l values.	$0, 1, 2, \ldots, n - 1$ for each n value
m_l	Magnetic	Integral values that depend on the value of l. For each l quantum number, there are $2l + 1$ possible m_l values.	$l, l - 1, l - 2, \ldots 0, \ldots,$ $-(l - 1), -l$ for each l value

Quantum number l	0	1	2	3
Designation of orbital	s	p	d	f

For example, the shorthand notation for a hydrogen-like orbital with quantum numbers $n = 2$ and $l = 1$ and having two electrons is

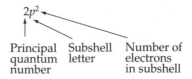

$2p^2$

Principal   Subshell   Number of
quantum   letter   electrons
number   in subshell

Note that the m_l value of a hydrogen-like orbital is not indicated.

- Note that all hydrogen-like orbitals with the same n and l values, such as the $2p$ orbitals, form a subshell.
- Carefully examine Table 6.2 in the text. Learn the subshells for each n level.
- Remember that each n level contains n^2 orbitals; each l subshell level contains $2l + 1$ orbitals. For example, the $n = 2$ principle quantum level contains a total of 2^2 or 4 orbitals: One $2s$ orbital and three $2p$ orbitals.

Graphical representations of the $1s$, $2s$, $2p$, and $3d$ orbitals provide you with an understanding of the properties of orbitals.

- You need to be familiar with the graphical representations of the following orbitals: $1s$, $2s$, $2p_x$, $2p_y$, $2p_z$, $3d_{z^2}$, $3d_{x^2-y^2}$, $3d_{xy}$, $3d_{xz}$, $3d_{yz}$.
- Be able to reproduce Figures 6.19, 6.22 and 6.23 your text. Note that the letter subscripts do not necessarily equal a given m_l value. They are only an aid in helping you know an orbital's orientation in space.
- Graphical representations of $1s$, $2s$, $2p$, and other orbitals provide you with an understanding of the properties of orbitals. In graphing wave functions, particularly ψ^2, it is important to remember that you cannot accurately and simultaneously measure or describe the position of an electron and its direction of movement (a component of velocity) as stated by Heisenberg's **uncertainty principle**. Since one cannot measure simultaneously all properties of an electron, we can only talk about its probability of being at a particular point in space. A common graphical representation is a 90-percent boundary plot, such as the one shown in Figure 6.2.

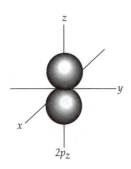

$2p_z$

▲ **FIGURE 6.2** A 90-percent boundary plot for a $2p_z$ orbital. The spherical surfaces enclose 90 percent of the total ψ^2 for the orbital. Note that there is no electron density in the xy plane. Such a surface is called a *node*.

EXERCISE 9   Identifying the significance of n, l, and m_l

Each wave function (ψ) describing an orbital has three quantum numbers, n, l, and m_l, associated with it. What do these quantum numbers determine in terms of orbital properties?

SOLUTION: The principal quantum number, n, defines the size of an orbital and its energy. As the value of n increases, so does the energy of the orbital with respect to the ground-state energy. That is, the energy of an orbital becomes a smaller negative quantity as the value of n increases. The shape of an orbital is related to the angular momentum quantum number, l. For a given l value for an orbital, the magnetic quantum number, m_l, determines the orientation of the orbital in space with respect to an axis in the presence of a magnetic field.

EXERCISE 10 Determining the number of hydrogen orbitals with same *n* value

How many hydrogen orbitals exist with the same $n = 3$ value but with differing l and m_l values?

SOLUTION: *Analyze*: We are asked to determine the total number of hydrogen orbitals that can have the same *n* value (3) but with differing l and m_l values. Thus, we need to know the relationships among n, l, and m_l.

Plan: Hydrogen orbitals are described by the quantum numbers n, l, and m_l: All three are needed to describe a single hydrogen orbital. Start with $n = 3$ and determine how many l states are possible. Then, for each l state, determine how many m_l states are possible. Each combination of the three quantum numbers describes a hydrogen orbital. Finally, sum the number of orbitals that are described to obtain the total.

Solve: For $n = 3$ there exists hydrogen orbitals with angular momentum quantum numbers $l = 0, 1, \ldots, n - 1$. In this case the values are 0, 1, and 2. For each l value, there exist hydrogen orbitals with the following range of m_l values: $l, l - 1, \ldots 0, \ldots, -(l - 1), -l$. These hydrogen orbitals are tabulated as follows:

l	m_l	Number of orbitals
0	0	1
1	1, 0, −1	3
2	2, 1, 0, −1, −2	5

The total number of hydrogen orbitals that can have an $n = 3$ value is therefore 9.

Check: The total number of orbitals for a given *n* level equals n^2. 3^2 orbitals = 9 orbitals which confirms the analysis in the solution.

EXERCISE 11 Writing shorthand notations for orbitals

Write the shorthand notation (for example, 2s) for the orbitals described by the following quantum numbers: (**a**) $n = 2, l = 1, m_l = 0$; (**b**) $n = 3, l = 2, m_l = 1$; (**c**) $n = 4, l = 3, m_l = 2$.

SOLUTION: *Analyze*: We are given n, l, and m_l values for three orbitals and asked to write the shorthand notation for each.

Plan: A shorthand notation has the format nl^x where x is the number of electrons in the subshell designated by its l and n values. A letter is used in place of the value of l (for example, s for $l = 1$ and p for $l = 1$). The value of m_l is not indicated.

Solve: For each case, determine the letter designation for the given l value. Remember that in the shorthand notation system the value of m_l is not indicated. (**a**) $l = 1$ corresponds to a p orbital; thus the orbital designation is 2p. (**b**) $l = 2$ corresponds to a d orbital; thus the orbital designation is 3d. (**c**) $l = 3$ corresponds to a f orbital; thus the orbital designation is 4f.

Check: In each case the number in front of the shorthand notation equals the value of *n* given in the question. A recheck of the relationship between l and its letter designation is confirmed by the information in the box on the previous page.

EXERCISE 12   Identifying a correct set of quantum numbers for an orbital

Why is it not possible for a hydrogen orbital to have the quantum numbers $n = 3, l = 2, m_l = 3$ associated with it?

SOLUTION: *Analyze:* We are asked to explain why a particular set of quantum numbers for a hydrogen orbital is not possible.

Plan: This type of question requires us to know the relationships among n, l, and m_l. Start with n and determine if the given value of l is possible; repeat this for l and m_l.

Solve: The possible l values for $n = 3$ are $l = 0, 1, 2$; thus the values of $n = 3$ and $l = 2$ are permitted. For a given l value, only certain m_l values are allowed: $m_l = \{-l \ldots 0 \ldots +l\}$. The maximum m_l value for $l = 2$ is 2. Thus, $m_l = 3$ is not possible for a hydrogen orbital with a $l = 2$ value.

Check: The range of l is $0, 1, 2, 3 \ldots n - 1$; thus the value of $l = 2$ is allowed because it is the maximum value permitted. The value of m_l can never be larger than the associated value of l; this is not the case in the question.

EXERCISE 13   Determining the number of orbitals from a shorthand notation

How many hydrogen orbitals are possible for the quantum state designated by the following shorthand notations: (**a**) 2s; (**b**) 3d?

SOLUTION: *Analyze:* We are given shorthand for two orbitals and asked to determine how many hydrogen orbitals are possible for each shorthand notation.

Plan: A shorthand notation describes a set of hydrogen orbitals with the same n and l values. To determine the maximum number of orbitals within that set (called a subshell) we must determine the possible m_l values. The number of m_l values for a subshell is $2l + 1$.

Solve: (**a**) The letter *s* corresponds to an orbital with $l = 0$. For $l = 0$, there is only *one* possible orbital with $m_l = 0$. (**b**) The letter *d* corresponds to an orbital with $l = 2$. For $l = 2$, there are *five* orbitals with five different orientations in space determined by the m_l values, which equal 2, 1, 0, −1, and −2.

Check: The total number of hydrogen orbitals associated with a given nl^x shorthand notation is only determined by the value of l: $2l + 1$. Use of this relationship confirms the number of hydrogen orbitals determined for each shorthand description of hydrogen orbitals.

EXERCISE 14   Characterizing a nodal surface

(**a**) What is a nodal surface? (**b**) What quantum number determines the number of nodal surfaces?

SOLUTION: (**a**) A nodal surface is a region in space where ψ^2 goes to zero. (**b**) The number of nodal surfaces, or nodes, depends on the principal quantum number, n. For an orbital with a given n value, there are $n - 1$ nodal surfaces. Thus a 2p orbital has one node; a 3d orbital has two nodes; and a 1s orbital has no nodes. For example, look at Figure 6.22 in your text and Figure 6.2 in the *Student's Guide*. The *xy* plane is a nodal surface for a $2p_z$ orbital.

EXERCISE 15   Identifying orbitals from their shapes

Label the following orbitals as to type:

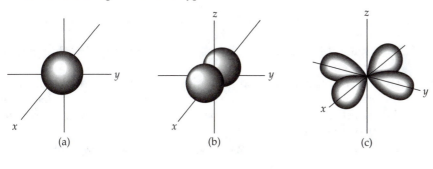

(a)   (b)   (c)

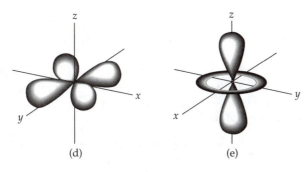

(d)   (e)

SOLUTION: *Analyze*: We are given pictures of different orbitals and asked to label them. By label we mean s or p_x, for example.

Plan: We either have to find similar orbital pictures in the text or we have memorized the various types and orientations with their labels.

Solve: (a) s; (b) p_x; (c) $d_{x^2-y^2}$; (d) d_{xy}; (e) d_{z^2}.

Check: A review of Figures 6.20, 6.23, and 6.24 in the text confirms the labels.

The electronic structure of the hydrogen atom forms the basis of electronic structure for atoms with two or more electrons.

- Orbitals are described like those for hydrogen, and have similar shapes.
- We can continue to use the orbital designations such as $1s$, $2s$, $2p_x$, $2p_y$, and $3d_{xy}$.

In the 1920s, S. A. Goudsmit and B. E. Uhlenbeck proposed that electrons possess intrinsic spin as well as orbital motion.

- Electrons in an atom show two intrinsic spin orientations in the presence of a magnetic field. The two spin orientations are in opposite directions and are quantized.
- Electron spin is defined by the **electron-spin quantum number**, m_s, which has two possible values, $m_s = +\frac{1}{2}$ and $m_s = -\frac{1}{2}$.

There is a further limitation placed on quantum numbers for electrons in an atom. In 1924, Wolfgang Pauli proposed what has become known as the **Pauli exclusion principle**.

- According to this principle, *no two electrons in an atom can have the same set of four quantum numbers, n, l, m_l, and m_s.*
- The effect of the Pauli exclusion principle is that an orbital can hold a maximum of two electrons and these electrons must have opposite spins.

ENERGIES OF ORBITALS IN MANY-ELECTRON ATOMS

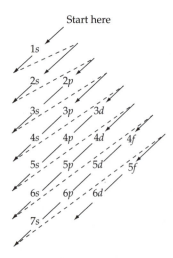

Start here

▲ FIGURE 6.3 A technique to aid memorization of the order of orbital energy levels in many-electron atoms. Each horizontal row in the diagram contains all the orbitals for a given n quantum number. The rows follow a sequential listing of n quantum numbers starting from $n = 1$. Parallel arrows are then drawn as shown.

The order of energies of orbitals in nonhydrogen atoms is shown in Figure 6.25 in the text.

- Note that subshells with the same n value (for example, 2s and 2p) do *not* have the same energies as they do in a hydrogen atom.
- Note that orbitals *within* a subshell (for example, $2p_x, 2p_y, 2p_z$) possess the same energy.
- Use the mnemonic method illustrated in Figure 6.3 of the *Student's Guide* to help you remember the relative order of energies. To determine the correct order of energies, simply follow each arrow from tail to head, and then go to the next arrow.
- Remember that this ordering of energies is only used to "construct" the electronic structure of atoms.

EXERCISE 16   Relating quantum numbers of orbitals to energy

In a many-electron atom, which of the following orbitals, described by their three quantum numbers, have the same energy in the absence of magnetic and electric fields: **(a)** $n = 1, l = 0, m_l = 0$; **(b)** $n = 2, l = 1, m_l = 1$; **(c)** $n = 2, l = 0, m_l = 0$; **(d)** $n = 3, l = 2, m_l = 1$; **(e)** $n = 3, l = 2, m_l = 0$?

SOLUTION: *Analyze:* We are given a set of quantum numbers for five different orbitals and asked to determine which of the five have the same energy when electric or magnetic fields are not present.

Plan: Orbitals with the same n and l values have the same energy in many-electron atoms when electric or magnetic fields are not present.

Solve: Orbitals **(d)** and **(e)**, which are 3d orbitals, are the only orbitals with the same n and l values and thus the same energy.

Comment: Note that the question is restrictive in that magnetic and electric fields are not present. If they are present, then the rule that orbitals with the same n and l values have the same energy does not necessarily apply.

EXERCISE 17   Determining the number of electrons occupying a subshell

What is the maximum number of electrons that can occupy each of the following subshells: 1s, 2p, 3d, and 4f?

SOLUTION: *Analyze:* We are given four different subshells indicated by their shorthand notations and asked to determine what is the maximum number of electrons that can occupy each.

Plan: A subshell consists of one or more orbitals; thus we first must determine the number of orbitals within each subshell. The number of orbitals within a subshell equals the total number of m_l values for the l value of the subshell, $2l + 1$. The second step is to recognize that a single orbital can hold a maximum of two electrons of opposite spin. Thus, the total number of electrons possible within a subshell is $2(2l + 1)$.

Solve: To determine the maximum electron occupancy for each subshell, you need to know how many degenerate (equal energy) orbitals exist for each subshell. The number of degenerate orbitals equals the total number of m_l values for the l value of the subshell. Each orbital can hold a maximum of two electrons. The maximum occupancy for each subshell is shown below:

Subshell	*l* value	m_l values	Number of m_l values	Maximum number of electrons in subshell
1s	0	0	1	$2 \times 1 = 2$
2p	1	1, 0, −1	3	$2 \times 3 = 6$
3d	2	2, 1, 0, −1, −2	5	$2 \times 5 = 10$
4f	3	3, 2, 1, 0, −1, −2, −3	7	$2 \times 7 = 14$

Check: A review of the maximum number of electrons for each subshell in the problem shows that the answer fits the general formula: maximum number of electrons in a subshell = $2(2l + 1)$. We can confirm the number of electrons in each subshell by calculating for each *l* this value. For example, if $l = 3$, then there are $2[2(3) + 1]$ electrons possible (14 electrons), in agreement with the solution. The other subshells are confirmed in a similar manner.

At this point, you should review the rules governing the possible values of the four quantum numbers for electrons:

1. The principal quantum number, *n*, has integral values of 1, 2, 3, . . .
2. The integral values of the *l* quantum number for a given *n* are $l = 0, 1, 2, 3, \ldots, n - 1$.
3. The integral values of the m_l quantum number for a given *l* are $m_l = l, l - 1, l - 2, \ldots, 0, \ldots, -1 + l, -l$.
4. The values of the m_s quantum number for a given m_l are $m_s = +\frac{1}{2}, -\frac{1}{2}$.

With this information you can now specify the four quantum numbers for all electrons in an atom. This description of how electrons are arranged in orbitals is called an **electron configuration**. The procedure for writing the electron configuration of an element is summarized as follows:

1. Electrons occupy orbitals in order of increasing energy. All orbitals that are equal in energy (degenerate) are filled with electrons first, before the next level of energy begins to fill up. (A few exceptions exist, for example, Cr and Cu.)
2. An orbital can have a maximum of two electrons with *opposite* spins. This is an application of the Pauli exclusion principle.
3. When several orbitals of equal energy exist (for example, $2p_x$, $2p_y$, $2p_z$), the filling of orbitals with electrons follows **Hund's rule**: *Electrons enter orbitals of equal energy singly and with the same spins, until all orbitals have one electron each.* Then electrons with opposite spins pair with the electrons in the half-filled set of equal-energy orbitals.

Three methods are used to describe the electron configuration for an element:

1. Assignment of the four quantum numbers (n, l, m_l, m_s) for each electron;
2. A shorthand using the orbital system notation described in Section 6.5 of the text with the number of electrons indicated (as a superscript) for each set of degenerate orbitals;
3. An orbital diagram (see Exercise 19).

These methods are described in Exercises 18 through 20.

ELECTRONIC STRUCTURE OF MANY-ELECTRON ATOMS

EXERCISE 18   Describing electron configurations for atoms I

Using the four quantum numbers, describe the electron configuration for each of the following atoms: (**a**) B; (**b**) N. Assume that an electron first enters an orbital in the spin state described by the spin quantum number $m_s = +\frac{1}{2}$

SOLUTION: *Analyze*: We are asked to write the complete electron configurations for B and N using the four quantum numbers.

Plan: First determine the number of electrons possessed by each atom. Then using the *Aufbau Principle, Hund's rule*, and *Pauli's Principle* construct a table that shows the four quantum numbers for each electron.

Solve: (**a**) Boron has five electrons. The first two electrons enter the lowest energy orbital, 1s, with opposite spins as per the Pauli exclusion principle. The next energy level, 2s, also has two electrons with opposite spins. The last electron enters the 2p energy level. The electron configuration for boron is as follows:

n	l	m_l	m_s	Orbital notation
1	0	0	$+\frac{1}{2}$	1s
1	0	0	$-\frac{1}{2}$	1s
2	0	0	$+\frac{1}{2}$	2s
2	0	0	$-\frac{1}{2}$	2s
2	1	1	$+\frac{1}{2}$	2p
2	1	0	$+\frac{1}{2}$]ᵃ	
2	1	−1	$+\frac{1}{2}$]	

ᵃ These two orbitals are equivalent to the one immediately above, and the fifth electron can be placed in any of the three 2p orbitals. The fifth electron occupies only one of them.

(**b**) Nitrogen has seven electrons. The first five electrons enter orbitals as for boron. The last two electrons enter 2p orbitals because they have a maximum occupancy of six electrons. According to Hund's rule, they enter singly until all degenerate orbitals are half filled:

n	l	m_l	m_s	Orbital notation
1	0	0	$+\frac{1}{2}$	1s
1	0	0	$-\frac{1}{2}$	1s
2	0	0	$+\frac{1}{2}$	2s
2	0	0	$-\frac{1}{2}$	2s
2ᵃ	1	1	$+\frac{1}{2}$	2p
2ᵃ	1	0	$+\frac{1}{2}$	2p
2ᵃ	1	−1	$+\frac{1}{2}$	2p

ᵃ These three 2p orbitals are referred to as a half-filled set of orbitals.

Check: A check of the total number of electrons in the summary of electron configurations for B and N confirms that five and seven electrons are identified. No two electrons have the same set of quantum numbers. If two electrons had the same set of quantum numbers this would tell us that we made an incorrect assignment. Electrons in the same orbital have opposite spins. If two electrons in the same orbital had the same spins this would tell us that we made an incorrect assignment.

EXERCISE 19   Describing electron configurations for atoms II

Another approach to describing the electron configuration of an atom is an orbital diagram, in which a box represents an orbital. Electrons are represented by arrows, with an arrow that points up ($\uparrow$) corresponding to the spin state $m_s = +\frac{1}{2}$ and an arrow that points down ($\downarrow$) corresponding to the spin state $m_s = -\frac{1}{2}$. For example, the electron configuration for boron can be represented as

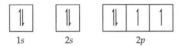

What are the orbital diagrams for: (**a**) O; (**b**) Sr; (**c**) V?

SOLUTION: *Analyze*: We are given information about how to construct electron box diagrams and are asked to construct electron box diagrams for O, Sr, and V.

Plan: First determine the number of electrons possessed by each atom. Then, using a box for each orbital within a subshell, construct the appropriate number of boxes, grouped as electron subshells, that permits a complete orbital description for the electrons in the atom.

Solve: (**a**) Oxygen has eight electrons:

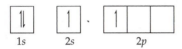

The electron configuration shown for the 2p orbitals conforms to Hund's rule, which requires that electrons remain unpaired until all degenerate (equal energy) orbitals are filled with one electron. The 2p orbitals are degenerate. Thus, the first three 2p electrons are placed singly in orbitals, and the fourth one is paired. (**b**) Strontium has 38 electrons:

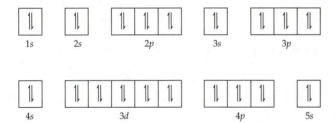

When *forming* an electron configuration, fill the 4s orbitals before the 3d orbitals. As a check, you should count the number of electrons. (**c**) Vanadium has 23 electrons:

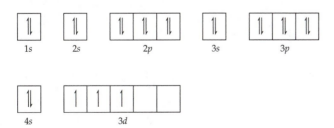

Check: Each box has no more than two electrons and if two electrons are in the same box they have opposite spins. The order of orbitals is reconfirmed by use of Figure 6.24 in the text.

EXERCISE 20   Describing electron configurations for atoms III

An orbital diagram is useful and informative, but cumbersome. More commonly, a shorthand notation system is also used to describe electron configurations. This notation uses the nl symbol for each subshell (such as $1s$, $2s$, $2p$, $3s$, $3d$), with the number of electrons occupying an orbital set indicated with a superscript. For example, boron has five electrons; its electron configuration is $1s^2 2s^2 2p^1$. Write the shorthand configuration for the atoms in Exercise 19.

SOLUTION: *Analyze:* We are given information on how to write the complete electron configuration for elements and asked to write the complete shorthand notation for all electrons possessed by O, Sr, and V.

Plan: First determine the number of electrons possessed by each atom. Then, use the principles of building electron configurations of atoms to place electrons in appropriate subshells. You have already done this in the previous exercise.

Solve: (**a**) O: $1s^2 2s^2 2p^4$; or $[\text{He}]2s^2 2p^4$, where [He] is used to describe the He core electron configuration $1s^2$. (**b**) Sr: $1s^2 2s^2 2p^6 3s^2 3p^6 4s^2 3d^{10} 4p^6 5s^2$; or $[\text{Kr}]5s^2$, where [Kr] is used to describe the Kr core electron configuration $1s^2 2s^2 2p^6 3s^2 3p^6 4s^2 3d^{10} 4p^6$. (**c**) V: $1s^2 2s^2 2p^6 3s^2 3p^6 4s^2 3d^3$ or $[\text{Ar}]4s^2 3d^3$, where [Ar] is used to describe the Ar core electron configuration $1s^2 2s^2 2p^6 3s^2 3p^6$.

Check: The number and types of electrons agree with the assignments in Exercise 19. Always count the total number of electrons in the shorthand notation to confirm that it agrees with the atomic number of the element.

THE PERIODIC TABLE: ELECTRON CONFIGURATIONS AND VALENCE ELECTRONS

The electronic configuration of an element can be constructed from its location in the periodic table.

*   Elements in **families** (*vertical columns*) of representative elements possess the same number and type of outer electrons beyond the core electron configuration. These outer-shell electrons are referred to as **valence electrons**. The distribution of valence electrons in transition elements can vary slightly within a family.
*   Valence electrons are involved in ion formation and chemical reactions.
*   If you know the valence electron configuration for the first member of a representative family of elements you then know the valence electron configuration for the other members. For example, the electron configuration for oxygen is

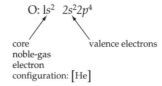

O: $1s^2$ $2s^2 2p^4$

core
noble-gas
electron
configuration: [He]   valence electrons

Any other member of the oxygen family will have a similar valence electron configuration:

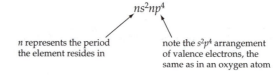

$ns^2 np^4$

n represents the period   note the $s^2 p^4$ arrangement
the element resides in   of valence electrons, the
                                   same as in an oxygen atom

A knowledge of the period in which an element resides will help you write complete electron configurations. Note the following key points shown in Figure 6.30 in the text:

- The first three periods consist only of **representative elements**: Only s and p orbitals are involved their valence electron configurations. The period number corresponds to the principal quantum number of the valence electrons.
- Periods 4–7 contain both representative elements and **transition elements**. The $(n - 1)d$ outer-orbitals are occupied in transition elements, where n is the period number. Thus, the valence orbitals of transition elements are the ns $(n - 1)d$ orbitals. Note that beyond the transition elements in a period, the d electrons form part of the inner-core of electrons and are no longer valence electrons.
- The **inner-transition elements** have f orbitals occupied.

EXERCISE 21 Writing valence-electron configurations for elements

(a) Write the valence electron configuration for the elements of the second period. (b) How, in general, do the valence-electron configurations of the other members of each family relate to the first member's valence-electron configuration? (c) How does the group number of a representative family relate to its valence-electron configuration?

SOLUTION: *Analyze*: We are asked to write the valence-electron configurations for the elements of the second row: Li, Be, B, C, N, O, F, and Ne. We are also asked to describe in what way these elements share similarities to the other elements within each family and how does the group number of the family help you construct valence-electron configurations?

Plan: The period number identifies the principal quantum number for the valence electrons in that period. (Note: This is only for the s and p electrons; if an atom possesses d valence electrons then for the d valence electrons the principal quantum number is $n - 1$). Starting from the left of the period, and as we move from element to element, we fill the s orbital first with electrons and then the p orbitals. This period does not contain transition metals and we do not have to use d orbitals. For the other two questions we need to look at the valence-electron configuration of each element and compare it to the other elements in the same family.

Solve: (a) The second period of elements begins after $H(1s^1)$ and $He(1s^2)$. The next quantum level to be filled is $n = 2$. Thus the valence shell of each element contains electrons in the energy level $n = 2$. Using Hund's rule and the Pauli exclusion principle, we obtain the following valence-electron configurations:

Family:	1A	2A	3A	4A
Element:	Li	Be	B	C
Electron configuration	$2s^1$	$2s^2$	$2s^2 2p^1$	$2s^2 2p^2$
Family:	5A	6A	7A	8A
Element:	N	O	F	Ne
Electron configuration:	$2s^2 2p^3$	$2s^2 2p^4$	$2s^2 2p^5$	$2s^2 2p^6$

(b) Any other element within a given family of representative elements will have the same number and type of valence electrons, except for the n quantum number, which will reflect the period the element is in. For example, in the halogen family, the next member after fluorine is chlorine, and its valence-electron configuration is $3s^23p^5$. Chlorine is in the third period, so $n = 3$. (c) The group number of a representative family equals the number of valence electrons for the family.

Check: The number of electrons in each valence-electron configuration agrees with the group number of the family in which the element resides. The valence electrons are assigned to the highest energy subshell. The period number corresponds to the n value in the shorthand notation, two.

EXERCISE 22   Writing complete electron configurations of elements

Use the periodic table to write the complete electron configurations of S and Ca.

SOLUTION: *Analyze*: We are to use the structure of the periodic table to help us write the complete electron configurations of S, a member of the oxygen family, and Ca, a member of the alkaline-earth family.

Plan: First determine the number of electrons possessed by each element from its atomic number. We can write the complete electron configurations by recognizing that each period has associated with it a set of valence-shell orbitals used by the atoms within the period. We saw this for the second period in the previous exercise. Fill the orbitals with electrons, period-by-period in a sequential manner, until the number of electrons possessed by the element is reached.

Solve: Sulfur is in the 6A [16] family and in the third period; it has 16 electrons. We can complete the electron configuration by writing the valence orbitals used by atoms within each period and adding the electrons, period by period, to fill the orbitals. Adding 16 electrons to valence orbitals requires using the first three periods.

Period 1: $1s^2$

Period 2: $2s^22p^6$

Period 3: $3s^23p^4$ (the $3p$ orbital has only four electrons because the total of 16 electrons has been reached)

Complete electron configuration: $1s^22s^22p^63s^23p^4$

Calcium is in the 2A [2] family and is in the fourth period; it has 20 electrons.

Period 1: $1s^2$

Period 2: $2s^22p^6$

Period 3: $3s^23p^6$

Period 4: $4s^2$ (the total number of electrons is now 20)

Complete electron configuration: $1s^22s^22p^63s^23p^64s^2$.

Check: We can confirm the answers by using procedures in Exercise 20.

EXERCISE 23   Writing valence-electron configurations for transition elements

Write the valence-electron configurations for the following transition elements: Ti, Cr, Cd, and Cu.

SOLUTION: *Analyze*: We are asked to write the valence-electron configurations of four transition elements. The periodic table can guide us in doing the problem.

Plan: We can answer this question by determining the electron configurations of the elements and then identify the valence-electrons for transition elements by recognizing that they are in the outer $(n − 1)$ d and ns orbitals. Alternatively we can use the periodic table directly to help us answer the question of the electron occupancy of the $(n - 1) d$ and ns valence orbitals. For most transition elements there are two ns valence electrons. Thus, the number of $(n - 1) d$ electrons is typically the group number of the transition metal family minus two. We also need to remember that there are some exceptions discussed in the text. We will use this approach.

Solve: Ti: $3d^2 4s^2$. Ti is in the fourth period and therefore n is four. Ti is in group 4; thus the number of $3d$ electrons is $4 - 2 = 2$.
Cr: $3d^5 4s^1$. Based on the previous example you might expect the following: $3d^2 4s^4$. However, Cr is an exception discussed in the text and the resulting valence-electron configuration has two half-filled subshells. We observe that this configuration of electrons is more stable.
Cd: $4d^{10} 5s^2$. Cd is the last transition element in the fifth period; thus its $4d$ orbitals are completely filled.
Cu: $3d^{10} 4s^1$. This situation is similar to that discussed in the answer for Cr. A filled d orbital and a half-filled s orbital are apparently more stable than $3d^9 4s^2$.

SELF-TEST QUESTIONS

Key Terms

Having reviewed key terms in Chapter 6, match key terms with phrases and identify statements as true or false, If a statement is, false indicate why it is incorrect.

Match each phrase with the best term:

6.1 In the expression $\Delta E = h\nu$ the name given to the term $h\nu$.

6.2 The name given to the pattern of energy emitted when a gas in placed under reduced pressure and high voltage is passed through it.

6.3 Radiation which is characterized by a wave with amplitude and frequency.

6.4 A spectrum characterized by a non-continuous pattern of energies.

6.5 The statement that tells us that if we know accurately the speed of an electron we can not measure accurately its position.

6.6 The term given to the collection of $2s$ and $2p$ set of orbitals.

6.7 A terms used to describe orbitals that have the same energy.

6.8 Mo is in this cluster of elements.

6.9 Eu is in this cluster of elements.

6.10 Mg and P belong to elements which are given this name.

6.11 It has two possible values, $+1/2$ or $-1/2$.

6.12 Particles with energy but no mass.

6.13 A mathematical equation that describes the periodic properties of an electron moving about a hydrogen nucleus.

6.14 A term used to describe the probability of finding an electron in space.

6.15 The allowed wave function for an electron.

6.16 The name given to the condition that a $2p$ hydrogen orbital possesses no electron density at the coordinates $x = y = z = 0$.

Key Terms:

- **(a)** degenerate
- **(b)** electromagnetic
- **(c)** electron density
- **(d)** electron-spin quantum number
- **(e)** Heisenberg's Uncertainty Principle
- **(f)** lanthanide
- **(g)** line spectrum
- **(h)** node
- **(i)** orbital
- **(j)** photons
- **(k)** representative
- **(l)** spectrum
- **(m)** subshells
- **(n)** transition
- **(o)** quantum of energy
- **(p)** wave function

True-False Statements:

6.17 According to Louis de Broglie's postulate, the wavelength of a *matter wave* associated with an electron increases with the velocity of an electron as it moves about a nucleus.

6.18 The $3s$ and $3p$ orbitals comprise the third *electron shell*.

6.19 Uranium is a *f-block element*.

6.20 The *electron configuration* for carbon is $1s^2 2s^2 2p^3$.

6.21 The *valence* electrons for fluorine are $2p^3$.

6.22 According to the *Pauli exclusion principle*, two electrons in an atom can have the following set of quantum numbers: $n = 2, l = 1, m_l = 0, m_s = +\frac{1}{2}$ and $n = 2, l = 1, m_l = 0, m_s = +\frac{1}{2}$.

6.23 In titanium, the two $3d$ electrons are predicted to be unpaired by application of *Hund's rule*.

6.24 The *electronic structure* of an atom refers to the number of electrons possessed by it.

6.25 Electromagnetic radiation moves at the speed of light in a vacuum.

6.26 The longer the *wavelength* of radiant energy, the higher the energy.

6.27 Frequency of a wave is directly proportional to wavelength.

6.28 When sunlight is passed through a prism, a *continuous spectrum* of all wavelengths is produced.

6.29 The *ground state* for a helium atom is $1s^1 2s^1$.

6.30 When a $2s$ electron of a carbon atom gains energy and occupies a $3s$ orbital, an *excited state* of carbon is produced.

6.31 According to de Broglie, the greater the *momentum* of a particle, the longer its wavelength.

6.32 All materials possess a *matter wave*.

6.33 The *probability density* for an electron in an atom is related directly to its wave function.

6.34 There are four orientations for *electron spin*.

6.35 *Core electrons* are those electrons in the shell closest to the nucleus.

6.36 The fourth row of the periodic table contains *transition metals*.

6.37 Calcium is an *active metal*.

Problems and Short-Answer Questions

6.38 Use the following figure to answer questions 6.38–6.39:

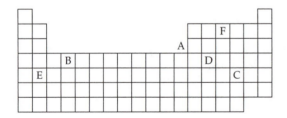

What is the ground-state valence-electron configuration of the elements in the family labeled by A?

6.39 Which of the elements identified by the letters B, C, D, E, and F have only four valence electrons? Explain your answer.

6.40 Calculate the energy of a photon with a wavelength of 420 nm.

6.41 Using the Bohr model, calculate the wavelength of radiant energy associated with the transition of an electron in the hydrogen atom from the $n = 4$ orbit to the $n = 2$ orbit. Is energy emitted or absorbed? Indicate the type of radiation emitted.

6.42 One of Bohr's postulates is that an electron moves in a well-defined path about the nucleus. What principle later negated this postulate? Why?

6.43 Differentiate between the functions ψ and ψ^2.

6.44 For each of the following pairs of hydrogen orbitals, indicate which orbital has the higher energy:

(a) $2s$ or $3s$ (b) $3p$ or $4d$ (c) $2s$ or $2p$

6.45 A hydrogen orbital is described by the quantum numbers $n = 4$, $l = 2$ and $m_l = 0$

(a) Write the shorthand notation for this orbital.

(b) Are there any other hydrogen orbitals that have the same shorthand notation and if so, which ones?

6.46 Write the shorthand notation for all hydrogen orbitals that belong to the $n = 4$ principal quantum number energy level. What property do they all have in common?

6.47 Give the values of n, l, and m_l for each hydrogen orbital in the $n = 2$ principal quantum number energy level.

6.48 An instructor shows you a three-dimensional picture of the surface of an orbital with a spherical shape. Is there sufficient information for you to conclude that the orbital is a $1s$ orbital? If there is insufficient information, what additional information do you need to determine the type of orbital?

6.49 An atom of an element has a valence-electron configuration that includes three $4p$ electrons. Is there sufficient information for you to determine which element it is? If there is insufficient information, what additional information do you need to determine the type of orbital?

6.50 Which orbital notation violates the Pauli Principle and why? (a) $2s^2$; (b) $3p_x^0$; (c) $4d_{xy}^2$; (d) $5d_{xz}^3$

6.51 What is a node? Is there a general relationship between the number of nodes possessed by a wave function and its energy? What differences in orbital size and number of nodes do you expect between a $3s$ and a $4s$ hydrogen orbital?

6.52 Draw the contour representations for the $3d_{z^2}$, $3d_{x^2-y^2}$, and $3d_{xy}$ hydrogen orbitals. In your sketches, not all of the orbitals will look identical. Does this indicate that these $3d$ orbitals are all different in energy?

6.53 Louis de Broglie showed that an electron's wavelength and momentum are related by the equation $\lambda = h/mv$. Why is the relationship $E = hc/\lambda$ not valid for matter typically used by humans.

6.54 Microwave ovens are now commonly used to heat foods. Most microwave ovens have warnings posted to the effect that microwave radiation is harmful to humans. A typical microwave frequency is 20,000 megacycles per second. Calculate the wavelength and energy of this radiation. Show that this energy is significantly less in magnitude than that of visible radiation with a frequency of 100,000,000 megacycles per second and, thus, the danger of microwave radiation cannot be attributed to its energy content.

6.55 (a) In what orbitals and in what order are electrons entering the atoms of the second and fifth periods?

(b) The valence orbitals of the elements of the fourth period are the $4s$, $3d$, and $4p$ orbitals. Why isn't the fourth period composed of elements with only outer $4s$ and $4p$ orbitals?

6.56 Write the shorthand notation and the box-diagram electron-configuration representation for each of the following elements and indicate the outer orbital(s):

(a) N   (c) Cu
(b) K   (d) Al

6.57 Each of the following electron configurations represents the valence-electron configuration for an atom of an element in the periodic table. For each, what is the element, and in what period does it belong?

(a) $4s^2 4p^2$ (d) $2s^2 2p^5$
(b) $3d^6 4s^2$ (e) $3d^{10} 4s^2$
(c) $6s^1$

6.58 An atom contains layers or shells of electrons. A completely filled layer or shell of electrons in an atom is a very stable electron configuration.

(a) Both Ne and Xe contain completely filled shells of electrons. What electrons comprise the shell that gets filled between Ne and Ar?
(b) When a representative element forms an ion, it usually forms one whose electron configuration corresponds to the electron configuration for the nearest inert gas. What electrons must be added to or removed from each of the following atoms to form an ion with the nearest inert gas electron configuration: Li; Sr; O?

6.59 Which of the following electron configurations for neutral atoms correspond to ground states, and which correspond to excited states?

(a) $1s^1 2s^1$ (c) $[Ar]3d^6$
(b) $[Ar]4s^1$ (d) $[Xe]6s^2 5d^{10} 6p^3$

6.60 Identify the groups of elements having outer electron configurations of (a) $ns^2 np^4$; (b) $ns^2 (n-1)d^1$; (c) ns^2; (d) $ns^2 np^1$.

Integrative Questions

6.61 Student (A) is asked to assign a set of n and l values to one of the valence electrons in a certain atom. The student reports that one of the valence electrons has the values of $n = 4$ and $l = 1$. Another student (B) looks at this information and concludes that the element must be arsenic.

(a) Is student (B)'s conclusion reasonable? Explain.
(b) If student (B)'s conclusion is not reasonable what additional information in terms of the electrons would the student need to correctly identify the element?
(c) Student (A) reports that the element is very stable and very unreactive. Does this information help student (B) identify the element? Explain.
(d) Student (A) reports the element is metallic. Does this help student (B) identify the element? Explain.

6.62 Sodium arc lamps emit an intense yellow color resulting from the emission spectrum of sodium atoms.

There are two closely spaced emission lines with a center at 589.3 *nm* that are primarily responsible for the yellow color.

(a) What is the energy of a photon with that wavelength?
(b) What is the energy emitted by a mole of sodium atoms for that wavelength?
(c) What is the electron configuration of sodium? What are its valence electrons?
(d) What is the electron configuration of Na^+?

6.63 Richard Feynman, a notable physicist, wrote in 1964 in *The Character of Physical Law*, The Messenger Lectures, 1964, MIT Press, pp. 127–128, that "Electrons behave ... in exactly the same ways as photons; they are both screwy, in exactly the same way."

(a) Based on your reading of chapter 6 why do you think Feynman characterized both electrons and photons as "screwy"?
(b) Feynman also said that our desire for familiarity will lead to perpetual torment as we learn about quantum mechanics. Does this make psychological sense? Explain.
(c) Thus, what must you do as you study the principles of quantum mechanics to avoid this torment?

Multiple-Choice Questions

6.64. A quantum of electromagnetic radiation has a wavelength equal to 7.52×10^6 Å. What is the frequency of this radiation in cycles/sec?

(a) 1.13×10^{-12} (d) 9.45×10^{13}
(b) 8.80×10^{-26} (e) 2.69×10^{-15}
(c) 3.99×10^{11}

6.65. What is the energy of radiation that has a frequency of 9.00×10^{11} cycles/sec? (Remember that Planck's constant, h, has a value of 6.63×10^{-34} J-sec.)

(a) 1.66×10^{-45} J   (d) 5.00×10^{-22} J
(b) 5.97×10^{-22} J   (e) 3.32×10^{-45} J
(c) 4.99×10^{-27} J

6.66. Calculate the wavelength of an electron traveling with a velocity of 4.0×10^9 cm/sec in an electron micro scope. The mass of an electron is 9.1×10^{-28} g; $h = 6.63 \times 10^{-34}$ J-sec; and 1 J $= 1$ kg-m^2/sec^2.

(a) 0.18 Å   (d) 1.5×10^8 cm
(b) $0.67^3 \, 10^{-8}$ cm   (e) 1.1×10^{-38} Å
(c) 1.5 Å

6.67. In the photoelectric effect, in order for an electron to be released from the surface of a clean metal, which one of the following conditions must exist?

(a) The metal must have a low temperature;
(b) the metal must have a high temperature;
(c) the kinetic energy of photons striking the metal's surface must equal that of the emitted electron;

(d) the kinetic energy of photons striking the metal's surface must be less than that of the emitted electrons;

(e) the kinetic energy of photons striking the metal's surface must be greater than or equal to that of the emitted electrons plus the binding energy holding the electron in the metal.

6.68. Passing an electrical charge through argon gas contained in a partially evacuated vessel yields which of the following?

(a) a continuous spectrum;

(b) a line spectrum;

(c) white light;

(d) no visible change;

(e) (a) and (c).

6.69 According to Bohr's model of the atom, which of the following characteristics of metallic elements explains the fact that many of these elements can easily form positively charged ions?

(a) few electrons;

(b) many electrons;

(c) electrons in lowest-energy n level having high energies;

(d) electrons having momentum;

(e) relatively small ionization energies in their ground states.

6.70 According to the Bohr model of the atom, emission of electromagnetic radiation by heated atoms in a vacuum is directly due to which of the following?

(a) photons absorbed by atoms;

(b) particle emission from the nucleus;

(c) momentum possessed by electrons;

(d) electrons being excited from an inner to outer orbit;

(e) electrons falling from an outer to inner orbit.

6.71 Which of the following is *not* true about the Bohr model of the hydrogen atom?

(a) Electrons decay into the nucleus;

(b) the model cannot account for the ionization of an electron;

(c) an electron in a stable Bohr orbit does not emit radiation continuously;

(d) an electron may remain in an orbit indefinitely;

(e) the hydrogen atom absorbs radiant energy in multiples of hv.

6.72 Which of the following statements is *not* true about the principal quantum number, n?

(a) It is related to the spin of an electron.

(b) It is related to the energy of an electron.

(c) The larger the n value of an electron, the higher its energy.

(d) The larger the n value of an electron, the larger the value of its Bohr radius.

(e) The lowest energy state for an electron in a Bohr atom corresponds to $n = 1$.

6.73 For an orbital with $l = 4$, what are the possible numerical values of m_l?

(a) 1, 0, −1

(b) 3, 2, 0, −1, −2, −3

(c) 3, 2, 1, 0

(d) 4, 3, 2, 1, 0, −1, −2, −3, −4

(e) 4, 3, 2, 1, 0

6.74 Which of the following combinations of quantum numbers for an electron is *not* permissible?

(a) $n = 5, l = 2, m_l = 0$

(b) $n = 3, l = 2, m_l = 3$

(c) $n = 4, l = 3, m_l = -2$

(d) $n = 1, l = 0, m_l = 0$

(e) $n = 2, l = 1, m_l = -1$

6.75 Which of the following statements is the most correct and complete with reference to an electron transition from a Bohr $n = 2$ to $n = 4$ hydrogen orbit?

(a) Energy is released during the electronic transition.

(b) Energy is absorbed during the electronic transition.

(c) The energy of the transition is proportional to $\frac{3}{16}$.

(d) Both (b) and (c).

6.76 Which set of hydrogen-like orbitals can contain a maximum of ten electrons?

(a) $3s$

(b) $3p$

(c) $4p$

(d) $4d$

(e) $4f$

6.77 For the principal quantum number 3 what are the allowed values of the l quantum number?

(a) 1, 2

(b) 0, 1, 2

(c) 1, 2, 3

(d) 1, 2, 3, 4

(e) −2, −1, 0, 1, 2

6.78 What designation is used to describe the following orbital?

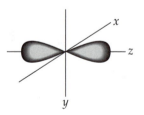

(a) s

(b) p_z

(c) p_z

(d) d_{z^2}

(e) d_{xz}

6.79 In iron, which of the following electrons, characterized by the four quantum numbers, has the lowest energy?

(a) $n = 4, l = 0, m_l = 0, m_s = +\frac{1}{2}$

(b) $n = 3, l = 2, m_l = 1, m_s = -\frac{1}{2}$

(c) $n = 3, l = 2, m_l = 0, m_s = +\frac{1}{2}$

(d) $n = 3, l = 1, m_l = 0, m_s = +\frac{1}{2}$

(e) Electrons (b), (c), and (d) are all degenerate and lowest in energy.

6.80 What is the electron configuration of the element Xe?

(a) $1s^2 2s^2 2p^6 3s^2 3p^6$

(b) $1s^2 2s^2 2p^6 3s^2 3p^6 3d^{10} 4s^1$

(c) $1s^2 2s^2 2p^6 3s^2 3p^6 3d^{10} 4s^2 4p^6$

(d) $1s^2 2s^2 2p^6 3s^2 3p^6 3d^{10} 4s^2 4p^6 5s^2$

(e) none of the above

6.81 How many electrons are there in the outermost shell of arsenic?

(a) 3   (d) 6

(b) 4   (e) 7

(c) 5

6.82 Which element possesses the electron configuration $1s^2 2s^2 2p^6 3s^2 3p^6 3d^6 4s^2$?

(a) V   (d) Ni

(b) Cr   (e) Fe

(c) Mn

6.83 How many electrons in an atom can possess an $n = 3$ principal quantum number?

(a) 3   (d) 10

(b) 6   (e) 18

(c) 8

6.84 Which of the following is the correct order of increasing energy of atomic orbitals when forming a many-electron atom?

(a) 1s2s2p3s3d   (d) 1s2s2p3d4s

(b) 1s2s3s3p4s   (e) none of the above

(c) 1s2s2p3s3p

SELF-TEST SOLUTIONS

6.1 (o). **6.2** (l). **6.3** (b). **6.4** (g). **6.5** (e). **6.6** (m). **6.7** (a). **6.8** (n). **6.9** (f). **6.10** (k). **6.11** (d). **6.12** (j). **6.13** (p). **6.14** (c). **6.15** (i). **6.16** (h). **6.17** False. $\lambda = h/mv$. The value of λ decreases with increasing velocity. **6.18** False. An electron shell consists of a complete collection of orbitals that have the same value of n. The 3s, 3p, and 3d comprise the third electron shell. **6.19** True. **6.20** False. The electron configuration for carbon is $1s^2 2s^2 2p^2$. **6.21** False. The valence-shell electrons for fluorine are $2s^2 2p^5$. **6.22** False. They must have different values for m_s because their values for n, l, and m_l are the same. **6.23** True. **6.24** False. Not only does it include the number of electrons, but also their energies. **6.25** True. **6.26** False. Energy and wavelength are inversely related. The longer the wavelength, the smaller the energy. **6.27** False. Frequency and wavelength are inversely related. **6.28** True. **6.29** False. It is $1s^2$. **6.30** True. **6.31** False. The wavelength associated with matter is inversely proportional to momentum. **6.32** True. **6.33** False. Probability density is proportional to the square of a wave function. **6.34** False. There are two spin orientations. **6.35** False. Core electrons are those that are not valence electrons. **6.36** True. **6.37** True. Group IA and IIA elements are active metals.

6.38 The column of elements belongs to the last group of the transition elements. Thus, the valence-shell orbitals are completely filled: $(n-1)d^{10}ns^2$.

6.39 Elements with the ground-state valence electron configuration ns^2np^2 or $(n-1)d^2ns^2$ have a total of four valence electrons. Only elements B and D meet one of these two requirements.

6.40 The relation between the energy of a photon and its wavelength is $E = hc/\lambda$. Substituting the appropriate values for h, c, and the wavelength (converted into the unit meter) yields

$$E = \frac{(6.63 \times 10^{-34}\,\text{J-sec})(3.00 \times 10^8\,\text{m/sec})}{(4.20 \times 10^2\,\text{nm})(1 \times 10^{-9}\,\text{m/1 nm})}$$

$$= 4.74 \times 10^{-19}\,\text{J}$$

6.41 The wavelength of radiation is related to the energy of the electron transition by the equation $\Delta E = hc/\lambda$, where ΔE is the energy difference between the $n = 4$ and $n = 2$ orbits. This energy difference is calculated as follows:

$$\Delta E = R_H\left(\frac{1}{n_i^2} - \frac{1}{n_f^2}\right) = (2.18 \times 10^{-10}\,\text{J})\left(\frac{1}{4^2} - \frac{1}{2^2}\right)$$

$$= -4.09 \times 10^{-19}\,\text{J}$$

Since the value of ΔE is negative, energy is released. The absolute value of ΔE is used in the calculation of the value of λ:

$$\lambda = \frac{hc}{\Delta E}$$

$$= \frac{(6.63 \times 10^{-34}\,\text{J-sec})(3.00 \times 10^8\,\text{m/sec})}{4.09 \times 10^{-19}\,\text{J}}$$

$$= 4.86 \times 10^{-7}\,\text{m} = 486\,\text{nm}$$

A wavelength of 4.86×10^{-7} m or 4860 Å belongs in the visible range of light.

6.42 Heisenberg's uncertainty principle negated this postulate. It states that it is impossible to measure the position and velocity of a small particle simultaneously and accurately. In order to describe an electron as having motion in a well-defined circular path, we would have to specify its velocity and position accurately at all times, which, according to the uncertainty principle, is impossible.

6.43 ψ is a mathematical expression that describes the amplitude and the motion of the matter wave of an electron as a function of time. ψ^2 is the probability of finding the electron at some point in space.

6.44 The energy of a hydrogen orbital increases as the value of the principal quantum number increases. Thus, among the pairs, the following have the higher energy: (a) 3s; (b) 4d; (c) both orbitals, because both have the same principal quantum number. Note: The answer to (c) is true only for the hydrogen atom.

6.45 (a) $4d$.

(b) Yes. Any orbital possessing the quantum numbers $n = 4$ and $l = 2$ has the same notation. The $l = 2$ quantum number has associated with it five m_1 quantum numbers, 2, 1, 0, -1, -2. Therefore a total of five orbitals have the designation $4d$.

6.46 For the $n = 4$ principal quantum energy level, hydrogen orbitals with $l = 0, 1, 2, 3$ values can exist. Thus, the $n = 4$ principal quantum energy level possesses the following subshells: $4s$, $4p$, $4d$, and $4f$. They all possess the same energy because they have the same principal quantum number, 4.

6.47 For the $n = 2$ principal quantum energy level, hydrogen orbitals with $l = 0, 1$ values can exist. For each l value, there exists a further subset of orbital types:

l	m_l	Shorthand notation
0	0	$2s$
1	+1	$2p$
1	0	$2p$
1	-1	$2p$

6.48 The fact that the surface is completely spherical tells you that it is an s type orbital; however, you cannot tell if it is a $1s$ orbital without knowing the total number of spherical nodes. If the electron density is only zero at the nucleus, then it is a $1s$ orbital. If it has another node at a distance from the nucleus it is a $2s$ orbital; if it has two nodes at a distance from the nucleus it is a $3s$ orbital.

6.49 There is sufficient information. The fact that the valence-shell electron configuration includes the $4p$ electrons tells you the element is beyond the transition elements of the fourth period. Only these elements in the fourth period have p electrons in the valence shell. Arsenic is the element because it possesses the valence electrons $4s^2 4p^3$ and it has a set of three $4p$ electrons.

6.50 (d) violates the Pauli Principle because a single orbital can possess no more than two electrons.

6.51 A node is a place in space where the amplitude of a wave function is zero. It is generally true that the more nodes a wave function possesses the higher its energy. The size of a hydrogen orbital increases as the value of the principal quantum number increases. A $4s$ orbital is thus larger than a $3s$ orbital. For each orbital with a given n principal quantum number, the number of nodes is the same—one less than the value of n. Therefore, the $4s$ orbital has three nodes, and the $3s$ orbital has two nodes.

6.52 See Figure 6.24 in the text for these contour representations. These orbitals all have the same energy in a hydrogen atom because they all have the same principal quantum number value, 3.

6.53 The expression $E = hc/\lambda$ is valid for electromagnetic radiation. Matter typically used by humans is not electromagnetic radiation and does not move at the speed of light.

6.54 First, convert megacycles per second into cycles per second using the equivalence 1 megacycle = 1×10^6 cycles

$$v_{microwave} = \left(20{,}000 \frac{\text{megacycles}}{\text{sec}}\right) \times \left(\frac{1 \times 10^6 \text{ cycles}}{1 \text{ megacycle}}\right)$$

$$= 2 \times 10^{10} \text{ cycles/sec}$$

$$v_{visible} = \left(100{,}000{,}000 \frac{\text{megacycles}}{\text{sec}}\right) \times \left(\frac{1 \times 10^6 \text{cycles}}{1 \text{ megacycle}}\right)$$

$$= 1 \times 10^{14} \text{ cycles/sec}$$

Substituting these values into the relation $E = hv$ yields

$$E_{microwave} = hv$$

$$= (6.63 \times 10^{-34} \text{ J-sec})(2 \times 10^{10}/\text{sec})$$

$$= 1.33 \times 10^{-23} \text{ J}$$

$$E_{visible} = hv$$

$$= (6.63 \times 10^{34} \text{ J-sec})(1 \times 10^{14}/\text{sec})$$

$$= 6.63 \times 10^{-20} \text{ J}$$

The value of the ratio $E_{microwave}/E_{visible}$ shows the relative magnitude of their energies:

$$\frac{E_{microwave}}{E_{visible}} = \frac{1.33 \times 10^{-23} \text{ J}}{6.63 \times 10^{-20} \text{ J}} = 2.01 \times 10^{-4}$$

From this we see that $E_{microwave} = 2.01 \times 10^{-3} E_{visible}$. Thus the energy content of microwave radiation is significantly less than that of visible radiation.

6.55 (a) Electrons enter the $2s$ and $2p$ orbitals in the atoms of the second-row elements and the $5s$, $4d$, and $5p$ orbitals in the atoms of the fifth period.

(b) The energy of a $3d$ orbital is lower than the energy of a $4p$ orbital. The $3d$ orbital is filled with electrons before the $4p$ orbital when atoms of the fourth period are formed. In order for the fourth row to be composed of elements sequentially increasing by one atomic number, the transition elements with their $3d$ valence electrons must be included.

6.56 (a) $1s^2 \,\widehat{2s^2 2p^3}$—outer orbital set

(b) $1s^2 2s^2 2p^6 3s^2 3p^6 \,\widehat{4s^1}$—outer orbital set

(c) $1s^2 2s^2 2p^6 3s^2 3p^6 \boxed{4s^1 3d^{10}}$—outer orbital set

The $4s$ and $3d$ orbitals are very similar in energy in the atom once the orbitals are filled with electrons. Thus, both orbitals comprise the outer set of orbitals. The outer orbital electron configuration is $4s^1 3d^{10}$ instead of the expected $4s^2 3d^9$ because a half-filled $4s$ orbital and a filled $3d$ orbital appear to be more stable than a partially filled $3d$ orbital.

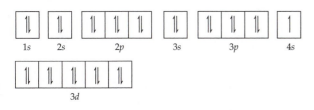

(d) $1s^2 2s^2 2p^6 \boxed{3s^2 3p^1}$—outer orbital set

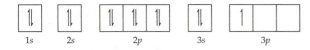

6.57 **(a)** The element with the highest-energy $4s^2 4p^2$ electron configuration belongs to the fourth period and the carbon family because $n = 4$ (fourth period) and the valence electron configuration of the carbon family is $s^2 p^2$. The element is germanium (Ge).

(b) The element with the highest-energy $3d^6 4s^2$ electron configuration belongs to the fourth period and to the transition element family because $n = 4$ for the s orbital and transition elements have incompletely filled d orbitals. The element is iron ($1s^2 2s^2 2p^6 3s^2 3p^6 3d^6 4s^2$): Fe.

(c) The only family that has an ns^1 electron configuration is the alkali metal family, group 1A. The element belongs to the sixth period because $n = 6$. It is cesium Cs.

(d) This element belongs to the second period, as $n = 2$ for the valence electrons: Fluorine (F).

(e) See (b) for rationale: Zinc (Zn).

6.58 **(a)** Neon has the electron configuration $1s^2 2s^2 2p^6$; argon's is $1s^2 2s^2 2p^6 3s^2 3p^6$. The electrons that comprise the shell from Ne to Ar are the $3s^2 3p^6$ electrons.

(b) Li has the electron configuration $1s^2 2s^1$. The nearest inert gas in the periodic table is He, with a $1s^2$ electron configuration. Li loses the $2s^1$ electron to form Li^+, with a $1s^2$ electron configuration. Sr has the electron configuration $[Kr]5s^2$. Sr loses the two $5s$ electrons to form Sr^{2+} with a $[Kr]$ electron configuration. Oxygen has the electron configuration $[He] 2s^2 2p^4$. The nearest inert gas in the periodic table is Ne. Oxygen adds two $2p$ electrons to form O^{2-} with the neon electron configuration $1s^2 2s^2 2p^6$.

6.59 The electron configurations shown in (a) and (c) correspond to excited states. The groundstate electron configuration for (a) is $1s^2$, and for (c), $[Ar] 3d^4 4s^2$. The electron configurations in (b) and (d) correspond to ground states.

6.60 **(a)** This group of elements has six valence electrons and is not a group of transition element. With six valence electrons the group is number 6A[16]. Oxygen is an example with its valence electron configuration of $2s^2 sp^4$. The other members of the oxygen family will have a similar electron configuration.

(b) This group must contain transition elements because the $(n - 1)d$ orbitals are involved in the outer electron configuration. Choose row four in the periodic table because it has transition elements and count three elements from the left side of the periodic table (the elements in the group possess three outer electrons, $2 + 1 = 3$): It is Sc and it has the valence electron configuration $3d^1 4s^2$. The other members of the group are Y, Lu and Lr.

(c) Using the approach in (a) and counting two elements from the left in row two you find the element Be. Thus, this group is the alkaline-earth elements.

(d) B, Al, Ga, In, and Tl.

6.61 **(a)** Student's (B) conclusion is not reasonable because other elements can also have a valence 4p orbital. The value of $n = 4$ for a valence orbital tells us that the element is in the fourth period. The value of $l = 1$ tells us that the electron is a p orbital; it is a $4p$ electron. Arsenic has the valence-electron configuration $[Ar]4s^2 3d^{10} 4p^3$. One of the valence electrons is a $4p$ electron; thus, As is a possibility. However the elements Ga, Ge, Se, Br, and Kr also possess $4p$ valence electrons and could also be the unknown element.

(b) To identify the unknown element student (B) needs either the total number of electrons or the total number of valence electrons possessed by the element.

(c) Yes. The noble or rare-gas elements have a filled valence electronic structure, $ns^2 np^6$, which is very stable. Thus these elements do not readily lose or gain electrons. This information helps identify the unknown element as krypton.

(d) Yes. The only metal in the fourth period with a valence 4p electron is gallium.

6.62 **(a)** $E = h\dfrac{c}{\lambda}$

$$E = \frac{(6.63 \times 10^{-34}\ \text{J-s})\left(3.00 \times 10^8\ \dfrac{\text{m}}{\text{s}}\right)}{(589.3\ \text{nm})\left(\dfrac{1\ \text{m}}{10^9\ \text{nm}}\right)} = 3.38 \times 10^{-19}\ \text{J}$$

(b) The energy calculated in (a) is for a photon emitted by a single sodium atom. To calculate the energy for a mole of sodium atoms you must use Avogadro's number:

$$E = \left(\frac{3.38 \times 10^{-19}\ \text{J}}{\text{Na atom}}\right)\left(\frac{6.022 \times 10^{23}\ \text{Na atoms}}{1\ \text{mole}}\right)$$

$$= 203{,}000\ \text{J} = 203\ \text{kJ}$$

(c) The electron configuration of sodium is $1s^2 2s^2 2p^6 3s^1$. The valence electron is $3s^1$.

(d) The formation of a positive ion occurs when an electron is removed. To form Na^+ the valence electron is removed: $1s^2 2s^2 2p^6$.

6.63 **(a)** Electrons and photons are "screwy" because if you say they behave like particles you give the wrong impression; similarly if you say they behave like waves you also give the wrong impression. They possess a duality of both. They can show both wave and particle properties depending on the experiment and circumstances. This duality behavior on the atomic scale is different and we do not observe it for particles in our macroscopic world.

(b) It is very difficult to find analogies between behaviors of particles in our familiar world and that of electrons and photons at the atomic level; the particles behave entirely different. They do not act in "normal" ways and we find this hard to model. If we find it hard to understand the behaviors of particles in the atomic world this can lead to frustration as we try to develop accurate models. It is likely whatever we visualize or model is not quite correct and the search for truth may never end, leading to our potential torment.

(c) Learn how electrons and photons in the atomic world behave; be able to describe their behaviors based on what we know today. Use the models that have been developed and be able to apply them. Recognize the limitations of the models. However, if you ask the question stated by Feynman, "But how can it be like that?" and demand an accurate and precise answer, then you will find yourself entering a world of possible torment.

6.64 **(c)** $v = \dfrac{c}{\lambda}$

$$= \frac{3.00 \times 10^8 \text{ m/sec}}{(7.52 \times 10^6 \text{ Å})(1 \times 10^{-10} \text{ m/Å})}$$

$$= 3.99 \times 10^{11}/\text{s}$$

6.65 **(b)** $E = hv$

$$= (6.63 \times 10^{-34} \text{ J-sec})(9.00 \times 10^{11}/\text{s})$$

$$= 5.97 \times 10^{-22} \text{ J}$$

6.66 **(a)** $\lambda = \dfrac{h}{mv}$

$$= \left(\frac{6.63 \times 10^{-34} \text{ J-sec}}{(9.1 \times 10^{-28} \text{ g})(4.00 \times 10^9 \text{ cm/sec})} \right)$$

$$\times \left(\frac{1 \text{ kg-m}^2/\text{sec}^2}{1 \text{ J}} \right)\left(\frac{100 \text{ cm}}{\text{m}} \right)^2\left(\frac{10^3 \text{ g}}{\text{kg}} \right)$$

$$= 1.8 \times 10^{-9} \text{ cm}$$

The answer can be converted to angstroms as follows:

$$\lambda = (1.8 \times 10^{-9} \text{ cm})\left(\frac{1 \text{ Å}}{1 \times 10^{-8} \text{ cm}} \right)$$

$$= 0.18$$

6.67 (e). **6.68 (b)**. **6.69 (e)**. **6.70 (e)**. **6.71 (b)**. **6.72 (a)**. **6.73 (d)**.

6.74   **(b)**. The value of m_l cannot be greater than the value of l.

6.75   **(d)**. $\Delta E \propto \left(\dfrac{1}{2^2} - \dfrac{1}{4^2} \right) = \dfrac{3}{16}$. Energy is absorbed because ΔE is positive.

6.76   **(d)**

6.77   **(b)**

6.78   **(c)**

6.79   **(d)**. In many-electron atoms, the energy of an electron depends on both n and l. The electron with the lowest n value along with the lowest l value has the lowest energy.

6.80   **(e)**. $1s^2 2s^2 2p^6 3s^2 3p^6 3d^{10} 4s^2 4p^6 4d^{10} 5s^2 5p^6$.

6.81   **(c)**

6.82   **(e)**

6.83   **(e)**. For $n = 3$, the maximum possible outer electron configuration is $3s^2 3p^6 3d^{10}$.

6.84   **(c)**

Sectional MCAT and DAT Practice Questions II

Our stomach secretes acids to help digest foods. If a hole in a mucosal lining exists in the stomach or excessive acid is secreted, a person may have either ulcers or gastric distress. Commercial antacids use a variety of acid-neutralizing agents to help reduce acidity in a stomach. Examples of these agents include $NaHCO_3$, $Al(OH)_3$, $CaCO_3$, and $Mg(OH)_2$.

Calcium carbonate can be formed in a laboratory by the following reaction:

$$CaCl_2(aq) + Na_2CO_3(aq) \rightarrow CaCO_3(s) + 2\,NaCl(aq)$$
Equation 1

Aluminum hydroxide can be formed in a laboratory by the following reaction:

$$Al(NO_3)_3(aq) + 3\,NaOH(aq) \rightarrow Al(OH_3)_3(s) + 3\,NaNO_3(aq)$$
Equation 2

The net ionic equation for Equation 2 is:

$$Al^{3+}(aq) + 3\,OH^-(aq) \rightarrow Al(OH)_3(s)$$
Equation 3

Standard enthalpies of formation for selected compounds, cations, and anions are given in Table 1. The standard enthalpies of formation of ions are calculated with respect to the $H^+(aq)$ ion and its standard enthalpy of formation is assigned a value of zero.

TABLE 1 Standard enthalpies of formation at 25 °C

Substance	ΔH_f^o (kJ mol^{-1})	Cation	ΔH_f^o (kJ mol^{-1})	Anion	ΔH_f^o (kJ mol^{-1})
$CaCO_3(s)$	−1207.1	Al^{3+}	−524.7	Cl^-	−167.4
$CaCl_2(s)$	−795.8	Na^+	−239.7	CO_3^{2-}	−676.3
$Na_2CO_3(s)$	−113.9	Ca^{2+}	−543.0	OH^-	−229.9
$NaCl(aq)$	−407.1	Mg^{2+}	−462.0	HCO_3^-	−691.1
$Al(OH)_3(s)$	−1276	—	—	—	—

1. Which is the net ionic equation for the reaction shown in Equation 1?
   (a) $CaCl_2(aq) + Na_2CO_3(aq) \rightarrow CaCO_3(s) + 2\,NaCl(aq)$
   (b) $CaCl_2(aq) + CO_3^{2-}(aq) \rightarrow CaCO_3(s) + 2\,Cl^-(aq)$
   (c) $Ca^{2+}(aq) + Na_2CO_3(aq) \rightarrow CaCO_3(s) + 2\,Na^+(aq)$
   (d) $Ca^{2+}(aq) + CO_3^{2-}(aq) \rightarrow CaCO_3(s)$

2.   If 0.0200 moles of $CaCl_2$ reacts with 0.0300 moles of Na_2CO_3 as shown in Equation 1, how many moles of NaCl form?
      (a) 0.0200 mol   (b) 0.0400 mol   (c) 0.0100 mol   (d) 0.0600 mol

3.   What is the molarity of a solution of calcium chloride $(110.98 \text{ g mol}^{-1})$ if 5.55 grams are dissolved in water to form 250.0 mL of solution?
      (a) 0.0500 M (b) 0.200 M (c) 0.0800 M (d) 3.20 M

4.   Which substance might be neutralized in a stomach by a commercial antacid?
      (a) NH_3 (b) NaOH   (c) HCl   (d) CH_4

5.   What is the enthalpy of reaction for the reaction given by Equation 3?
      (a) −2490 kJ   (b) −1276 kJ   (c) −521 kJ   (d) −62 kJ

6.   What is the electron configuration of Al^{3+}?
      (a) $1s^2 2s^2 2p^3$ (b) $1s^2 2s^2 2p^6$
      (c) $1s^2 2s^2 2p^6 3s^2 3p^1$ (d) $3s^2 3p^1$

7.   Which substance at 25 °C should be the most thermally stable with respect to decomposition into its elements?
      (a) $Al(OH)_3(s)$ (b) $CaCl_2(s)$
      (c) $CaCO_3(s)$ (d) $Na_2CO_3(s)$

| Questions 8 through 12 are **not** based on a descriptive passage. |

8.   What is the molarity of a $NaNO_3$ solution if 25.0 mL of a 0.200 M $NaNO_3$ solution is diluted to 100.0 mL?
      (a) 0.0500 M (b) 0.100 M (c) 0.150 M (d) 0.250 M

9.   Which of the following conditions in a gaseous system *always* results in a decrease in the internal energy?
      (a) An endothermic process and a corresponding decrease in total volume.
      (b) An endothermic process and a corresponding increase in total volume.
      (c) An exothermic process and a corresponding increase in total volume.
      (d) An exothermic process and a corresponding decrease in total volume.

10.   Which set of quantum numbers for an electron in an atom is *not* allowed?

	n	l	m_l	m_s
(a)	2	1	0	−1/2
(b)	3	2	1	+1/2
(c)	3	3	2	+1/2
(d)	1	0	0	+1/2

11.   Which element has the electron configuration $1s^2 2s^2 2p^6 3s^2 3p^6 3d^{10} 4s^2 4p^3$?
      (a) P   (b) Si   (c) Ge   (d) As

12.   How many unpaired electrons does Co possess in its ground state?
      (a) 0   (b) 1   (c) 2   (d) 3

ANSWERS

1. (d) The ion equation before eliminating spectator ions is
$$Ca^{2+}(aq) + 2\cancel{Na^+}(aq) + 2\cancel{Cl^-}(aq) + CO_3^{2-}(aq) \rightarrow CaCO_3(s) + 2\cancel{Na^+}(aq) + 2\cancel{Cl^-}(aq).$$

The spectator ions have a line through their symbols. The net result when spectator ions are removed is the equation given in the correct response.

2. (b) $CaCl_2$ and Na_2CO_3 react in a 1:1 mole ratio. The given ratio of moles in the question is 2:3 or 1/1.33. This shows that $CaCl_2$ is the limiting reactant. The stoichiometry of the reaction shows a stoichiometric mole equivalence for $CaCl_2/NaCl$ of 1:2. Therefore, the number of moles of NaCl formed is:
(0.0200 mol $CaCl_2$)(2 mol NaCl/1 mol $CaCl_2$) = 0.0400 mol NaCl

3. (b) M = moles $CaCl_2$/volume solution in liters
   = [(5.55 g $CaCl_2$)(1 mol/ 110.98 g)][/0.2500 L
   = 0.200 moles $CaCl_2$/L = 0.200 M.

4. (c) HCl is the only acid listed and our stomachs secrete acids as stated in the passage.

5. (d) $\Delta H^o = \sum \Delta H_f^o(P) - \sum \Delta H_f^o(R)$
   = -1276 kJ $- [-524.7$ kJ $-3(-229.9$ kJ$)]$ = -62 kJ.

6. (b) Al has a total of 13 electrons: $1s^2 2s^2 2p^6 3s^2 3p^1$. When it forms a 3+ ion it loses three electrons from the valence shell. The remaining electrons are $1s^2 2s^2 2p^6$.

7. (a) The standard enthalpy of formation refers to the enthalpy change accompanying the formation of one mole of a substance from its elements at standard conditions. All enthalpies of formation of the substances in the question are exothermic as shown in Table 1. Thus, the decomposition of each substance to its elements is endothermic. The substance with the largest negative standard enthalpy of formation requires the largest amount of energy to decompose it and therefore it is the most thermodynamically stable. $Al(OH)_3(s)$ meets this requirement.

8. (a) Dilution results in a solution with a lower molarity than the initial concentration and therefore, **(d)** cannot be correct. Calculate the new molarity using the relationship:$M_i V_i = M_f V_f$ or $M_f = \dfrac{M_i V_i}{V_f}$.

$$M_f = \frac{[0.200\ M][25.0\ \text{mL}]}{100.0\ \text{mL}} = 0.0500\ M$$

9. (c) An exothermic process releases energy and this results in a decrease in the internal energy of a system. In addition, when a system does work on its surroundings by expanding its volume, a decrease in internal energy occurs.

10. (c) The quantum number l has integral values from 0 to $n - 1$ for each value of n. In answer (c) both n and l have the same value, 3 , which is not permissible.

11. (d) The neutral element has 33 electrons and As has this number of electrons.

12. (d) Cobalt has 27 electrons and the electron configuration $1s^2 2s^2 2p^6 3s^2 3p^6 3d^7 4s^2$ or [Ar] $3d^7 4s^2$. All orbitals are filled with electrons except for the 3d orbitals. Using a box diagram and distributing the electrons in the set of five 3d orbitals using Pauli's principle and Hund's rule shows three unpaired electrons:

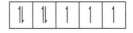

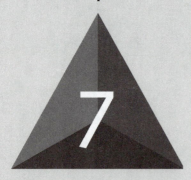

Chapter 7

Periodic Properties of the Elements

OVERVIEW OF THE CHAPTER

7.1, 7.2, 7.3 PERIODIC PROPERTIES, EFFECTIVE NUCLEAR CHARGE, ATOMIC SIZE, AND ION SIZE

Review: Bohr radius (6.3)

Learning Goals: You should be able to:

1. Explain why the periodic table can be viewed as a classification scheme.
2. Describe the contributions of Dmitri Mendeleev, Lothar Meyer, and Henry Mosely to the development of the modern periodic table.
3. Define effective nuclear charge, Z_{eff}, and explain how it is determined.
4. Describe the periodic trends in Z_{eff}.
5. Explain the effect of Z_{eff} on the radial density function.
6. Explain the variations in bonding atomic radii among the elements and predict the relative sizes of atoms based on their positions in the periodic table.
7. Explain the variations in ion size among the elements, the change in size when an atom gains or loses electrons, and predict the relative sizes of ions based on their charge and positions in the periodic table.

7.4, 7.5 PERIODIC PROPERTIES: IONIZATION ENERGY AND ELECTRON AFFINITY

Learning Goals: You should be able to:

1. Explain the observed changes in values of successive ionization energies for a given atom.
2. Explain the general variations in first ionization energies among the elements and predict the periodic trends within a family and period.
3. Explain the variations in electron affinities among the elements and predict the periodic trends within a family and period.

7.6 OVERVIEW: METALS AND NONMETALS

Review: Anions (2.7); cations (2.7); family (2.5); period (2.5); metals (2.5); nonmetals (2.5); semimetals (2.5).

Learning Goals: You should be able to:

1. Describe the periodic trends in metallic and nonmetallic behavior.
2. Describe the general differences in chemical reactivity between metals and nonmetals.

Review: Alkali and alkaline-earth metals (2.5); concept of family (2.5).

Learning Goals: You should be able to:

1. Describe the general physical and chemical behavior of the alkali metals and alkaline-earth metals, and explain how their chemistry relates to their position in the periodic table.
2. Write balanced equations for the reaction of hydrogen with metals to form metal hydrides.
3. Write balanced equations for simple reactions between the active metals (groups 1A and 2A) and the nonmetals in groups 6A and 7A.

Review: Nonmetals (2.5).

Learning Goals: You should be able to:

1. Write balanced equations for the reaction of hydrogen with non-metals such as oxygen and chlorine.
2. Describe the allotropy of oxygen.
3. Explain the dominant chemical reactions of oxygen and relate this behavior to its position in the periodic table.
4. Describe the physical states and colors of the halogens, and explain the trends in reactivity with increasing atomic number in the family.
5. Explain the very low chemical reactivity of the noble gas elements.

TOPIC SUMMARIES AND EXERCISES

The periodic table shows an arrangement of elements based on increasing atomic numbers and grouping of elements within periods and families. Dmitri Mendeleev is given primary credit for the development of a modern periodic table. Knowledge of the periodic table and the electronic structure of the elements helps us to explain and predict trends in their physical and chemical properties.

Energies of electrons in many-electron atoms differ from those in a hydrogen atom because the nucleus contains more than one proton and the additional electrons repel one another. The **effective nuclear charge**, Z_{eff}, experienced by valence electrons determines many atomic properties. It is a measure of the average electrical environment experienced by an electron in an atom and it is created by all other electrons in the atom and the nucleus: $Z_{eff} = Z - S$.

- Z equals the number of protons in the nucleus.
- S is the shielding constant and represents an average electron density created by the electrons between the electron of interest and the nucleus. The electron density between the nucleus and the electron of interest decreases the nuclear charge and is said to shield or screen the electron from the full nuclear charge.
- Electrons in the same shell only minimally screen each other from the nuclear charge.
- Z_{eff} for outermost electrons tends to increase left to right across a period and to decrease down a family.

7.7 GROUP TRENDS EXEMPLIFIED: THE ACTIVE METALS

7.8 GROUP TRENDS EXEMPLIFIED: SELECTED NON-METALS

PERIODIC PROPERTIES, EFFECTIVE NUCLEAR CHARGE, ATOMIC SIZE, AND ION SIZE

Accurate calculations of electron-charge distributions in many-electron atoms have shown that the distribution of electronic charge in an atom is not continuous but rather occurs in layers or shells. Thus, atomic size is only an approximation of the radius of an atom.

- Each shell corresponds to a collection of electrons with the same principal quantum number; for example, $1s^2$ forms a shell that is different from a $2s^2 2p^6$ shell.
- The distribution of electron density as a function of the radial distance of the electron from the nucleus is termed radial electron density. The size of an atom is approximated by its **bonding atomic radius**.
- There is a decrease in bonding atomic radii from left to right across a period and an increase in bonding atomic radii with increasing atomic number within a family. Be sure you can use the concepts of effective nuclear charge and energy levels of electrons to explain periodic trends in bonding atomic radii.

When atoms form ions in a crystal, they tend to achieve an octet by forming a valence-electron configuration equal to that of the *nearest* noble gas in the periodic table.

- Atoms form positive ions (**cations**) by losing electrons. Elements that typically form cations in compounds are found on the left side and in the middle of the periodic table: These are the metallic elements.
- Atoms form negative ions (**anions**) by gaining electrons. Elements that typically form anions in compounds are found on the right side: These are the nonmetallic elements.
- Transition elements typically form cations. The outer electron configurations of their atoms are either $(n-1)d^x ns^2$ or $(n-1)d^x ns^1$ ($n = 4, 5, 6$ and $x = 1 - 10$). When they form ions, the ns electrons are lost *first*, and then the necessary number of $(n-1)d$ electrons are lost to form an ion of a particular charge. For example,

$$Fe(3d^6 4s^2) \longrightarrow Fe^{3+}(3d^5) + 3e^-$$
$$Cu(3d^{10} 4s^1) \longrightarrow Cu^+(3d^{10}) + e^-$$

It is possible to measure the distance between ions in a crystal. By appropriate calculation methods, an average ion size can be assigned to cations and anions.

- Removing electrons from an atom forms a cation that is *smaller* than the corresponding neutral atom from which it is derived.
- Adding electrons to an atom forms an anion that is *larger* than the corresponding neutral atom. You should note the trends in ion size across a row and within a family of the periodic table.
- Ions in an **isoelectronic series** have the same number of electrons. The size of an ion in an isoelectronic series decreases with increasing atomic number. You should be able to explain this trend based on the attraction of an increasing number of protons in a nucleus for the same number of valence electrons.

EXERCISE 1   Writing electron configurations for ions

Write the electron configuration for each of the following ions and indicate which ions possess a noble-gas electron configuration: (a) S^{2-}; (b) Be^{2+}; (c) Cl^-; (d) Fe^{3+}.

SOLUTION: *Analyze*: We are asked to write the electron configuration for four ions and to identify which have electron configurations equivalent to one of the noble-gas elements.

Plan: First write the complete electron configuration of each element. Second, remove electrons (one for each positive charge) or add electrons (one for each negative charge) to form the appropriate ion. Finally compare the electron configurations to those of the noble-gas elements. Electron configurations that reflect a noble-gas electron configuration can be written in a shorthand notation format, for example; $1s^2 2s^2 2p^6$ can be written as [Ne].

Solve: (a) Sulfur has the electron configuration $[Ne]3s^2 3p^4$. Two electrons are added to the $3p$ orbitals to form the S^{2-} ion, whichhas the noble-gas electron configuration $[Ne]3s^2 3p^6$ or [Ar]. (b) Be has the electron configuration $[He]2s^2$. The two $2s$ electrons are lost to form the Be^{2+} ion, which has the noble-gas electron configuration $1s^2$ or [He]. (c) Cl has the electron configuration $[Ne]3s^2 3p^5$. One electron is added to a $3p$ orbital to form the Cl^- ion, which has the noble-gas electron configuration $[Ne]3s^2 3p^6$ or [Ar]. (d) Fe has the electron configuration $[Ar]3d^6 4s^2$. We find experimentally that when a transition element loses electrons to form an ion, the ns electrons are lost before the $(n-1)d$ electrons. The two $4s$ electrons and one $3d$ electron are lost to form the Fe^{3+} ion, which has the electron configuration $[Ar]3d^5$; this is not a noble-gas electron configuration.

Check: S^{2-} and Cl^- show addition of electrons and Be^{2+} and Fe^{3+} show loss of electrons to reflect their charges. Each element, before its electron configuration is altered, shows the correct number of outer electrons for its group number.

EXERCISE 2   Determining trends in radii for ions

In each of the following pairs of series, which series is arranged in order of *decreasing* radius? In each case state the general rule that the series illustrates.

(a) Mg^{2+}, Na^+, F^-, O^{2-} or O^{2-}, F^-, Na^+, Mg^{2+}
(b) F^-, Cl^-, Br^-, I^- or I^-, Br^-, Cl^-, F^-
(c) Mn^{2+}, Mn^{3+} or Mn^{3+}, Mn^{2+}
(d) Ca^{2+}, Cu^{2+} or Cu^{2+}, Ca^{2+}

SOLUTION: *Analyze*: We are asked to determine which series of ions is arranged in order of decreasing radius. We are also asked to deduce a general rule that each series illustrates.

Plan: Consider each series and determine what relationships may exist. Are the elements in the same family? Are the elements in similar groups? Are the elements adjacent to one another in a series? Are the charges the same or different? Recognize that adding electrons to an element increases its radius and removing electrons decreases its radius.

Solve: (a) O^{2-}, F^-, Na^+, Mg^{2+}. These five ions form an **isoelectronic series** (a series of ions with the same number of electrons). The radii of ions in an isoelectronic series decrease with increasing atomic number. (b) I^-, Br^-, Cl^-, F^-. These five ions all belong to the halogen family. The radii of ions of the same charge in a family increase with increasing atomic number. (c) Mn^{2+}, Mn^{3+}. The radii of cations of the same element decrease with increasing charge. (d) Ca^{2+}, Cu^{2+}. The radii of Group 1B or 2B ions are smaller than the radii of 1A or 2A ions if the ions have the same charge and are in the same period.

EXERCISE 3 Understanding limitations in the concept of bonding atomic radius

Describe the limitations that must be placed on the concept of bonding atomic radii.

SOLUTION: The electronic-charge distribution in an atom is not sharply defined because the electron cloud extends through all space. We can define an approximate atomic radius in terms of bond distances by apportioning the bond distance between two atoms of a compound to two atomic radii. The atomic radius of an atom in a homonuclear diatomic molecule, a molecule that consists of two atoms of the same element, such as Br_2, is approximated as one-half the distance between the two bonded atoms.

EXERCISE 4 Determining the relative order of atomic sizes for a series of atoms

Arrange the following atoms in order of increasing atomic size: (**a**) Mg, Ca, Sr; (**b**) B, F, Ge, Pb.

SOLUTION: (**a**) Within a family, atomic size increases with increasing atomic number: Mg < Ca < Sr. (**b**) Boron and fluorine are smaller than germanium and lead because they are in the second period, whereas the other two are in the fourth and sixth periods, respectively. Boron is larger than fluorine because atomic size tends to decrease across a period. Lead is larger than germanium because it is in a period with a higher number: F < B < Ge < Pb.

PERIODIC PROPERTIES: IONIZATION ENERGY AND ELECTRON AFFINITY

The minimum energy required to remove an electron from an isolated gaseous atom or ion so that the electron and resulting species are an infinite distance apart is termed **ionization energy**.

• An example of this process is the loss of one electron by a sodium atom:

$$Na(g) \longrightarrow Na^+(g) + e^-(g) \qquad I_1 = 496 \text{ kJ/mol}$$

Energy is always added because an electron is being removed from a positively charged nucleus.

• The symbol I is used for ionization energy, and the subscript 1 in I_1 means that the first electron in a gaseous atom is removed.
• Within each period there is a gradual increase in first ionization energy from left to right. Within a family there is a decrease in ionization energy with increasing atomic number. Be sure that you can use the concepts of effective nuclear charge and atomic radii trends to explain periodic trends in ionization energies. Also, you should note the irregularities in the general trend of first ionization energies across a period.

When an electron is added to an isolated gaseous atom, energy is usually released. For a few metals and rare gases energy is absorbed. The energy change during the process of adding an electron to an atom or ion is termed **electron affinity** (We use the symbol ΔE for this energy change).

• An example of such a process is the addition of one electron to iodine:

$$I(g) + e^-(g) \longrightarrow I^-(g) \qquad \Delta E = -295 \text{ kJ/mol}$$

Energy is released during this process; the sign is negative.

- There is a rough trend of electron affinities becoming more negative from left to right across a period; in going down a family only small changes are usually observed. Carefully note rationales given in the text for trends and irregularities in trends.

EXERCISE 5   Characterizing the first and second ionization energies for calcium

Energy is required for the following reaction and it is the first ionization energy (I_1) for calcium:

$$Ca(g) \longrightarrow Ca^+(g) + e^-(g)$$

What ionization reaction is associated with the second ionization energy (I_2) for calcium? Will it be endothermic or exothermic?

SOLUTION: *Analyze*: We are asked to identify the reaction associated with the second ionization energy of calcium and to state if the energy change is exothermic or endothermic.

Plan: The term "second ionization energy," refers to the energy required to remove the second highest-energy electron from calcium, after the first has been removed. We should also recognize that it will take work (energy) to remove an electron from an atom.

Solve: The reaction associated with the second ionization energy of calcium is

$$Ca^+(g) \longrightarrow Ca^{2+}(g) + e^-(g)$$

It is an endothermic process because energy is required.

EXERCISE 6   Comparing the magnitudes of the first and second ionization energies for silicon

The first ionization energy for Si(g) is 780 kJ/mol; the second ionization energy is 1575 kJ/mol. Why is the second ionization energy greater than the first?

SOLUTION: *Analyze*: We are given the first and second ionization energies for Si(g), 780 kJ and 1575 kJ, respectively, and asked to explain why the second ionization energy is greater.

Plan: We will write the reactions describing the removal of the first and second electrons from Si(g) and consider why removing electrons changes the charges of the atoms and the ionization energies. Coulomb's law is useful when charged particles exist and explanation of physical properties is required.

Solve: The first ionization energy is the energy required for the ionization reaction

$$Si(g) \longrightarrow Si^+(g) + e^-(g)$$

The second ionization energy is the energy required for the ionization reaction

$$Si^+(g) \longrightarrow Si^{2+}(g) + e^-(g)$$

Note that in the first reaction the electron is removed from a noncharged Si(g) atom, whereas in the second reaction the electron is removed from a positive Si$^+$(g) ion. According to Coulomb's law, more energy is required to remove an electron from a positive ion than from a noncharged particle; thus the second ionization energy is greater than the first.

Comment: A general trend exists: It is energetically more difficult to remove an electron as the magnitude of the charge of an atom increases.

EXERCISE 7   Comparing the ionization energy and electron affinity of fluorine

What is the ionization energy for $F^-(g)$ if the electron affinity for $F(g)$ is -328 kJ/mol?

SOLUTION: *Analyze*: We are given the electron affinity for $F(g)$, -328 kJ/mol, and asked what is the ionization energy for $F^-(g)$.

Plan: The question suggests that there is some simple relationship between $F(g)$ and $F^-(g)$ and we need to determine it. We should first write the reaction describing the addition of an electron to $F(g)$ and relate it to the reaction describing the removal of an electron from $F^-(g)$.

Solve: When $F(g)$ accepts an electron according to the reaction

$$F(g) + e^-(g) \longrightarrow F^-(g)$$

328 kJ/mol of energy is released. The ionization energy for $F^-(g)$ is the energy required for the reaction

$$F^-(g) \longrightarrow F(g) + e^-(g)$$

Inspection of these two reactions shows that the reaction associated with the ionization energy of $F^-(g)$ is the reverse of the reaction associated with the electron affinity of $F(g)$. Reversing the direction of either reaction causes the heat for the reversed reaction to be the negative of the heat of the original reaction on the basis of the first law of thermodynamics. Thus the ionization energy of $F^-(g)$ is $+328$ kJ/mol.

Check: Note that when the two energies are added the sum is zero. If we add the two reactions together there is no net reaction. This suggests that we have appropriately discovered a valid relationship.

EXERCISE 8   Predicting relative magnitudes of ionization energies

Predict which member of each pair has the higher ionization energy and provide a brief explanation: (**a**) Na or Rb; (**b**) Si or P?

SOLUTION: *Analyze*: We are given pairs of atoms and asked which member of each pair has the higher ionization energy and why.

Plan: When determining trends in ionization energy it is useful to look at the following: (1) The location of each element in the periodic table; (2) the relative atomic sizes of the atoms; (3) the electron configurations of the atoms; and (4) the effective nuclear charge experienced by the outer electrons.

Solve: (**a**) Na. Rb is a larger atom than Na; therefore the outermost electron in Na is closer to the nucleus and thus requires a greater expenditure of energy to remove it compared to the one in Rb. (**b**) P. The electron configuration for P is $[Ne]3s^23p^3$ and it shows that the three electrons in the $3p$ subshell are all unpaired in different $3p$ orbitals. A half-filled $3p$ subshell has associated with it an extra stabilization energy that is not present in the same subshell for silicon, $[Ne]3s^23p^2$. Removing an electron from a half-filled $3p$ subshell thus requires more energy than removing an electron from one that has only two electrons in it. Also, the effective nuclear charge experienced by the outer electrons increases from Si to P. This results in P having a higher ionization energy than Si.

Check: Data in Figure 7.10 in the text confirms which element of each pair has the higher ionization energy.

EXERCISE 9   Predicting relative magnitudes of electron affinities

Which element has the greater electron affinity, sulfur or chlorine? Why?

SOLUTION: Chlorine. More energy is released when chlorine accepts an electron than when sulfur accepts one. This occurs because the effective nuclear charge experienced by the electron being added to a chlorine atom is greater than in the case of an electron that is added to a sulfur atom.

SOLUTION: *Analyze*: We are asked whether sulfur or chlorine has the greater electron affinity and why.

Plan: Electron affinity is the energy change that occurs when an electron is added to a gaseous atom. The greater that attraction of an atom for an electron the more exothermic the process. The text discusses trends in electron affinities which are less consistent than those of ionization energies. We need to review this discussion to answer the question.

Solve: Chlorine. Both elements are in the third row of the periodic table and in general electron affinity becomes more exothermic as the atomic number increases in the period. Primarily this occurs because the effective nuclear charge experienced by the electron being added to chlorine is greater than in the case of an electron added to a sulfur atom.

Check: Data in Figure 7.12 in the text confirms that chlorine has a greater electron affinity than sulfur.

Metals comprise roughly 70 percent of the known elements, and are situated to the left of the dark line in Figure 7.15 of the text.

OVERVIEW: METALS AND NONMETALS

- All except mercury, a liquid, are solids at room temperature and pressure.
- They exhibit good electrical and thermal conductivity.
- Many metals melt only at high temperatures.
- Metals tend to lose electrons in chemical reactions and become positively charged ions, called **cations**.

$$2\ Ca(s) + O_2(g) \longrightarrow 2\ CaO(s)$$

no ions   2+ ion   2− ion

- Compounds consisting of a metal and an element from the right side of The periodic table tend to be **ionic**: the metal has a positive charge and the other element a negative charge.

Oxides of metals form an important class of metallic compounds.

- The **oxide ion** is O^{2-}.
- *Most metal oxides are basic:* They dissolve in water to form metal hydroxides (containing a OH^- unit) or react with acids to form salts and water.

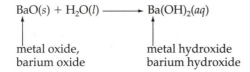

$$BaO(s) + H_2O(l) \longrightarrow Ba(OH)_2(aq)$$

metal oxide,   metal hydroxide
barium oxide   barium hydroxide

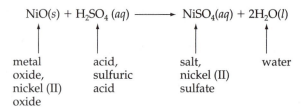

$$NiO(s) + H_2SO_4\ (aq) \longrightarrow NiSO_4(aq) + 2H_2O(l)$$

| metal oxide, nickel (II) oxide | acid, sulfuric acid | salt, nickel (II) sulfate | water |

Important trends in metallic character are

- Metallic character is strongest on the left side of a row (period) and tends to decrease to the right.
- Metallic character in a family tends to increase from top to bottom.
- Group 1A and 2A elements tend to lose electrons very readily.

Nonmetals show a variety of physical and chemical properties.

- They are found to the right of the dark line in Figure 7.15 of the text.
- H_2, N_2, O_2, F_2, Cl_2, Br_2, and I_2 exist as diatomic molecules.
- They are generally poor conductors of heat and electricity.
- When they react with metals, they tend to gain electrons and become negatively charged (**anions**).
- Compounds consisting only of nonmetals are **molecular substances**.

Most oxides of nonmetals are acidic.

- Some dissolve in water, for example CO_2, P_4O_{10}, and SO_3, to form acidic solutions:

$$CO_2(g) + H_2O(l) \longrightarrow H_2CO_3(aq)$$

- They may also dissolve in base to form salts and water.

$$CO_2(g) + 2\,KOH(s) \longrightarrow K_2CO_3(aq) + H_2O(l)$$

- CO and N_2O do not react with water or bases.

EXERCISE 10   Characterizing properties of metals

Which of the following are characteristic properties of metals: low luster, malleable; poor conductor of heat; form basic oxides; form positive ions?

SOLUTION: Refer to Table 7.3 in the text. The following are characteristic of metals: malleable, form basic oxides, and form positive ions.

EXERCISE 11   Identifying relative metallic character

Which member of each pair is expected to have the more metallic character: (**a**) Na or Rb; (**b**) K or Ca; (**c**) Pb or F; (**d**) C or Sn?

SOLUTION: *Analyze*: We are given pairs of elements and asked to identify which element in each pair is more metallic.

Plan: We can use the periodic table to help us answer this question. Metals are found in the left and mid regions of the periodic table and nonmetals are found in the right region. Metallic character tends to increase within a family with increasing atomic number.

Solve: (a) Rb. Metallic character increases down a family. (b) K. Metallic character tends to decrease from left to right across a periodic table. (c) Pb. Metallic character increases to the left and bottom of the periodic table. (d) Sn. Metallic character increases down a family.

Check: Data in Figure 7.13 in the text confirm the choices made in the solution.

EXERCISE 12   Identifying which oxides are basic

Which of the following are basic oxides: (a) P_2O_5; (b) MgO; (c) CO_2; (d) Na_2O?

SOLUTION: *Analyze*: We are given four oxides and asked to determine which are basic oxides.

Plan: Questions of this type require us to understand what types of elements combine to form basic oxides. Review the discussion of metal and nonmetal oxides in Section 7.6 of the text.

Solve: Basic oxides consist of metals combined with oxygen: (b) MgO and (d) Na_2O are basic oxides. The others do not contain metals.

Comment: For a substance to be a basic oxide it must react in one of two ways: Dissolve in water to form a basic solution or react with an acid.

EXERCISE 13   Identifying which oxides are acidic

Which oxide would you expect to be more acidic, Al_2O_3 or MgO? Explain.

SOLUTION: *Analyze*: We are given two oxides and asked which one is more acidic.

Plan: Questions of this type require us to understand what types of elements combine to form acidic oxides and relative acidity of acidic oxides. Review the discussion of metal and nonmetal oxides in Section 7.6 of the text.

Solve: Both aluminum and magnesium have metallic properties and thus we would expect them to be basic oxides. However, aluminum has significantly less metallic character than magnesium and it is adjacent to metalloids and nonmetals in the periodic table. This causes it to have far less basic character; in fact it possesses properties of both basic and acidic oxides. Therefore, Al_2O_3 is more acidic.

Comment: For a substance to be an acidic oxide it must react in one of two ways: Dissolve in water to form an acidic solution or react with a base. The larger the ratio $\dfrac{Z_+}{r_+}$ for a cation, the more acidic it tends to be.

EXERCISE 14   Writing reactions of oxides with water to form acids

Which oxides react with water to form the acids H_3PO_3, H_2CO_3, and H_2SO_4? Write reactions showing how they are formed.

SOLUTION: *Analyze*: We are given three acids, phosphorous, carbonic, and sulfuric, and asked which oxides react with water to form them and to write the reactions.

Plan: We need to write the reaction of a nonmetal oxide with water because acidic oxides react with water to form acids. It is useful to remove one or two water molecules from the formula of an acid to help determine the parent nonmetal oxide. If the result is not familiar then we have to review the material in Sections 7.6 and 7.8 to find an appropriate example.

Solve: The removal of one water molecule from each gives: HPO, CO_2, and SO_3. The last two are familiar nonmetal oxides and should be the ones we require. The first one is not familiar and we need to look at examples in the text. Equation 7.15 in the text gives us the reaction that forms phosphoric acid, H_3PO_4, which is related to phosphorous acid, H_3PO_3. This leads to the following reactions and acidic oxides:

$$P_4O_6(s) + 6\,H_2O(l) \longrightarrow 4\,H_3PO_3(aq)$$
$$(\textbf{Note:}\ P_4O_{10}\ \text{forms}\ H_3PO_4.)$$
$$CO_2(g) + H_2O(l) \longrightarrow H_2CO_3(aq)$$
$$SO_3(g) + H_2O(l) \longrightarrow H_2SO_4(aq)$$

Check: Equation 7.14 in the text confirms the reaction of carbon dioxide. Sulfur dioxide would be expected to react in a similar manner. Equation 7.15 supports our reaction to form H_3PO_3.

GROUP TRENDS EXEMPLIFIED: THE ACTIVE METALS

The text uses the principles of periodic properties to examine the chemistry of **alkali metals** (Group 1A) and the **alkaline-earth metals** (Group 2A). Also, the chemistry of the groups 1B and 2B is compared to those of groups 1A and 2B. Both group 1A and 1B elements form ions that have a noble-gas electron configuration. Thus, their compounds are colorless or white unless the anion has its own characteristic color.

Be sure you can explain the following general observations about the physical and chemical properties of group 1A and 2A elements. Pertinent concepts that can help us are noted:

- Alkali metals readily form 1+ cations: The result is explained by their ns^1 valence electron configuration and the low first ionization energies of the active metals.
- Alkali metals combine directly with most nonmetals and water: Alkali metals are high in the activity series and have low first ionization energies.
- Alkali metals react with oxygen to form three different types of compounds: *Oxides* (O^{2-} ion); *Peroxides* (O_2^{2-} ion); and *Superoxides* (O_2^- ion). Note the differences among the alkali metals as shown in equations (7.20–7.22) in the text.
- Lithium frequently shows significantly different properties than Na, K, Rb, or Cs: It has an extremely small size compared to the other alkali metals.
- Alkaline-earth metals readily lose two electrons to form 2+ cations, explained by their ns^2 valence electron configuration and relatively low first and second ionization energies.
- Alkaline-earth metals are less reactive than alkali metals: A ns^2 outer electron configuration is more stable than ns^1 and alkaline-earth metals have relatively higher ionization energies than alkali metals.
- Beryllium, like lithium, shows properties frequently different from those shown by other alkaline-earth metals: Note the extremely small size of Be compared to other group 2A elements.

Group 1A elements, the **alkali metals**, consist of Li (lithium), Na (sodium), K (potassium), Rb (rubidium), Cs (cesium), and Fr (francium).

- The elements are soft, have relatively low densities and melting points, and readily lose one electron in chemical reactions.

- Metallic hydrides contain the H^- ion:

$$2\,Na(s) + H_2(g) \longrightarrow 2\,NaH(s)$$

　　　　　1+ ion　　　　　　1− hydride ion

- Halogens (F_2, Cl_2, Br_2, I_2) readily react with alkali metals to form halide-ion–containing salts.

$$2\,Na(s) + Cl_2(g) \longrightarrow 2\,NaCl(s)$$

　　　　　1+ ion　　　　　　1− ion: a halide ior

- Alkali metals are usually kept under a hydrocarbon solvent such as kerosene to prevent their reaction with oxygen or water.

$$2\,Cs(s) + 2\,H_2O(l) \longrightarrow 2\,CsOH(aq) + H_2(g)$$

　　　　　　Note: Hydrogen gas is produce

Group 2A metals, the alkaline-earth metals, consist of Be (beryllium), Mg (magnesium), Ca (calcium), Sr (strontium), Ba (barium), and Ra (radium).

- They are harder, more dense, and melt at higher temperatures than Group 1A elements.
- They tend to undergo the same reactions as Group 1A, but are less reactive. For example, Mg will not react with cold H_2O, but Na, which is adjacent to it in the periodic table, will.
- They react to form 2+ ions.

EXERCISE 15　Writing combination reactions for forming alkali metal salts

Write balanced chemical reactions for the preparation of the following alkali metal compounds: (**a**) KOH; (**b**) Cs_2O; (**c**) NaH; (**d**) KBr.

SOLUTION: *Analyze*: We are given four alkali metal compounds and asked to write a balanced chemical reaction that describes their preparation.

Plan: We can answer problems of this type by studying analogous chemical reactions in Sections 7.6 and 7.7 in the text. Also, as we read these sections, we should look for general types of chemical reactions for specific classes of elements or compounds.

Solve: By referring to the discussion of chemical reactions of alkali metals in Section 7.6, we can write the following by analogy. (**a**) To form a hydroxide, combine an alkali metal with water:

$$2\,K(s) + 2\,H_2O(l) \longrightarrow 2\,KOH(aq) + H_2(g)$$

(**b**) To form an oxide, combine an alkali metal with oxygen:

$$4\,Cs(s) + O_2(g) \longrightarrow 2\,Cs_2O(s)$$

(**c**) To form a hydride, combine an alkali metal with hydrogen:

$$2\,Na(s) + H_2(g) \longrightarrow 2\,NaH(s)$$

(**d**) To form a bromide salt, combine an alkali metal with liquid bromine:

$$2\,K(s) + Br_2(l) \longrightarrow 2\,KBr(s)$$

Check: Chemical reactions described in equations 7.9, 7.13, 7.17, and 7.19 in the text support the chemical reactions given in the solution.

EXERCISE 16   Comparing properties of two metals

How does magnesium compare with sodium in terms of the following properties: (**a**) atomic size; (**b**) number of outer-shell electrons; (**c**) ionization energy; (**d**) formula of bromide salt?

SOLUTION: *Analyze*: We are asked to compare the atomic size, number of valence-shell electrons, ionization energy, and formula of bromide salt for Mg and Na.

Plan: Na is an alkali metal and Mg is an alkaline-earth metal, thus they will have different physical and chemical properties. We can use the ideas developed in prior exercise to compare their properties.

Solve: (**a**) Mg is smaller; size decreases across a period. (**b**) Mg has two $3s$ electrons, whereas Na has one $3s$ electron. (**c**) The first ionization energy of Mg is greater than that of Na. (**d**) The formula of magnesium bromide is $MgBr_2$, and that of sodium bromide is NaBr. Note that magnesium exists as a 2+ ion, whereas sodium exists as a 1+ ion.

Comment: The alkaline-earth metals are slightly harder, more dense, and melt at higher temperatures than alkali metals. They are generally less reactive than their alkali metal neighbors.

GROUP TRENDS EXEMPLIFIED: SELECTED NON-METALS

Hydrogen is a unique element and doesn't belong to any family. It is discussed in this section for convenience.

- The majority of its compounds contain hydrogen in the 1+ oxidation state.
- When it combines with very active metals, such as Na and Ca, a salt containing the hydride (H^-) ion is formed.
- Small quantities can be prepared in the lab by the reaction between zinc and HCl to form $ZnCl_2(aq)$ and $H_2(g)$.

The **oxygen family** (Group 6A) consists of O (oxygen), S (sulfur), Se (selenium), Te (tellurium), and Po (polonium).

- O_2 is a colorless gas at room temperature; the others are solids.
- Oxygen exists in two allotropic forms, O_2 and O_3 (ozone). Allotropes are different chemical forms of the same element.
- Review combustion reactions involving oxygen (Section 3.2 of the text).
- Small amounts of oxygen are produced by heating $KClO_3(s)$ in the presence of a small amount of $MnO_2(s)$.

$$2\,KClO_3(s) \xrightarrow[\Delta]{MnO_2} 2\,KCl(s) + 3\,O_2(g)$$

The **halogen family** (Group 7A) consists of F (fluorine), Cl (chlorine), Br (bromine), I (iodine), and At (astatine).

- All form diatomic molecules.
- Remember their physical states: $F_2(g)$, $Cl_2(g)$, $Br_2(l)$, and $I_2(s)$.

- All react with metals to form ionic halide (X^-) salts.
- They react with hydrogen gas to form gaseous HX compounds.

The **noble-gas family** (Group 8A) consists of He (helium), Ne (neon), Ar (argon), Kr (krypton), Xe (xenon), and Rn (radon).

- All are monatomic gases at room temperature.
- They have completely filled s and p subshells.
- They have low reactivities. Kr, Xe, and Rn react directly with fluorine to form fluorides such as XeF_2, XeF_4, and XeF_6.
- Radon is radioactive.

EXERCISE 17   Identifying an unknown substance I

A flask contains a colorless gas at room temperature. The gas extinguishes a flame. When the gas dissolves in water it forms an acidic solution. Which of the following substances could the unknown gas be? O_2, NaCl, CO_2, I_2. Explain.

SOLUTION: *Analyze*: We are given that a colorless gas extinguishes a flame and forms an acidic solution in water. From a list of substances we are asked to identify the substance and explain our reasoning.

Plan: We will first characterize each of the given substances: oxygen gas, sodium chloride, carbon dioxide, and molecular iodine. Then we will compare their properties to those given in the question. We have encountered these substances in the text.

Solve: NaCl and I_2 are solids at room temperature and can be eliminated. $O_2(g)$ would not extinguish a flame as it is a reactant in a combustion reaction. Also, it does not react with water to form an acidic solution. $CO_2(g)$ dissolves in water to form $H_2CO_3(aq)$ and extinguishes a flame.

Comment: This type of question requires that you have some knowledge of common substances and their physical and chemical properties. Chemistry does require that you develop a collection of key facts that you can use to solve problems.

EXERCISE 18   Identifying an unknown substance II

Identify each of the following unknown elements from the information given: (a) This gas is produced when an active metal is added to a strong acid. The gas is colorless and reacts with active metals to form a salt in which it is an anion. (b) This gas is highly reactive and has a light blue color and pungent smell. It is an allotrope of a group 6A element. (c) It is a dark violet solid that readily sublimes to form vapors. When it reacts with hydrogen gas, it forms an acid. This element is added to alcohol to form an antiseptic. (d) It is a monatomic gas that is radioactive and chemically unreactive. (e) It is a pale yellow diatomic gas that reacts with almost all substances. It has a lower electron affinity than chlorine. (f) It is a solid that exhibits several allotropic forms; the most common and stable one is a yellow solid with molecular formula X_8. (f) This gas is produced in a laboratory by heating a mixture of potassium chlorate and manganese dioxide. (g) This element is a brown liquid that contains the element in a diatomic state and is highly corrosive. Upon contact with the skin it forms severe blisters and burns. (h) This gas was once thought to be unreactive; however it is now known to form several compounds with fluorine and oxygen.

SOLUTION: *Analyze*: We are given information about eight elements and are asked to identify each element.

Plan: This type of question requires that we have learned specific facts about a variety of elements that we have encountered in the text. If we do not recognize the element we will have to review appropriate sections in the text.

Solve: (a) $H_2(g)$. Active metals react with strong acids to form this gas. It also reacts with active metals to form a hydride ion, H^-. (b) $O_3(g)$. Ozone is an allotrope of oxygen. (c) $I_2(s)$. When iodine reacts with hydrogen gas it forms $HI(g)$. (d) Rn. Radon is a noble gas that is radioactive. (e) $F_2(g)$. Fluorine is one of the most reactive elements; it is diatomic in its naturally occurring state. It has a smaller electron affinity than chlorine because of its extremely small size. (f) $S(s)$. Sulfur occurs in several structural forms; its most common one is yellow S_8. (f) $O_2(g)$. (g) $Br_2(l)$. (h) $Xe(g)$. At one time it was believed that all noble gases were inert. However, Xe was the first noble gas shown to be chemically reactive.

Comment: This question examines how effectively you have either learned specific information about elements or your ability to find the information in the text to identify each element. As you continue your learning of chemistry it is important that you develop a collection of facts about key elements. It will help you answer other problems in future chapters.

EXERCISE 19   Writing combination reactions

Write the expected formula when the following elements combine to form compounds: (a) Al and O; (b) B and Cl; (c) Ca and F; (d) Na and I; (e) S and O; and (f) H and Ca.

SOLUTION: *Analyze*: You are given six pairs of atoms and asked to write the expected chemical formula when each pair combines to form a substance.

Plan: We will determine if the elements in each pair are metals or nonmetals. If one is a metal and the other is a nonmetal then we can assume that they form an ionic substance. We can use our knowledge of the most common charge of each metal and nonmetal as an ion to predict the chemical formula. If both are nonmetals we have to consider several molecular (covalent) combinations as possibilities.

Solve: (a) Al forms the 3+ state in salts. Oxygen forms the oxide ion, O^{2-}. Al_2O_3. (b) Boron is a member of the family containing aluminum; thus, you should expect it to behave in a similar manner to aluminum. BCl_3. (c) Calcium forms Ca^{2+} and fluorine forms F^-. CaF_2. (d) Na forms Na^+ and iodine in the presence of metals forms I^-. NaI. (e) Sulfur and oxygen, both nonmetals, form nonmetal oxides: For example, $SO_2(g)$ and $SO_3(g)$. (f) Ca is an active metal. In the presence of an active metal hydrogen forms the hydride ion, H^-: CaH_2.

Comment: This question examines how effectively we have either learned specific information about elements or our ability to find the information in the text to identify each element. We can verify the answers by referring to appropriate chapters or the index in the text.

SELF-TEST QUESTIONS

Key Terms

Having reviewed key terms in Chapter 7, match key terms with phrases and identify statements as true or false. If a statement is false, indicate why it is incorrect.

Match each phrase with the best term:

7.1   A term used for elements in group 7A(17).

7.2   Its value for an atom in a family increases with increasing atomic mass.

7.3   Ozone is a form of this class of elements.

7.4   A term used for elements in group 1A(1).

7.5   The minimum energy change associated with adding an electron to an atom.

7.6   A family of nonmetals that has eight valence-shell electrons.

7.7   An endothermic energy associated with removing an electron from an atom

7.8   These elements are harder and more dense than their nearest group 1A(1) elements.

Terms:

(a) alkaline-earth metals

(b) alkali metals

(c) allotrope

(d) bonding atomic radius

(e) electron affinity

(f) halogens

(g) ionization energy

(h) noble gases

True-False Statements:

7.9 The formula of *ozone* is O_4.

7.10 Potassium is an element of group 1A and thus it is an *alkali metal*.

7.11 A *valence orbital* of sulfur is the $2p$.

7.12 In metallic hydrides, the *hydride* ion carries a negative one charge.

7.13 Metallic character is more predominant on the left bottom section of the periodic table than the right top section.

7.14 In iodine the *effective nuclear charge* experienced by electrons increases with increasing value of their n quantum number.

7.15 A $2s$ electron in Mg experiences a smaller effective nuclear charge than a $1s$ electron because of the *screening effect*.

7.16 An *isoelectronic* series of ions possess the same number of electrons.

Problems and Short-Answer Questions

7.17 Which element in the following periodic table has the largest atomic radius? Justify your response.

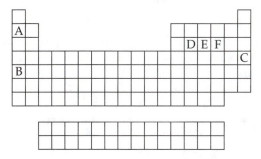

7.18 Which element in the previous periodic table has the largest first ionization energy? Justify your response.

7.19 The bonding atomic radii for the halogen family are as follows:

Element	Atomic radius (Å)	Atomic number
F	0.64	9
Cl	0.99	17
Br	1.14	35
I	1.33	53
At	1.4	85

Explain the observed trend in bonding atomic radii.

7.20 The first ionization energy for a particular element is 578 kJ/mol, the second ionization energy is 1817 kJ/mol, the third ionization energy is 2745 kJ/mol, and the fourth

ionization energy is 11,577 kJ/mol. Is this sufficient information to conclude that the M^{3+} ion has a rare-gas electron configuration? If there is insufficient information, what additional information is needed and why?

7.21 Does the addition of an electron to $O^-(g)$ to form $O^{2-}(g)$ require or evolve energy (that is, is it an endothermic or exothermic process)?

7.22 Based on their positions in the periodic table, select the atom of each pair that has the larger value of the indicated atomic property:

(a) ionization energy, Na or Mg

(b) ionization energy, Mg or Cl

(c) electron affinity, Cl or Br

(d) atomic radius, K or Cs

(e) atomic radius, Se or Br

7.23 Fe_2O_3 is an insoluble substance in water. Is this sufficient information to conclude that it is *not* a basic oxide? If there is insufficient information, what additional information is needed and why?

7.24 Identify each of the following compounds as a halide, oxide, peroxide, superoxide, carbonate, or hydroxide:

(a) $Ca(OH)_2$

(b) CsCl

(c) K_2O

(d) Rb_2O_2

(e) CsO_2

(f) K_2CO_3

7.25 List three properties that distinguish nonmetals from metals.

7.26 In what way does Na(s) metal differ from transition metals, such as iron?

7.27 Classify each as a metal, metalloid, or nonmetal:

(a) Si

(b) Ba

(c) Au

(d) Xe

(e) I

7.28 Identify the element in each series with the greatest value or property indicated.

(a) Metallic character: Mg, Ca, Sr, Ba

(b) Nonmetallic character: K, Fe, C, Sn

(c) Atomic radius: B, C, O, F

(d) Ionization energy: Mg, Ca, Sr, Ba

(e) Electron affinity: Be, O, S, Sn

7.29 Write chemical reactions to describe the reaction between:

(a) Na(s) and $H_2(g)$

(b) Ca(s) and HCl(g)

(c) $Al_2O_3(s)$ and HCl(aq)

(d) Al(s) and NaOH(aq)

(e) $H_2(g)$ and $Cl_2(g)$

(f) Ba(s) and $Br_2(l)$

(g) Li(s) and $H_2O(l)$

7.30 What is the charge of each of the following ions:

(a) hydride

(b) calcium

(c) chloride

(d) oxide

(e) potassium

7.31 A student had three test tubes containing the gases O_2, CO, and Cl_2. The gas in test tube *A* extinguishes a flame, does not react with aluminum, and when dissolved in water forms neither an acidic nor basic solution. A burning splint is inserted into test tube *B*. The splint bursts into flame. The gas is colorless. The gas in test tube *C* has a yellow-green color and reacts with hydrogen gas to form an acid. Identify the gas in each test tube.

7.32 Which of the following are solids at room temperature, and which are gases?

(a) CO_2 (d) F_2
(b) BaO (e) NO
(c) CuO

7.33 Which substances are ionic and which are covalent?

(a) Br_2 (d) SO_2
(b) KO_2 (e) $Ca(ClO_4)_2$
(c) AsH_3

Integrative Questions

7.34 An element has the following successive ionization energies(*kJ*): 737, 1450, 7731, 10,545, ... The element directly above it in the periodic table has the following successive ionization energies(*kJ*): 899, 1757, 14,845, and 21,000 (complete).

(a) In which family in the periodic table does the first element belong? Explain.
(b) What is the second element? What is the first element? Explain.
(c) Why is the first ionization energy of the first element less than that of the second element?
(d) Will an ion with one electron removed be larger or smaller than the parent atom? Explain.

7.35 A certain diatomic element is a yellowish-green gas and is highly reactive. It has two isotopes with mass number 35 and 37.

(a) What are the naturally occurring diatomic elements and which one is identified by the given information?
(b) Write the isotopic symbols.
(c) How many neutrons does each isotope possess?
(d) What is the electron configuration of each isotope?
(e) What are the expected chemical formulas of the compounds formed between this element and sodium and barium? What type of compounds are they?

7.36 Xenon hexafluoride, a solid, can be prepared by combining xenon gas with fluorine gas. It cannot be prepared in a glass or quartz container because it readily reacts with silicon dioxide to form $XeOF_4(1)$ and silicon tetrafluoride solid. It also reacts readily with water to form xenon trioxide, a solid, and hydrogen fluoride gas.

(a) What is the oxidation number of xenon in $XeOF_4$?
(b) Write the balanced chemical equation describing the reaction between xenon hexafluoride and water.

(c) What are the grams of hydrogen fluoride gas produced if in a 1.00 L container at 25.0°C 1.00 g of xenon hexafluoride reacts with sufficient water?

7.37 An element is monatomic gas, is not radioactive, reacts with fluorine gas and has an ionization energy between 1000–1500 kJ/mol. Can you determine if it is a member of the 1A, 5A, or 8A family? Explain.

Multiple-Choice Questions

7.38 Which elements have the lowest ionization energies?

(a) those on the right side of the periodic table;
(b) nonmetals;
(c) those on the left side of the periodic table;
(d) halogens;
(e) (a) and (b).

7.39 Which element has the smallest atomic radius?

(a) Mg   (d) Ba
(b) Ca   (e) K
(c) Sr

7.40 Which element has the largest atomic radius?

(a) Ge   (d) Br
(b) As   (e) Kr
(c) Se

7.41 Why doesn't sodium ordinarily occur in the 2+ ion state?

(a) low electron affinity;
(b) high second ionization energy;
(c) large atomic radius;
(d) small density.

7.42 Copper has one valence electron in its $4s$ orbital and the same occurs for potassium. In spite of this similarity, we find copper is less reactive than potassium. What is the most likely cause for this difference in reactivity?

(a) The presence and effect of $3d$ electrons in copper;
(b) the higher ionization energy of copper compared to that of potassium;
(c) the smaller size of the potassium atom compared to that of copper;
(d) (a) and (b);
(e) (b) and (c).

7.43 Based on its position in the periodic table, which atom has the largest ionization energy?

(a) Li   (d) N
(b) B   (e) Ne
(c) C

7.44 Which of the following represents the chemical reaction for the heating of Bi_2O_3 in a hydrogen environment?

(a) $Bi_2O_3 \longrightarrow 2\ Bi + 3/2\ O_2$
(b) $Bi_2O_3 + 3\ H_2 \longrightarrow 2\ Bi + 3\ H_2O$
(c) $2\ Bi_2O_3 + 6\ H_2 \longrightarrow 4\ Bi + 3\ OH^-$
(d) $H_2O + Bi_2O_3 + H_2 \longrightarrow 2\ Bi(OH)_2$

7.45 What are the expected ions when $K_2O(s)$ is dissolved in water?

(a) K^+, H^+ (c) K^+, OH^-
(b) K^+, O^{2-} (d)   No reaction occurs

7.46 Which of the following elements exhibit allotropic forms?

(a) hydrogen   (c)   sodium
(b) oxygen   (d)   xenon

7.47 SeO_2 is

(a) an ionic substance
(b) a basic oxide
(c) an ionic oxide
(d) a covalent oxide

7.48 Which of the following is true for Group 4A?

(a) consists only of nonmetals
(b) contains no metalloids
(c) contains an active metal
(d) metallic character increases from top to bottom

7.49 Which of the following descriptions is *not* true?

(a) Rb, an active metal that forms a +1 cation in aqueous solution
(b) I_2, a grayish-black solid that readily forms a purple vapor
(c) Pb, a bluish-gray, soft and dense metal
(d) H, shows only nonmetallic properties

7.50 Which element has its most common ion state as 2+ and its physical properties are partially determined by d electrons?

(a) Ca   (c)   Cu
(b) K   (d)   Al

7.51 Which substance is not expected to be a gas at room temperature and pressure?

(a) NO   (c) $ZnCl_2$
(b) PH_3 (d) F_2

7.52 Which substance should form an acidic solution in water?

(a) Na_2O (c)   CO
(b) SO_2 (d)   BaO

7.53 Which has the lowest *second* ionization energy?

(a) Ca   (c)   Ar
(b) K   (d)   P

7.54 Typically an alkaline-earth metal has a density significantly greater than that of an alkali metal in the same period. A primary reason is that for alkaline-earth metals

(a) The second ionization energy is smaller.
(b) The atomic number is larger.
(c) The atomic size is smaller.
(d) The electron affinity is more positive.

7.55 Which of the following is the largest in size?

(a) Cl^- (c) I^+
(b) Cl   (d) I^-

7.56 A property of a metal that relates to its ability to form a wire is its

(a) luster   (c)   ionic character
(b) ductility   (d)   conductance

7.57 When an alkali metal reacts with water, a metal hydroxide and a gas form. Which gas forms?

(a) carbon dioxide   (c)   ozone
(b) oxygen   (d)   hydrogen

7.58 What is the valence-electron configuration for Se^{2-}?

(a) $2s^2 2p^4$ (d) $4s^2 4p^6$
(b) $2s^2 2p^6$ (e) $3s^2 3p^2$
(c) $4s^2 4p^4$

7.59 What is the valence-electron configuration for Ni^{2+}?

(a) $3d^6 4s^2$ (d) $3d^5 4s^2$
(b) $3d^6 4s^1$ (e) $3d^4 4s^2$
(c) $3d^8$

7.60 What is the correct order of increasing radii for the isoelectronic series $Rb^+, Sr^{2+}, Se^{2-}, Br^-$?

(a) $Rb^+ < Sr^{2+} < Se^{2-} < Br^-$
(b) $Br^- < Se^{2+} < Sr^{2+} < Rb^+$
(c) $Se^{2-} < Br^- < Rb^+ < Sr^{2+}$
(d) $Sr^{2+} < Rb^+ < Br^- < Se^{2-}$
(e) none of the above

7.61 Which of the following has the largest ionic radius?

(a) Be^{2+} (d) Sr^{2+}
(b) Mg^{2+} (e) Ba^{2+}
(c) Ca^{2+}

SELF-TEST SOLUTIONS

7.1 (f). **7.2** (d). **7.3** (c). **7.4** (b). **7.5** (e). **7.6** (h). **7.7** (g). **7.8** (a). **7.9** False. It is O_3. **7.10** True. **7.11** False. The valence orbitals of sulfur lie in the third shell; therefore a $3p$ orbital is a valence orbital. **7.12** True. **7.13** True. **7.14** False. As the value of n increases there are more inner electrons that shield the outer electrons, and the effective nuclear charge is thereby reduced. **7.15** True. **7.16** True.

7.17 Element B has the largest atomic radius. Atomic size increases down a family and decreases across a period.

7.18 Element C has the greatest ionization energy. Element C is a rare gas, Kr. The rare gases have significantly larger first ionization energies than other elements. Although Kr is in a different period than elements D, E, and F and it has valence electrons further from the nucleus, its first ionization energy is sufficiently high that it is still higher than those of D, E, and F. This is shown in Figure 7.12 in the text.

7.19 The larger the value of the n quantum number for the highest-energy orbital(s) of an atom, the larger the size of the atom. Since the value of the n quantum number of the outermost orbital increases with increasing atomic number in a family of elements, the atomic radii of the halogen family increase in the order F < Cl < Br < I < At.

7.20 Note that the difference between the first and third ionization energies is 2167 kJ/mol but the difference between the third and fourth ionization energies is 8832 kJ/mol. Removing the first three electrons is relatively easy compared to the fourth electron, which is very difficult to remove from an energy viewpoint. Obviously M^{3+} is an extremely stable ion. This strongly supports the idea that M^{3+} ion has a rare-gas electron configuration and removing an electron from it is very difficult because it a very stable energy state. Note: The element is aluminum.

7.21 According to Coulomb's law, the addition of a negatively charged electron to a negatively charged $O^-(g)$ ion requires overcoming the repulsion forces between the electron and O^- ion. This requires energy. The process

$$O^-(g) + e^- \longrightarrow O^{2-}(g)$$

is thus endothermic.

(a) Mg. The outer electron configuration of Mg is $3s^2$, whereas the outer electron configuration of Na is $3s^1$. Because it is more difficult to remove an electron from $3s^2$ than $3s^1$, Mg has the higher ionization energy.

(b) Cl. Ionization energy generally increases across a period.

(c) Cl. Within a family, electron affinity generally decreases with increasing atomic number.

(d) Cs. Atomic radii within a family increase with increasing atomic number.

(e) Se. Atomic radii decrease across a period of representative elements.

7.23 There is insufficient information. A basic oxide is a metal oxide that either forms a basic solution (forms hydroxide ion) or reacts with acid. Fe_2O_3 is insoluble in water and thus should not increase the basicity of water. However, if it reacts with acid, it then is a basic oxide; thus, additional information is needed. It indeed reacts with acid as follows: $Fe_2O_3(s) + 6 H^+(aq) \rightarrow 2 Fe^{3+}(aq) + 3 H_2O(l)$.

7.24 (a) hydroxide;   (d) peroxide;
(b) halide;   (e) superoxide;
(c) oxide;   (f) carbonate.

7.25 Nonmetals tend to be less lustrous, are poorer conductors of heat and electricity, and form covalent instead of ionic substances.

7.26 Na is a soft, highly reactive metal that exists only in the 1+ ion state. Transition metals tend to be more dense, less reactive, and exhibit more than one ion state.

7.27 (a) metalloid: (d) nonmetal;
(b) metal;   (e) nonmetal.
(c) (transition) metal;

7.28 (a) Mg—metallic character tends to decrease down a family;

(b) C—nonmetallic character tends to increase left to right in a period and near the top of a family;

(c) B—atomic radius tends to decrease from left to right in a group of representative elements;

(d) Mg—ionization energy tends to decrease down a family;

(e) O—elements with greatest electronegativity values are nonmetals at the top right section of the periodic table.

7.29 (a) $2 Na(s) + H_2(g) \rightarrow 2 NaH(s)$;
(b) $Ca(s) + 2 HCl(g) \rightarrow CaCl_2(s) + H_2(g)$;
(c) $Al_2O_3(s) + 6 HCl(aq) \rightarrow 2 AlCl_3(s) + 3 H_2O(l)$;
(d) $2 Al(s) + 6 NaOH(aq) \rightarrow 2 Na_3AlO_3(aq) + 3 H_2O$;
(e) $H_2(g) + Cl_2(g) \rightarrow 2 HCl(g)$;
(f) $Ba(s) + Br_2(l) \rightarrow BaBr_2(s)$;
(g) $2 Li(s) + 2 H_2O(l) \rightarrow 2 LiOH(aq) + H_2(g)$.

7.30 (a) H^-;   (d) O^{2-};
(b) Ca^{2+};   (e) K^+.
(c) Cl^-;

7.31 CO is in test tube A. It is a relatively nonreactive nonmetal oxide. When placed in water it does not act as an acidic oxide. O_2 is in test tube B. It supports combustion and is colorless. Cl_2 is in test tube C. It has a yellow-green color and forms HCl in the presence of hydrogen gas.

7.32 In general, metallic compounds are solids whereas nonmetallic compounds occur commonly in the gaseous or liquid form.

(a) gas;   (d) gas;
(b) solid;   (e) gas.
(c) solid;

7.33 (a) covalent;   (d) covalent;
(b) ionic;   (e) ionic
(c) covalent;

7.34 (a) Note that there is an extreme increase in ionization energy when the third electron is removed, 7731 kJ versus 1450 kJ and 737 kJ. This suggests that forming a 3+ ion is extremely difficult whereas forming a 1+ or 2+ ion is energetically feasible. This dramatic difference should occur in the alkaline-earth metal family, group 2A. These elements form 2+ ions but not 3+ ions.

(b) The data for the second element includes all possible ionization energies, and since there are four values there are four electrons that can be removed; this is the total number of electrons possessed by the element. The element with four electrons is Be. Thus, the element directly below it in the periodic table is Mg.

(c) The outer electrons in Mg are further from the nucleus than those in Be. Thus, the outer electrons in Mg experience a lesser effective nuclear charge and are less strongly held; this results in a smaller first ionization energy.

(d) The removal of one electron to produce a positive ion reduces electron–electron repulsions within an atom, thereby increasing the effective nuclear charge experienced by the electrons. This increased attraction between the nucleus and electrons causes a contraction in the average distance of the electrons from the nucleus and thus a smaller size than the parent atom.

7.35 (a) The naturally occurring diatomic elements are: H_2, N_2, O_2, F_2, Cl_2, Br_2, and I_2. Chlorine is a yellowish-green gas that is highly reactive. The molar mass of chlorine is 35.45 g/mol and this number is close to the mass numbers of the two isotopes, further supporting the conclusion that chlorine is the element identified.

(b) $^{35}_{17}$Cl and $^{37}_{17}$Cl.

(c) Chlorine-35 possess 18 neutrons and chlorine-37 possess 20 neutrons.

(d) The electron configurations of all isotopes are the same, only the number of neutrons differ: $1s^2 2s^2 2p^6 3s^2 3p^5$.

(e) When chlorine combines with active metals it is in a 1^- ion state: KCl and BaCl$_2$. These are ionic compounds because they contain metal ions (K$^+$ and Ba^{2+}) and the nonmetal Cl$^-$ ion.

7.36 (a) $+6$. Each fluorine atom is assigned a -1 oxidation number and the oxygen atom is assigned a -2; thus, $0 = +6 - 2 + 4(-1)$.

(b) $XeF_6(s) + 3 H_2O(l) \rightarrow XeO_3(s) + 6 HF(g)$

(c) Calculate the number of grams of HF gas produced based on the stoichiometry in (b).

$$\text{mol HF} = (\text{mol XeF}_6)\left(\frac{6 \text{ mol HF}}{1 \text{ mol XeF}_6}\right)$$

$$= (1.00 \text{ g})\left(\frac{1 \text{ mol XeF}_6}{245.29 \text{ g}}\right)\left(\frac{6 \text{ mol HF}}{1 \text{ mol XeF}_6}\right)$$

$$= 0.0245 \text{ mol HF}$$

$$\text{g HF} = (0.0245 \text{ mol HF})\left(\frac{20.01 \text{ g HF}}{1 \text{ mol HF}}\right) = 0.490 \text{ g HF}$$

7.37 The element cannot be a member of the 1A family because it is a gas and members of this family are solids (metals). Group 5A elements are N$_2$(g) and solids (P, As, Sb, and Bi). The unknown element is monatomic and cannot be nitrogen and not P, As, Sb, or Bi because it is a gas. Members of group 8A are monatomic and have high ionization energies, which is confirmed by the magnitude of ionization energy. If we look at the data in Figure 7.7 in the text our conclusion is confirmed; the element is either Xe(g) or Kr(g). Rn(g) is radioactive and it cannot be the element.

7.38 (c) Ionization energies increase across periods from left to right.

7.39 (a) **7.40** (a)

7.41 (b) Formation of Na^{2+} from Na$^+$ requires removal of an electron from a $1+$ ion, which according to Coulomb's law, requires a large expenditure of energy.

7.42 (d) In Cu, the $3d$ electrons do not effectively shield the outer $4s$ electrons, and the effective nuclear charge experienced by them is thereby increased. This effect leads to a higher ionization energy for copper, which reduces the reactivity of copper compared to potassium.

7.43 (e). **7.44** (b). **7.45** (c). **7.46** (b). **7.47** (d). **7.48** (d). **7.49** (d). **7.50** (c). **7.51** (c). **7.52** (b). CO is a nonmetallic oxide but it is not acidic.

7.53 (a) Calcium is the only atom listed that forms a cation with a positive two charge; thus, it has a reasonable second ionization energy compared to the others which do not form $2+$ ions. Their second ionization energies are too large.

7.54 (c) Alkaline-earth metals have atomic sizes appreciably smaller than those of the alkali metals in the same period. This permits the atoms to pack closer and the metals to have a larger density.

7.55 (d) Iodine is a larger atom than Cl. The addition of an electron causes a further increase in size.

7.56 (b)

7.57 (d) The general reaction is $2 M(s) + 2 H_2O(l) \rightarrow 2 MOH(aq) + H_2(g)$.

7.58 (d)

7.59 (c) $4s$ electrons are lost before $3d$ electrons are lost when a transition metal loses electrons.

7.60 (d) **7.61** (e) Within a family of elements and of the same charge, size increases down the family.

Chapter

8

Basic Concepts of Chemical Bonding

OVERVIEW OF THE CHAPTER

8.1 LEWIS SYMBOLS: OCTET RULE

Review: Concept of outermost electron shell (6.7, 6.8, 6.9); electron configurations (6.8).

Learning Goals: You should be able to:

1. Determine the number of valence electrons for any atom, and write its Lewis symbol.
2. Recognize when the octet rule applies to the arrangement of electrons in the valence shell for an atom.

8.2 IONIC BONDING: LATTICE ENERGY

Review: Atomic radii (7.3); electron configurations (6.8, 6.9); ions (2.7); periodic table (2.5, 6.9).

Learning Goals: You should be able to:

1. Describe the origin of the energy terms that lead to stabilization of ionic lattices.
2. Predict on the basis of the periodic table the probable formulas of ionic substances formed between common metals and nonmetals.

8.3 THE LEWIS MODEL FOR COVALENT BONDING

Review: Electron configurations (6.8); periodic table (2.3, 6.9).

Learning Goals: You should be able to:

1. Describe the basis of the Lewis theory, and predict the valence of common nonmetallic elements from their positions in the periodic table.
2. Be able to describe a covalent bond in terms of sharing of electron density between bonded atoms.
3. Describe the formation of a covalent compound using Lewis symbols.
4. Be able to look at a Lewis structure and determine if it properly fits the Lewis model.
5. Describe a single, double, and triple covalent bond.

8.4 BOND POLARITY, ELECTRO-NEGATIVITY, AND NOMENCLATURE

Review: Oxidation numbers (2.8).

Learning Goals: You should be able to:

1. Explain the significance of electronegativity and in a general way relate the electronegativity of an element to its position in the periodic table.

2. Predict the relative polarities of bonds using either the periodic table or electronegativity values.
3. Name a binary compound given its chemical formula or write the chemical formula given its chemical name.

Review: Electron configurations (6.8); periodic table (2.5, 6.9).

Learning Goals: You should be able to:

1. Write the Lewis structures for molecules and ions containing covalent bonds, using the periodic table.
2. Write resonance forms for molecules or polyatomic ions that are not adequately described by a single Lewis structure.

Learning Goals: You should be able to write the Lewis structures for molecules and ions containing covalent bonds that have an odd number of electrons, a deficiency of electrons, or an expanded octet.

Review: Energy changes in chemical reactions (5.2, 5.4); enthalpy (5.3, 5.4); Hess's law (5.6).

Learning Goals: You should be able to relate bond enthalpies to bond strengths and use bond enthalpies to estimate ΔH for reactions.

TOPIC SUMMARIES AND EXERCISES

Valence electrons are the electrons involved in forming chemical bonds. If valence electrons are transferred between atoms, an ionic bond forms; if they are shared between atoms, a covalent bond forms.

The number of valence electrons for an atom can be shown by its **Lewis symbol**, or electron-dot formula.

*   The Lewis symbol for an element consists of the chemical symbol for the element plus a dot for each valence electron. For example, the Lewis symbol for H ($1s^1$) is H· and for F ($2s^2 2p^5$) it is $:\ddot{\text{F}}\cdot$
*   Note that pairs of electrons are kept together and that each quadrant about the element's symbol can hold up to two electrons. Also, the placement of electrons can be in any of the quadrants. For example, equally valid Lewis structures are H· or $\dot{\text{H}}$ or ·H

In chemical reactions we find that atoms of representative elements atoms gain, lose or share valence electrons. The atoms in their new bonding state are surrounded by a noble gas electron configuration equivalent to that possessed by the nearest noble gas in the periodic table.

*   The **octet rule** is a statement of the observation that atoms of representative elements in their compounds often possess eight electrons in their valence shell.
*   Many representative elements in their compounds do not obey the octet rule. Some elements show two electrons or six electrons in their valence shell. Examples of these are hydrogen (two valence electrons) and boron (six valence electrons). Others can expand their valence shell to 10 or 12 electrons. The latter commonly occurs with the heavier, non-metallic, representative elements.

8.5, 8.6 DRAWING LEWIS STRUCTURES

8.7 EXCEPTIONS TO THE OCTET RULE

8.8 STRENGTHS OF COVALENT BONDS

LEWIS SYMBOLS: OCTET RULE

EXERCISE 1   Characterizing ionic, covalent, and metallic bonds

Three types of chemical bonds are summarized in Section 8.1: Ionic bond, covalent bond, and metallic bond. Characterize for each chemical bond what types of elements typically combine to form the bond.

SOLUTION: *Analyze*: We are asked to characterize three types of chemical bonds discussed in the text.

Plan: Read Section 8.1 and note carefully the characteristics of ionic, covalent, and metallic bonds.

Solve: An *ionic bond* occurs when electrostatic forces exist between ions of opposite charges. Ions occur when valence electrons are transferred from one atom to another. Metals gain electrons, whereas nonmetals prefer to lose electrons if electron transfer occurs. A *covalent bond* occurs when valence electrons are shared between two atoms. This commonly occurs when the two atoms are nonmetals. A *metallic bond* occurs when a metal atom bonds to several neighboring metal atoms. Some of the valence electrons are relatively free to move throughout the three-dimensional structure of the metallic substance.

Check: A review of the material in Section 8.1 shows the main characteristics have been identified.

EXERCISE 2   Writing Lewis symbols for elements

Write the Lewis symbols for the third-row representative elements (Na, Mg, Al, Si, P, S, Cl, and Ar).

SOLUTION: *Analyze*: We are asked to write Lewis symbols for non-transition elements in the third row of the periodic table.

Plan: In order to write the Lewis symbol for any element, we must first determine how many valence electrons are associated with it. To do this, we write the shorthand notation description of the element's valence-electron configuration (see Section 6.9 of the text). For the third-row elements, the valence electrons are those beyond the neon electron configuration ($1s^2 2s^2 2p^6$):

Solve:

Group number	Valence-electron configuration	Lewis symbol
1A	$3s^1$	Na·
2A	$3s^2$	Mg:
3A	$3s^2 3p^1$	$\overset{\cdot\cdot}{\text{Al}}$·
4A	$3s^2 3p^2$	$\overset{\cdot\cdot}{\text{Si}}$·
5A	$3s^2 3p^3$	·$\overset{\cdot\cdot}{\text{P}}$·
6A	$3s^2 3p^4$	·$\overset{\cdot\cdot}{\text{S}}$:
7A	$3s^2 3p^5$	:$\overset{\cdot\cdot}{\text{Cl}}$·
8A	$3s^2 3p^6$	:$\overset{\cdot\cdot}{\underset{\cdot\cdot}{\text{Ar}}}$:

Check: If we look at the group number of each element we see that it corresponds to the number of valence electrons shown for each element. Also, no more than two electrons are paired in any location about an element.

IONIC BONDING: LATTICE ENERGY

Many inorganic substances, such as table salts (NaCl), are composed of a three-dimensional array of ions. The forces holding the ions together are largely electrostatic (coulombic attractions between positive and negative ions).

- A measure of the stability of a solid ionic substance is its **lattice energy.** This quantity is the energy required for one mole of a solid ionic substance to be separated completely into ions far removed from one another:

$$MgCl_2(s) \longrightarrow Mg^{2+}(g) + 2\,Cl^-(g) \qquad \Delta H_{lattice} = 2326 \text{ kJ/mol}$$

- Lattice energy is approximated by the product of ion charges $(Q_1 Q_2)$ and the reciprocal of the distance between two adjacent ions of opposite charge $(1/d)$:

$$E_{lattice} \propto \frac{Q_1 Q_2}{d}$$

- Note that *large* ion charges and *small* distances between oppositely charged ions yield large crystal lattice energies and the most stable ionic solids.

When atoms form ions in a crystal, they tend to achieve an octet by forming a valence-electron configuration equal to that of the *nearest* noble gas in the periodic table.

- Atoms that lose electrons form positive ions (**cations**) and are found on the left side and in the middle of the periodic table: These are the metallic elements.
- Atoms that gain electrons form negative ions (**anions**) and are found on the right side: These are the nonmetallic elements.
- Transition elements usually do not obey the octet rule when they form ions. The outer electron configurations of these elements are either $(n-1)d^x ns^2$ or $(n-1)d^x ns^1$ ($n = 4, 5, 6$ and $x = 1 - 10$). When they form ions, the ns electrons are lost *first*, and then the necessary number of $(n-1)d$ electrons are lost to form an ion of a particular charge. For example,

$$Fe(3d^6 4s^2) \longrightarrow Fe^{3+}(3d^5) + 3\,e^-$$
$$Cu(3d^{10} 4s^1) \longrightarrow Cu^+(3d^{10}) + e^-$$

EXERCISE 3 Explaining why a transition element may have more than one ion state

Explain why titanium has two stable ions, 2+ and 4+, rather than just one.

SOLUTION: *Analyze:* We are asked to explain why titanium has two stable ions.

Plan: We should first determine in what family or region of the periodic table titanium occurs. If it is a member of the transition elements it is likely to show more than one ion state because of the presence of ns and $(n-1)d$ valence electrons. We need to analyze the valence electron configuration for the possible loss of electrons in different ways.

Solve: By referring to a periodic table we see that titanium is a transition element, atomic number 22. The complete electron configuration for titanium is: $1s^2 2s^2 2p^6 3s^2 3p^6 4s^2 3d^2$. The valence electrons are the outer s and d electrons: $4s^2 3d^2$. The Ti^{2+} ion can form by the loss of the two $4s$ electrons. Remember that the outer s electrons in a transition element are lost before the outer d electrons. The Ti^{4+} ion can form by the loss of all four valence electrons.

Comment: Contrast this situation with the ion state of calcium, Ca. Calcium has the valence electron configuration: $4s^2$. Therefore it can lose two valence electrons to form Ca^{2+} but after it loses two electrons the ion has the electron configuration of Ar, which is a stable electron configuration. It does not lose more electrons to achieve a greater ion state.

EXERCISE 4   Identifying trends in lattice energies

Arrange the oxides of magnesium, calcium, and barium in order of increasing lattice energies, without referring to a table of data.

SOLUTION: *Analyze*: We asked to arrange MgO, CaO, and BaO in order of increasing lattice energies.

Plan: The trend in magnitude of the lattice energies can be estimated by using the relationship $Q(+)Q(-)/d$, where $Q(+)$ is the charge of the cation, $Q(-)$ is the charge of the anion, and d is the closest distance between a cation and anion.

Solve: For the oxides of magnesium, calcium, and barium, the product of charges is a constant because the cations are all 2+, and the oxide ion is 2−. Thus, the factor that causes the lattice energies to be different is the distance term, d. The trend in sizes of the 2+ cations is Mg < Ca < Ba which means that the trend in Metal—O distances is Mg—O < Ca—O < Ba—O. The trend in lattice energies should be BaO < CaO < MgO because lattice energy increases in magnitude with decreasing distance between the ions.

Check: Table 8.2 in the text gives the lattice energies of MgO and CaO. This table confirms the order for these two and BaO would fit the trend based on the given arguments.

THE LEWIS MODEL FOR COVALENT BONDING

Most compounds do not show ionic properties; instead the combined atoms act as a unit and are not easily separated. These compounds have bonds between atoms, referred to as **covalent bonds**, which result from the *sharing* of one or more pairs of electrons.

The Lewis model for covalent bonds emphasizes the idea that atoms share valence electrons with other atoms in order to achieve a noble-gas electron configuration.

- **Lewis structures** show the arrangement of covalent bonds and unshared valence electron pairs about each atom in a compound.
- An example of the formation of a Lewis structure is shown for water:

$$2\,H\cdot + \cdot \ddot{O}\cdot \longrightarrow H - \ddot{O} - H$$

- A solid line between adjacent atoms represents a pair of bonding electrons in a covalent bond, and each dot represents an unshared valence electron.

In some molecules, the only way bonded atoms can attain an octet is by sharing more than one pair of electrons between them: Two and three electron pairs may be shared.

- In the example of H_2O, the H—O bond is referred to as a **single bond** because one pair of bonding electrons holds the two atoms together.
- **Multiple bonds** occur when there is an insufficient number of valence electrons for bonded atoms to achieve an octet using single bonds only. Consider the formation of ethylene (C_2H_4) from its atoms:

When atoms form bonds, valence electrons can unpair as shown in structure (1). Structure (2) shows one C—C bond and four C—H bonds with one unshared valence electron for each carbon atom. Each carbon atom shows only seven valence electrons about it, not eight. To achieve an octet about each carbon atom, the electrons on the carbon atoms pair [see structure (3)] to form an *additional* C—C bond as shown in structure (4). This multiple bond between the carbon atoms in C_2H_4 is termed a **double bond**.

- Acetylene is an example of a compound in which a **triple bond** binds two atoms together. The formation of multiple bonds occurs when atoms need to share extra electrons to form an octet.

$$H—C\equiv C—H$$

- The relative strengths of bonds in a series of related compounds is: single < double < triple.

EXERCISE 5   Identifying covalent bonds

Which of the following bonds should be covalent: C—Cl; Mg—Cl; and S—N? Explain.

SOLUTION: *Analyze*: We are given three pairs of bonded atoms and asked to determine which form covalent bonds.

Plan: A covalent bond typically forms when two nonmetallic elements are bonded.

Solve: C—Cl and S—N each contain a pair of nonmetallic elements; thus, the bonds should be covalent. Mg—Cl contains a metal and nonmetal, which typically leads to an ionic bond.

Check: C, S, and N are in the nonmetallic section of the periodic table whereas Mg is shown as a metal.

EXERCISE 6   Determining which covalent bond is shorter

The Lewis structure for carbon monoxide is :C $\equiv$ O: and carbon dioxide is Which structure should have the shorter C—O covalent bond distance?
:Ö $=$ C $=$ Ö:

SOLUTION: *Analyze*: We are given two Lewis structures and asked which should have the shorter C—O covalent bond.

Plan: Determine which C—O bond has the greater number of covalent bonds. The greater the number of covalent bonds between two atoms the stronger the covalent bond and the shorter the covalent bond distance.

Solve: Carbon monoxide contains a triple bond and carbon dioxide contains a double bond. Therefore, the C—O bond distance in carbon monoxide should be the shorter one.

Check: Table 8.5 in the text confirms this conclusion.

EXERCISE 7   Recognizing correctly written Lewis structures

Which Lewis structure does not properly represent a Lewis structure? Why?

$$\begin{array}{ccc} & H & H \\ & | & | \\ H—C=N: \qquad H—C—H \qquad H—C=\ddot{O} \\ & | & \\ & H & \end{array}$$

SOLUTION: *Analyze*: We are given three Lewis structures and asked to determine which one is not valid.

Plan: We need to carefully count bonding and nonbonding electrons about each atom to ensure that each carbon and nitrogen atom has eight and each hydrogen atom has two. A line represents two electrons. We need to count the total number of valence electrons in the molecule to ensure that it equals the sum of valence electrons for each atom in the substance. If there is an excess or shortage, then it cannot be a correct Lewis structure.

Solve: The first Lewis structure does not properly represent a Lewis structure. A hydrogen atom contributes one valence electron, a carbon atom contributes four valence electrons, and a nitrogen atom contributes a total of five valence electrons for a total number of 10 valence electrons to form the covalent bonds. (Remember that a line in a Lewis structure represents a single covalent bond and contains two electrons.) If you count the number of electrons in H—C≡N: you find eight valence electrons, which is two valence electrons deficient. Furthermore, carbon and nitrogen atoms do not show an octet of electrons. If you repeat this process for CH_4 you find eight total electrons and for H_2CO you find 12 total electrons, which in both cases equals the number of electrons used in the Lewis structures. Also, both carbon and oxygen atoms show an octet of electrons.

Check: Carbon is observed to have typically 4 bonds about it. A recheck confirms that the first structure does not meet this observation whereas the other structures do.

BOND POLARITY, ELECTRONEGATIVITY, AND NOMENCLATURE

Covalent bonds are classified as polar covalent or nonpolar covalent. A **nonpolar covalent bond** involves equal sharing of bonding electrons between the bonded atoms. A **polar covalent bond** involves unequal sharing of the bonding electrons between the bonded atoms. In a polar bond, one atom more strongly attracts the electrons in the bond than the other one. In an extreme case of unequal sharing, an ionic bond is formed.

Electronegativity is a quantity that can help you determine whether a covalent bond is polar or nonpolar.

- **Electronegativity** (*EN*) is the relative ability of an atom in a molecule to attract electrons to itself.
- If electrons are shared *equally* between two covalently bonded atoms, the atoms have equal values of electronegativity, and the covalent bond is said to be **nonpolar**.
- When electrons are *unequally* shared between two covalently bonded atoms, the atoms have differing electronegativity values, and the covalent bond is said to be **polar**.
- A large difference in electronegativities between two bonded atoms leads to the formation of an ionic bond.
- Electronegativity tends to increase from left to right in a period (row) of the periodic table. Thus, a metal has a lower electronegativity than a nonmetal in the same period. Electronegativity tends to decrease with increasing atomic number in a family or group. Transition metal ions will show exceptions to these general trends.
- Figure 8.6 in your text should be reviewed to observe general trends.

A polar covalent bond results from unequal sharing of bonding electrons between two atoms; thus, the bonded atoms become partially charged with respect to one another. You can represent the polarity of a bond by using the symbols δ^+ and δ^- where δ is the Greek symbol "delta." It represents a partial charge, a value less than one. The more electronegative atom in a covalent bond is assigned δ^- and the one with a lesser electronegativity is assigned δ^+.

- For example, the polarity of a H—Cl bond is described as follows:

$$\overset{\delta^+}{H}—\overset{\delta^-}{Cl}$$

Hydrogen has electronegativity value of 2.1 and chlorine has a value of 3.0.
- A **dipole** exists when two electrical charges of the same magnitude and of opposite charge are separated by a distance. The quantitative measure of the magnitude of a dipole is called a **dipole moment**, μ. If Q represents the magnitude of each of the equal and opposite electrical charges and r is the distance between the two electrical charges, then the dipole moment is

$$\mu = Qr$$

The unit of dipole moment is debyes (D), a unit that equals 3.34×19^{-30} coulomb-meter (C-m).

The naming of ionic and covalent compounds is discussed in Section 2.8 and the approach used depended on whether a substance is ionic or covalent. Common to both is that the *less* electronegative element is named first and the more electronegative element is modified to have an *-ide* ending. For ionic compounds the oxidation number of the element is given in Roman numerals and enclosed in parentheses following the name of the element; for example, $NiCl_2$ is nickel(II) chloride. If the compound is covalent, the prefixes in Table 2.6 of the text are used to indicate the number of each kind of element in the substance; for example, N_2O_4 is *di*nitrogen *tetra*oxide.

EXERCISE 8 Determining trends in electronegativities of elements

For each of the following pairs, which series shows the correct trend of electronegativity?

(a) $EN_B > EN_C > EN_N > EN_O > EN_F$ or $EN_F > EN_O > EN_N > EN_C > EN_B$
(b) $EN_C > EN_{Si} \approx EN_{Ge}$ or $EN_{Ge} \approx EN_{Si} > EN_C$
(c) $EN_{Cs} > EN_{Ge} > EN_F$ or $EN_F > EN_{Ge} > EN_{Cs}$

SOLUTION: *Analyze*: We are asked to determine which electronegativity series is correct in three different pairs.

Plan: We need to determine what relationships exist among the elements in each series. Are they in the same family? Are they in the same period? Are some metals whereas others are nonmetals? We then apply the principles of trends in electronegativity values to each series.

Solve: (a) $EN_F > EN_O > EN_N > EN_C > EN_B$. Electronegativity generally increases from left to right across a period. (b) $EN_C > EN_{Si} \approx EN_{Ge}$. The electronegativities of second-row elements are greater than those of elements in the third and fourth rows. (c) $EN_F > EN_{Ge} > EN_{Cs}$. The atoms of highest electronegativity are found in the upper right-hand section of the periodic table; the next highest in electronegativity are found in the middle and the lower right-hand sections; and the lowest are found in the left-hand section of the periodic table.

Check: By referring to Figure 8.6 in the text we can verify the validity of each conclusion.

EXERCISE 9   Determining the relative polarities of chemical bonds

(a) For each of the following bonds indicate its polarity: H—C; H—O; and C—N.

(b) Which bond is the most polar?

SOLUTION: *Analyze*: We are given three pairs of bonded atoms and asked to determine the direction of the polarity and to determine which bond is the most polar.

Plan: We will use the data in Figure 8.6 of the text to find the electronegativity of each atom. The more electronegative atom of a pair of bonded atoms is assigned the partial negative charge. The most polar bond will be the one with the two atoms having the largest difference in electronegativities.

Solve: (a)

	δ^+ δ^- H—C	δ^+ δ^- H—O	δ^+ δ^- C—N
Electronegativity values	↑ 2.1 ↑ 2.5	↑ 2.1 ↑ 3.5	↑ 2.5 ↑ 3.0
Difference	0.4	1.4	0.5

(b) Calculate the difference in electronegativity values for each pair of atoms. The pair with the largest difference will be the most polar: H—O.

Check: Note that the most electronegative atom is indeed carrying the δ^- charge. A recheck of the data in Figure 8.9 confirms the assignment of electronegativity values.

EXERCISE 10   Writing chemical formulas

Write chemical formulas for the following compounds: (a) nitrogen(III) fluoride (b) arsenic(III) oxide (c) manganese(IV) oxide.

SOLUTION: *Analyze*: We are given three chemical names and asked to write their chemical formulas.

Plan: The least electronegative atom is written first in the chemical formula, and it is normally the one listed first in the chemical name. The oxidation state of the first element is listed and by determining the oxidation state of the second element we can then write the chemical formula. Remember that the sum of oxidation numbers must equal the charge of the substance.

Solve: (a) NF_3. Nitrogen has an oxidation state of +3 and because fluorine has an oxidation state of -1 there must be three fluorine atoms: $+3 + 3(-1) = 0$. (b) As_2O_3. Note that oxygen has an oxidation number of -2; thus three oxygen atoms are required ($3 \times -2 = -6$) to balance the total oxidation number of arsenic atoms ($2 \times +3 = +6$). (c) MnO_2. Note that two oxygen atoms are required to balance the total oxidation number of manganese with an oxidation number of +4. By now you may have noted the following for binary compounds containing atom A with an oxidation number of $+y$ and atom B with oxidation number of $-x$: The chemical formula of the compound can be generated by using the absolute value of the oxidation number as the subscript of the other atom, with integers reduced to their smallest ratio of whole numbers:

$$\mathbf{A}^{+\textcircled{y}}\!\!\mathbf{B}^{-\textcircled{x}} \text{ has the chemical formula } A_yB_x$$

$Mn^{+4}O^{-2}$ becomes Mn_2O_4; we must reduce the subscripts to the simplest ratio of whole numbers $-Mn_{2/2}O_{4/2}$ or MnO_2.

Comment: There is no table that permits you to confirm your chemical formula. We should take the chemical formulas in the solution and independently write the chemical names without referring to the question. If the chemical names agree with the chemical names in the question, this helps confirm that you did the problem correctly.

EXERCISE 11   Naming chemical substances

Name the following substances: (a) SF_6; (b) NO; (c) N_2O; (d) P_2O_5.

SOLUTION: *Analyze:* We are given chemical formulas of substances and asked to name them.

Plan: Review the rules of nomenclature. The least electronegative atom is normally written first. If the elements are all nonmetals, as in the problem, we can use the prefixes in Table 2.6 of the text to name the substances.

Solve: (a) sulfur hexafluoride; (b) nitrogen oxide; (c) dinitrogen oxide; (d) diphosphorus pentaoxide.

Comment: Use the same approach to checking as in the previous problem: Write the chemical formulas independently and compare to the ones given in the question.

The following rules summarize the procedures for drawing Lewis structures:

DRAWING LEWIS STRUCTURES

1. Determine the arrangement of the atoms in the compound with respect to each other. In simple molecules or ions, such as CCl_4, NO_3^-, HCN, or SO_3^{2-}, there is usually one central atom to which all other atoms are bonded. If the compound is binary (a compound containing only two kinds of elements), the first element given, such as C in CCl_4, is the central one. When a ternary compound (one with three kinds of elements), such as HCN, is written, the middle atom in the formula is usually the central one. If the compound is ionic, for example, NH_4Cl, the cation is separated from the anion and a Lewis structure is written for each ion.
2. Determine the total number of valence electrons available for bonding by adding the number of valence electrons for each atom. Also, if the species has a negative charge, add one electron for each negative charge to the previous number of valence electrons. If the species has a positive charge, subtract one electron for each positive charge.
3. Place the valence electrons about the atoms so that each atom has an octet (8) of electrons. Common exceptions to atoms achieving an octet are H (2 electrons), Be (4 electrons), B (6 electrons), and elements in the second, third, and fourth rows, such as P, S, and Si, which *may* have 10 to 12 electrons.
4. If there is an insufficient number of valence electrons for atoms in the compound to have their required number (see 3), form double or triple bonds between appropriate atoms. Multiple bonds often occur among the atoms C, N, and O.
5. If there are more than the required number of valence electrons for atoms to achieve their required number, expand the octets of the appropriate atoms (see 3).
6. If there is more than one way of arranging bonds and unshared electrons pairs to form equivalent Lewis structures, write all resonance forms.

There are molecules or ions for which drawing one Lewis structure provides an incomplete picture of bonding. An example of this situation is the NO_3^- ion:

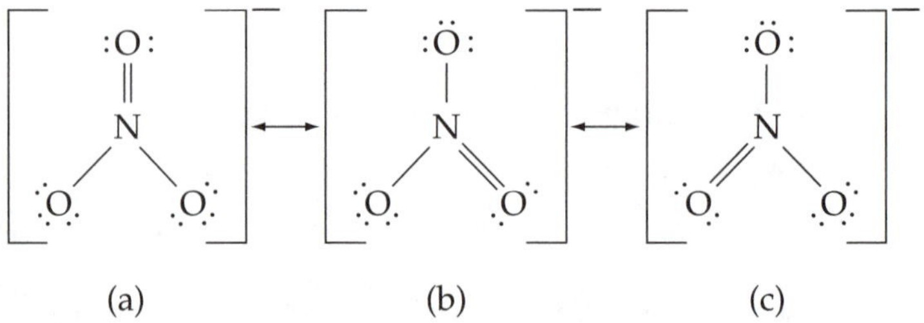

(a)   (b)   (c)

- We can write three Lewis structures, with the double bond between the N—O atoms placed differently in each structure. The positions of the nuclei remain fixed and only the placement of electrons varies. Each structure shows all atoms following the octet rule.
- According to experimental evidence, all bond lengths between N and O atoms in the NO_3^- ion are equivalent. But any single Lewis structure for NO_3^-—for example, Lewis structure (a) above—shows two kinds of bonds, single and double. As we learned earlier, double bonds are shorter than single bonds. To describe the structure of NO_3^- properly, we write all three Lewis structures and indicate that the real molecule is described by a blend of the three.
- Equivalent Lewis structures of this sort are called **resonance structures** or **resonance forms**. Double-headed arrows are used to indicate resonance forms.

The text presents the concept of **formal charge** as an aid in determining which of several different Lewis resonance structures is the most reasonable. You must remember that formal charges of atoms do not represent real charges. The formal charge of any atom in a Lewis structure can be calculated as follows:

- Formal charge = (number of valence electrons of isolated atom) $- \frac{1}{2}$ (number of bonding electrons shared by the atom) $-$ (number of unshared valence electrons)

Usually the best Lewis structure is the one that minimizes formal charges of atoms. The more electronegative atoms should also carry the negative formal charges. A caution: Recent quantum mechanical calculations suggest that Lewis structures which obey the octet rule are the best Lewis structures, irrespective of formal charge distributions. You should consult your instructor for latest information and applications.

EXERCISE 12   Drawing Lewis structures

Draw the Lewis structures for: (a) HOCl; (b) NH_4Cl; (c) SiH_4; (d) BH_4^-; (e) H_2CO

SOLUTION: *Analyze*: We are given the chemical formulas for five substances and asked to write their Lewis structures.

Plan: We will apply the rules for drawing Lewis structures to each substance.

Solve: (**a**) When hydrogen and a halide such as chlorine occur with oxygen, oxygen is the central element. The total number of valence electrons for HOCl is calculated as follows:

	No. of atoms	×	No. of valence electrons	=	Total no. of valence electrons
O:	1	×	6	=	6
H:	1	×	1	=	1
Cl:	1	×	7	=	7
			Total	=	14

Place the 14 electrons (seven pairs) about H, O, and Cl so that O and Cl each have an octet and H has two electrons:

$$\text{H} - \overset{\cdot\cdot}{\underset{\cdot\cdot}{\text{O}}} - \overset{\cdot\cdot}{\underset{\cdot\cdot}{\text{Cl}}}\colon$$

Finally, check your work to make sure that the total number of electrons used in your Lewis structure equals the total number available. (**b**) NH_4Cl is an ionic compound containing the cation NH_4^+ and the anion Cl^-. The single atom in compounds of type AB_x is usually the central atom. The total number of valence electrons for NH_4^+ is calculated as follows:

	No. of atoms	×	No. of valence electrons	=	Total no. of valence electrons
N:	1	×	5	=	5
H:	4	×	1	=	4
Subtract one electron for positive charge of ion				=	−1
			Total	=	8

For Cl^-, the number of valence electrons is seven ($3s^2 3p^5$) plus the addition of one electron for the negative charge. The total number is eight electrons (or 4 pairs). Place an octet of electrons about N and Cl and two electrons about each H. Separate the cation from the anion:

$$\left[\begin{array}{c} \text{H} \\ | \\ \text{H} - \text{N} - \text{H} \\ | \\ \text{H} \end{array}\right]^+ \qquad \left[\colon\overset{\cdot\cdot}{\underset{\cdot\cdot}{\text{Cl}}}\colon\right]^-$$

(**c**) Silicon is the central atom because it is the single atom in the type AB_x compound. The total number of valence electrons is

	No. of atoms	×	No. of valence electrons	=	Total no. of valence electrons
Si:	1	×	4	=	4
H:	4	×	1	=	4
			Total	=	8

Eight valence electrons represents an octet. Place four single bonds (four pairs of electrons) about silicon.

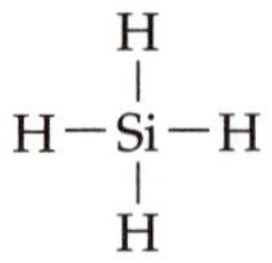

$$H-\underset{\underset{\displaystyle H}{|}}{\overset{\overset{\displaystyle H}{|}}{Si}}-H$$

Note: SiH_4 has the same Lewis structure as CH_4 which is shown in the text. This is another example of periodic relationships. (**d**) Boron is the central atom because it is the single atom in the type AB_x compound. The total number of valence electrons is

	No. of atoms	×	No. of valence electrons	=	Total no. of valence electrons
B:	1	×	3	=	3
H:	4	×	1	=	4
Add one electron for the negative charge of BH_4^-				=	1
			Total	=	8

As in (c) this represents an octet of electrons.

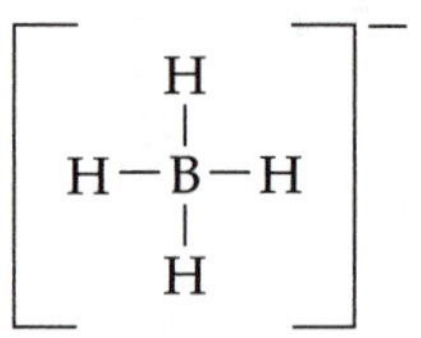

$$\left[\, H-\underset{\underset{\displaystyle H}{|}}{\overset{\overset{\displaystyle H}{|}}{B}}-H \,\right]^{-}$$

(**e**) In simple carbon-containing compounds, such as H_2CO, carbon is the central element. The total number of valence electrons for H_2CO is calculated as follows:

	No. of atoms	×	No. of valence electrons	=	Total no. of valence electrons
H:	2	×	1	=	2
C:	1	×	4	=	4
O:	1	×	6	=	6
			Total	=	12

The following Lewis structures use 12 electrons each, but in both cases the octet rule is not followed for the atoms circled.

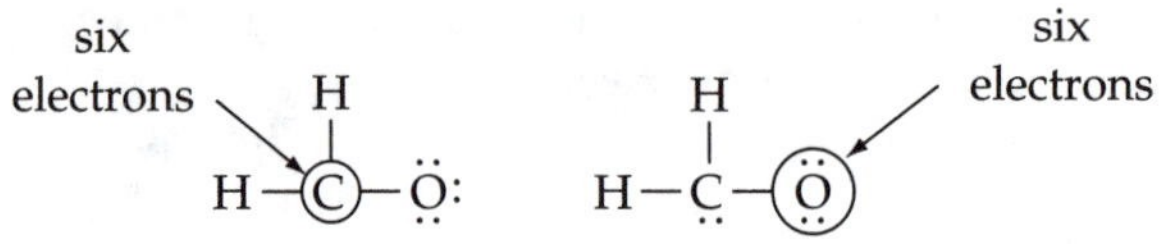

six electrons

six electrons

$$H-\overset{\overset{\displaystyle H}{|}}{\textcircled{C}}-\ddot{\text{O}}: \qquad H-\overset{\overset{\displaystyle H}{|}}{\underset{\displaystyle ..}{C}}-\textcircled{\ddot{\text{O}}}$$

In a case like this, we need to form a multiple bond between carbon and oxygen atoms so both possess an octet of electrons:

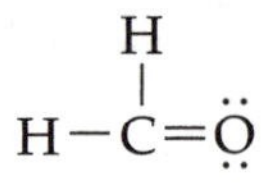

$$H-\overset{\overset{\displaystyle H}{|}}{C}=\ddot{\text{O}}$$

Check: For each substance check its Lewis structure to ensure that the total number of valence electrons shown equals the total number of valence electrons available. Check that the octet rule has been correctly applied to second row elements. Check that hydrogen has no more than two valence electrons about it. A recheck confirms our Lewis structures.

EXERCISE 13 Drawing resonance structures

From experiments it is known that the S—O bonds in SO_2 are equivalent. A Lewis structure for SO_2 is

$$:\ddot{O}-\ddot{S}=\ddot{O}$$

It shows that one S—O bond is a single bond and that the other S—O bond is a double bond. How does the Lewis model for covalent bonding explain this discrepancy?

SOLUTION: *Analyze*: We are given a Lewis structure and it appears that one bond is single and the other double, yet the experimental bond lengths are equivalent. We are asked to explain this observation.

Plan: When we observe structures with double bonds we need to consider whether resonance forms exist. If resonance forms exist then this can explain equivalency of bond lengths.

Solve: There are, in fact, two Lewis structures for SO_2 that can be written:

$$:\ddot{O}-\ddot{S}=\ddot{O} \longleftrightarrow \ddot{O}=\ddot{S}-\ddot{O}:$$

not just one. The true structure of SO_2 is not represented by either structure, but has the character of both structures. Each S—O bond has some single-bond and some double-bond character. The two Lewis structures are called resonance structures. The overall effect of more than one valid Lewis structure is called resonance.

Check: In Section 8.6 of the text we observe other examples in which resonance forms result in equal bonds within a structure, such as in NO_3^-. An alternation of single and double covalent bonds also gives rise to the phenomenon of resonance.

EXERCISE 14 Assigning formal charges to atoms and determining plausible Lewis structures

Two possible arrangements of atoms for the empirical formula HCN are HCN and HNC. Draw the Lewis structure for each, assign formal charges to the atoms, and determine which one is more plausible.

SOLUTION: *Analyze*: We are given two arrangements of atoms for HCN and asked to determine which one is more likely based on their Lewis structures and formal charges of atoms.

Plan: First we need to draw the Lewis structures of HCN and HNC. Next we need to assign formal charges to atoms to help determine which structure is most likely. We will use the rule that the structure having its atoms with smallest formal charges is the most likely one to exist.

Solve: Exercise 8.7 in the text shows us how to write the Lewis structure of HCN. The Lewis structure of HNC is equivalent except the atoms N and C are reversed. Therefore the two Lewis structures are:

$$H-C\equiv N: \quad \text{and} \quad H-N\equiv C:$$

The formal charges(FC) of the atoms in HCN are calculated as follows:

H: FC $= 1 - \frac{1}{2}$(1 bond $\times$ 2 electrons) $-$ 0 unshared electrons $= 0$

C: FC $= 4 - \frac{1}{2}$(4 bonds $\times$ 2 electrons) $-$ 0 unshared electrons $= 4 - 4 - 0 = 0$

N: FC $= 5 - \frac{1}{2}$(3 bonds $\times$ 2 electrons) $-$ 2 unshared electrons $= 5 - 3 - 2 = 0$

The formal charges of the atoms in HNC are calculated as follows:

H: FC $= 1 - \frac{1}{2}$(1 bond $\times$ 2 electrons) $-$ 0 unshared electrons $= 0$

N: FC $= 5 - \frac{1}{2}$(4 bonds $\times$ 2 electrons) $-$ 0 unshared electrons $= 5 - 4 - 0 = +1$

C: FC $= 4 - \frac{1}{2}$(3 bonds $\times$ 2 electrons) $-$ 2 unshared electrons $= 4 - 3 - 2 = -1$

HCN is more plausible than HNC because all atoms in the Lewis structure for HCN have zero formal charges.

Comment: The formal charge is calculated by taking the number of valence electrons in the isolated atom less all of the unshared electrons and half of the bonding electrons. This requires that a Lewis structure be correctly written if formal charges are to be correctly calculated. We need to not only check the validity of our Lewis structures as we have done previously but also to double check our math in terms of counting and subtracting electrons. A recheck confirms our conclusions.

EXCEPTIONS TO THE OCTET RULE

Many molecules and ions contain atoms which do not obey the octet rule. The text discusses three exceptions to the octet rule.

- A few molecules and ions contain atoms with an *odd* number of electrons in their valence shell: ClO_2, NO, and NO_2 are three examples you should know.
- A few molecules and ions contain atoms with *fewer* than eight electrons in their valence shell: H, Be, B, and Al are the primary examples of elements which exhibit this behavior.
- The largest class of exceptions consists of molecules and ions which contain atoms having *more* than eight electrons in their valence shell. These atoms are from the third period and beyond. They can use their empty *d* orbitals to accept more than eight valence electrons. In general, expansion of octet occurs when the central atom is large and the surrounding atoms are small and strongly electron-attracting, such as F, Cl, and O.

EXERCISE 15   Drawing a Lewis structure for an odd-electron molecule

Draw the Lewis structure for the odd-electron molecule NO.

SOLUTION: *Analyze*: We are given the odd-electron species NO and asked to draw its Lewis structure.

Plan: NO must possess at least one single electron since it has an unpaired electron. We will draw its Lewis structure using procedures we learned in the prior section.

Solve: We first determine the number of valence electrons:

	No. of atoms	×	No. of valence electrons	=	Total no. of valence electrons
N:	1	×	5	=	5
O:	1	×	6	=	6
			Total	=	11

NO possesses an odd number of valence electrons. The first 10 electrons can be used to form five pairs of electrons. Five pairs of electrons are insufficient to form an octet about each atom in NO. The extra (11th) electron is added in such a manner as to best agree with experimental data.

$$\ddot{N}\!=\!\ddot{O}$$

Check: Note that the sum of valence electrons confirms that it is an odd-electron molecule. The sum of valence electrons in the Lewis structure equals this sum.

EXERCISE 16   Drawing Lewis structures: Exceptions to the octet rule

Write the Lewis structures for (**a**) BI_3 and (**b**) PF_5.

SOLUTION: *Analyze*: We are given two chemical formulas and asked to write their Lewis structures.

Plan: We will write their Lewis structures using procedures we learned in the prior section. We should note that compounds of boron and aluminum can be electron deficient and elements in the third and higher periods may expand their octets.

Solve: (**a**) Boron is the single atom in the type - AB_x compound and should be the central atom. The total number of valence electrons for BI_3 is calculated as follows:

	No. of atoms	×	No. of valence electrons	=	Total no. of valence electrons
B:	1	×	3	=	3
I:	3	×	7	=	21
			Total	=	24

Place an octet of electrons about each I and six electrons about B (an exception to the octet rule):

$$:\!\ddot{I}\!-\!B\!-\!\ddot{I}\!:$$
$$|$$
$$:\!\ddot{I}\!:$$

(**b**) Phosphorus is the single atom and therefore should be the central one. The total number of valence electrons for PF_5 is calculated as follows:

	No. of atoms	×	No. of valence electrons	=	Total no. of valence electrons
P:	1	×	5	=	5
F:	5	×	7	=	35
			Total	=	40

Phosphorus must expand its octet to accommodate five fluorine atoms. Place an octet of electrons about each F. This results in the presence of 10 electrons about P:

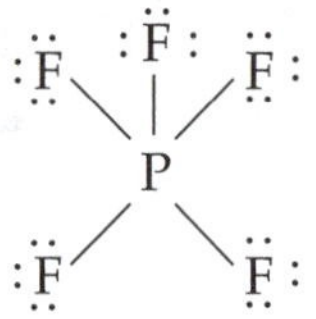

Check: The sum of valence electrons in each Lewis structure equals the sum of valence electrons. BI_3 is an electron-deficient molecule as noted in the *Plan* and phosphorus did expand its octet as an element of the third period.

STRENGTHS OF COVALENT BONDS

Section 8.8 focuses on the relationship between bond enthalpies and the strengths of covalent bonds and estimating enthalpies of reactions using bond enthalpies.

- Bond enthalpy is defined as the enthalpy change required to break one mole of bonds in the gas state. When gaseous H_2 molecules dissociate to form gaseous hydrogen atoms, energy is required:

$$H_2(g) \longrightarrow 2\,H(g) \qquad \Delta H = 436\ kJ/mol$$

- In the previous example, 436 kJ of energy is required when one mole of gaseous H—H bonds is broken.
- Another way of representing this energy change is $\Delta H = D(H—H) = 436\ kJ/mol$ where $D(H—H)$ means the energy required to break or dissociate one mole of H—H bonds.
- In most molecules the bond enthalpy of a particular bond depends on its environment; so when we speak of bond enthalpy we are usually referring to an average bond enthalpy. However, you should note that some bond enthalpies, such as H—H and Cl—Cl, can be precisely determined. Tables of bond enthalpies enable us to calculate approximate enthalpies of reactions using the relation

$$\Delta H_{rxn} = \Sigma D(\text{bonds } broken) - \Sigma D(\text{bonds } formed)$$

- The above equation is another example of the application of Hess's law of heat summation. However, heats of reaction are seldom calculated in this manner since bond enthalpies usually do not represent exact values, only averages for a particular bond in several environments.
- Bond enthalpies represent thermodynamic quantities that can be measured or calculated. They also tell us about the strength of a bond: The larger the bond enthalpy, the stronger the bond. For example, bond strength and bond enthalpy increase in the following order of bond types between atoms: single < double < triple.

An average **bond enthalpy** for bond types can be defined. These are shown in Table 8.4 in your text in Table 8.5 average bond lengths for some single, double and triple bonds are provided. You should note the following:

- There is a relationship between bond enthalpy and bond length. As bond enthalpy for a particular type of bond increases, the bond length shortens.
- Also note that a bond between two atoms grows shorter and stronger as the number of bonds between the two atoms increases.

Another quantity that can tell us something about the nature of a bond and its strength is the electronegativity difference between bonded atoms.

EXERCISE 17 Estimating ΔH_{rxn} from bond enthalpies

Estimate ΔH for the reaction

$$H_2(g) + Cl_2(g) \longrightarrow 2\,HCl(g)$$

using the following bond enthalpies:

$$D(H\text{—}H) = 436 \text{ kJ/mol}$$
$$D(Cl\text{—}Cl) = 242 \text{ kJ/mol}$$
$$D(H\text{—}Cl) = 431 \text{ kJ/mol}.$$

SOLUTION: *Analyze*: We are asked to calculate the enthalpy of reaction given bond enthalpies.

Plan: We can calculate ΔH for the reaction by using the relation:

$$\Delta H_{rxn} = \Sigma D(\text{bonds } broken) - \Sigma D(\text{bonds } formed)$$

or, for the reaction being considered:

$$\Delta H_{rxn} = D(H\text{—}H) + D(Cl\text{—}Cl) - 2D(H\text{—}Cl)$$

The two in front of the term $D(H\text{—}Cl)$ is required because 2 mol of H—Cl bonds are formed.

Solve:

$$\Delta H_{rxn} = (1 \text{ mol})(436 \text{ kJ/mol}) + (1 \text{ mol})(242 \text{ kJ/mol}) - (2 \text{ mol})(431 \text{ kJ/mol})$$

$$= -184 \text{ kJ}$$

Check: An estimate of the calculation is $(1)(450) + (1)(250) - (2)(450) = -150$, which is close to the calculated value. The unit is appropriately kJ. A recheck that the data is correctly substituted will confirm the format of the calculation.

SELF-TEST QUESTIONS

Key Terms

Having reviewed key terms in Chapter 8, match key terms with phrases and identify statements as true or false. If a statement is false, indicate why it is incorrect.

Match each phrase with the best term:

8.1 It shows the arrangements of valence electrons about atoms in a chemical species.

8.2 This bond is weaker than a multiple bond in a similar bonding environment.

8.3 This type of bond exists between N and O in NO_3^- and between C and H in CH_4.

8.4 A measure of the distribution of electron density in a covalent chemical bond.

8.5 The outer electron shell of a K atom.

8.6 The type of bond that exists in MgO.

8.7 The value of this term is inversely proportional to the distance between adjacent particles.

8.8 The type of covalent bond that exists in H_2CCH_2.

8.9 The type of covalent bond that exists in HCCH.

8.10 The name given to the $2s^1$ electron in sodium.

8.11 A measure of the relative attraction an atom has for electrons in a chemical bond.

8.12 When a chemical bond is broken this term is used to describe the energy change.

8.13 HCl has this type of covalent bond but Cl_2 does not have it.

Key Terms:

(a)	bond enthalpy	**(h)**	Lewis structure
(b)	bond polarity	**(i)**	polar covalent bond
(c)	covalent	**(j)**	single covalent bond
(d)	double bond	**(k)**	triple bond
(e)	electronegativity	**(l)**	valence-shell
(f)	ionic	**(m)**	valence electron
(g)	lattice energy		

True-False Statements

8.14   The reaction of calcium with chlorine to form $CaCl_2$ illustrates the *octet rule*.

8.15   The *oxidation state* of iodine in ICl is −1.

8.16   One of the three *resonance forms* of SO_3 is

8.17   The form of the equation for lattice energy is

$$E_{\text{lattice}} \propto \frac{Q_+Q}{d^2}$$

where the symbol $\propto$ means "proportional to."

8.18   The *formal charge* of an atom in a binary ionic compound is simply its ionic charge.

8.19   An *ionic bond* is characterized by a small difference in electronegativity values between the bonded atoms.

8.20 $FeCl_3$ contains a *metallic bond*.

8.21   The *Lewis symbol* for arsenic is $\cdot\ddot{A}s\cdot$

8.22 CO_2 contains *covalent bonds*.

8.23   *Multiple bonding* occurs among the atoms carbon, nitrogen, and oxygen.

8.24   The *bond enthalpy* of a C≡N bond is larger than that for a C≡N bond.

8.25 Na_2 contains a *nonpolar covalent bond*.

Problems and Short-Answer Questions

8.26   Two atoms react to form an ionic bond. This chemical reaction is shown by the figure below. Given this figure, which atom is a metal and which is a nonmetal? Explain.

$$\text{(A)} + \text{(B)} \longrightarrow \left[\text{(A)}\ \text{(B)}\right]$$

8.27   Elements A and B are nonmetals. The following figure shows the distribution of electron density in a covalent bond between atoms A and B. Identify the polarity of the chemical bond and the most electronegative atom. Explain.

8.28   Critique the following statement: The driving force for the formation of $KCl(s)$ is the ionization of potassium to form K^+ and the addition of an electron to chlorine to form Cl^-.

8.29   Which should have the larger lattice energy, MgO or SrO? Why?

8.30   Why doesn't $K_2Cl(s)$ form?

8.31   Rank the following in order of increasing radii: Li^+, K, F, and I^-.

8.32   Differentiate between electronegativity and electron affinity.

8.33   Which of the following atoms are found to form covalent compounds that are either deficient or expansive in terms of the octet rule: B, H, O, and S? Explain.

8.34   In Sample Exercise 8.9 in the text you found that the nitrogen atom in the Lewis resonance form $:\ddot{N}-C\equiv S^-$ has a −2 formal nitrogen, the carbon atom a zero formal charge, and the sulfur atom a +1 formal charge. What does the formal charge of nitrogen mean in terms of the charge of nitrogen in the Lewis structure. What do the formal charges tell us about the relative contribution of the resonance form to the overall Lewis structure?

8.35   Critique the following statement: If a molecule is described using several Lewis resonance structures this means that at some point in time the molecule will exist in one of these forms.

8.36   Answer the following questions with respect to bond enthalpy:

   (a) Why is it always a positive quantity?
   (b) In a molecule such as H_2O does each O—H bond have the same bond enthalpy as each O—H bond is broken stepwise?
   (c) The average bond enthalpy for 8.9 is 614 kJ/mol and 418 kJ/mol for S=S. Indicate which double bond is stronger and suggest a reason for the difference.

8.37   Element A has a slightly larger ionization energy than element B. Is there sufficient information to conclude that element B is in the same family as A and above A in the family? If there is insufficient information, what other information is needed and why?

8.38   Mn has two $4s$ valence electrons, just as Ca does. We know that the chemical formula of calcium chloride is $CaCl_2$. Is there sufficient information to write the chemical formula of manganese chloride? If there is insufficient information, what other information is needed and why?

8.39   Iodine reacts with bromine to form IBr. Is there sufficient information to conclude that IBr_3 may also form under the appropriate reaction conditions? If there is insufficient information, what other information is needed and why?

8.40   Methane and the ammonium ion are isoelectronic. What does this mean in terms of these two species? If they are isoelectronic why do they have dissimilar chemical properties?

8.41   When Li(s) reacts with $F_2(g)$, LiF(s) is produced and 610 kJ/mol of heat is released. Since it requires energy to break the F—F bond in F_2 and to form Li^+ from Li(s), why should the formation of LiF(s) from Li(s) and $F_2(g)$ release energy?

8.42   Predict the formula of the solid ionic compound formed when each of the following pairs of atoms are combined:

   (a) Na and I   (c)   Ca and Br
   (b) Ba and S   (d)   Ca and H

8.43 Which element or ion in each pair is larger?

(a) Mg or Mg^{2+} (c) S or S^{2-}
(b) Co^{2+} or Co^{3+} (d) Rb^+ or Br^-

8.44 Draw the Lewis structures for:

(a) $HCCCH_3$ (d) H_3O^+
(b) ICl (e) $CS_2(S-C-S$
(c) $CaBr_2$ *arrangement of atoms)*

8.45 (a) Draw the Lewis structures for the following series of related compounds: N_2, HN_2H, and $H_2N_2H_2$.

(b) How do the N—N bond distances and bond enthalpies vary in this series?

8.46 (a) Arrange the following pairs of covalently bonded atoms in expected order of increasing bond polarity: C—N, P—Cl, I—Cl, and C—O.

(b) Indicate the direction of the polarity for each bond using the symbols δ^+ and δ^-.

8.47 Estimate the standard enthalpy of formation of ammonia given the following average bond enthalpies in kilojoules: D(N—H), 391; D(N—N), 163; D(N=N), 418; D(N≡N), 941; D(H—H), 436.

8.48 Name the following compounds:

(a) SO_3 (f) $Fe(NO_3)_2$
(b) Ag_2S (g) AuCl
(c) $AsCl_3$ (h) $AuCl_3$
(d) ICl (i) $HgCl_2$
(e) $Fe(NO_3)_3$

8.49 Write chemical formulas for the following:

(a) mercury(II) (d) hydrogen sulfide
   cyanide (e) sulfur
(b) iron(II) sulfate tetrafluoride
(c) sodium carbonate (f) dinitrogen oxide

Integrative Questions

8.50 Calcium carbide, $CaC_2(s)$, is made by reacting calcium oxide (lime) with carbon at high temperatures; a toxic, carbon-containing gas is also produced. CaC_2 reacts readily with water to form acetylene, $C_2H_2(g)$. This reaction also produces a white solid which is basic.

(a) Write balanced chemical reactions for both reactions.

(b) What type of compound is calcium carbide? Explain.

(c) Write the Lewis structure for calcium carbide.

8.51 (a) Draw all appropriate Lewis structures for $ClNO_2$.

(b) What are the formal charges of Cl, N, and O in any structures you wrote?

(c) Are any of the oxygen atoms equivalent?

(d) What nitrogen- and oxygen-containing ion is isoelectronic with $ClNO_2$?

8.52 Pentane is a carbon–hydrogen-containing substance (referred to as a hydrocarbon) with a chain of five carbon atoms. When all carbon—carbon bonds in a hydrocarbon

are single it is said to be saturated; if a double or triple bond exists it is unsaturated.

(a) Draw the Lewis structures of a saturated and unsaturated five-carbon hydrocarbon.

(b) Estimate the enthalpy of reaction for the conversion of the unsaturated to a saturated form by adding H_2 given the following bond enthalpies (kJ/mol): C—H, 413; C—C, 348; C=C, 614; and H—H, 436.

(c) Given that the C=C bond is rigid and rotation about it does not occur, what two Lewis structures can exist for $CH_3CH=CHCH_3$? Would you expect their heats of formation to be different in the gas phase? Which structure is energetically more stable? Why?

8.53 Can more than one Lewis structure be written for a compound which contains 52.2% C, 13.1% H, and 34.7% O and has an approximate molar mass of 46 g? Explain.

Multiple-Choice Questions

8.54 What is the valence-electron configuration for Se^{2-}?

(a) $2s^2 2p^4$ (d) $4s^2 4p^6$
(b) $2s^2 2p^6$ (e) $3s^2 3p^2$
(c) $4s^2 4p^4$

8.55 Which of the following ionic crystals would you expect to have the largest lattice energy?

(a) LiCl (d) SrO
(b) LiBr (e) $BaSO_4$
(c) CaO

8.56 Which is the valence-electron configuration for Co^{2+}?

(a) $3d^7 4s^2$ (d) $3d^6 4s^1$
(b) $3d^5 4s^2$ (e) $3d^9$
(c) $3d^7$

8.57 What is the correct order of increasing radii for the isoelectronic series Rb^+, Sr^{2+}, Se^{2-}, Br^-?

(a) $Rb^+ < Sr^{2+} < Se^{2-}$, Br^-
(b) $Br^- < Se^{2+} < Sr^{2+} < Rb^+$
(c) $Se^{2-} < Br^- < Rb^+ < Sr^{2+}$
(d) $Sr^{2+} < Rb^+ < Br^-$, Se^{2-}
(e) none of the above

8.58 Which has the largest ion radius?

(a) N^{3-} (d) S^{2-}
(b) P^{3-} (e) F^-
(c) O^{2-}

8.59 Which of the following compounds contains a multiple bond in its Lewis dot structure?

(a) ICl (d) NH_4^+
(b) SO_2 (e) SO_4^{2-}
(c) Cl_2

8.60 Given the following average bond enthalpies:

$$C—C = 348 \text{ kJ/mol}$$
$$C=C = 614 \text{ kJ/mol}$$
$$C≡C = 839 \text{ kJ/mol}$$

which of the following is most likely to contain a $C-C$ bond that has a bond enthalpy of 820 kJ/mol?

(a) C_2H_6

(b) C_3H_8

(c) C_4H_{10}

(d) C_2H_4

(e) C_2H_2

8.61 In which of the following compounds is the direction of the symbol $=C$ *not* indicated correctly?

(a) HBr: $H \longmapsto Br$

(b) SO_2: $S \longmapsto O$

(c) BI_3: $I \longmapsto B$

(d) H_2S: $H \longmapsto S$

(e) none of the above

8.62 Which of the following is a valid Lewis structure for CO?

(a) $:\ddot{C}-\ddot{O}:$

(b) $:C=O:$

(c) $:C\equiv\ddot{O}:$

(d) $C\equiv O:$

(e) $:C\equiv O:$

8.63 Which of the following is a valid Lewis structure for C_2Cl_4?

(a) $:\ddot{C}l-\ddot{C}-\ddot{C}-\ddot{C}l:$ with $:\ddot{C}l:$ $:\ddot{C}l:$

(c) $:\ddot{C}l-C=C-\ddot{C}l:$ with $:\ddot{C}l:$ $:\ddot{C}l:$

(b) $:\ddot{C}l-\ddot{C}-C-Cl:$ with $:\ddot{C}l:$ $:\ddot{C}l:$

(d) $:\ddot{C}l-C\equiv C-\ddot{C}l:$ with $:\ddot{C}l:$ $:\ddot{C}l:$

(e) none of the above

8.64 Which of the following are valid resonance structures for HN_3?

(a) $H-N\equiv N-\ddot{N}:$

(b) $H-\ddot{N}-N\equiv N:$

(c) $H-\ddot{N}=N=\ddot{N}:$

(d) (a) and (b)

(e) (a), (b), and (c)

8.65 Why is the Lewis structure (1) $:\ddot{C}l-\overset{\overset{\displaystyle :\ddot{C}l:}{|}}{B}-\ddot{C}l:$ a more reasonable Lewis structure than (2)

$:\ddot{C}l=\overset{\overset{\displaystyle :\ddot{C}l:}{|}}{B}-\ddot{C}l:$?

(a) Structure **2** does not have the correct number of electrons.

(b) In structure **1** chlorine has a formal charge of +1 which is preferred.

(c) In structure **2** boron has a +1 formal charge which is preferred.

(d) In structure **2** the sum of formal charges is not zero.

(e) In structure **1** all formal charges of the atoms are zero.

SELF-TEST SOLUTIONS

8.1 (h) **8.2** (j) **8.3** (c) **8.4** (b) **8.5** (l) **8.6** (f) **8.7** (g) **8.8** (d) **8.9** (k) **8.10** (m) **8.11** (e) **8.12** (a) **8.13** (i) **8.14** True. In $CaCl_2$ both atoms achieve an octet of valence electrons. **8.15** False. Because chlorine is more electronegative than iodine, it is assigned a -1 oxidation state value; therefore iodine must have a $+1$ oxidation state so that the sum of oxidation states will equal zero. **8.16** True. **8.17** False. The denominator of the expression should be d not d^2. **8.18** True. Note: This will not be true for an atom contained within a polyatomic ion, but for simple salts such as $MgCl_2$ the statement is true. **8.19** False. An ionic bond is characterized by a large difference in electronegativity values. **8.20** False. $FeCl_3$ is an ionic compound. A metallic bond occurs when two metals are bonded together. **8.21** True. **8.22** True. **8.23** True. Most occurrences observed involve these three atoms. **8.24** False. More energy is required to break a triple bond than a double bond; therefore the bond-enthalpy of a triple bond is larger than that of a double bond. **8.25** False. It contains a metallic bond. A nonpolar covalent bond involves two nonmetals possessing equivalent electronegativity value.

8.26 The cation is atom B, a metal, because the final figure shows its size diminishing when the ionic substance is formed. The formation of a cation from an atom results occurs with the loss of electrons and results in a smaller sized species; cations are formed by metals. The anion is atom A, a nonmetal, because its size increases. The formation of an anion from an atom occurs with the addition of electrons; this causes an expansion of the electron cloud and a larger sized species. Nonmetallic elements typically form anions in ionic compounds.

8.27 The figure shows the electron density in the chemical bond centered more on A than on B. Thus, A is more electronegative than B. This means that the polarity of the chemical bond is $\overset{\delta^- \quad \delta^+}{A\text{-}B}$.

8.28 The formation of Na^+ requires energy (ionization energy) and the addition of an electron to chlorine to form Cl^- is exothermic (electron affinity); however, the sum of the two is endothermic, which is typical for many similar salts. The driving force for the formation of salts is the electrostatic interaction of the cations and negative ions; this energy is reflected in the lattice energy.

8.29 Lattice energy is the energy required to separate a mole of an ionic compound into its gaseous ion. This energy is dependent on the following ratio: $\dfrac{Q_1Q_2}{d}$. In this ratio Q_1 and Q_2 are the charges of the cation and anion and d is the distance between their centers. For MgO and SrO, the charges are a constant factor, 2+ and 2−. The value that changes is d. Sr is a larger atom than Mg; thus Sr^{2+} is a larger ion than Mg^{2+}, and the $Sr-O$ distance is larger than the $Mg-O$ distance. Since d is in the denominator of the ratio, the larger the distance the smaller the value of the lattice energy. Therefore, MgO has the larger lattice energy (3795 kJ/mol versus 3217 kJ/mol).

8.30 Potassium is always a K$^+$ ion in ionic substances and from this we conclude that the chlorine atom would have a 2− charge. We know that the formation of Cl$^-(g)$ from Cl(g) is an exothermic process; however, the addition of the second electron would be highly endothermic because an electron would have to be added to a negatively charged chlorine atom. Coulomb's law tells you that two negative charges separated by a distance undergo repulsion and energy is required to force them together. Also, the addition of an electron to Cl$^-$ would require that the electron be placed in a $n = 4$ energy level, which is beyond the valence shell of chlorine; this is unfavorable.

8.31 Li$^+ <$ F $<$ K $<$ I$^-$. See Figure 8.5 in your text. The removal of an electron from a lithium atom significantly reduces its size. Adding an electron to an iodine atom significantly increases its size.

8.32 Electronegativity refers to the tendency of an atom in a bonding situation to attract electrons to itself. Electron affinity is a thermodynamic quantity: It is the energy associated with adding a gaseous electron to a gaseous atom. It is a property of an isolated atom. Electronegativity is related to both the ionization energy and electron affinity of an atom.

8.33 H has a $1s^1$ electron configuration and only needs one more electron to complete its valence shell and does not need an octet. The valence electron configuration of B is $2s^2 2p^1$. We find boron is able to form covalent substances in which it has only six valence electrons in the chemical bonds (for example, BCl$_3$). This suggests that boron readily forms the electron configuration $2s^1 2p^2$ when it forms covalent bonds. Sulfur can form substances in which it obeys the octet rule (H$_2$S, for example). It also forms substances in which it expands its octet (SF$_6$, for example).

8.34 Formal charge is a way of "bookkeeping" the number of valence electrons and the charge that an atom would have in a molecule if the atoms had all the same electronegativity. It is not a "real" charge for atoms in molecular substances because ions do not exist as they do in ionic substances. Thus, nitrogen is not a −2 ion; it simply is carrying an "excess" of valence electrons in the molecule compared to its valence state as an atom. The Lewis structures that typically contribute significantly to the overall molecule are those in which the atoms have formal charges with the smallest magnitudes. Since this Lewis resonance form has a nitrogen with a −2 formal charge it should not contribute as effectively as the other Lewis resonance forms with atoms possessing lower formal charges.

8.35 There are molecules in which one Lewis structure does not represent accurately the molecular arrangement, including bond angle and distances between atoms. In such cases it is often possible to construct two or more

Lewis structures that are equivalent except for the placement of the electron domains. The actual molecule can be considered a "blend" of all the contributing Lewis structures. The molecule is not oscillating between each Lewis structure, thus it does not exist in each Lewis structure form. Resonance is a way of trying to describe these situations and reflects a deficiency of a single Lewis structure.

(**a**) Bond enthalpy is associated with the breaking on one mole of a particular bond in substance. Energy is always required to break a chemical bond.

(**b**) No. When the first bond is broken the fragment O—H remains. The O—H bond experiences a different environment than in the case of water, H—O—H. The lack of a second hydrogen atom influences the strength of the OH bond in the hydroxyl radical and it will have a different bond energy. If we were to measure the total energy to break two O—H bonds in water and divide by two we then have an average bond enthalpy.

(**c**) The larger the bond enthalpy the stronger the chemical bond; thus, C=C is a stronger bond than S=S. Sulfur is a larger atom than carbon and this limits the effectiveness of the *p-p* atomic orbital overlap between two sulfur atoms.

8.37 You cannot make the conclusion stated in the problem. Element B might also be an element in the same period as A but to its right because there is a general increase in ionization energy across a period from left to right. We need information about other physical and chemical properties of A and B so that we can properly place the elements in perspective. If the information provided lets you conclude that elements A and B are in the same family, then the statement that B is above A in the family would be correct.

8.38 There is insufficient information. Manganese is a member of the transition elements and exhibits a variety of oxidation states. The valence electrons of Mn are not only the $4s$ electrons but also the $3d$ electrons; thus, it has the ability to lose more than two electrons. Knowledge of the oxidation state of manganese in a given context is needed to write the correct chemical formula.

8.39 Iodine is a member of the fifth row of the periodic table. Nonmetallic elements in the fifth row may expand their octet when compounds form. IBr$_3$ has the potential to form because we can write a valid Lewis structure for it with expansion of the octet of electrons for iodine:

$$:\ddot{Br} - \dot{\ddot{I}} - \ddot{Br}:$$
$$|$$
$$:\ddot{Br}:$$

However, without doing an experiment to determine if IBr$_3$ actually forms, we cannot conclude with absolute certainty that it will form.

8.40 The term isoelectronic means that the species have the same number of electrons. Both CH$_4$ and NH$_4^+$ have

10 electrons. They have dissimilar chemical properties because the nitrogen atom has one more proton in the nucleus than does carbon. This one extra proton causes a significant change in the overall properties of the nitrogen atom compared to the carbon atom. This results in ammonium ion having a positive charge and acting as a cation in ionic substances. Methane is a molecular compound.

8.41 Although the two processes $F_2(g) \rightarrow 2\,F^-(g) + 2\,e^-$ and $Li(s) \rightarrow Li^+(g) + e^-$ require energy, the energy released in $Li^+(g) + F^-(g) \rightarrow LiF(s)$ more than compensates for this required energy. Thus, the formation of LiF(s) from Li(s) and $F_2(g)$ releases energy.

8.42 The sum of ion charges in a formula must equal zero for an ionic compound.

(a) Na belongs to group 1A and thus has a 1+ charge; I has a 1− charge in ionic compounds. Therefore, they occur in a 1:1 ratio in an ionic compound because they have the same charge. The formula is NaI.

(b) Ba belongs to group 2A and thus has a 2+ charge; and S, like O, normally has a 2− charge. The formula is BaS.

(c) Ca, like Ba, has a 2+ charge; and Br, like I, has a 1− charge. Two bromine ions are needed to equal the 2+ charge of Ca. The formula is $CaBr_2$.

(d) Ca has a 2+ charge. Since H cannot also have a positive charge, it must have a negative charge, in the form of a hydride ion (H^-). Two hydride ions are needed to balance the 2+ charge of Ca. The formula is CaH_2.

8.43 (a) Mg. The removal of two electrons forms Mg^{2+}. This reduces electron–electron repulsions resulting in a decreased size of Mg^{2+} compared to Mg.

(b) Co^{2+}. The larger ion charge of Co^{3+} causes the electron cloud to contract more, resulting in a decreased size compared to Co^{2+}.

(c) S^{2-}. The addition of two electrons to S causes an increase in repulsions among electrons resulting in an expansion of the size of the electron cloud and the size of the ion compared to S.

(d) Br^-. Both ions are isoelectronic. Thus, the larger nuclear charge of Rb^+ compared to Br^- results in a reduction in the size of the electron cloud and in a smaller size compared to Br^-.

8.44 (a) The structural formula is

$$H-C-C-C-H$$

with H atoms completing the structure:

$$\begin{array}{c} \quad\quad H \\ \quad\quad | \\ H-C-C-C-H \\ \quad\quad | \\ \quad\quad H \end{array}$$

The number of valence electrons is $(4\,H\,atoms)(1$ *valence electron*$) + (3\,C\,atoms)(4\,valence\,electrons) = 16$ electrons or 8 pairs. Note that the Lewis structure

$$H-\ddot{C}-\ddot{C}-\ddot{C}-H$$

requires 20 electrons and is incorrect; thus, multiple bonds are required and the correct Lewis structure is

$$\begin{array}{c} \quad\quad\quad\quad\quad H \\ \quad\quad\quad\quad\quad | \\ H-C\equiv C-C-H \\ \quad\quad\quad\quad\quad | \\ \quad\quad\quad\quad\quad H \end{array}$$

(b) For ICl, the number of valence electrons is (1 I atom)(7 valence electrons) + (1 Cl atom) × (7 valence electrons) = 14 electrons or 7 pairs. or 7 pairs.

The Lewis structure $:\ddot{I}-\ddot{C}l:$ fits this requirement.

(c) $CaBr_2$ is an ionic compound, therefore, the ions must be separated. All of calcium's valence electrons are removed to form a 2^+ ion. Each bromine atom adds one electron to achieve a rare gas electron configuration. The Lewis structure is $[:\ddot{B}r:]^- \; [:\ddot{C}a:]^{2-} \; [:\ddot{B}r:]^-$

(d) In H_3O^+, all of the hydrogen atoms are bonded to oxygen. The number of valence electrons is (3 H atoms) (1 valence electron) + (1 O atom)(6 valence electrons) − 1 electron for positive charge = 8 valence electrons or 4 pairs. The Lewis structure is

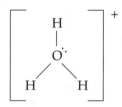

(e) Carbon is the central element. The number of valence electrons is (1 C atom)(4 valence electrons + (2 S atoms) (6 valence electrons) = 16 valence electrons, or 8 pairs. In order to achieve an octet of electrons about each atom, multiple bonds must be used and the Lewis structure is: $:\ddot{S}=C=\ddot{S}:$

8.45 (a) Using the procedures for determining the number of valence electrons and Lewis structures yields

$$:N\equiv N: \quad H-\ddot{N}=\ddot{N}-H \quad \begin{array}{c} N-\ddot{N}-\ddot{N}-H \\ | \quad | \\ H \quad H \end{array}$$

(b) Bond strengths and bond enthalpies increase with an increasing number of bonds between atoms. With increasing bond strength, bond distances shorten. Thus, for these three molecules N—N bond distances increase in the order $N_2 < HN_2H < H_2N_2H_2$. The N—N bond enthalpies increase in the following order: $H_2N_2H_2 < HN_2H < N_2$ (that is, single bond, double bond, triple bond).

8.46 (a) $I-Cl \approx C-N < P-Cl < C-O$. The order is determined by differences in electronegativity values between the pairs. The greater the difference in electronegativity, the more polar the bond. Electronegativity values are as follows: I = 2.5, Cl = 3.0, difference = 0.5; C = 2.5, N = 3.0, difference = 0.5; C = 2.5, O = 3.5, difference = 1.0; P = 2.1, Cl = 3.0, difference = 0.9.

(b) $I^{\delta+}-Cl^{\delta-}$; $C^{\delta+}-N^{\delta-}$; $C^{\delta+}-O^{\delta-}$; $P^{\delta+}-Cl^{\delta-}$

8.47 The reaction for the formation of ammonia from its constituent elements is $N_2(g) + 3 H_2(g) \rightarrow 2 NH_3(g)$. Thus, the standard enthalpy of formation is estimated by using the relationship ΔH_{rxn} = sum bond energies reactants − sum bond energies of products = $D(N\equiv N)$ + $3D(H-H) - 2 \times 3 D(N-H)$ = (941 kJ) + 3(436 kJ) − 6(391 kJ) = − 97 kJ. ΔH_f is for *one* mole; thus, $\Delta H_f = \frac{1}{2}(-97$ kJ) = −48.5 kJ. $D(N\equiv N)$ is used in the calculation because the nitrogen atoms in N_2 are bonded together by a triple bond. $2 \times 3 D(N-H)$ is used in the calculation because there are two ammonia molecules and each ammonia molecule contains three $N-H$ bonds; therefore there are a total of six $N-H$ bonds.

8.48 (a) sulfur(VI) oxide or sulfur trioxide;
(b) silver(I) sulfide;
(c) arsenic(III) chloride or arsenic trichloride;
(d) iodine(I) chloride or iodine monochloride;
(e) iron(III) nitrate;
(f) iron(II) nitrate;
(g) gold(I) chloride;
(h) gold(III) chloride;
(i) mercury(II) chloride.

8.49 (a) $Hg(CN)_2$;   (d) H_2S;
(b) $FeSO_4$;   (e) SF_4;
(c) Na_2CO_3;   (f) N_2O_4.

8.50 (a) $CaO(s) + 3 C(s) \longrightarrow CaC_2(s) + CO(g)$
$CaC_2(s) + 2 H_2O(l) \longrightarrow C_2H_2(g) + Ca(OH)_2(s)$

(b) Calcium carbide is an ionic compound consisting of the metal ion Ca^{2+} and the nonmetal ion C_2^{2-}.

(c) When writing the Lewis structure for an ionic compound the metal ion is separated from the nonmetal ion. This results in:

$$[Ca^{2+}][:C\equiv C:^{2-}]$$

8.51 (a)

$$:\overset{..}{O}: \quad\quad\quad :\overset{..}{\overset{|}{O}}:$$
$$:\overset{..}{\underset{..}{C}l}-N-\overset{..}{\underset{..}{O}}: \longleftrightarrow :\overset{..}{\underset{..}{C}l}-N=\overset{..}{\underset{..}{O}}$$

(b) The formal charge of Cl is: Number of valence electrons of atom − $\frac{1}{2}$(number of bonding electrons)− (number of nonbonding electrons)

$$= 7 - \frac{1}{2}(2) - 6 = 0$$

The formal charge of the nitrogen atom is: $5 - \frac{1}{2}(8) - 0 = +1$

The formal charge of the doubly bonded O in one resonance form is: $6 - \frac{1}{2}(4) - 4 = 0$

The formal charge of the singly bonded O in the same resonance form is: $6 - \frac{1}{2}(2) - 6 = -1$

(c) The two oxygen atoms are equivalent. Since there are two resonance forms, each oxygen atom experiences an average $1\frac{1}{2}$ bonds; neither has a single nor a double bond.

(d) Chlorine is a member of the halogen family which is one family to the right of the oxygen family and an atom of it has one more valence electron than an oxygen atom. Thus, if an oxygen atom is substituted for a chlorine atom, an extra electron must be added to maintain isoelectronic conditions. The ion which meets this requirement is NO_3^-.

8.52 (a) An example of a saturated five-carbon atom hydrocarbon with a linear chain is

$$H-\overset{\overset{\displaystyle H}{|}}{\underset{\underset{\displaystyle H}{|}}{C}}-\overset{\overset{\displaystyle H}{|}}{\underset{\underset{\displaystyle H}{|}}{C}}-\overset{\overset{\displaystyle H}{|}}{\underset{\underset{\displaystyle H}{|}}{C}}=\overset{|}{\underset{\underset{\displaystyle H}{|}}{C}}-\overset{\overset{\displaystyle H}{|}}{\underset{\underset{\displaystyle H}{|}}{C}}-H$$

An example of an unsaturated five-carbon atom hydrocarbon with a linear chain is

$$H-\overset{\overset{\displaystyle H}{|}}{\underset{\underset{\displaystyle H}{|}}{C}}-\overset{\overset{\displaystyle H}{|}}{\underset{\underset{\displaystyle H}{|}}{C}}-\overset{\overset{\displaystyle H}{|}}{\underset{\underset{\displaystyle H}{|}}{C}}-\overset{\overset{\displaystyle H}{|}}{\underset{\underset{\displaystyle H}{|}}{C}}-\overset{\overset{\displaystyle H}{|}}{\underset{\underset{\displaystyle H}{|}}{C}}-H$$

Note that the existence of the carbon–carbon double bond has resulted in two less hydrogen atoms.

(b) Conversion of the unsaturated hydrocarbon to a saturated hydrocarbon requires the addition of two hydrogen atoms. This can be accomplished by reacting it with a species that has two hydrogen atoms, $H_2(g)$. To estimate the enthalpy of reaction you first have to write the chemical reaction and then determine the number and types of bonds that are broken and formed.

$$C_5H_{10}(g) + H_2(g) \longrightarrow C_5H_{12}(g)$$

$\Delta H_{rxn} = \Sigma D(\text{bonds broken}) - \Sigma D(\text{bonds formed})$

$= [3 D(C-C) + 1 D(C=C) + 10 D(C-H)$
$+ 1 D(H-H)] - [4 D(C-C) + 12 D(C-H)]$

$= [3 \text{ mol}(348 \text{ kJ/mol}) + 1 \text{ mol}(614 \text{ kJ/mol})$
$+ 10 \text{ mol}(413 \text{ kJ/mol}) + 1 \text{ mol}(436 \text{ kJ/mol})]$

$- [4 \text{ mol}(348 \text{ kJ/mol}) + 12 \text{ mol}(413 \text{ kJ/mol})]$

$= -124 \text{ kJ}$

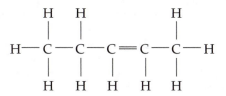

(A)   (B)

The heats of formation should be very similar since they have very similar structures and the same number and types of atoms. However, structure B should have a slightly greater heat of formation and thus be energetically more stable (the sign of the heats of formation will be negative; an exothermic process). This occurs because in structure A the two methyl groups, CH_3, are next to each other and undergo electronic repulsion. In structure B the methyl groups are adjacent to H atoms and H atoms are smaller than methyl groups, resulting in less electronic repulsion. With less electronic repulsion a more stable structure occurs. The difference in kJ is not large.

8.53 Using the percent composition data, assuming a 100 g sample, calculate the empirical formula. C: 52.2 g/12.01 g/mol; H: 13.1 g/1.008 g/mol; O: 34.6 g/16.0 g/mol or 4.35 mol C; 13.0 mol H and 2.16 mol O. Dividing by the smallest number gives: C_2H_6O. The molar mass of this substance is approximately 46 g, thus, it is also the molecular formula. Possible Lewis structures include

$$
\begin{array}{ccccc}
& H & & H & \\
& | & .. & | & \\
H - & C - & \overset{..}{O} - & C - & H \\
& | & & | & \\
& H & & H &
\end{array}
\qquad \text{and}
$$

$$
\begin{array}{cccc}
& H & H & \\
& | & | & \\
H - & C - & C - & \overset{..}{\underset{..}{O}} - H \\
& | & | & \\
& H & H &
\end{array}
$$

8.54 (d)

8.55 (c) Lattice energy is proportional to Q_1Q_2/d. For CaO, $Q_1Q_2 = 4$ (same as for SrO and $BaSO_4$). And CaO has the smallest value of d. Thus it has the largest lattice energy.

8.56 (c) The 4s electrons in transition elements are lost first when ions form.

8.57 (d) In an isoelectronic series size decreases with increasing atomic number.

8.58 (b) P^{3-}. P^{3-} has more electrons than the ions of nitrogen, oxygen, and fluorine because it is in a higher numbered period and the outermost electrons are in a higher principal quantum number state; this results in a larger ion size. P^{3-} and S^{2-} are isoelectronic and because S^{2-} has more protons in its nucleus this will result in a smaller ion size compared to P^{3-}.

8.59 (b) Draw the Lewis structures for all. Only SO_2 has a multiple bond. One resonance form is

$$
:\overset{..}{\underset{..}{O}} - \overset{..}{S} = \overset{..}{\underset{..}{O}}:
$$

8.60 (e) Value of bond enthalpy corresponds to that of a $C=C$ bond. Only C_2H_2 has a triple bond. C_2H_4 has a double bond, and all others have a single bond.

8.61 (c)

8.62 (e) The octet rule is obeyed, and the total number of valence electrons is correct.

8.63 (c)

8.64 (e)

8.65 (e) The sum of formal charges is zero for both Lewis structures; however in the first Lewis structure the formal charges of all atoms are zero, whereas in the second structure the chlorine with a double bond has a positive formal charge (not favorable with an electronegative atom) and boron has a negative formal charge. The rules of formal charge suggest that the Lewis structure in which all atoms have zero formal charges is the most stable form.

Molecular Geometry and Bonding Theories

Chapter

9

OVERVIEW OF THE CHAPTER

Review: Lewis structures (8.1, 8.5); bond polarity (8.4).

Learning Goals: You should be able to:

1. Relate the number of electron domains in the valence shell of an atom in a molecule to the geometrical arrangement around that atom.
2. Explain why nonbonding electron domains exert a greater repulsive interaction on other domains than do bonding electron domains.
3. Predict the geometrical structure of a molecule or ion from its Lewis structure.
4. Predict, from the molecular shape and the electronegativities of the atoms involved, whether a molecule can have a dipole moment.

Review: Orbital diagrams (6.6); orbital types and shapes (6.6); concept of valence orbitals (6.8).

Learning Goals: You should be able to:

1. Explain the concept of hybridization and its relationship to geometrical structure.
2. Assign a hybridization to the valence orbitals of an atom in a molecule, knowing the number and geometrical arrangement of the atoms to which it is bonded.

Review: Lewis structures containing multiple bonds (8.5, 8.6); resonance (8.7).

Learning Goals: You should be able to:

1. Formulate the bonding in a molecule in terms of σ bonds and π bonds, from its Lewis structure.
2. Explain the concept of delocalization in π bonds.

Review: Hund's Rule (6.8); the Pauli principle (6.7); orbital shapes (6.6).

Learning Goals: You should be able to:

1. Explain the concept of orbital overlap and the reason why overlap may in some cases be zero because of symmetry.
2. Describe how molecular orbitals are formed by overlap of atomic orbitals.
3. Explain the relationship between bonding and antibonding molecular orbitals.
4. Construct the molecular-orbital energy-level diagram for a diatomic molecule or ion built from elements of the first or second row and predict the bond order and number of unpaired electrons.

TOPIC SUMMARIES AND EXERCISES

VSEPR MODEL: A TOOL FOR QUALITATIVELY PREDICTING MOLECULAR STRUCTURE AND MOLECULAR POLARITY

The three-dimensional array of atoms in a molecule, referred to as the geometrical structure of the molecule, depends on many factors. The valence-shell electron-pair (**VSEPR**) model contains concepts that help us to predict the geometrical shapes of simple molecules.

- An essential idea of the VSEPR model is that repulsions among electron domains about a central atom in a molecule are minimized, and that this minimization of repulsions is primarily responsible for the geometrical arrangement of bonded atoms about the central atom.
- An electron domain is a region of space in which electrons of an atom will most likely exist. The electron domains responsible for the geometrical structure of a molecule are of two types: nonbonding pair of electrons (also called unshared pairs or lone pairs) and bonding pair of electrons.
- The geometrical structure of a AB_n molecule (A and B are representative elements) can be predicted using the following steps:

1. Draw the Lewis structure of the molecule or ion, showing all bonding and nonbonding pairs of valence electrons.
2. Count the total number of electron domains about the central atom. (*A multiple bond is counted as a single electron domain.*)
3. Use Table 9.1 on the next page (or Tables 9.1, 9.2, and 9.3 of the text) to determine the arrangement of atoms about the central atom. *Note that the structure of a molecule is described in terms of arrangement of atoms, not electron domains.* For example, consider molecules where the total number of electron domains is the same: In the series CH_4, :NH_3, and HÖH, each central atom is surrounded by four electron domains. The geometrical arrangement of electron domains is the same for all in the series (tetrahedral). However, the structures of the molecules (in terms of arrangement of atoms) differ because of differing numbers of atoms about the central atom (CH_4, tetrahedral; NH_3, trigonal pyramidal; H_2O, nonlinear).
4. In a molecule or ion, atoms bonded to the central atom arrange themselves so as to minimize repulsions among electron domains. The strength of this repulsion is in the following order:

$$\text{nonbonding domains–nonbonding domains} >$$
$$\text{nonbonding domains–bonding domains} >$$
$$\text{bonding domains–bonding domains}$$

These repulsions cause distortions from the idealized geometry.

In Chapter 8 we learned about polarity of bonds, dipoles, and dipole moments. Molecules can also be polar and have dipole moments. The dipole

TABLE 9.1 Relationship between Numbers of Bonding and Nonbonding Electron Domains and Molecular Structure

Molecular type[a]	Total number of electron domains about central atom	Number of bonding electron domains	Number of nonbonding electron domains (E)	Arrangement of atoms about central atom[d]	Example
AX_2	2	2	0	Linear	BeH_2
AX_3	3	3	0	Trigonal planar	BCl_3
AX_2E	3	2	1	Nonlinear	SO_2
AX_4	4	4	0	Tetrahedral	CH_4
AX_3E	4	3	1	Trigonal pyramidal	NH_3
AX_2E_2	4	2	2	Nonlinear	H_2O
AX_5	5	5	0	Trigonal bipyramidal	PF_5
AX_4E	5	4	1	Seesaw[b]	$TeCl_4$
AX_3E_2	5	3	2	T-shaped[b]	ClF_3
AX_2E_3	5	2	3	Linear[b]	XeF_2
AX_6	6	6	0	Octahedral	SF_6
AX_5E	6	5	1	Square pyramidal	BrF_5
AX_4E_2	6	4	2	Square planar[c]	BrF_4^-

[a] A represents the central atom. X represents the atoms bonded to the central atom. E represents the number of nonbonding electron domains possessed by the central atom.
[b] Nonbonding electron domains are in the triangular plane.
[c] Nonbonding electron domains are 180° apart.
[d] These structures are diagrammed in Tables 9.1, 9.2, and 9.3 of the text.

moment of a molecule depends on both the magnitudes of the bond dipoles and the geometry of the bond dipoles.

- If a polyatomic molecule contains equivalent polar bonds that form a totally symmetric arrangement about the central atom, then the bond dipoles will cancel and the polyatomic molecule is not polar. Examples of such cases are CH_4 (tetrahedral), BF_3 (trigonal planar), and CO_2 (linear). A molecule is usually polar if the bond dipoles are not all equivalent, such as in CH_3Br, or if the bond dipoles are equivalent but do not cancel, such as in H_2O.
- Predicting whether or not a molecule is polar is not always straightforward. However, you can do this for simple molecules by first determining the geometrical shape of the molecule using the VSEPR model. Then, determine which bonds in the molecule are polar. Finally, analyze the geometrical arrangement of the bond dipoles to see if they overall cancel or not. If they do not cancel then the molecule is polar and has a dipole moment.

EXERCISE 1 Writing Lewis structures and predicting molecular structures

For each of the following species, draw its Lewis structure: (a) NO_3^-; (b) XeO_3; (c) H_2S; (d) F_2SO_2; (e) $SnCl_2$ (covalent); (f) PCl_3; (g) IF_5; and (h) SF_4. State the total number of bonding electron domains and the number of bonding and non-bonding electron domains about the central atom for each structure. Finally, using Table 9.1, predict the structure about the central atom.

SOLUTION: *Analyze*: We are asked to write the Lewis structure for eight species and to identify the number of bonding and nonbonding electron domains bonding about the central atom. The sum of these two numbers is the total number of electron domains. We are then asked to predict the molecular shape for each species.

Plan: All species are of the type AB_n or AB_xX_y, where A is the central atom and B and X are bonded to it. We first will use the procedures developed in Sections 8.1 and 8.5 of the text to draw the Lewis structures. Then we will carefully inspect each structure to determine the number of bonding and nonbonding electron domains about the central element A. Finally, we will use the information in Table 9.1 to help us predict the molecular structure.

Solve: Following the *Plan*, we write the following Lewis structures: Use the procedures developed in Sections 8.1 and 8.6 of the text to draw the Lewis structures. Your efforts should result in the following Lewis structures:

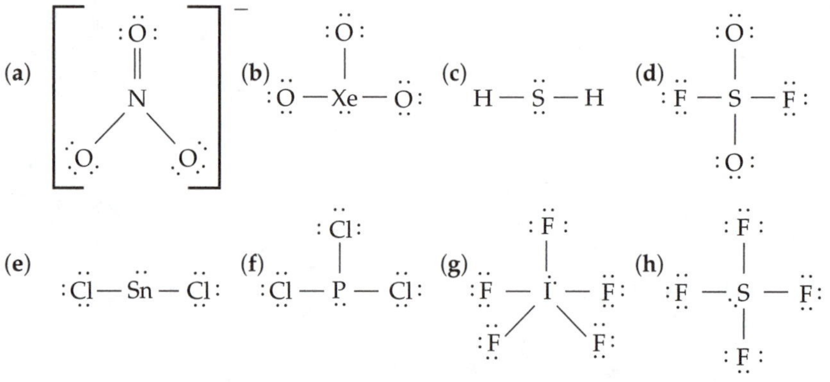

We will use H_2S as an example of how to answer the question. The Lewis structure for H_2S shows two bonding and two nonbonding electron domains for a total number of four electron domains. Thus, H_2S is a AX_2E_2, (E is an nonbonding electron pair) molecular type as shown in Table 9.1. This molecular type has a nonlinear structure about the central atom, S. A nonlinear structure involving H—S—H bonds is named V-shaped or bent. Another view is that H_2S has four electron domains about sulfur and this results in the electron domains forming a tetrahedron. If we place two hydrogen atoms at vertices of the tetrahedron a bent or V-shaped molecular shape results. The other molecular shapes are summarized in the following table:

Compound	Total number of electron domains about central atom	Total number of bonding electron domains about central atom	Total number of nonbonding electron domains about central atom	Structure
(a) NO_3^-	3[a]	3	0	Triangular
(b) XeO_3	4	3	1	Trigonal pyramidal
(c) H_2S	4	2	2	Bent
(d) SO_2F_2	4	4	0	Tetrahedral
(e) $SnCl_2$	3	2	1	Bent
(f) PCl_3	4	3	1	Trigonal pyramidal
(g) IF_5	6	5	1	Square pyramidal
(h) SF_4	5	4	1	Seesaw

[a] Remember that a multiple bond counts as one electron domain in VSEPR theory. Thus, only three electron domains are used in structure prediction.

EXERCISE 2   Predicting the molecular structure of SO_2 and the O—S—O bond angle

Predict the structure of SO_2. Do you expect the OSO bond angle to expand or contract compared to the idealized geometry?

SOLUTION: *Analyze*: We are asked to predict the molecular structure of sulfur dioxide and to predict whether the OSO bond angle expands or contracts compared to the idealized geometry.

Plan: We will use the procedure described in Exercise 1 to predict the molecular shape of sulfur dioxide. We will then consider the effect of nonbonding electron domains interacting with bonding electron domains to predict what happens to the OSO bond angle.

Solve: Each sulfur and oxygen atom contributes six valence electrons to the electron domains. Thus, there are 18 valence electrons distributed in the electron domains about the central atom, sulfur. This results in the following Lewis structures (resonance forms exist):

$$:\ddot{O}-\ddot{S}=\ddot{O}: \longleftrightarrow :\ddot{O}=\ddot{S}-\ddot{O}:$$

Using Table 9.1, we predict that the SO_2 molecule has the following nonlinear structure (V-shaped or bent):

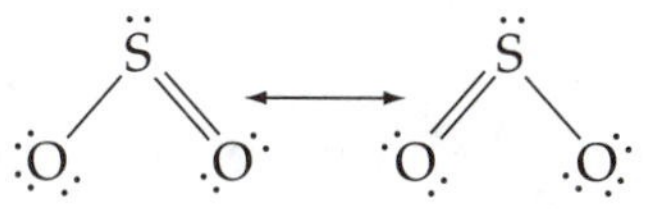

The repulsive forces existing between the nonbonding electron domain on sulfur and the bonding electron domains on sulfur and oxygen atoms cause the O—S—O bond angle to contract. The repulsive forces operating between a nonbonding electron domain and a bonding electron domain are greater than those between a bonding electron domain and another bonding electron domain. Thus, the O—S—O bond angle is not 120°, but less than this.

EXERCISE 3   Predicting whether SO_2 and SO_3 have dipole moments

Do SO_2 and SO_3 possess dipole moments?

SOLUTION: *Analyze*: We are asked to determine whether sulfur dioxide and sulfur trioxide possess molecular dipole moments.

Plan: We will draw the Lewis structure of each and apply VSEPR rules.

Solve:

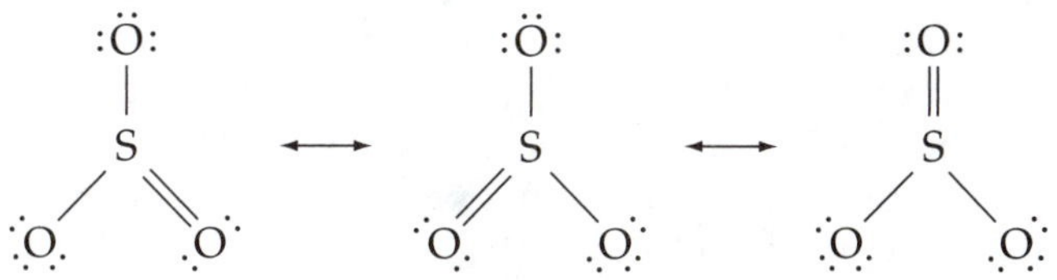

In SO_3, there are three bonding electron domains about the sulfur atom, resulting in a trigonal-planar structure. The existence of resonance structures results in three equal bond dipoles in a symmetrical arrangement. Therefore there is no molecular dipole. The Lewis structure for SO_2 is shown in Exercise 2 and is not reproduced here. The S—O bond dipoles do not cancel because they do not form a totally symmetric arrangement. Therefore it is a nonpolar molecule. The bond dipoles are shown below. Remember that the tip of the dipole arrow points toward the more electronegative atom.

Dipole moment exists because centers of charge are not symmetrically arranged about sulfur. A lone pair dipole also exists, but it does not cancel the bond dipoles.

There is no net dipole moment because all three bond dipoles are equivalent and symmetrically arranged.

COVALENT BONDING, HYBRID ORBITALS, AND MOLECULAR STRUCTURE

Covalent bonding involves sharing of electrons between atoms to form a stable molecular structure. In section 9.4 of the text a theory of covalent bonding is introduced; it is called the valence-bond model. The key idea is that a valence atomic orbital of one atom overlaps with a valence atomic orbital of another atom resulting in electron density shared between two nuclei. This buildup of electron density holds the two positive nuclei together.

- Note in Figure 9.14 in the text the overlap of two $1s$ atomic orbitals to form an overlap region. This leads to a stable bond. In Figure 9.15 in the text you should observe that there is a minimum in the potential energy diagram corresponding to the observed bond length for H_2. Atoms cannot approach extremely close together because nuclei will begin to repel each other.
- Sigma (σ) bonds are shown in Figure 9.14 of the text. Note that electron density is concentrated between nuclei along the internuclear axis.
- Pi (π) bonds are shown in Figure 9.22 of the text. Note that the electron density parallels the internuclear axis. A pi bond is weaker than a sigma bond. Also observe that a nodal plane exists in a pi bond. This is a region in which there is no electron density.

The geometrical structures of many molecules cannot be explained by considering the geometrical structure formed by the overlap of "pure" valence-shell atomic orbitals in bond formation. For example, consider the formation of a compound from beryllium and hydrogen atoms, BeH_2.

- Beryllium has no unpaired electrons available for bonding because the valence-shell electron configuration for beryllium is $2s^2$ and the shell is completely filled. Therefore, we might predict that beryllium does not form a compound with hydrogen.
- The compound BeH_2 in the gas state is known to have a linear structure and equivalent beryllium-hydrogen bonds: H—Be—H.
- The existence of two linear Be—H bonds in BeH_2 implies that beryllium is using two new equivalent atomic orbitals, each with one electron, when it bonds with hydrogen.
- These new atomic orbitals are called **hybrid orbitals**. Hybrid orbitals are formed from linear combinations of "pure" atomic orbitals and have

their own individual physical characteristics. They provide for better atomic orbital overlap, stronger bonds, and more stable structures.

Given the structure of a covalent molecule, you can predict the type of hybrid orbitals used by the central atom in bonding. The type of hybrid orbitals used by an atom in its bond formations is related to the geometry of the bonding and nonbonding electron domains about the atom. Use the following procedure to determine the type of hybrid orbitals used by a central atom in a covalent compound.

1. Draw the Lewis structure for the compound.
2. Use the VSEPR model to predict the arrangement of electron domains about the atom of interest. (Remember that a multiple bond is treated as one electron domain.)
3. Use Table 9.2 to predict the type of hybrid orbitals used by the central atom. Remember that nonbonding electron domains can be hybrid orbitals containing electron pairs.

The use of hybrid atomic orbitals to explain observed geometries of molecules provides an easy-to-use model at this stage of your learning of general chemistry. The question of whether or not an atom actually uses hybrid orbitals in bonding to other atoms is a more complex issue. For example, the bent angle in H_2S can also be explained by using pure atomic orbitals and then invoking electron domain repulsions to cause the angle to change. However, at this stage of your learning, use the hybrid orbital model when molecular geometries cannot be explained by overlap of pure atomic orbitals.

TABLE 9.2 Types of Hybrid Orbitals

Number of electron domains about the atom	Geometry of the electron domains[a] (not atoms)	Types of hybrid orbitals used by the atom (types of atomic orbitals combined indicated in parentheses)	Example
2	Linear	sp $(s + p)$	BeH_2
3	Trigonal planar	sp^2 $(s + p + p)$	BI_3
4	Tetrahedral	sp^3 $(s + p + p + p)$	CH_4
4	Square planar	dsp^2 $(d + s + p + p)$	$PtCl_4^{2-}$
5	Trigonal bipyramidal	dsp^3 $(d + s + p + p + p)$	PF_5
6	Octahedral	d^2sp^3 $(d + d + s + p + p + p)$	SF_6

[a] This is also the geometry of hybrid orbitals about the central atom. Note that a nonbonding electron domain can be a hybrid orbital.

Exercise 4   Explaining the molecular structure of BCl_3

Boron trichloride has a trigonal-planar structure with the three bonds about the central atom at angles of 120°. Can its structure be explained using only the $2s$ and $2p$ valence-shell orbitals of boron? If not, suggest an alternative explanation using the concepts developed in Section 9.5 of the text.

SOLUTION: *Analyze*: We are told that boron trichloride has a trigonal-planar structure and bond angles of 120°. We are then asked to explain if the $2s$ and $2p$

valence atomic orbitals on boron can be used to form covalent bonds with chlorine atomic orbitals to give the known molecular structure. If not, we are asked to provide an alternative model.

Plan: We will have to analyze the atomic orbitals on boron and chlorine and determine how many covalent bonds can form and what molecular structure results. If we cannot predict the given molecular structure then we will have to invoke the concept of hybridization, the mixing of atomic orbitals to form new atomic orbitals with the necessary geometrical orientation.

Solve: The valence orbital diagrams for boron and chlorine are

A covalent bond between two atoms occurs when a half-filled valence orbital on one atom combines (overlaps) with an appropriate half-filled valence orbital on the other. Boron has only one $2p$ atomic orbital with one electron that can overlap with a chlorine $2p$ atomic orbital containing one electron. Thus, we should predict the formula BCl, not BCl_3. To form three B—Cl bonds, the boron atom must use three hybrid orbitals with one electron in each. Also, these three hybrid orbitals of boron must be 120° apart in a trigonal plane. The nature of the three hybrid orbitals can be determined by visualizing the following process. First, a $2s$ electron in boron is promoted to obtain three unpaired electrons:

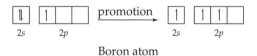

Second, the $2s$ and $2p$ atomic orbitals with one electron in each are hybridized, leaving one "pure" $2p$ atomic orbital in the boron atom:

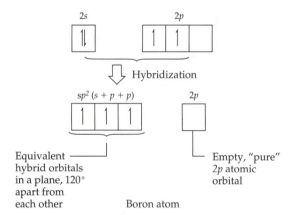

Three chlorine atoms, each with a half-filled $2p$ atomic orbital, combine with the three half-filled sp^2 hybrid orbitals on boron to yield three B—Cl bonds and the compound BCl_3.

Comment: Note that it is the geometry of the electron domains about the central atom that helps us understand what types of hybrid orbitals may be necessary in explaining a molecular structure.

EXERCISE 5

Predict the types of hybrid orbitals used by the central atom in the compounds in Exercises 1 and 2.

SOLUTION: *Analyze*: We are asked to predict the types of hybrid orbitals used by the central atom in each structure in Exercises 1 and 2.

Plan: For each structure determine the geometry of the electron domains and then determine the type of hybrid orbitals that fit the geometry.

Solve: See Exercises 1 and 2 for the Lewis structures. In the following table, the central atom is underlined, and the total number of electron domains about it are indicated:

Compound	Total number of electron domains	Geometry of electron domains	Hybrid orbitals used by central atom
$\underline{N}O_3^-$	3^a	Trigonal planar	sp^2
$\underline{Xe}O_3$	4	Tetrahedral	sp^3
$H_2\underline{S}$	4	Tetrahedral	sp^3
$\underline{S}O_2F_2$	4	Tetrahedral	sp^3
$\underline{Sn}Cl_2$	3	Trigonal planar	sp^2
$\underline{P}Cl_3$	4	Tetrahedral	sp^3
$\underline{I}F_5$	6	Octahedral	d^2sp^3
$\underline{S}F_4$	5	Trigonal bipyramidal	dsp^3
$\underline{S}O_2$	3	Trigonal planar	sp^2

^{a} Remember that a multiple bond counts as one electron domain in the VSEPR model.

Note: The geometry of electron domains is different in some cases from the geometry of the molecule. H_2S is a bent molecule, but the geometry of the electron domains is approximately tetrahedral. Two of the sp^3 hybrids on sulfur are used in the bonds to the hydrogens. The other two sp^3 hybrids contain the unshared electron domains. The experimentally observed structure of a molecule is determined by the locations of the atoms in the molecule and we can only infer the locations of nonbonding electron domains from its structure.

HYBRIDIZATION IN MOLECULES CONTAINING BOTH SIGMA AND PI BONDS

The Lewis structures of some molecules show atoms sharing more than one pair of electrons; that is, multiple bonds exist between the atoms. For example, molecular nitrogen contains a triple bond:

$$:N\equiv N:$$

The nature and origin of multiple bonds in molecules is discussed in Section 9.6 of the text. In the case of N_2, the multiple bond consists of one sigma (σ) bond and two pi (π) bonds.

- A **sigma bond** is directed along the internuclear axis. Electron density is concentrated directly between the bonded atoms. Sigma bonds can form from:

  1. The overlap of two s atomic orbitals
  2. The overlap of an s orbital with a p orbital that is directed along the internuclear axis

3. The overlap of two *p* orbitals, both directed along the internuclear axis; or the overlap of two hybrid orbitals, both directed along the internuclear axis. For example, the overlap of two *sp* hybrid orbitals to form a σ bond is shown in Figure 9.1 above.

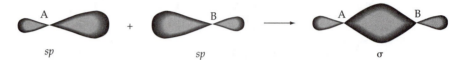

▲ **FIGURE 9.1** Two *sp* hybrid orbitals overlap to form a σ bond.

•   **Pi bonds** can occur with a σ bond if there are parallel *p* orbitals oriented perpendicularly to the sigma bond on the atoms bonded together and if these *p* orbitals contain two electrons between them. They can overlap and share the two valence electrons. A π bond consists of two distinct regions of electron density. One region is on one side of the plane containing the internuclear axis, and the other region is on the opposite side. An example of two $2p_x$ orbitals overlapping to form a π bond is pictured in Figure 9.2. The sigma bond is along the *y* axis and is not shown.

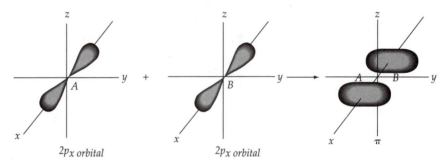

▲ **FIGURE 9.2** Two $2p_x$ hybrid orbitals overlap to form a π bond.

•   In some molecules, the overlap of *p* orbitals among two or more neighboring atoms leads to a delocalized π bond. When such a condition exists, you write more than one Lewis structure to show resonance forms. The delocalization of π electrons in benzene is an important example discussed in the text.

EXERCISE 6   Describing the C—C bonds in C_2^{2-}

Describe the C—C bonding in the acetylide ion, C_2^{2-}.

SOLUTION: *Analyze:* We are asked to describe the types of bonds (may involve hybrids) and electron domains in the acetylide ion.

Plan: We will first write the Lewis structure of the acetylide ion and then determine the types of electron domains and types of bonds.

Solve: **(a)** The Lewis structure of the acetylide ion is:

$$[:C \equiv C:]^{2-}$$

All diatomic species are linear. The Lewis structure shows one σ and two π bonds between the bonded carbon atoms. The presence of only one σ bond in a plane containing the bonded atoms suggests that the two carbon atoms each use one of their two sp hybrid orbitals to form the σ bond. The two $2p$ orbitals remaining on each carbon atom are used to form two pi bonds. The nonbonding electron domain on each carbon atom is contained within the other sp hybrid orbital.

EXERCISE 7 Determining the number of electrons in pi orbitals of compounds

How many electrons occupy π orbitals in: (**a**) CO_2; (**b**) HCN?

SOLUTION: *Analyze*: We are asked to determine the number of pi electrons in carbon dioxide and hydrogen cyanide. We will need to determine the number of pi orbitals in each.

Plan: We can write the Lewis structures for the two substances and count the number of pi bonds. Each pi bond holds two electrons.

Solve: (**a**) The Lewis structure for CO_2 is:

$$\ddot{O} = C = \ddot{O}$$

It shows that in CO_2 there are two σ bonds (one between each C—O bond) and two π bonds (one between each C—O bond). Since each π bond contains two electrons, there is a total of four electrons occupying the π bonds. (**b**) The Lewis structure for HCN is

$$H - C \equiv N:$$

Since there are two π bonds between the carbon and nitrogen atoms, there is a total of four electrons occupying the π bonds.

EXERCISE 8 Applying the concept of delocalization of pi electrons to NO_3^-

Show how the concept of delocalization of π electrons applies to NO_3^-.

SOLUTION: *Analyze*: We are asked to apply the concept of delocalization of π electron to the nitrate ion.

Plan: We can write the Lewis structures for the nitrate ion. If delocalization of electrons exists, then it occurs because of resonance.

Solve: The Lewis structures for NO_3^- involve resonance as shown below:

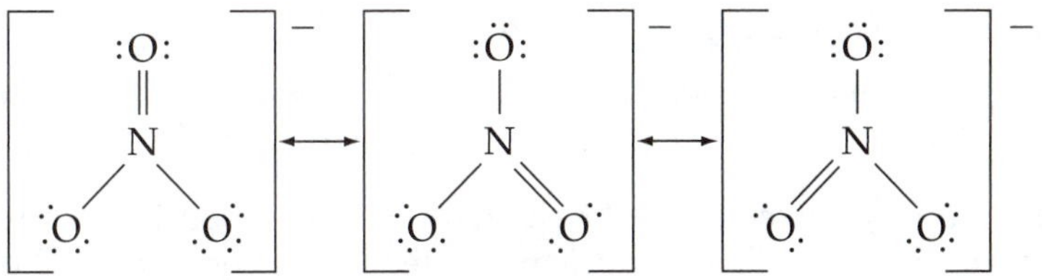

The π electrons are not localized in any one nitrogen-oxygen electron domain, but rather are associated with all three N—O bonds. That is, they are equally delocalized over the three N—O bonds. The following Lewis structure is one way of qualitatively representing the delocalization of the π electrons:

The dashed lines represent delocalized π electrons.

Comment: Delocalization of electrons within a molecular structure creates a more stable substance and it tends to be less reactive than if the substance existed as a single resonance form.

MOLECULAR ORBITALS

The Lewis model for covalent bonding, including the use of hybrid orbitals, provides us with a simple picture of bonding in which only valence orbitals are used to construct localized bonds between atoms. However, this model does not always correctly predict the electronic structure of a molecule. For example, from the Lewis structure of O_2,

$$\ddot{O}=\ddot{O}$$

we would predict molecular oxygen to be **diamagnetic** (containing no unpaired electrons); however, it is observed experimentally to be **paramagnetic** (containing unpaired electrons).

The molecular orbital model for bonding is an alternative model. The molecular orbital model is contrasted to the Lewis model by the following:

- Electrons are not necessarily confined to bonds localized between two atoms or to lone pair orbitals.
- Atomic orbitals combine to form **molecular orbitals** (**MOs**) that can spread or delocalize over an entire molecule or portions of it.
- Atomic orbitals combine to form **molecular orbitals** that can have bonding or antibonding properties.

The essential principles of the molecular orbital model are:

- Atomic orbitals on two (or more) different atoms combine to form molecular orbitals.
- The number of molecular orbitals formed equals the number of atomic orbitals combined.
- When two atomic orbitals are linearly combined, a bonding and anti-bonding molecular orbital are formed.
- A **bonding molecular orbital** has electron density occurring primarily between the two nuclei, which holds the nuclei together.
- An **antibonding molecular orbital** has electron density not occurring primarily between the nuclei. A node occurs, thereby destabilizing the nuclei. The nuclei repel one another.

Molecular orbitals formed from two atomic orbitals are classified as sigma (σ) molecular orbitals or pi (π) molecular orbitals.

- **Sigma (σ) molecular orbitals** have electron density symmetrical about and along the internuclear bonding axis. A bonding MO concentrates electron density between the nuclei whereas an antibonding MO concentrates electron density away from the region between the nuclei. Be sure you know how σ_s and σ_p MOs are formed and their shapes.
- **Pi (π) molecular orbitals** have electron density above and below the internuclear bonding axis. Bonding π MOs concentrate electron density

between nuclei whereas antibonding MOs concentrate electron density away from the region between the nuclei. Note in Figure 9.38 of the text that π_{2p_x} and π_{2p_y} MOs are equivalent, except that they are at 90° angles to one another.

You should memorize the energy-level diagrams for MOs of second row diatomic elements as shown in Figures 9.42 and 9.45 of the text.

- Note that there are two different MO orbital energy-level diagrams for the second-row diatomic molecules. For the diatomic boron, carbon, and nitrogen molecules the π_{2p} MO is lower in energy than the σ_{2p} MO because of the large 2s–2p orbital interactions. When these interactions are small, the σ_{2p} MO is lower in energy than the π_{2p} MO and this ordering is observed for the oxygen, fluorine, and neon diatomic molecules.
- Electrons enter MOs according to Hund's rule and Pauli's principle.
- The subscripts associated with σ and π MOs indicate the kind of atomic orbitals used to form a molecular orbital.
- Note that a bonding MO is of lower energy than either parent atomic orbital whereas an antibonding MO is of higher energy.

A useful concept for determining the stability of a diatomic molecule is bond order.

- Bond order is the net number of bonding electrons.
- Bond order

$$= \frac{\text{number of bonding electrons} - \text{number of antibonding electrons}}{2}$$

- A bond order equal to zero means the combination of nuclei is not stable and the molecule will not permanently exist.
- The greater the bond order, the greater the bond strength.

EXERCISE 9   Identifying types of molecular orbitals

Match the following labels with the correct MO:

$$\sigma_{2s}^*, \ \pi_{2p}^*, \ \pi_{2p}, \ \sigma_{1s}$$

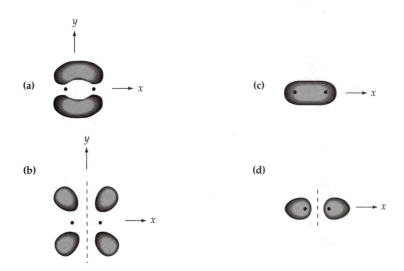

SOLUTION: *Analyze*: We are given the symbols for four molecular orbitals and asked to match them to the appropriate figure.

Plan: If a symbol has a * as a superscript it is an antibonding molecular orbital (MO) and the electron density is not concentrated between the nuclei; if a symbol does *not* have a * it is a bonding MO and the electron density is concentrated between the nuclei. A sigma (σ) MO has electron density symmetrical about the bonding axis and a pi (π) MO has electron density above and below the bonding axis but not on the bonding axis.

Solve: (a) π_{2p}; (b) π_{2p}^*; (c) σ_{1s}; (d) σ_{2s}^*.

EXERCISE 10 Determining bond orders from MO energy-level diagrams

What are the bond orders for N_2, N_2^+, and N_2^-? Which one of the three species is the most stable?

SOLUTION: *Analyze*: We are given three forms of N_2 and asked to predict which one is the most stable based on the magnitude of bond orders.

Plan: To calculate bond orders we must determine the number of bonding and antibonding electrons in each form of N_2. We must construct the MO energy-level diagram for each form of N_2 to obtain the necessary information. The bond order is calculated by the relation:

$$\text{Bond order} = \frac{(\text{no. bonding electrons}) - (\text{no. antibonding electrons})}{2}$$

The larger the bond order, the more stable the bond.

Solve: First, determine the number of electrons each species has: N_2 has 14 electrons, N_2^+ has 13 electrons, and N_2^- has 15 electrons. Then, for each nitrogen species, place the electrons in MOs by using the energy-level diagrams given in Figure 9.42 and 9.45 in the text. Electrons are placed in MOs using Hund's rule and the Pauli exclusion principle: Electrons occupy lower energy levels before a higher one is filled, and if an energy level is degenerate (more than one MO of the same energy), all orbitals must contain one electron before pairing occurs. The electron configurations for N_2, N_2^+, and N_2^- are given in Table 9.3.

TABLE 9.3 Electron configurations of N_2, N_2^+, and N_2^-

	N₂ (14 electrons)		N₂⁺ (13 electrons)		N₂⁻ (15 electrons)	
σ_{2p}^*	□		□		□	
π_{2p}^*	□	□	□	□	↑	□
σ_{2p}	↑↓		↑		↑↓	
π_{2p}	↑↓	↑↓	↑↓	↑↓	↑↓	↑↓
σ_{2s}^*	↑↓		↑↓		↑↓	
σ_{2s}	↑↓		↑↓		↑↓	
σ_{1s}^*	↑↓		↑↓		↑↓	
σ_{1s}	↑↓		↑↓		↑↓	

Next calculate the bond order for each:

$$N_2: \text{Bond order} = \frac{10 - 4}{2} = 3$$

$$N_2^+: \text{Bond order} = \frac{9 - 4}{2} = 2.5$$

$$N_2^-: \text{Bond order} = \frac{10 - 5}{2} = 2.5$$

N_2 is the most stable of the three species because it has the largest bond order. That is, N_2 has more net bonding electrons than either N_2^+ or N_2^-

The diagram on the previous page shows the σ_{2p} MO above the π_{2p} MOs. This occurs because there are significant interactions between the $2s$ and $2p$ atomic orbitals. You need to remember that the above diagram is only good for B_2, C_2, and N_2. For O_2, F_2, and Ne_2 the σ_{2p} MO is below the π_{2p} MO.

SELF-TEST QUESTIONS

Key Terms

Having reviewed key terms in Chapter 9, match key terms with phrases and identify statements as true or false. If a statement is false, indicate why it is incorrect.

Match each phrase with the best term:

9.1   The name and symbol given to the type of bond formed between C and H in CH_4.

9.2   The name and symbol given to the weaker bond existing between C and O in CO.

9.3   This type of molecular orbital results in destabilization of a chemical bond.

9.4   This type of molecular orbital is symmetrical about the internuclear axis in a homonuclear diatomic molecular compound.

9.5   Graphite and diamond are examples of this form of an element.

9.6   What must happen to two atomic orbitals on different atoms for a covalent bond to form.

9.7   The property of a substance being attracted into a magnetic field.

9.8   A measure of the net number of covalent bonds connecting two atoms in a covalent substance.

9.9   A model that helps us determine the molecular structure of simple covalent compounds.

9.10   The formation of an atomic orbital from the blending of two or more atomic orbitals.

9.11   A wavefunction describing the energy state of an electron distributed over all the atoms in a covalent substance.

9.12   The term given to the characteristic of a wavefunction of an electron being distributed throughout a molecule rather than between a pair of atoms.

9.13   The property of a substance being weakly repelled by a magnetic field.

9.14   Two electrons localized between two nuclei resulting in a chemical bond.

9.15   The regions in a molecular substance containing bonding or nonbonding electrons.

Terms:

(a)	allotropic	**(i)**	molecular orbital
(b)	antibonding	**(j)**	overlap
(c)	bond order	**(k)**	paramagnetism
(d)	bonding pair	**(l)**	pi(π)
(e)	delocalized	**(m)**	sigma(σ)
(f)	diamagnetism	**(n)**	sigma bonding molecular orbital
(g)	electron domains		
(h)	hybridization	**(o)**	VSEPR

True-False Statements:

9.16 N_2 contains three *bonding pairs* of electrons between the nitrogen atoms.

9.17   A *molecular orbital diagram (energy-level diagram)* for homonuclear second-row elements as found in Figure 9.45 in the text shows the σ_{2p} molecular orbital lower in energy than a π_{2p} orbital for C_2.

9.18   The H—C—H *bond angle* in H_2CO equals the ideal bond angle of 120 degrees.

9.19   The *electron-pair geometry* in NH_3 is tetrahedral.

9.20   The *molecular geometry* of NH_3 is tetrahedral.

9.21   *Dipole moments* are vector quantities.

9.22   The tin atom in $SnCl_3^-$ possesses one *nonbonding pair* of electrons.

9.23   ICl is a *polar* molecule whereas I_2 is not.

9.24   A *bond dipole* exists when two bonded atoms have the same electronegativity.

9.25 According to *valence bond theory*, a chemical bond between two atoms occurs when a valence orbital on each atom overlaps and four electrons are shared.

9.26 The oxygen atom in water uses sp^3 *hybrid orbitals*.

9.27 According to *MO theory*, a sigma molecular orbital is never higher in energy than a pi molecular orbital.

9.28 A *bonding MO* concentrates electron density between two bonded atoms.

9.29 A *MO energy-level diagram* for F_2 shows the highest molecular orbital is empty of electrons.

9.30 A *pi molecular orbital* concentrates electron density along the bonding axis.

Problems and Short-Answer Questions

9.31 Which of the following patterns of bonding and nonbonding electrons about a central atom show the octet rule?

9.32 Estimating the enthalpy of reaction using bond energies requires an ability to analyze a chemical reaction to determine which bonds are broken and formed. From the following diagram determine which bonds, and how many of each, are broken and formed.

9.33 A molecule has the formula AB_3X. What physical measurements tell you the general shape of the molecule? To more precisely define the molecular structure what additional information do you need?

9.34 What are common possible molecular shapes for a molecule with the general formula AB_3? Draw the three structures.

9.35 What is the difference between electron-domain geometry and molecular geometry?

9.36 What is the role of double and triple bonds in determining molecular geometry?

9.37 To predict the dipole of a simple molecule what information do you need and why?

9.38 You are told that atom A in the molecule $:AX_3$ uses sp^3 hybrid atomic orbitals in bonding to atom X. What does the term sp^3 mean? How many electron domains exist? What does this information tell you about the X—A—X bond angles?

9.39 Differentiate between an atomic orbital and a molecular orbital (MO). What are two conditions necessary for a bonding MO to form?

9.40 The bond order of He_2 is zero. What is bond order and what does a value of zero mean?

9.41 Draw the Lewis structures for the following molecules and predict their molecular structures and the types of hybrid orbitals used by the central atoms:

    (a) XeO_4 (d) ICl_2^-

    (b) PO_4^{3-} (e) SbF_5

    (c) TeF_6

9.42 Draw the Lewis structures for CO, CO_2, and CO_3^{2-} and predict the trend in C—O bond distances.

9.43 Write the MO electron description for F_2 and F_2^+. Is either species paramagnetic?

9.44 In each of the following molecules, indicate the directions of the individual bond dipoles and predict the direction of the overall molecular dipole, if any:

    (a) ClCN   (c) BrCCBr

    (b) NO   (d) CF_4

9.45 Give the formula of a molecule or ion in which

    (a) diatomic oxygen has a bond order of $1\frac{1}{2}$ (also construct the molecular-orbital energy levels for this oxygen species);

    (b) carbon uses sp^2 hybrid orbitals;

    (c) nitrogen forms three σ bonds;

    (d) phosphorus has five σ bonds.

9.46 Using NO_3^-, which contains delocalized π electrons, describe the hybrid atomic orbitals used by nitrogen. Would you expect NO_3^- to have a dipole moment?

9.47 Suggest a reason for the fact that at room temperature CO_2 is a gas but SiO_2 is a solid. (Hint: Silicon forms four σ bonds to oxygen, not two as might be indicated by the formula SiO_2.)

9.48 Would you expect N_2 or O_3 to be more reactive? Why?

9.49 A substance has the formula AB_3 and it is determined that the A—B bond is polar. Is this sufficient information to conclude that AB_3 has a molecular dipole? If not, what additional information is needed and why?

9.50 A molecular orbital is symmetrical about the bonding axis between two atoms and has electron density concentrated along the axis. Is this sufficient information to conclude that it is a sigma bonding molecular orbital and that it forms from the combination of s atomic orbitals? If not, what additional information is needed and why?

9.51 A molecule has the formula AX_4 and there are four bonding electron domains about the central atom A. Is this sufficient information to predict that it has a tetrahedral molecular structure? If not, what additional information is needed and why?

Integrative Exercises

9.52 Ethylene gas, C_2H_2, is extensively used in the petroleum and polymer industries. Conceptually one might prepare ethylene from ethane, C_2H_6, by a reaction involving the loss of two hydrogen atoms:

$$C_2H_6(g) \longrightarrow C_2H_4(g) + H_2(g)$$

(a) What are the Lewis structures of ethylene and ethane?

(b) What is the structure about each carbon atom? What types of hybrid orbitals does each carbon atom use in bonding to hydrogen atoms?

(c) Given the following average bond enthalpies (BE),

	kJ/mol
C=C	614
C—C	348
C—H	413
H—H	436

calculate an estimated heat of reaction for the proposed synthesis of ethylene.

(d) Does it seem likely that ethylene gas can be made commercially from the proposed reaction? Explain.

9.53 The molecules SF_2, SF_4, and SF_6 are known.

(a) Draw the Lewis structure for each.

(b) Predict the structure of each.

(c) What type of hybrid atomic orbitals are used by each sulfur atom in bonding to fluorine atoms?

(d) Which are polar?

(e) Why can sulfur exhibit a variety of sulfur-fluorine compounds but oxygen does not?

9.54 Acetic acid contains 39.9% C, 6.7% H, and 53.4% O and has a molar mass of 60.0 g/mol.

(a) What are the empirical and molecular formulas for acetic acid?

(b) The empirical formula for acetic acid is identical to the molecular formula for formaldehyde. Draw the Lewis structures for both.

(c) What is the molecular geometry about each carbon atom in formaldehyde and acetic acid?

(d) What hybrid orbitals does each carbon atom utilize in the two structures?

Multiple-Choice Questions

9.55 Which of the following possess delocalized π electrons?

(a) CH_2Cl_2
(b) OCN^-
(c) N_2O
(d) (a) and (b)
(e) (b) and (c)

9.56 N_2O has several resonance forms, one of which is

$$\overset{a \quad b \quad c}{:\ddot{N}-N\equiv O:}$$

What are the formal charges, a, b, and c?

	a	b	c
(a)	+1	−1	0
(b)	−1	+1	0
(c)	0	+1	−1
(d)	−2	+1	+1
(e)	−2	0	+2

9.57 Which molecule possesses a trigonal-bipyramidal structure?

(a) ICl_2
(b) BrF_3
(c) SF_4
(d) PF_5
(e) PF_6^-

9.58 Which has a trigonal-pyramidal structure?

(a) BF_3
(b) NH_3
(c) $SOCl_2$
(d) (a) and (b)
(e) (b) and (c)

9.59 In which molecule would you expect the central atom to use sp^3d^2 hybrid orbitals?

(a) PF_5
(b) SF_6
(c) CO_2
(d) SiO_2
(e) SO_3

9.60 In a diatomic molecule of the second row, which MO has the highest energy?

(a) σ_{1s}
(b) σ_{2s}^*
(c) σ_{2p}
(d) π_{2p}
(e) π_{2p}^*

9.61 Which molecule has a bond order of two based on MO theory?

(a) Li_2
(b) B_2
(c) C_2
(d) N_2
(e) F_2

9.62 Which molecule does *not* have a dipole moment?

(a) CO
(b) HBr
(c) CH_3Cl
(d) XeF_4
(e) NH_3

9.63 Which statement about BrF_5 is correct?

(a) It has a trigonal bipyramidal structure.

(b) Bromine uses sp^3d hybrid orbitals in bonding to fluorine.

(c) It should have the same molecular structure as PF_5.

(d) The F—Br—F bond angles should be about 85°.

(e) Delocalization of electrons occurs.

SELF-TEST SOLUTIONS

9.1 (m). **9.2** (l). **9.3** (b). **9.4** (n). **9.5** (a). **9.6** (j). **9.7** (k). **9.8** (c). **9.9** (o). **9.10** (h). **9.11** (i). **9.12** (e). **9.13** (f). **9.14** (d). **9.15** (g). **9.16** True. **9.17** False. The reverse is true. **9.18** False. The bond angle is less than the ideal bond angle of 120 degrees. The electron domains of the double bond of C=O in the molecule causes the H—C—H bond angle to contract because of electron–electron repulsions. **9.19** True. **9.20** False. The molecular geometry is trigonal pyramid. Note that the electron-domain geometry is not the same as the molecular geometry. **9.21** True. **9.22** True. **9.23** True. **9.24** False. A bond dipole exists when electronegativity values are different. **9.25** False. Two electrons are shared in a single chemical bond. **9.26** True. **9.27** False. For certain elements, a 2s sigma molecular orbital is higher in energy than a 2p sigma molecular orbital. **9.28** True. **9.29** True. **9.30** False. A sigma MO concentrates electron density along the bonding axis. Electron density in a pi molecular orbital lies between the bonded atoms, but above and below the internuclear bonding axis. There is a node along this axis.

9.31 Figures **a**, **d**, **e**, and **f** show central atoms which obey the octet rule. Each atom has a combination of lines (a pair of electrons) and pairs of electrons (two dots) adding to a total of eight electrons.

9.32 The following bonds are broken: 1 C=C 4 C—H; and 1 H—Br. The following bonds are formed: 1 C—C; 5 C—H; and 1 C—Br.

9.33 If you know the bond angles between atoms, B—A—B and B—A—X, you can define the general shape of the molecule. If you also know the bond distances A—B and A—X you also can define the size of the molecule.

9.34

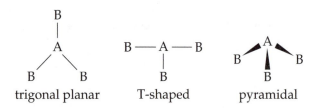

| trigonal planar | T-shaped | pyramidal |

9.35 The geometrical arrangement of electron domains about the central atom in a molecule of type AB_n is referred to as electron-domain geometry. The arrangement of atoms in three-dimensional space is the molecular geometry. We predict the molecular geometry from its electron-domain geometry and arrangement of atoms.

9.36 A double or triple bond is considered as one electron domain when determining molecular geometry. Pi bonds parallel a sigma bond and thus they do not alter the basic directional character of a sigma bond. However, there will be a greater concentration of electron density in a double or triple bond and this will cause the electron domain to inter-

act more strongly with other electron domains—that is, a greater degree of electron–electron repulsions.

9.37 To predict the dipole of a simple molecule you need two key pieces of information: The nature of each bond dipole and the molecular structure. The former is determined by looking at the difference in electronegativity values for the bonded atoms. The second is predicted using VSEPR theory. Both the magnitude and directions of the bond dipoles must be considered. If the bond dipoles are symmetrical about the central atom and equivalent, then they will cancel and the molecule will not have a dipole. The molecule should have a dipole if they are not equivalent and/or not symmetrical.

9.38 The term sp^3 means there are four hybrid atomic orbitals formed from a blending of one s atomic orbital and three p atomic orbitals. There are four sp^3 hybrid atomic orbitals, and each, when occupied by one or two electrons, describes an electron domain. The electron domains form a tetrahedral geometry. In this question the formula is: AB_3. Three of the sp^3 hybrid atomic orbitals are used by A in bonding to B; the fourth one contains a pair of nonbonding electrons. This gives rise to bonding-nonbonding electron repulsions and would cause the B—A—B bond angle to contract to somewhat less than 109.5°.

9.39 An atomic orbital is an energy state for an electron in an atom; it is often represented graphically by a 90% contour plot. One to two electrons can be in an atomic orbital and they are localized on the atom. A molecular orbital can also hold one to two electrons but it is associated with the entire molecule or region within a molecule. A molecular orbital forms when two atomic arbitals overlap.

9.40 Bond order is related to the stability of a covalent bond in molecule. It is defined as: 1/2 (no. of bonding MO electrons − no. of antibonding MO electrons). A bond order of zero means that no bond exists; thus He_2 should be an unstable molecule, which it is.

9.41 (a)

:Ö:
|
:Ö—Xe—Ö:
|
:O:

Tetrahedral—
sp^3

(b)

$$\left[\begin{array}{c} :\ddot{O}: \\ | \\ :\ddot{O}—P—\ddot{O}: \\ | \\ :\ddot{O}: \end{array} \right]^{3-}$$

Tetrahedral—
sp^3 hybrids

(c)

:F:
:F.͵ | .͵F:
Te
:F: | ͵F:
:F:

Octahedral—
d^2sp^3 hybrids

(d) $[:\ddot{Cl}—\dot{I}—\ddot{Cl}:]^-$

Linear, dsp^3 hybrids. The unshared electron domains on iodine are in the radial (equatorial) plane of the trigonal bipyramidal arrangement of electron domains.

(e) The structure is trigonal bipyramidal and the central atom uses dsp^3 hybrid orbitals.

$$:\ddot{F}:$$
$$| \quad \cdot\ddot{F}\cdot$$
$$:\ddot{F} - Sb$$
$$| \quad \cdot\ddot{F}\cdot$$
$$:\ddot{F}:$$

9.42 The Lewis structures for CO, CO_2, and CO_3^{2-} respectively are:

$$:C=O:, \ddot{O}=C=\ddot{O}$$

$$\left[\begin{array}{c} :O: \\ || \\ C \\ :\ddot{O} \quad \ddot{O}: \end{array}\right]^{2-} \longleftrightarrow \left[\begin{array}{c} :\ddot{O}: \\ | \\ C \\ :\ddot{O} \quad \ddot{O}: \end{array}\right]^{2-} \longleftrightarrow \left[\begin{array}{c} :\ddot{O}: \\ | \\ C \\ :\ddot{O} \quad \ddot{O}: \end{array}\right]^{2-}$$

The C—O bond types are: CO, a triple bond; CO_2, double bonds; and CO_3^{2-}, a bond that is between a single and a double bond. The greater the number of bonds between the C and O atoms, the shorter the bond distance will be. Thus we predict that the C—O bond distances decrease in the order $CO_3^{2-} > CO_2 > CO$.

9.43 F_2 (18 electrons): $(\sigma_{1s})^2 (\sigma_{1s}^*)^2 (\sigma_{2s})^2 (\sigma_{2s}^*)^2 (\sigma_{2p})^2$ $(\pi_{2p})^2 (\pi_{2p})^2 (\pi_{2p}^*)^2 (\pi_{2p}^*)^2$; F_2^+ (17 electrons): $(\sigma_{1s})^2 (\sigma_{1s}^*)^2$ $(\sigma_{2s})^2 (\sigma_{2s}^*)^2 (\sigma_{2p})^2 (\pi_{2p})^2 (\pi_{2p})^2 (\pi_{2p}^*)^2 (\pi_{2p}^*)^1$. F_2^+ is paramagnetic because one of its π_{2p}^* molecular orbitals has an unpaired electron. F_2 is diamagnetic because all of its electrons are paired.

9.44 (a) $Cl \longleftrightarrow C \longleftrightarrow N$
   linear molecule

Since $EN_{Cl} = EN_N$, we predict no molecular dipole.

(b) $N \longleftrightarrow O$ linear molecule
The bond dipole and molecular dipole are coincident.

(c) $Br \leftarrow C - C \rightarrow Br$ linear molecule
No molecular dipole exists as the bond dipoles cancel because of their equivalence in magnitude and symmetrical arrangement.

(d)   F   tetrahedral molecule
$$F \longleftrightarrow C \longleftrightarrow F$$
$$F$$

CF_4 is a tetrahedral molecule. No net molecular dipole exists.

9.45 (a) O_2^-: The MO energy levels are $(\sigma_{1s})^2 (\sigma_{1s}^*)^2$ $(\sigma_{2s})^2 (\sigma_{2s}^*)^2 (\sigma_{2p})^2 (\pi_{2p})^2 (\pi_{2p})^2 (\pi_{2p}^*)^2 (\pi_{2p}^*)^1$.

(b) In $H_2C_2H_2$, the arrangement of atoms around each carbon atom is trigonal planar. Therefore, each car-

$$\begin{array}{cc} H\diagdown & \diagup H \\ C=C \\ H\diagup & \diagdown H \end{array}$$

bon is utilizing sp^2 hybrids.

(c) NH_3 contains three single bonds that are σ bonds.

(d) PCl_5 contains five single bonds that are σ bonds.

9.46 As explained in Exercise 8, all N—O bonds are

equivalent because of π-electron delocalization:
Since NO_3 has a trigonal-planar structure, the bond dipoles cancel, and there is no net molecular dipole. For a trigonal-planar structure to occur, nitrogen must use sp^2 hybrid orbitals for bonding to the three oxygen atoms. The remaining unhybridized p orbital on nitrogen is involved in the delocalized π bond.

9.47 CO_2 is a nonpolar molecular compound; thus electrostatic forces holding CO_2 molecules together are minimal, and it therefore exists as a gas. In SiO_2, a diamond-like lattice of Si—O bonds exists, with each silicon atom bonded to four oxygen atoms. Therefore, the total forces holding Si and O atoms together are more numerous than in the case of CO_2, and SiO_2 exists as a solid.

9.48 The Lewis structure of N_2 is

$$:N \equiv N:$$

and the Lewis structure of O_3 is

$$\begin{array}{cc} \ddot{O} & \ddot{O} \\ \diagup \diagdown & \diagup \diagdown \\ :\ddot{O} \quad \ddot{O}: & :\ddot{O} \quad \ddot{O}: \end{array}$$

As shown by these structures, there is a triple bond between the nitrogen atoms, and the bond between oxygen atoms in ozone is between a single and a double bond. Since there are fewer net bonds between O atoms in O_3 than between N atoms in N_2, we expect the O—O bond in O_3 to be weaker than the N—N bond in N_2. Thus we would expect O_3 to be more reactive. This is verified by experimental evidence.

9.49 There is insufficient information. We also need the molecular structure of AB_3.

9.50 There is sufficient information to conclude that the bonding molecular orbital is a sigma type. A bonding sigma molecular orbital is symmetrical about the bonding axis and has electron density concentrated along the bonding axis. There is insufficient information to conclude that the molecular orbital forms from combining s atomic orbitals. The molecular orbital could also, for example, form from the combination of $2s$ and a $2p_z$ atomic orbitals if the z axis is the bonding axis. We need additional information about the energy of the atomic orbitals on both atoms and how they might interact (this will require complex calculations that are beyond the scope of this text).

9.51 There is insufficient information. We also need to know the number of nonbonding electron domains. For example, if it has two nonbonding electron domains then we would predict a square planar structure. If it has one nonbonding electron domain then we would predict a seesaw structure; and if it has zero nonbonding electron domains then we would predict a tetrahedral structure.

9.52 (a)

H — C — C — H   ethane

$C=C$ ethylene

(b) The structure about each carbon atom in ethane is tetrahedal and about each carbon atom in ethylene it is trigonal planar. If the carbon atoms in ethane are using sp^3 hybrid atomic orbitals a tetrahedral geometry about each carbon atom results. If the carbon atoms in ethylene are using sp^2 hybrid atomic orbitals a trigonal planar geometry about each carbon atom results.

(c) The heat of reaction is calculated using the equation

$$\Delta H_{rxn} = \Sigma(\text{Bond Enthalpies of Reactants}) - \Sigma(\text{Bond Enthalpies of Products})$$

$$\Delta H_{rxn} = BE(C-C) + 6[BE(C-H)] - BE(C=C) - 4[BE(C-H)] - BE(H-H)$$

$$\Delta H_{rxn} = 348\,kJ + 6(413\,kJ) - 614\,kJ - 4(413\,kJ) - 436\,kJ = 124\,kJ$$

(d) The proposed synthesis of ethylene requires a significant amount of heat input which means a commercial process would require an expensive energy source. Another method of preparing ethylene gas would be likely.

9.53 (a), (b), (c), and (d)

bent sp^3 polar

seesaw dsp^3 polar

octahedral d^2sp^3 nonpolar

(e) Sulfur is a larger atom than oxygen and can accommodate more than four atoms about it. Furthermore, sulfur has accessible empty d orbitals which it can use to participate in more bond formation.

9.54 (a) A 100 g sample of acetic acid contains 39.9 g of carbon, 6.7 g of hydrogen, and 53.4 g of oxygen. The number of moles of each element in that sample is calculated by the following approach: number of moles = (number of grams) $\left(\dfrac{1 \text{ mol of element}}{\text{molar mass of element}}\right)$. Using this expression results in the following number of moles: 3.32 mol C; 6.6 mol H; and 3.34 mol O. Dividing by the smallest number of moles gives the following ratio of moles: CH_2O. This is the empirical formula and it has an empirical mass of 30.0 g. Thus, the number of empirical formula units in the molecular formula $(CH_2O)_x$ is x = 60.0/30.0 = 2. The molecular formula is $C_2H_4O_2$.

(b)

formaldehyde   acetic acid

(c) Trigonal planar about carbon in formaldehyde. Tetrahedral about the carbon attached to three hydrogen atoms and one carbon atom, and trigonal planar about the carbon atom attached to two oxygen atoms and a carbon atom in acetic acid.

(d) sp^2, sp^3, and sp^2 in the same order of carbon atoms as in part (c). **9.55** (e) **9.56** (d) **9.57** (d) **9.58** (e) NH_3 and $SOCl_2$ are trigonal pyramidal; BF_3 is trigonal planar. **9.59** (b) **9.60** (e) **9.61** (c) **9.62** (d) Square planar: All bond dipoles and nonbonding-electron domain dipoles cancel because of the symmetrical structure. **9.63** (d) BrF_5 is a AB_5E molecule where E means one nonbonding electron domain (see Table 9.3 in the text). Thus, it has a square pyramidal geometry. The nonbonding electron domain causes the bonding electron domains to compress from the idealized geometry of 90° to a lesser value. The actual bond angle is 83.5°.

Sectional MCAT and DAT Practice Questions III

Many important small molecules and ions contain either carbon–oxygen or sulfur–oxygen bonds. Among these are the following: CO, CO_2 H_2CO, CH_3CO_2H, CO_3^{2-}, SO_2, SO_3, and SO_4^{2-}. Carbon monoxide and carbon dioxide are both found in the atmosphere and are toxic to humans in sufficient concentrations. Vinegar is a solution of approximately 5% by mass of CH_3CO_2H (acetic acid) in water. Scientists believe water exists on Mars because they have found sulfate salts and hematite, an iron–oxygen compound.

Combustion reactions are reactions that produce heat and a flame. Most combustion reactions involve oxygen as a reactant; however, this is not a requirement for this class of reactions. Hydrocarbons are compounds containing carbon and hydrogen atoms only. In a sufficient amount of oxygen hydrocarbons burn to form water and carbon dioxide.

Average bond enthalpies are given in Table 1 for several combinations of elements.

TABLE 1 Average bond enthalpies in kJ/mol

X—A	kJ/mol	X=A	kJ/mol	X≡A	kJ/mol
C—H	413	O=O	495	C≡C	839
C—O	358	C=O	799	C≡O	1072
C—C	348	S=O	523	C≡N	891
H—H	436	N=O	607	N≡N	941
H—O	463	P=O	545		
P—O	335				

Electronegativity is a relative measure of the ability of an atom in a molecule to attract electrons to itself. Several scales of electronegativities exist. Pauling's electronegativity values are often used to analyze the polarity of chemical bonds. Examples of Pauling's electronegativity values are H, 2.1; B, 2.0; C, 2.5; N, 3.0; O, 3.5; Si, 1.8; and P, 2.1.

1. Which bond is the most polar?
   (a) C—H (b) C—C (c) C—O (d) N—O
2. A 125.0 g solution contains 10.0% by mass of acetic acid. What is the mass of the solvent?
   (a) 12.5 g (b) 112.5 g (c) 115.0 g (d) 125.0 g

3.   Which bond releases the most energy when it is formed from its atoms?
      (a) $C{=}O$ (b) $N{\equiv}N$ (c) $S{=}O$ (d) $H{-}O$

4.   Which is the least likely substance to form when H_2CNH is combusted in a sufficient amount of oxygen?
      (a) N_2 (b) H_2O (c) CO_2 (d) CO

5.   Which has a trigonal planar structure about the central atom?
      (a) $SO_4{}^{2-}$ (b) H_2S (c) H_2CO (d) NH_3

6.   Which set of hybrid orbitals does boron use in bonding to fluorine in BF_3?
      (a) sp (b) sp^2 (c) sp^3 (d) sp^2d

7.   Which is the correct chemical formula for aluminum sulfate?
      (a) $Al(SO_4)_3$ (b) $Al_2(SO_4)_3$ (c) Al_3SO_4 (d) $AlSO_4$

Questions 8 through 12 are **not** based on a descriptive passage.

8.   Which element has the largest first ionization energy?
      (a) S   (b) P   (c) Si   (d) Na

9.   Which order of ion sizes is correct?
      (a) $Mg^{2+} > Na^+$ (b) $F^- > O^{2-}$ (c) $Rb^+ > K^+$ (d) $Ca^{2+} > Sr^{2+}$

10.   How many nonbonding valence electrons are possessed by the bromine atom in BrF_3?
      (a) 0   (b) 2   (c) 4   (d) 6

11.   Which type of hybridization is associated with a central atom surrounded by bonded atoms in an octahedral structure?
      (a) sp^2d (b) sp^3d (c) sp^3d^2 (d) sp^3d^4

12.   What is the total number of pi electrons in H_2CNH?
      (a) 0   (b) 2   (c) 4   (d) 6

ANSWERS

1. (c) The larger the difference in electronegativity values between bonded atoms, the more polar the bond. The $C{-}O$ bond has the largest difference, $|2.5{-}3.5|$ or 1.0. The other pairs of atoms have differences less than 1.0.

2. (b) The solution has a mass of 125.0 g and 10.0% of it is acetic acid by mass. This means $(0.10)(125.0$ g) or 12.50 g of the solution is acetic acid. Thus the mass of solvent is 125.0 g $-$ 12.5 g $=$ 112.5 g.

3. (a) An average bond enthalpy for a two bonded atoms represents the average enthalpy change for breaking the bond to form the separated atoms. Bond breaking is an endothemic process. The pair of atoms with the largest average bond enthalpy is the one which shows the largest amount of energy released when the bond is formed, the reverse of breaking the bond.

4. (d) The passage states that carbon dioxide and water form when a hydrocarbon burns in sufficient oxygen. H_2CNH contains carbon, hydrogen, and nitrogen atoms. Thus, because it contains carbon and hydrogen atoms, it should combust to form carbon dioxide and water. The nitrogen atom must form a new substance after combustion. Nitrogen should readily form N_2 because of the strong triple bond between the two nitrogen atoms and the large average bond enthalpy. CO readily forms in a limited quantity of oxygen. In a sufficient quantity of oxygen it forms CO_2 as stated in the passage. Thus, CO is the best answer.

5. (c) To determine the type of hybrid atomic orbitals used by an atom draw the Lewis structure of each molecule and determine the number of bonded and nonbonded pairs of electrons about the central atom. Apply the VSEPR model to determine the type of hybrid orbitals that might be used by the central atom. The Lewis structure for H_2CO is

About carbon, the central atom, there are two single bonds, one double bond, and no nonbonded electron pairs. In the VSEPR model this type of bonding arrangement leads to a trigonal planar structure about the carbon atom. The sulfur atom in the sulfate ion has four single bonds and no nonbonded electron pairs resulting in a tetrahedral structure. H_2S has two S—H bonds and two sets of nonbonding electron pairs on sulfur which results in a bent structure. NH_3 has a nitrogen atom with three single bonds and one nonbonded electron pair resulting in a trigonal pyramid structure.

6. (b) The Lewis structure for BF_3 is

$$:\ddot{F}: \\ | \\ B \\ \diagdown \diagup \\ :\ddot{F}: \quad :\ddot{F}:$$

Boron is as an exception to the octet rule in this structure. The structure is trigonal planar and the associated hybrid orbitals used by boron are sp^2.

7. (b) The sulfate ion has a 2− charge and an aluminum ion has a 3+ charge. Two aluminum ions combine with three sulfate ions, resulting in the sum of charges equaling zero in the compound.

8. (b) Within each row of the periodic table of representative elements the first ionization energy generally increases with increasing atomic number. There are a few significant exceptions such as N and P having higher first ionization energies compared to their adjacent atoms in the same period. Both Si and S are adjacent to P and Na is the first member of the same period; therefore, P should have the largest first ionization energy.

9. (c) The size of an ion depends on the nuclear charge of the atom, the number of electrons it possesses and the types of orbitals in the valence shell. Removing an electron from an atom creates a positive ion which results in a smaller sized particle. Adding an electron to an atom creates a negative ion which increases the size of the particle. In (a) both atoms have the same valence shell and removing two electrons from Mg, along with its larger nuclear charge, creates a smaller sized ion compared to Na^+. In (b) an isoelectronic series of ions occurs with the same number of electrons for each element. The nuclear charge increases from O to F and this results in F^- being smaller in size compared to O^{2-}. In (c) both Rb and K are in the same family and size increases with increasing atomic number. Thus Rb^+ should be larger than K^+. In (**d**) Sr^{2+} is larger than Ca^{2+} for reasons given in (**c**).

10. (**c**) The Lewis structure for BrF_3 is
Br has two sets of nonbonding electrons about it, or four electrons.

11. (**c**) An octahedral structure about a central atom is shaped by six hybrid orbitals using the types and numbers of atomic orbitals indicated in the answer.

12. (**b**) The Lewis structure for H_2CNH is

$$\overset{\displaystyle H}{\underset{\displaystyle H}{N}}=C\overset{\displaystyle H}{\underset{\displaystyle H}{\diagup\diagdown}}$$

The $C=N$ bond contains a sigma and one pi bond. A pi bond contains two electrons; thus, the total number of pi electrons is two.

Chapter 10

Gases

OVERVIEW OF THE CHAPTER

Review: States of matter (1.2); dimensional analysis (1.6).

Learning Goals: You should be able to:

1. Describe the general characteristics of gases as compared to other states of matter, and list the ways in which gases are distinct.
2. Define atmosphere, torr, and pascal, the most important units in which pressure is expressed. Also describe how a barometer and manometer work.

Review: Density (1.4); temperature (1.4); mole (3.4); stoichiometry (3.1, 3.6).

Learning Goals: You should be able to:

1. Describe how a gas responds to changes in pressure, volume, temperature, and quantity of gas.
2. Use the ideal-gas equation to solve for one variable (P, V, n, or T), given the other three variables or information from which they can be determined.
3. Use the gas laws, including the combined gas law, to calculate how one variable of a gas (P, V, n, or T) responds to changes in one or more of the other variables.
4. Calculate the molar mass of a gas, given gas density under specified conditions of temperature and pressure. Also calculate gas density under stated conditions, knowing molar mass.

Learning Goals: You should be able to:

1. Calculate the partial pressure of any gas in a mixture, given the composition of that mixture.
2. Calculate the mole fraction of a gas in a mixture, given its partial pressure and the total pressure of the system.

10.1, 10.2 PROPERTIES AND CHARACTERISTICS OF GASES

10.3, 10.4, 10.5 USING THE GAS LAWS: SOLVING PROBLEMS

10.6 DALTON'S LAW OF PARTIAL PRESSURES: MIXTURE OF GASES

10.7, 10.8 THE KINETIC-MOLECULAR THEORY OF GASES: MOLECULAR SPEEDS AND EFFUSION

Review: Molecules (2.6); kinetic energy (6.1).

1. Describe how the distribution of speeds and the average speed of gas molecules changes with temperature.
2. Describe how the relative rates of effusion and diffusion of two gases depend on their molar masses (Graham's law).
3. Use the principles of the kinetic-molecular theory of gases to explain the nature of gas pressure and temperature at the molecular level.

Learning Goals: You should be able to:

10.9 DEPARTURES FROM IDEAL-GAS BEHAVIOR

1. Explain the origin of deviations shown by real gases from the relationship $PV/RT = 1$ for a mole of an ideal gas.
2. Cite the general conditions of P and T under which real gases most closely approximate ideal-gas behavior.
3. Explain the origins of the correction terms to P and V that appear in the van der Waals equation.

TOPIC SUMMARIES AND EXERCISES

PROPERTIES AND CHARACTERISTICS OF GASES

Gases play an important role in the support of human life. If we lose contact with the atmosphere—which is a mixture of gases—for more than 4–5 min, permanent brain damage occurs; longer loss of contact eventually results in death. This is just one example of why we need to fully understand the physical properties and characteristics of gases and the physical laws that govern their behavior.

From observing the behavior of gases we find that:

- They expand to fill the container in which they are enclosed.
- They are compressible.
- They readily flow.
- They form homogenous mixtures with other gases.
- The volume of the gas molecules themselves is only a small portion of the total volume in which they are contained at room temperature and pressure.

The state (condition) of a gas is defined by giving the following properties:

- Temperature (in K) − T
- Volume (usually liters) − V
- Quantity (usually moles) − n
- Pressure [chemists commonly use atmosphere (atm) as a unit, but the SI unit is pascal (Pa)] − P

You need to remember the following about **gas pressure**:

- It results from gas molecules striking a surface with which they are in contact. Pressure is the force exerted by gas molecules striking a given surface area: $P = \frac{F}{A}$
- A **barometer** contains a column of mercury whose height is directly related to the pressure exerted by the atmosphere. Thus atmospheric pressure may be reported in units related to the height of the mercury column: mm Hg (or torr).

- A **standard atmospheric pressure** represents 1 atm of pressure. This corresponds to the amount of pressure that supports a column of Hg 760 mm in height in a barometer.
- A **manometer** is used to measure the pressure of enclosed gases, usually below atmospheric pressure. The pressure exerted is determined from the difference in heights of mercury levels in a U-tube.
- Units of pressure and conversion factors you should be familiar with are:

$$1 \text{ atm} = 760 \text{ mm Hg} = 760 \text{ torr} = 101.3 \text{ kPa}$$

$$1 \text{ atm} = 29.9 \text{ in Hg} = 33.9 \text{ ft H}_2\text{O} = 10.3 \text{ m H}_2\text{O}$$

EXERCISE 1   Choosing a liquid for a barometer

(**a**) Suppose a barometer is constructed using methanol as the liquid. Given that the density of Hg is 13.6 g/mL and that of methanol is 0.791 g/mL, and disregarding the effect of methanol gas molecules above methanol liquid, calculate the height of the methanol column at 1 atm of pressure at 0 °C. (**b**) Why do we use Hg rather than methanol as a liquid in a barometer?

SOLUTION: *Analyze*: In (**a**) we are given the density of mercury (13.6 g/mL) and methanol (0.791 g/mL) and are asked to calculate the height of a methanol column equivalent to 1 atm of pressure at 0 °C. In (**b**) we are asked to explain why mercury rather than methanol is used in a barometer.

Plan: (**a**) At 1 atm of pressure a mercury column has a height of 760 mm. Since the methanol is placed in a column of the same diameter, the force exerted by a column of mercury of 760 mm in length would be equivalent to a column of methanol of the same mass. The density of Hg is 17.2 times that of methanol per unit volume; so it will take 17.2 mm of methanol to equal the same mass of 1 mm of mercury.

(**b**) We compare the heights of mercury and methanol columns at 1 atm pressure and 0 °C to determine which is more practical to use.

Solve: (**a**) We convert 760 mm Hg to mm methanol:

$$(760 \text{ mm Hg})\left(\frac{17.2 \text{ mm methanol}}{1 \text{ mm Hg}}\right) = 13{,}000 \text{ mm methanol}$$

(**b**) From this calculation you can see that mercury is an excellent choice for a liquid in a barometer because its column height at 1 atm is reasonable.

Check: An estimate of the calculation in (**a**) gives $(750)(20) = 15{,}000$, which is of the same magnitude as the exact solution. The answer should be larger than 760 mm because it takes significantly more methanol in a fixed diameter tube to equal the same mass as mercury.

EXERCISE 2   Calculating pressure

A diver is 261 ft below sea level in pure water. What pressure, in atmospheres, is exerted upon the diver?

SOLUTION: *Analyze*: A diver is 261 ft below sea level and we are asked to find the total pressure in atmospheres exerted upon the diver.

Plan: The total pressure exerted upon the diver is the atmospheric pressure above the water plus the pressure exerted by the water above the diver. We can use dimensional analysis and the relationship 1 atm = 33.9 ft H_2O to calculate the pressure exerted upon the diver.

Solve: Using dimensional analysis, the pressure exerted upon the diver is calculated as follows:

Pressure = atmospheric pressure at sea level + pressure exerted by water

$$\text{Pressure} = 1.00 \text{ atm (at sea level)} + (261 \text{ ft})\left(\frac{1 \text{ atm}}{33.9 \text{ ft}}\right)$$

Pressure = 1.00 atm + 7.70 atm = 8.70 atm

Check: An estimate of the calculation gives $1 + (300)/30 = 11$, which is of the same magnitude as the exact solution. The final answer is also in agreement with the concept that the value should be greater than the pressure of the atmosphere.

USING THE GAS LAWS: SOLVING PROBLEMS

Problems involving application of the gas laws require an understanding of the quantitative relationships involving pressure (P), volume (V), quantity (n), and temperature (T in the unit of Kelvin).

Gas laws discovered before the twentieth century that you need to remember are

- **Boyle's law:** For a given amount of a gas, volume is inversely proportional to pressure at constant temperature:

$$V = \frac{\text{constant}}{P} \qquad (T, n \text{ constant})$$

- **Charles's law:** For a given amount of a gas, volume is directly proportional to temperature at constant pressure.

$$V = \text{constant} \times T \qquad (P, n \text{ constant})$$

- **Avogadro's hypothesis:** Equal volumes of gases at the same temperature and pressure contain equal numbers of molecules.
- **Avogadro's law:** At constant pressure and temperature, the volume of a gas is directly proportional to the moles of gas:

$$V = \text{constant} \times n \qquad (P, T \text{ constant})$$

Eventually the four variables pressure (P), volume (V), moles of gas (n), and temperature (T) were tied together into one equation called the **ideal-gas equation:**

$$PV = nRT$$

- T must be expressed in degrees Kelvin.
- R is a constant, called the **gas constant**, whose value depends on the units chosen for pressure and temperature: $R = 0.0821$ L-atm/mol-K if pressure is measured in atmospheres and volume in liters.

Relationships that are useful for solving gas-law problems are in Table 10.1 on the next page.

Exercises 3–6 demonstrate the methodology of solving several kinds of problems involving gases. When in doubt about how to solve problems begin with the ideal-gas equation. It is the most general statement of the quantitative behavior of an ideal gas.

TABLE 10.1 Useful Relationships for Solving Gas Problems

Equation	Comment
Ideal-gas equation: $PV = nRT$	The most general description of the quantitative relationships among P, V, T, and n. It can be used as the starting point to solve most gas-law problems. Typically you use it when you are asked to calculate P, V, T, or n given the other three. See Exercise 4.
Same gas at two different states: $\dfrac{P_1V_1}{T_1} = \dfrac{P_2V_2}{T_2} \ (n \text{ constant})$	This equation is used when the same gas is compared under two different conditions (state 1 and state 2) with n constant. Typically you are asked to calculate P_2, V_2, or T_2 given two of the values and also P_1, V_1, and T_1. See Exercise 4.
Density of gas (g/L): $d = \dfrac{P(\mathcal{M}\text{ of gas})}{RT}$	This is derived from $PV = nRT$. Rather than memorize the equation you should be able to derive it knowing that $\dfrac{n}{V} = \dfrac{m/\mathcal{M}}{V} = \dfrac{g/V}{\mathcal{M}} = \dfrac{d}{\mathcal{M}}$. See Exercise 5.
Molar mass of a gas ($\mathcal{M}$): $(\mathcal{M}) = \dfrac{(\text{grams})(R)(T)}{PV}$	This is also derived from $PV = nRT$ using the relationship $n = m/\mathcal{M}$

EXERCISE 3   Using standard temperature and pressure for gases

There is defined for gases a standard state, given the symbol STP, which is the standard reference condition for the properties of a gas. What is the standard state for a gas? What is the volume of one mole of an ideal gas at STP?

SOLUTION: *The standard state for a gas corresponds to 0 °C (273 K) and 1 atm of pressure.* The letters in STP stand for standard temperature and pressure. The volume of one mole of an ideal gas at STP is 22.4 L.

EXERCISE 4   Solving a gas-law problem in three different ways

A 5.00-L container is filled with $N_2(g)$ to a pressure of 3.00 atm at 250 °C. What is the volume of a container that is used to store the same gas at STP? Solve this problem using each of the following approaches: (1) gas-law equation; (2) $\frac{PV}{T}$ = constant at different conditions if the number of moles is constant; (3) "common sense" and Charles's and Boyle's laws.

SOLUTION: *Analyze*: We are given 5.00 L of nitrogen gas in a container at a pressure of 3.00 atm at 250 °C and asked to determine the volume of a container needed to store the same amount of gas at STP conditions using three different methods: (1) Applying the ideal-gas law equation; (2) recognizing that the amount of gas does not change and thus $\frac{PV}{T}$ = constant; and (3) using "common sense" and Charles's and Boyle's gas laws.

Plan: The question identifies the three different methods to use. First we will use the ideal-gas law $PV = nRT$ to solve for n and then V at STP conditions. Second we will use the relation $\dfrac{P_1V_1}{T_1}$ and the condition that it has a fixed value at STP conditions to solve for V. Third, we will use the concepts of Charles's and Boyle's gas laws to determine how volume changes from the initial conditions to the final conditions.

Solve: We are given that the initial conditions: $P = 3.00$ atm, $V = 5.00$ L, and $T = 523$ K (*remember*: when doing gas problems we must convert degrees Celsius to Kelvin; K $= 250 \, °C + 273 = 523$ K). The problem tells us that the final conditions are at STP (1.00 atm and 273 K) and that we are to calculate the new volume of the gas.

Approach 1: The words *same gas* in the problem tell us that the number of moles of gas does not change; therefore n is constant. If we are not sure how to solve gas-law problems, always start with the ideal-gas law. Let's solve for n using the initial conditions and then use this value of n to solve for the new volume.

$$PV = nRT$$

or rearranging the equation to solve for n

$$n = \frac{PV}{RT} = \frac{(3.00 \text{ atm})(5.00 \text{ L})}{(0.0821 \text{ L-atm/K-mol})(523 \text{ K})} = 0.349 \text{ mol}$$

Use this value of n to solve for the new volume at STP:

$$V = \frac{nRT}{P} = \frac{(0.349 \text{ mol})(0.0821 \text{ L-atm/K-mol})(273 \text{ K})}{1.00 \text{ atm}} = 7.83 \text{ L}$$

Approach 2: Instead of using the previous approach we can shorten the number of steps involved in the calculation if we use the following equation: $P_1V_1/T_1 = P_2V_2/T_2$. It can be used whenever the number of moles of gas does not change.

Substituting the quantities $P_1 = 3.00$ atm, $V_1 = 5.00$ L, $T_1 = (250 + 273)$ K, $P_2 = 1.00$ atm, and $T_2 = 273$ K into the relation $P_1V_1/T_1 = P_2V_2/T_2$ gives

$$\frac{(3.00 \text{ atm})(5.00 \text{ L})}{523 \text{ K}} = \frac{(1.00 \text{ atm})(V_2)}{273 \text{ K}}$$

Solving for V_2 gives a volume of 7.83 L.

Approach 3: When the number of moles of gas remains constant we can use a "common sense" method of solving for the new volume. Boyle's law tells us that volume varies inversely with pressure and Charles's law tells us that it varies directly with absolute temperature. Thus it makes sense to say

New volume = old volume × pressure correction ratio × temperature correction ratio

In this problem the pressure decreases from 3.00 atm to 1.00 atm. A decrease in the pressure increases the volume according to Boyle's law. Thus the pressure correction ratio must increase the volume of the gas. This ratio must be 3.00 atm/1.00 atm. The temperature decreases from 523 K to 273 K. A decrease in the temperature decreases the volume of a gas according to Charles's law. Thus the temperature correction ratio must be less than one. This ratio must be 273 K/523 K. We can now solve for the new volume

$$\text{New Volume} = 5.00 \text{ L} \times \frac{3.00 \text{ atm}}{1.00 \text{ atm}} \times \frac{273 \text{ K}}{523 \text{ K}} = 7.83 \text{ L}$$

This approach is no different from that in Approach 2 except that instead of using an equation we memorized we used our knowledge of the behavior of gases to solve for an unknown quantity.

Check: All three answers agree with each other and this suggests that the calculations were done correctly. Another check is that the units properly cancel to give the desired unit of liters. Estimates of the calculated values in the first method gives

$(3)(5)/(0.1)(500) = 0.3$ and $(0.3)(0.1)(300)/1 = 9$, which have the same magnitudes as the exact solution. This supports the answer for all three methods.

EXERCISE 5 Calculating the density of a gas

Cyclopropane is used as a general anesthetic. It has a molar mass of 42.0 g. What is the density of cyclopropane gas at 25 °C and 1.02 atm?

SOLUTION: *Analyze*: We are asked to determine the density of cyclopropane gas at 1.02 atm and 25 °C and given its molar mass of 42.0 g/mol.

Plan: We will use the ideal-gas law equation to relate density to molar mass, volume, pressure, and temperature. Density is m/V and the number of moles of gas is $n = \left(\frac{m}{\mathcal{M}}\right)$, where $\mathcal{M}$ is the molar mass of the gas.

Solve: In this problem we are given P, T, and the molar mass ($\mathcal{M}$) of the gas. We are asked to calculate density $(d = m/V)$. The ratio m/V for a gas can be related to its pressure, temperature, and ($\mathcal{M}$) by using the ideal-gas equation:

$$PV = nRT = \left(\frac{m}{\mathcal{M}}\right)RT$$

Rearranging the equation so the ratio m/V is on the left side of the equal sign yields:

$$\frac{m}{V} = \frac{(\mathcal{M})P}{RT} = d$$

The gas density of cyclopropane is therefore

$$d = \frac{(\mathcal{M})P}{RT}$$

$$= \frac{\left(42.0\dfrac{\text{g}}{\text{mol}}\right)(1.02 \text{ atm})}{(0.0821 \text{ L-atm/K-mol})(298 \text{ K})}$$

$$= 1.75 \text{ g/L}$$

Converting the unit of density to g/cm^3,

$$d = \left(1.75\frac{\text{g}}{\text{L}}\right)\left(\frac{1 \text{ L}}{10^3 \text{ cm}}\right) = 1.75 \times 10^{-3} \text{ g/cm}^3$$

Check: An estimate of the first calculation gives $(40)(1)/(0.1)(300) = 1.4$, which is in close agreement with the calculated value. Also the answer, 1.75×10^{-3} g/cm^3, is a small number in agreement with the idea that gas molecules are not densely packed.

EXERCISE 6 Using volumes of gases in a stoichiometry problem

What volume of $N_2(g)$ at 720 torr and at 23 °C is required to react with 7.35 L of $H_2(g)$ at the same temperature and pressure?

$$N_2(g) + 3 H_2(g) \longrightarrow 2 NH_3(g)$$

SOLUTION: *Analyze*: We are given the reaction describing how nitrogen gas reacts with hydrogen gas to form ammonia gas and asked what volume of nitrogen at 720 torr and 23 °C is required to react with 7.35 L of ammonia.

Plan: This is a stoichiometry problem and we will need a stoichiometric equivalence involving the number of moles or volume between nitrogen and ammonia to calculate the volume of nitrogen gas. Avogadro's hypothesis is useful when we are given volumes of gases and a chemical reaction.

Solve: Avogadro's hypothesis tells us that equal volumes of gases at the same temperature and pressure contain equal numbers of moles of gas. Therefore, the number of moles of gas and volumes of gas at the same temperature and pressure are directly proportional. This means that the volumes of gases, measured at the same temperature and pressure, are in the same ratio as the coefficients in the balanced chemical equation. From the balanced chemical equation,

$$1 \text{ mole } N_2 \simeq 3 \text{ moles } H_2$$

or since the volumes of gases and the number of moles are directly related at the same temperature and pressure

$$1 \text{ liter } N_2 \simeq 3 \text{ liters } H_2$$

$$\text{Volume } N_2 = 7.35 \text{ liters } H_2 \times \frac{1 \text{ L } N_2}{3 \text{ L } H_2} = 2.45 \text{ liters } N_2$$

Check: An estimate of the calculation gives $7/3 = 2$, which is close to the calculated value. Also the value is in agreement with the principles of Avogadro's hypothesis since three volumes of hydrogen gas are needed to react with one volume of nitrogen gas—the volume of nitrogen must be smaller than 7.35 L, which it is.

DALTON'S LAW OF PARTIAL PRESSURES: MIXTURE OF GASES

The relationship describing the behavior of gases in a mixture was originally proposed by John Dalton in the nineteenth century.

- **Dalton's Law of partial pressure states:** The total pressure (P_t) of a mixture of ideal gases is the sum of the individual pressures (partial pressures) each ideal gas would exert if it were the only gas in the container:

$$P_t = P_1 + P_3 + \cdots P_n \quad (n \text{ gases in mixture})$$

- $P_i = \dfrac{n_i RT}{V}$ (P_i = partial pressure of an individual gas, n_i = moles of an individual gas, V = total volume of container)

- $P_t = (n_1 + n_2 + n_3 + \cdots n_n)\dfrac{RT}{V}$ (Use this equation to calculate P_t when you know the number of moles of each gas)

Dalton's law of partial pressures is applied when measuring the pressure of a gas stored over water.

- A gas stored over water also contains water vapor. At a given temperature the partial pressure of water vapor is known.
- P_t = partial pressure of dry gas + partial pressure of water vapor.

$$P_t = P_{gas} + P_{H_2O}$$

- Therefore, the partial pressure of the dry gas is calculated by

$$P_{gas} = P_t - P_{H_2O}$$

Mole fraction and volume percentage of each gas in a mixture also describe the composition of gas mixtures. Note that mole fraction and volume percentage expressed as a decimal are the same numerically. Key concepts are

- **Mole fraction** (X) is the ratio of the number of moles of a component in a mixture to the total number of moles present.

$$X_i = \frac{\text{number of moles of component } i}{\text{total number of moles in mixture}}$$

- The partial pressure of gas A (P_A) in a mixture can be calculated from its mole fraction and the total gas pressure (P_t) of the mixture:

$$P_A = X_A P_t$$

- The volume percent of a gas in a mixture is equal to the mole fraction of that gas. Thus, a mixture of a gas containing 11.1% of oxygen has a mole fraction of oxygen of 0.111.

EXERCISE 7   Using Dalton's law to calculate the partial pressure of a gas over water

Oxygen gas evolved from heating $KClO_3(s)$ is collected over water in a bottle. The total volume of gas in the bottle is 150.0 mL at 27.0 °C and 810.0 torr. Calculate the partial pressure of the collected $O_2(g)$. The partial pressure of pure water at 27.0 °C is 22.4 torr.

SOLUTION: *Analyze*: We are given the volume (150.0 mL) and temperature (27.0 °C) of oxygen gas in a bottle containing water and the total pressure in the container is 810.0 torr. We are asked to calculate the partial pressure of oxygen gas in the bottle. In addition we are told that the partial pressure of pure water vapor at this temperature is 22.4 torr.

Plan: A gas collected over water contains both water vapor and the gas. Thus, the measured pressure is a sum of the partial pressures of oxygen gas and water vapor. The partial pressure of water vapor is given in the question. We can use Dalton's law of partial pressures to solve for the partial pressure of oxygen gas.

Solve: A gas collected over water contains water vapor. Thus the measured pressure is a sum of the partial pressures of oxygen and water above the water:

$$P_t = P_{O_2} + P_{H_2O}$$

Solving for P_{O_2} gives:

$$P_{O_2} = P_t - P_{H_2O} = 810.0 \text{ torr} - 22.4 \text{ torr} = 787.6 \text{ torr}$$

Check: The partial pressure of oxygen gas must be less than the total pressure, which is confirmed by the calculation.

EXERCISE 8   Using mole fraction and Dalton's law

Answer the following questions concerning a gas mixture that is held in a 1.00-L container at 25.0 °C and contains 0.0200 mol of nitrogen and 0.0300 mol of ammonia at a total pressure of 1.22 atm. (**a**) What is the mole fraction of ammonia gas? (**b**) What is the partial pressure of ammonia gas?

SOLUTION: *Analyze*: We are asked to calculate the mole fraction of ammonia gas and its partial pressure given that a 1.00 L container at 25 °C contains a mixture of gases, 0.0200 mole of nitrogen and 0.0300 mole of ammonia.

Plan: We can use the definition of mole fraction to calculate the mole fraction of

ammonia: $X_{\text{substance}} = \dfrac{\text{moles of substance of interest}}{\text{sum of moles of all substances in mixture}}$. If we know the mole fraction of a substance and the total pressure of a system we can calculate the partial pressure of the substance using the relation:

$$P_{substance} = X_{substance}P_{total}.$$

Solve:

(a) The mole fraction of ammonia is calculated from the relationship

$$X_{NH_3} = \frac{\text{moles of NH}_3}{\text{moles of NH}_3 + \text{moles N}_2}$$

$$= \frac{0.0300 \text{ moles NH}_3}{0.0300 \text{ moles NH}_3 + 0.0200 \text{ moles O}_2} = \frac{0.0300 \text{ moles}}{0.0500 \text{ moles}}$$

$$= 0.600 \quad \text{(note that there are no units)}$$

(b) Substitute the mole fraction of ammonia, X_{NH_3}, and the total pressure, P_t, into the equation

$$P_{NH_3} = X_{NH_3}P_t = (0.600)(1.22 \text{ atm}) = 0.732 \text{ atm}$$

The partial pressure of ammonia is 60% of the total gas pressure, 0.732 atm.

Check: The calculated value of the mole fraction is reasonable because the mole fraction of any substance is less than one. The calculated pressure of ammonia is less than the total pressure which is expected based on Dalton's law of partial pressures.

EXERCISE 9   Calculating total gas pressure using Dalton's law

A 0.50-L container holds 0.25 mol of N_2 and 0.15 mol of He at 25 °C. What is the total pressure exerted by the mixture of gases?

SOLUTION: *Analyze*: We are asked to calculate the total pressure exerted by 0.25 mol of nitrogen and 0.15 mol of helium in a 0.50 L container 25 °C.

Plan: We can use Dalton's law of partial pressures to calculate the total pressure exerted by the gases. The partial pressure exerted by each gas can be calculated using the ideal-gas equation: $PV = nRT$:

$$P_{O_2} = \frac{n_{O_2}RT}{V} \quad \text{and} \quad P_{He} = \frac{n_{He}RT}{V}$$

Rather than substitute the known values into these equations to calculate the partial pressures, let us see if there is a way to reduce the number of calculations. We know from Dalton's law of partial pressures that the total pressure is a sum of partial pressures:

$$P_t = P_{O_2} + P_{He} \quad \text{or} \quad P_t = \frac{n_{O_2}RT}{V} + \frac{n_{He}RT}{V}$$

Rearranging the right-hand side of the equation gives the following:

$$P_t = (n_{O_2} + n_{He})\frac{RT}{V}$$

Solve: We can substitute the number of moles of each gas, the gas-law constant $\left(R, 0.0821 \frac{\text{L-atm}}{\text{mol-K}}\right)$, the temperature in Kelvin (298 K), and the volume in liters (0.50 L) into the previous equation to solve for the total pressure:

$$P_t = \left[\frac{\left(0.0821\dfrac{\text{L-atm}}{\text{mol-K}}\right)(298 \text{ K})}{0.50 \text{ L}}\right] \times (0.25 \text{ mol} + 0.15 \text{ mol}) = 20 \text{ atm}$$

Note that the previous equation can be generalized to:

$$P_t = (n_a + n_b + n_c + \cdots + n_z)\frac{RT}{V}$$

where n_a, n_b, etc. are the number of moles of each gas in the mixture.

Check: An estimate of the total pressure gives $[(0.1)(300)/0.50][0.0.50] = 30$, which is approximately the calculated value. The units also properly cancel to give the desire unit of atm.

Application of the ideal-gas law in solving gas-law problems does not require any knowledge of the behavioral nature of gas molecules. By using some simple assumptions about the behavioral nature of gas molecules, scientists in the nineteenth century formulated the kinetic-molecular theory (KMT) of gases, from which the ideal-gas law can be derived. Section 10.7 of the text has a detailed discussion of the assumptions used in developing the KMT.

| **THE KINETIC-MOLECULAR THEORY OF GASES: MOLECULAR SPEEDS AND EFFUSION** |

The KMT explains pressure and temperature at a molecular level. Perhaps the main idea to grasp from the kinetic-molecular theory of gases is that *at the same temperature, molecules of all gases have the same average translational kinetic energy.* Two types of molecular speeds are discussed in the text: root-mean-square (rms) and average.

- Rms speed, μ, is not the same as average speed. The rms speed is the square root of the average of the squared velocities of the gas molecules:
  $\left[\frac{1}{N}(v_1^2 + v_2^2 + \cdots v_N^2)\right]^{\frac{1}{2}}$ where N is the number of particles and v_i is the velocity of a particular particle.

- The average speed is the average of the velocities: $\frac{1}{N}(v_1 + v_2 + \cdots v_N)$
- The rms speed of gas molecules is slightly larger than the average speed.

- Note in Figure 10.18 of the text that speeds increase with increasing temperature, and also study the shape of the curve.

- The rms speed of a gas is *inversely* proportional to the molar mass of the gas at a given temperature:

$$\mu = \sqrt{\frac{3\,RT}{M}}$$

Thus lighter molecules move faster than heavier molecules at a particular temperature. For example, H_2 molecules move faster than N_2 molecules at 25 °C. See Figure 10.19 in the text.

Effusion is the flow of gas molecules through a small pinhole or small opening into a vacuum. **Graham's law of effusion** relates the rates of two gases escaping through the same pinhole and their molar masses at constant temperature.

- $\dfrac{\text{rate of effusion of } A}{\text{rate of effusion of } B} = \sqrt{\dfrac{M_B}{M_A}}$

- Note that the rate of effusion for substance A is inversely proportional to the mass of one mole of substance A.

- Usually time of effusion, rather than rate of effusion, is measured. Measured times are *inversely* proportional to rates

$$\frac{\text{time}_B}{\text{time}_A} = \frac{\text{rate}_A}{\text{rate}_B} = \sqrt{\frac{\mathcal{M}_B}{\mathcal{M}_A}}$$

Lighter substances effuse faster than heavier substances.

• Do not confuse diffusion with effusion. **Diffusion** is a process in which a substance gradually mixes with another. An example of this is the fragrance of a perfume smelled after the bottle is opened.

Mean free path is the average distance gas molecules travel between collisions. In general, as the molecular size of a gas particle increases, the mean free path decreases. The mean free path also increases with decreasing density of a gas.

EXERCISE 10   Understanding the kinetic-molecular theory of gases

Some of the following ideas of the kinetic-molecular theory of gases are stated incorrectly. Where this is the case, change the statement so it is correct. (a) A gas consists of molecules in ceaseless, ordered motion. (b) These molecules occupy a large percentage of the volume of the container. (c) The time between the collision of two gas molecules is long compared to the time that the molecules are in contact during their collision. (d) Attractive and repulsive forces act between the molecules. (e) The average kinetic energy of the molecules is proportional to the absolute temperature.

SOLUTION: *Analyze:* We are given statements relating to the kinetic-molecular theory of gases and asked to determine which are stated incorrectly and to write them so they are correct.

Plan: Take each statement and compare it to information provided in the text re the kinetic-molecular theory of gases. If the statement does not agree with the information in the text we will correct it.

Solve: (a) A gas consists of molecules in ceaseless, *chaotic* (random) motion. (b) The gas molecules themselves occupy a *small, almost negligible* percentage of the volume of the container. (c) Correct. (d) The attractive or repulsive forces acting between the molecules, including gravitational forces, are negligible. (e) Correct.

EXERCISE 11   Applying Graham's law of effusion

How does the rate of effusion of $SO_2(g)$ compare to that of $O_2(g)$ through the same pinhole at the same temperature and pressure?

SOLUTION: *Analyze:* We are asked to compare the rate of effusion of $SO_2(g)$ to that of $O_2(g)$ through the same pinhole at the same temperature and pressure. We can use Graham's law to make the comparison on a relative basis.

Plan: Graham's law relates the rates of effusion of two gases to their molar masses:

$$\frac{r_1}{r_2} = \sqrt{\frac{\mathcal{M}_2}{\mathcal{M}_1}}$$

Solve: Substituting the molar masses of sulfur dioxide and oxygen into this equation gives:

$$\frac{r_{SO_2}}{r_{O_2}} = \sqrt{\frac{32.0 \text{ g/mol}}{64.1 \text{ g/mol}}} = 0.707$$

The rate of effusion of $SO_2(g)$ molecules is 0.707 of the rate of effusion of $O_2(g)$ molecules. Thus SO_2 molecules move through the hole at a slower rate.

Check: We learned in the text that the greater the molar mass of a gas the slower it moves compared to another gas of lesser molar mass at the same conditions. The calculated answer is in agreement with this concept.

EXERCISE 12 Comparing the average speed of a gas to its root-mean-square speed

The average speed of Ar(*g*) at 273.15 K is 380.8 m/s. How does the value of root-mean-square speed compare to 380.8 m/s?

SOLUTION: *Analyze*: We are asked to compare the root-mean-square velocity of Ar(*g*) to its average speed, 380.8 m/s, at 273.15 K.

Plan: μ is the root-mean-square speed and is calculated using the relationship:

$$\mu = \sqrt{\frac{3\,RT}{\mathcal{M}}}$$

Solve:

Substituting for *R*, *T*, and $\mathcal{M}$ gives:

$$\mu = \sqrt{\frac{3(8.314\ \text{J/K-mol})(273.15\ \text{K})}{39.95\ \text{g/mol}} \times \frac{10^3\ \text{g}}{1\ \text{kg}}}$$

The conversion of g to kg is necessary so that the units can be converted to m^2/s^2 using the equivalence: $1\ \text{J} = 1\ \text{kg}-m^2/s^2$.

$$\mu = \sqrt{1.705 \times 10^5 \frac{\text{J}}{\text{kg}} \times \frac{1\ \text{kg-}m^2/s^2}{1\ \text{J}}} = 413.0\ \text{m/s}$$

This calculation shows that μ (root-mean-square speed) is greater than the average speed for Ar. This is also true for other gases.

Check: An estimate of the calculated value gives

$$\sqrt{\frac{1(10)(300)}{30} \times \frac{10^3}{1}} = \sqrt{10^5} \approx 320$$ which is approximately the answer

calculated in the problem. As mentioned in the problem μ is also slightly greater than the average speed, which is in agreement with the discussion in the text.

The behavior of gases departs from the ideal-gas equation primarily because real molecules have finite volume and they do repel one another.

DEPARTURES FROM IDEAL-GAS BEHAVIOR

- Gas molecules are attracted to one another at short distances. Therefore, gas molecules do not collide with a wall of a container as often as is expected for an ideal gas. This causes the observed (experimental) gas pressure to be less than the ideal-gas pressure.
- The volume available for a gas to move about in a container is less than the total volume of the container. Gas molecules possess finite volume and exclude part of the volume of the container from one another. Therefore, the experimental volume is greater than the ideal volume, the volume actually available to the gas molecules.

The **van der Waals equation** accounts for the above two factors:

$$\left(P_{\text{exp}} + \frac{an^2}{V^2}\right)(V_{\text{exp}} - nb) = nRT$$

The term an^2/V^2 in $\left(P_{exp} + \dfrac{an^2}{V^2}\right)$ corrects for the intermolecular attractions that cause the observed pressure to be lower than what would be expected for an ideal gas. Typically, the more polar a substance is, the larger the value of it will be. The nb term in $(V_{exp} - nb)$ corrects for the volume excluded to gas molecules because of their size. The value of b increases with increasing molecular size.

EXERCISE 13   Using *PV/RT* to understand when a gas behaves as an ideal gas

The quantity PV/RT can be used to show whether 1 mol of a gas acts like an ideal gas as the pressure is varied over a wide range. (**a**) What is the value of PV/RT for 1 mol of an ideal gas? (**b**) When will PV/RT for a gas be greater than 1? (**c**) When will PV/RT for a gas be less than 1?

SOLUTION: *Analyze*: We are told that the ratio PV/RT can be used to demonstrate whether one mole of a gas acts like an ideal gas. We are asked what is the ratio for one mole of an ideal gas and when is the value greater than one or less than one.

Plan: The ideal-gas law equation is $PV = nRT$ and can be rearranged to $n = PV/RT$. For a given temperature the value of the ratio depends on how the experimental values of P and V compare to those expected for an ideal gas. We can analyze how the factors of the size of particles and intermolecular forces affect P and V.

Solve: (*a*) The ratio PV/RT equals n, which has the value of one for 1 mol of an ideal gas. (**b**) The observed value of PV is greater than the ideal value of PV for a gas when the observed volume—that is, the measured volume—is larger than the ideal volume. This happens when the gas molecules occupy a significant portion of the volume of the container. Thus the gas molecules have an available ideal volume that is less than the observed volume. This behavior of gases occurs at very high pressures. (**c**) A gas shows this behavior if it is not an ideal gas and if the following relation exists:

$$PV_{experimental} < PV_{ideal}$$

The observed pressure will be less than the ideal pressure when gas molecules are attracted to each other and do not collide with the walls as frequently as in the ideal case.

Check: The discussion of Figure 10.23 in the text confirms the statements in the solution.

EXERCISE 14   Understanding van der Waals equation

(**a**) Which term in the van der Waals equation accounts for the fact that gas molecules have a finite volume and exclude one another from space? (**b**) Under what conditions does a real gas show ideal-gas behavior?

SOLUTION: *Analyze*: We are asked to identify the term in van der Waals equation that accounts for the sizes of particles and their impact on the behavior of gases and what conditions result in a gas behaving as an ideal gas.

Plan: Van der Waals equation is $\left(P + \dfrac{n^2a}{V^2}\right)(V - nb) = nRT$. We need to determine whether a or b accounts for the sizes of particles. We also need to determine under what conditions the left side of the equation simplifies to PV.

Solve: (**a**) The term nb accounts for the excluded volume. The term $V - nb$ corresponds to the ideal volume. (**b**) When volume becomes very large, the term n^2a/V^2 becomes very small compared to P, and the term $P + n^2a/V^2$ reduces to P. Also, the value of nb in the term $V - nb$ becomes very small compared to V, and the term reduces to V. The equation $(P + n^2a/V^2)(V - nb) = nRT$ then reduces to $PV = nRT$. Note that the volume becomes large when the pressure is low. Also, at high temperature the motional energies are larger relative to intermolecular forces between gas molecules, and this leads to more ideal behavior.

Check: The discussion of van der Waals constants for gas molecules and Table 10.3 in the text confirms the answer to (**a**). The solution to (**b**) is confirmed by the fact that when the values of a and b become zero the left side of the equation becomes PV.

EXERCISE 15   Using van der Waals equation

What is the pressure exerted by one mole of Ar gas at a volume of 2.00 L and at 300 K when it acts as an ideal gas and as a nonideal gas? $a = 1.34$ L^2-atm/mol^2 and $b = 0.0322$ L/mol for Ar(g). Explain any significant departure from ideality. Qualitatively predict if $CO_2(g)$ would show greater or lesser departure from ideality. $a = 3.59$ L^2-atm/mol^2 and $b = 0.427$ L/mol for $CO_2(g)$.

SOLUTION: Analyze: We are given one mole of Ar gas with a volume of 2.00 L at 300 K and its van der Waals constants and asked what pressure does the gas exert. We are also given van der Waals constants for carbon dioxide gas and asked to qualitatively predict whether it departs from ideal-gas behavior to a greater or lesser extent than argon gas at the same conditions.

Plan: We will use van der Waals equation to answer both questions.

Solve: First, rearrange the van der Waals equation so that pressure is only on one side of the equal sign:

$$P + \frac{n^2a}{V^2} = \frac{nRT}{V - nb} \quad \text{or} \quad P = \frac{nRT}{V - nb} - \frac{n^2a}{V^2}$$

Substitute the following values into the van der Waals equation:

$n = 1.00; V = 2.00$ L; $T = 300$ K; $R = 0.0821$ L-atm/K-mol;
$n^2a = (1.00 \text{ mol})^2(1.34 \text{ L}^2\text{-atm/mol}^2) = 1.34$ L^2-atm;
$nb = (1.00 \text{ mol})(0.0322 \text{ L/mol}) = 0.0322$ L

$$P = \frac{(1.00 \text{ mol})(0.0821 \text{ L-atm/mol-K})(300 \text{ K})}{(2.00 \text{ L} - 0.0322 \text{ L})} - \frac{1.34 \text{ L}^2\text{-atm}}{(2.00 \text{ L})^2}$$

$$P = 12.5 \text{ atm} - 0.335 \text{ atm} = 12.2 \text{ atm}$$

The pressure calculated from the ideal-gas equation is

$$P = \frac{nRT}{V} = \frac{(1.00 \text{ mol})(0.0821 \text{ L-atm/mol-K})(300 \text{ K})}{2.00 \text{ L}}$$

$$P = 12.3 \text{ atm}$$

The calculated pressures are almost equal. The slight reduction in pressure from the ideal pressure is caused by the a term. If CO_2 had been chosen as the gas, the departure from ideality would have been greater because its a value is significantly larger.

Check: An estimate of the calculated gas pressure gives $(1)(0.08)(300)/2 = 12$, which is close to the calculated value in the solution. The units properly cancel. The value of P when van der Waals constants are used in the calculation is indeed lower because internal molecular forces reduce the pressure from the ideal value. For the second part of the question, the value of n^2a/V^2 indeed increases in magnitude as the value of a becomes larger.

SELF-TEST QUESTIONS

Key Terms

Having reviewed key terms in Chapter 10, match key terms with phrases and identify statements as true or false. If a statement is false, indicate why it is incorrect.

Match each phrase with the best term:

10.1 This quantity depends on the height of a liquid in a column and not the diameter of the column.

10.2 101.325 kPa equals this quantity.

10.3 Conditions equal to 0 °C and 1 atm pressure.

10.4 According to this relationship, the volume of a flexible container holding a gas is inversely proportional to the pressure of the gas.

10.5 The name given to a law that shows this relationship: $P = constant/V$.

10.6 The following figure demonstrates this relationship:

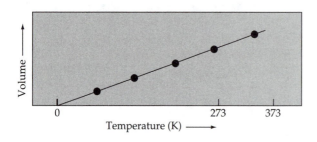

10.7 This hypothesis states equal volumes of gases maintained at the same temperature and pressure contain the same number of molecules of gases.

10.8 Another name given to a mm of mercury.

10.9 This distance for a gas increases with increasing altitude.

10.10 According to this law ammonia gas effuses through a tiny hole faster than carbon monoxide gas.

10.11 In this relationship the term an^2/V^2 corrects the observed pressure for attractive forces between molecules.

10.12 The movement of gas particles through a tiny hole.

Key Terms:

(a) Avogadro's
(b) Boyle's law
(c) Charles's law
(d) effusion
(e) Graham's law of effusion
(f) ideal-gas equation
(g) mean free path
(h) pressure
(i) R
(j) standard atmosphere pressure
(k) STP
(l) torr
(m) Van der Waals equation

True-False Statements:

10.13 When you measure the rate of NH_3 spreading throughout a long tube, you are measuring the rate of *diffusion* of NH_3.

10.14 According to the postulates of the *kinetic-molecular theory* of gases, gases have differing average kinetic energies at the same temperature.

10.15 In the ideal-gas equation, the *gas constant R* equals PV/nT.

10.16 In a mixture of two gases, A and B, A has a partial pressure of 20 mm Hg, and B has a partial pressure of 30 mm Hg at 25 °C in a total volume of 1 L. According to *Dalton's law of partial pressures*, if these two gases were allowed to expand to 2 L at 25 °C, the new total pressure of the gas mixture would be 100 torr, double the total pressure before expansion occurred.

10.17 The vapor pressure of ether is greater than that of ethanol. Therefore you should expect more *vapor* above ether than ethanol in a closed system.

10.18 A *Pascal* is a unit of pressure which equals $1 \ N/m^2$.

10.19 The units *atmosphere* and torr are equivalent.

10.20 According to *Avogadro's law*, the pressure of a gas maintained at constant temperature and volume is directly proportional to the number of moles of the gas.

10.21 You would expect a gas at high pressures to behave like an *ideal gas*.

10.22 If the mole fraction of a gas increases in a mixture of gases, the *partial pressure* of the gas increases.

10.23 The *mole fraction* of 0.02 mol HCl in 10 mol H_2O is 0.02/10.

10.24 The *root-mean-square speed* of a gas will increase as the molar mass of the gas increases at constant temperature.

Problems and Short-Answer Questions

10.25 The following three figures represent different mixtures of gases in containers of the same volume and temperature. The unshaded spheres represent particles of a gas with a smaller molar mass than the particles of a gas represented by the shaded spheres. How do the pressures of the gases compare? Explain.

10.26 A mixture of three gases, Ar, Ne, and He, has a total pressure of 1000 torr in a 1 liter container at 25 °C. Given the following representation of the mixture, what is the partial pressure of Ar? Explain.

10.27 The following figure represents the condition of two gases effusing through a pinhole from container 1 to container 2 after 10 minutes. Container 2 has no particles of gas initially present. Which gas has a smaller molar mass? Explain.

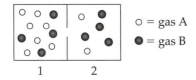

10.28 What is the source of pressure in a container holding a gas?

10.29 Charles's law does not accurately describe gas behavior as temperature decreases. Why?

10.30 Why do gases behave more like ideal gases at higher temperatures than at lower temperatures?

10.31 An unknown gas has a pressure of 1.0 atm in a 25.0 mL container that weighs 100.0 g and is at 25.0 °C. Without doing a calculation can you determine whether there is sufficient information to determine the mass of the unknown gas? If not, explain what additional information is needed and why.

10.32 A container holds 10.0 g Ar and 5.0 g Ne at 25.0 °C and the total pressure is 15.0 atm. Without doing a calculation can you determine whether there is sufficient information to determine the volume of the container? If not, explain what additional information is needed and why.

10.33 A tire pressure gauge measures the pressure exerted by the air within the tire above atmospheric pressure. Given that atmospheric pressure at sea level is about 14.5 lb/in.^{2}, answer the following questions:

   (a) What is the total pressure of air in a tire when the gauge reads 28.0 lb/in.^{2}, at 25 °C?
   (b) Suppose the temperature of the tire described in (a) rises to 35 °C with no volume change; what is the new gauge reading?

10.34 A gas with a pressure of 652 torr at 25.0 °C occupies a volume of 455 mL.

   (a) If the temperature is changed to 262 °C while the pressure remains constant, what will be the new volume?

   (b) If the gas is compressed to 25.5 mL at constant temperature, what will be the new pressure?

10.35 How many molecules of $CH_4(g)$ are contained in 1.00 mL at STP?

10.36 A mixture of 2.00 g of $O_2(g)$ and 3.25 g of $SO_2(g)$ exerts how many atmospheres of pressure inside a 2.00-L container at 300 K?

10.37 How many liters of $O_2(g)$ at STP are evolved when 3.25 g of $KNO_3(s)$ decomposes to $KNO_2(s)$ and $O_2(g)$?

10.38 What volume of oxygen is required for the complete combustion of 10.0 L of $CH_4(g)$ if the gases are all at 600 K and 1.00 atm?

10.39 Calculate the rate of flow of $O_2(g)$ through the walls of a porous tube if $H_2(g)$ flows at the rate of 3.95 × 10^{-3} mL/sec under identical conditions.

10.40 Helium is a gas at the boiling point of liquid nitrogen, 77.0 K. If 2.00 g of He is placed inside a 1.50-L container immersed in liquid nitrogen at 77 K, what is the pressure exerted by the helium gas?

10.41 A mixture of gases at 2.00 atm pressure and at 273 K contains 0.70 mol of N_2 and 0.30 mol of O_2. Calculate the partial pressure of each gas in the mixture.

10.42 A glass bulb was filled with the vapors of a volatile liquid at 100 °C, and 750 torr pressure. The net weight of the sample in the bulb under these conditions was 0.124 g. When the bulb was filled with water at 25 °C, the volume of the bulb was found to be 120.2 cm^3. Calculate the molar mass of the unknown gas.

10.43 A mixture containing 0.250 mol $N_2(g)$, 0.150 mol $CH_4(g)$, and 0.100 mol $O_2(g)$ is confined in a 1.00 L vessel at 35 °C.

   (a) Calculate the partial pressure of $O_2(g)$ in the mixture.
   (b) Calculate the total pressure of the mixture.
   (c) If the temperature is raised to 50 °C, what is the new partial pressure of $O_2(g)$, its mole fraction and its percentage in the gas mixture?

10.44 How do the average kinetic energies and rms speeds of samples of N_2 and CO compare at 25 °C?

Integrative Exercises

10.45 (a) Palmitic acid, $CH_3(CH_2)_{14}COOH$, is an example of a fatty acid found in the human body. The heat of combustion of palmitic acid is −9977 kJ/mol. If the standard heat of formation of $H_2O(l)$ at 25 °C is −285.8 kJ/mol, what volume of hydrogen gas is required to react with oxygen gas at 25 °C and 1.00 atm to produce an amount of energy equivalent to combusting 100.0 g of palmitic acid?

   (b) How many molecules of hydrogen are required?

10.46 A 0.1530 *g* sample of a gas that reacts with ozone in the atmosphere exerts a pressure of 470.6 torr in a 50.00 mL vessel at 298.0 *K*. This gas has the following elemental composition: 9.934% C; 58.64% Cl; and 31.43% F. What is the molecular formula of the gas?

10.47 Xenon gas reacts with fluorine gas to form xenon tetrafluoride solid, which sublimes slightly. This can be accomplished by heating in a nickel can at 400 °C xenon and fluorine in a one-to-five volume ratio and at an initial pressure of 6 atm. After a few hours the reaction is cooled and colorless crystals form.

    (a) If 10 mL of xenon reacts, what volume of fluorine reacts?

    (b) What are the partial presures of xenon and fluorine in the container at the start of the reaction?

    (c) Look up the word "sublime." What does it mean? Describe the contents of the container after the reaction is over.

    (d) Which gas should deviate the most from ideal behavior at high pressures? Why?

Multiple-Choice Questions

10.48 According to the ideal-gas equation, which of the following is true?

    (a) If gases are mixed, the partial pressure of each lowers the partial pressure of the others.

    (b) For Boyle's law to apply, a gas must be kept at constant pressure.

    (c) The volume of a gas is not changed if it is heated from 0 °C to 100 °C and at the same time the pressure is increased from 750 torr to 850 torr.

    (d) The volume of a gas doubles when the centigrade temperature doubles if all other variables are held constant.

    (e) The volume of a gas decreases by a factor of 2 when the pressure is doubled if all other variables are held constant.

10.49 Which statement about the kinetic-molecular theory of gases is true?

    (a) It does not give a satisfactory explanation of Boyle's law.

    (b) It can be applied to crystals.

    (c) It explains why small molecules exert a smaller pressure than do the same number of larger molecules at the same temperature and volume.

    (d) It explains why gas molecules consume a finite portion of the container in which they are contained.

    (e) It assumes that small and large molecules have the same average kinetic energy at the same temperature.

10.50 Deviations in the behavior of gases from the ideal-gas equation: (1) Occur because gas molecules occupy a finite volume in a container; (2) occur because attractions between gas molecules exist; (3) decrease with increasing pressure; (4) decrease with increasing temperature; (5) decrease with decreasing volume.

    (a) (1) and (2)

    (b) (1), (2), and (3)

    (c) (1), (2), and (4)

    (d) (1), (2), (4), and (5)

    (e) (1), (2), (3), (4), and (5)

10.51 A 15.0-L gas sample is heated from 29 °C to 67 °C at constant-pressure conditions. Under these conditions, which of the following statements is correct?

    (a) The number of moles of gas increases.

    (b) The number of moles of gas decreases.

    (c) The volume decreases.

    (d) The volume increases.

    (e) None of the above is correct.

10.52 If the density of a gas is 0.08987 g/L at STP conditions, what is its molar mass?

    (a) 0.08987 g   **(d)** 89.87 g

    (b) 2.01 g   **(e)** 100 g

    (c) 249 g

10.53 A sample of Ne(*g*) occupies 4.0 L at 35 °C at 2.0 atm. What is its new volume if the temperature and pressure are changed to 48 °C and 1.0 atm?

    (a) 0.48 L   **(d)** 7.7 L

    (b) 2.1 L   **(e)** 8.3 L

    (c) 6.2 L

10.54 What is the density of Ne gas in grams per liter at STP conditions and when the temperature is raised from STP conditions to 1500 °C and all other conditions remain constant?

    (a) 0.0446 g/L; 0.139 g/L

    (b) 0.901 g/L; 0.139 g/L

    (c) 0.901 g/L; 0.0446 g/l

    (d) 0.139 g/L; 1.25 g/L

    (e) insufficient information to answer question

10.55 Several commercial drain cleaners contain NaOH(*s*) and small amounts of Al(*s*). When one of these cleaners is added to water a reaction occurs resulting in the formation of H_2 bubbles:

$$2\,Al(s) + 2\,OH^-(aq) + 2\,H_2O \longrightarrow 3\,H_2(g) + 2\,AlO_2^-(aq)$$

The purpose of the H_2 bubbles is to agitate the solution and thereby increase the cleansing action. According to this equation, how many liters of $H_2(g)$ are released when 0.200 g of Al are dissolved in an excess of OH^- at 25 °C and 1.00 atm pressure?

    (a) 1.00 L   **(d)** 2.72 L

    (b) 0.111 L   **(e)** 0.272 L

    (c) 0.0111 L

10.56 Which of the following statements about the properties of a gas are true when the temperature is increased at constant volume? (1) Pressure increases; (2) pressure decreases; (3) average molecular speed increases; (4) average molecular speed decreases; (5) ideality of gas does not change.

    (**a**) (1), (2), (3), (4), and (5)
    (**b**) (2) and (4)
    (**c**) (2), (4), and (5)
    (**d**) (1) and (3)
    (**e**) (1), (3), and (5)

10.57 Which statement about the density of a gas is correct?

    (**a**) It is independent of volume.
    (**b**) It decreases with increasing temperature at constant temperature.
    (**c**) It is independent of pressure.
    (**d**) It doubles when pressure and kelvin temperature double.
    (**e**) It does not depend on molar mass.

10.58 Acetylene gas reacts with oxygen gas as follows:

$$2\,C_2H_2(g) + 5\,O_2(g) \longrightarrow 4\,CO_2(g) + 2\,H_2O(g)$$

What quantity of $CO_2(g)$ is formed when 6.0 L of acetylene gas is completely combusted?

    (**a**) 12 mol   (**d**) 12 L
    (**b**) (12×44) g   (**e**) (12×22.4) L
    (**c**) 6.0 L

SELF-TEST SOLUTIONS

10.1 (h). **10.2** (j). **10.3** (k). **10.4** (f). **10.5** (b). **10.6** (c). **10.7** (a). **10.8** (l). **10.9** (g). **10.10** (e). **10.11** (m). **10.12** (d). **10.13** True. **10.14** False. At the same temperature, gases have the same average kinetic energies. **10.15** True. **10.16** False. Initially the total pressure is 50 mm Hg: $P = P_1 + P_2 = 20$ mm $P = P_1 + P_2 = 20$ mm Hg + 30 mm Hg = 50 mm Hg. When the volume of the container is doubled at constant temperature the partial pressure of each gas must decrease by a factor of 2. Thus the new total pressure is 10 mm Hg + 15 mm Hg = 25 mm Hg. **10.17** True. **10.18** True. **10.19** False. A torr is equivalent to 1 mm Hg. **10.20** False. The volume of a gas maintained at constant temperature and pressure is.... **10.21** False. At high pressures, particles of gas are close together and thus experience strong interactions. At low pressures, they tend to behave as ideal gases. **10.22** True. **10.23** False. The mole fraction is $0.02/(0.02 + 10)$. The denominator is the sum of all moles of components in the mixture. **10.24** False. It will decrease with increasing molar mass. rms is proportional to the square root of the inverse of molar mass.

10.25 The pressure exerted by the gases is the same in all three containers. Each container shows eight particles of gas. Although the distribution of gases differs in each container the total number of particles is constant. According to Dalton's law the total pressure is the sum of the partial pressures of each gas and also is proportional to the total number of moles of all gases in the mixture.

The total number of moles is simply the total number of gas particles divided by Avogadro's number; it is a constant for all three containers.

10.26 The total pressure is a sum of the partial pressures of the gases. The partial pressure of a gas is proportional to the number of moles, which is related to the number of particles of the gas, for a fixed volume and temperature. In the figure for the mixture there are ten gas particles, of which Ar shows four particles, or 40% of the mixture; thus, the mole fraction of Ar is 0.40. The partial pressure of Ar is $(0.40)(1000 \text{ torr}) = 400$ torr.

10.27 Gas B has a smaller molar mass. According to Graham's law of effusion the effusion rate of a gas is inversely proportional to the square root of its molar mass. The gas with the smaller molar mass escapes from the initial container through the tiny hole more rapidly than the gas with a larger molar mass. More molecules of gas B have effused through the pinhole after ten minutes than gas A; thus gas B has effused at the faster rate.

10.28 Molecules of a gas are in ceaseless, chaotic motion and collide with the walls of a container. The impact of gas molecules with the walls of a container causes gas pressure.

10.29 Charles's law states that at constant volume and mass the volume of a gas is directly proportional to absolute temperature. As temperature is lowered the average kinetic energy of the gas particles decreases and real forces of attraction between particles increases; this causes an increase in non-ideality. Furthermore, as the temperature is lowered the gas particles become sufficiently attracted to one another that the gas becomes a liquid or solid. Charles's law does not apply when a solid or liquid forms.

10.30 At higher temperatures the velocities of molecules significantly increase as compared with the forces of attraction between molecules. Because the motional energies of gas molecules cause a reduction in the effect of intermolecular attractions at higher temperatures, the molecules stick together less upon collision. This results in a decrease in the value of the an^2/V^2 term in the $(P + an^2/V)$ component of van der Waals equation. Thus at high temperatures the "experimental" pressure becomes more like the "ideal-gas" pressure.

10.31 There is insufficient information. The number of moles of a gas can be determined from the ideal-gas law, $PV = nRT$. P, V, T are given in the problem and R is a known constant. To determine the mass of the substance from the number of moles, the molar mass must be known: $n = m/M$. The molar mass is not given in the problem and thus the problem cannot be solved.

10.32 There is sufficient information. The volume of the container holding the two gases can be determined from the ideal-gas law, $P_{total}V_{total} = n_{total}RT$. $P_{total}(15.0 \text{ atm})$ and T (298.15 K) are given in the question, and n_{total} is the sum of the number of moles of Ar and Ne (calculated from the number of grams of each gas) and R is the gas-law constant. V_{total} can thus be calculated.

10.33 (a) $P_t = P_{gauge} + P_{atmosphere} = 28.0$ lb/in.^{2}
$+ 14.5$ lb/in.$^2 = 42.5$ lb/in.^{2}
(b) $T_1 = (25 + 273)$ K $= 298$ K; $T_2 = (35 + 273)$ K $= 308$ K. Since the number of moles of gas remains constant, we can use the relation of $P_1V_1/T_1 = P_2V_2/T_2$, and substitute into it the values of P_1, T_1, and T_2. V_1 and V_2 are the same; thus $P_1/T_1 = P_2/T_2$ or $P_2 = P_1(T_2/T_1)$: $P_2 =$ $(42.5$ lb/in.$^2)(308$ K$)/(298$ K$) = 43.9$ lb/in.^{2}. The gauge pressure is 43.9 lb/in.$^2 - 14.5$ lb/in.$^2 = 29.4$ lb/in.^{2}.

10.34 The number of moles gas does not change; thus we can use the following relationship to solve the two problems:

$$\frac{P_1V_1}{T_1} = \frac{P_2V_2}{T_2}$$

(a) In this problem the pressure does not change and therefore the pressures cancel. We now have

$$V_2 = \frac{V_1T_2}{T_1} = \frac{(455 \text{ mL})(535 \text{ K})}{298 \text{ K}} = 817 \text{ mL}$$

(b) Similarly, with temperature constant:

$$P_2 = \frac{P_1V_1}{V_2} = \frac{(652 \text{ torr})(455 \text{ mL})}{25.5 \text{ mL}} = 11,200 \text{ torr}$$

10.35 One milliliter of $CH_4(g)$ at STP contains the following number of moles:

$$n = \frac{PV}{RT} = \frac{(1 \text{ atm})(0.00100 \text{ L})}{\left(0.08206\dfrac{\text{L-atm}}{\text{K-mol}}\right)(273 \text{ K})}$$

$$= 4.46 \times 10^{-5} \text{ mol}$$

The number of molecules of $CH_4(g)$ is its number of moles times Avogadro's number:

$$\text{Molecules} = (4.46 \times 10^{-5} \text{ mol})$$

$$\times \left(6.022 \times 10^{23}\frac{\text{molecules}}{\text{mol}}\right)$$

$$\text{Molecules} = 2.69 \times 10^{19} \text{ molecules}$$

10.36 The total pressure is the sum of partial pressures: $P_t = P_{O_2} + P_{SO_2} = (n_{O_2}RT/V) + (n_{SO_2}RT/V)$ Since R, T, and V are known quantities, we need only calculate the number of moles of each gas in order to calculate P_t:

$$n_{O_2} = \frac{2.00 \text{ g}}{32.00 \text{ g/mol}} = 0.0625 \text{ mol}$$

$$n_{SO_2} = \frac{3.25 \text{ g}}{64.00 \text{ g/mol}} = 0.0507 \text{ mol}$$

$$P_t = \left(0.08206\frac{\text{L-atm}}{\text{K-mol}}\right)$$

$$\times \frac{(300 \text{ K})(0.0625 \text{ mol} + 0.0507 \text{ mol})}{2.00 \text{ L}} = 1.40 \text{ atm}$$

10.37 The chemical equation describing the reaction is $2 \text{ KNO}_3(s) \rightarrow 2 \text{ KNO}_2(s) + O_2(g)$. The mole equivalence is 2 mol $KNO_3 \simeq 2$ mol $KNO_2 \simeq 1$ mol O_2. At STP, 1 mol of an ideal gas has a volume of 22.4 L (see Exercise 3). The number of moles of KNO_3 is $(3.25$ g$/(101.1$ g/mol$))$ $= 0.0321$ *mol*. The volume of O_2 evolved at STP is

$$V = \left(22.4\frac{\text{L}}{\text{mol}}\right)(0.0321 \text{ mol KNO}_3)$$

$$\times \left(\frac{1 \text{ mol O}_2}{2 \text{ mol KNO}_3}\right) = 0.360 \text{ L}$$

10.38 The balanced chemical equation describing the combustion is $CH_4(g) + 2 O_2(g) \rightarrow CO_2(g) + 2 H_2O(l)$. Equal volumes of gases at the same temperature and pressure contain an equal number of moles. One mole of $CH_4(g)$ requires 2 mol of $O_2(g)$ to complete the combustion. Thus, 1 volume of $CH_4(g)$ requires 2 volumes of $O_2(g)$, or a volume ratio of 2:1. Therefore, $V_{O_2} = 2V_{CH_4} = (2)(10.0 \text{ L}) = 20.0 \text{ L}$.

10.39 The ratio of flow rates is

$$r_{O_2}/r_{H_2} = \sqrt{M_{H_2}/M_{O_2}} \text{ where r is the flow rate.}$$

Substituting the appropriate values into the equation gives

$$\frac{r_{O_2}}{3.95 \times 10^{-3} \text{ mL/sec}} = \sqrt{\frac{2.02 \text{ g/mol}}{32.00 \text{ g/mol}}}$$

Solving for the rate of flow of $O_2(g)$ yields $r_{O_2} = 9.92 \times 10^{-4}$ mL/sec.

10.40 $PV = nRT$

$$P = \frac{nRT}{V}$$

$$= \frac{(2.00 \text{ g He})\left(\dfrac{1 \text{ mol He}}{4.00 \text{ g He}}\right)\left(0.08206\dfrac{\text{L-atm}}{\text{K-mol}}\right)(77.0 \text{ K})}{1.50 \text{ L}}$$

$$= 2.11 \text{ atm}$$

10.41 Let n_t equal total number of moles of gases. Then the unknown volume of the system is

$$P_tV_t = n_tRT \quad \text{or} \quad V_t = \frac{n_tRT}{P_t}$$

The partial pressures are calculated as follows:

$$P_{N_2} = \frac{n_{N_2}RT}{V_t} = \frac{n_{N_2}RT}{\left(\dfrac{n_tRT}{P_t}\right)} = \left(\frac{n_{N_2}}{n_t}\right)P_t = X_{N_2}P_t$$

$$= \left(\frac{0.70 \text{ mol}}{0.70 \text{ mol} + 0.30 \text{ mol}}\right)(2.00 \text{ atm})$$

$$= 1.4 \text{ atm}$$

$$P_{O_2} = 2.00 \text{ atm} - 1.4 \text{ atm} = 0.6 \text{ atm}$$

10.42 $PV = nRT = (g/\mathcal{M})$ RT. Solving for $\mathcal{M}$ gives

$$\mathcal{M} = \frac{gRT}{PV}$$

$$= \frac{(0.124\ g)\left(0.0821\dfrac{\text{L-atm}}{\text{K-mol}}\right)(373\ K)}{(750\ \text{torr})\left(\dfrac{1\ \text{atm}}{760\ \text{torr}}\right)(120.0\ cm^3)\left(\dfrac{1\ L}{10^3\ cm^3}\right)}$$

$$= 32.1\ g/mol$$

10.43 (a)

$$P_{O_2} = \frac{(\text{moles oxygen})RT}{V}$$

$$= \frac{(0.100\ mol)\left(\dfrac{0.0821\ \text{L-atm}}{\text{mol-K}}\right)(308\ K)}{1.00\ L}$$

$$P_{O_2} = 2.53\ atm$$

(b) $P_{O_2} = X_{O_2}P_t$ or $P_t = P_{O_2}/X_{O_2}$

$$P_t = \frac{2.53\ atm}{\left(\dfrac{0.100\ mol}{(0.250\ mol\ +\ 0.150\ mol\ +\ 0.100\ mol)}\right)}$$

$$P_t = 2.53\ atm/0.200 = 12.7\ atm$$

(c) Use $PV = nRT$ to calculate the new total pressure, where n is the total number of moles in the mixture. Then use $P_i = X_i P_t$ to calculate the new partial pressure.

$$P_t = \frac{nRT}{V} = \frac{(0.500\ mol)\left(0.0821\dfrac{\text{L-atm}}{\text{mol-K}}\right)(323\ K)}{1.00\ L}$$

$$= 13.3\ atm$$

$$P_{O_2} = X_{O_2}P_t = \left(\frac{0.100\ mol}{0.500\ mol}\right)(13.3\ atm) = 2.66\ atm$$

The mole fraction does not change unless more material is added to the mixture. The percentage of oxygen in the mixture is the same as its mole fraction multiplied by 100: 20.0%.

10.44 The average kinetic energies of N_2 and CO are exactly the same at 25 °C because all gaseous substances have the same average kinetic energies at the same temperature. To compare the rms speeds of the gas molecules at 25 °C, use the following relation:

$$\frac{u_{N_2}}{u_{CO}} = \sqrt{\frac{\mathcal{M}_{CO}}{\mathcal{M}_{N_2}}} = \sqrt{\frac{28\ g\ mol}{28\ g\ mol}} = 1.0$$

Therefore $u_{N_2} = 1.0(u_{CO})$; N_2 molecules move with an average speed that is equal to the value for CO.

10.45 (a) First calculate the quantity of heat produced when 100.0 g of palmitic acid is combusted in oxygen.

This requires that the number of moles of palmitic acid also be calculated:

$$\text{moles palmitic acid} = (100.0\ g)(1\ mol/256.43\ g)$$

$$= 0.3900\ \text{mol palmitic acid}$$

The heat produced when this quantity of palmitic acid is combusted is:

$$(0.3900\ mol)(-9977\ kJ/mol) = -3891\ kJ.$$

The heat of formation reaction for $H_2O(l)$ is:

$$H_2(g) + \tfrac{1}{2}\,O_2(g) \longrightarrow H_2O(l)$$

We need the number of moles of hydrogen gas reacting with oxygen gas to give -3891 kJ. The enthalpy of reaction for x moles of hydrogen reacting to give -3891 kJ is related to the heat of formation for the reaction, -285.5 kJ, as follows:

$$\Delta H°_{rxn} = (x\ \text{moles}\ H_2O)(\Delta H°_f)$$

$$-3891\ kJ = (x\ \text{moles}\ H_2O)(-285.5\ kJ/mol)$$

$$(x\ \text{moles}\ H_2O) = 13.61\ mol$$

From the number of moles of water formed, which is stoichiometrically equivalent to the number of moles of hydrogen gas reacted, we can conclude that the number of moles of hydrogen gas required is 13.61 mol. The volume is calculated from the ideal-gas law:

$$PV = nRT$$

$$V = \frac{nRT}{P} = \frac{(13.61\ mol)\left(0.0821\dfrac{\text{L-atm}}{\text{mol-K}}\right)(298\ K)}{1.00\ atm}$$

$$V = 333\ L$$

(b) The number of molecules of hydrogen gas required is calculated using Avogadro's number:

$$\text{Number of molecules} = (13.61\ \text{mol hydrogen gas})$$

$$(6.022 \times 10^{23}\ \text{molecules/mol})$$

$$\text{Number of molecules}$$

$$= 8.20 \times 10^{24}\ \text{molecules hydrogen gas}$$

10.46 To solve for the molecular formula you need the empirical formula and the molar mass of the gas. The molar mass of the gas can be calculated from the mass-pressure-volume-temperature data using the ideal-gas law:

$$PV = nRT = \left(\frac{g}{\text{molar mass}}\right)RT$$

$$\text{Molar mass} = \frac{gRT}{PV} = \frac{(0.1530\ g)\left(0.0821\dfrac{\text{L-atm}}{\text{mol}}\right)(298.0\ K)}{(470.6\ \text{torr})\left(\dfrac{1\ \text{atm}}{760\ \text{torr}}\right)(0.05000\ L)}$$

$$\text{Molar mass} = 120.9\frac{g}{mol}$$

The empirical formula is calculated from the elemental composition data and assuming a 100.0 g sample:

$$mol\ C = (100.0\ g)(1\ mol/12.011\ g)(0.09934)$$
$$= 0.8271\ mol\ C$$

$$mol\ Cl = (100.0\ g)(1\ mol/35.45\ g)(0.5864)$$
$$= 1.654\ mol\ C$$

$$mol\ F = (100.0\ g)(1\ mol/19.00\ g)(0.3143)$$
$$= 1.654\ mol\ F$$

The empirical formula is $C_{0.8271}Cl_{1.654}F_{1.654}$ or CCl_2F_2. The empirical formula mass is 120.9 g, the same as the calculated molar mass. Thus the molecular formula is the same as the empirical formula.

10.47 **(a)** First write the chemical equation to determine the stoichiometry of the reaction.

$$Xe(g) + 2\ F_2(g) \longrightarrow XeF_4(s)$$

Gay-Lussac's law of combining volumes says that the volumes of gases react in small, whole numbers when they are at the same temperature. Avogadro's hypothesis says that equal volumes of gases contain equal numbers of molecules; thus, the volumes reacting must be in the same ratio as the stoichiometric coefficients in the balanced chemical reaction. Since fluorine reacts with xenon in a two-to-one ratio, the volume of fluorine reacting must be 2(10 mL) = 20 mL.

(b) We are told in the question that xenon and fluorine are placed in the container in a one-to-five volume ratio. Using the same logic as in **(b)**, this means the partial pressures of the two gases must also be in a one-to-five ratio. The total pressure is 6 atm; thus we can deduce that the pressure of xenon is 1 atm and that of fluorine is 5 atm, since they are in a one-to-five ratio and the sum of partial pressures must be 6 atm. Both conditions are met.

(c) The term sublime means a solid transforms to a gas without going through a liquid phase. Thus, some xenon tetrafluoride gas will form when the reaction is completed. The container will also contain residual fluorine gas and a small amount of solid xenon tetrafluoride. This is a limiting reactant problem because the ratio of partial pressures in which xenon and fluorine react is 1:2 yet the initial ratio is 1:5 and fluorine is in excess.

(d) At high pressures the volume of the particles of gas cause significant deviations from ideal-gas pressure; the larger the particle the greater the deviation. A look at a table of bonding atomic radii should convince you that xenon tetrafluoride molecules are larger than those of fluorine gas and therefore xenon tetrafluoride should exhibit the greater deviations.

10.48 **(e)**

10.49 **(e)**

10.50 **(c)**

10.51 **(d)** $V_2 = V_1(T_2/T_1)$. Since $T_2/T_1 > 1$, volume increases.

10.52 **(b)** At STP conditions, 1 mol of gas occupies 22.4 L. Therefore the mass of the gas in 222.4 L is the molar mass. This mass is calculated as follows: $(0.08987\ g/L)(22.4\ L) = 2.01\ g$.

10.53 **(e)**

$$V_2 = \left(\frac{P_1}{P_2}\right)\left(\frac{T_2}{T_1}\right)(V_1)$$

$$V_2 = \left(\frac{2.0\ atm}{1.0\ atm}\right)\left(\frac{321\ K}{308\ K}\right)(4.0\ L) = 8.3\ L$$

10.54 **(b)** At STP conditions, 22.4 L contains 1 mol of Ne. Therefore, because we know the molar mass,

$$d = \frac{1\ mol}{22.4\ L} = \frac{20.18\ g}{22.4\ L} = 0.901\ g/L$$

At 1500 °C we can use the relation $V_1/T_1 = V_2T_2$ because the pressure and number of moles do not change. $V_2 = V_1$
$x\frac{T_2}{T_1} = 22.4\ L \times \frac{1773\ K}{273\ K} = 145\ L$ and $d = \frac{20.18\ g}{145\ L}$
$= 0.139\ g/L$

10.55 **(e)** To calculate the number of moles of H_2 produced you need to make the conversion: grams $Al \rightarrow$ moles $Al \rightarrow$ moles H_2.

$$Moles\ H_2 = (0.200\ g\ Al)\left(\frac{1\ mol\ Al}{26.98\ g\ Al}\right)\left(\frac{3\ mol\ H_2}{2\ mol\ Al}\right)$$

$$= 0.0111\ mol\ H_2$$

Using the ideal-gas equation, calculate the volume of gas produced:

$$V = \frac{nRT}{P}$$

$$V = \frac{(0.0111\ mol)\left(0.0821\ \frac{L\text{-}atm}{K\text{-}mol}\right)(298\ K)}{1\ atm}$$

$$= 0.272\ L\ or\ 272\ mL$$

10.56 **(d)**

10.57 **(b)**

10.58 **(d)** This is an application of Avogadro's law using the stoichiometry of the chemical reaction.

Chapter

11

Intermolecular Forces, Liquids and Solids

OVERVIEW OF THE CHAPTER

Review: Enthalpy (5.3); kinetic-molecular theory of gases (10.7); nature of matter (1.2, 8.2).

Learning Goal: You should be able to: Employ the kinetic-molecular model to explain the differences in motion of particles in gases, liquids, and solids and how these relate to their states.

11.1 KINETIC MOLECULAR THEORY FOR LIQUIDS AND SOLIDS

Review: Electronegativity (8.4); bond dipole (8.4); dipole moment (8.4, 9.3).

Learning Goal: You should be able to: Describe the various types of intermolecular attractive forces, and state the kinds of intermolecular forces expected for a substance given its molecular structure.

11.2 INTERMOLECULAR FORCES: HYDROGEN BONDING

Review: Nature of gases (10.1); pressure exerted by gases (10.2); X-radiation (6.1).

Learning Goal: You should be able to:

1. Explain the meaning of the terms viscosity, surface tension, critical temperature, and critical pressure, and account for the variations in these properties in terms of intermolecular forces and temperature.
2. Explain the way in which the vapor pressure of a substance changes with intermolecular forces and temperature.
3. Describe the relationship between the pressure on the surface of a liquid and the boiling point of the liquid.
4. Given the needed heat capacities and enthalpies for phase changes, calculate the heat absorbed or evolved when a given quantity of a substance changes from one condition to another.

11.3, 11.4, 11.5 PHYSICAL PROPERTIES OF LIQUIDS AND SOLIDS: CHANGES OF STATE

Learning Goal: You should be able to: Draw a phase diagram of a substance given appropriate data, and use a phase diagram to predict which phases are present at any given temperature and pressure.

11.6 PHASE DIAGRAMS

11.7, 11.8 SOLIDS | **Review:**   Density (1.4); covalent bonds (8.3); ionic bonds (8.2).

Learning Goals: You should be able to:

1. Distinguish between crystalline and amorphous solids.
2. Determine the net contents in a cubic unit cell, given a drawing or verbal description of the cell. Use this information, together with the atomic weights of the atoms in the cell and the cell dimensions, to calculate the density of the substance.
3. Describe the most efficient packing patterns of equal-sized spheres.
4. Predict the type of solid (molecular, covalent network, ionic, or metallic) formed by a substance, and predict its general properties.

TOPIC SUMMARIES AND EXERCISES

KINETIC-MOLECULAR THEORY FOR LIQUIDS AND SOLIDS

According to the kinetic-molecular theory, a gas consists of a collection of molecules in constant, turbulent motion.

- Gas particles move about in the total volume of the container they occupy; however, the volume occupied by the gas particles themselves is small compared to the total volume.
- Gas particles remain largely separated from one another because their average kinetic energy is considerably greater than the average energy of forces between the gas particles.

Why is a liquid formed when a gas is cooled at an appropriate temperature and pressure? The kinetic-molecular theory helps you understand the physical processes that are occurring.

- As the temperature of a gas is lowered the average speed of particles of a gas decreases and the attractive forces between gas particles begin to dominate. With sufficient cooling, gas particles cluster together and form a liquid.
- In the condensed state of a liquid, the particles are in close proximity to one another and undergo frequent collisions.

Why is a solid formed when a liquid is sufficiently cooled?

- With additional lowering of temperature, attractive forces between particles become very large compared to the average kinetic energy of the particles and a solid forms.
- Particles in a solid have minimal or no translational (kinetic) energy, but they do have vibrational energy.
- Particles in a solid occupy certain positions relative to one another and this arrangement extends throughout the solid.
- When a solid and a liquid exist in a state of dynamic equilibrium, the rate of freezing equals the rate of melting. Similarly, when a liquid and a gas exist in a state of dynamic equilibrium, the rate of condensation equals the rate of vaporization.

EXERCISE 1   Identifying characteristics of a liquid or gas

Which of these physical properties is characteristic of a liquid, a gas, or both: (a) flows; (b) virtually incompressible; (c) slow diffusion of molecules; (d) always occupies entire volume of container?

SOLUTION: *Analyze*: We are asked to match a given physical property to the state (gas or liquid) it describes.

Plan: We can review the characteristics of the states of matter given in the chapter and use this information to answer the question.

Solve: (a) Both liquids and gases flow. (b) This is characteristic of a liquid; a gas is compressible. (c) This is characteristic of a liquid; molecules in a gas diffuse rapidly because of their high kinetic energies. (d) This is true of a gas; it occupies the entire volume in which it is contained.

EXERCISE 2   Applying the principles of kinetic theory to liquids

Using the kinetic theory of liquids, explain why the temperature of a volatile liquid decreases when it is poured into a shallow, insulated dish.

SOLUTION: *Analyze*: Using the kinetic theory of liquids we are asked to explain why the temperature of a liquid decreases when it is in a shallow, insulated dish.

Plan: We will consider four factors when answering questions using the kinetic theory of liquids: (1) The temperature of a liquid decreases when the average kinetic energy of the particles decreases; (2) liquid particles undergo frequent collisions; (3) liquid particles aggregate when their average kinetic energy becomes smaller than attractive forces between them; and (4) particles at the surface of a liquid will escape if their kinetic energy is sufficient to overcome attractive forces.

Solve: At the surface of the liquid in the shallow, insulated dish, molecules with sufficient kinetic energy can escape from the liquid to form gas molecules. Not all molecules have sufficient kinetic energy because a range of molecular speeds occurs. Therefore, when some of the liquid molecules escape, there is a loss of average kinetic energy within the liquid. Because the dish is insulated, energy from the surroundings except at the surface is not available to raise the average kinetic energy of the remaining liquid molecules. With a decrease in the average kinetic energy of the molecules in the liquid the temperature of the liquid decreases.

Comment: As the liquid cools, the rate of evaporation slows. There is some energy absorbed by the liquid at its surface and this will enable slow evaporation over a particular time period.

Attractive forces between neutral molecules are referred to as intermolecular forces. In Section 11.2 of the text two primary types of intermolecular forces are characterized and distinguished. The properties and characteristics of these intermolecular forces, and forces between ions and dipoles, are summarized in Table 11.1 on page 226.

A special kind of electrostatic attraction between molecules occurs when hydrogen is bonded to *nitrogen, oxygen,* or *fluorine* atoms.

- A hydrogen atom attached to one of these three elements possesses a substantial partial positive charge. The positive end of each of the following bond dipoles (H atom)

$$\overset{\leftarrow}{N}{-}\overset{+}{H} \qquad \overset{\leftarrow}{O}{-}\overset{+}{H} \qquad \overset{\leftarrow}{F}{-}\overset{+}{H}$$

INTERMOLECULAR FORCES: HYDROGEN BONDING

is capable of strongly interacting with an unshared electron pair possessed by a nitrogen, oxygen, or fluorine atom of an adjacent molecule.

The intermolecular attraction that results is called a *hydrogen bond*. For example, the hydrogen bonds between HF molecules can be pictured as

$$\cdots H-F\cdots H-F\cdots H-F\cdots$$

The dashes between HF molecules represent hydrogen bonds.

- Hydrogen-bond energies are a small percentage of the energies of ordinary covalent bonds; however, they have important consequences for the properties of many chemical substances.
- You should learn to identify molecules that have the capacity to undergo hydrogen bonding with another like or differing molecule: Look for the presence of a N—H, O—H, or F—H bond in a molecule.

TABLE 11.1 The Various Types of Intermolecular Forces

Type of intermolecular attraction	Characteristics of the interaction	Examples
Ion–dipole	Attractive force between an ion and an oppositely charged end of a permanent dipole possessed by a neutral molecule	KCl dissolved in water to give $K^+\cdots OH_2$ and $Cl^-\cdots H_2O$
Dipole–dipole[*]	Positive end of permanent dipole on one molecule aligns itself with negative end of permanent dipole on another molecule	$\overset{\delta^+\delta^-}{HCl}\cdots\overset{\delta^+\delta^-}{HCl}$
	Only significant in effect when molecules are close together	
London dispersion[*]	Short-range attractive forces between molecules resulting from momentary mutual distortion (polarization) of electron clouds	$Ne(g)$
	Magnitude increases with increasing molecular volume and molecular mass	$N_2(g)$ $C_8H_8(l)$

[*]Together known as van der Waals forces

EXERCISE 3   Assigning boiling points of nonpolar compounds

London dispersion forces allow nonpolar compounds to become liquefied. Assign the boiling points $-162\ ^\circ C$, $-88\ ^\circ C$, $-42\ ^\circ C$, and $0\ ^\circ C$ to the following nonpolar compounds: C_4H_{10}, CH_4, C_2H_6, and C_3H_8.

SOLUTION: *Analyze:* We are given four nonpolar C_aH_b compounds and four boiling points and are asked to assign the boiling points to the compounds.

Plan: Boiling a liquid requires overcoming internal attractions between molecules in the liquid. We will consider six factors when answering questions involving the physical properties of molecular substances: (1) Are the molecules polar or nonpolar? (2) If the molecules are polar what is the magnitude of dipole–dipole forces?

(3) What is the magnitude of London dispersion forces? (4) If any molecule contains a nitrogen, oxygen, or fluorine atom attached to a hydrogen atom does hydrogen bonding exist? (5) The greater the internal attractions between molecules in the liquid, the greater the boiling point. (6) The greater the internal attractions between molecules in a solid, the greater the melting point.

Solve: The magnitude of London dispersion forces between nonpolar molecules is related to the extent to which electron clouds in molecules can be polarized. In general, the larger the size of an atom or molecule, the greater the magnitude of London dispersion forces. The boiling points of a series of related nonpolar compounds should increase with increasing molecular mass (increasing size of molecule), because the boiling points of such substances depend on the London dispersion forces between molecules.

Compound	Molecular mass (amu)	Number of electrons	b.p. (°C)
CH_4	16.05	10	−162
C_2H_6	30.08	18	−88
C_3H_8	44.11	26	−42
C_4H_{10}	58.14	34	0

Check: The trend in boiling points is confirmed by looking at the section on Physical Constants of Organic Compounds in the *Handbook of Chemistry and Physics*.

EXERCISE 4   Relating trends in boiling points to molecular compositions

Explain why the replacement of a hydrogen atom in CH_4 by a chlorine atom to form CH_3Cl causes an increase in boiling point from −164 °C to −24.2 °C.

SOLUTION: *Analyze*: We are asked to explain why changing a hydrogen atom in CH_4 to a chlorine atom increases the boiling point.

Plan: We will use the same plan as in Exercise 3.

Solve: The difference in boiling-point temperatures exists because CH_4 molecules are nonpolar, whereas CH_3Cl molecules are polar. The additional dipole–dipole forces among CH_3Cl molecules as compared to only London dispersion forces among CH_4 molecules result in the increase in boiling point from −164 °C for CH_4 to −24.2 °C for CH_3Cl.

In Sections 11.3–11.5 of the text, many new terms and concepts relating to the nature of liquids and solids are introduced. You should study these carefully.

- The physical properties of a liquid are described by equilibrium vapor pressure, boiling point, melting point, viscosity, and surface tension.
- Table 11.2 on page 228 summarizes some of the characteristics of these physical properties.

A relative measure of the forces holding particles of a substance together (intermolecular forces) is given by either critical temperature or critical pressure.

- **Critical temperature** is the highest temperature at which a liquid can exist; at any higher temperature, the substance exists as a gas.
- The pressure required to liquify a gas at its critical temperature is termed **critical pressure**.

PHYSICAL PROPERTIES OF LIQUIDS AND SOLIDS: CHANGES OF STATE

TABLE 11.2 **Physical Properties of Liquids**

Physical property	Definition	Characteristics
Vapor pressure	Pressure exerted by a gas (vapor) in equilibrium with its liquid	Increases nonlinearly with temperature Volatile liquids have high vapor pressures
Normal boiling point	The temperature at which the vapor pressure of a liquid equals *1 atm pressure*	The higher a liquid's vapor pressure, the lower is its normal boiling temperature
Boiling point	The temperature at which the vapor pressure of a liquid equals the external pressure	Boiling point increases with increasing external pressure
Viscosity	A measure of the resistance of liquids to flow under a standard applied force	The larger the viscosity of a liquid, the more slowly it flows Viscosity of a liquid decreases with increasing temperature The SI unit for viscosity is N-sec/m^2
Surface tension	The energy required to increase the surface of a liquid by a unit amount	Cohesive intermolecular forces cause a liquid surface to become as small as possible A liquid exhibits a meniscus in a glass tube when cohesive forces do not equal adhesive forces The SI unit for surface tension is J/m^2

- The larger the values of critical temperature and critical pressure, the stronger the intermolecular forces holding the particles together.

A **phase change** occurs when matter is transformed from one physical state to another.

- The phase changes you should know are shown in Figure 11.1.
- During a phase change temperature remains constant.

Use the following information when you calculate enthalpy changes associated with phase changes.

- **Enthalpy of fusion**, ΔH_{fus}, is the enthalpy change required to melt one mole of a substance at its melting point. Energy is required to disrupt the regularity of the solid state in order for particles in the liquid state to form.

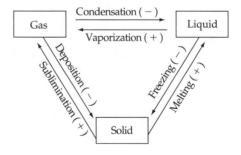

▲ **FIGURE 11.1** Phase changes. A plus sign indicates an endothermic process, and a minus sign indicates an exothermic process.

- **Enthalpy of vaporization**, ΔH_{vap}, is enthalpy change required to vaporize one mole of a substance at its boiling point. Energy must be supplied so that liquid particles gain sufficient kinetic energy to escape from one another at the surface of the liquid.
- To calculate the enthalpy change associated with a change in temperature of a substance at constant pressure, use specific heat capacity, C_s, or molar heat capacity, C_m.

$$q_p = mC_s\Delta T$$

A **dynamic equilibrium** exists when two or more forms of the same matter exist together and there is no apparent change in the quantity of matter, temperature, or pressure.

- At the microscopic level, particles are passing from one phase or form to another and back again at equal rates.
- When a solid and a liquid exist in a state of dynamic equilibrium, the rate of freezing equals the rate of melting; when a liquid and a gas exist in a state of dynamic equilibrium, the rate of condensation equals the rate of vaporization.

EXERCISE 5   Relating temperature and viscosity of a liquid

The viscosity of a liquid decreases with increasing temperature. Why?

SOLUTION: *Analyze*: We are asked to explain why the viscosity of a liquid decreases with increasing temperature.

Plan: Viscosity is a measure of the resistance of a liquid to flow under a standard applied force. The larger the value of viscosity for a liquid the more resistance to flow. Since we are dealing with a liquid, we should consider the kinetic-molecular theory of liquids.

Solve: Increasing the temperature of a liquid causes the average kinetic energy of the molecules to increase. This increase in the average kinetic energy of molecules in the liquid overcomes some of the attractive forces existing between liquid molecules, and thus the molecules are better able to move with respect to each other. That is, their resistance to motion or flow is decreased, resulting in decreased viscosity.

EXERCISE 6   Predicting changes in the boiling point of a liquid when temperature, pressure, and volume are changed

Explain what happens to the boiling point of a liquid in an open container when the following occur: (**a**) The temperature of the surroundings increases without a change in atmospheric pressure. (**b**) The pressure above the liquid increases. (**c**) The volume of the liquid is doubled.

SOLUTION: *Analyze*: We are asked to predict what happens to the boiling point of a liquid when (**a**) the temperature of the surroundings increases at constant atmospheric pressure, (**b**) the external pressure above the liquid increases, and (**c**) the volume of the liquid is doubled.

Plan: For question (**a**) we will apply the kinetic-molecular theory of liquids. For (**b**) and (**c**) we will use the principle that the boiling point of a liquid is the temperature when the vapor pressure of the liquid equals atmospheric pressure. Notice that the volume or mass of a liquid is not part of the principle.

Solve: (**a**) The boiling point does not change as long as the atmospheric pressure does not change. As the temperature of the surroundings increases, the rate of evaporation or boiling increases. (**b**) The increased pressure above the liquid requires an increase in the vapor pressure of the solution to cause boiling. Thus the boiling point increases with an increase in pressure. (**c**) The boiling point does not change because the vapor pressure has not changed. Vapor pressure does not depend on the volume of a liquid.

Comment: The normal boiling point is the temperature when atmospheric pressure is 1 atm. When the external pressure is not 1 atm, it is referred to simply as the boiling point. Also, what happens to a liquid when it is placed in a vacuum?

EXERCISE 7   Calculating the enthalpy change for water undergoing phase and temperature changes

Calculate the enthalpy change associated with converting 100 g of water at $-10.0\ °C$ to steam at $110.0\ °C$. $\Delta H_{fus} = 6.02$ kJ/mol, $\Delta H_{vap} = 40.67$ kJ/mol, C_m of ice $= 37.78$ J/mol$-°C$, C_m of $H_2O(l) = 75.3$ J/mol$-°C$, and C_m of steam $= 33.1$ J/mol$-°C$.

SOLUTION: *Analyze*: We are asked to calculate the enthalpy change required to convert 100 g of ice at $-10\ °C$ to steam at $110.0\ °C$. We are given the molar heat capacities of ice, liquid water, and steam and the enthalpies of fusion and vaporization for water.

Plan: When you encounter enthalpy problems involving a substance changing temperature and physical state it is very useful to qualitatively describe the changes in the physical state of the substance as temperature changes. For this problem,

$$\text{Ice } (-10.0\ °C) \xrightarrow{\ 1\ } \text{ice } (0.0\ °C) \xrightarrow{\ 2\ } \text{liquid } (0.0\ °C) \xrightarrow{\ 3\ }$$

$$\text{liquid } (100.0\ °C) \xrightarrow{\ 4\ } \text{steam } (100.0\ °C) \xrightarrow{\ 5\ } \text{steam } (110.0\ °C)$$

Solve: Calculate the enthalpy change associated with each physical change described above and then add the values to calculate the total enthalpy change.

Heating of ice from $-10.0\ °C$ to $0.0\ °C$:

$$\Delta H_1 = n_{ice}C_{m_{ice}}\Delta T$$
$$= (100\ g)\left(\frac{1\ mol}{18.0\ g}\right)\left(\frac{37.78\ J}{mol-°C}\right)(0\ °C - (-10.0\ °C))$$
$$= 2100\ J = 2.10\ kJ$$

Melting of ice to liquid at constant $0.0\ °C$:

$$\Delta H_2 = n\Delta H_{fus} = (100\ g)\left(\frac{1\ mol}{18.0\ g}\right)\left(\frac{6.02\ kJ}{1\ mol}\right) = 33.4\ kJ$$

Heating of liquid from $0.0\ °C$ to $100.0\ °C$:

$$\Delta H_3 = n_{liquid}C_{m_{liquid}}\Delta T$$
$$= (100\ g)\left(\frac{1\ mol}{18.0\ g}\right)\left(\frac{75.3\ J}{mol-°C}\right)(100.0\ °C - 0\ °C)$$
$$= 41,8000\ J = 41.8\ kJ$$

Vaporization of liquid to steam at constant $100.0\ °C$:

$$\Delta H_4 = n\Delta H_{vap} = (100\ g)\left(\frac{1\ mol}{18.0\ g}\right)\left(\frac{40.67\ kJ}{1\ mol}\right) = 226\ kJ$$

Heating of steam from 100.0 °C to 110.0 °C:

$$\Delta H_5 = n_{steam}C_{m_{steam}}\Delta T$$

$$= (100 \text{ g})\left(\frac{1 \text{ mol}}{18.0 \text{ g}}\right)\left(\frac{33.1 \text{ J}}{\text{mol}-°C}\right)(110.0 °C - 100 °C)$$

$$= 1840 \text{ J} = 1.84 \text{ kJ}$$

The total enthalpy change (ΔH) is

$$\Delta H = \Delta H_1 + \Delta H_2 + \Delta H_3 + \Delta H_4 + \Delta H_5$$

$$= 2.10 \text{ kJ} + 33.4 \text{ kJ} + 41.8 \text{ kJ} + 226 \text{ kJ} + 1.84 \text{ kJ}$$

$$= 305 \text{ kJ}$$

Check: An estimate of the enthalpy change associated with each step gives $(100)(1/20)(40)(10) = 2000$; $(100)(1/20)(6) = 30$; $(100)(1/20)(80)(100) = 40,000$; $(100)(1/20)(40) = 200$; and $(100)(1/20)(30)(10) = 1500$. All are close in value to the values calculated in the solution. An estimate of the sum gives $2 + 35 + 40 + 200 + 2 = 279$, which also approximates the calculated value in the solution. Also, the sign of ΔH is positive, indicating that energy is added to change ice to steam.

PHASE DIAGRAMS

Depending on pressure and temperature, different phases of a particular substance may exist in equilibrium. A **phase diagram** concisely shows in a graphical presentation how the phases of a substance change with pressure and temperature.

- Lines are used to represent a condition of equilibrium between two phases.
- A single point connecting three lines on a phase diagram shows the pressure and temperature at which all three phases are in equilibrium. The condition at which all three phases of a substance are in equilibrium is called a **triple point**.
- Regions between lines represent conditions when a single phase exists. These ideas are briefly illustrated in Exercises 9 and 10.

EXERCISE 8 Identifying phases of CO_2 using a phase diagram

The following figure shows a phase diagram for CO_2. (a) What phases exist in the regions labeled 1, 2, and 3? (b) What conditions exist at the points labeled A and B?

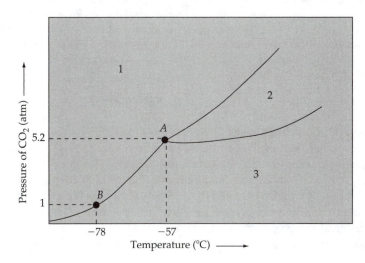

SOLUTION: *Analyze*: We are given the phase diagram for carbon dioxide and asked to identify which phases exist in three regions and what conditions exist at two points.

Plan: Typically three phases are shown in a phase diagram, solid, liquid, and gas. The *y* axis is pressure and the *x* axis is temperature; a line between phases represents a condition of equilibrium between the two phases. An intersection of three lines is the triple point, a condition when all three phases are in equilibrium.

Solve: **(a)** The regions labeled 1, 2, and 3 correspond respectively to the pressure–temperature conditions at which the phases solid, liquid, and gas exist. **(b)** Point *A* corresponds to the triple point of CO_2, the pressure and temperature at which all three phases coexist. Point *B* corresponds to the temperature at 1 atm at which CO_2 sublimes—that is, an equilibrium exists between the solid and gas phases.

Comment: Carbon dioxide is an interesting substance in that its triple point occurs at a pressure greater than one atmosphere. This means that at one atmosphere and room temperature carbon dioxide will change from a solid to a gas without an intermediate liquid state.

EXERCISE 9   Identifying phase changes of CO_2 using a phase diagram when temperature and pressure are changed

Refer to the figure in Exercise 9 to answer the following questions. **(a)** What happens to the state of CO_2 when the temperature is increased at a *constant* pressure of 5.2 atm starting at −70 °C? **(b)** Repeat question (a) when the pressure is reduced from 7 atm pressure to 1 atm pressure at a *constant* temperature of −60 °C.

SOLUTION: *Analyze*: Using the figure in Exercise 9 we are asked to predict what happens to carbon dioxide under two different starting conditions (pressure and temperature) when one of the physical properties is changed and the other is held constant.

Plan: First we need to identify the phase of carbon dioxide at each of the given starting pressure/temperature conditions. Then for each situation determine which physical property is held constant and which changes. Finally, construct a line on the figure that shows the physical property changing while the other is held constant (that is, a linear line). Analyze the changes in properties of carbon dioxide along the line from starting to final conditions.

Solve: **(a)** At the given initial conditions, CO_2 is in region 1 and is a solid. With increasing temperature, CO_2 reaches the triple point at A and all three phases exist; then it becomes a liquid; next an equilibrium between liquid and gas phases occurs; and finally CO_2 becomes a gas in region 3. **(b)** Initially CO_2 is a solid. As the pressure is lowered it remains a solid until the pressure–temperature condition on the solid–gas phase line is reached and both phases occur. With a further lowering in pressure a gas forms.

SOLIDS | Two types of solids are discussed: crystalline solids and amorphous solids.

- **Crystalline solids** consist of an ordered three-dimensional arrangement of particles. Section 11.7 emphasizes in its coverage this type of solid.
- **Amorphous solids** do not form a regular arrangement of particles.
- Crystalline solids can be classified in terms of the types of particles and the attractive forces between them. Study carefully Table 11.7 in the text. Note particularly the crystal classifications, the form of unit particles and the forces between particles. Use your knowledge of intermolecular forces to understand the characteristics of each type of crystal classification.

The three-dimensional arrangement of atoms or molecules in a crystal—called crystal lattice or space lattice—is often shown using a unit-cell representation.

- A **unit cell** is the smallest repeating unit of the three-dimensional crystal lattice. The entire space lattice can be considered an extension of the unit cell in three dimensions.
- You should look at Figures 11.31 through 11.37 in the text and be able to recognize or reproduce the three cubic unit cells (simple cubic, body-centered cubic, and face-centered cubic) and the NaCl unit cell shown.

You can also analyze crystal structures by considering how spherical particles in a crystal lattice are packed together. Close-packed structures contain the most efficient packing arrangements of spheres.

- **Hexagonal close-packed structures** result from an ABABAB... arrangement of layers: atom A in a plane of atoms sits over itself in every other layer. See Figure 11.37 in the text.
- **Cubic close-packed structures** result from an *ABCABCABC...* arrangement of layers: atom A in a plane of atoms sits over itself in every third layer below it.
- In both types of close-packed structure, each sphere has 12 nearest neighbors: The number of nearest neighbors is termed **coordination number**.

Solids can be classified according to the types of forces between particles. Four types of solids are presented in Table 11.7 in the text. You should focus on the types of forces between particles and the properties of the solids. Key relationships to note are:

- Intermolecular forces such as London forces, dipole–dipole, and hydrogen bonds are associated with solids which are soft, have low to moderately high melting points, and have poor thermal and electrical conduction.
- Atoms bonded by covalent bonds are usually very hard and have high melting points.
- Electrostatic attractions are dominant in ionic solids and give rise to brittle or hard materials and poor thermal or electrical conduction.
- Metallic elements are bonded by metallic bonds, which involve mobile valence electrons. This produces solids which are soft to very hard, and have excellent thermal and electrical conduction.

EXERCISE 10   Identifying a unit cell

Identify the following unit cell as to type:

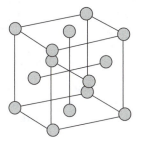

SOLUTION: *Analyze*: We are given a figure of a unit cell and are asked to identify it.

Plan: Unit cells differ by arrangement of atoms at corners, faces, and sides of a three-dimensional structure. We can identify the basic shape of the cell and then focus on how many atoms are at corners, faces, and sides and compare this information to what we know about unit cells.

Solve: By referring to the figure, we see immediately that it belongs to a cubic unit cell group because eight particles are located at the corners of the cell and are spatially arranged to form a cube. Since particles are also located at the center of each face of the cube, this unit cell is known specifically as a face-centered cubic cell.

Check: We can confirm our conclusion by looking at Figure 11.33 in the text.

EXERCISE 11   Determining the number of chemical formula units per unit cell

The following figure represents the cubic unit cell for CsCl:

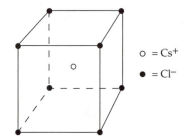

Calculate the number of formula units per unit cell for CsCl.

SOLUTION: *Analyze*: We are given a cubic unit cell and asked to determine the number of formula units of CsCl represented in the unit cell.

Plan: To determine the number of formula units in a unit cell of CsCl we need to identify how each atom is shared with other unit cells. This will enable us to calculate how many atoms belong to the unit cell. Atoms at corners and sides are shared with other unit cells but an atom in the center of the unit cell is not shared.

Solve: There are eight Cl^- ions at the corners of the cube and one Cs^+ ion at the center. An ion on a corner is shared by eight other unit cells and contributes only $\frac{1}{8}$ of itself to the unit cell. An ion at the center of a cube contributes all of itself to the unit cell. Therefore:

$$\text{Number of } Cl^- \text{ per unit cell } = (8)\left(\tfrac{1}{8}\right) = 1$$
$$\text{Number of } Cs^+ \text{ per unit cell } = (1)(1) = 1$$

The ratio of Cs^+ to Cl^- is 1:1; thus the number of formula units per unit cell is 1.

EXERCISE 12   Calculating the mass of an atom from its density and dimensions of its unit cell

At room temperature iron crystallizes in a body-centered cubic close-packed structure. The length of an edge of the cubic cell is 287 pm. A body-centered cubic structure has two atoms per unit cell. Given that the density of Fe is 7.86 g/cm^3, calculate the mass of an iron atom.

SOLUTION: *Analyze*: We are told that Fe crystallizes in a body-centered close-packed structure with an edge of 287 pm, it has two atoms per cubic unit cell and it has a density of 7.86 g/cm^3. Given this information we are asked to calculate the mass of a single iron atom.

Plan: If we can calculate the mass of a single cubic unit cell of Fe we can then calculate the mass of a single Fe atom because the cubic unit cell contains two iron atoms. We can calculate the mass of a single unit cell by calculating the volume of a single unit cell: density = mass/volume.

Solve: The length of an edge of the cubic unit cell of iron is 287 pm, which is equivalent to 287×10^{-12} m, or 2.87×10^{-8} cm. From this length, the volume of the cubic unit cell of iron is calculated:

$$V = l \times l \times l = l^3 = (2.87 \times 10^{-8} \text{ cm})^3 = 2.36 \times 10^{-23} \text{ cm}^3$$

From the given density, and the calculated volume, the mass of the unit cell is

$$d = \frac{m}{V}$$

or

$$m = d \times V = \left(\frac{7.86 \text{ g}}{1 \text{ cm}^3}\right)(2.36 \times 10^{-23} \text{ cm}^3) = 1.85 \times 10^{-22} \text{ g}$$

Note that the unit cell contains two iron atoms; thus, if you divide the mass of the unit cell by two, you determine the mass of a single iron atom:

$$\text{mass Fe atom} = \left(\frac{1 \text{ unit cell}}{2 \text{ Fe atoms}}\right)\left(\frac{1.85 \times 10^{-22} \text{ g}}{1 \text{ unit cell}}\right) = 9.25 \times 10^{-23} \frac{\text{g}}{\text{Fe atom}}$$

Check: We can check the calculated mass of an iron atom by taking its molar mass and dividing by Avogadro's number:

$$\frac{55.845 \text{ g/mol}}{6.02 \times 10^{23} \text{ atoms/mol}} = 9.28 \times 10^{-23} \text{ g/atom.}$$

This is excellent agreement with the mass calculated in the problem.

SELF-TEST QUESTIONS

Key Terms

Having reviewed key terms in Chapter 11, match key terms with phrases and identify statements as true or false. If a statement is false, indicate why it is incorrect.

Match each phrase with the best term:

11.1 This condition eventually exists between the particles of water in the liquid and gas states when liquid water is placed in an evacuated container and sealed.

11.2 At an atmospheric pressure of 700 mm Hg water boils at 97.7 °C.

11.3 The temperature at which the vapor pressure of a gas above the liquid is one atmosphere.

11.4 The pressure exerted by the particles of a liquid in the gas phase.

11.5 An oil that has a low rating of this term is best used at low temperatures in an automobile engine, thereby permitting pistons to slide easily against cylinder walls.

11.6 A physical property exhibited by alcohol at 20 °C when it spreads out on a piece of wax paper more readily than water.

11.7 A liquid is drawn up into a thin tube by this action.

11.8 Type of forces holding Ar atoms together in the liquid state.

11.9 The existence of this type of intermolecular force causes water to have a higher normal boiling point than hydrogen sulfide.

11.10 These types of forces permit molecular compounds to exist as liquids and solids.

11.11 A solid whose arrangement of particles does not exhibit a regular pattern throughout the crystal.

11.12 A solid whose arrangement of particles shows a regular pattern in all directions.

11.13 The smallest portion of a crystal that can be used to reproduce the entire structure of the crystal.

11.14 The temperature at which the solid and liquid phases of a substance are at equilibrium and the pressure is one atmosphere.

11.15 $CO_2(s)$ cannot sublime to form $CO_2(g)$ above this temperature.

11.16 In this type of packing of particles in a solid the atoms are adjacent to six other atoms in the same plane.

11.17 The highest temperature at which CH_4 can be converted from a gas to a liquid.

11.18 The number of nearest-neighbor anions surrounding a cation in a crystal.

Terms:

(a) amorphous
(b) boiling point
(c) capillary
(d) coordination number
(e) critical temperature
(f) crystalline
(g) cubic closest-packing
(h) dynamic equilibrium
(i) hydrogen bonds
(j) intermolecular
(k) London dispersion forces
(l) normal boiling point
(m) normal melting point
(n) surface tension
(o) triple point
(p) unit cell
(q) vapor pressure
(r) viscosity

True-False Statements:

11.19 As critical temperature increases for a series of related substances, *critical pressure* decreases.

11.20 Figure 11.2 could represent the *phase diagram* for a pure substance.

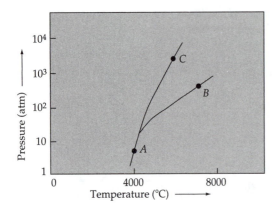

▲ **FIGURE 11.2** Phase diagram.

11.21 According to Figure 11.2, it is possible for the substance to undergo a *phase change* from solid to liquid at 9 atm of pressure (point *A*).

11.22 The *ion–dipole forces* between Ca^{2+} and water should be greater than those between Na^+ and water.

11.23 Significant *dipole–dipole forces* should exist between CCl_4 and SO_2.

11.24 The *polarizability* of CCl_4 should be larger than that of CH_4.

11.25 Water collected on a well waxed surface tends to form a spherical bead. This shows that the *surface tension* of water is high relative to the intermolecular forces between water and wax.

11.26 Oxygen freezes at a lower temperature than carbon dioxide. Thus, you should expect the *heat of fusion* of oxygen to be greater than that of carbon dioxide.

11.27 Heat of *vaporization* for a liquid is always smaller than its heat of fusion.

11.28 A *volatile* substance has a high heat of vaporization.

11.29 A crystal lattice represents a two-dimensional array of points describing a *crystalline solid*.

11.30 A *primitive cubic* unit cell shows lattice points only at the corners of a cube.

11.31 A *body-centered cubic* unit cell is a primitive cubic unit cell with the addition of a lattice point in the center of the cube.

11.32 A *face-centered cubic* unit cell is a body-centered unit cell with the addition of lattice points at the center of each side (face) of the cube.

11.33 *X-ray diffraction* techniques are used to determine the spacing of layers of atoms in a solid crystal.

11.34 *Molecular solids* tend to be rigid solids.

11.35 An example of a *covalent (network) solid* is diamond.

11.36 *Ionic solids* show good thermal conduction.

11.37 *Metallic solids* consist of arrays of atoms of metallic elements and show good conductivity of heat.

Problems and Short-Answer Questions

11.38 The figure below shows a graph of the vapor pressure versus temperature for three different molecular substances. At the temperature of 30°, shown by a vertical line in the figure, which substance has the greatest relative strength of intermolecular forces? Explain.

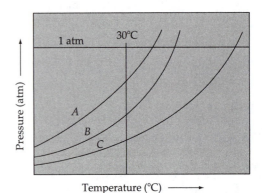

11.39 In the previous figure a horizontal line shows the pressure of one atmosphere. Which substance has the highest normal boiling point? Explain. Assume that sodium chloride is soluble in substance (B). How would the curve for substance (B) change if sodium chloride is added to it and how would this affect its boiling point? Explain.

11.40 Which figure shows hydrogen bonding?

(a) H—H

(b) H—NH$_2$

(c) H—F---H—F

(d)
$$
\begin{array}{cccc}
 & O & & O \\
 & / \ \backslash & & / \ \backslash \\
H & & H\text{- - -}H & & H
\end{array}
$$

11.41 In which phase(s) of a substance are:

(a) collisions between particles not occurring;

(b) particles able to slide past one another;

(c) average kinetic energies of the particles are the greatest?

11.42 Compare the diffusion rates of the particles in a gas, liquid, and solid.

11.43 Which of the following exhibit ion–dipole forces between the particles:

(a) H_2O and CH_3OH;

(b) $CaCl_2$ and benzene;

(c) NaCl and H_2O? Explain.

11.44 Compare the magnitude of dipole–dipole forces between particles of a substance in the liquid state to the gaseous state.

11.45 How does the polarizability of a molecule relate to the magnitude of London dispersion forces? What property shown on the periodic table should help you predict the relative magnitude of London dispersion forces between chlorine, iodine, and bromine molecules?

11.46 How does the shape of a molecule play a role in the magnitude of London dispersion forces for nonpolar molecules? Will a linear-shaped molecule of the same substance have a higher or lower boiling point than a spherical-shaped sample of the same substance?

11.47 Critique the following statement: London dispersion forces operate between nonpolar molecules but not between polar molecules.

11.48 H_2Te has a higher boiling point than SnH_4. Why? Do any of these substances exhibit hydrogen bonding? Explain.

11.49 Rank the expected trend in hydrogen-bonding forces between the nitrogen atom of $NH_3(l)$ and the following molecules: HF, H_2O, and NH_3.

11.50 Rank KCl, F_2, CO_2, and He in expected order of increasing boiling points. Explain.

11.51 Rank the following in expected order of increasing critical temperatures: NH_3, Ar, and CO_2. Explain.

11.52 Water boils at 71 °C on top of Mt. Everest (8848 m above sea level) and at 86 °C on top of Mt. Whitney (4418 m above sea level). Why does water boil at a higher temperature on Mt. Whitney than on Mt. Everest?

11.53 Should ionic salts in water increase the surface tension of aqueous solutions above the value for pure water?

11.54 Explain why methanol (CH_3OH) has a higher normal boiling point and a higher normal melting point than methylfluoride (CH_3F), given the following data:

Compound	Molecular weight amu	b.p. (°C)	m.p. (°C)
CH_3OH	32.04	65	−94
CH_3F	34.03	−78	−142

11.55 A heating curve shows the change in temperature of a substance as a function of the time it is heated. It is like a phase diagram, but instead of pressure as one of the axes, time is plotted. Figure 11.3 shows an idealized heating curve for a substance at a constant pressure of 1 atm beginning with a solid at a low temperature. Explain what physical changes in the substance are occurring in the regions from points A to B, B to C, C to D, D to E, and E to F. What phases are present at temperatures x and y?

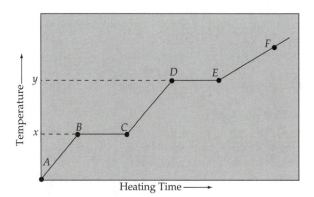

▲ **FIGURE 11.3** Heating curve.

11.56 The cooling curve of 1 mol of a substance is shown in Figure 11.4 below.

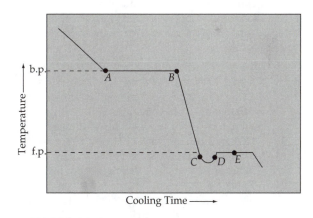

▲ **FIGURE 11.4** Cooling curve.

(a) Why is the curve flat and horizontal between points A and B?

(b) The curve between points C and D results from a phenomenon known as supercooling. At points on this curve, a liquid has a temperature below its freezing point. Suggest a reason for the existence of this phenomenon.

(c) What are the phases of the substances at points C, D, and E?

11.57 The length of a CsCl cubic unit cell is 4.123 Å. Calculate the density of CsCl in g/cm³. See Exercise 12 for a diagram of the cubic unit cell.

11.58 Indicate the type of crystal (molecular, metallic, covalent-network, or ionic) each of the following would form upon solidification:

(a) SO_3;   (d) $MgCl_2$;
(b) Ne;   (e) SiO_2.
(c) Cs;

11.59 The critical temperature of ammonia is 405.6 K and its critical pressure is 111.5 atm. Is this sufficient information to conclude that at a temperature of 500 K and a pressure of 111.5 atm ammonia is a gas? If there is insufficient information, what additional information is needed and why?

11.60 A substance is known to boil at 293 K whereas Ne boils at 27 K. Is this sufficient information to conclude that the unknown substance must be polar? If there is insufficient information, what additional information is needed and why?

11.61 The triple point temperature for a substance is $-56\,°C$. Is this sufficient information to conclude that the unknown substance is a solid at sea level and a temperature of $-65\,°C$? If there is insufficient information, what additional information is needed and why?

Integrative Exercises

11.62 The vapor pressure of water at 60.00 °C is 149.4 torr. A 0.100 g sample of water is introduced into a 525 mL vessel and the temperature is maintained at 60.0 °C. Will the water be present as a mixture of liquid and vapor or vapor only? Explain.

11.63 Pentane, C_5H_{12}, has three isomers [substances having the same formula but the atoms are arranged differently].

(a) Draw the three isomeric forms showing the arrangement of atoms given that a hydrogen atom forms only one bond to a carbon atom and the carbon atoms can bond to each other to form a chain of carbon–carbon bonds.

(b) Which form will have the highest boiling point? Explain.

11.64 Ethane, CH_3CH_3, dimethyl ether, CH_3OCH_3, and ethyl alcohol, CH_3CH_2OH, are each a liquid at $-100\,°C$.

(a) Draw their Lewis structures and determine which are polar.

(b) Rank each in expected order of increasing vapor pressures.

(c) Which can exhibit hydrogen bonding? Explain.

11.65 The text in Section 11.2 focuses on how intermolecular forces affect boiling and melting points of substances. Application of the concept of intermolecular forces to two different substances interacting can help explain solubility. Decide whether each of the following substances is more likely to be soluble in water or *n*-hexane (C_6H_{14}—carbon atoms consecutively connected to one another): CCl_4; $Ca(NO_3)_2$; and $[Ag(NH_3)_2]Cl$.

Multiple-Choice Questions

11.66 What type of cubic crystal is represented by the following diagram?

(a) Normal   (d) body-centered
(b) cubic   (e) simple
(c) face-centered

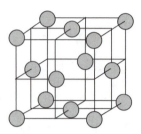

11.67 Which type of crystalline solid is likely to consist of NO particles?

(a) ionic   (d) metallic
(b) molecular   (e) covalent
(c) atomic

11.68 Which of the following substances should have the highest normal boiling point?

(a) CH_3Cl (d) CH_4
(b) CH_3Br (e) CH_3OH
(c) CH_3I

11.69 In liquid SO_2 which of the following intermolecular forces operate in the liquid phase: (1) ion–ion; (2) ion–dipole; (3) dipole–dipole; (4) London dispersion?

(a) (1) and (2)   (d) (2) and (4)
(b) (3) and (4)   (e) all of them
(c) (1) and (3)

11.70 Which substance crystallizes as a molecular solid?

(a) Ar   (d) phosphorus
(b) Ni   (e) NO
(c) NaCl

11.71 Which substance should have the highest boiling point?

(a) Ar   (d) CH_2Cl_2
(b) SO_2 (e) H_2O
(c) KCl

11.72 Which of the following represents the correct order of increasing boiling points for CCl_4, Cl_2, $ClNO$, N_2?

(a) $CCl_4 < ClNO < Cl_2 < N_2$
(b) $N_2 < Cl_2 < ClNO < CCl_4$
(c) $Cl_2 < N_2 < CCl_4 < ClNO$
(d) $CCl_4 < N_2 < Cl_2 < ClNO$
(e) $ClNO < Cl_2 < N_2 < CCl_4$

11.73 Which substance should exhibit hydrogen bonding in the liquid phase?

(a) CF_4 (d) HI
(b) CHF_3 (e) $CH_3CH_2CH_2CH_2NH_2$
(c) CsH

11.74 The vapor pressure of a liquid will decrease if

(a) the volume of the vapor above the liquid is increased.

(b) the volume of the liquid is decreased.

(c) the temperature is decreased.

(d) the surface area of the liquid is decreased.

(e) a more volatile liquid is added.

11.75 Which statement about HCl and ICl is true?

(a) Both are ionic substances.

(b) HCl is ionic and ICl is covalent.

(c) Dipole–dipole forces exist between particles in both.

(d) ICl exhibits hydrogen bonding.

(e) London forces are relatively not important for ICl.

11.76 Which substance should exhibit hydrogen bonding in the liquid phase?

(a) SF_4 (d) HBr

(b) CH_3CH_3 (e) $CH_3CH_2NH_2$

(c) NaH

11.77 TiO_2 does not form a cubic crystal lattice, but a tetragonal one. How many oxide ions are there per unit cell if two of the oxide ions are within the interior of the cell, four oxide ions are in the faces of the cell and one Ti^{4+} ion is at the center of the cell and each corner of the cell has a Ti^{4-} ion?

(a) 2   (d) 5

(b) 3   (e) 6

(c) 4

SELF-TEST SOLUTIONS

11.1 (h). **11.2** (b). **11.3** (l). **11.4** (q). **11.5** (r). **11.6** (n). **11.7** (c). **11.8** (k). **11.9** (i). **11.10** (j). **11.11** (a). **11.12** (f). **11.13** (p). **11.14** (m). **11.15** (o). **11.16** (g). **11.17** (e). **11.18** (d). **11.19** False. Pressure is directly related to temperature; thus critical pressure tends to increase with increasing critical temperature. **11.20** True. **11.21** False. At 9 atm of pressure, only solid and gas exist; thus only a transformation from solid to gas is possible. **11.22** True. The higher charge of the calcium ion will more strongly interact with the negative end of the water dipole. **11.23** False. CCl_4 is a nonpolar molecule; therefore London forces will exist, but not dipole–dipole. **11.24** True. CCl_4 has more electrons and is a larger molecule. **11.25** True. **11.26** False. It takes less energy to melt oxygen; thus, it should have a lower heat of fusion. **11.27** False. Vaporization requires the particles to be separated, which requires more energy than simply moving particles from a rigid lattice to a liquid state. **11.28** False. A high heat of vaporization implies strong intermolecular forces in the liquid. This would result in a lower vapor pressure and a higher boiling point. **11.29** False. It is a three-dimensional array. **11.30** True. **11.31** True. **11.32** False. It is a primitive cubic cell with the addition of lattice points to the faces of the cube.

11.33 True. **11.34** False. They tend to be soft solids because intermolecular forces between particles are somewhat weak. **11.35** True. **11.36** False. They are poor conductors of heat.

11.37 True.

11.38 For a given temperature we find the following: The lower the vapor pressure of a liquid the greater the magnitude of intermolecular forces. In the figure, we see that the vapor pressure line of substance (C) intersects the line for temperature at the lowest vapor pressure. Substance (C) has the greatest relative strength of intermolecular forces.

11.39 The normal boiling point of a substance can be determined by the intersection between the lines for one atmosphere pressure and the vapor pressure for the substance. We see for substance (C) that its vapor pressure line intersects the one atmosphere line at the highest temperature (furthest to the right). Substance (C) has the highest boiling point. This is also in agreement with the idea that the greater the relative magnitude of intermolecular forces the greater the boiling point. If sodium chloride is present in a solution with substance (B), this would cause an increase in the intermolecular forces experienced by substance (B) (ion–dipole and dipole–dipole) resulting in a lower vapor pressure for the solution. This would cause the vapor pressure line of the solution to be lower at any temperature compared to the vapor pressure line of pure substance (B). The vapor pressure curve shifts to look more like the vapor pressure line for substance (C). This would cause an increase in the boiling point.

11.40 Figure (c). Hydrogen bonding is between a H atom attached to generally F, O, or N atom in a substance and a different F, O, N atom.

(a) Solid state.

(b) Gas and liquid states.

(c) Gas state.

11.42 Diffusion is rapid for gas particles, slow for liquid particles, and nonexistent for solid particles.

11.43 Ion–dipole forces require the presence of an ion and a polar molecule.

(a) has polar water and moderately polar methanol. No ions are present from a salt.

(b) has an ionic calcium chloride but benzene is a nonpolar molecule.

(c) has ionic sodium chloride and a polar water molecule; thus this pair exhibits ion–dipole forces between particles.

11.44 Intermolecular forces between particles in the liquid state are greater than between particles of the same type in the gas state. The average kinetic energy of particles in the gas state overcomes the intermolecular forces found within the liquid state and this results in gas particles that are not close to one another.

11.45 Polarizability is a measure of the ease with which an electron cloud of an atom can be changed in the presence of an electric charge or field. The greater the polarizability the more easily the electron cloud can be distorted from its equilibrium condition. London dispersion forces result from temporary dipoles that exist during the oscillation of electrons in the electron cloud. This oscillation causes a temporary separation of charge in the molecule and can also induce a temporary dipole in a nearby molecule. The more easily the electron cloud can be distorted, the greater the extent of the temporary dipole. Thus, the greater the polarizability of a molecule, the greater the extent of London dispersion forces. Generally as the number of electrons in a molecule increases the greater the polarizability. This property is reflected by molecular size and volume. If you look at the atomic masses of chlorine, iodine, and bromine you see that the trend is chlorine < bromine < iodine. Thus, the expected trend in London dispersion forces is the same.

11.46 The magnitude of intermolecular forces partially depends on how close molecules can come together and the extent of potential interaction along the surfaces of their structures. Linear molecules permit a greater extent of contact along the surfaces of the molecules than a spherical surface. Thus the linear form of a substance should exhibit a higher boiling point than a spherical form.

11.47 Electron clouds in both polar and nonpolar molecules oscillate and thus form temporary dipoles. Thus London dispersion forces exist in both types of molecules.

11.48 H_2Te is a bent molecule and polar. SnH_4 is a tetrahedral molecule and thus is nonpolar. Neither exhibits hydrogen bonding because there are no $O—H, H—F$, and $N—H$ bonds.

11.49 The extent of the hydrogen bond between $A—H \cdots NH_3$ will depend of the extent of the bond dipole in the AH bond and the polarization of the electrons in the bond towards atom A. The expected trend is $N—H \cdots NH_3 < O—H \cdots NH_3 < H—F \cdots NH_3$.

11.50 The boiling point partially depends on the attractive forces within the liquid phase. These should be extremely strong in a molten liquid ionic salt such as KCl. For the other substances you have to look at molecular weights (London dispersion forces) and potential for polarity and hydrogen bonding. The boiling point of He should be the lowest because it is nonpolar and has the lowest molecular weight. Fluorine and CO_2 are nonpolar molecules and thus their relative boiling points should be reflected in their molecular weights. Carbon dioxide is a more massive substance and thus should experience greater London dispersion forces. The expected trend is $He < F_2 < CO_2 < KCl$.

11.51 The magnitude of a substance's critical temperature reflects intermolecular forces. Ammonia should have the greatest critical temperature because it is polar and exhibits hydrogen bonding. Ar and CO_2 only exhibit London dispersion forces, and CO_2 should have greater intermolecular forces because it is slightly more massive. The expected trend is $Ar < CO_2 < NH_3$.

11.52 At higher altitudes; atmospheric pressure is less than at lower altitudes. Boiling occurs when the vapor pressure of the liquid equals the surrounding atmospheric pressure. Thus water has a lower vapor pressure and boiling point on Mt. Everest than on Mt. Whitney.

11.53 Ions in solution tend to pull water molecules into the interior of the solution because of ion–dipole attractions. The liquid surface in an aqueous ionic solution therefore contracts to a smaller area than is the case for pure water under identical conditions. More energy is required to increase the surface area of the aqueous ionic solution than is required to increase that of pure water; thus the addition of ionic salts to water increases the surface tension of the aqueous solution above the value for pure water. Note that the increasing of the surface tension of water by addition of salts is of a far smaller magnitude than the decreasing of surface tension by addition of detergent.

11.54 Both CH_3OH and CH_3F are polar and have similar molecular weights. If dipole–dipole forces and London dispersion forces were the only intermolecular forces, we would expect the compounds to have similar melting and boiling points. However, CH_3OH molecules are also attracted to each other by hydrogen-bonding forces, which must be overcome in the melting and boiling processes. The additional energy required to break hydrogen bonds between methanol molecules increases the melting and boiling points of CH_3OH compared to those of CH_3F.

11.55 The physical changes occurring in the specified regions are as follows: A to B corresponds to heating the solid to its melting point; B to C corresponds to the melting of the solid at constant temperature; C to D corresponds to heating the liquid to its boiling point; D to E corresponds to the boiling of the liquid at constant temperature and pressure; E to F corresponds to heating the gas from the boiling point of the liquid to higher temperatures. Temperature x corresponds to the melting point of the solid. Thus both solid and liquid phases are present. Temperature y corresponds to the boiling point. At this temperature both liquid and gas phases may exist.

11.56 (a) Between points A and B, heat is being removed from the substance at its boiling point with no change in the substance's temperature. During this time, condensation of the substance from gas to liquid occurs. The average kinetic energies of gas molecules are lowered to the point that attractive forces dominate and a liquid forms. As long as the kinetic energy removed is exactly equal to the energy released during the formation of intermolecular bonds between liquid particles, the temperature of the substance remains constant. This temperature is the boiling point of the liquid or condensation temperature of the gas.

(b) Supercooling represents the time lag associated with the change from the random organization of liquid molecules to their regular and orderly arrangement in a crystal. This time lag often occurs when cooling is too rapid for a true dynamic equilibrium between liquid and solid to occur. Until the lowest point on the curve between

C and *D* is reached, there are no crystal nuclei in the solution upon which other particles can be deposited to form larger crystals.

(c) Point *C* corresponds to the supercooled liquid phase only. Point *D* corresponds to a mixture of tiny crystals and liquid. At this point crystals are forming rapidly. At point *E* the rate of crystal formation has "caught up" with the cooling rate, and the solid and liquid phases are in equilibrium.

11.57 CsCl has a cubic unit cell with one formula of CsCl per unit cell. The volume of the unit cell is $V = (4.123 \text{ Å})^3 = 70.08 \text{ Å}^3$. The mass of the unit cell is the mass of one Cs^+ ion and one Cl^- ion: $m = 132.90 \text{ amu} + 35.45 \text{ amu}$ $m = 132.90 \text{ amu} + 35.45 \text{ amu} = 168.35 \text{ amu}$. The density (m/V) of one unit cell is:

$$d = \left(\frac{168.35 \text{ amu}}{70.08 \text{ Å}^3}\right)\left(\frac{1 \text{ Å}}{10^{-8} \text{ cm}}\right)^3 = 2.402 \times 10^{24} \text{ amu/cm}^3$$

One gram is equivalent to 6.022×10^{23} amu. The density of CsCl expressed in g/cm^3 is

$$\left(2.402 \times 10^{24} \frac{\text{amu}}{\text{cm}^3}\right)\left(\frac{1 \text{ g}}{6.022 \times 10^{23} \text{ amu}}\right) = 3.989 \text{ g/cm}^3$$

11.58 (a) SO_3 is a covalent molecule; thus it is a molecular type.

(b) Ne would possess London dispersion forces only between atoms; thus it is a molecular type.

(c) Cs is a metal; thus it is a metallic type.

(d) $MgCl_2$ consists of a metal and nonmetal; thus it is an ionic type.

(e) SiO_2 contains a network of covalent Si—O linkages creating a three-dimensional array of interconnected silicon and oxygen atoms; thus it is a covalent-network type.

11.59 There is sufficient information because the critical temperature represents the highest temperature at which a substance can exist as a liquid. Since ammonia is above this temperature it cannot exist as a liquid, irrespective of the pressure.

11.60 There is insufficient information. The data suggests that nonpolar substances such as Ne boil at low temperatures and given the significantly higher b.p. of the unknown substance it might be polar. However, nonpolar substances with significant London dispersion forces can exhibit high b.p.s. We need to measure its dipole moment to determine if the substance is polar.

11.61 There is insufficient information because we also need to know the pressure at the triple point temperature. If the triple point pressure is above the pressure at sea level, then it is possible that the substance is either a solid or gas, depending on the phase line for the equilibrium. If the triple point pressure is below the pressure at sea level then it is either a solid or liquid depending on the slope of the phase line for the solid–liquid equilibrium. We need a

phase diagram so that we can determine the phase at the given temperature and pressure.

11.62 First, calculate the vapor pressure of water as if it existed only as a gas:

$$P = \frac{nRT}{V} = \frac{(0.100 \text{ g})\left(\frac{1 \text{ mol}}{18.02 \text{ g}}\right)\left(0.0821\frac{L\text{-atm}}{\text{mol-}K}\right)(333 \text{ K})}{0.525 \text{ L}}$$

$$P = 0.289 \text{ atm} = (0.289 \text{ atm})\left(\frac{760 \text{ torr}}{1 \text{ atm}}\right) = 220 \text{ torr}$$

The calculated vapor pressure exceeds the vapor pressure of water that can exist at 60.0 °C, 149.9 torr; therefore some of the water must exist in a liquid phase. Both the liquid and vapor forms exist together.

11.63 (a) Carbon can form a maximum of four bonds to other atoms. A C— is an abbreviation for a C—H bond.

(pentane, b.p. 36 °C)

(2-methylbutane, b.p. 28 °C)

(2,2-dimethylpropane, b.p. −10 °C)

The intermolecular interactions between molecules in all three isomers are London dispersion forces. The magnitude of dispersion forces primarily depends on the polarizability of the electron clouds and the ability of the molecules to approach one another and interact. Electrons in elongated molecules are more easily distorted than those in compact molecules. Pentane is the highest boiling point substance because it is more elongated. Also, branching of carbon atoms in a chain of carbon atoms tends to prevent a close approach of electron clouds compared to elongated molecules.

11.64 (a) Ethane is a hydrocarbon and is nonpolar. VSEPR analysis of the structure of dimethyl ether tells you that the CH_3 groups are bent about the central oxygen atom and thus it is a slightly polar molecule. Ethyl alcohol is polar around the OH group and thus it is a polar molecule.

ethane   dimethyl ether

ethyl alcohol

(b) The predicted order is developed by analyzing the various intermolecular forces that exist between the particles. The greater the intermolecular forces within the liquid, the smaller the vapor pressure. Ethane is nonpolar and thus should exhibit the weakest intermolecular forces and the greatest vapor pressure. Both dimethyl ether and ethyl alcohol are polar and have the same molecular weight; thus they should have similar dispersion forces. However, ethyl alcohol is more polar about the oxygen atom and thus it should exhibit the greatest intermolecular forces. Ethyl alcohol < dimethyl ether < ethane.

(c) Look for the existence of a N—H or F—H or O—H bond to indicate the potential existence of a hydrogen bond. Only ethyl alcohol has this type of bond and will undergo hydrogen bonding. This also contributes to a lower vapor pressure.

11.65 If a substance is to be soluble in another substance there has to be sufficient attractive forces between them. Thus, polar substances are likely to be attracted to one another and nonpolar substances are likely to be attracted to one another and not a polar substance. The ions of an ionic substance will be attracted to the appropriate positive or negative dipole regions of a polar substance. Nonpolar substances are attracted to one another through London forces. CCl_4 is a symmetrical, nonpolar substance; thus it should be soluble in *n*-hexane, which is nonpolar (hydrocarbons, C—H containing compounds,

consist of nonpolar regions and thus are nonpolar). $Ca(NO_3)_2$ is ionic as is $[Ag(NH_3)_2]Cl$ and thus both should be soluble in water. The later ionic compound consists of $Ag(NH_3)^+$ and Cl^- ions.

11.66 (c)

11.67 (b) Polar-covalent molecules such as tend to crystallize in a molecular crystal lattice.

11.68 (e) Normally, the heaviest compound in a related series will have the highest boiling point because London dispersion forces increase with increasing weight. However, CH_3OH has the highest boiling point because hydrogen bonds also exist between CH_3OH molecules.

11.69 (b)

11.70 (e) Molecular solids usually form from polar or nonpolar *molecular* compounds. NO is the only case listed that fits this requirement.

11.71 (c) In general, ionic substances have higher boiling points than the atomic or molecular substances listed here.

11.72 (b) The following three molecules are easily ordered based on increasing molecular volume and therefore increasing London forces: $N_2 < Cl_2 < CCl_4$. ClNO needs to be placed. ClNO is polar and should have a higher boiling point than Cl_2 CCl_4 is *far* more massive than ClNO and thus should have a higher boiling point, even though CCl_4 is nonpolar. London forces are very significant with large molecules.

11.73 (e)

11.74 (c) Lowering the temperature of a liquid decreases its vapor pressure because intermolecular forces become stronger relative to kinetic energy of particles.

11.75 (c) Both substances are molecular and polar.

11.76 (e) Hydrogen bonding will occur between the same molecules if a hydrogen atom is covalently bonded to one of the highly electronegative atoms: N, O, or F. Inspection of the five substances shows that only (e) has hydrogen bonded to one of these atoms, that is, a nitrogen atom.

11.77   (c) The oxide ions at the center of the unit cell count fully and thus contribute two ions. The ions at the faces are shared with another unit cell and thus count only $\frac{1}{2}$ for each oxide ion; thus there is a net of $4\left(\frac{1}{2}\right)$ ions, or two. The total is thus four oxide ions per unit cell.

Sectional MCAT and DAT Practice Questions IV

Many instruments have been developed to measure the pressures of gases. One type is a manometer which is normally used to measure pressures at less than atmospheric. When a manometer uses mercury as the liquid, the pressure of a gas is commonly reported in mm Hg or torr (1 mm Hg = 1 torr). The SI unit of pressure is the Pascal (Pa) and one torr is equivalent to 133.32 Pa. One atm pressure is equivalent to 760 torr.

An example of one type of manometer is shown in Figure 1.

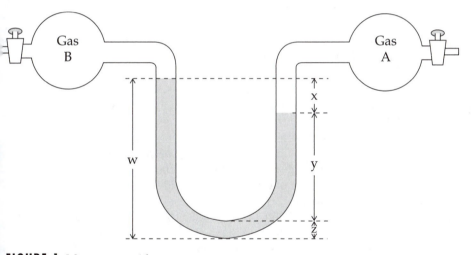

FIGURE 1 Manometer with two gases

Different gases at different pressures can be placed in the two chambers, which are separated by a liquid. Mercury is typically the chosen liquid because it has a low vapor pressure. Mercury also flows readily and can be found in tiny small spheres in pores of rocks in which it is typically found. The volume of a gas in one of the chambers can be determined along with its pressure if the nature and amount of gas is known.

Daniel Fahrenheit designed the first thermometer using mercury as the liquid. Temperature is correlated to the length of a mercury column contained in a narrow bore within a glass tube. Mercury thermometers cannot be used at low temperatures because the freezing point of mercury is 38.83 °C. Listed in Table 1 are the temperatures at which several liquids have vapor pressures of 100 mm Hg.

TABLE 1 Temperatures of selected liquids when their vapor pressures are 100 mm Hg

Substance	Temperature (°C) when vapor pressure is 100 mm Hg
Benzene, C_6H_6	26.1
Ethyl alcohol (ethanol), C_2H_5OH	34.9
Propane, C_3H_8	−79.6
Water	50.2

The ideal gas equation is $PV = nRT$ where R is 0.0821 L-atm/mol-K.

1. Which measurement in Figure 1, w, x, y, or z, represents the pressure difference between the two gases?
    (a) w   (b) x   (c) y   (d) z

2. If the partial pressure of gas A is 6.00 atm and the values of x, y, and z are 825 mm Hg, 15 mm Hg, and 12 mm Hg, respectively, what is the partial pressure of gas B at the same temperature?
    (a) 4.91 atm   (b) 5.98 atm   (c) 6.12 atm   (d) 7.09 atm

3. Which best characterizes mercury at room temperature and pressure?
    (a) high vapor pressure
    (b) typical metallic properties
    (c) valence electrons are localized on mercury atoms
    (d) high surface tension

4. Which substances in Table 1 possess London forces between particles?
    (a) benzene   (b) ethanol and water
    (c) propane and benzene   (d) All

5. Which substances in Table 1 exhibit hydrogen bonding?
    (a) H_2O only   (b) propane and benzene
    (c) ethanol and water   (d) benzene only

6. What is the density of carbon tetrachloride gas (molar mass, 154.0 g/mol) at 755 torr and 155 °C?
    (a) 0.0833 g/L (b) 0.230 g/L   (c) 4.35 g/L   (d) 12.0 g/L

7. Which substance in Table 1 has the lowest boiling point when the external pressure is 100 mm Hg?
    (a) benzene   (b) ethanol   (c) propane   (d) water

Questions 8 through 12 are **not** based on a descriptive passage.

---

8. When a particular substance forms a solid that is hard, has a high melting point, and is a good electrical conductor, we classify its crystalline state as
    (a) molecular.   (b) covalent network.
    (c) ionic.   (d) metallic.

9. A real gas typically exhibits behavior that is closest to an ideal gas at
    (a) low pressure and high temperature.
    (b) high pressure and high temperature.
    (c) low pressure and low temperature.
    (d) high pressure and low temperature.

10. The viscosity of a liquid
    (a) increases with increasing temperature.
    (b) decreases with increasing temperature.
    (c) increases with decreasing volume.
    (d) decreases with decreasing volume.
11. Which solvent should best dissolve $I_2(s)$ to form a homogeneous solution?
    (a) $H_2O(l)$ (b) $KI(s)$ (c) $CCl_4(l)$ (d) $NO(g)$
12. For a particular substance which enthalpy change has the largest value?
    (a) ΔH_{fusion} (b) $\Delta H_{vaporization}$
    (c) $\Delta H_{melting}$ (d) $\Delta H_{sublimation}$

ANSWERS

1. (b) The difference in heights of the two mercury columns represents the pressure difference between the two gases.

2. (a) $P_{gas\ B} = P_{gas\ A} - P_{difference} = 6.00$ atm $-$ (825 mm Hg)
(1 atm/760 mm Hg) = 4.91 atm

3. (d) The passage states that Hg can occur in tiny drops in pores of rocks. Tiny drops of a liquid form when it has a high surface tension. The passage also states that Hg at room temperature is a liquid and has a low vapor pressure. Typical metallic properties include "hardness." Mercury is a shiny liquid, and it is not a "hard" metal. The bonding electrons (involves valence electrons) in a metal are relatively free to move throughout the elemental structure.

4. (d) All molecular substances have London forces between particles. Other intermolecular forces may also exist such as in water and ethanol.

5. (c) Both water and ethanol possess an —OH group. The hydrogen atom in one —OH group can hydrogen bond to an oxygen atom of an —OH group in a different molecule. Propane and benzene molecules do not possess a hydrogen atom attached to N, F, or O and thus do not have the ability to hydrogen bond to one another.

6. (c) Use the ideal-gas law to calculate density and write n as g/(Molar Mass).

$$d = \frac{P(\text{Molar Mass})}{RT} = \frac{(755\ \text{torr})\left(\dfrac{1\ \text{atm}}{760\ \text{torr}}\right)\left(154.0\,\dfrac{g}{\text{mol}}\right)}{\left(0.0821\dfrac{\text{L-atm}}{\text{mol-K}}\right)(428\ \text{K})} = 4.35\ \frac{g}{L}$$

Note: the answer can be obtained by approximation and compared to given answers.

$$d = \frac{(760\ \text{torr})\left(\dfrac{1\ \text{atm}}{760\ \text{torr}}\right)\left(155\,\dfrac{g}{\text{mol}}\right)}{\left(0.1\dfrac{\text{L-atm}}{\text{mol-K}}\right)(400\ \text{K})} \approx 4\ \frac{g}{L}$$

Only answer (c) is close to the approximate answer. Use an approximation method to solve a math problem when you do not have a calculator.

7. (c) The boiling point is the temperature when the vapor pressure of a liquid equals the external pressure. Thus, the liquid in Table 1 with the lowest temperature has the lowest boiling point at an external pressure of 100 mm Hg.

8. (d) Metallic bonds occur between atoms of metals and the properties described fit this type of bonding situation.

9. (a) Ideal gases do not interact with one another and have negligible volumes compared to the volume of the container. At low pressures the volume of a container is large compared to the volume occupied by the molecules themselves. At high temperatures, which correspond to higher average kinetic energies, molecules readily overcome intermolecular attractions. Both lead to real gases behaving more like ideal gases.

10. (b) Viscosity is a measure of the resistance of liquids to flow. The greater the viscosity of a liquid, the more slowly it flows. At higher temperatures the liquid particles have greater kinetic energies and decreasing attractive forces between particles. Particles of a liquid are able to flow more readily with decreasing attractive forces and this leads to a lower viscosity. The amount of a substance does not change the property of viscosity.

11. (c) I_2 is a nonpolar substance and will dissolve in a nonpolar solvent, CCl_4. Particles with similar intermolecular forces typically dissolve in one another.

12. (d) Sublimation involves a solid converting directly to a gas. The enthalpy change for this transition is the sum of the enthalpies of fusion and vaporization.

Modern Materials

Chapter 12

OVERVIEW OF THE CHAPTER

Review: Metal (2.5); nonmetals (2.5); metallic oxides (7.6, 7.7); molecular orbitals (9.7); valence electrons (6.8, 8.1).

Learning Goals: You should be able to:

1. Characterize the differences among metals, semiconductors, and insulators.
2. Describe bands of energy, conduction bands, valence bands, and band gap, and their origins.

Review: Metallic oxides (7.6, 7.7).

Learning Goals: You should be able to:

1. Describe the properties and types of semiconductors.
2. Determine the types of dopants that can be used to create *n*-type and *p*-type semiconductors.
3. Explain how solar energy is converted to electrical energy using semiconductors.
4. Describe ceramics and their characteristics.
5. Indicate the advantages and disadvantages of engineering ceramics as compared with other materials, and give examples of ceramic composites.
6. Describe the sol-gel process for forming ceramic materials.
7. Define superconductivity and give several examples of superconducting materials.
8. Define superconducting temperature, and explain why it is an important property of superconducting materials.

Review: Chemical formulas (3.3, 3.5); covalent structures (8.3, 8.5, 11.); intermolecular forces (11.2).

Learning Goals: You should be able to:

1. Describe what is meant by the terms polymer and monomer, give some examples, and describe how certain types of polymers react to changes in temperature.

12.1, 12.2 MODERN MATERIALS: CHARACTERIZATION AND ELECTRONIC STRUCTURES

12.3, 12.4, 12.5 SEMICONDUCTORS, CERAMICS, AND SUPERCONDUCTORS

12.6, 12.7 POLYMERS AND BIOMATERIALS

2. Explain how a polymer is formed from monomers via (**a**) addition polymerization and (**b**) condensation polymerization.
3. Explain what is meant by the term crystallinity in polymers, and indicate how polymer properties generally vary with degree of crystallinity.
4. Describe the process of cross-linking in polymers, and explain how it affects polymer properties.
5. Describe the necessary properties of polymeric biomaterials to make them biocompatible with body tissues and fluids and to provide long life expectancy.

12.8 LIQUID CRYSTALS

Review: Liquids (11.1).

Learning Goals   You should be able to:

1. Describe how a liquid crystalline phase differs from an ordinary (isotropic) liquid phase.
2. Distinguish among the major classes of liquid crystalline phases.
3. Describe how liquid-crystalline materials are employed in liquid crystal displays for electronics equipment.

12.9 NANO-TECHNOLOGY

Review: Graphite (11.8); hybrid orbitals (9.5); mean-free path (10.8); nano- (1.4); photon (6.2).

Learning Goals: You should be able to:

1. Describe why particles in the 1–100 nm size scale have electronic properties important to technology.
2. Explain what quantum dots are and describe how they are used as nanomaterials.
3. Describe why the nanoscale properties of materials do not always have to be in three dimensions.
4. Explain how nanoscale dispersions of certain metals can exhibit differing colors depending on the size of the particles.
5. Describe the characteristics of a carbon nanotube and how it can be viewed as a form of graphite rolled up into a tube.

TOPIC SUMMARIES AND EXERCISES

MODERN MATERIALS: CHARACTERIZATION AND ELECTRONIC STRUCTURES

In this chapter we learn about the types, properties, and applications of several classes of modern materials. Modern materials are composed of substances, which can be pure or mixtures, developed through human invention or design to perform specific functions. Scientists are working to improve the technology of modern materials for specialized applications.

Section 12.1 introduces general concepts about the characteristics of modern materials based on their electronic structures. One class of materials consists of metals, semiconductors, and insulators. The three types differ in their electronic conduction properties: metals, which readily conduct electrical current; semiconductors, which have limited electrical conductivity; and insulators, which are nonconductive.

Materials can also be classified by the type of bonding that holds the atoms together. Our modern understanding of the electronic structures and the conductivity and bonding of crystalline solids is based on molecular orbital theory.

- In large aggregates of metal atoms, atomic orbitals combine to form numerous molecular orbitals whose energies are very close to one another. Groups of molecular orbitals with energies close to one another form a continuous band of energy states. Bands are separated by gaps in energies.
- The highest band filled with electrons is the valence band. The next higher empty energy band, if it is reasonably close in energy to the valence band, is a conduction band. When this situation exists an electron in the valence band can gain energy and jump to the conduction band. Alternatively, an electron can move to a higher energy level in a partially filled valence band and conduction of current can occur.
- In an insulator we find that the energy of the conduction band is much higher than the valence band and the large band gap effectively prevents electrons from jumping.
- In a semiconductor we find that the band gap is smaller than in an insulator and some electrons are available to jump to the conduction band.
- In a conductive metal we find that there is an overlap between the valence band and the conduction band. Electrons in the valence band can readily jump to higher energy states and then move through the structure.

EXERCISE 1 Identifying the electrical properties of sodium chloride

Which form of sodium chloride has a high resistance to the flow of electric current, NaCl(s) or NaCl(l)? What name do we give to this type of material?

SOLUTION: *Analyze:* We are asked which phase of sodium chloride has a high resistance to the flow of electric current.

Plan: Sodium chloride is an ionic substance containing Na^+ and Cl^- ions. The phase which best prevents the ions from being mobile will have a higher resistance to the flow of electric current.

Solution: In a solid phase the sodium and chloride ions are in fixed positions and they are not in a mobile situation. In a liquid phase the sodium and chloride ions are in motion. For a substance to be a conductor the ions must be in motion or the valence electrons must be mobile in the band structure. In contrast an insulator does not have ions in motion or mobile valence electrons in the band structure. Thus NaCl(s) has a high resistance to the flow of electric current and is an insulator.

SEMICONDUCTORS, CERAMICS, AND SUPERCONDUCTORS

Elemental semiconductors are silicon, germanium, grey tin and carbon in the form of graphite. They are all members of group 4A with four valence electrons. Compound semiconductors exist when the average of valence electrons of the atoms in the empirical formula is four: For example, in the compound semiconductor GaAs, gallium contributes three valence electrons and arsenic contributes five and the average is four valence electrons.

The conductivity of a semiconductor can be increased by doping it with a small amount of impurity that either adds electrons to the semiconductor (*n*-type semiconductor) or has empty orbitals to accept electrons from the semiconductor (*p*-type semiconductor). An *n*-type semiconductor requires the donor electrons of the dopant to be close in energy to the conduction band of the semiconductor, and a *p*-type semiconductor requires the acceptor orbitals of the dopant to be close in energy to the valence band.

Inorganic materials that are high-melting, normally hard or brittle, and stable to very high temperatures, are termed **ceramics**. Bonding can involve a covalent network structure, ionic structure, or a combination of the two.

- Ceramic materials are based on a variety of chemical forms; the major ones are silicates (Si and O), oxides of metals, carbides (metals and carbon), nitrides (metals and nitrogen), and aluminates (Al_2O_3 with other metallic oxides).
- Ceramic materials are used to replace other engineering materials such as wood, metals, or plastics and to provide heat and corrosion resistance, low deformity properties, lighter weight materials, or lower cost materials. A major disadvantage of ceramics is their brittleness, a tendency to shatter.

The method of processing ceramics to form parts is an important focus for research. Unless the process can prevent random, extremely tiny micro cracks or voids in the structure from forming, the material is subject to stress fractures during use.

- One method for improving fracture resistance is to produce extremely pure, uniform, sub-micron ceramic particles, and then sinter them at high temperature under pressure to create the desired material.
- The **sol-gel process** is used to make extremely fine ceramic particles of uniform size. A gel is produced from a sol (a suspension of a metallic hydroxide) whose pH is adjusted to cause a condensation reaction involving splitting water out from two of the reacting particles. The resulting gel is heated, driving off all liquid, and fine metallic oxide powder is produced.

Superconductors are substances showing no resistance to the flow of electrical current. *Superconductivity* is the property of exhibiting a "frictionless" flow of electrons within a material. Materials that show this property do so only below a certain temperature, called the *superconducting transition* temperature, T_c. Superconductors showing a T_c above 77 K are desirable because relatively inexpensive liquid nitrogen instead of expensive liquid helium can be used to cool the material.

Superconductors are typically mixed metallic oxides such as $YBa_2Cu_3O_7$. Copper and oxygen are presently the most important constituents of superconducting ceramics. Yttrium and barium, for example, can be replaced by other metals without losing superconducting properties.

EXERCISE 2   Identifying the type of semiconductor

Classify the following semiconductor as to type: silicon doped with boron.

SOLUTION: *Analyze*: We are asked to determine the type of semiconductor formed when boron is added as a doping atom to silicon.

Plan: We need to determine if boron has more or fewer valence electrons than silicon. We can obtain this information from the periodic table and then determine if it is an *n*-type semiconductor or a *p*-type semiconductor.

Solve: If the dopant has more electrons than silicon the semiconductor is an *n*-type semiconductor. If it has fewer electrons then it is a *p*-type semiconductor. Silicon has four valence electron and boron has three valence electrons. Therefore, boron has fewer valence electrons than silicon and boron thus acts as an acceptor dopant. The semiconductor is a *p*-type semiconductor.

EXERCISE 3   Identifying direction of electron flow in a photovoltaic cell

A photovoltaic cell converts sunlight to electricity and consists of a thin layer of a *p*-type semiconductor in contact with an *n*-type semiconductor. Charged particles do not cross the boundary unless sunlight shines on the *p*-type semiconductor. Suggest a reason why electrons can flow across the boundary when sunlight is present.

SOLUTION: *Analyze*: We are asked to explain how a photovoltaic cell generates electrical current when sunlight strikes the *p*-type semiconductor.

Plan: We can consider the characteristics of a *p*-type semiconductor and an *n*-type semiconductor and develop a model that uses the properties of each to permit electron flow across the boundary between the two.

Solve: A *p*-type semiconductor has a dopant with acceptor orbitals near the valence band of the semiconductor. The sunlight provides energy for electrons in the valence band to jump to the acceptor orbitals, leaving positive holes in the valence band. The electrons in the acceptor orbitals of the dopant are mobile and can then move across the boundary to the empty conduction band in the *n*-type semiconductor. The excess electrons in the *n*-type semiconductor move through the electrical circuit and eventually return to neutralize the positive holes in the *p*-type semiconductor.

EXERCISE 4   Using superconducting transition temperature

Using the data in Table 12.4 in the text, explain the significant interest in $HgBa_2Ca_2Cu_3O_{8+x}$ compared to $La(Ba)_2CuO_4$.

SOLUTION: A key feature of superconductors is the superconducting transition temperature. The higher the temperature the more accessible the superconducting property. $HgBa_2Ca_2Cu_3O_{8+x}$ has a superconducting transition temperature of 133 K which is above 77 K whereas $La(Ba)_2CuO_4$ has a superconducting transition temperature of 35 K. This means $HgBa_2Ca_2Cu_3O_{8+x}$ can be studied using inexpensive liquid nitrogen whereas $La(Ba)_2CuO_4$ requires expensive liquid helium.

POLYMERS AND BIOMASS MATERIALS

Polymers form the basis of a huge industry involving such things as plastics, resins, lacquers, and synthetics. **Polymers** are macromolecules, of high molecular weights, that are formed from smaller repeating units called **monomers**.

- The coupling of monomers to form polymers is termed **polymerization**. Equation 12.1 in the text shows the polymerization reaction for polyethylene. A carbon–carbon double bond opens to form carbon atoms possessing a single electron. These carbon atoms join together into long chains by **addition polymerization**.

- A chemical formula such as $\displaystyle \begin{array}{cc} H & H \\ | & | \\ {+}\!\!-\!C\!-\!C\!-\!\!{+}_n \\ | & | \\ H & Cl \end{array}$ shows the repeating unit in the polymer. The subscript n is used to denote that a large number of these units are bonded in a chain. This type of formula is characteristic of addition polymers.

* **Condensation polymerization** is shown in Equation 12.2 in the text. In the reaction shown, a hydrogen attached to a nitrogen reacts with a OH group of adipic acid to form water which splits out. Look at the product (nylon-6,6) and observe that a N—C bond (thickened bond below) has formed between the diamine and adipic acid. Water may be a product in a condensation reaction because of the reaction between a hydrogen atom and a hydroxyl group:

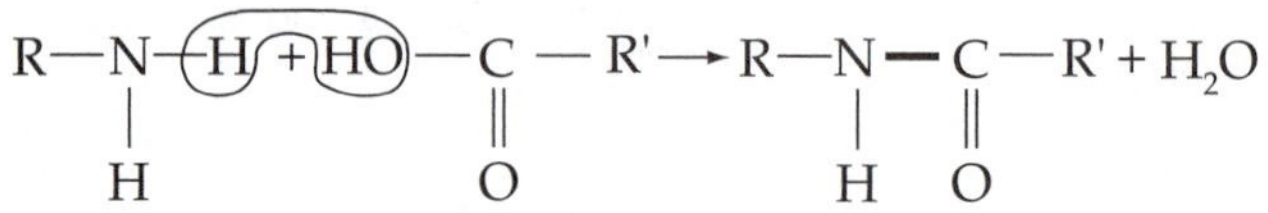

$$R-N-(H + HO)-C-R' \rightarrow R-N-C-R' + H_2O$$

Polymers are typically amorphous substances, with no definite crystalline order. However, there are local regions of ordering termed **crystallinity** that arise from interactions between one polymer chain and another.

* **Cross-linking** is a process of forming chemical bonds between chains in polymers. When cross-linking occurs, the polymer becomes more dense and rigid. Polymers are cross-linked to maintain their shapes. The text shows in Equation 12.4 the vulcanization of rubber as an example of cross-linking. Note how sulfur is used to cross-link polymer chains by reactions at carbon–carbon double bonds to form a thermosetting polymer.
* **Plasticizers** are low molecular weight materials that are added to polymers to minimize inter-chain interactions. This makes polymers more flexible.

Biomaterials constructed of polymers must meet many requirements before they can be used in human biological systems. Our bodies readily reject foreign substances and to integrate biomaterials in the body requires careful consideration of factors such as strength, reliability, purity, flexibility, hardness, and chemical nonreactivity. For example, pyrolytic carbon can be used to develop a smooth, chemically inert surface, thus minimizing the coagulation of blood; or Dacron can be formed in a polymeric mesh permitting a support for the growing of body tissue. Successful artificial tissues usually contain a copolymer that has an abundance of polar carbon–oxygen functional groups, thereby enhancing hydrogen bonding interactions.

EXERCISE 5   Writing a chemical reaction for polymerization

Write a reaction showing the formation of polystyrene $-(CHCH_2)-n$
$\quad\quad\quad\quad\quad\quad\quad\quad\quad\quad\quad\quad\quad\quad\quad\quad C_6H_5$

Is the reaction addition or condensation polymerization?

SOLUTION: *Analyze*: We are given the repeating unit in polystyrene and asked to write a reaction describing the formation of polystyrene and to identify the type of polymerization reaction.

Plan: Two types of polymerization reactions are discussed in the text: condensation and addition. The former involves repeating units coming together and splitting out small molecules, whereas the latter involves repeating units attaching together without splitting out other particles. By writing the reaction for the formation of the polymer we can identify the type of polymerization reaction.

Solve: To write the reaction describing the formation of polystyrene we first determine the monomer. Rewrite the chemical formula to show the arrangement of bonds in the carbon backbone:

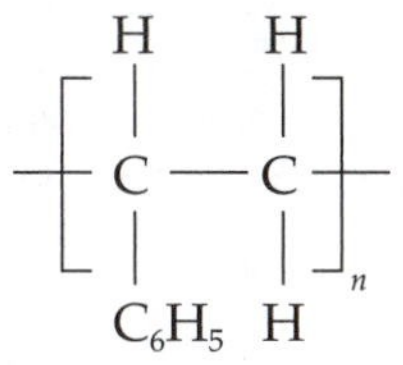

$$\left[\begin{array}{cc} \overset{\displaystyle H}{\underset{\displaystyle C_6H_5}{\overset{\displaystyle |}{\underset{\displaystyle |}{C}}}} & \overset{\displaystyle H}{\underset{\displaystyle H}{\overset{\displaystyle |}{\underset{\displaystyle |}{C}}}} \end{array}\right]_n$$

From this structure, the monomer can be deduced by recognizing that a carbon–carbon double bond in a monomer is typically opened up to permit carbon atoms of different monomers to chain together. Thus, the monomer unit used to form polystyrene should be
$$\underset{\displaystyle C_6H_5}{CH}=CH_2$$

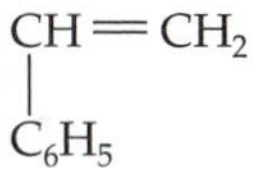

Therefore, the reaction describing the formation of polystyrene is

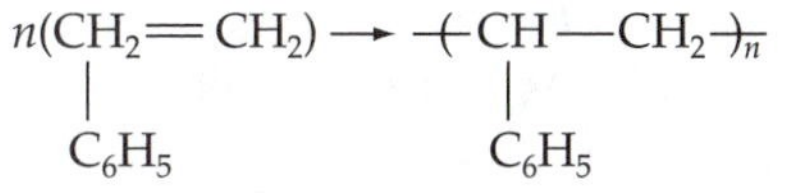

$$n(\underset{\displaystyle C_6H_5}{CH_2}=CH_2) \longrightarrow (\underset{\displaystyle C_6H_5}{CH}-CH_2)_n$$

This reaction is an example of addition polymerization because the polymer backbone is formed by carbon atoms adding together without another molecule splitting out.

Comment: If a compound contains only multiple bonds, it is likely to react by addition polymerization; if it contains $-OH$ or $-NH_2$ and $-COOH$ groups it is likely to react by condensation polymerization.

EXERCISE 6   Writing the simplest formula for the formation of a condensation polymer

A condensation polymer is formed from the following two monomers:

$$\underset{\displaystyle A}{HOCH_2CH_2COOH} \qquad \underset{\displaystyle B}{HOCH_2CH_2CH_2COOH}$$

Given that a carboxylic acid group ($-COOH$) reacts with an alcohol group ($-OH$) to split out water and form a $-C-O-C-$ internal linkage, write the product formed when two molecules of A react with one molecule of B to initiate a condensation polymerization reaction.

SOLUTION: *Analyze:* To write the formula of the product when two molecules of A react with one molecule of B we need to analyze where the condensation reactions occur.

Plan: We are told that the condensation reaction occurs between a carboxylic acid and an alcohol group. We see that each substance A and B contains these two groups. Thus each molecule has the capacity to react at two sites along its carbon chain. This is characteristic of monomers involved in condensation reactions. We can write the beginning of the polymerization reaction:

$$HOCH_2CH_2COOH + HOCH_2CH_2CH_2COOH +$$
$$\text{A} \qquad\qquad\qquad \text{B}$$
$$HOCH_2CH_2COOH \rightarrow$$
$$\text{A}$$

Note that when molecules A and B are next to each other a carboxylic acid group is oriented toward an alcohol group to establish the correct orientation of groups to permit a condensation reaction to occur. We are told that a water molecule splits out in a condensation reaction. If we remove a water molecule (—OH and —H) at both reactive sites (highlighted in bold below) we can write the following:

$$\overset{\text{O}}{\overset{\|}{\text{HO-CH}_2\text{CH}_2\text{C}}}\text{-OH} + \overset{\text{O}}{\overset{\|}{\text{HO-CH}_2\text{CH}_2\text{CH}_2\text{C}}}\text{-OH} + \overset{\text{O}}{\overset{\|}{\text{HO-CH}_2\text{CH}_2\text{C}}}\text{-OH} \rightarrow$$
$$\text{A} \qquad\qquad\qquad \text{B} \qquad\qquad\qquad \text{A}$$

$$\overset{\text{O}}{\overset{\|}{\text{HO-CH}_2\text{CH}_2\text{C}}}\text{-O-CH}_2\text{CH}_2\text{CH}_2\overset{\text{O}}{\overset{\|}{\text{C}}}\text{-O-CH}_2\text{CH}_2\overset{\text{O}}{\overset{\|}{\text{C}}}\text{-OH} + 2H_2O$$

LIQUID CRYSTALS

Liquid crystals are substances that are above their melting points and exhibit properties intermediate between those of the solid and liquid phases. The intermediate phase, or phases, is described as liquid crystalline.

- Typical liquid crystalline substances show a molecular shape that is long and rod-like, with some rigidity. These rod-like substances can maintain parallel ordering in the liquid crystal state.
- Another molecular shape is flat, similar to a pancake. These molecules stack on top of each other.
- Ionic substances and flexible molecules usually don't show this behavior.

Liquid crystalline phases are characterized by the arrangements of the molecules.

- In the **nematic** phase, the simplest type, molecules are aligned along their long axis, but there is no ordering with respect to the alignment of their ends.
- In the **smectic** phase, the rods are aligned parallel and their ends are aligned together. The ordering that persists is due primarily to London dispersion forces and dipole–dipole interactions.
- In the **cholesteric** phase, characteristic of derivatives of the cholesterol molecule, flat, disc-shaped molecules align themselves in a stacking arrangement. Twisting of the stacks may occur and this property may be useful in specific applications.
- Commercial applications of liquid crystals are numerous, with the most well-known in the area of electrically controlled liquid crystal displays (*LCD*).

- By application of an electrical potential in an appropriate direction in a liquid crystalline substance, the orientation of the liquid crystal molecules can be altered. Depending on the orientation of the molecules, polarized light may pass through the liquid crystal medium, or not.
- Some cholesteric substances change color as a function of temperature.
- Liquid crystal behavior is seen in certain types of biological structures, such as in tissue structures.

EXERCISE 7　Characterizing linkages in liquid crystals

Which of the following central linkages would you expect to find in liquid crystal chemistry? Which liquid crystalline phases would primarily involve these linkages?

$$-C-C-　　-CH=CH-　　-N-N-$$
$$-CH=N-　　-CH_2-NH-　　-N=N-$$

SOLUTION: *Analyze*: We are given six types of linkages and are asked which we would expect to observe in liquid crystals and in which types of liquid crystals.

Plan: We need to carefully compare the linkages in the question to the types of linkages commonly found in different liquid crystals. Compounds that form the nematic and smectic liquid crystal phase are rod-shaped and fairly rigid about the long axis of the molecule. The presence of pi bonds prevents rotation of bonded atoms about the bond, and a series of pi bonds near one another in the molecule will give it rigidity. Thus, the linkages that show pi bonds should form a central linkage in a liquid crystal system.

Solve: As noted in the *Plan* we expect linkages that have pi bonds to be in nematic or smectic liquid crystal phases: $-CH=CH-$; $-CH=N-$; $-N=N-$.

Comment: It is helpful to remember that the following atoms may be involved in pi bonding among themselves: C, N, O, and S.

Molecular nanotechnology is a manufacturing technology which uses particles in the 1–100 nm size scale. It involves building things from the atom up and assembling them into a structure with useful properties. Nanomaterials have dimensions in this scale.

NANO-TECHNOLOGY

- The prefix "nano-" means 10^{-9}.
- Semiconductor particles in the 1–10 nm range are called quantum dots.
- As the size of semiconductor nanoparticles become smaller the size of the band gap increases. Thus, quantum dots can be manufactured with various band gaps and with different colors.
- Nanosize particles of metals are being studied because the behavior of electrons can be different at this scale than in larger size particles.
- Carbon nanotubes are being intensely studied today. These are thin, long tubes of carbon atoms. Carbon nanotubes have a very broad range of electronic, thermal, and structural properties that change depending on the diameter of the tube, the thickness of the carbon wall, and the length of the nanotube.

EXERCISE 8　Identifying carbon hybrid orbitals in carbon nanotubes

Carbon nanotubes are often viewed as graphite rolled into tubes. What type of hybridization is used by carbon atoms in carbon nanotubes if this view is correct?

SOLUTION: *Analyze*: We are asked to identify the type of hybrid atomic orbitals used by carbon atoms in carbon nanotubes assuming a graphite model.

Plan: We can review the properties of graphite in Chapter 11 (covalent-network solid) and hybrid atomic orbitals in Chapter 9. This information will help us determine the type of hybrid atomic orbitals used by carbon.

Solve: In graphite the carbon atoms are in layers of hexagonal rings of carbon atoms. The bonding distance between adjacent carbon atoms in the hexagonal ring is very similar to that found in benzene, which has delocalized π electrons, and each carbon atom is bonded to three other atoms. Thus, we can conceive of graphite as a benzene-like carbon structure extended over a layer. Each carbon atom in benzene uses a p orbital to form a π bond and three sp^2 hybrid orbitals to form three sigma bonds. This suggests that each carbon atom in a carbon nanotube uses three sp^2 hybrid orbitals to form sigma bonds to the other three carbon atoms and a p orbital to form a π bond.

SELF-TEST QUESTIONS

Key Terms

Having reviewed key terms in Chapter 12, match key terms with phrases and identify statements as true or false. If a statement is false, indicate why it is incorrect.

Match each phrase with the best term:

12.1 C_2H_2 is the building block for polyethylene and thus is one of these.

12.2 Glass is an amorphous substance, an insulator, and is also in this class of materials.

12.3 In this class of materials the band gap between the valence band and the next higher band is very large.

12.4 In this type of liquid crystal the molecules are aligned end to end.

12.5 In this type of liquid crystal the molecules are aligned with their long axes parallel.

12.6 In this type of liquid crystal the molecules form long fibers that align through stacking in layers of fibers.

12.7 This class of materials has dimensions in the 10^{-9} m range.

12.8 This class contains materials which have relatively homogeneous properties and are above their melting points, yet show some properties of both solid and liquid phases.

12.9 Rubber is formed in this process.

12.10 This class of polymers can be shaped into a wide variety of materials.

12.11 In this type of reaction a polymer forms by the coupling of monomers with the splitting out of small molecules.

12.12 In this type of polymer two different monomers are used to form it.

12.13 Monomers containing carbon–carbon π bonds are often used to form this type of polymer.

12.14 This class of materials is composed of many repeating units.

12.15 This class of materials can be a ceramic and shows the property of exhibiting zero electrical resistance to current flow below a certain temperature.

(a) addition polymerization reaction
(b) ceramic
(c) cholesteric liquid crystal phase
(d) condensation polymerization reaction
(e) copolymer
(f) insulator
(g) liquid crystal
(h) monomer
(i) nanomaterial
(j) nematic liquid crystal phase
(k) plastic
(l) polymer
(m) smetic liquid crystal phase
(n) superconductor
(o) vulcanization

True-False Statements:

12.16 Some advantages of *ceramics* compared to natural materials are high resistance to heat and corrosion, lower weight in some cases, and potentially lower cost to use.

12.17 A *sol* consists of large particles dispersed in a medium.

12.18 Jell-O is comprised of primarily protein molecules dispersed in water, and is an example of a *gel*.

12.19 In the *sol-gel* process a gel is heated to a high temperature to drive off the sol.

12.20 Vulcanization of rubber involves *cross-linking* of polymers using nitrogen.

12.21 One of the most widely studied *superconductors* is $YBa_2Cu_3O_x$ which shows a T_c near 95 K.

12.22 A *superconducting ceramic* shows the property of exhibiting zero electrical resistance above a certain temperature.

12.23 A *biomaterial* that is used to form a disposable contact lens might be expected to have a number of hydrophobic groups on its surface.

12.24 MgB_2 has a *superconducting transition temperature* of 39 K; this means the superconductivity effect can be obtained using liquid nitrogen.

12.25 *Metals* are materials that are shiny, ductile, and malleable.

12.26 A *semiconductor* can have a large band gap.

12.27 The models for conductivity of metals show significant *band gaps*.

12.28 A *valence-band* in a material consists of a collection of highest-energy occupied molecular orbitals which appear as a region of continuous energy.

12.29 A valence band may also be a *conduction band* if some of the molecular orbitals are not occupied.

12.30 The process of adding impurities to a semiconductor to increase its strength is referred to as *doping*.

12.31 An *n-type semiconductor* has excess electrons in the conduction band.

12.32 A *p-type semiconductor* has fewer holes in the valence band.

12.33 The *crystallinity* of a polymer can sometimes be decreased by mechanical stretching or drawing the molten form through a small hole.

12.34 A key feature in determining the *biocompatibility* of a substance is the chemical nature and physical form of its surface.

12.35 A *liquid crystal* has no solid state, only a liquid form that appears to have the features of a solid.

12.36 *Light-emitting diodes* generally consist of two semiconductors, both of the same type.

Problems and Short-Answer Questions

12.37 What type of material is suggested by the following figure? The spheres represent atoms and the dots electrons. Explain.

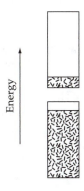

12.38 What type of material is suggested by the following energy picture of bands? The darkened region represents MOs occupied by electrons.

12.39 Ceramics are strong in resisting compression, but are brittle in bending. Explain how the presence of pre-existing defects in a ceramic creates brittleness.

12.40 The hardness of substances is provided by a scale known as Knoop values. Ag has a Knoop value of 60, Cr a value of 935, and Al_2O_3 a value of 2100. Is there sufficient

information to conclude that Cr and Al_2O_3 are ceramics whereas Ag is not based on their Knoop hardness? If not, what additional information is needed and why?

12.41 A metallic conductor is a substance with a resistance that increases with increasing temperature. How does a superconducting substance compare in its resistance properties? Why are we interested in superconducting substances with a T_c above 77 K?

12.42 Show how the polymer polyvinyl chloride can be formed from its monomer cholorethene:

$$CH_2 = CH_2$$
$$|$$
$$Cl$$

12.43 Does a condensation chemical reaction lead to smaller or larger molecules? For the following show how a condensation reaction can occur between the two substances:

$$\begin{array}{c} O \\ \| \\ CH_3-C-OH \end{array} \text{ and } CH_3-CH_2-CH_2-NH_2$$

12.44 Why must the production of plastics used in living systems such as humans be carefully monitored for impurities? What term is used to describe this requirement?

12.45 Why would a manufacturer of a heart valve desire the surface of the material to be smooth and not rough?

12.46 In problem 12.37 you wrote the repeating unit that forms the polymer polyvinyl chloride. Is it possible that this repeating unit could be used to form a ceramic polymer? Explain.

12.47 *Kevlar* is a commercial product used in bullet-proof vests. It is a polymer. Would you expect it to be a thermoplastic or thermosetting plastic? Explain.

12.48 The formula of a particular amino acid is

$$\begin{array}{c} H \\ | \\ H \quad O \\ | \quad | \\ H-C-C=O \\ | \\ N-H \\ | \\ H \end{array}$$

Is there sufficient information given by the chemical formula to permit you to conclude it could participate in a polymerization reaction? If not, what additional information is needed and why?

12.49 Does a liquid crystal become more or less viscous than the normal liquid in its transition temperature range? Explain your answer.

12.50 Which of the three forms of liquid crystals discussed in the text is the most ordered? What characteristics does it possess that permit this ordering?

12.51 Would you expect the following substance to exhibit the smectic liquid-crystalline phase? Explain.

$$CH_3 - CH_2 - \overset{\overset{\displaystyle O}{\|}}{C} - CH_3$$
$$|$$
$$CH_2$$
$$|$$
$$CH_3$$

12.52 Suggest three reasons why the length of the nanoscale is important.

12.53 Nanotubes of carbon are being manipulated to form different arrangements of the tubes. It is important to be able to manipulate nanotubes for developing commercial applications. One method is to use an atomic force microscope to move a nanotube on a surface, like a plow moves dirt, to change the shape of the original nanotube. Suggest two possible reasons why the nanotube can maintain its new shape on the surface of the material on which it rests. What intermolecular forces primarily exist?

Integrative Questions

12.54 Alumina, Al_2O_3, is the most common non-silicate ceramic and it has a density of 3.97 g/cm^3. Another important non-silicate ceramic is beryllia, Be_2O_3, and it has a density of 3.00 g/cm^3.

   (a) How do the ionic distances between aluminum and oxygen atoms and beryllium and oxygen atoms compare? Answer this question using only the periodic table.
   (b) Why is beryllia used in the space and aviation industry?

12.55 When HCl(s) is slowly heated it forms a phase which is a solid termed a plastic crystal. With further heating it becomes a normal liquid. Terephthal-*bis*-(4-*n*-butylaniline) (TBBA) is an organic compound that has a rigid backone of three benzene rings and two N=C bonds with a short chain of four carbon atoms at each end. When solid TBBA is heated a liquid crystalline state is formed before the ordinary liquid state is achieved.

   (a) Why doesn't HCl form a liquid crystalline phase as does TBBA? Explain.
   (b) What is the expected difference between a plastic crystal and liquid crystal? Explain using HCl and TBBA as examples.

12.56 Egyptian paste—consisting of sand, clay, potash feldspar, and soda—is used to make a ceramic coating. It which was developed by the ancient Egyptians approximately 7000 years ago. The paste can be cured at moderately low-firing temperatures (around 900 °C) and produces its own glaze. Copper carbonate is added to give the characteristic turquoise color of ancient Egyptian art.

   (a) How can this salt-containing paste produce a glazed surface during heating?
   (b) Why is the ceramic surface subject to damage if disturbed?
   (c) Use the Internet to find some other salts that could be added to produce different colors.

12.57 Liquid-crystal displays (LCD) are used in a wide variety of devices including laptop computers, airplane instruments, and wristwatches.

   (a) What is the basis for the technology of flat, high-resolution laptop computer screens in terms of the materials making up the LCD?
   (b) Early LCD screens had poor display resolutions. They consisted of segments of the substance arranged into electrically insulated segments. Each segment can be addressed and made to appear black. Early LCD screens consisted of a single layer of liquid-crystal materials with poor resolution. Today, screens can show sufficient picture elements (pixels) to display both numbers and letters. How might this increase in resolution have been obtained?
   (c) If sufficient liquid-crystal materials are used to create a high number of pixels, what might happen if they become too close?

Multiple-Choice Questions

12.58 Which type of material is used in measuring temperature changes based on color?

   (a) nematic liquid crystal
   (b) smectic liquid crystal
   (c) cholesteric liquid crystal
   (d) thermistor ceramic

12.59 Which is a material made by an addition polymerization reaction?

   (a) metal oxide
   (b) superconducting material
   (c) PVC
   (d) grey tin

12.60 A fiber must have the following property if it is added to a polymer to form an improved product.

   (a) low tensile strength
   (b) small length, large breadth
   (c) very high crystalline melting point
   (d) high cohesive energy with polymer substrate

12.61 Which statement is *not* true about most ceramics?

   (a) strong
   (b) low cost
   (c) abrasion resistant
   (d) withstand wide temperature range

12.62 What type of sol is typically formed in the sol-gel process?

   (a) alkoxide   (c) metal amide
   (b) metal   (d) metal hydroxide

12.63 Which type of ceramic generates an electrical potential when subjected to mechanical stress?

(a) thermistor
(b) LCD
(c) piezoelectric
(d) superconductor

12.64 What coolant used in applications of superconductors is readily available and cools below 77 K?

(a) N_2
(b) He
(c) H_2
(d) Alcohol

12.65 What is 100 nm?

(a) 10^{-11} m
(b) 10^{-9} m
(c) 10^{-7} m
(d) 10^{-6} m
(e) 10^{-3} m

12.66 When an applied electrical field to a liquid crystal display, what happens to change the optical properties?

(a) The crystal structure changes in chemical composition.
(b) The orientation of molecules change with the applied electrical field.
(c) The liquid crystal becomes a semiconductor.
(d) The molecules align themselves against the direction of the applied voltage.
(e) The distances between molecules increase significantly.

SELF-TEST SOLUTIONS

12.1 (h). **12.2** (b). **12.3** (f). **12.4** (m). **12.5** (j). **12.6** (c). **12.7** (i). **12.8** (g). **12.9** (o). **12.10** (k). **12.11** (d). **12.12** (e). **12.13** (a). **12.14** (l). **12.15** (n). **12.16** True. **12.17** False. The particles are extremely tiny. **12.18** True. **12.19** False. Water, not sol, is driven off. **12.20** False. Sulfur is the crosslinker, not nitrogen. **12.21** True. **12.22** False. Superconductivity occurs *below* a certain temperature. **12.23** False. A contact lens must have a wettable surface. The presence of hydrophobic groups would cause water to be repelled and thus the lens surface would not "wet." **12.24** False. The superconducting temperature of superconductors must be above 77 K for the superconducting effect to be accessible by cooling with liquid nitrogen. **12.25** True. **12.26** False. The band gap is small. **12.27** False. The valence and conduction bands overlap and the electrons move throughout the metallic structure. **12.28** True. **12.29** True **12.30** False. Doping is used to add excess electrons or create positive holes in energy bands. **12.31** True. **12.32** False. More holes in the valence band are created. **12.33** False. The processes normally increase the degree of crystallinity. **12.34** True. **12.35** False. It is a phase above the melting point of the solid and it exhibits a liquid form which has a high degree of crystallinity. **12.36** False. One is a *p*-type semiconductor and the other is an *n*-type.

12.37 Electrons are separated from the atoms throughout the material. This suggests that the atoms must possess a positive charge to maintain charge neutrality. This figure is a representation of a mobile sea of electrons around positive atoms: conductive metal.

12.38 A semiconductor with some of the valence electrons in the higher energy conduction band.

12.39 The term brittle means to be easily broken or shattered. Consider two types of ceramics. The structure with voids has regions of space and the structure when bent cracks more readily around the solid region surrounding voids. Compare this to the structure with no voids, which is solid throughout, and structurally it will resist bending.

12.40 There is insufficient information. Ceramics are hard, but hardness alone does not prove a substance is a ceramic. Ag and Cr are both elements and metals, yet Cr has a high Knoop value. To determine if Al_2O_3 is a ceramic we would have to study its resistance to heat, corrosion, and wear and its deformation under stress.

12.41 A superconducting substance exhibits a more complicated resistance pattern. Above a certain temperature, the superconducting transition temperature (T_c), resistance increases with increasing temperature, just as a metallic conductor. However, at T_c and below the resistance becomes zero. N_2 boils at 77 K and can be used to cool a superconducting substance below its T_c if its T_c is above 77 K. N_2 is relatively cheap as a low-temperature liquid compared to He.

12.42 $n(CH_2 \!=\! CH_2) \xrightarrow[\text{initiator}]{} n[\dot{C}H_2 \!-\! \dot{C}H_2] \longrightarrow$

12.43 A condensation reaction forms a larger molecule by combining two substances with the release of a smaller particle, usually water.

12.44 Plastics used in human systems cannot contain contaminants that might leak into the living environment; this can be harmful. Also, contaminants might alter the surface of the plastic and this can also cause problems such as blood clots. The term is biocompatibility.

12.45 Surface roughness facilitates hemolysis, the breakdown of red blood cells. The red blood cells can become entangled in the rough surface and promote coagulation.

12.46 Ceramics contain a network structure usually with cross-linked structures that leads to rigidity and hardness. Polyvinyl chloride does not have the appropriate groups attached to the carbon backbone chain to prepare a highly rigid, hard structure. Furthermore it could not be resistant to very high temperatures, which is a characteristic of ceramics. It is possible to use the monomer to prepare a rigid type of plastic but it still does not have the correct thermal properties.

12.47 To form a bulletproof vest the plastic must have extensive cross-linking between polymer chains to form a rigid framework so a bullet cannot penetrate it. The extensive cross-linking will give it properties of a thermosetting plastic.

12.48 There is sufficient information. The amino acid contains both a carboxylic acid ($-$COOH group) and basic amine ($-NH_2$ group). A condensation reaction (acid–base reaction) occurs when an amine of one molecule reacts with a carboxylic acid of a different molecule. This produces a new substance which also contains amine and carboxylic acid groups. These can continue to react by a condensation process to produce a polymer.

12.49 A characteristic of liquid crystals is their tendency to become partially organized compared to the normal liquid phase. This organization leads to greater intermolecular attractions between particles, less random motions of particles, and thus a greater viscosity.

12.50 Smectic liquid-crystalline phase. These molecules tend to be long in relation to their thickness. Usually they contain $C=N$ or $N=N$ somewhere in the backbone of the chain and this leads to rigidity. Also, polar groups in appropriate locations give rise to intermolecular dipole–dipole attractions that enhance these alignments.

12.51 No. The molecule lacks rigidity and does not possess a long chain structure relative to its thickness.

12.52 (1) Wavelength properties of electrons contained within atoms can be changed by variations of atomic structure on the nanometer scale. This can influence changes in the fundamental properties of the material. (2) As with all very small particles, a nanoscale substance will have a very high surface area. This will provide means for new reactive surfaces, energy changes, and storage of energy. (3) The wavelength scales of external energy applications can become comparable to the nanoscale, and this can permit the development of opto-electromechanical applications.

12.53 The nanotube will maintain its new shape if (1) the intermolecular forces between the different nearby surfaces of the nanotube are more effective than in the original structure or (2) there are sufficient intermolecular forces between the surface of the carbon nanotube and the surface of the material upon which it rests. Reason (2) is believed to be the key reason why a nanotube will maintain its new structure when using atomic force microscopy. Carbon nanotubes have surfaces containing carbon atoms and these surfaces are nonpolar. Therefore, the types of forces between surfaces should be van der Waals forces.

12.54 (a) Beryllium is a smaller atom than aluminum (111 pm versus 143 pm). This is expected since beryllium is a member of the second row in the periodic table and is in group 2A whereas aluminum is in the third row and is a member of group 3A. The ions show a similar trend (Be^{2+}, 31 pm versus Al^{3+}, 50 pm). Cations are always smaller than the parent atom. Thus the Be–O distance is less than the Al–O distance.

(b) Ceramics used in the space and aviation industry must exhibit both good mechanical strength and low mass per unit volume. Beryllia has a significantly lower density than alumina and thus is preferred as a ceramic. However, special care must be taken with beryllia in production as beryllium dust is toxic and beryllia forms a toxic vapor when heated.

12.55 To answer both questions, you need to analyze the expected structures of each solid. HCl is a small, somewhat spherical molecule, whereas TBBA has a rod-like structure because of the rigid chain described.

(a) When HCl is heated the small, spherical particles begin to tumble and this disruption of the solid structure creates an intermediate phase known as a plastic crystal. Upon further heating the particles move so rapidly that the solid structure is converted completely to a liquid phase. To form a liquid crystalline phase some degree of orientational order must be kept as a solid is converted to a liquid. TBBA contains long rod-like molecules that are not able to easily tumble but are able to retain partial organization as the particles begin to move upon heating. HCl molecules are too small to prevent tumbling and thus they cannot form a liquid crystal phase.

12.56 (a) The soluble salts in the Egyptian paste can rise to the surface during heating and can form a crystalline film that gives a glaze.

(b) The saltlike surface is thin and doesn't have an extensive networking of bonds; thus if the surface is disturbed, the ceramic film is broken.

(c) Various metal chromates, iron oxides, and lead oxide are some examples that can be found.

12.57 (a) Nematic liquid crystals are the basis for LCD screens. They can be divided into cells of long, rodlike molecules with their axes aligned parallel to one another and each cell electrically segmented from the others. Electrical switches can be used to control each cell and thus whether it is black or not. Today, twisted nematic liquid crystals are used in the construction of LCD displays.

(**b**) Increased resolution requires an ability to control pixels in the x,y,z planes, not just the x,y plane. By forming nematic liquid crystals on substrates which are at right angles (row/column) to each other and in two different planes, the intersection of a row and column creates a controllable pixel and more control of density of pixels.

(**c**) If the number of nematic liquid crystals becomes too large, then it is likely that cross-talk between them will occur; that is, they are no longer separated electrical cells. The switching of one cell might cause an adjacent one to switch.

12.58 (**c**). **12.59** (**c**). **12.60** (**d**). **12.61** (**b**). **12.62** (**d**). **12.63** (**c**). **12.64** (**a**). **12.65** (**c**). **12.66** (**b**).

Chapter

13

Properties of Solutions

OVERVIEW OF THE CHAPTER

13.1, 13.2, 13.3
SOLUTIONS

Review: Energy (5.1, 5.2); pressure (10.2); intermolecular forces (11.2).

Learning Goals: You should be able to:

1. Describe the energy changes that occur in the solution process in terms of the solute–solute, solvent–solvent, and solute–solvent attractive forces; describe the role of disorder in the solution process.
2. Rationalize the solubilities of substances in various solvents in terms of their molecular structures and intermolecular forces.
3. Describe the effects of pressure and temperature on solubilities.

13.4 CONCENTRA-
TION EXPRESSIONS

Review: Molarity (4.5); solution (4.5); dimensional analysis (1.6).

Learning Goals: You should be able to:

1. Define mass percentage, parts per million, mole fraction, molarity and molality and calculate concentrations in any of these concentration units.
2. Convert concentration in one concentration unit into any other (given density of the solution where necessary).

13.5 COLLIGATIVE
PROPERTIES

Review: Boiling point (11.4); melting point (11.4); vapor pressure (11.5); gas-law constant (10.4); pressure units (10.2).

Learning Goals:

1. Determine the concentration and molar mass of a nonvolatile non-electrolyte from its effect on the colligative properties of a solution.
2. Explain the difference between the magnitude of changes in colligative properties caused by electrolytes compared to those caused by nonelectrolytes.
3. Describe the effects of solute concentration on the vapor pressure, boiling point, freezing point, and osmotic pressure of a solution, and calculate any of these properties given appropriate concentration data.

13.6 COLLOIDS

Objective: You should be able to describe how a colloid differs from a true solution.

TOPIC SUMMARIES AND EXERCISES

A solution is a homogeneous mixture of two or more substances, one of | **SOLUTIONS**
which is the solvent and the others are solutes. The solvent is the dissolving
medium and is in the largest quantity.

- When a solute dissolves in a solute three distinct processes occur:
  (1) The solvent particles separate sufficiently from one another to create space (a "hole") for the solute particles;
  (2) the solute particles separate from one another in order to enter the
  "holes"; and
  (3) the solvent particles are attracted to the solute particles and surround them to form aggregates.
- The process whereby solute particles are attracted and surrounded by
  the solvent particles is referred to as **solvation**. When water is the solvent, the interaction is called **hydration**. Some aggregates of solvent and
  solute particles contain a definite ratio of solute to solvent particles.
- The quantity of solute dissolved in a solvent depends on:
  (1) the energetics of interactions between particles;
  (2) the extent to which particles mix and spread out in a larger volume;
  (3) temperature; and
  (4) pressure.
- These factors are discussed in this chapter and you should be familiar
  with each.

The formation of a solution is favored by two general factors:

- The dissolving of solute particles in the solvent releases heat (exothermic). This occurs when the energies of solvent–solute interactions are
  greater than the sum of energies of solute–solute and solvent–solvent
  interactions. Separating solute particles from each other and solvent
  particles from each other requires energy. Mixing may not occur if insufficient attractive forces exist between solvent–solute particles. However, endothermic solution processes can occur spontaneously if the
  next factor operates to a sufficient extent.
- The solution exhibits a greater dispersal of energy compared to the
  original solute and solvent. This is often shown by particles in the solution being dispersed in a less organized, or more disordered way compared to their separated states. A measure of this dispersal of energy, or
  change in disorder, is a thermodynamic quantity called **entropy**. If entropy increases then the dispersal of energy increases and entropy has a
  positive sign. In most cases the formation of a solution is accompanied
  by an increase in entropy.
- In general, we find that a solute is likely to dissolve in a solvent when
  the solute and solvent particles have similar structures and similar
  types of polarity. This is summarized by the saying: "Like dissolves
  like." For example, nonpolar benzene dissolves readily in nonpolar
  carbon tetrachloride but not in polar water.

When excess, undissolved solid exists in a solution two dynamic oppos-
ing processes exist: (1) dissolving of solid particles and their solvation; and
(2) coalescence of dissolved solute particles and reforming the solid, termed

crystallization. When the rates of these two processes are equal the concentration of dissolved solid particles is constant and a state of equilibrium exists.

- A **saturated solution** exists when a solution of a dissolved substance can no longer dissolve the solid. The maximum amount of solute dissolved under this condition is termed **solubility** and is commonly expressed in grams of solute per 100 g or 100 cm^3 of solvent. A solution with a relatively large quantity of solute is termed concentrated and one with relatively little solute is dilute.
- An **unsaturated solution** contains the solute in a concentration less than needed to form a saturated solution. No solid will be apparent.
- A **supersaturated solution** is a solution in which the concentration of the dissolved solid is greater than in a saturated solution. Such solutions are not stable and under the right conditions, such as adding solute crystals, the solid will crystallize and a saturated solution forms.
- When liquids mix and form a solution, they are said to be **miscible**. If they form a heterogeneous mixture (more than a single phase) they are **immiscible**. For example, oil and water are immiscible: They do not mix to form a homogeneous mixture.

Increasing the pressure of a gas over a solvent increases the solubility of the gas, whereas pressure has minimal effect on the solubility of solids and liquids.

- The relationship between pressure and solubility of a gas is expressed by **Henry's law**:

$$S_g = kP_g$$

where k is a proportionality constant known as the Henry's law constant, S_g is the solubility of the gas in the solvent, and P_g is the pressure of the gas over the solution.
- Note the linear relationship between solubility and pressure.
- The solubility of a gas in general decreases with increasing temperature.

EXERCISE 1   Applying concepts of intermolecular forces and entropy changes to solutions

Hexane and heptane are nonpolar liquids. (**a**) In a solution containing only hexane and heptane, what kinds of forces exist between molecules? (**b**) Why does a solution containing equal quantities of hexane and heptane have a greater disorder of particles than do pure hexane and pure heptane alone? Is entropy increasing or decreasing? (**c**) Would you expect NaCl(s) to dissolve in a solution containing only heptane and hexane?

SOLUTION: *Analyze*: We are told that heptane and hexane are nonpolar liquids and asked four questions involving them. These questions pertain to the types of interactions between solute and solvent particles, the arrangement of particles in the pure and solution phases, and whether NaCl, an ionic compound, will dissolve in them.

Plan: We will have to determine the types of intermolecular forces in hexane and heptane, and analyze how mixing the two will change the arrangement of particles. For NaCl to dissolve in a solution of hexane and heptane there must be sufficient ion–solvent interactions. We have to determine if sufficient interactions will exist.

Solve: (**a**) London dispersion forces—that is, instantaneous induced-dipole attractions (see Section 11.2)—exist between nonpolar molecules such as heptane and hexane.

(**b**) A solution containing hexane and heptane has a larger volume than either the separate hexane or heptane liquids from which the solution is formed. Thus, in the solution state, there exists more possible kinds of arrangements of molecules than in the separated hexane and heptane liquids. This results in a more disordered state for the solution particles compared to that for the separated liquids. This is also reflected in an increase in entropy. (**c**) No. An ionic salt such as NaCl dissolves in polar solvents, not in a nonpolar liquid containing only heptane and hexane.

EXERCISE 2   Analyzing why a solution forms when the entropy change is negative

Mixing HCl(g) and H_2O(l) leads to a solution with a more ordered state of particles than exists for the separated HCl(g) and H_2O(l) particles. Explain why the particles of the solution exist in a more ordered state than the particles in the two separated phases, and why, in spite of this condition, HCl(g) dissolves in H_2O(l).

SOLUTION: *Analyze*: In the text we are told that an increase in entropy favors the formation of a solution. In this problem we are told that a mixture of HCl(g) dissolved in H_2O(g) contains particles in a more ordered state than the particles in the separated phases and are asked why the solution still forms.

Plan: We will have to carefully analyze both the initial state (separated components) and the solution state (the mixture) for changes in disorder and the energetics of the process when HCl(g) and H_2O(l) mix.

Solve: The initial system consists of two phases, gas and liquid. The gaseous state is a highly disordered state because of the large volume available for HCl molecules to move about randomly. HCl(g) and H_2O(l) form a solution, in which HCl molecules are considerably more restricted in their motions than in the gaseous state. Therefore, the solution possesses a more ordered arrangement of particles compared to the separated states of a gas and a liquid. Entropy has decreased in forming the solution; this does not favor the formation of a solution. Nevertheless, HCl and H_2O readily form a solution because strong attractive forces exist between polar HCl and polar H_2O molecules. This results in a large exothermic heat of solution that enables the solution process to occur.

Comment: When you are asked any questions about why substances may or may not dissolve in each other you must consider two factors: energies of interactions and changes in disorder.

EXERCISE 3   Determining which solvent is likely to dissolve a solute

For each solute listed in the left-hand column of the following table, select a solvent in the right-hand column that will dissolve it:

Solute	Solvent
Sodium	Water
Gasoline	Mercury
KCl	Hexane

SOLUTION: *Analyze*: We are given a set of solutes and asked to determine in which solvent each will dissolve.

Plan: Solvents usually dissolve solutes that have internal intermolecular forces similar to those in the solvent. Thus, we will have to analyze which types of intermolecular forces exist in each solute and solvent and match them.

Solve: The results of this analysis are summarized in the following table. The types of intermolecular forces are given in parentheses.

Solute	Appropriate solvent
Sodium (metallic)	Mercury (metallic)
Gasoline (van der Waals)	Hexane (van der Waals)
KCl (ionic)	Water (dipole–dipole, hydrogen bonds)

Comment: This problem is an example of the application of the general rule: "Like dissolves like."

EXERCISE 4   Using Henry's law

When we open a bottle of carbonated soft drink, bubbles of carbon dioxide gas escape from the solution. Explain this phenomenon using Henry's law.

SOLUTION: *Analyze*: We are asked to use Henry's law to explain why carbon dioxide escapes from a bottle of carbonated soft drink when opened.

Plan: Henry's law states that the mass of a gas dissolved in a definite volume of a liquid is directly proportional to the pressure of the gas above the liquid. We can use Henry's law to help us explain the observed phenomenon.

Solve: During the bottling of the beverage, carbon dioxide is forced into the solution by application of pressure. The bottle is tightly capped to ensure the carbon dioxide stays dissolved. When we open the bottle, the pressure above the solution is decreased to atmospheric pressure and thus the solubility of the carbon dioxide decreases. This results in $CO_2(g)$ escaping from solution, a process known as effervescence.

CONCENTRATION EXPRESSIONS

Molarity and mole fraction are two ways of expressing concentration. In Section 13.4 of the text, we are introduced to the following new concentration expressions: parts per million (ppm); molality (m); and percentage. Concentration expressions are defined and expressed in equation form in Table 13.1 on page 267. Concentration expressions can be used as conversion factors when you do problems.

EXERCISE 5   Calculating mass from parts per million (ppm)

A 3.0 g sample of ground water contains 3.5 ppm of arsenic ion. How many grams of arsenic are in this sample?

SOLUTION: *Analyze*: We are asked to calculate the grams of arsenic in a 3.0 g of ground water containing 3.5 ppm of arsenic.

Plan: We can use the definition of parts per million (ppm) to solve for the mass of arsenic in the sample.

$$\text{ppm of a component} = \frac{\text{mass of component in solution}}{\text{total mass of solution}} \times 10^6$$

Solve: Let x equal the number of grams of arsenic in the solution. Substituting the given values into this equation yields:

$$3.5 \text{ ppm} = \frac{x}{3.0 \text{ g}} \times 10^6$$

Solving for the number of grams, x, gives:

$$x = \frac{(3.5)(3.0 \text{ g})}{1 \times 10^6} = 1.1 \times 10^{-5} \text{ g}$$

TABLE 13.1 **Units of Concentration**

Concentration unit	Definition	Equation
Molarity (M)	Number of moles of solute in a liter of solution	$M = \dfrac{\text{moles solute}}{\text{volume of solution in } \textit{liters}}$
Mass percentage (wt %)	The percentage of mass of a component of a solution in a given mass of the solution	$\text{Mass \%} = \dfrac{\text{mass of solution component}}{\text{total mass of solution}} \times 100$
Parts per million (ppm)	The grams of a solute in a million (10^6) grams of solution. This is equivalent to 1 mg of solute per kg of solution	$\text{ppm} = \dfrac{\text{mass of solute}}{\text{mass of solution}} \times 10^6$
Mole fraction (X)	Ratio of the number of moles of a component in a mixture to total number of moles of all components in solution	$X = \dfrac{\text{moles of component}}{\text{total moles of all components}}$
Molality (m)	Number of moles of solute in a *kilogram of solvent*	$m = \dfrac{\text{moles of solute}}{\text{mass of } \textit{solvent} \text{ in } \textit{kilograms}}$

Check: We can check the final answer by estimating it in the calculation:

$$x = \frac{(4)(3)}{10^6} = 1.2 \times 10^{-5},$$

which is close to the calculated value.

EXERCISE 6   Calculating mass of a solution from its molality

An antifreeze mixture is an 11.0 molal ethylene glycol solution. Water is the solvent and it has a mass of exactly 800 g. The molar mass of ethylene glycol is 62.0 g. What is the mass of the solution?

SOLUTION: *Analyze*: We are given a 11.0 molal solution of ethylene glycol and asked to calculate the mass of the solution given that the solvent, water, has a mass of 800 g.

Plan: We need the mass of ethylene glycol to calculate the mass of the solution. The mass of ethylene glycol can be determined if we know how many moles of it are present in the solution. We can use the definition of molality to calculate the number of moles of ethylene glycol:

$$\text{molality} = \frac{\text{moles of solute}}{\text{kg of solvent}}$$

We know both molality and the mass of water, the solvent, and therefore we can calculate the number of moles of solute.

Solve: Rearrange the previous equation to calculate the number of moles of ethylene glycol:

Moles ethylene glycol = molality $\times$ mass of water in kg

$$= \left(\frac{11.0 \text{ mol ethylene glycol}}{1 \text{ kg H}_2\text{O}} \right) (800 \text{ g H}_2\text{O}) \left(\frac{1 \text{ kg H}_2\text{O}}{1000 \text{ g H}_2\text{O}} \right)$$

$$= 8.80 \text{ mol ethylene glycol}$$

The mass of ethylene glycol is calculated using its molar mass:

$$\text{Mass ethylene glycol} = (8.80 \text{ mol ethylene glycol})\left(\frac{62.0 \text{ g ethylene glycol}}{1 \text{ mol}}\right)$$

$$= 546 \text{ g ethylene glycol}$$

The mass of the solution is:

$$\text{Mass solution} = 800 \text{ g } H_2O + 546 \text{ g ethylene glycol} = 1346 \text{ g}$$

Check: An estimate of the number of moles of ethylene glycol is $(10)(800)(10^{-3}) = 8$, which is approximately the value calculated exactly. The mass is estimated as $(9)(60) = 540$, which is essentially the same as the mass calculated exactly.

EXERCISE 7   Calculating different concentrations

A solution contains 10.0 g of KOH and 900.0 mL of water at 20 °C. The volume of the solution is 904.7 mL. (**a**) Which component of the solution is the solute? Which is the solvent? (**b**) What is the molarity of the solution? (**c**) What is the mass percentage of KOH? What is the mass percentage of H_2O? (**d**) What is the mole fraction of KOH? (**e**) What is the molality of the solution?

SOLUTION: *Analyze*: We are given 904.7 mL of a solution containing 10.0 g of KOH and 900.0 mL of water. We are asked several questions about concentrations. Note that the volume of water is not the volume of the solution.

Plan: (**a**) This question requires us to apply the definitions of solvent and solute. (**b**)–(**d**) We will have to use the definitions of molarity, mass percentage, mole fraction, and molality to answer each question.

Solve: (**a**) KOH is the solute because it is the component of the solution that occurs in the smallest amount; water is the solvent. Also, water is the solvent because KOH(*s*) dissolves in it to form a *liquid* solution—the same phase as water. (**b**) The molarity (*M*) of the solution is calculated from the relation

$$\text{Molarity} = \frac{\text{moles KOH}}{\text{volume of solution in liters}}$$

Substituting the appropriate values into this equation and solving for molarity of the solution yields

$$\text{Molarity} = \frac{(10.0 \text{ g KOH})\left(\dfrac{1 \text{ mol KOH}}{56.1 \text{ g KOH}}\right)}{0.9047 \text{ L}} = \frac{0.178 \text{ mol}}{0.9047 \text{ L}} = 0.197 \text{ mol/L}$$

(**c**) The mass percentage of KOH is calculated from the relation

$$\text{Mass \% KOH} = \left(\frac{\text{mass KOH}}{\text{mass of solution}}\right)(100)$$

The mass of water is calculated from its density and volume:

$$\text{Mass } H_2O = (\text{volume})(\text{density}) = (900.0 \text{ mL})\left(1.00\frac{\text{g}}{\text{mL}}\right) = 900 \text{ g}$$

The mass percentage of KOH is

$$\text{Mass \% KOH} = \left(\frac{10.0 \text{ g KOH}}{10.0 \text{ g KOH} + 900 \text{ g } H_2O}\right)(100) = 1.10\%$$

The mass percentage of water is

$$100\% - \text{mass \% KOH} = 100\% - 1.10\% = 98.90\%$$

because the sum of mass percentages of all components in solution equals 100 percent. (**d**) The mole fraction of KOH (X_{KOH}) is calculated as follows:

$$X_{KOH} = \frac{\text{moles KOH}}{\text{moles KOH + moles } H_2O}$$

First calculate the number of moles of KOH and H_2O:

$$\text{Moles KOH} = (10.0 \text{ g KOH})\left(\frac{1 \text{ mol KOH}}{56.1 \text{ g KOH}}\right) = 0.178 \text{ mol}$$

$$\text{Moles } H_2O = (900.0 \text{ g } H_2O)\left(\frac{1 \text{ mol } H_2O}{18.0 \text{ g } H_2O}\right) = 50.0 \text{ mol}$$

Then substitute the two values into expression for X_{KOH}

$$X_{KOH} = \frac{0.178 \text{ mol KOH}}{0.178 \text{ mol KOH + 50.0 mol } H_2O} = 3.55 \times 10^{-3}$$

(**e**) The molality of the solution is calculated as follows:

$$\text{Molality} = \frac{\text{moles KOH}}{\text{mass of } \textit{solvent} \text{ in } \textit{kilograms}}$$

$$\text{Molality} = \frac{0.178 \text{ mol KOH}}{(900 \text{ g})\left(\dfrac{1 \text{ kg}}{1000 \text{ g}}\right)} = 0.198 \, m$$

Check: This problem has several calculations. Each can be checked for errors in calculation by substituting approximate values and checking that this answer is close in value to the one calculated exactly. For example in (**b**) we can approximate the calculation by $\frac{\frac{10}{50}}{1} = 0.2$, which is approximately the answer calculated exactly. We should check again that all units properly cancel to give the desired unit. We should also check that we have substituted values (mass, volume, moles) for the solvent and solute appropriately into equations.

EXERCISE 8   Calculating molarity

Calculate the molarity of a solution that contains 18.0 percent HCl by mass and has a density of 1.05 g/mL.

SOLUTION: *Analyze*: We are given an 18.0 percent by mass solution of HCl with a density of 1.05 g/mL and asked to calculate the molarity of the solution.

Plan: To calculate the molarity of the solution we need the number of moles of HCl in 1.00 L of solution. We can use density and the mass percentage to determine the number of grams of HCl in a 1.00 L of solution and thus the number of moles of HCl.

Solve: The mass of 1.00 L of solution is calculated from its density:

$$\text{Mass of solution} = \left(1.05 \frac{\text{g}}{\text{mL}}\right)(1.00 \text{ L})\left(\frac{1000 \text{ mL}}{1 \text{ L}}\right) = 1050 \text{ g}$$

The number of grams of HCl in this solution is calculated from the given mass percentage of HCl:

$$\text{Mass \% HCl} = \left(\frac{\text{Mass HCl}}{\text{Mass solution}}\right)(100)$$

Or, rearranging the equation, we have

$$\text{Mass HCl} = \frac{\text{Mass \% HCl}}{100}(\text{Mass solution})$$

$$\text{Mass HCl} = \left(\frac{18.0}{100}\right)(1050 \text{ g}) = 189 \text{ g}$$

The molarity of the solution is calculated by knowing that in 1.00 L of solution there are 189 g of HCl.

$$\text{Molarity} = \frac{\text{Moles HCl}}{1.00 \text{ L}} = \frac{(189 \text{ g HCl})\left(\dfrac{1 \text{ mol HCl}}{36.5 \text{ g HCl}}\right)}{1.00 \text{ L}} = 5.18 \text{ } M$$

Check: The molarity is estimated by $\dfrac{200\left(\dfrac{1}{40}\right)}{1} = 5$, which is approximately the *value calculated exactly.*

COLLIGATIVE PROPERTIES

Those properties of a solution that depend on the *number* of solute particles are called **colligative properties**. In Section 13.5 of the text, the following colligative properties of a solution are discussed: vapor-pressure lowering, boiling-point elevation, freezing-point depression, and osmotic pressure. The text discussion is limited to the effect of nonvolatile solutes on the colligative properties of a solution, assuming ideal conditions.

- Useful relations for calculating colligative properties of solutions are given in Table 13.2 on page 271.
- Remember to account for all forms of the solute when calculating X, m, or M. For example, when calculating the osmotic pressure of a 0.10 M Na$_2$SO$_4$ solution, you must use $M = 3 \times 0.10M = 0.30M$ because Na$_2$SO$_4$ ionizes to form one sodium ion and two sulfate ions.
- Another useful relation is **Raoult's law**, which enables you to calculate the vapor pressure of a solvent when a solute is present:

$$P_A = X_A P_A^\circ$$

In this equation, P_A° is the vapor pressure of pure solvent; X_A is the mole fraction of solvent; and P_A is the vapor pressure of solvent with a solute present.

EXERCISE 9   Calculating osmotic pressure

Compare the osmotic pressures of 0.10 M sucrose and 0.10 M NaCl aqueous solutions at 25 °C.

SOLUTION: *Analyze*: We are given aqueous solutions of NaCl and sucrose of equal concentration and asked to compare their osmotic pressures.

Plan: Osmotic pressure is a colligative property; thus, we need to determine how NaCl and sucrose behave in an aqueous solution. Sucrose is a nonelectrolyte; therefore it does not dissociate and the concentration of sucrose molecules is 0.10 M. NaCl is ionic and when it dissolves in water it forms sodium and chloride ions; that is, two moles of ions form for every mole of NaCl that dissociates. Therefore, the molarity of all ions in solution is twice the molarity of NaCl,

$$M = M_{\text{Na}^+} + M_{\text{Cl}^-} = 0.10M + 0.10M = 0.20M$$

TABLE 13.2 Relations for Calculating Colligative Properties of Solutions

Colligative property	Quantitative relation	Comments
Vapor-pressure lowering (*VPL*)	$VPL = X_B P_A^\circ$	A nonvolatile solute lowers the vapor pressure of a solvent by an amount *VPL* X_B = mole fraction of solute particles P_A° = vapor pressure of pure solvent
Boiling-point elevation (ΔT_b)	$\Delta T_b = K_b m$	A nonvolatile solute raises the boiling point of a solution by an amount ΔT_b K_b = molal boiling-point elevation constant, which is unique for a given solvent m = molality of solute
Freezing-point depression (ΔT_f)	$\Delta T_f = K_f m$	A nonvolatile solute lowers the freezing point of a solution by an amount ΔT_f K_f = molal freezing-point depression constant, which is unique for a given solvent m = molality of solute
Osmotic pressure (π)	$\pi = MRT$	Osmosis is the net movement of solvent particles across a semipermeable membrane from a more dilute solution into a more concentrated one; osmotic pressure (π) of a solution is the applied pressure required to prevent osmosis from pure solvent into a solution. *Isotonic* solutions have the same osmotic pressure. M = total molarity of all ionized and unionized solute species R = ideal-gas constant T = absolute temperature

The equation for osmotic pressure is $\pi = MRT$.

Solve: The osmotic pressure for the sucrose solution is calculated by substituting the known values of *M*, *R*, and *T*:

$$\pi = \left(\frac{0.10 \text{ mol}}{1 \text{ L}}\right)\left(0.082 \frac{\text{L-atm}}{\text{mol-K}}\right)(298 \text{ K}) = 2.4 \text{ atm}$$

In a similar manner, the osmotic pressure of NaCl can be calculated. However, since *R* and *T* are the same, the only difference is *M*, which is twice the magnitude. Therefore, the osmotic pressure for the NaCl solution is

$$\pi = 2 \times 2.4 \text{ atm} = 4.8 \text{ atm}.$$

EXERCISE 10   Calculating vapor pressure, boiling point elevation, freezing-point depression, and osmotic pressure

A solution is formed by dissolving 10.0 g of KCl in 500.0 g of H_2O. (a) What is the vapor pressure of the solution at 25 °C if the vapor pressure of pure H_2O is 23.8 torr at 25 °C? (b) What is the boiling-point elevation? K_b for H_2O is 0.52 °C/*m*. (c) What is the freezing-point depression? K_f for H_2O is 1.86 °C/*m*. (d) What is the osmotic pressure of the solution? Assume that the volume of the solution is 0.500 L.

SOLUTION: *Analyze*: We are given a solution that consists of 10.0 g of KCl dissolved in 500.0 g of H_2O and are asked four different questions that involve colligative properties.

Plan: First we need to determine the behavior of KCl in an aqueous solution. It is an ionic substance and therefore it will dissociate to form potassium and chloride ions: $KCl(s) \rightarrow K^+(aq) + Cl^-(aq)$. Thus, one mole of KCl will form two moles of ions. For each question we will use the appropriate relation in Table 13.2 and substitute the given or calculated values for the parameters in each equation.

Solve: (**a**) The vapor pressure of a volatile component in a solution, P_A, is related to its vapor pressure as a pure liquid, P_A°, and its mole fraction, X_A, as expressed by Raoult's law:

$$P_A = X_A P_A^\circ$$

The mole fraction of water in the solution is

$$X_{H_2O} = \frac{\text{moles } H_2O}{\text{moles } H_2O + \text{moles } K^+ + \text{moles } Cl^-}$$

$$= \frac{\text{moles } H_2O}{\text{moles } H_2O + (2)(\text{moles KCl})}$$

The mole fraction of water is calculated as follows:

$$\text{Moles } H_2O = (500.0 \text{ g } H_2O)\left(\frac{1 \text{ mol } H_2O}{18.0 \text{ g } H_2O}\right) = 27.8 \text{ mol}$$

$$\text{Moles KCl} = (10.0 \text{ g KCl})\left(\frac{1 \text{ mol KCl}}{74.55 \text{ g KCl}}\right) = 0.134 \text{ mol}$$

$$X_{H_2O} = \frac{27.8 \text{ mol } H_2O}{27.8 \text{ mol } H_2O + (2)(0.134 \text{ mol KCl})} = 0.990$$

The partial pressure of the water when 10.0 g of KCl is present is thus

$$P_{H_2O} = X_{H_2O}P_{H_2O}^\circ = (0.990)(23.8 \text{ torr}) = 23.6 \text{ torr}$$

(**b**) The boiling-point elevation is found using the equation $\Delta T_b = K_b m$, where m is the molality of the solute and K_b is the boiling-point elevation constant for the pure solvent:

$$m = \frac{\text{moles of solute}}{\text{kg of solvent}} = \frac{\text{moles } K^+ \text{ ions} + \text{moles } Cl^- \text{ ions}}{\text{kg of solvent}}$$

$$m = \frac{0.268 \text{ mol ions}}{(500.0 \text{ g})\left(\frac{1 \text{ kg}}{1000 \text{ g}}\right)} = 0.536 \ m$$

Substituting the values of m and K_b into $\Delta T_b = K_b m$ gives

$$\Delta T_b = (0.51 \text{ °C}/m)(0.536 \ m) = 0.27 \text{ °C}$$

(**c**) Similarly, the freezing-point depression is found by substituting the values of m and K_f into

$$\Delta T_f = K_f m$$

$$\Delta T_f = (1.86 \text{ °C}/m)(0.536 \ m) = 1.0 \text{ °C}$$

(**d**) Osmotic pressure is calculated using the equation $\pi = MRT$. First calculate the value of M

$$M = \frac{\text{moles of all solute species}}{\text{volume in L}}$$

$$= \frac{(0.134 \text{ mol KCl})(2 \text{ mol ions}/1 \text{ mol KCl})}{0.500 \text{ L}}$$

$$= 0.536 \, M$$

Then calculate π:

$$\pi = (0.536 \, M)\left(0.0821 \frac{\text{L-atm}}{\text{mol-K}}\right)(298 \text{ K}) = 13.1 \text{ atm}$$

Check: This problem has several calculations. Each can be checked for errors in calculation by substituting approximate values and checking that this answer is close in value to the one calculated exactly. For example, in (**b**) we can approximate the calculation by

$$m = \frac{0.3}{(500)\left(\dfrac{1}{1000}\right)} = 0.60 \text{ and } \Delta T = (0.5)(0.6) = 0.3,$$

which is approximately the answer calculated exactly. We should check again that all units properly cancel to give the desired unit. We should also check that we have substituted values (mass, volume, moles, mole fraction) for the solvent and solute appropriately into equations.

A **colloid** contains gaseous, liquid, or solid particles with diameters approximately 10 Å to 2000 Å dispersed in a solvent-like substance.

COLLOIDS

- Colloidal dispersions (colloids) scatter light very effectively and thus appear cloudy or opaque. The scattering of a light beam that is passed through a colloid is known as the **Tyndall effect**.
- Colloid particles do not settle under the influence of gravity as do larger particles.
- A **hydrophilic colloid** contains colloidal particles that are attracted to water molecules and thereby form a single-phase suspension. Often such colloidal solutions show a gelatinlike character and are known as gels.
- A **hydrophobic colloid** contains colloidal particles that are not attracted to water molecules; they do not form gels.
- Colloidal particles can be removed from a dispersing medium by coagulation, which involves increasing the size of colloidal particles until they settle out of solution. Dialysis involves using a semipermeable membrane to separate colloidal particles from ions in solution.

EXERCISE 11 Determining colloidal properties of a solution

A colloidal suspension of As_2S_3 can be prepared by reacting As_2O_3 with H_2S in water. (**a**) How could you show that the As_2S_3 solution is colloidal? (**b**) The colloidal solution appears gelatinous. Is it a hydrophilic or hydrophobic colloid?

SOLUTION: (**a**) The solution will appear cloudy, and the suspended particles will not settle under the influence of gravity. Also, the solution will scatter a beam of light (Tyndall effect). (**b**) Since the As_2S_3 particles do not separate from H_2O (as evidenced by the gelatinlike character of the solution), the colloid must be hydrophilic.

EXERCISE 12   Identifying dispersed and dispersing phases of colloids

Identify the phase of the dispersed substance, the phase of the dispersing substance, and the colloid type for each of the following: beer foam, clouds, milk, smoke, and paints. Refer to Table 13.6 of the text for descriptions of different kinds of colloids.

SOLUTION:

Example	Dispersed phase	Dispersing phase	Type
Beer foam	Gas	Liquid	Foam
Clouds	Liquid	Gas	Aerosol
Milk	Liquid	Liquid	Emulsion
Smoke	Solid	Gas	Aerosol
Paint	Solid	Liquid	Sol

SELF-TEST QUESTIONS

Key Terms

Having reviewed key terms in Chapter 13, match key terms with phrases and identify statements as true or false. If a statement is false, indicate why it is incorrect.

Match each phrase with the best term:

13.1 The concentration expression used when calculating freezing-point depression.

13.2 The concentration expression expressed by 10 g NaOH/(10 g NaOH and 190 g H_2O).

13.3 The amount of a solute present in a solution when it is saturated under the given conditions.

13.4 The process of surrounding a solute by water molecules.

13.5 The type of solution formed when an equilibrium exists between a solid solute and its dissolved ions.

13.6 The condition of a solution when 3.0×10^{-3} g/mL of $CdCO_3$ is completely dissolved at 25 °C and the solubility of cadmium carbonate is 4.48×10^{-4} g/mL.

13.7 The condition of a solution when 5.00×10^{-5} g of $CdCO_3$ is completely dissolved at 25 °C.

13.8 The type of solution that can form in a mixture with water and particles of 500 nm.

13.9 The name given to the scattering of light that occurs when a beam of light at right angles to the line of sight passes through a suspension.

13.10 When two solutions are connected by a semipermeable membrane and are isotonic they have the same physical property which has this name.

13.11 The process of solvent molecules passing through a semipermeable membrane from a more dilute solution to a more concentrated one.

13.12 A property of a solution that depends on the numbers of solute and solvent particles.

13.13 When a gas condenses the value of this term decreases.

13.14 A statement that the pressure of a gas over a solvent in which it is dissolved depends on the solubility of the gas in the solvent.

Key terms:

(a) colligative	**(h)** osmosis
(b) colloid	**(i)** osmotic pressure
(c) entropy	**(j)** saturated
(d) Henry's law	**(k)** solubility
(e) hydration	**(l)** supersaturated
(f) mass percent	**(m)** Tyndall effect
(g) molality	**(n)** unsaturated

True-False Statements:

13.15 A solution contains 0.30 mg of sodium ion per liter of solution: This corresponds to a concentration of sodium ion of 3.0 *ppm*.

13.16 *Crystallization* occurs more readily when a seed crystal is present.

13.17 The *molal boiling-point-elevation constant* for carbon tetrachloride is 2.53 °C/*m*. This means that 1 *m* nonvolatile solute particles in carbon tetrachloride will cause the solution to boil 2.53 °C higher than pure carbon tetrachloride.

13.18 In general, *molal freezing-point-depression constants* for solvents are smaller in magnitude than their molal boiling-point-elevation constants.

13.19 The presence of *ion pairs* in a NaCl solution will cause a further increase in osmotic pressure.

13.20 CH$_4$ should have *hydrophilic* properties.

13.21 A CH$_3$(CH$_2$)$_{20}$ group in a molecule should cause the molecule to be *hydrophobic*.

13.22 Water and CH$_3$(CH$_2$)$_7$OH are both liquids and *miscible*.

13.23 Dissolving NaCl in water should increase *entropy*.

13.24 Copper metal is *miscible* in water.

13.25 Hydrocarbons are *immiscible* in water.

13.26 A 0.50 *M* solution of KCl and a 0.10 *M* solution of sucrose are *isotonic*.

13.27 A *hypotonic* solution has an osmotic pressure that is higher than another solution.

13.28 A 0.20 *M* NaCl solution is *hypertonic* with respect to a 0.20 *M* CaCl$_2$ solution.

Problems and Short-Answer Questions

13.29 The following figure represents enthalpy changes accompanying a solution process. One letter refers to the overall process of dissolving the solute in the solvent. Other letters refer to different steps in the solution process but viewed in a step-wise fashion. Use it to answer the following questions:

   (a) Which letter represents the enthalpy change between separated solute and separated solvent particles to form a solution?
   (b) Which letter represents the enthalpy change for the heat of solution?
   (c) What does the sum of enthalpy changes A and B represent?
   (d) Is the solution process exothermic or endothermic? Explain.

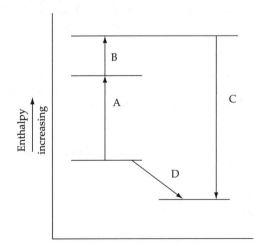

13.30 The following figure shows two solutions separated by a semipermeable membrane. Construct a figure that shows the situation after some time has passed.

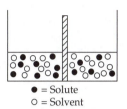

● = Solute
○ = Solvent

13.31 (a) When considering whether a solution will form when a substance is added to a solvent, what three energetic factors must be considered? Identify each factor as endothermic or exothermic.
   (b) Why doesn't KBr dissolve in a nonpolar liquid like CCl$_4$?

13.32 The heat of solution of RbCl in water at 25 °C is 16.7 kJ/mol. Given that the process is endothermic, how can RbCl dissolve in H$_2$O?

13.33 A sample of copper(II) sulfate is dried and weighed. It is left overnight in a damp environment. When it is placed in a dry environment and weighed again it has gained mass. Explain this observation.

13.34 Which is more likely to be soluble in water, CH$_3$CH$_3$ or HOCH$_2$CH$_2$OH? Explain.

13.35 When a sweetened soft drink is made, the water is saturated with carbon dioxide at several atmospheres pressure.

   (a) Why is the carbon dioxide gas pressure increased to several atmospheres during the manufacturing process?
   (b) At 25 °C the partial pressure of carbon dioxide gas in the atmosphere is about 4.0×10^{-4} atm. Why does carbon dioxide gas escape when the soft drink can is opened? Use Henry's law in your analysis.

13.36 What are two significant advantages of using the concentration unit molality compared to using molarity? When will the two terms have similar values for an aqueous solution?

13.37 Which acid should give a greater freezing point depression, a 0.002 *m* HF solution or a 0.002 *m* HCl solution? Explain.

13.38 Contrast an aerosol to an emulsion. How are they different and similar? Give an example of each.

13.39 Iron(III) hydroxide can form a sol because excess iron(III) ions form on the surface of a particle. What is a sol? How can it be made to aggregate?

13.40 A cucumber is placed into a salt brine. Explain what will happen to it and the principles involved.

13.41 Explain how you could prepare 50.0 mL of a 1.00 *M* KI solution from a bottle containing 1.00 L of 5.00 *M* KI.

Assume you have a balance, a beaker, pipets of different volumes, burets, and volumetric flasks of different volumes available.

13.42 Calculate the concentrations asked for in each problem.

(a) The molality of 142 g of Na_2CO_3 in 2.00 kg of water at 0 °C (a saturated solution).

(b) The molality of 24.50 g of codeine, $C_{18}H_{21}NO_3$ in 150.5 g of ethanol, C_2H_5OH.

(c) The mole fraction of 32.3 g of NaCl in 265.0 g of H_2O.

(d) The mass percentage of HCl in a tile cleaner containing 140 g of HCl and 800 g of H_2O.

13.43 The freezing-point depressions for 0.01 m solutions of $Co(NH_3)_6Cl_3$, $MgSO_4$, NH_4Cl, and CH_3COOH are 0.0643 °C, 0.0308 °C, 0.0358 °C, and 0.0193 °C, respectively. K_f for H_2O is 1.86 °C/m. Which compounds are strong electrolytes and which are weak? Which compound forms the greatest number of ions in a 0.01 m solution?

13.44 A 0.157 M NaCl solution is to be used to replace lost blood. The average osmotic pressure of blood is 7.7 atm at 25 °C. Is the NaCl solution isotonic with blood?

13.45 When 20.00 g of sucrose is dissolved in 100.00 g of water at 20 °C, a vapor-pressure lowering of 0.185 torr Hg is observed. The vapor pressure of pure H_2O at 20 °C is 17.54 torr. Determine the mass of one mole of sucrose from the vapor-pressure-lowering data.

13.46 Which substance will be more soluble in liquid NH_3, $CCl_4(l)$ or $CH_3OH(l)$?

13.47 10.0 g of $MgCl_2$ is dissolved in water to form 1525 g of solution. Without doing a calculation, is there sufficient information to determine the molality of the solution? If not, what additional information is needed and why?

13.48 The concentration of a KI solution is 2.25 M. Without doing a calculation, is there sufficient information to determine the molality of the solution? If not, what additional information is needed and why?

Integrative Exercises

13.49 Sucrose, $C_{12}H_{22}O_{11}$, is found in maple syrup. Maple syrup is produced by boiling the sap from a maple tree thereby increasing the sucrose concentration.

(a) What is the mass of sucrose in 100.0 mL of sap if the density of sap is 1.01 g/cm^3 and the approximate concentration of sucrose is 3.0%?

(b) Sap rises in trees largely because of osmosis. What is the osmotic pressure at 25 °C of sap? R = 0.0821 L-atm/mol-K.

(c) What is the boiling point of sap? K_b(water) = 0.512 K-kg/mol.

13.50 (a) Two solutions are prepared: 0.10 M Na_3PO_4 and 0.10 M H_3PO_4. The vapor pressure of the sodium phosphate solution is less than that of the phosphoric acid solution. Given that both have a similar formula, X_3PO_4, explain this observation.

(b) If X = Fe^{3+} how would the vapor pressure of this solution compare to the vapor pressures of the original solutions? Explain.

13.51 50.0 mL of a 0.500 M $BaCl_2$ solution is reacted with 50.0 mL of a 0.300 M Na_2SO_4 solution. Assuming the reaction goes to completion and the volumes are additive and reflect the volume of the final solution:

(a) write the chemical reaction;

(b) calculate the molarities of $BaCl_2$ and NaCl in the solution after the reaction is completed.

13.52 Assume that we live on a different planet in which carbon dioxide is toxic and hydrogen chloride is not and water is still a safe solvent. Could we use hydrogen chloride gas to form carbonated beverages? Explain.

Multiple-Choice Questions

13.53 What is the molality of ethylene glycol, $C_2H_4(OH)_2$, in a solution prepared by mixing 5.00 g of ethylene glycol in 125 g of water?

(a) 0.644   (d) 0.000619
(b) 0.000644   (e) none of these
(c) 0.619

13.54 What is the mole fraction of water in a solution prepared by mixing 12.5 g of H_2O with 220 g of acetone, C_3H_6O?

(a) 0.817   (d) 0.155
(b) 0.845   (e) none of these
(c) 0.183

13.55 In which of the following solvents would you expect the solubility of $CaCl_2$ to be greatest?

(a) CH_3OH (d) H_2O
(b) C_6H_6 (benzene)   (e) insufficient information
(c) CCl_4 to answer question

13.56 Given that K_b = 2.53 °C/m for benzene, which mass of acetone (CH_3COCH_3) must be dissolved in 200 g of benzene to raise the boiling point of benzene by 3.00 °C?

(a) 7.26 g   (d) 0.138 g
(b) 0.0726 g   (e) 13.8 g
(c) 2.56 g

13.57 Which set of conditions is true when a nonvolatile solute is added to a volatile solvent and an ideal solution forms? The symbols I and D mean increase and decrease, respectively, for the given properties of the volatile solvent when the solution forms.

	Freezing Pt.	Boiling Pt.	Vapor Pressure
(a)	I	I	I
(b)	D	D	D
(c)	I	D	D
(d)	D	I	D
(e)	I	D	I

13.58 Which of the following substances might stabilize a colloidal suspension of oil in water?

(a) octane, C_8H_{18}

(b) sodium bicarbonate, $NaHCO_3$

(c) sodium stearate, $NaCO_2(CH_2)_{16}CH_3$

(d) HCl

(e) $CaCl_2$

13.59 Two solutions of sugar with concentrations of 0.2 *M* and 0.5 *M* are separated by a semipermeable membrane. The 0.2 *M* solution is labeled A and the other one is labeled B. During osmosis:

(a) Sugar molecules move from solution A to B.

(b) Water molecules move from solution A to B.

(c) Sugar molecules move from solution B to A.

(d) Water molecules move from solution B to A.

(e) Osmosis does not occur.

13.60 When 0.200 g of a high-molecular-weight compound is dissolved in water to form 12.5 mL of solution at 25 °C, the osmotic pressure of the solution is found to be 1.10×10^{-3} atm. What is the molar mass of the compound?

(a) 3.56×10^5 g

(b) 3.56×10^4 g

(c) 2.98×10^4 g

(d) 2.98×10^3 g

(e) 3.00×10^4 g

SELF-TEST SOLUTIONS

13.1 (g). **13.2** (f). **13.3** (k). **13.4** (e). **13.5** (j). **13.6** (l). **13.7** (n). **13.8** (b). **13.9** (m). **13.10** (i). **13.11** (h). **13.12** (a). **13.13** (c). **13.14** (d). **13.15** False. 1 *ppm* corresponds to 1 mg of solute per liter of solution. Thus the concentration should be 0.30 ppm. **13.16** True. **13.17** True. **13.18** False. Inspect Table 13.5 in the text. You will see the opposite is true. This results from the slope of the solid–gas equilibrium line in a phase diagram being less than that of the liquid–gas equilibrium line. **13.19** False. Ion pairs reduce the total number of solute species in solution; therefore, osmotic pressure decreases. **13.20** False. A hydrophilic species will have a highly charged region or a number of very polar regions. This is not the case for the given molecule. **13.21** True. A hydrophobic group will be non-polar and molecular hydrophobic properties are enhanced by a large non-polar group. **13.22** False. Although both are liquids, they are immiscible, not miscible. The long nonpolar carbon chain is the dominant feature of the alcohol (octanol) and the polar OH group plays a minimal role in interacting with water. **13.23** True. The solution has a greater degree of disorder than the separated substances; particularly because NaCl is a highly organized crystalline solid. **13.24** False. Copper does not dissolve in water. The term miscible means substances mix in all proportions. **13.25** True. Hydrocarbons are nonpolar and do not dissolve (immiscible) in water. **13.26** False. Isotonic solutions have the same osmotic pressure. The NaCl solution is more concentrated and therefore it has a higher osmotic pressure.

13.27 False. A hypotonic solution has a lower osmotic pressure.

13.28 False. A hypertonic solution is more concentrated in terms of total number of solute particles than the other one. The NaCl solution has $2 \times 0.20M$ in solute particles whereas $CaCl_2$ has 3×0.20 *M* in solute particles.

13.29 (a) C. (b) D. (c) The enthalpy change required to separate the solute particles from each other and to separate the solvent particles from each other. (d) Exothermic. A process is exothermic when the final enthalpy state is at a lower value than the initial enthalpy state.

13.30 There is a net movement of solvent particles from the less concentrated solution (fewer dark spheres) to the more concentrated solution (more dark spheres) until the solutions are isotonic. This movement of solvent particles will change the volumes and molarities of both solutions.

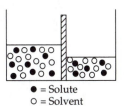

● = Solute
○ = Solvent

13.31 (a) The three factors are: (1) separation of solute particles from one another (endothermic); (2) separation of solvent particles from one another (endothermic); and (3) attraction of solute particles to solvent particles (exothermic).

(b) The sum of energetics of each step leads to the overall heat of solution. If it is too endothermic a solution will not form. There is insufficient energy associated with the attractive forces between the ions of KBr(*s*) and the nonpolar $CCl_4(l)$ molecules to overcome the sum of energies required to separate the ions of KBr(*s*) from one another and the molecules of $CCl_4(l)$ from one another.

13.32 The energy associated with the ion–water attractions is not sufficient to overcome the endothermic energetics of solute and solvent particle separation. Another principle exists and operates in this case: Processes that lead to an increase in the disorder of the system tend to occur spontaneously. The dissolving of a highly ordered crystalline RbCl in water leads to solvated ions that now are able to move. This results in a net increase in disorder in the system and permits the dissolution process to occur.

13.33 Refer to *A Closer Look: Hydrates* in the text. Copper(II) sulfate is known to form hydrates in which water occupies positions in the crystal lattice. Thus, when a dried sample of the anhydrous form is left in a wet environment, some water takes up locations in the crystal lattice structure and gains mass.

13.34 CH_3CH_3 is a hydrocarbon and is nonpolar. No polar groups are attracted to the dipole of water; thus it should not be soluble in water. $HOCH_2CH_2OH$ has two —OH groups that can hydrogen bond to water molecules; thus, it should be soluble.

13.35 (a) Henry's law states that the concentration of a gas dissolved in a solution is directly proportional to the partial pressure of the gas above the solution. When the partial pressure of carbon dioxide is increased to several atmospheres above sweetened water, the quantity of carbon dioxide dissolved increases. This will increase the amount of carbonic acid that also forms and gives the solution a more "acidic" taste.

(**b**) When the soft drink can is opened, the carbon dioxide that can remain in a dissolved form is significantly reduced because the partial pressure of the gas in the atmosphere is much less than one atmosphere. This results in the carbon dioxide escaping from the sweetened water and eventually resulting in a "flat" soft drink.

13.36 Molality is defined as

$$\text{Molality} = \frac{\text{number of moles of solute}}{\text{kilograms of solvent}}$$

and molarity is defined as

$$\text{Molarity} = \frac{\text{number of moles of solute}}{\text{number of liters of solution}}$$

The key difference is the term in the denominator. First, when you work with volume (liters) you are working with a measurement unit that is temperature dependent. Thus, molarity varies with temperature, whereas molality does not. Second, one can usually measure mass more accurately than volume. When an aqueous solution is very dilute the mass of the solute contributes very little to the density; thus there is essentially no difference between the number of kilograms and the number of liters.

13.37 The freezing-point depression is proportional to the molality of the solution. The two acids have the same molality, m, and thus one might expect the freezing-point depression to be the same for both. However, HF is a weak acid, whereas HCl is a strong acid. A weak acid ionizes to a small extent, whereas a strong acid ionizes completely. The number of solute particles produced in the ionization of HCl is greater than that for HF; thus the molality of all solute species is greater for HCl than HF. This results in a greater freezing-point depression for HCl.

13.38 Both are colloids. An aerosol contains liquid drops or solid particles dispersed in a gas. An emulsion contains liquid drops dispersed through a different liquid. An example of an aerosol is fog, and an example of an emulsion is mayonnaise (oil dispersed in water).

13.39 A sol is a solid particle dispersed in another liquid. The iron(III) particles on the surface of the solid particles have to be removed or the charge neutralized so that repulsion of solid particles is minimized and the particles can aggregate. Addition of a salt with a highly charged anion, such as the phosphate ion, can assist in this neutralization of charge.

13.40 A cucumber placed in a salt brine will eventually shrink and shrivel to form a pickle. The skin of the cucumber acts like a semipermeable membrane. The liquid inside the cucumber has a lower concentration of salts than in the salt brine. By osmosis water leaves the cucumber to make the solution inside the cucumber more concentrated in an attempt to equalize concentration. As the water leaves the cucumber it shrivels.

13.41 We will assume that we have a 50.0 mL volumetric flask available so that we can dilute a portion (an aliquot) of the more concentrated 5.00 M KI solution to 50.0 mL and in the process make the desired 50.0 mL of 1.00 M KI solution. You can always take a concentrated solution and prepare a more dilute solution from it. What aliquot of 5.00 M KI is needed to add to the 50.0 mL volumetric flask? The number of moles of KI in the 50.0 mL solution of 1.00 M KI solution can be calculated as follows: Moles = Molarity × Volume = $1.00 M \times 0.500$ L = 0.0500 mol KI. We need to take out of the 5.00 M KI solution exactly this number of moles of KI. We do this by taking a portion of the solution by taking an aliquot (a volume). The volume needed is again calculated from the definition of molarity: Volume = Moles of KI required/Molarity = 0.0500 $mol/5.00$ M = 0.01000 L. If we use a pipet or buret to accurately dispense 0.0100 L (10.0 mL) of the 5.00 M KI solution into the 50.0 mL volumetric flask we will have placed 0.0500 mol of KI into it. We then carefully add deionized water until the flask is about half filled and gently swirl to mix. Continue adding water until the bottom of the meniscus of the solution is at the 50.0 mL mark. Carefully swirl and we have the required 1.00 M KI solution.

13.42 (**a**) $m = \dfrac{\text{moles of Na}_2\text{CO}_3}{\text{kg of solvent}}$

$$= \frac{(142 \text{ g Na}_2\text{CO}_3)\left(\dfrac{1 \text{ mol}}{106.0 \text{ g}}\right)}{2.00 \text{ kg H}_2\text{O}}$$

$$= 0.670 \text{ molal}$$

(**b**) $m = \dfrac{\text{moles codeine}}{\text{kg of solvent}}$

$$= \frac{(24.50 \text{ g codeine})\left(\dfrac{1 \text{ mol}}{299.4 \text{ g}}\right)}{(150.5 \text{ g ethanol})\left(\dfrac{1 \text{ kg}}{1000 \text{ g}}\right)}$$

$$= 0.5437 \text{ molal}$$

(**c**) $X(\text{NaCl}) = \dfrac{\text{mol NaCl}}{\text{mol NaCl} + \text{mol H}_2\text{O}}$

$$= \frac{(32.3 \text{ g NaCl})\left(\dfrac{1 \text{ mol}}{58.4 \text{ g}}\right)}{\left[(32.3 \text{ g NaCl})\left(\dfrac{1 \text{ mol}}{58.5 \text{ g}}\right) + (265.0 \text{ g H}_2\text{O})\left(\dfrac{1 \text{ mol}}{18.01 \text{ g}}\right)\right]}$$

$$= \frac{0.533 \text{ mol}}{0.552 \text{ mol} + 14.71 \text{ mol}} = 0.0349$$

(**d**) % HCl $= \dfrac{140 \text{ g HCl}}{140 \text{ g HCl} + 800 \text{ g H}_2\text{O}} \times 100$

$$= 14.9\% \text{ HCl}$$

13.43 The ΔT lowering for a 0.01 m weak electrolyte solution is approximately $\Delta T_f = K_f m = (1.86\ °C/m)(0.01m)$ = 0.0186 °C. Only the 0.01m CH$_3$COOH solution has a

ΔT_f approximately equal to 0.0186 °C; it is the only weak electrolyte. The other salt solutions produce freezing-point depressions significantly greater than that of acetic acid; thus they are all strong electrolytes. Since the 0.01 m solution of $Co(NH_3)_6Cl_2$ produces the greatest ΔT_f lowering, the molality of all species in the solution is the largest; it forms the greatest number of ions in solution.

13.44 The solutions are isotonic if they have the same osmotic pressure. The osmotic pressure of 0.157 M NaCl is calculated using the expression $\pi = MRT$. The value of M is 2(0.157 M) because there are two ions per NaCl formula:

$$\pi = (0.314M)\left(0.0821\frac{\text{L-atm}}{\text{K-mol}}\right)(298\text{ K}) = 7.68\text{ atm}$$

The solutions are isotonic.

13.45 VPL = $X_{\text{sucrose}}P_{H_2O}$; X_{sucrose} = mol sucrose/(mol sucrose + mol H_2O). For very dilute solutions, such as this one in which the number of moles of H_2O is far greater in magnitude than number of moles of sucrose, we can assume that (mol sucrose + mol H_2O) $\approx$ mol H_2O

$$VLP = X_{\text{sucrose}}P^\circ_{H_2O} = \left(\frac{\text{mol sucrose}}{\text{mol } H_2O}\right)P^\circ_{H_2O}$$

$$VLP = \left[\frac{(\text{grams sucrose})\left(\dfrac{1\text{ mol sucrose}}{\mathcal{M}\text{ sucrose}}\right)}{(\text{grams } H_2O)\left(\dfrac{1\text{ mol } H_2O}{\mathcal{M}\ H_2O}\right)}\right]P^\circ_{H_2O}$$

Rearranging the equation gives

$$\mathcal{M}\text{ sucrose} = \left(\frac{\text{grams sucrose}}{\text{grams } H_2O}\right)\left(\frac{1\text{ mol sucrose}}{1\text{ mol } H_2O}\right)$$

$$\times \left(\frac{P^\circ_{H_2O}}{VLP}\right)(\mathcal{M}\ H_2O)$$

$$= \left(\frac{20.00\text{ g}}{100.00\text{ g}}\right)\left(\frac{1\text{ mol}}{1\text{ mol}}\right)$$

$$\times \left(\frac{17.54\text{ mm Hg}}{0.185\text{ mm Hg}}\right)\left(\frac{18.0\text{ g}}{1\text{ mol}}\right)$$

$$= 341\text{ g/mol}$$

13.46 $NH_3(l)$ is a polar solvent. The most polar substance the most soluble. $CH_3OH(l)$ is more soluble because it is polar; $CCl_4(l)$ is nonpolar.

13.47 There is sufficient information. Molality = moles solute/kg of *solvent*. We can calculate the number of moles of $MgCl_2$ from the given mass and its molar mass. We can calculate the mass of solvent from the relation: mass solution = mass $MgCl_2$ + mass water. We know the mass of solution and mass of $MgCl_2$; thus we can calculate the mass of water and convert it to kg.

13.48 There is insufficient information. We need both the number of moles of KI and kg of solvent. We are not given any quantity of the solution, thus we can assume an arbitrary value: 1.00 L. Because molarity = number of moles

of KI/volume of solution in liters, we can then calculate the number of moles of KI in 1.00 L: number of moles of KI = $V \times M$. From the number of moles of KI and its molar mass we can calculate the number of grams of KI in 1.00 L of solution. To determine the number of grams of solvent we need the number of grams of solution representing 1.00 L. We need the density of solution to determine the mass of solution: Mass of solution = Volume $\times$ *density*. Density is not given in the problem; thus we cannot calculate the mass of solvent.

13.49 (a) The mass of the solution (sap) is (100.0 mL) (1.01 g/mL) = 101 g. The mass of sucrose is (101 g) (0.030) = 3.0 g

$$\pi = MRT = \frac{(\text{moles of solute particles})RT}{\text{volume in liters}}$$

$$\pi = \frac{(3.0\text{ g})\left(\dfrac{1\text{ mol}}{342.3\text{ g}}\right)\left(0.0821\dfrac{\text{L-atm}}{\text{mol-}K}\right)(298\text{ K})}{0.1000L}$$

$$\pi = 2.2\text{ atm}$$

(b) $\qquad\qquad \Delta T_b = K_b m$

$$\Delta T_b = \left(0.512\frac{\text{K-kg}}{\text{mol}}\right)\left(\frac{(3.0\text{ g})\left(\dfrac{1\text{ mol}}{342.3\text{ g}}\right)}{(101\text{ g} - 3.0\text{ g})\left(\dfrac{1\text{ kg}}{1000\text{ g}}\right)}\right)$$

(Note: the deonominator in m is the kg of solvent only)

$$\Delta T_b = 0.044K$$

The normal boiling point of water is 100.00 °C. Thus the boiling point of the sap solution is increased to 100.04 °C.

13.50 (a) The vapor pressure of an ideal solution is $P_{\text{solution}} = X_{\text{solvent}}P^\circ_{\text{solvent}}$ where

$$X_{\text{solvent}} = \frac{\text{moles solvent}}{\Sigma(\text{moles of all particles in solution})}$$

Na_3PO_4 is a strong electrolyte in water and completely ionizes to three Na^+ and one PO_4^{3-} ions per formula unit, or a total of four ions per formula unit. H_3PO_4 is a weak acid and only partially ionizes, thus forming fewer particles per formula unit than sodium phosphate. In the formula X_{solvent} the denominator is larger for the sodium phosphate solution than for phosphoric acid solution; this results in a smaller mole fraction and therefore a smaller vapor pressure. The number of moles of solvent is very large and essentially a constant.

(b) $FePO_4$ is insoluble in water, according to the solubility rules learned earlier. Thus, $FePO_4$ does not appreciably dissolve and the solution is essentialy water. The mole fraction of this solution approaches one and thus this solution should have the greatest vapor pressure.

13.51 (a) $BaCl_2(aq) + Na_2SO_4(aq) \rightarrow BaSO_4(s)$
$$+ 2\text{ NaCl}(aq)$$

(b) mol $BaCl_2$ = MV_{liters} = (0.500 M)(0.0500 L)
$$= 0.0250\text{ mol}$$

Barium chloride and sodium sulfate react in a 1:1 mole ratio; however, the mole ratio of the actual reactants is not 1:1, thus it is a limiting reactant situation. After the reaction, barium chloride, a reactant, will be left over since it is present in a greater amount. The number of moles of materials present is calculated as follows:

$$BaCl_2 + Na_2SO_4 \longrightarrow BaSO_4 + 2\,NaCl$$

Initial(mol)	0.0250	0.0150	0	0
Change(mol)	−0.0150	−0.0150	+0.0150	+2(0.0150)
End(mol)	0.0100	0	0.0150	0.0300

$$\text{molarity of } BaCl_2 = \frac{mol}{V_L} = \frac{0.0100 \text{ mole}}{0.100 \text{ L}} = 0.100 \ M$$

$$\text{molarity of } NaCl = \frac{0.0300 \text{ mol}}{0.100 \text{ L}} = 0.300 \ M$$

Barium sulfate is insoluble and thus is not a dissolved substance.

13.52 Hydrogen chloride would not make a suitable gas for forming carbonated beverages. HCl is polar and water is polar; thus HCl would readily dissolve in the water. Therefore, it would not readily escape as a gas unless the solution was supersaturated with it at a very high concentration. In a concentrated solution it is likely the HCl would also affect the taste of the beverage. It is also a corrosive gas.

13.53 (a)

$$m = \frac{(5.00 \text{ g } C_2H_4(OH)_2)\left(\dfrac{1 \text{ mol } C_2H_4(OH)_2}{62.07 \text{ g } C_2H_4(OH)_2}\right)}{(125 \text{ g } H_2O)\left(\dfrac{1 \text{ kg}}{1000 \text{ g}}\right)}$$

$$= 0.644$$

13.54 (d)

$$X_{H_2O} = \frac{(12.5 \text{ g } H_2O)\left(\dfrac{1 \text{ mol } H_2O}{18.0 \text{ g } H_2O}\right)}{\left[(12.5 \text{ g } H_2O)\left(\dfrac{1 \text{ mol } H_2O}{18.0 \text{ g } H_2O}\right)\right.}$$

$$\left. + (220 \text{ g } C_3H_6O)\left(\dfrac{1 \text{ mol } C_3H_6O}{58.09 \text{ g } C_3H_6O}\right)\right]$$

$$= 0.155$$

13.55 (d) $CaCl_2$ is an ionic substance and will dissolve best in the most polar solvent, water.

13.56 (e) $\Delta T = K_b m = K_b\left(\dfrac{\text{moles acetone}}{1 \text{ kg benzene}}\right)$

$$\text{Moles acetone} = \left(\frac{\Delta T}{K_b}\right)(1 \text{ kg benzene})$$

$$= \left(\frac{3.00 \ ^\circ C}{2.53 \ ^\circ C/m}\right)(1 \text{ kg benzene})$$

$$= \left(\frac{1.19 \text{ mol acetone}}{1 \text{ kg benzene}}\right)(1 \text{ kg benzene})$$

$$= 1.19 \text{ mol}$$

The number of moles of acetone in 200 g of benzene is calculated as follows:

$$\frac{1.19 \text{ mol acetone}}{\text{moles acetone}} = \frac{1000 \text{ g benzene}}{200 \text{ g benzene}}$$

$$\text{Moles acetone} = 0.238 \text{ mol}$$

$$\text{Mass acetone} = (0.238 \text{ mol acetone})$$

$$\times \left(\frac{58.09 \text{ g acetone}}{1 \text{ mol acetone}}\right)$$

$$= 13.8 \text{ g}$$

13.57 (d). 13.58 (c).

13.59 (b) The solution with the largest number of solute particles has largest osmotic pressure.

13.60 (a) $M = \dfrac{\pi}{RT} = \dfrac{1.10 \times 10^{-3} \text{ atm}}{\left(0.0821 \dfrac{\text{L-atm}}{\text{K-mol}}\right)(298 \text{ K})}$

$$= 4.50 \times 10^{-5} \text{ mol/L}$$

The number of moles in 0.200 g of the compound in 12.5 mL is:

$$\text{Moles} = \left(4.50 \times 10^{-5}\frac{\text{mol}}{\text{L}}\right)\left(\frac{1 \text{ L}}{1000 \text{ mL}}\right)(12.5 \text{ mL})$$

$$= 5.62 \times 10^{-7} \text{ mol}$$

The molar mass of the compound is calculated as follows:

$$\frac{0.200 \text{ g}}{\text{Molar mass}} = \frac{5.62 \times 10^{-7} \text{ mol}}{1 \text{ mol}}$$

$$\text{Molar mass} = (0.200 \text{ g})\left(\frac{1 \text{ mol}}{5.62 \times 10^{-7} \text{ mol}}\right)$$

$$= 3.56 \times 10^5 \text{ g}$$

Chemical Kinetics

Chapter

14

OVERVIEW OF THE CHAPTER

Review: Concentration units (13.4); graphing techniques (see Appendix in the text).

14.1, 14.2, 14.3 REACTION RATES: GENERAL CONSIDERATIONS

Learning Goals: You should be able to:

1. Express the rate of a given reaction in terms of the variation in concentration of a reactant or product substance with time.
2. Calculate the average rate over an interval of time, given the concentrations of a reactant or product at the beginning and end of that interval.
3. Calculate instantaneous rates from a graph of reactant or product concentrations as a function of time.
4. Explain the meaning of the term *rate constant* and state the units associated with rate constants.
5. Calculate rate, rate constants, or reactant concentration, given two of these together with the rate law.
6. Determine the rate law from experimental results that show how concentration affects rate.

Review: Calculation of the slope of a straight line (see Appendix A in the text); use of logarithms (see Appendix A in the text).

14.4 REACTION RATES: FIRST AND SECOND ORDER

Learning Goals: You should be able to:

1. Use the equations

$$\ln\left(\frac{[A]_t}{[A]_0}\right) = -kt$$

$$\ln[A]_t = -kt + \ln[A]_0$$

to determine (**a**) the concentration of a reactant or product at any time after a reaction has started, (**b**) the time required for a given fraction of sample to react, or (**c**) the time required for a reactant concentration to reach a certain level.

2. Explain the concept of reaction half-life and describe the relationship between half-life and rate constant for a first-order reaction.

3. Use the equations

$$\ln[A]_t = -kt + \ln[A]_0$$

$$\frac{1}{[A]_t} = \frac{1}{[A]_0} + kt$$

to determine graphically whether the rate law for a reaction is first or second order.

4. Determine the rate law for a given higher-order reaction from the appropriate data.

14.5 REACTION RATES: TEMPERA-TURE AND ACTIVATION ENERGY

Review: Energy (5.1, 5.2); endothermic and exothermic processes (5.3).

Learning Goals: You should be able to:

1. Explain the concept of activation energy and how it relates to the variation of reaction rate with temperature.
2. Determine the activation energy for a reaction from a knowledge of how the rate constant varies with temperatures (the Arrhenius equation).
3. Use the collision model of chemical reactions to explain how reactions occur at the molecular level.

14.6 REACTION MECHANISMS

Learning Goals: You should be able to:

1. Explain what is meant by the mechanism of a reaction using the terms elementary steps, rate-determining step, and intermediate.
2. Derive the rate law for a reaction that has a rate-determining step, given the elementary steps and their relative speeds; or, conversely, choose a plausible mechanism for a reaction given the rate law.

14.7 CATALYSIS

Learning Goals: You should be able to:

1. Describe the effect of a catalyst on the energy requirements for a reaction.
2. Relate the factors that are important in determining the activity of a heterogeneous catalyst.
3. Explain how enzymes act as biological catalysts using the lock-and-key model.

TOPIC SUMMARIES AND EXERCISES

REACTION RATES: GENERAL CONSIDERATIONS

The study of chemical kinetics is concerned with (1) the rate, or speed, at which reactants form products; (2) what factors influence the rates of reactions; and (3) how reactions occur. To describe the **rate** (or speed) **of a chemical reaction**, we use the ratio of the change in concentration of a reactant or product to the time interval required for the observed concentration change.

• For example, the rate at which N_2O_5 disappears in the chemical reaction

$$N_2O_5(g) \longrightarrow 2\,NO_2 + \tfrac{1}{2}O_2(g)$$

is written as

$$\text{Rate} = -\frac{\Delta[N_2O_5]}{\Delta t}$$

$$= -\left(\frac{\begin{array}{l}\text{concentration of } N_2O_5 \text{ at } t_{\text{final}} \\ - \text{ concentration of } N_2O_5 \text{ at } t_{\text{initial}}\end{array}}{t_{\text{final}} - t_{\text{initial}}}\right)$$

- The Greek symbol Δ means "change in" or "difference in." Thus, $\Delta[N_2O_5]$ means the change in concentration of N_2O_5 at time t_{final} minus the concentration of N_2O_5 at time t_{initial}.
- Rates are assigned a positive quantity. The negative sign in front of $\Delta[N_2O_5]/\Delta t$ is necessary because $\Delta[N_2O_5]$ is a negative number (the concentration of N_2O_5, a reactant, decreases with time).

We need to differentiate between average rates and instantaneous rates of reactions.

- An **average rate** is calculated by taking the ratio of the change in concentration to change in time for a particular set of data points chosen from rate data. Note in Table 14.1 of the text that average rates vary depending on the pair of data chosen.
- An **instantaneous rate** is the slope of a tangent drawn at any point along a graph of concentration versus time. See Figure 14.1 below. Note that instantaneous rates are rates at a particular time and thus vary with time. Unless otherwise stated, the rate of a reaction will refer to an instantaneous rate.
- **Initial rates** are instantaneous rates, evaluated at $t = 0$.

Another way of describing the velocity of a chemical reaction is by a rate law.

- A **rate law** is an *experimentally* derived expression that relates the rate of a chemical reaction to the concentration of one or more of the species in

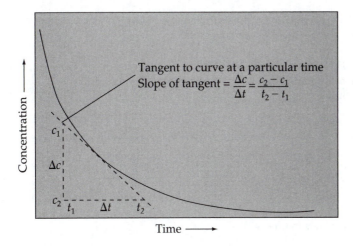

Tangent to curve at a particular time
Slope of tangent $= \dfrac{\Delta c}{\Delta t} = \dfrac{c_2 - c_1}{t_2 - t_1}$

▲ **FIGURE 14.1** Graphing technique to determine an instantaneous rate.

the chemical reaction. For example, the rate law for the previously described decomposition of $N_2O_5(g)$ is

$$\text{Rate} = k[N_2O_5]$$

This expression tells us that the speed of the reaction is directly proportional to the concentration of N_2O_5. If we double the concentration of N_2O_5, the rate of the reaction doubles.

- The constant k, a proportionality constant called the **rate constant**, is determined experimentally and is temperature dependent and concentration independent.
- A general form of the rate law for simple kinds of reactions is

$$\text{Rate} = k[\text{reactant 1}]^n[\text{reactant 2}]^m \ldots$$

The exponent associated with the concentration of a species in the rate law is called the **reaction order** for that species. If we add all the exponents ($n + m \ldots$), we obtain what is called the **overall reaction order**. Reaction orders do *not* have to be integral.

We use units with these terms: Concentrations, time, rates, and rate constants.

- Units of concentration are molarity for solutions and atmospheres of pressure for gases.
- Units of time are seconds, minutes, hours, or any appropriate time period. Typically seconds or minutes are used.
- Units of rate are [concentration]/time; for example, Ms^{-1}. Note that the unit expression Ms^{-1} is equivalent to the expression M/s. Ms^{-1} can also be expressed as moles $L^{-1}s^{-1}$ because molarity, M, is equivalent to moles/L or moles L^{-1}.
- Units of a rate constant depend on the rate law.

EXERCISE 1   Writing rate expressions for chemical reactions

(**a**) Write three different expressions which might be used to describe rates of reaction for

$$2\,N_2O(g) \longrightarrow 2\,N_2(g) + O_2(g)$$

using the changes in the concentrations of $N_2O(g)$, $N_2(g)$, $O_2(g)$ per unit time. (**b**) Relate the three to each other.

SOLUTION: *Analyze*: We are given the reaction for the decomposition of dinitrogen oxide and asked to write three different rate expressions and to relate them to each other.

Plan: A rate expression involving changes in concentration takes the form: $\text{Rate} = \dfrac{\Delta C}{\Delta t}$, where C refers to concentration and t refers to time. We will write this type of expression for each substance in the reaction.

Solve: (**a**) The rate of the change of concentration of $N_2O(g)$ with time will be determined first. Assume that the $N_2O(g)$ concentration at time t_{initial} is $[N_2O]_{\text{initial}}$ and that at time t_{final} it is $[N_2O]_{\text{final}}$. The decrease in the concentration of $N_2O(g)$ with a change in time is

$$\text{Rate}_1 = -\frac{1}{2}\frac{[N_2O]_{\text{final}} - [N_2O]_{\text{initial}}}{t_{\text{final}} - t_{\text{initial}}} = -\frac{1}{2}\frac{\Delta[N_2O]}{\Delta t}$$

The negative sign is required because $\Delta[N_2O]$ is a negative number; the N_2O concentration is decreasing with time, and a *measured rate is by convention a positive quantity*. The other average rates are

$$\text{Rate}_2 = \frac{1}{2}\frac{\Delta[N_2]}{\Delta t} = \frac{1}{2}\frac{[N_2]_{\text{final}} - [N_2]_{\text{initial}}}{t_{\text{final}} - t_{\text{initial}}}$$

$$\text{Rate}_3 = \frac{\Delta[O_2]}{\Delta t} = \frac{[O_2]_{\text{final}} - [O_2]_{\text{initial}}}{t_{\text{final}} - t_{\text{initial}}}$$

The latter two rates do not require a negative sign, because the concentrations of O_2 and N_2 increase with time. (**b**) The three rates are related to each other as follows:

$$\text{Rate} = -\frac{1}{2}\frac{\Delta[N_2O]}{\Delta t} = \frac{1}{2}\frac{\Delta[N_2]}{\Delta t} = \frac{\Delta[O_2]}{\Delta t}$$

The factor of $1/2$ before $\dfrac{\Delta[N_2O]}{\Delta t}$ and $\dfrac{[N_2]}{\Delta t}$ is needed because the rate of disappearance of $N_2O(g)$ and the rate of appearance of $N_2(g)$ are each twice the rate of appearance of $O_2(g)$. That is, for every two molecules of $N_2O(g)$ that react, two molecules of $N_2(g)$ and one molecule of O_2 appear. Thus, $N_2O(g)$ must disappear more rapidly than $O_2(g)$ appears.

EXERCISE 2   Explaining why initial rates are commonly used to determine rate laws

Why are initial rates commonly used to determine rate laws?

SOLUTION: An initial rate is the slope of the tangent, measured at time zero, to a curve such as shown in Figure 14.1. Take a ruler and use it as a tangent to the curve. Move the ruler along the curve until you reach time equals zero. At this point, you should observe that the slope of the curve has its greatest value. During early changes in a reaction, the rates of greatest change are measured. Experimental errors associated with rate data are less when the measured rates are large.

EXERCISE 3   Interpreting a rate law

For the reaction

$$2\,NO(g) + 2\,H_2(g) \longrightarrow N_2(g) + 2\,H_2O(g)$$

the rate law is

$$\text{Rate} = k[NO]^2[H_2]$$

(**a**) Identify the rate constant in the rate law and indicate the overall reaction order. (**b**) What are the units of k if the unit for concentration is atmospheres? (**c**) How does the initial rate change if the initial concentration of $NO(g)$ is doubled and that of $H_2(g)$ remains constant?

SOLUTION: *Analyze*: We are given the rate law for the reaction of nitrogen oxide with hydrogen and asked three questions relating to the rate law.

Plan: (**a**) We need to identify which term in the rate law is the rate constant and recognize that the overall reaction order is the sum of exponents for each concentration term. (**b**) We can rearrange the rate law expression and solve for k. Then we can substitute the appropriate units for each term. (**c**). We can determine how the value of $[NO]^2$ changes when the concentration of NO is doubled.

Solve: (**a**) The rate constant is k. The overall order of the reaction is calculated by adding the powers to which each term is raised in the rate law. In this problem, the overall order of the reaction is 2 + 1, or 3. (**b**) To solve for the units of k, you must relate k to all of the other terms in the rate law:

$$\text{Rate} = k[NO]^2[H_2]$$

Rearrange the equation to obtain k:

$$k = (\text{Rate})\left(\frac{1}{[NO]^2[H_2]}\right)$$

Substituting the appropriate units for each term in the expression yields the following:

$$k = \left(\frac{\text{atm}}{\text{s}}\right)\left(\frac{1}{(\text{atm})^2(\text{atm})}\right) = \frac{1}{(\text{s})(\text{atm})^2} = \text{atm}^{-2}\,\text{s}^{-1}$$

(**c**) The rate is proportional to the *square* of [NO] if the concentration of H_2 is held constant. Thus, doubling the concentration of NO causes the rate to increase by a factor of four.

EXERCISE 4   Writing a rate law for a chemical reaction

The rate law for the reaction

$$O_3(g) + 2\,NO(g) \longrightarrow N_2O_5(g) + O_2(g)$$

contains $NO(g)$ and $O_3(g)$ with orders of $\frac{2}{3}$ each. Write the rate law.

SOLUTION: *Analyze*: We are given a reaction between ozone and nitrogen oxide and the order for each reactant in the rate law and asked to write the rate law.

Plan: An order for a concentration of a particular substance in the rate law is its exponent. We can write a rate law with a rate constant and each reactant with its given order.

Solve: The order with respect to a reactant is the power to which the concentration of that reactant is raised. The rate law is

$$-\frac{\Delta[O_3]}{\Delta t} = k[NO]^{2/3}[O_3]^{2/3}$$

Comment: Note that orders for concentrations of substances in a rate law do not have to be whole numbers; the same is true for overall order of the reaction. Furthermore, the orders are not equal to the stoichiometric coefficients for the reactants. For example, the stoichiometric coefficient for NO in the chemical reaction is 2 but its order is 2/3. A rate law cannot be determined from inspection of the balanced chemical reaction.

EXERCISE 5   Determining a rate law for a chemical reaction

In three different experiments, the following data were obtained for the reaction $A + B \rightarrow C$:

Experiment number	Initial concentration of A, M	Initial concentration of B, M	Initial rate, Ms^{-1}
1	0.20	1.00	5.0×10^{-2}
2	0.20	2.00	2.0×10^{-1}
3	0.40	2.00	4.0×10^{-1}

Determine the rate law for the reaction.

SOLUTION: *Analyze*: We are given a hypothetical reaction with two reactants, A and B, and three sets of experimental data. Each set gives the initial concentrations of A and B and the initial rate. We are asked to determine the rate law.

Plan: The rate law may have the form

$$Rate = k[A]^n[B]^m$$

where n and m may be zero or some other number. One approach to solving this type of problem is to evaluate the value of n at two different rates when [B] is constant and then to evaluate the value of m when [A] is constant. For example, to solve for the value of n, use the ratio of rates for two differing initial concentrations of A but with the same initial concentrations of B (Experiments 2 and 3):

$$\frac{Rate_2}{Rate_3} = \frac{k[A]^n_{expt2}[B]^m_{expt2}}{k[A]^n_{expt3}[B]^m_{expt3}}$$

Since $[B]^m_{expt2} = [B]^m_{expt3}$, you can write

$$\frac{Rate_2}{Rate_3} = \frac{k[A]^n_{expt2}[B]^m_{expt2}}{k[A]^n_{expt3}B]^m_{expt3}} = \frac{[A]^n_{expt2}}{[A]^n_{expt3}} = \left(\frac{[A]_{expt2}}{[A]_{expt3}}\right)^n$$

We can then solve for the value of n. A similar approach is used to solve for m.

Solve: Using the experimental data,

$$\frac{Rate_2}{Rate_3} = \frac{2.0 \times 10^{-1}\ Ms^{-1}}{4.0 \times 10^{-1}\ Ms^{-1}} = \left(\frac{0.20M}{0.40M}\right)^n$$

or

$$0.50 = (0.50)^n$$

The latter relation is valid only if the value of n is equal to 1. Repeating this procedure to evaluate the value of m when the concentration of A is kept constant (Experiments 1 and 2):

$$\frac{Rate_1}{Rate_2} = \left(\frac{[B]_{expt1}}{[B]_{expt2}}\right)^m$$

or

$$\frac{5.0 \times 10^{-2}\ Ms^{-1}}{2.0 \times 10^{-1}\ Ms^{-1}} = \left(\frac{1.00M}{2.00M}\right)^m$$

or

$$0.25 = (0.500)^m$$

For this relationship to be valid, the value of m must be 2. Therefore, the rate law is

$$Rate = -\frac{\Delta[A]}{\Delta t} = k[A][B]^2$$

The experimental rate law for the reaction

$$C_2H_6(g) \longrightarrow C_2H_4(g) + H_2(g)$$

is

$$Rate = -\frac{\Delta[C_2H_6]}{\Delta t} = k[C_2H_6]$$

REACTION RATES: FIRST AND SECOND ORDER

- This is a **first-order reaction** because there is only one reactant concentration in the rate law and its concentration is raised to the first power—that is, it has an exponent of one. Only a few reactions are first order; most reactions have more complex rate laws.

- An important relationship that first-order reactions obey is

$$\ln[A]_t = -kt + \ln[A]_0$$

  where $[A]_t$ is the concentration at time t of species A and $[A]_0$ is the initial concentration of species A at the start of the reaction ($t = 0$). It is important to remember that the two concentrations $[A]_t$ and $[A]_0$ must have the same concentration units.

- An important characteristic of a first-order reaction is its **half-life**. The half-life, $t_{1/2}$, is the time required for a concentration of a reactant to decrease to one-half of its original value—that is, $[A]_{t_{1/2}} = \frac{1}{2}[A]_0$. For a first-order reaction, the half-life depends only on the rate constant and is given by the relationship

$$t_{1/2} = \frac{0.693}{k}$$

- Note that the value of $t_{1/2}$ for a first-order reaction does not depend on concentrations. This is an important property of first-order reactions.

Most reactions have rate laws with overall orders of two or more. The rate law of a higher-order reaction is often determined by proposing a rate law and then determining whether the experimental data fit it.

- A simple **second-order rate law** is a rate law with one reactant that has a reaction order of two. The general rate law for this type of second-order reaction is

$$\text{Rate} = k[A]^2$$

- A second-order reaction of this type obeys the following relationship

$$\frac{1}{[A]_t} = \frac{1}{[A]_0} + kt$$

- The half-life for a second-order reaction is

$$t_{1/2} = \frac{1}{k[A]_0}$$

Note that the half-life is concentration dependent.

EXERCISE 6   Determining the order of a chemical reaction and its half-life

The following data were obtained for the decomposition of N_2O_5 in CCl_4 at 45 °C:

Initial concentration of $[N_2O_5]$, M	Initial rate, Ms^{-1}
0.50	1.55×10^{-4}
1.0	3.10×10^{-4}
1.5	4.65×10^{-4}

(a) Show that the reaction is first order. (b) What is the value of k? (c) What is the value of the half-life for the reaction?

SOLUTION: *Analyze:* We are given data for the decomposition of N_2O_5 and asked to show that it is a first-order reaction and to determine the value of the rate constant and the half-life for the reaction.

Plan: To demonstrate that it is a first-order reaction we have to show that it has the following form: rate $= k[N_2O_5]$. We can do this by demonstrating that the rate changes linearly with the concentration of N_2O_5. The value of k is determined by averaging the values of k calculated from the initial rate and initial concentration for each set of data. The value of k for a set of data is obtained from: $k = \dfrac{\text{rate}}{[N_2O_5]}$. The half-life for a first-order reaction is calculated using: $t_{1/2} = \dfrac{0.693}{k}$.

Solve: (a) If it is a first-order reaction, the rate law can be written as

$$\text{rate} = k[N_2O_5]$$

The initial rate should double if the concentration of N_2O_5 is doubled. Doubling the concentration of N_2O_5 from 0.50 M to 1.0 M doubles the rate from 1.55×10^{-4} M to 3.10×10^{-4} M/sec; thus the reaction is first order. (b)

$$k = \frac{\text{initial rate}}{[N_2O_5]_0}$$

where $[N_2O_5]_0$ is the initial concentration of N_2O_5. The calculated values of k are

$$k = \frac{1.55 \times 10^{-4}\ Ms^{-1}}{0.50M} = 3.1 \times 10^{-4}\ s^{-1}$$

$$k = \frac{3.10 \times 10^{-4}\ Ms^{-1}}{1.0M} = 3.1 \times 10^{-4}\ s^{-1}$$

$$k = \frac{4.65 \times 10^{-4}\ Ms^{-1}}{1.5M} = 3.1 \times 10^{-4}\ s^{-1}$$

Obviously, the average value of k is $3.1 \times 10^{-4}\ s^{-1}$. (c) The half-life of the reaction is calculated using the relation

$$t_{1/2} = \frac{0.693}{k} = \frac{0.693}{3.1 \times 10^{-4}\ s^{-1}} = 2.2 \times 10^3\ s$$

Exercise 7 Determining a rate constant from the half-life for a first-order chemical reaction

The half-life of a first-order reaction is 6.00×10^{-2} s. What is the rate constant for the reaction?

SOLUTION: *Analyze:* We are given the half-life for a first-order reaction and asked to determine the rate constant.

Plan: We can rearrange the equation $t_{1/2} = \dfrac{0.693}{k}$ to solve for the rate constant, k.

Solve: Rearranging the previous equation gives:

$$k = \frac{0.693}{t_{1/2}} = \frac{0.693}{6.00 \times 10^{-2}\ s} = 11.6\ s^{-1}$$

Check: An estimate of the value for rate constant is $1/0.1 = 10$, which is approximately the value of the calculated number in the solution.

EXERCISE 8   Using the specific rate constant for a first-order chemical reaction to determine concentration and time

N_2O_5 decomposes in the solvent CCl_4 as follows:

$$2\,N_2O_5 \longrightarrow 4\,NO_2 + O_2$$

The specific rate constant for the first-order decomposition at 45 °C is $6.32 \times 10^{-4}\,s^{-1}$. (a) What is the concentration of N_2O_5 remaining after 2.00 hr if the initial concentration of N_2O_5 was 0.500 M? (b) How much time is required for 90% of the N_2O_5 to decompose?

SOLUTION: *Analyze*: We are given the rate constant for the decomposition of N_2O_5 with an initial concentration of 0.500 M and are asked what is the concentration of N_2O_5 after two hours. Then we are asked how much time it takes for 90% of N_2O_5 to disappear.

Plan: (a) We need a relationship between time and concentration. This relationship for first-order reactions is:

$$\ln[A]_t = -kt + \ln[A]_0$$

In this problem A represents N_2O_5, k is the specific rate constant, t is the time for the reaction, $\ln[A]_t$ is the natural log of the concentration of N_2O_5 after 2.00 hr, and $\ln[A]_0$ is the natural log of the initial concentration of N_2O_5. From the given information, the only unknown in the equation is the concentration of N_2O_5 after 2.00 hr, $[A]_t$.

(b) Note that the initial concentration is not given. This quantity is not required because we are told that we want the time required for 90% of N_2O_5 to disappear. You can use any arbitrary initial concentration.

Solve: (a) We can substitute the given data in the equation relating time and concentration to give:

$$\ln[N_2O_5]_t = -kt + \ln[N_2O_5]_0$$

or

$$\ln[N_2O_5]_t = -\left(6.32 \times 10^{-4}\frac{1}{s}\right)\left(2.00\text{ hr} \times \frac{3600\text{ s}}{1\text{ hr}}\right) + \ln(0.500)$$

$$\ln[N_2O_5]_t = -4.55 + (-0.693) = -5.24$$

Taking the antiln of $\ln[N_2O_5]$ gives

$$[N_2O_5] = \text{antiln}(-5.24) = 5.30 \times 10^{-3}M$$

This represents the concentration of N_2O_5 after 2.00 hr has elapsed and when the initial concentration is 0.500 M. (b) The initial concentration is not stated and you can assume it is as in problem (a). The information that 90.0% of N_2O_5 has decomposed means that 10.0% remains, or $(0.100)(0.500M\ N_2O_5) = 0.0500M\ N_2O_5$ remains. You can substitute this data into the equation relating time and concentration and solve for t:

$$\ln[0.0500] = -(6.32 \times 10^{-4}\,s^{-1})t + \ln[0.500]$$

$$-2.996 = -(6.32 \times 10^{-4}\,s^{-1})t + (-0.693)$$

Solving for t gives:

$$t = \frac{-2.996 + 0.693}{-6.32 \times 10^{-4}\,s^{-1}} = 3.64 \times 10^3\text{ s}$$

Thus, the time it takes for 90.0% of N_2O_5 to disappear is 3640 s or 1.01 hours.

Comment: The statement in (**b**) that it is not necessary to know the initial concentration can be shown by the following analysis. We can rearrange the equation for first-order reactions to the following form:

$$\ln\left[\frac{[N_2O_5]}{[N_2O_5]_0}\right] = -kt$$

If the final concentration is 10% of the initial concentration we can then write:

$$\ln\left[\frac{(0.10)[N_2O_5]_0}{[N_2O_5]_0}\right] = -kt$$

or, canceling the initial concentration of N_2O_5 gives $\ln[0.10] = -kt$. We see that the initial concentration of N_2O_5 cancels irrespective of the percent of N_2O_5 that decomposes.

EXERCISE 9   Determining whether a chemical reaction is first-order or second-order

A particular reaction is either first-order or second-order. How can you determine which one it is?

SOLUTION: Measure the half-life of the reaction. If the reaction is first-order, its half-life will be independent of concentration; if its half-life decreases with increasing concentration, it is second-order. Alternately, you can use a graphing procedure to distinguish between the two. To determine if the rate-law expression is a first- or second-order reaction, first graph both $\ln[A]_t$ and $1/[A]_t$ against t. If the reaction is first-order, the graph of $\ln[A]_t$ versus t will be a straight line, with a slope of $-k$ and an intercept of $\ln[A]_0$. On the other hand, if it is a second-order reaction, the graph of $1/[A]_t$ against t will be a straight line with a slope of k and an intercept of $1/[A]_0$.

Energy must be available to the reactants in a chemical reaction so that their chemical bonds can undergo rearrangements to form products. There will be no reaction unless a minimum amount of energy, referred to as **activation energy** (E_a), is provided to the reactants to overcome the energy barrier between reactants and products.

REACTION RATES: TEMPERATURE AND ACTIVATION ENERGY

- Reactant molecules gain energy from their collisions with other molecules. During collisions between molecules, kinetic energy is transferred to potential energy in chemical bonds. If sufficient kinetic energy is transferred to the chemical bonds of reactant molecules—that is, an amount of potential energy equal to or greater than the activation energy for the reaction—products may form.
- Even if the reactant molecules have gained sufficient activation energy, there may be no reaction unless the molecules are properly aligned during their collisions.
- If a collision between reacting molecules is successful in initiating the reaction, the reacting molecules form a transitory intermediate termed an **activated complex** or **transition state**. An activated complex has the maximum energy of any species formed during the reaction pathway of reactants going to products.
- The rate at which an activated complex forms (which determines the rate of the reaction) depends on both the activation energy for the reaction and the number of collisions that are effective in producing products.

Svante Arrhenius determined a relationship between the rate constant for a reaction and the activation energy required for the reaction to occur. Most reaction-rate data obey the **Arrhenius equation**:

$$\ln k = \ln A - \frac{E_a}{RT}$$

where A is a constant termed the frequency factor, R is the gas constant, T is absolute temperature, k is the rate constant, and E_a is the activation energy.

- Note that as the value of activation energy increases, the value of $\ln k$ decreases, and therefore, the value of k decreases. A lower value of k corresponds to a smaller rate of reaction.
- E_a is a positive quantity (except for some free-radical reactions). Therefore, with increasing temperature the term $\frac{E_a}{RT}$ decreases in magnitude and the value of k increases. This means that the rate of a reaction increases with increasing temperature.
- Another form of the Arrhenius equation relates two rate constants at two different temperatures, T_1 and T_2:

$$\ln\frac{k_1}{k_2} = \frac{E_a}{R}\left(\frac{1}{T_2} - \frac{1}{T_1}\right)$$

This is the most useful form when working problems.

Note: E_a and R must have the same units—convert E_a to joules because $R = 8.314\ \text{J/K-mol}$.

EXERCISE 10   Determining the Arrhenius activation energy for a chemical reaction

What is the Arrhenius activation energy for the decomposition of N_2O_5 given the following data: at 35 °C, $k_1 = 6.60 \times 10^{-5}\ \text{s}^{-1}$, and at 65 °C, $k_2 = 2.40 \times 10^{-3}\ \text{s}^{-1}$.

SOLUTION: *Analyze*: We are given two rate constants at different temperatures for N_2O_5 and asked to calculate the Arrhenius activation energy.

Plan: We are given rate-constant data at two different temperatures, therefore we can use the following equation to solve for the activation energy:

$$\ln\frac{k_1}{k_2} = \frac{E_a}{R}\left(\frac{1}{T_2} - \frac{1}{T_1}\right)$$

Solve: We can solve for E_a by substituting the given data

$$k_1 = 6.60 \times 10^{-5}\ \text{s}^{-1}\quad T_1 = 308\ \text{K}$$
$$k_2 = 2.40 \times 10^{-3}\ \text{s}^{-1}\quad T_2 = 338\ \text{K}$$
$$R = 8.314\ \text{J/K-mol}$$

into the Arrhenius equation:

$$\ln\frac{6.60 \times 10^{-5}\ \text{s}^{-1}}{2.40 \times 10^{-3}\ \text{s}^{-1}} = \frac{E_a}{8.314\ \text{J/K-mol}}\left(\frac{1}{338\ \text{K}} - \frac{1}{308\ \text{K}}\right)$$

Solving for E_a gives

$$E_a = \frac{(8.314\ \text{J/K-mol})(\ln 2.75 \times 10^{-2})}{\left(\dfrac{1}{338\ \text{K}} - \dfrac{1}{308\ \text{K}}\right)}$$

$$= 1.04 \times 10^5\ \text{J/mol} = 104\ \text{kJ/mol}$$

Comment: When you work with the Arrhenius equation the following relation may be useful when solving problems:

$$\frac{1}{T_2} - \frac{1}{T_1} = \frac{T_1 - T_2}{T_1 T_2}.$$

EXERCISE 11 Interpreting a graph of energy change versus progress of a chemical reaction

Figure 14.2 below represents energy changes that might accompany a particular chemical reaction. Referring to this figure, answer the following questions: **(a)** Points 1, 2, and 3 correspond to the energy of which species in the reaction? **(b)** To what energy changes do *A* and *B* correspond? **(c)** Is the reaction exothermic or endothermic?

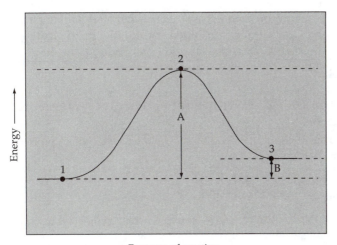

▲ **FIGURE 14.2** Graph of energy change during a chemical reaction.

SOLUTION: *Analyze*: We are given a graph of energy change versus progress of a chemical reaction. Using the graph we are asked to characterize species at points 1, 2, and 3; the types of energy changes associated with A and B; and the type of energy change for the overall chemical reaction.

Plan: We can characterize species 1, 2, and 3 by recognizing species 1 exists at the start of the chemical reaction, species 2 is at the maximum energy in the reaction pathway, and species 3 is formed at the completion of the chemical reaction. We can characterize the types of energy change associated with A by recognizing that energy is required to reach species 2 and species 2 is at the maximum energy in the pathway. That energy change associated with B is simply the difference between the energies of the product and reactant. An exothermic energy change occurs if the energy state of the final state is lower than that of the initial state, and endothermic if the opposite is true.

Solve: **(a)** Point 1 corresponds to the energy of the reactants. Point 2 corresponds to the energy of the activated complex with an activation energy equal to E_a. Point 3 corresponds to the energy of the products of the reaction. **(b)** Energy change A corresponds to the activation energy, E_a. Energy change B corresponds to the net energy absorbed during the reaction. **(c)** Since energy is absorbed (as shown by the products being at a higher energy level than the reactants), this particular reaction is endothermic.

REACTION MECHANISMS

The rate law of a reaction does not depend on the stoichiometry of the reaction, but rather on the reaction mechanism. A **reaction mechanism** is a proposed sequence of elementary reactions which, when added together, result in the overall chemical equation. An example of a proposed reaction mechanism containing two elementary reactions is that for the conversion of ozone into oxygen:

$$O_3(g) \longrightarrow O_2(g) + O(g) \tag{1}$$

$$\underline{O_3(g) + O(g) \longrightarrow 2\,O_2(g)} \tag{2}$$

$$2\,O_3(g) \longrightarrow 3\,O_2(g) \tag{3}$$

- Step 1 is **unimolecular** because the elementary reaction involves only a single reactant.
- Step 2 is **bimolecular** because the elementary reaction involves two reactant molecules.
- $O(g)$ is an **intermediate** because it is neither a reactant nor a product of the overall reaction but is consumed and formed in an elementary reaction.
- Note that when reactions (1) and (2) are added, the overall reaction (3) is obtained.

EXERCISE 12   Writing rate laws given elementary processes

What is the molecularity of each of the following elementary processes? Write the rate law for each.

(a) $CO(g) + Cl_2 \longrightarrow COCl_2(g)$
(b) $HClI(g) \longrightarrow HCl(g) + I(g)$

SOLUTION: *Analyze*: We are given two different elementary reactions and asked to determine the molecularity and the rate law for each.

Plan: The molecularity of an elementary process is determined by the number of molecules that are reactants. A rate law for an elementary reaction is dependent only on the concentrations of reactants, each with an order reflecting its stoichiometry in the reaction.

Solve: (a) The molecularity is two because two reactants are present. The rate law is: Rate $= k[CO][Cl_2]$

(b) The molecularity is one because only one reactant is present. The rate law is: Rate $= k[HClI]$

EXERCISE 13   Relating a reaction mechanism to the rate law

For the reaction

$$2\,H_2(g) + 2\,NO(g) \longrightarrow N_2(g) + 2\,H_2O(g)$$

show that the proposed mechanism

(1) $\qquad\qquad 2\,NO(g) \underset{k_{-1}}{\overset{k_1}{\rightleftharpoons}} N_2O_2(g)$ (fast)

(2) $\quad N_2O_2(g) + H_2 \overset{k_2}{\longrightarrow} H_2O_2(g) + N_2(g)$ (slow, rate determining)

(3) $\qquad\quad H_2(g) + H_2O_2(g) \underset{k_{-3}}{\overset{k_3}{\rightleftharpoons}} 2\,H_2O$ (fast)

is consistent with the experimental rate law

(4) $\qquad\qquad\qquad$ Rate $= k[H_2][NO]^2$

SOLUTION: *Analyze*: We are given a three-step mechanism for a reaction and asked to show that the mechanism is consistent with Rate $= k[H_2][NO]^2$.

Plan: We first examine the rate-determining step and write a rate law based on it. It should be consistent with the given rate law. If the rate law contains an intermediate we can use prior fast steps in the mechanism to develop a relationship between the concentration of the intermediate and that of a reactant. Fast steps in a mechanism are assumed to reach equilibrium quickly.

Solve: If we write the rate law based only on the rate-determining step, the result is

(5) $$\text{Rate} = k_2[N_2O_2][H_2]$$

This rate law is not acceptable because N_2O_2 is an intermediate—it does not appear in the overall reaction. Therefore we need to look at one of the fast steps in the mechanism that will enable us to represent the concentration of N_2O_2 in terms of an original reactant. If we assume that the forward and reverse rates in step 1 are both fast and are approximately equal in value, then we can write

$$\text{Rate}_1 = k_1[NO]^2 = \text{rate}_{-1} = k_{-1}[N_2O_2]$$

Solving for $[N_2O_2]$ yields

(6) $$[N_2O_2] = \frac{k_1[NO]^2}{k_{-1}}$$

Substituting the expression for $[N_2O_2]$ in equation (6) into equation (5) gives

$$\text{Rate} = k_2\left(\frac{k_1[NO]^2}{k_{-1}}\right)[H_2] = \frac{k_2k_1}{k_{-1}}[H_2][NO]^2$$

If we define

$$k = \frac{k_2k_1}{k_{-1}}$$

then we have the experimentally observed rate law

$$\text{Rate} = k[H_2][NO]^2$$

CATALYSIS

Wilhelm Ostwald was the first to define a **catalyst** in terms of its influence on the rate of a chemical reaction: "A catalyst is a substance that changes the velocity of a chemical reaction without itself appearing in the end products."

- A catalyst increases the rate of a chemical reaction by providing a new reaction path with a lower activation energy.
- The concentration of a catalyst may appear in the rate law for a reaction.
- **Homogeneous catalysts** have the same phase as the reaction medium.
- A catalyst in a different phase than the reaction mixture is a **heterogeneous catalyst**. Heterogeneous catalysis usually involves the adsorption of a reactant molecule onto the surface of the heterogeneous catalyst at an active site.

EXERCISE 14 Determining if a catalytic process is homogeneous or heterogeneous

The primary process for the manufacturing of H_2SO_4 is the contact process. Sulfur is oxidized to $SO_2(g)$ in the presence of $O_2(g)$. $SO_2(g)$ is then oxidized to $SO_3(g)$ in the presence of $O_2(g)$ and the catalyst $V_2O_5(s)$ at 400 °C. $SO_3(g)$ is then reacted with $H_2O(l)$ to form $H_2SO_4(l)$. Is the oxidation of SO_2 to SO_3 an example of homogeneous or heterogeneous catalysis?

SOLUTION: *Analyze*: We are given the phases of the reactants, intermediates, and products in the commercial synthesis of sulfuric acid. Using this information we are asked to determine if the reaction involves homogeneous or heterogeneous catalysis.

Plan: We can recognize the type of catalysis using the definitions of homogeneous and heterogeneous.

Solve: In homogeneous catalysis, the catalyzed reaction occurs in one phase; in heterogeneous catalysis, the catalyzed reaction occurs at a phase interface. The catalyzed reaction in this problem involves $SO_2(g)$ and the solid catalyst V_2O_5. The catalysis occurs at the surface of the solid V_2O_5 and is thus an example of heterogeneous catalysis.

EXERCISE 15   Explaining why a species may appear in a rate law when it is not a reactant in the overall chemical reaction

The hydrolysis of sucrose (table sugar) in an acid solution is described by the chemical equation

$$C_{12}H_{22}O_{11}(aq) + H_2O(l) \longrightarrow C_6H_{12}O_6(aq) + C_6H_{12}O_6(aq)$$
$$\text{Sucrose} \qquad\qquad\qquad \text{Glucose} \qquad \text{Fructose}$$

The rate law for the reaction is

$$-\frac{\Delta[C_{12}H_{22}O_{11}]}{\Delta t} = k[C_{12}H_{22}O_{11}][H^+]$$

Why does $[H^+]$ appear in the rate law?

SOLUTION: *Analyze*: We are given a rate law with $[H^+]$ in the expression and asked to explain why it appears in the expression when H^+ is not a reactant in the overall chemical reaction.

Plan: A species appearing in a rate law is involved in a step in the reaction mechanism. We need to consider how H^+ could appear in the reaction mechanism.

Solve: An intermediate species in a reaction mechanism does not appear in the overall chemical reaction; however, it may appear in the rate expression if it is a catalyst. Therefore, H^+ must be a catalyst.

SELF-TEST QUESTIONS

Key Terms

Having reviewed key terms in Chapter 14, match key terms with phrases and identify statements as true or false. If a statement is false, indicate why it is incorrect.

Match each phrase with the best term:

14.1 The term given to an expression having the form: rate = $k[A]^n$.

14.2 The name given to the exponent in the expression: rate = $k[HI]^2$.

14.3 The name given to k in the previous expression.

14.4 The name given to an expression such as $-\dfrac{\Delta[Br_2]}{\Delta t}$.

14.5 The type of reaction describing $2\,N_2O_5(g) \longrightarrow 4\,NO_2(g) + O_2(g)$ given that Rate = $k[N_2O_5]$.

14.6 The value of this term is 6.9×10^2/s if the rate constant for a first-order chemical reaction is 1.0×10^{-3} s.

14.7 The highest energy form of a species in the chemical reaction between $NO(g)$ and $O_3(g)$.

14.8 A chemical reaction that has a very large rate constant should be expected to have a relatively low value of this energy.

14.9 This equation can be used to show that as the temperature of a chemical reaction increases as the value of the rate constant increases.

14.10 If this information is known it is possible to develop a rate law expression for a chemical reaction.

14.11 The term given to the slowest step in the proposed reaction mechanism for a chemical reaction.

14.12 This substance affects the rate of a chemical reaction by lowering the activation energy.

14.13 A substance that speeds up a biochemical reaction but is not shown in the overall chemical reaction.

14.14 The type of catalyst HCl(*aq*) is in the hydrolysis of sucrose(*aq*).

14.15 The type of catalyst $MnO_2(s)$ is in the decomposition of molten $KClO_3$.

Terms:

(a)	activated complex	(i)	homogeneous
(b)	activation energy	(j)	rate constant
(c)	Arrhenius	(k)	rate determining
(d)	catalyst	(l)	rate law
(e)	enzyme	(m)	reaction mechanism
(f)	first-order	(n)	reaction order
(g)	half-life	(o)	reaction rate
(h)	heterogeneous		

True-False Statements:

14.16 The rate of decomposition of NO(*g*) to yield $N_2O(g)$ and $NO_2(g)$ at 100 atm of pressure follows a third-order rate law:

$$\text{Rate} = k[\text{NO}]^3$$

From the rate law we can conclude that the decomposition of NO(*g*) requires a *termolecular* reaction.

14.17 A plausible reaction mechanism for the decomposition of $N_2O(g)$ to $N_2(g)$ and $O_2(g)$ is

(a) $N_2O(g) \longrightarrow N_2(g) + O(g)$
(b) $N_2O(g) + O(g) \longrightarrow N_2(g) + O_2(g)$

Given this plausible mechanism we can conclude that $N_2(g)$ is an *intermediate*.

14.18 Reaction (a) shown in problem 14.17 is an example of an *elementary* reaction.

14.19 Reaction (a) shown in problem 14.17 is a *bimolecular* reaction.

14.20 Neither reaction (a) nor reaction (b) in problem 14.17 is an example of a *unimolecular* reaction.

14.21 The *chemical kinetics* of a reaction gives us information about the spontaneous direction of a reaction.

14.22 For a first-order reaction, the *instantaneous rate* increases with time of reaction.

14.23 The rate law for the reaction of NH_4^+ with NO_2^- is first order in each reactant. Therefore, the *overall order* is two.

14.24 The magnitude of the *frequency factor* in the Arrhenius equation is related to the probability that collisions are favorably oriented for reaction.

14.25 According to the *collision model*, reaction rates do not vary significantly with changes in the number of reactant molecules.

14.26 The *transition state* for a chemical reaction is the energy barrier between the starting molecule and the highest energy along the reaction pathway.

14.27 The *molecularity* for the following elementary step in a reaction mechanism

$$NO_2(g) + NO_2(g) \longrightarrow NO_3(g) + NO(g)$$

is four, the number of molecules reacting and formed.

14.28 *Adsorption* refers to the uptake of molecules into the interior of another substance.

14.29 A particular catalyst with a large number of *active sites* will be more effective than one with fewer.

14.30 A *substrate* may change the structural features around an active site when it binds.

14.31 In the *lock-and-key model* for the specificity of enzymes, the substrate is the key.

Problems and Short-Answer Questions

14.32 The following figure shows plots for two first-order chemical reactions. Each is a decomposition reaction, one involving species A and the other species B, at the same temperature. The initial concentrations of A and B are the same. Which reaction has the greater average rate? Explain.

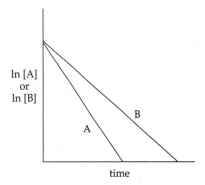

14.33 Critique the following statement: A rate for a chemical reaction is constant over time.

14.34 The following figure shows a plot of *k*, the rate constant, versus temperature, T(K), for a chemical reaction. What relationship does this curve show? What does it tell you about the rate of the reaction? Explain.

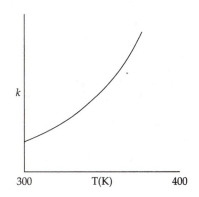

14.35 For the reaction

$$2\,H_2(g) + 2\,NO(g) \longrightarrow N_2(g) + 2\,H_2O(g)$$

(a) Write a relationship between the rate of the disappearance of $NO(g)$ and the appearance of $N_2(g)$.

(b) What is the rate law if the reaction is first order with respect to H_2 and second order with respect to NO?

14.36 There exists a proportionality constant between a reaction rate and concentrations.

(a) What is its name and symbol?

(b) Is its set of units constant?

(c) How does its value change with temperature?

14.37 Using the rate law expression you found in problem 14.35 predict the effect if

(a) the concentration of NO is doubled;

(b) the concentration of NO is reduced by two-thirds;

(c) the temperature is reduced.

14.38 (a) Does the half-life of a first-order chemical reaction depend on concentrations? Explain.

(b) By what factor does the concentration of a reactant in a first-order reaction decrease if it goes through three half-lives?

14.39 What leads to effective collisions between reacting molecules in the gas phase and thus leads to products? Are most collisions effective at room temperature?

14.40 A free radical is a species that contains an unpaired electron. For example, a free hydrogen atom contains one electron and thus is a free radical. Why do you think a number of free-radical reactions have activation energies that are very small and in a few cases zero? What is the expected product when two radicals react?

14.41 What is the role of a catalyst in a chemical reaction? How does it affect the reaction mechanism? How is the equilibrium state affected by a catalyst? If a reaction is thermodynamically unfavorable, can a catalyst change this condition? How might a rate law give you a clue that a catalyst is operating?

14.42 Compare the likelihood of a bimolecular elementary step to a termolecular elementary step occuring in a reaction mechanism. If the following elementary reaction is the rate-determining step, what is the rate expression for it: $A + B \longrightarrow C$? Is this rate expression the same as for the overall chemical reaction?

14.43 Carbon tetrachloride can be obtained by adding chlorine to chloroform, $CHCl_3$. The gas-phase reaction mechanism involves first the formation of a chlorine radical from chlorine followed by radical abstraction of a hydrogen radical from $CHCl_3$. The final step in the reaction mechanism is a reaction of two radicals to form carbon tetrachloride. From this information write the mechanism and the overall chemical reaction.

14.44 Given the following kinetic data for the reaction

$$2\,A + B + C \longrightarrow 2\,D$$

$[A]_{initial}(M)$	$[B]_{initial}(M)$	$[C]_{initial}(M)$	Initial rate (Ms^{-1})
0.1	0.1	0.1	x
0.2	0.1	0.1	$4x$
0.2	0.2	0.1	$32x$
0.2	0.1	0.2	$8x$

(a) Determine the rate law as a function of the concentrations of A, B, and C.

(b) What is the overall order of the reaction?

14.45 Given the following data for the reaction

$$CF_4(g) + H_2(g) \longrightarrow CHF_3(g) + HF(g)$$

determine the rate law for the reaction.

$[CF_4]_{initial}(M)$	$[H_2]_{initial}(M)$	Rate (Ms^{-1})
0.10	0.10	45
0.15	0.10	67.5
0.20	0.20	180
0.20	0.30	270

14.46 An alteration in the structure of a certain virus follows first-order kinetics with an activation energy of 587 kJ/mol. The half-life of the reaction at 29.6 °C is 1.62×10^4 s (1 yr = 3.154×10^7 s). What are the rate constant at 29.6 °C and the half-life at 32.0 °C for the alteration of structure of the virus?

14.47 The decomposition of $CrCl^{2+}$ in water occurs as follows:

$$CrCl^{2+} \longrightarrow Cr^{3+} + Cl^{-}$$

The process is first order, and the rate constant for the reaction is $2.85 \times 10^{-7}\,s^{-1}$. If Cl^{-} is removed from the reaction as it forms so that it cannot recombine with Cr^{3+} and if 10 g of $CrCl^{2+}$ is present initially in 1 L of solution, how many grams of $CrCl^{2+}$ are left after 10 hr?

14.48 The rate law for the reaction

$$2\,NO(g) + O_2(g) \longrightarrow 2\,NO_2(g)$$

is

$$-\frac{\Delta[O_2]}{\Delta t} = k[NO]^2[O_2]$$

The proposed mechanism is

$$NO(g) + O_2(g) \longrightarrow NO_3(g)$$
$$NO(g) + NO_3(g) \longrightarrow 2\,NO_2(g)$$

(a) The reaction mechanism does not include a collision among three molecules. Since the reaction order is three, shouldn't the mechanism include such a collision? Explain.

(b) Which step in the proposed mechanism is the slowest step?

(c) What is the molecularity of each step?

14.49 The decomposition of HI(g) to H$_2$(g) and I$_2$(g) is catalyzed by platinum. The observed rate equation is

$$\text{Rate} = k[\text{HI}][\text{surface area of Pt}]$$

(a) Is this an example of homogeneous or heterogeneous catalysis?

(b) Why is the surface area of Pt in the rate law?

14.50 The decomposition of nitrogen gas is catalyzed by metallic iron. It is believed that nitrogen undergoes chemisorption at the surface of iron. Is this sufficient information to conclude that the chemisorption on the catalytic surface most likely involves splitting the nitrogen molecule as follows:

$$:N::::::N:$$
$$\vdots \qquad \vdots$$
$$\text{----Fe----Fe----}$$

If not, what additional information is needed and why?

14.51 The rate constant for the reaction of NO gas with oxygen gas to form nitrogen dioxide gas at 80 K has a rate constant that is less than one. Is this sufficient information to conclude that the reaction is exothermic? If not, what additional information is needed and why?

Integrative Questions

14.52 (a) Calculate the activation energy for the decomposition reaction of NO$_2$:

$$2\,\text{NO}_2(g) \longrightarrow 2\,\text{NO}(g) + \text{O}_2(g)$$

given that the rate constant at 300 °C is $5.7 \times 10^{-1}/M\text{-s}$. and at 230 °C it is $2.7 \times 10^{-2}/M\text{-s}$.

(b) NO$_2$ can also form a bond between two nitrogen atoms to form N$_2$O$_4$:

$$2\,\text{NO}_2(g) \longrightarrow \text{N}_2\text{O}_4(g)$$

For this reaction, $k = 5.2 \times 10^8/M\text{-s}$ at both 300 K and 325 K. What is the activation energy?

(c) Explain using Lewis structures the difference in magnitude between the two activation energies.

14.53 (a) In this chapter rates are usually expressed in units of M/s. The rates of gaseous reactions can also be expressed in units of atm/s. Using the ideal-gas law show why atm/s is a valid unit for rate.

(b) The half-life for the first-order decomposition of *di-t*-butylperoxide(DTBP), a catalyst used in the manufacturing of polymers, is $4.8 \times 10^2\,s$ at 147 °C. If a reaction vessel contains pure DTBP at a pressure of 925 torr at 147 °C, how long will it take for the pressure to decrease to 115 torr?

14.54 Answer the following questions:

(a) What relationship exists between the enthalpy of an endothermic chemical reaction and the activation energy? Which energy of the two will be a larger quantity and why?

(b) The rate law for the dimerization of NO$_2$(g) is rate $= k[\text{NO}_2]^2$. What effect will doubling the volume of the container in which the reaction is occurring have upon the value of the specific rate constant, k?

(c) Will running the dimerization reaction of NO$_2$ in a CCl$_4$ solution affect the value of k?

(d) A single-step reaction has an activation energy that is the same for both the forward and reverse directions. What conclusion can you make about ΔH if the reaction is done under constant temperature and pressure conditions?

14.55 Construct an approximate energy-profile diagram for the decomposition of HI(g) to its elements in the gas state at 25 °C. The heat of formation of HI(g) is 26.5 kJ/mol at 25 °C. What assumption must you make when you construct the diagram?

Multiple-Choice Questions

14.56 Given that a reaction is exothermic and has an activation energy of 50 kJ/mol, which of the following statements are correct: (1) the reverse reaction has an activation energy equal to 50 kJ/mol; (2) the reverse reaction has an activation energy less than 50 kJ/mol; (3) the reverse reaction has an activation energy greater than 50 kJ/mol; (4) the reaction rate increases with temperature; (5) the reaction rate decreases with temperature?

(a) (1) and (4)   (d) (2) and (5)

(b) (2) and (4)   (e) (3) and (5)

(c) (3) and (4)

14.57 A catalyst increases the rate of a reaction by doing which of the following?

(a) increasing reactant concentrations;

(b) increasing temperature;

(c) decreasing temperature;

(d) increasing activation energy of reaction;

(e) decreasing activation energy of reaction.

14.58 Given the following mechanism for a reaction:

$$\text{Cl}_2(g) \longrightarrow 2\,\text{Cl}(g)$$
$$2\,\text{NO}(g) + 2\,\text{Cl}(g) \longrightarrow \text{N}_2(g) + 2\,\text{ClO}(g)$$
$$2\,\text{ClO}(g) \longrightarrow \text{Cl}_2(g) + \text{O}_2(g)$$

which of the following is a catalyst in the reaction?

(a) Cl$_2$ (d) ClO

(b) N$_2$O$_5$ (e) O$_2$

(c) N$_2$

14.59 What can we say about catalysts, given the following information?

$$\text{CO}(g) + 3\,\text{H}_2(g) \xrightarrow{\text{Ni catalyst}} \text{CH}_4(g) + \text{H}_2\text{O}(g)$$

$$\text{CO}(g) + 2\,\text{H}_2(g) \xrightarrow[\text{catalyst}]{\text{ZnO Cr}_2\text{O}_3} \text{CH}_3\text{OH}(g)$$

(a) Catalysts are nonspecific in their activity.
(b) Catalysts are highly specific in their activity.
(c) Ni is a better catalyst than ZnO/Cr_2O_3.
(d) ZnO/Cr_2O_3 is a better catalyst than Ni.
(e) Most metals can act as catalysts.

14.60 What are the units of k for the rate law

$$Rate = k[A][B]^2$$

when the concentration unit is mol L^{-1}?

(a) s^{-1} (d) $L^2\,mol^{-2}\,s^{-1}$
(b) s   (e) $L^2 - s^2\,mol^{-2}$
(c) $L\,mol^{-1}\,s^{-1}$

14.61 Given the energy-profile diagram in Figure 14.3 below which of the following phrases could correctly describe the kind of reaction shown: (1) an exothermic reaction; (2) an endothermic reaction; (3) a reaction with two steps in its reaction mechanism; (4) a reaction with three steps in its reaction mechanism; (5) a catalyzed reaction involving two successive steps?

(a) (1) and (3)   (d) (2), (4), and (5)
(b) (2) and (4)   (e) (1), (2), (3), (4), and (5)
(c) (1), (3), and (5)

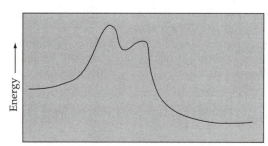

▲ **FIGURE 14.3** An energy-profile diagram.

14.62 Figure 14.4 is an energy-profile diagram for two reactions, A and B. Which of the following statements about the reactions are correct: (1) reaction B is slower than

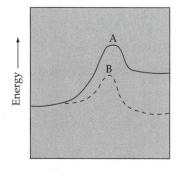

Reaction pathway

▲ **FIGURE 14.4** An energy-profile diagram for two reactions.

reaction A; (2) reaction A is slower than reaction B; (3) the reverse reaction of B has a higher activation energy than that for the reverse reaction of A; (4) the reverse reaction of A has a higher activation energy than that for the reverse reaction of B; (5) the energy-profile diagram for reaction B could represent a catalyzed reaction A?

(a) (2) and (3)   (d) (1), (4), and (5)
(b) (2) and (4)   (e) (2), (3), and (5)
(c) (1) and (3)

14.63 The rate law for the reaction between $NO(g)$ and $H_2(g)$ is rate $= k[H_2][NO]^2$. Which statement is true based on the rate law?

(a) $NO(g)$ is consumed twice as fast as $H_2(g)$
(b) The rate determining step involves a three-body collision
(c) If the concentration of NO is doubled the rate doubles
(d) The reaction is third order
(e) The chemical reaction is
$$H_2(g) + 2\,NO(g) \longrightarrow 2\,HO(g) + N_2(g)$$

SELF-TEST SOLUTIONS

14.1 (l). **14.2** (n). **14.3** (j). **14.4** (o). **14.5** (f). **14.6** (g). **14.7** (a). **14.8** (b). **14.9** (c). **14.10** (m). **14.11** (k). **14.12** (d). **14.13** (e). **14.14** (i). **14.15** (h). **14.16** False. A third-order rate law by itself does not prove that the reaction is termolecular. In fact, reaction among three NO molecules in the gas phase is highly unlikely. **14.17** False. An intermediate is a species in the reaction mechanism that is neither a reactant nor a product of the overall reaction. Thus $N_2(g)$ is not an intermediate because it is a product of the overall reaction. $O(g)$ is an example of an intermediate. **14.18** True. **14.19** False. It is a unimolecular reaction because in the reaction only one molecule of reactant is shown to form products. **14.20** False. See answer to problem 14.19. **14.21** False. Chemical kinetics gives us information about the speeds or rates of chemical reactions, not the most likely direction. **14.22** False. Instantaneous rate is greatest at the start of a chemical reaction. **14.23** True. **14.24** True. **14.25** False. Reaction rates increase with increasing number of reactant molecules. **14.26** False. Transition state refers to the chemical species that forms at the highest entry along the reaction pathway. **14.27** False. It is two, the number of molecules reacting in the elementary step. **14.28** False. It refers to the binding of molecules to the surface of another species. **14.29** True. **14.30** True. **14.31** True.

14.32 The average rate of each reaction is $-\dfrac{\Delta[A]}{\Delta t}$ or $-\dfrac{\Delta[B]}{\Delta t}$.

We can see from the graph that for the same time interval the change in concentration of A is greater than for B. Thus, the average rate for A is greater.

14.33 The shape of the curve is a hyperbola. A hyperbola is represented by the equation: $y = e^x$. The Arrhenius equation has this same form: $k = Ae^{-E_a/RT}$. As the temperature increases the negative exponent becomes a smaller number. This results in a larger value of k. The larger the value of k, the greater the rate.

14.34 The statement is true only for zero-order reactions. The reaction rate decreases as the chemical reaction proceeds for all other cases studied in the text.

14.35 (a) A relationship between rates requires the rates be equal:

$$\text{rate} = -\frac{1}{2}\frac{\Delta[NO]}{\Delta t} = \frac{\Delta[N_2]}{\Delta t}$$

The factor of $\frac{1}{2}$ for NO is necessary to make the rates equal. For every two molecules of NO that disappears, only one molecule of N_2 appears.

(b) rate $= k[NO]^m[H_2]^n = k[NO]^2[H_2]$. m and n represent the orders with respect to each reactant.

14.36 (a) k: It is the specific rate constant.

(b) The general form of a reaction rate law expression is rate $= k[A]^a[B]^b\ldots$ in which the rate has units of concentration/time and the concentrations have units of molarity (solution) or atmospheres (gas). Thus,

$$k = \frac{\text{rate}}{[A]^a[B]^b\ldots}.$$

The units of the denominator will change depending on the number of different species and their orders. Thus, the set of units for k depends on the form of the rate law expressions.

(c) $k = Ae^{-Ea/RT}$ is the Arrhenius relationship. Note that as the value of the temperature increases, the value of k also increases; if the temperature decreases, the value of k decreases. E_a is a positive quantity or it is essentially zero for a very few reactions.

14.37 (a) The rate expressions is $= k[NO]^2[H_2]$. Note that doubling the concentration results in quadrupling the rate: $[2 \times \text{concentration}]^2 = 4 \times \text{concentration}^2$.

(b) If the concentration is reduced by two-thirds this means only one-third of the original amount exists. Thus, $[1/3 \times \text{concentration}]^2 = 1/9 \times \text{concentration}^2$. The rate is reduced by 8/9ths.

(c) The reaction rate is reduced as discussed in the solution to the previous problem.

14.38 (a) The half-life of a first-order reaction does not depend on concentration, it depends on the specific rate constant:

$$t_{1/2} = \frac{0.693}{k}$$

(b) The concentration decreases by one-half for every half-life step. Therefore, if the reaction goes through three half-lives, the reduction factor is $\frac{1}{2} \times \frac{1}{2} \times \frac{1}{2}$ or $(1/2)^3$.

14.39 Molecules in the gas phase are in constant motion. The average speed is a constant at a fixed temperature but there is a distribution of speeds. To form products the molecules must collide and undergo energy changes. The frequency of collisions depends on their relative concentrations. The colliding molecules must approach each other in a proper orientation (e.g., head-on versus side-on). Some orientations require more energy than others. The molecules must have sufficient kinetic energies to overcome repulsions of electrons in the molecules. If the repulsions are too large, the molecules will fly away from one another. When the molecules collide and form a transition state they are able to redistribute potential and kinetic energy. Finally, as the new molecules separate, the distribution of electrons and bonds must result in a stable molecular structure. Most collisions are not effective at room temperature.

14.40 Activation energy reflects the energy required to break and rearrange bonds between colliding molecules before any products form. The free electron of a radical is readily available and need not undergo significant changes in its energy state in order for it to pair with another electron. This results in a lower activation energy. When two radicals react, no bond breaking is necessary and the energy of activation approaches zero. The product of two radicals reacting is nonradical—the two electrons pair to form a bond.

14.41 A catalyst functions by providing an alternative mechanism that has a lower activation energy. This permits the forward and reverse reactions to become equally faster, but it does not change the position of equilibrium. A catalyst cannot cause a thermodynamically unfavorable reaction to become favorable. If you know the reaction rate law for the uncatalyzed reaction, then the one for the catalyzed reaction usually has a different form and the concentration of the catalyst can appear in the rate expression.

14.42 A bimolecular step is relatively common; the collision of two particles is not an unexpected event. A termolecular step requires three particles to simultaneously collide and form a transition state. This is highly unlikely. The rate expression is rate $= k[A][B]$. This expression may be the overall rate expression, but not necessarily. If there are other elementary reactions preceding it in the proposed reaction mechanism, then the form of the overall rate expression will be different.

14.43 The first step in the gas-phase reaction involves forming a radical from chlorine gas

$$Cl_2 \rightleftharpoons 2\,Cl\cdot$$

The next step is removal of a hydrogen radical using the chlorine radical

$$Cl\cdot + CHCl_3\cdot \longrightarrow HCl + CCl_3\cdot$$

The last step is the reaction of two radicals:

$$Cl\cdot + CCl_3\cdot \longrightarrow CCl_4$$

If you add the reactions together, the chlorine radicals and the CCl_3 radicals act as intermediates and disappear. The overall reaction in the gas-phase is

$$Cl_2 + CHCl_3 \longrightarrow HCl + CCl_4$$

14.44 (a) Doubling the concentration of A increases the rate by 4; thus, in the rate law [A] must be squared. Doubling the concentration of B increases the rate by a factor of 8; thus, [B] must have a power of 3($2^3 = 8$) in the rate law. Doubling the concentration of C doubles the rate; thus, [C] must have a power of 1 in the rate law. The rate law is $-\Delta[A]/\Delta t = k[A]^2[B]^3[C]$.

(b) The overall order of the reaction is 6(2 + 3 + 1).
14.45 Assuming that the rate law has the form Rate = $k[CF_4]^n[H_2]^m$, we will use the approach in Exercise 6 to solve this problem:

$$\frac{Rate_1}{Rate_2} = \frac{k[0.10]^n[0.10]^m}{k[0.15]^n[0.10]^m} = \left(\frac{0.10}{0.15}\right)^n$$

$$\frac{45\ Ms^{-1}}{67.5\ Ms^{-1}} = 0.667 = (0.667)$$

Therefore $n = 1$.

$$\frac{Rate_3}{Rate_4} = \frac{k[0.20][0.20]^m}{k[0.20][0.30]^m}$$

$$= \frac{180\ Ms^{-1}}{270\ Ms^{-1}} = 0.667 = (0.667)^m$$

Therefore $m = 1$. The rate law is Rate = $k[CF_4][H_2]$.

14.46 The rate constant at 29.6 °C is calculated using the relation $k_1 = 0.693/t_{1/2} = 0.693/1.62 \times 10^4\ s = 4.28 \times 10^{-5}\ s^{-1}$. The half-life at 32.0 °C is calculated from the rate constant at 32.0 °C, which can be calculated using the Arrhenius equation. E_a is given as 587 kJ/mol, but A is not given. However, k_2 at 32.0 °C can be calculated without knowledge of the value of A using the equation

$$\ln\left(\frac{k_1}{k_2}\right) = \frac{E_a}{R}\left(\frac{1}{T_2} - \frac{1}{T_1}\right)$$

$$\ln\left(\frac{4.28 \times 10^{-5}/sec}{k_2}\right) = \frac{(587\ kJ/mol)(1000\ J/kJ)}{(8.314\ J/K\text{-}mol)}$$
$$\times \left(\frac{1}{305.0\ K} - \frac{1}{302.6\ K}\right)$$

Solving for k_2 gives $k_2 = 2.68 \times 10^{-4}\ s^{-1}$. Thus, the half-life for the alteration of the virus at 32.0 °C is $t_{1/2} = 0.693/k_2$

$$= 0.693/(2.68 \times 10^{-4}\ s^{-1}) = 2.58 \times 10^3\ s.$$

14.47 Use the relation $\ln([A]_t/[A]_0) = -kt$. In this problem we know $[A]_0$, k, and t, are asked to solve for $[A]_t$. Since the ln expression contains a ratio of concentrations, you can use a ratio of gram quantities because the volume and molar mass of $CrCl^{2+}$ cancel.

$$\ln\left(\frac{[A_t]}{10.00\ g}\right) = (-2.85 \times 10^{-7}\ s^{-1})(10\ hr)$$

$$\times \left(\frac{60\ min}{1\ hr}\right)\left(\frac{60\ sec}{1\ min}\right)$$

$$= -0.0103$$

Solving for $[A_t]$ yields $[A_t] = 9.90\ g$. After 10 hr, only 0.10 g of $CrCl^{2+}$ has decomposed.

14.48 (a) No, A collision among three molecules is not likely. The $[NO]^2$ term results because the slowest step in the mechanism depends on [NO] and some other concentration term that is directly related to $[NO][O_2]$.

(b) If the first elementary step were the rate determining step, the rate law would be $k[NO][O_2]$. This does not agree with the observed rate law. Therefore, the second elementary step must be the rate determining step. You can show this by using the approach in Exercise 15.

(c) Both steps are bimolecular.

14.49 (a) It is an example of heterogeneous catalysis, because the catalysis of HI(g) occurs at a solid–gas interface—the surface of solid platinum and the gases.

(b) A surface Pt atom absorbs an HI molecule, enabling the formation of $H_2 + I_2$. The rate of the reaction depends on the number of surface Pt atoms, that is, the surface area of Pt available for adsorbing HI molecules. Thus, the surface area of Pt is involved in the slowest step and occurs in the rate law.

14.50 There is insufficient information. The structure shown is a hypothesis. Unless we can show by experimental methods the actual existence of the adsorbed species we can only state it as a hypothesis.

14.51 We cannot tell anything about the exothermic or endothermic character of a reaction from the rate constant. We can calculate the activation energy if we know two rate constants at different temperatures. To determine if the reaction is exothermic we need to know the enthalpy of the products and reactants: $\Delta H_{rxn} = H_P - H_R$. We can also use a table of enthalpies of formation to calculate the heat of reaction.

14.52 (a) $\ln\dfrac{k_2}{k_1} = \dfrac{Ea}{R}\left(\dfrac{1}{T_1} - \dfrac{1}{T_2}\right)$

$$E_a = \frac{R\ln\left(\dfrac{k_2}{k_1}\right)}{\left(\dfrac{1}{T_1} - \dfrac{1}{T_2}\right)}$$

$$= \frac{8.314\dfrac{J}{mol\text{-}K}\left[\ln\left(\dfrac{5.7 \times 10^{-1}/M\text{-}s}{2.7 \times 10^{-2}/M\text{-}s}\right)\right]}{\left(\dfrac{1}{503K} - \dfrac{1}{573K}\right)}$$

$$E_a = \frac{8.314\dfrac{J}{mol\text{-}K}(\ln 21)}{2.43 \times 10^{-4}/K}$$

$$E_a = 1.0 \times 10^5\ J/mol = 100\ kJ/mol$$

(b) Note that $k_1 = k_2$ at the two temperatures. This means that $\ln(k_2/k_1) = \ln 1 = 0$. Therefore, E_a must equal 0 kJ/mol.

(c) The Lewis structure for NO_2 is

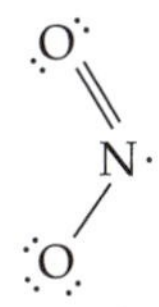

.O.
‖
N·
/
:O.

It is an odd-electron molecule with a single electron that can readily combine with the single electron on another NO_2 molecule to form N_2O_4:

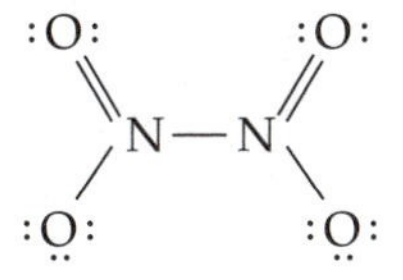

:O: :O:
‖   ‖
N—N
/   \
:O: :O:

Since no bonds are broken to form N_2O_4 the activation energy is zero. When NO_2 decomposes, bonds are broken and made, which requires activation energy.

14.53 (a) The ideal-gas law is $PV = nRT$. The ratio n/V is the same as for molarity. Thus,

$$\frac{n}{V} = \frac{P}{RT}$$

For a given reaction at constant temperature the ratio n/V will vary only with pressure. Therefore, pressure can be used instead of molarity for a concentration unit when gases are involved in a chemical reaction.

(b)

$$k = \frac{0.693}{t_{1/2}} = \frac{0.693}{4.8 \times 10^2/s} = 1.4 \times 10^{-3}/s$$

$$\ln\frac{(A_t)}{(A_0)} = -kt$$

$$\ln\left(\frac{115 \text{ torr}}{925 \text{ torr}}\right) = (-1.4 \times 10^{-3}/s)t$$

$$t = \frac{\ln(0.124)}{-1.4 \times 10^{-3}/s}$$

$$t = 1400 \text{ s}$$

14.54 (a) In an endothermic chemical reaction the products exist in a higher energy state than the reactants. This means that the energy state of the products is closer in energy to the transition state than the reactants. Therefore,

the activation energy will always be a larger quantity than the enthalpy of reaction. See Figure 14.2.

(b) Doubling the volume will change the partial pressure of $NO_2(g)$ and this will change the rate of the chemical reaction; however, k, which is temperature dependent (Arrhenius equation), should not change because the temperature is held constant.

(c) Running the reaction in a solvent rather than in the gas phase is likely to change the mechanism or alter the activation energy. The solvent will interact with the NO_2 and alter such things as the collision rate, the number of effective collisions, and activation energy.

(d) If the activation energy is the same for both the forward and reverse directions this means that there is no energy difference between the reactants and products and $\Delta H = 0$.

14.55 To construct the energy diagram we need to know the sign of the enthalpy of reaction. We can determine this by using the equation: $\Delta H^\circ = \Sigma \Delta H^\circ_f(\text{Products}) - \Sigma \Delta H^\circ_f(\text{Reactants})$. The reaction is $2\,HI(g) \longrightarrow H_2(g) + I_2(s)$. Therefore the standard heat of reaction is $\Delta H = 0\,kJ + 0\,kJ - 2\,\text{mol}(26.5\,kJ/\text{mol}) = -53\,kJ$. The standard enthalpies of formation of elements are zero. If we assume that $\Delta H^\circ \simeq \Delta E^\circ$ then the energy diagram is

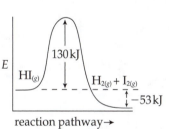

reaction pathway→

14.56 (c) 14.57 (e) 14.58 (a) A catalyst is regenerated during a reaction. $Cl_2(g)$ catalyzes the decomposition of $NO(g)$.

14.59 (b) Changing the catalyst changed the nature of products; thus catalysts are specific in the kinds of products they help form.

14.60 (d)

$$k = \frac{(Ms^{-1})}{(M)(M)^2} = \frac{1}{(M)^2(s)}$$

$$= \frac{L^2}{(\text{mol})^2(s)} = L^2\,\text{mol}^{-2}\,s^{-1}$$

14.61 (c) 14.62 (a) 14.63 (d)

Chapter

15

15.1, 15.2, 15.3, 15.4, 15.5 THE EQUILIB-RIUM STATE: K_p AND K_c

15.6 CALCULATING EQUILIBRIUM CONCENTRATIONS

15.6, 15.7 REACTION QUOTIENT AND LE CHATELIER'S PRINCIPLE

Chemical Equilibrium

OVERVIEW OF THE CHAPTER

Review: Gas laws (10.3, 10.4); gas-law constant (10.4); stoichiometric equivalences (3.6); concept of dynamic equilibrium (11.4).

Learning Goals: You should be able to:

1. Write the equilibrium-constant expression for a chemical system at equilibrium, whether heterogeneous or homogeneous, using the law of mass action.
2. Numerically evaluate K_c from a knowledge of the equilibrium concentrations (or pressure) of reactants or products, or from initial concentrations and the equilibrium concentration of at least one substance in the equilibrium.
3. Interpret the magnitude of K_c and what this tells us about the composition of an equilibrium mixture.
4. Determine the equilibrium constant for a chemical system when it is reversed having been given its equilibrium constant.

Review: Stoichiometric relationships (3.6).

Learning Goal: You should be able to use the equilibrium constant to calculate equilibrium concentrations.

Review: Enthalpy (5.3, 5.4); Catalysis (15.7).

Learning Goal: You should be able to:

1. Calculate the reaction quotient, Q, and by comparison with the value of K_c determine whether a reaction is at equilibrium. If it is not at equilibrium, you should be able to predict in which direction it will shift to reach equilibrium.
2. Explain how the relative equilibrium quantities of reactants and products are shifted by changes in temperature, pressure, or the concentrations of substances in the equilibrium reaction.
3. Explain how the change in equilibrium constant with change in temperature is related to the enthalpy change in the reaction.
4. Describe the effect of a catalyst on a system as it approaches equilibrium.

TOPIC SUMMARIES AND EXERCISES

At constant temperature, a mixture of $H_2(g)$ and $I_2(g)$ in a closed container reacts to form $HI(g)$ until the partial pressures of $H_2(g)$, $I_2(g)$, and $HI(g)$ are constant with time. If the container originally contains $HI(g)$ instead of $H_2(g)$ and $I_2(g)$, $HI(g)$ decomposes until the partial pressures of $H_2(g)$, $I_2(g)$, and $HI(g)$ are constant with time. This reaction is an example of a reversible reaction that leads to an equilibrium state.

- A state of **chemical equilibrium** exists when both forward and reverse reactions are proceeding at equal rates and concentrations of reactants and products, pressure, volume, and temperature do not change with time.
- Double arrows between reactants and products in a chemical reaction show that the reaction proceeds in both a forward and reverse direction. A single arrow is used to indicate a reaction that proceeds essentially to completion.

Another statement of the equilibrium condition is the **law of mass action.**

- This law states that if an equilibrium condition exists for a reaction then there also exists for the reaction a constant K_c called the **equilibrium constant.** Its value depends on a ratio of concentrations of products to concentrations of reactants at equilibrium. More specifically, the equilibrium constant for the general reaction

$$jA + kB \rightleftharpoons pR + qS$$

has the form

Stoichiometric coefficient in chemical equation

$$\text{Equilibrium-constant expression} = K_c = \frac{[R]^p[S]^q}{[A]^j[B]^k} \leftarrow \text{products}$$
$$\leftarrow \text{reactants}$$

- In the equilibrium-constant expression, the concentration of each substance is raised to a power equal to its coefficient in the balanced chemical equation.

K_c is used as the equilibrium constant when concentrations are expressed in units of molarity. When the reactants and products are gases there are two ways of expressing concentration: molarity and partial pressure. If partial pressure is used to express the concentration of a gas then the equilibrium constant is denoted as K_p. In general, the values of K_c and K_c for the same equilibrium involving gases are different. The relationship between the two is:

$$K_p = K_c(RT)^{\Delta n}$$

Δn equals the number of moles of gas products minus the number of moles of gas reactants for the chemical reaction being considered. The number of moles of liquid and solid present in the equilibrium are not included in the calculation of Δn. Only when Δn equals zero does K_p equal K_c.

The phase change of solid water to liquid water at constant temperature and pressure is an example of **heterogeneous equilibrium**—that is, a reversible reaction involving substances in two or more phases.

- At constant temperature, the value of the equilibrium constant for a heterogeneous system depends only on the equilibrium concentrations of solutes and gases. The concentrations of pure liquids and pure solids are constant and do not affect the position of equilibrium.

*   *Thus, the concentration of a pure solid or liquid is not shown in the equilibrium-constant expression for a heterogeneous equilibrium.* For example, the equilibrium-constant expression for

$$PbBr_2(s) \rightleftharpoons Pb^{2+}(aq) + 2\, Br^-(aq)$$

is

$$K_c = [Pb^{2+}][Br^-]^2$$

Reactions that do not go to completion achieve an equilibrium state. The extent to which a reaction proceeds is indicated by the value of K_c.

* $K_c \gg 1$: The reaction tends to proceed far to the right; mostly product forms at equilibrium.
* $K_c \ll 1$: The reaction lies far to the left; very little product forms at equilibrium.
* $K_c \approx 1$: Both the products and the reactants exist with neither highly favored versus the other.

EXERCISE 1   Writing equilibrium constant expressions

Write equilibrium-constant expressions using K_p and K_c for the following reversible reactions:

(a) $H_2(g) + I_2(g) \rightleftharpoons 2\, HI(g)$
(b) $2\, Cl_2(g) + 2\, H_2O(g) \rightleftharpoons 4\, HCl(g) + O_2(g)$

SOLUTION: *Analyze*: We are given two reactions describing systems at equilibrium and asked to write the equilibrium-constant expressions K_c and K_p.

Plan: According to the law of mass action the equilibrium-constant expression involves a ratio of the concentrations of products to reactants. The concentration of each substance is raised to a power equal to the stoichiometric coefficient in front of it in the chemical reaction. If a substance is in solution and its concentration is in the unit of molarity it is shown in a bracket $[X]$. The symbol P is used for the substance when the unit of concentration is atmospheres.

Solve:

(a) $K_c = \dfrac{[HI]^2}{[H_2][I_2]}$ or $K_p = \dfrac{P_{HI}^2}{P_{H_2}P_{I_2}}$

The concentration of HI is raised to the second power because there is a two in front of HI(g) in the balanced equation. Similarly, $[H_2]$ and $[I_2]$ both have exponents of one.

(b) $K_c = \dfrac{[HCl]^4[O_2]}{[Cl_2]^2[H_2O]^2}$ or $K_p = \dfrac{P_{HCl}^4 P_{O_2}}{P_{Cl_2}^2 P_{H_2O}^2}$

Again, the exponent of each substance in the equilibrium expression is raised to the same as the coefficient in front of the substance in the balanced equation.

Check: A review of the equilibrium constant expressions confirms the products of each equilibrium are the numerators and the reactants are the denominators.

Also, the exponent of each substance does equal the appropriate stoichiometric coefficient. For example, the partial pressure of $HCl(g)$ is raised to the power of 4 and the stoichiometric coefficient in front of it in (**b**) is also 4.

EXERCISE 2 Writing an equilibrium constant expression when a chemical reaction is reversed

How is the equilibrium-constant expression (K_c) for the reaction

$$2\,HI(g) \rightleftharpoons H_2(g) + I_2(g)$$

related to the one for the reverse reaction:

$$H_2(g) + I_2(g) \rightleftharpoons 2\,HI(g)$$

SOLUTION: *Analyze*: We are given two equilibriums, one the reverse of the other, and asked to relate their equilibrium-constant expressions.

Plan: The equilibrium-constant expression for a reversed reaction is simply the reciprocal of the equilibrium-constant expression for the original reaction.

Solve: For the reaction

$$2\,HI(g) \rightleftharpoons H_2(g) + I_2(g)$$

the associated equilibrium constant is

$$K_c = \frac{[H_2][I_2]}{[HI]^2}$$

The equilibrium constant for the reverse reaction is

$$K_c = \frac{[HI]^2}{[H_2][I_2]}$$

Note that the second equilibrium-constant expression is the reciprocal (inverted form) of the first one.

EXERCISE 3 Writing equilibrium-constant expressions for heterogeneous equilibria

Write the equilibrium-constant expression (K_c) for each of the following heterogeneous equilibria:

(**a**) $H_2(g) + I_2(s) \rightleftharpoons 2\,HI(g)$
(**b**) $H_2(g) + Br_2(l) \rightleftharpoons 2\,HBr(g)$
(**c**) $(CH_3)_4Sn(s) \rightleftharpoons (CH_3)_4Sn(g)$
(**d**) $Ag(CH)_2^-(aq) + AgI(s) \rightleftharpoons I^-(aq) + 2\,AgCN(s)$

SOLUTION: *Analyze*: We are given four reactions describing different equilibrium conditions and are asked to write the expressions for K_cs.

Plan: We will follow the same procedure as in Exercise 1 but we must remember that only gases and solutes are shown in an equilibrium-constant expression.

Solve:

(**a**) $K_c = \dfrac{[HI]^2}{[H_2]}$

(b) $K_c = \dfrac{[\text{HBr}]^2}{[\text{H}_2]}$

(c) $K_c = [(\text{CH}_3)_4\text{Sn}(g)]$

(d) $K_c = \dfrac{[\text{I}^-]}{[\text{Ag}(\text{CN}_2)^-]}$

Comment: Pure liquids and pure solids are not shown in an equilibrium-constant expression because their concentrations do not depend on the amount of substance present. That is, at a specified temperature and pressure, density is a constant.

EXERCISE 4   Calculating K_c and K_p for a chemical reaction

A 1.000-L container contains an equilibrium mixture of $\text{H}_2(g)$, $\text{I}_2(g)$, and $\text{HI}(g)$ at 721 K. The number of moles of each substance present is 1.843×10^{-3}, 1.843×10^{-3}, and 1.310×10^{-2}, respectively. Calculate the values of K_c and K_p for the reaction

$$\text{H}_2(g) + \text{I}_2(g) \rightleftharpoons 2\,\text{HI}(g)$$

SOLUTION: *Analyze*: We are given a 1.000-L container with hydrogen, iodine, and hydrogen iodide gases and the number of moles of each at equilibrium and are asked to calculate K_c and K_p.

Plan: Since the container has a volume of 1.000 L, the molarity of each species is equivalent to the number of moles of each species present. Thus the molarity of H_2 is 1.843×10^{-3}; the molarity of I_2 is 1.843×10^{-3}; and the molarity of HI is 1.310×10^{-2}. Substituting these values into the equilibrium-constant expression yields the value of K_c. We can calculate K_p using the relationship: $K_p = K_c(RT)^{\Delta n}$.

Solve: Substituting the molarity of each gas at equilibrium into K_c gives:

$$K_c = \frac{[\text{HI}]^2}{[\text{H}_2][\text{I}_2]}$$

$$= \frac{(1.310 \times 10^{-2})^2}{(1.843 \times 10^{-3})(1.843 \times 10^{-3})}$$

$$= 50.52$$

The value of Δn, the number of moles of gas products minus the number of moles of gas reactants, is: $2 - (1 + 1) = 0$. Therefore,

$$K_p = K_c(RT)^0 = K_c(1) = K_c = 50.52$$

Check: An estimate of K_c gives $\dfrac{(10^{-2})^2}{(2 \times 10^{-3})(2 \times 10^{-3})} = 2.5 \times 10^1$. This estimate is the same order of magnitude as the exact solution.

EXERCISE 5   Calculating K_p

N_2O_4, a colorless gas, and NO_2, a brown gas, exist in equilibrium as follows:

$$2\,\text{NO}_2(g) \rightleftharpoons \text{N}_2\text{O}_4(g)$$

A closed container at 25 °C is charged with N_2O_4 at a partial pressure of 0.51 atm and NO_2 at a partial pressure of 0.56 atm. At equilibrium the partial pressure of N_2O_4 is found to be 0.54 atm. What is K_p?

SOLUTION: *Analyze*: We are given an equilibrium between N_2O_4 and NO_2, the initial partial pressures of the two gases and the equilibrium partial pressure of N_2O_4. We are asked to determine K_p for the equilibrium.

Plan: We need to determine the equilibrium partial pressure of NO_2 to calculate K_p; we already know the equilibrium partial pressure of N_2O_4. We will have to use the stoichiometry of the reaction to determine the equilibrium partial pressure of N_2O_4. We know that NO_2 reacts to form more N_2O_4 because the equilibrium concentration of N_2O_4 is greater than its initial concentration; thus N_2O_4 had to gain concentration when the reaction achieved equilibrium.

Solve: First, write the equilibrium expression for the reaction:

$$K_p = \frac{P_{N_2O_4}}{P_{NO_2}^2}$$

To solve for the equilibrium constant, we need to determine the equilibrium partial pressures of both gases. We can determine the equilibrium partial pressures from the stoichiometry of the equilibrium as follows: (1) Write the equilibrium and underneath it develop a table of concentrations that shows (2) initial conditions, (3) the change that occurs when the reaction proceeds to equilibrium, and (4) the equilibrium concentrations determined from either the given information or from the changes in partial pressures.

(1)		$2 NO_2 \rightleftharpoons N_2O_4$		
(2)	Initial pressure (atm):	0.56		0.51
(3)	Change in pressure (atm):	?		?
(4)	Equilibrium pressure (atm):	?		0.54

At this point, we do not know the information for the second line and the equilibrium partial pressure of NO_2. However, we can determine the change in pressure from the data in the N_2O_4 column. We can calculate the change in N_2O_4 pressure from the difference in equilibrium pressure and initial pressure: 0.54 atm $-$ 0.51 atm $=$ +0.03 atm. The change in NO_2 pressure is a negative quantity because N_2O_4 gained pressure; it is calculated from this change and the stoichiometry:

$$\text{Change in pressure of } NO_2 = -(0.03 \text{ atm } N_2O_4)\left(\frac{2 \text{ mol } NO_2}{1 \text{ mol } N_2O_4}\right) = -0.06 \text{ atm}$$

Thus, the equilibrium partial pressure of NO_2 is the sum of the initial partial pressure and the change in partial pressure, 0.56 atm $+$ (-0.06 atm) $=$ 0.50 atm. The table of concentrations now becomes:

	$2 NO_2$	$\rightleftharpoons$	N_2O_4
Initial pressure (atm)	0.56		0.51
Change in pressure (atm)	-0.06		$+0.03$
Equilibrium pressure (atm)	0.50		0.54

We can now solve for K_p by substituting the equilibrium partial pressures in the third line of the previous table into the equilibrium constant expression:

$$K_p = \frac{P_{N_2O_4}}{P_{NO_2}^2} = \frac{0.54}{(0.50)^2} = 2.2$$

Comment: Note that the value of K_p does not have units; therefore we do not show concentration units in the equilibrium-constant expression. An estimate of the value of K_p gives $\dfrac{0.5}{(0.5)^2} = 2$ which is in close agreement with the exact solution.

CALCULATING EQUILIBRIUM CONCENTRATIONS

The large majority of chemical equilibrium problems require calculating equilibrium concentrations. Equilibrium concentrations are calculated from the relationship between K and equilibrium concentrations. The next series of sample exercises explore how to solve for equilibrium concentrations. At the end of this section is a summary of the steps that were used in solving the sample exercises.

EXERCISE 6   Calculating equilibrium concentrations I

A 1.0-L container initially holds 0.015 mol of and 0.020 mole of I_2 at 721 K. What are the concentrations of H_2, I_2, and HI after the system has achieved a state of equilibrium? The value of K_c is 50.5 for the reaction $H_2(g) + I_2(g) \rightleftharpoons 2\,HI(g)$.

SOLUTION: *Analyze*: We are given a 1.0 L container that holds a given number of moles of hydrogen and iodine gas at 721 K and K_c is 50.5. We are asked to calculate the equilibrium concentrations of H_2, I_2, and HI.

Plan: We know the initial concentrations of hydrogen and iodine gas because we are given the number of moles of each and the volume of the container:

$$M = \frac{\text{moles solute}}{\text{volume in liters}}$$

We will use an approach similar to Sample Exercise 15.12 in the text. When equilibrium is achieved, hydrogen iodide forms; thus the concentrations of hydrogen and iodine gas must decrease. We will develop a concentration table that helps us determine the amounts of hydrogen and iodine gas that disappear and the quantity of hydrogen iodide that forms.

Solve: According to the stoichiometry of the balanced chemical equation, if we let x equal the molarity of H_2 that reacts, then x also equals the molarity of I_2 that reacts, and $2x$ equals the molarity of HI that forms. The last equivalence is derived by the following conversions:

$$1 \text{ mol } H_2 \simeq 1 \text{ mol } I_2 \simeq 2 \text{ mol HI}$$

$$[HI] = \left(x\frac{\text{mol } H_2}{L}\right)\left(\frac{2 \text{ mol HI}}{1 \text{ mol } H_2}\right) = 2x \text{ mol HI/L}$$

All of this information is summarized by the following table of concentrations:

	$H_2(g)$	$+$	$I_2(g)$	$\rightleftharpoons$	$2\,HI(g)$
Initial (*M*):	0.015		0.020		0
Change (*M*):	$-x$		$-x$		$+2x$
Equilibrium (*M*):	$(0.015 - x)$		$(0.020 + x)$		$2x$

Substituting the equilibrium concentrations into the K_c expression for the reaction $H_2(g) + I_2(g) \rightleftharpoons 2\,HI(g)$ yields

$$K_c = \frac{[HI]^2}{[H_2][I_2]} = \frac{(2x)^2}{(0.015 - x)(0.020 - x)} = \frac{4x^2}{3.0 \times 10^{-4} - 0.035x + x^2} = 50.5$$

Rearranging this equation into quadratic form $(ax^2 + bx + c = 0)$ gives

$$46.5x^2 - 1.77x + 0.0152 = 0$$

Dividing through by 46.5 yields

$$x^2 - 0.0381x + 3.27 \times 10^{-4} = 0$$

To solve this equation for the value of x, we must use the quadratic formula (see Appendix A.3 in the text):

$$x = \frac{-b \pm \sqrt{(b^2) - 4ac}}{2a}$$

$$= \frac{-(-0.0381)}{2(1)} \pm \frac{\sqrt{(-0.0381)^2 - 4(1)(3.27 \times 10^{-4})}}{2(1)}$$

$$= \frac{0.0381 \pm 0.0119}{2}$$

$$= 0.025 \text{ and } 0.0131$$

One solution, 0.025, is not valid because it is larger in value than either initial concentration. Thus $x = 0.0131$ is the only valid (physically significant) answer. Therefore

$$[H_2] = 0.015 - x = 0.015 - 0.013 = 0.002 \ M$$

$$[I_2] = 0.020 - x = 0.020 - 0.013 = 0.007 \ M$$

$$[HI] = 2x = 2(0.013) = 0.026 \ M$$

Check: An estimate of the value of x gives

$$\frac{-(-0.04) \pm \sqrt{(-.04)^2 - (4)(1)(3 \times 10^{-4})}}{2} = \frac{0.04 \pm 0.02}{2} = 0.01$$

or 0.03; both are close in value to the two exact solutions. Note that indeed the equilibrium concentrations of the reactants, hydrogen and iodine, did decrease and that of hydrogen iodide did increase.

EXERCISE 7   Calculating equilibrium concentrations II

A container initially holds 1.00 atm of NO_2 and 2.00 atm of NO. The equilibrium that is established is

$$2 \, NO_2(g) \rightleftharpoons 2 \, NO(g) + O_2(g)$$

All of the species are gases. K_p is 6.76×10^{-5}. What is the equilibrium partial pressure of O_2?

SOLUTION: *Analyze*: We are given an equilibrium involving NO_2, NO, and O_2 as gases, the initial partial pressures of NO_2 and NO and the value of K_{eq}. We are asked to determine the equilibrium concentration of O_2.

Plan: We will develop a concentration as we did in Exercise 6. The concentration table will show that NO_2 decreases in concentration and both NO and O_2 increase in concentration when equilibrium is reached. We conclude this because there is no oxygen initially present; it cannot have a zero concentration at equilibrium and therefore it must form.

Solve: The given quantities are the initial concentrations of NO_2 and NO; the questions asks for the equilibrium partial pressure of O_2. Let x equal the equilibrium partial pressure of O_2.

	$2 \, NO_2(g)$	$\rightleftharpoons$	$2 \, NO(g)$	$+O_2(g)$
Initial (atm):	1.00		2.00	0
Change (atm):	$-2x$		$+2x$	$+x$
Equilibrium (atm):	$(1.00 - 2x)$		$(2.00 - 2x)$	x

The change in pressure of NO is $+2x$ because of the stoichiometry of the reaction—two moles of NO form for every mole of O_2 that forms. Using a similar analysis, the change in pressure of NO_2 is $-2x$. For NO and O_2 to form, NO_2 must dissociate; two moles of NO_2 dissociated for every mole of O_2 that forms. The number of moles of gas and atmospheres of pressure are directly related at constant temperature and volume because of the ideal-gas law, $PV = nRT$.

Note that the value of K_p is small; therefore, you should expect the amount of NO_2 that dissociates, x, to be very small, particularly when compared to 2.00 and 1.00. Thus we can approximate the equilibrium concentration of NO_2 and NO as follows:

$$(1.00 - 2x) \text{ atm} \simeq 1.00 \text{ atm} \quad \text{and} \quad (2.00 - 2x) \text{ atm} \simeq 2.00 \text{ atm.}$$

$$K_p = \frac{[NO]^2[O_2]}{[NO_2]^2} \quad \text{or} \quad 6.76 \times 10^{-5} = \frac{(2.00)^2(x)}{(1.00)}$$

Solving for x, the equilibrium partial pressure of O_2, gives $x = 1.69 \times 10^{-5}$ atm.

Comment: The value of x is indeed much smaller than 2.00 or 1.00. For example, the percent x is of 2.00 is $x/2.00 \times 100 = 1.69 \times 10^{-5}/2 \times 100 = 8.5 \times 10^{-4}\%$. Any time this percent is less than 5%, the approximation will be considered valid.

SUMMARY OF STEPS INVOLVED IN CALCULATING EQUILIBRIUM CONCENTRATIONS

1. Write a balanced chemical equation describing the chemical changes that occur.
2. Identify knowns and the unknown asked for in the question.
3. Assign a variable, such as x, to the unknown. This variable usually represents a change in concentration.
4. Prepare a table of concentrations, using the substances involved in the chemical equilibrium:
     Row 1   chemicals involved in equilibrium
     Row 2   initial concentrations
     Row 3   changes in concentration (x usually first appears here)
     Row 4   equilibrium concentrations (add rows 2 + 3)
5. Write the equilibrium constant expression.
6. Solve for the unknown variable by using K and the equilibrium concentrations in row 4. Determine if any approximations can be made to simplify the problem.
7. Answer the question.
8. Check the approximation for validity: For expressions of type $C + x$ or $C - x$, x can be neglected if

$$\frac{x}{C} \times 100 < 5\%$$

Up to a 5% error caused by the approximation is acceptable.

REACTION QUOTIENT AND LE CHATELIER'S PRINCIPLE

A reaction that is not at equilibrium proceeds to a state of equilibrium. The direction of change in approaching equilibrium is determined by the **reaction quotient**, Q.

- A reaction-quotient expression for a given reaction has the same form as the reaction's equilibrium-constant expression, but it is numerically evaluated using the given reactant and product concentrations.
- The relationships between numerical values of K_c and Q and the direction of a reaction are summarized in Table 15.1 on the next page.

TABLE 15.1 Relationships among Numerical Values of K_c and Q

Relationship	Interpretation	Direction of change
$Q = K_c$	System is at equilibrium	none
$Q > K_c$	System is not at equilibrium; reaction proceeds to form more reactants until $Q = K$.	$\leftarrow$
$Q < K_c$	System is not at equilibrium; reaction proceeds to form more products until $Q = K$.	$\rightarrow$

The composition of a reaction at equilibrium may be altered by changing the concentration of a product or reactant, or by changing the pressure, volume, or temperature of the system. The question of what effect a change in one of these variables has on the extent of a reaction can be qualitatively answered by using **LeChatelier's principle:**

- *When a stress is applied to a system at equilibrium the system will change, if possible, in the direction of the reaction that will relieve the applied stress.*
- Alternatively, the principle may also be stated as follows: *When a system at equilibrium is disturbed, the system adjusts so as to restore the equilibrium state.*
- The effects of changing concentration, pressure, or temperature on a system at equilibrium are summarized in Table 15.2. Note that changes in concentration and pressure do not change the value of K_c for a reaction.

TABLE 15.2 Effect of Changing Concentration, Pressure, or Temperature on a System at Equilibrium

Change	Effect
Concentration	
(1) Increasing the concentration of a substance	Reaction shifts to *remove* some of the added substance
(2) Decreasing the concentration of a substance	Reaction shifts to produce *more* of the substance
Pressure	
(1) Increasing total gas pressure	Reaction shifts in a direction that causes the total number of moles of *gases* to *decrease*
(2) Decreasing total gas pressure	Reaction shifts in a direction that causes the total number of moles of *gases* to *increase*
Temperature	
(1) Increase	*Exothermic reaction: K_c decreases—reaction shifts toward reactants*
	Endothermic reaction: K_c increases—reaction shifts toward products
(2) Decrease	*Exothermic reaction: K_c increases—reaction shifts toward products*
	Endothermic reaction: K_c decreases—reaction shifts toward reactants

Note: In Table 15.2 the value of K_c changes when temperature changes. Thus, when heat is added or removed from a system at equilibrium (i.e., a temperature change), the position of equilibrium is modified because the value of K_c is altered.

A chemical system at equilibrium is dynamic at the molecular level. At equilibrium, the forward reaction is occurring at the same rate as the reverse reaction. Although the concentrations of reactant and product molecules remain constant, they are continuously forming and disappearing. By speeding up both forward and reverse reactions, a **catalyst** decreases the time required for a reaction to reach equilibrium. *A catalyst does not change the value of the equilibrium constant.*

EXERCISE 8   Applying LeChatelier's Principle I

What changes in the equilibrium composition of the reaction

$$N_2(g) + O_2(g) \rightleftharpoons 2\,NO(g)$$

occur at constant temperature if: (**a**) the partial pressure of $N_2(g)$ is increased; (**b**) the pressure of $NO(g)$ is decreased; (**c**) the total pressure of the system is increased; (**d**) the total volume of the system is increased?

SOLUTION: *Analyze*: We are given a reaction at equilibrium and asked to predict what happens at constant temperature if four different changes occur.

Plan: The direction of change can be qualitatively answered by application of LeChatelier's principle. We can use Table 15.2 to help us answer this question.

Solve: (**a**) Increasing the concentration (partial pressure) of $N_2(g)$ shifts the equilibrium to the right, removing added $N_2(g)$ and also some $O_2(g)$, and thereby forming more $NO(g)$. (**b**) Decreasing the concentration (partial pressure) of $NO(g)$ shifts the equilibrium to the right, forming more $NO(g)$ and consuming $N_2(g)$ and $O_2(g)$. (**c**) The reaction proceeds in a direction that reduces the total number of moles of gas. Since the sum of moles of N_2 and O_2, two, equals the number of moles of $NO(g)$ formed, two, there is no way for the system to change to reduce the number of moles of gas. Thus, the composition of the equilibrium mixture does not change. (**d**) Pressure is inversely proportional to volume at constant temperature. Although the total pressure is decreased, no change in the equilibrium mixture occurs as explained in (**c**).

Check: The answer agrees with the general idea from LeChatelier's principle that adding a substance involved in an equilibrium shifts the reaction away from the substance and removing a substance shifts the reaction towards the substance. Pressure is affected by the total number of moles of substances; thus, increasing the total pressure causes a change to reduce the total pressure to restore equilibrium, that is, fewer moles of gas are needed.

EXERCISE 9   Applying LeChatelier's Principle II

What change in the equilibrium composition of the reaction

$$2\,CO(g) + O_2(g) \rightleftharpoons 2\,CO_2(g)$$

occurs if the pressure of the system is increased?

SOLUTION: *Analyze*: We are given a chemical reaction and asked what happens at equilibrium if the pressure of the system is increased.

Plan: The system will respond to an increase in the pressure of the system by reducing the total number of moles of gas until the equilibrium condition is again established. We need to analyze the chemical reaction in this manner.

Solve: There are three moles of gaseous reactants and two moles of gaseous product in the balanced equation. Formation of more product, $CO_2(g)$, and consumption of reactants, $CO(g)$ and $O_2(g)$, result in a net decrease in the total number of moles of gas. Thus the reaction shifts to the right.

EXERCISE 10   Applying LeChatelier's Principle III

In which direction do the following reactions at equilibrium proceed when temperature is increased?

(a) $CH_4(g) + 2\,O_2(g) \rightleftharpoons CO_2(g) + 2\,H_2O(l)$ (exothermic)
(b) $H_2(g) + I_2(g) \rightleftharpoons 2\,HI(g)$ (endothermic)

SOLUTION: *Analyze*: We are given two reactions at equilibrium, one exothermic and one endothermic. We are asked to predict what happens when temperature is increased.

Plan: An increase in temperature is caused by adding heat to the system. A change in temperature is also accompanied by a change in the value of K_c. We will apply LeChatelier's principle to determine how the value of K_c and thus the direction of a reaction is affected by a change in temperature.

Solve: (a) Since the reaction is exothermic, the chemical equation may be rewritten so heat is included as a product:

$$CH_4(g) + 2\,O_2(g) \rightleftharpoons CO_2(g) + 2\,H_2O(l) + \text{heat}$$

According to LeChatelier's principle, the position of equilibrium shifts to the left in order to remove the added heat. This results in a smaller value of K_c for the equilibrium. (b) The reaction is endothermic; heat may be considered as a reactant:

$$\text{Heat} + H_2(g) + I_2(g) \rightleftharpoons 2\,HI(g)$$

According to LeChatelier's principle, the position of equilibrium shifts to the right in order to remove the added heat. This results in a larger value of K_c for the equilibrium.

EXERCISE 11   Using Q to predict the direction of a chemical reaction

The value of K_c for the reaction

$$H_2(g) + I_2(g) \rightleftharpoons 2\,HI(g)$$

is 50.5. In what direction will the reaction proceed if the initial concentrations are: $[H_2] = 0.015\ M$; $[I_2] = 0.0012\ M$; $[HI] = 0.025\ M$?

SOLUTION: *Analyze*: We are given a reaction of hydrogen and iodine gases to form hydrogen iodide gas and the value of K_c. We are also given the initial concentration of all three substances and asked to predict in which direction the reaction proceeds.

Plan: The reaction quotient, Q, helps us answer this question. Q has the same form as the equilibrium-constant expression and is evaluated with the given conditions. The value is compared to K_c to determine the direction of the reaction.

Solve: Write the expression for Q and then substitute the given starting concentrations:

$$Q = \frac{C_{HI}^2}{C_{H_2}C_{I_2}} = \frac{(0.025)^2}{(0.015)(0.0012)} = 35$$

Next, compare the values of Q and K_c ($K_c = 50.5$). In this problem, $Q < K_c$. Therefore, the reaction proceeds to form more products (to the right) until the value of Q increases to equal the value of K_c.

EXERCISE 12   Applying LeChatelier's Principle IV

According to LeChatelier's principle, what changes in the composition of the equilibrium system

$$H_2(g) + Br_2(l) \rightleftharpoons 2\,HBr(g)$$

occur when: (a) $Br_2(l)$ is added; (b) the partial pressure of $H_2(g)$ is increased; (c) the volume of the container that contains the equilibrium system is increased at constant temperature?

SOLUTION: *Analyze*: We are given a reaction at equilibrium involving hydrogen gas, bromine liquid, and hydrogen bromide gas and are asked how the composition changes when there is a change in the amount of bromine, the volume of the container, and the partial pressure of hydrogen gas.

Plan: This problem will be answered using the same approach as in Exercise 8. There is one new feature in this problem: bromine, a reactant, is a liquid. We must remember that a pure liquid or solid is not shown in the equilibrium-constant expression, and thus when their quantities are changed at constant temperature they do not affect the equilibrium.

Solve: (a) No change occurs because the equilibrium composition is independent of the total quantity of *liquid* Br_2. (b) The increased partial pressure of $H_2(g)$ is partly offset by reaction of $H_2(g)$ with $Br_2(l)$ to form more $HBr(g)$. (c) The increase in volume is accompanied by a decrease in the total pressure of the system. (Remember that $P \propto 1/V$.) The system reacts to a decrease in the total pressure by forming more moles of gas. The partial pressure of a pure liquid or solid in a system at equilibrium is constant and thus does not change so long as the temperature is unchanged. The consumption of $H_2(g)$ and $Br_2(l)$ to form $HBr(g)$ produces more moles of *gas*. Thus the reaction changes by forming more product.

EXERCISE 13   Using Q and K_c to calculate equilibrium concentrations

At 25 °C the equilibrium constant, K_c, for the reaction

$$\alpha\text{-D-glucose}(aq) \rightleftharpoons \beta\text{-D-glucose}(aq)$$

is 1.75. What are their equilibrium concentrations if initial concentrations are 1.00 mol/liter for both?

SOLUTION: *Analyze*: We are given the equilibrium constant for a reaction involving a change in the structure of glucose and the initial concentrations each are 1.00 *M*. We are asked to determine the equilibrium concentrations.

Plan: We can use an approach similar to that used in Exercise 6 to calculate the equilibrium concentrations. However, there is one difference: One of the substances does not have a zero concentration and thus we cannot qualitatively predict whether product or reactant forms. To determine the direction of the reaction when it reaches equilibrium we can use the reaction quotient.

Solve: The first step in this problem is to determine the direction of the reaction when it changes from its initial state to the equilibrium state by using the reaction quotient. We need this information because we do not know if the concentration of β-D-glucose increases or decreases at equilibrium.

$$Q = \frac{C_{\beta\text{-D-glucose}}}{C_{\alpha\text{-D-glucose}}} = \frac{1.00}{1.00} = 1.00$$

Because $Q < K_c$, the reaction shifts to form more product. Therefore, let x equal the concentration of β-D-glucose formed during the change.

α-D-glucose(aq) ⇌ β-D-glucose(aq)

	α-D-glucose	β-D-glucose
Initial (*M*):	1.00	1.00
Change (*M*):	$-x$	$+x$
Equilibrium (*M*):	$(1.00 - x)$	$(1.00 + x)$

The value of K_c is large; thus, *no* approximations should be made.

$$K_c = \frac{[\beta\text{-D-glucose}]}{[\alpha\text{-D-glucose}]} = 1.75$$

$$\frac{(1.00 + x)}{(1.00 - x)} = 1.75$$

Rearranging the above equation gives

$$1.00 + x = 1.75(1.00 - x) = 1.75 - 1.75x$$

or

$$2.75x = 0.75$$

$$x = \frac{0.75}{2.75} = 0.27 \ M$$

The equilibrium concentrations are therefore

$$[\beta\text{-D-glucose}] = (1.00 + x) \ M = 1.27 \ M$$
$$[\alpha\text{-D-glucose}] = (1.00 - x) \ M = 0.73 \ M$$

Check: The calculated values of the reactant and product can be substituted in the equilibrium-constant expression to check that they give a value of 1.75:
$\frac{1.27}{0.73} = 1.74$, which is the same value within ±0.01.

EXERCISE 14   Relating K_c to chemical kinetics

The equilibrium constant for the reaction

$$2\,SO_2(g) + O_2(g) \rightleftharpoons 2\,SO_3(g)$$

is 279. (a) Is the reaction fast or slow? (b) Nitrogen oxide catalyzes the reaction. How does the value of the equilibrium constant change when nitrogen oxide is added?

SOLUTION: (a) The equilibrium constant provides no information about the rate of reaction. **(b)** The equilibrium constant does not change in value because a catalyst affects the forward and reverse reaction rates equally.

EXERCISE 15   Relating enthalpy change and a catalyst in a chemical reaction

Does a catalyst affect the enthalpy change in a reaction?

SOLUTION: The enthalpy change for a reaction is independent of the way products are formed because it is a state function (see Section 5.2 in the text). However, a catalyst, by speeding up the reaction, causes the rate of heat absorption or evolution to increase. For example, in an exothermic reaction the catalyst causes all the heat to be released in a short time as compared to a much longer time when the catalyst is absent. However, the position of equilibrium is unchanged, and the total quantity of heat evolved is the same in each case.

SELF-TEST QUESTIONS

Having reviewed key terms in Chapter 15, match key terms with phrases and identify statements as true or false. If a statement is false, indicate why it is incorrect.

Match each phrase with the best term:

15.1 The type of chemical equilibrium exemplified by the following:

$$BaSO(s) \rightleftharpoons Ba^{2+}(aq) + SO_4^{2-}(aq)$$

15.2 We say that this condition exists when a closed system contains reactants and over time products form and concentrations are constant.

15.3 The name given to the commercial process when ammonia is formed from its elements.

15.4 The name given to the following expression, $\dfrac{[CO]^2[O_2]}{[CO_2]}$, which is associated with the following reaction:

$$2 CO_2(g) \rightleftharpoons 2 CO(g) + O_2(g)$$

15.5 The type of equilibrium exemplified by the following:

$$HCN(aq) \rightleftharpoons H^+(aq) + CN^-(aq)$$

Terms:

    (a) chemical equilibrium   **(d)** homogeneous
    (b) Haber   **(e)** mass action
    (c) heterogeneous

True-False Statements:

15.6 An equilibrium exists between $CO_2(g)$ in human lungs and $CO_2(aq)$ on the blood system at the interface between the lung and blood tissues. According to *LeChatelier's principle*, an increase in the CO_2 concentration in the air a person breathes should cause a reduction in the CO_2 concentration in the person's blood system.

15.7 Suppose that during the Haber process we find $[NH_3] = 1.20\ M$, $[N_2] = 0.80\ M$, and $[H_2] = 0.50\ M$. By comparing the value of the *reaction quotient Q*, to the value of K_c, which is 0.105 for the Haber reaction, we know that the system is at equilibrium.

15.8 The *equilibrium expression* for the reaction $N_2(g) + O_2(g) \rightleftharpoons 2\ NO(g)$ is

$$K_c = \frac{[N_2][O_2]}{[NO]^2}$$

15.9 The *equilibrium constant* for a particular reaction, K_c, is equal to K_c^{-1} for the reverse reaction.

Problems and Short-Answer Questions

15.10 The reaction $A(g) + 2\ B(g) \rightleftharpoons AB_2(g)$ has an equilibrium constant of $K_p = 0.5$. The diagram below shows a mixture containing A atoms (unshaded), B atoms (partially shaded), and AB_2 molecules. Is the reaction at equilibrium? Explain. If the system is not at equilibrium, how many B atoms needed to be added or removed to form an equilibrium condition, assuming the numbers of the other species do not change? Explain.

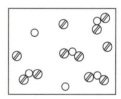

15.11 The following diagrams represent equilibrium mixtures for the reaction $A(g) + X(g) \rightleftharpoons AX_2(g)$ at (1) 300 K and (2) when heat is added until a temperature of 450 K is achieved. Each diagram shows A atoms (unshaded), X atoms (partially shaded), and AX_2 molecules. Is the reaction exothermic or endothermic? Explain.

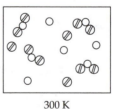 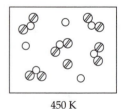

      300 K   450 K

15.12 Critique the following statement: When a system has reached equilibrium the reactants and products have reached a constant condition in concentrations and the reactants have stopped reacting.

15.13 Is there any relationship between an equilibrium-constant expression and the proposed reaction mechanism for the reaction? Would a catalyst change the equilibrium-constant expression? Explain.

15.14 Write K_c expressions for the following reactions:
    (a) $Cl_2(g) \rightleftharpoons Cl_2(l)$
    (b) $Ag_2SO_4(s) \rightleftharpoons 2\ Ag^+(aq) + SO_4^{2-}(aq)$

(c) $C_3H_6(g) + H_2(g) \rightleftharpoons C_3H_8(g)$

(d) $2 C_3H_6(g) + 9 O_2(g) \rightleftharpoons 6 CO_2(g) + 6 H_2O(l)$

15.15 At 740 °C the equilibrium constant K_c is 6×10^{-3} for the reaction

$$CaCO_3(s) \rightleftharpoons CaO(s) + CO_2(g)$$

What does this tell you about the extent of the reaction at equilibrium? What is the equilibrium constant for the reverse reaction and what is the extent of the reaction?

15.16 Refer to Problem 15.15 to answer these questions. Each of the following mixtures was placed in a sealed container and allowed to react for several days at 740 °C. Which systems will reach a state of equilibrium?

(a) carbon dioxide gas only with a concentration of one molar

(b) calcium carbonate only

(c) calcium carbonate and calcium oxide

(d) calcium oxide and carbon dioxide with a concentration of 0.5 molar

15.17

(a) Critique the following statement: When more silver chloride is added to the following reaction at equilibrium, $AgCl(s) \rightleftharpoons Ag^+(aq) + Cl^-(aq)$, the system responds by forming more silver and chloride ions.

(b) At 25 °C, $K_c = 4.7 \times 10^{-3}$ for the reaction $N_2O_4(g) \rightleftharpoons 2 NO_2(g)$. If the two gases are each 1.0×10^{-3} molar, is the system at equilibrium? If the system is not at equilibrium in which direction does it proceed?

15.18 Given the following information

$$2 NO(g) \rightleftharpoons NO(g) + Cl_2(g) \quad \Delta H^0 = 75 \text{ kJ}$$

what change occurs in a system at equilibrium (a) if the temperature is decreased; (b) if the partial pressure of chlorine gas is increased; and (c) if the total volume is increased.

15.19 Suppose you are given only the initial concentrations of reactants and the value of K_c for the following equilibrium

$$CO_2(g) + H_2(g) \rightleftharpoons CO(g) + H_2O(g)$$

and are asked to calculate the equilibrium concentrations. What relationships in changes in concentrations can you determine using the stoichiometry of the reaction? Use the expression Δ (chemical formula of reactant or product) to help you express these relationships. The term Δ means "a change in" and in this problem it means "a change in concentration." Thus, you want to write an expression such as the following: $\Delta(X) = \Delta(Y)$. The interpretation of this relationship is that a change in the concentration of species X is also accompanied by an equal change in species Y.

15.20 The equilibrium in the previous problem involved stoichiometric coefficients that are 1:1 throughout. Repeat the question using the following equilibrium:

$$CO_2(g) + 3 H_2(g) \rightleftharpoons CH_4(g) + H_2O(g)$$

15.21 A tank is pressurized with carbon dioxide gas at a concentration of 3.0 M and oxygen gas at concentration of 1.0 M at a high temperature. The reaction that occurs is

$$2 CO_2(g) \rightleftharpoons 2 CO(g) + O_2(g)$$

Fill in the following table:

	CO_2	CO	O_2
Initial(M)	What are the initial concentrations for each species?		
ΔConc.(M)	Use x as a variable to represent the change in concentration of each species.		
Equil.(M)	What are the equilibrium concentrations represented using the above?		

15.22 At 603 K, K_c for the reaction

$$NH_4Cl(g) \rightleftharpoons NH_3(g) + HCl(g)$$

is 1.1×10^{-2}. What is the equilibrium concentration of $NH_3(g)$ if the initial concentration of $NH_4Cl(g)$ is 0.500 M?

15.23 A 1.00 L container initially contains 0.476 mol of $SO_3(g)$ at 1105 K. What is K_c for the reaction

$$2 SO_3(g) \rightleftharpoons 2 SO_2(g) + O_2(g)$$

if 50.0 percent of the $SO_3(g)$ has decomposed at equilibrium?

15.24 For the equilibrium

$$C(s) + CO_2(g) \rightleftharpoons 2 CO(g)$$

the value of K_p is 167.5 at 1000 °C. If a 5.00-L container initially contains 2.00 atm pressure of $CO_2(g)$ and $C(s)$ at 1000 °C, what are the partial pressures of $CO_2(g)$ and $CO(g)$ at equilibrium?

15.25 At 25 °C carbon dioxide decomposes as follows: $2 CO_2(g) \rightleftharpoons CO(g) + O_2(g)$. A 2.0 L vessel initially contains 0.50 mol of each gas. Is there sufficient information to determine the equilibrium concentrations of the gases? If there is information, what data is needed and why?

15.26 At 1000 K, K_c is 2×10^{15} for the reaction $W(s) + 2 Cl_2(g) \rightleftharpoons WCl_4(g)$. A 1.0 L vessel contains all three substances at equilibrium. Is there sufficient information to determine what the value of Q will be if the amount of $W(s)$ is tripled? If there is insufficient information, what data is needed and why?

Integrative Exercises

15.27 An equilibrium involving N_2O_4 is established at 25 °C in a 1.00 L vessel. The initial concentrations are 0.0500 mol NO_2 and 0.100 mol N_2O_4:

$$N_2O_4(g) \rightleftharpoons 2 NO_2(g) \qquad K_c = 4.61 \times 10^{-3} \text{ at 25 °C}$$

This vessel is connected to another vessel which has a volume of 3.00 L. If the system at equilibrium is permitted to expand into the 3.00 L vessel, (a) in what direction does the reaction proceed when equilibrium is established in the smaller volume before expansion and (b) what are the equilibrium concentrations of N_2O_4 and NO_2 in the total volume after expansion?

15.28 An application of equilibrium principles not discussed in the text, but important in laboratory work involving separating chemical substances, occurs in the partitioning of a solute species between two mutually insoluble liquids. Iodine can be partitioned between water and carbon tetrachloride, both of which are immiscible (mutually insoluble). If iodine is added to this two phase system in a container and the system is thoroughly shaken, iodine distributes itself between water and carbon tetrachloride. The upper phase is aqueous iodine and the lower phase is carbon tetrachloride and iodine. The ratio of concentrations of iodine in the two phases is a constant at constant temperature and is called the partition coefficient:

$$K_c = \frac{[I_2]_{CCl_4}}{[I_2]_{aq}} = 85 \quad \text{(at 25 °C)}$$

(a) In which phase is iodine more soluble? Write the chemical equilibrium.
(b) An aqueous solution has an iodine concentration of 3.00×10^{-2} M. How much iodine remains after 0.500 L of this solution is mixed and shaken with 0.100 L of CCl_4 at 25 °C?
(c) How could the amount of iodine in the aqueous phase be further reduced?

15.29 A chemical system at 800 K initially contains 0.0729 M $CO(g)$ and 0.1687 M $H_2(g)$. The following equilibrium is established:

$$CO(g) + 3 H_2(g) \rightleftharpoons CH_4(g) + H_2O(g)$$

(a) At equilibrium it is found that 72.6% of the CO reacts. What is the value of the equilibrium constant?
(b) Another study at 1000 K finds K_c is 2.5×10^2. Is the reaction exothermic or endothermic? Explain.

Multiple-Choice Questions

15.30 Which of the following factors affect equilibrium concentrations: (1) magnitude of equilibrium constant; (2) catalyst; (3) initial concentrations of products and reactants; (4) temperature; (5) overall reaction rate?

(a) (1), (2), (3)
(b) (1), (3), (4)
(c) (1), (2), (3), (4)
(d) (1), (3), (4), (5)
(e) all of them

15.31 The position of equilibrium for the reaction

$$ZnO(s) + H_2(g) \rightleftharpoons Zn(s) + H_2O(g)$$

does *not* depend upon which of the following: (1) concentration of $ZnO(s)$; (2) concentration of $H_2(g)$; (3) concentration of $Zn(s)$; (4) concentration of $H_2O(g)$; (5) the value of K_c?

(a) (1), (2), (5)
(b) (2), (3), (5)
(c) (1), (3)
(d) (2), (4)
(e) (5)

15.32 Which of the following reactions, initially at equilibrium, will form more product when the total pressure is increased at constant temperature?

(1) $2 CO_2(g) \rightleftharpoons 2 CO(g) + O_2(g)$
(2) $N_2(g) + 3 H_2(g) \rightleftharpoons 2 NH_3(g)$
(3) $SO_2(g) \rightleftharpoons S(s) + O_2(g)$

(a) (1)
(b) (2)
(c) (3)
(d) all of them
(e) none of them

15.33 Which of the following reactions, initially at equilibrium, will shift to the left when the temperature is decreased at constant pressure?

(1) $2 C_2H_2(g) + 5 O_2(g) \rightleftharpoons$
$\qquad 4 CO_2(g) + 2 H_2O(l) \Delta H = -1297$ kJ
(2) $CO_2(g) \rightleftharpoons O_2(g) + C(s) \qquad \Delta H = 393$ kJ
(3) $4 Fe(s) + 3 O_2(g) \rightleftharpoons 2 Fe_2O_3(s)$
$\qquad\qquad\qquad\qquad \Delta H = -1644$ kJ

(a) (1)
(b) (2)
(c) (3)
(d) all of them
(e) none of them

15.34 Which of the following is *not* affected when a catalyst influences a gaseous chemical reaction?

(a) forward and reverse reaction rates
(b) initial reaction rate
(c) value of activation energy
(d) reaction mechanism pathway
(e) value of equilibrium constant

15.35 At 240 °C, the value of K_c is 20 for the reaction

$$PCl_3(g) + Cl_2(g) \rightleftharpoons PCl_5(g)$$

If a 1.0-L container contains 0.25 mol of PCl_5 at 240 °C, how many moles of it dissociate to form PCl_3 and Cl_2? (You'll need to use the quadratic formula to solve this problem.)

(a) 0.0081 mol
(b) 0.090 mol
(c) 0.013 mol
(d) 0.11 mol
(e) 0.25 mol

15.36 What is K_c for the following reaction, $2 SO_3(g) \rightleftharpoons 2 SO_2(g) + O_2(g)$, if K_p is 0.76 at 900 K?

(a) 1.3
(b) 0.76
(c) 0.56
(d) 0.050
(e) 0.010

15.37 Which statement about catalysts and equilibrium is correct?

(a) A catalyst does not appear in a rate equation; thus it does not affect the equilibrium.

(**b**) A catalyst finds a new mechanistic pathway that affects concentrations differently.

(**c**) A catalyst affects the rates of the forward and reverse reactions equally; thus, there is no net change in concentrations.

(**d**) Solid catalysts, which are common in several industrial applications, would not affect the equilibrium constant since solids do not appear in the equilibrium-constant expression.

(**e**) A different and unique catalyst is needed for every equilibrium at different temperatures.

SELF-TEST SOLUTIONS

15.1 (**c**) **15.2** (**a**) **15.3** (**b**) **15.4** (**e**) **15.5** (**d**) **15.6** False. CO_2(lung) $\rightleftharpoons CO_2$ (blood). According to LeChatelier's principle, increasing $[CO_2 \text{ (lung)}]$ causes more CO_2 (blood) to form. **15.7** False. The reaction is $N_2 + 3\,H_2 \rightleftharpoons 2\,NH_3$. Consequently: $Q = [NH_3]^2/[N_2]$ $[H]^3 = (1.20)^2/(0.80)(0.50)^3 = 14$. Since $Q > K_c$, the reaction is not at equilibrium and reacts to form more reactants until $Q = K_{eq}$. **15.8** False. The equilibrium expression is related to the concentration of products/reactants. The given equilibrium expression is reactants/products. The correct equilibrium expression is the reciprocal of the one shown. **15.9** True.

15.10 Partial pressure of a gas is directly proportional to the number of gas particles. If the number of particles of a gas is used to represent partial pressure and used in the K_p expression, the value of 0.5 is not obtained. To obtain a value of 0.5 for K_p, one particle of B needs to be removed:

$K_p = [4]/[2][3-1]^2 = [4]/[2][2]^2 = 0.5$.

15.11 When heat is added more product is formed at equilibrium. This occurs because the value of K_p is larger at 450 K. Refer to Table 15.2 in the *Student's Guide*. Thus the reaction is endothermic.

15.12 The first part of the statement is true: One characteristic of an equilibrium system is that the concentrations of reactants and products have reached steady-state values, a constant condition. The second part of the statement is incorrect. At equilibrium there is a dynamic condition at the microscopic level. The reactant molecules are changing to product molecules and vice versa. However, the forward and reverse rates are equal and this leads to constant concentrations.

15.13 No, the equilibrium-constant expression depends only on the stoichiometry of the given reaction. If a catalyst is used, the reaction mechanism is altered; however, the equilibrium state is not changed. Thus, the equilibrium-constant expression would not change.

15.14 (**a**) $K_c = \dfrac{1}{[Cl_2]}$ The concentration of a pure solid or liquid is not included in the equilibrium-constant expression.

(**b**) $K_c = [Ag1]^2[SO_4^{2-}]$

(**c**) $K_c = \dfrac{[C_3H_8]}{[C_3H_6][H_2]}$

(**d**) $K_c = \dfrac{[CO_2]^6}{[C_3H_6]^2[O_2]^9}$ Note that pure water, a liquid, is not shown.

15.15 The equilibrium-constant value is much less than one; this means that the equilibrium strongly favors the reactant side of the reaction. The equilibrium constant value for the reverse reaction

$$CaO(s) + CO_2(g) \rightleftharpoons CaCO_3(s)$$

is simply the inverse of the original reaction

$$K_{rev} = \frac{1}{K_c} = \frac{1}{6 \times 10^{-3}} = 2 \times 10^2$$

The formation of calcium carbonate is strongly favored at equilibrium.

15.16 (**a**) This system cannot reach a state of equilibrium because no calcium oxide is present to react with the carbon dioxide. Although the calcium oxide does not appear in the equilibrium-constant expression, it must be present for equilibrium to exist.

(**b**) This system can achieve a state of equilibrium by the decomposition of calcium carbonate to the products.

(**c**) This system can also achieve a state of equilibrium. The calcium carbonate can decompose to form both calcium oxide and carbon dioxide. The key is that calcium carbonate is the only reactant present.

(**d**) The system can achieve a state of equilibrium because the concentration of carbon dioxide is greater than $K_c = 6 \times 10^{-3} = [CO_2]$. The carbon dioxide will react with the calcium oxide and form calcium carbonate until its concentration reaches the value of K_c.

15.17 (**a**) The equilibrium-constant expression does not include a pure solid, in this case silver chloride. Thus, changes in the amount of silver chloride do not affect the position of equilibrium. The concentrations of the silver and chloride ions do not change.

(**b**) First, calculate the value of Q.

$$Q = \frac{C_{NO_2}^2}{C_{N_2O_4}} = \frac{(1 \times 10^{-3})^2}{1 \times 10^{-3}} = 1 \times 10^{-3}$$

The given concentrations are used to calculate the value of Q and are not necessarily the equilibrium concentrations. Compare the values of Q and K_c: $1 \times 10^{-3} < 4.7 \times 10^{-3}$. This means that the system is not at equilibrium and the value of Q must increase until it equals the value of K_c. The reaction continues in the forward direction until equilibrium exists.

15.18 (**a**) The reaction is endothermic and the reaction can be viewed as follows:

$$\text{heat} + 2\,NOCl(g) \rightleftharpoons 2\,NO(g) + Cl_2(g)$$

If the temperature is decreased, heat is removed from the

system and the quantity of reactant decreases. The system responds by changing the position of equilibrium so that more reactant is formed.

(b) If the partial pressure of chlorine gas is increased, the system responds by removing some of the added chlorine gas; the system responds by forming more reactant.

(c) The total pressure of the system is inversely proportional to the total volume ($P = nRT/V$). If the total volume is increased this results in a reduction in pressure. The system responds by trying to increase the pressure, which can occur if the total number of gas particles increases: The system responds by forming more product because three moles of products are gases compared to two moles of reactant.

15.19 An inspection of the chemical reaction

$$CO_2(g) + H_2(g) \rightleftharpoons CO(g) + H_2O(g)$$

shows that here is a 1:1 stoichiometric relationship between all species. Thus, a decrease in the concentration of carbon dioxide is also accompanied by an equal decrease in the concentration of hydrogen gas; also, there is a simultaneous increase in the concentrations of carbon monoxide and water. These changes can be expressed as follows: $\Delta(CO_2) = \Delta(H_2)$; and $\Delta(CO) = \Delta(H_2O)$. Since Δ (reactant) must be a negative number and Δ (product) must be a positive number, you can write the following relationships: $-\Delta(CO_2) = \Delta(CO) = \Delta(H_2O)$ and similarly $-\Delta(H_2) = \Delta(CO) = \Delta(H_2O)$.

15.20 In this reaction, $CO(g) + 3 H_2(g) \rightleftharpoons CH_4(g) + H_2O(g)$, an increase or decrease in the concentration of CO is accompanied by a change in the hydrogen gas concentration that is three times that for CO: Note that there is a 3:1 stoichiometric ratio between H_2 and CO. Thus we can write $\Delta(H_2) = 3\Delta(CO)$ or $\Delta(CO) = 1/3\Delta(H_2)$. Relationships can be written between reactants and products through a similar analysis of stoichiometric relationships. Note, for example, that water forms in a 1:3 ratio when compared to the change in concentration of hydrogen gas. $\Delta(H_2O) = -1/3\Delta(H_2)$. The negative sign is needed because the change in the concentration of hydrogen gas is a negative number, it is a reduction in concentration.

15.21 The table is completed based on the stoichiometry of the reaction,

$$2 CO_2(g) \rightleftharpoons 2 CO(g) + O_2(g)$$

	CO_2	CO	O_2
Initial(M)	3.0	0	1.0
ΔConc.(M)	$-2x$	$+2x$	$+x$
Equil.(M)	$3.0 - 2x$	$2x$	$1.0 + x$

The first line is based on the given starting concentrations in the problem. There is no CO initially present so it is assigned a value of zero. One of the substances must be assigned a change in concentration based on the variable x. Since oxygen has a stoichiometric coefficient of one it is arbitrarily assigned the variable. Why is it positive? The answer is provided by the first line: the number

under CO is zero. No CO is initially present, thus it *must* form when the system proceeds to equilibrium. If it forms, then oxygen must form and carbon dioxide must disappear. If instead you wanted to assign the variable x to CO, what would the table look like? Here is the result:

	CO_2	CO	O_2
Initial(M)	3.0	0	1.0
ΔConc.(M)	$-x$	$+x$	$+1/2x$
Equil(M)	$3.0 - x$	x	$1.0 + 1/2x$

15.22 Let x be the molar concentration of $NH_3(g)$ formed. This gives us the following:

	$NH_4Cl(g) \rightleftharpoons$	$NH_3(g) +$	$HCl(g)$
Initial (M):	0.500	0	0
Change (M):	$-x$	$+x$	$+x$
Equilibrium (M):	$(0.500 - x)$	x	x

Thus, $K_c = [NH_3][HCl]/[NH_4Cl] = (x)(x)/(0.500 - x)$

$= 1.1 \times 10^{-2}$. Solving for x using the quadratic formula yields $x = [NH_3] = 6.9 \times 10^{-2}$ mol/L.

15.23 According to the stoichiometry of the reaction, when 50.0% of the $SO_3(g)$ decomposes at equilibrium, the following equilibrium concentrations result: 0.238 M $SO_3(g)$, $\frac{1}{2}(0.238\ M)$ $O_2(g)$, and 0.238 M $SO_2(g)$. Solving for K_c

$$K_c = \frac{[SO_2][O_2]}{[SO_3]} = \frac{(0.238)^2(0.119)}{(0.238)^2} = 0.119$$

15.24 Using the stoichiometry of the equilibrium $C(s) + CO_2(g) \rightleftharpoons 2 CO(g)$, we can tabulate concentrations of CO_2 and CO. We need not consider $C(s)$ because

	CO_2	CO
Initial (atm):	2.00	0
Change (atm):	$-x$	$+2x$
Equilibrium (atm):	$(2.00 - x)$	$2x$

it is a solid

(The expression $2x$ is required for CO because for every 1 mol of CO_2 that reacts 2 mol of CO forms.) Thus, $K_p = \dfrac{P_{co}^2}{P_{co_2}}$

$= (2x)^2/(2.00 - x) = 167.5$ or $4x^2 = (2.00 - x)(167.5)$, $= 335 - 167.5x$ or $4x^2 + 167.5x - 335 = 0$. Using the quadratic formula to solve for x yields

$$x = \frac{-(167.5) \pm \sqrt{(167.5)^2 - 4(4)(-335)}}{(2)(4)} = 1.91$$

Thus, the partial pressure of CO_2 is 2.00 atm $- x = 2.00$ atm $- 1.91$ atm $= 0.09$ atm and for CO it is $2x = 3.82$ atm.

15.25 There is insufficient information because K_c for the reaction is not given. K_c is necessary to calculate equilibrium concentrations when initial concentrations are provided. However, we may be able to find the value of K_c in a reference.

15.26 There is sufficient information. The position of equilibrium is not affected with the addition of more tungsten (W), a solid. A solid does not appear in an equilibrium-constant expression; thus, $Q = K_c$.

15.27 (a) To answer this question we need to calculate the value of the reaction quotient and compare it to the value of the equilibrium concentration. The initial molarities of N_2O_4 and NO_2 are 0.100 M and 0.0500 M, respectively, because the volume is one liter. The reaction quotient, Q, is calculated as follows:

$$Q = \frac{C_{NO_2}^2}{C_{N_2O_4}} = \frac{(0.0500)^2}{0.100} = 0.0250$$

Since $Q > K_c$ the direction of the reaction is towards the reactants as equilibrium is established

(b) The equilibrium concentrations can be calculated as if the two gases were originally in a 3.00 L vessel and then they are permitted to reach equilibrium. The initial concentrations in a 3.00 L vessel are:

$$C_{NO_2} = \frac{0.0500 \text{ mol}}{3.00 \text{ L}} = 0.0167 \text{ M}$$

$$C_{N_2O_4} = \frac{0.100 \text{ mol}}{3.00 \text{ L}} = 0.0333 \text{ M}$$

	N_2O_4	NO_2
Initial(M)	0.0333	0.0167
Change(M)	$+x$	$-2x$
Equilibrium(M)	$0.0333 + x$	$0.0167 - 2x$

Next, develop a concentration table:
The value of K_c is reasonably large and it is likely that we cannot neglect the value of x.

$$K_c = \frac{(0.0167 - 2x)^2}{0.0333 + x} = 4.61 \times 10^{-3}$$

$$4x^2 - 0.0378x + 1.26 \times 10^{-4} = 0$$

$$x = 0.00197$$

The equilibrium concentrations are:

$$[NO_2] = (0.0167 - 2x)M = (0.0167 - 2(0.00197))$$
$$M = 0.0128 \text{ M}$$

$$[N_2O_4] = (0.0333 + x)M = (0.0333 + 0.00197)$$
$$M = 0.0353 \text{ M}$$

15.28 (a) The carbon tetrachloride layer as $K_c \gg 1$. The chemical equilibrium is:

$$I_2(aq) \rightleftharpoons I_2(CCl_4)$$

(b) Since there are two independent solutions involved it is necessary to calculate the number of moles of iodine in each solution at equilibrium in order to determine the molarity of iodine in each phase. Note that the volumes of the phases are different. The number of moles of iodine initially present is: $(3.00 \times 10^{-2} \text{ mol/L})(0.500 \text{ L}) = 0.0150 \text{ mol } I_2$. Let x = the number of moles of iodine extracted from water and dissolved in the carbon tetrachloride layer. This permits

	$I_2(aq)$	$I_2(CCl_4)$
Initial (M)	3.00×10^{-2}	0
Change (M)	$\dfrac{-x}{0.500 \text{ L}}$	$\dfrac{+x}{0.100 \text{ L}}$
Equil. (M)	$3.00 \times 10^{-2} - \dfrac{x}{0.500 \text{ L}}$	$\dfrac{x}{0.100 \text{ L}}$

$$K_c = \frac{\left(\dfrac{x}{0.100}\right)}{\left(3.00 \times 10^{-2} - \dfrac{x}{0.500}\right)} = 85$$

$$K_c = \frac{5x}{(3.00 \times 10^{-2})(0.500) - x} = 85$$

Solving for x gives:

$$x = 0.0142 \text{ mol } I_2$$

The amount of iodine in the aqueous layer after extraction is

$$3.00 \times 10^{-2} \text{ M} - \frac{0.0142 \text{ mol}}{0.500 \text{ L}} = 0.17 \times 10^{-2} \text{ M}$$

This represents about 6% of the original amount.

(c) Extracting the aqueous layer again with pure carbon tetrachloride will further reduce the concentration of iodine in the aqueous phase.

15.29 (a) To develop a concentration table for the equilibrium we will have to calculate the changes in concentration for each species. You are told 72.6% of the CO disappears. This means that $(0.726)(0.0729 \text{ M}) = 0.0529 \text{ M}$ of the CO reacts. Based on the stoichiometry of the reaction we can then calculate the change in the concentrations of the other species:

$$CO(g) + 3H_2(g) \rightleftharpoons CH_4(g) + H_2O(g)$$

	0.0729	0.1687	0	0
Initial (M)	0.0729	0.1687	0	0
Change (M)	-0.0529	$-3(0.0529)$	$+0.0529$	$+0.0529$
Equil. (M)	0.0200	0.0100	0.0529	0.0529

$$K_c = \frac{[CH_4][H_2O]}{[CO][H_2]^3} = \frac{(0.0529)(0.0529)}{(0.0200)(0.0100)^3} = 1.40 \times 10^5$$

(b) The value of the equilibrium constant has decreased but the temperature is higher. Thus the input of heat is driving the reaction to the left, the reactant side. This will happen if the reaction is exothermic. The system absorbs energy to counteract the effect of increasing temperature; it does this by shifting its position of equilibrium toward the reactants:

$$CO(g) + 3H_2(g) \rightleftharpoons CH_4(g) + H_2O(g) + heat$$

15.30 (b)

15.31 (c) Pure solids do not affect the position of equilibrium.

15.32 (b) The system restores equilibrium by changing so that fewer total moles of gas exist, that is, so there is less total pressure.

15.33 (b) Since in an endothermic reaction heat can be considered as a reactant, the removal of heat by lowering the temperature will cause the equilibrium to shift to the left to produce more heat.

15.34 (e). 15.35 (b).

$$K_c = [PCl_5]/[PCl_3][Cl_2] = (0.25 - x)/(x)(x)$$
$$= 20; x = 0.090 \text{ M}$$

15.36 (e) $\Delta n = 1$ in the relationship $K_p = K_c(RT)^{\Delta n}$.

15.37 (c)

Sectional MCAT and DAT Practice Questions V

The solubility of a gas in water is determined by the partial pressure of the gas above the solution. The relationship between the concentration of a dissolved gas and its partial pressure is given by Henry's law:

$$C_{gas} = kP_{gas}$$
Equation 1

C_{gas} is the concentration of the dissolved gas, P_{gas} is the partial pressure of the gas above the solution, and k is Henry's constant. Table 1 lists Henry's law constants for several gases dissolved in water.

TABLE 1 Henry's law constants, k

Gas	k (mol/L-atm)
He	3.9×10^{-4}
Ne	4.7×10^{-4}
Ar	1.5×10^{-3}
N_2	7.1×10^{-4}
O_2	1.4×10^{-3}

When $CO_2(g)$ dissolves in water it forms carbonic acid. Carbonic acid ionizes in water to form hydrogen, bicarbonate, and carbonate ions. The following three equations show the formation of carbonic acid and its step-wise ionization in water

$$H_2O(l) + CO_2(aq) \rightleftarrows H_2CO_3(aq) \quad K = 1.7 \times 10^{-3} \text{ at } 25 \text{ °C}$$
Equation 2

$$H_2CO_3(aq) \rightleftarrows H^+(aq) + HCO_3^-(aq) \quad K = 2.5 \times 10^{-4} \text{ at } 25 \text{ °C}$$
Equation 3

$$HCO_3^-(aq) \rightleftarrows H^+(aq) + CO_3^{2-}(aq) \quad K = 5.6 \times 10^{-11} \text{ at } 25 \text{ °C}$$
Equation 4

Calcium carbonate (calcite) is only slightly soluble in water as shown by the equilibrium

$$CaCO_3(s) \rightleftarrows Ca^{2+}(aq) + CO_3^{2-}(aq) \quad K = 4.5 \times 10^{-9} \text{ at } 25 \text{ °C}$$
Equation 5

Calcium carbonate forms in natural water if the concentrations of calcium and carbonate ions are sufficiently high. Monuments made of calcite degrade over time because of the following reaction in acid environments.

$$CaCO_3(s) + 2\,H^+(aq) \rightarrow Ca^{2+}(aq) + CO_2(g) + H_2O(l)$$
Equation 6

The reaction is first order in hydrogen ion and also in calcium carbonate.

1. Which has the greatest concentration when $CO_2(g)$ is in contact with water at normal atmospheric conditions when its partial pressure is 3.5×10^{-4} atm?
   (a) $CO_2(aq)$ (b) $H_2CO_3(aq)$
   (c) $HCO_3^-(aq)$ (d) $H^+(aq)$

2. Which change in the condition of a saturated solution of calcium carbonate containing excess solid at 25 °C will increase the concentration of calcium ions?
   (a) Making the solution more acidic.
   (b) Adding more solid calcium carbonate.
   (c) Adding KCl.
   (d) Adding K_2CO_3.

3. The equilibrium constant expression for Equation 5 is
   (a) $K = [Ca^{2+}][CO_3^{2-}]$ (b) $K = \dfrac{[Ca^{2+}][CO_3^{2-}]}{[CaCO_3]}$
   (c) $K = \dfrac{1}{[Ca^{2+}][CO_3^{2-}]}$ (d) $K = \dfrac{[CaCO_3]}{[Ca^{2+}][CO_3^{2-}]}$

4. Water seeping through the ground to a deep underground layer of water experiences an increasing partial pressure of carbon dioxide. When this water rises to the surface at ambient conditions, which occurs?
   (a) Carbonic acid concentration increases
   (b) Calcium ion concentration remains unchanged
   (c) Carbon dioxide gas is released
   (d) Hydrogen ion concentration increases

5. Which of the following gases has a Henry's law constant in Table 1 that appears inconsistent with its expected solubility in water based on London dispersion forces?
   (a) Ar (b) He
   (c) N_2 (d) O_2

6. What is the rate expression for the reaction of calcium carbonate with hydrogen ion?
   (a) rate $= k[H^+]$ (b) rate $= k[CaCO_3][H^+]$
   (c) rate $= k[CaCO_3][H^+]^2$ (d) rate $= k[H^+]^2$

7. If the concentration of hydrogen ions is doubled in an experimental study of the chemical kinetics of Equation 6, with all other conditions held constant, the rate of the reaction
   (a) does not change. (b) doubles.
   (c) quadruples. (d) decreases.

Questions 8 through 12 are **not** based on a descriptive passage.

8. When NaCl dissolves in water,
    (a) energy is released when NaCl bonds are broken.
    (b) Na^+ ions align themselves with the positive regions of dipoles of water.
    (c) the positive regions of dipoles of water interact with Cl^- ions.
    (d) sodium and chloride ions move as independent entities in water.

9. Which aqueous solution should have the largest increase in boiling point compared to the boiling point of pure water?
    (a) 0.10 m $CaCl_2$ (b) 0.10 m HCl
    (c) 0.10 m KNO_3 (d) 0.10 m NH_3

10. Which rate law is consistent with the following mechanism in the solvent water?

    $H_2O_2 + HI \rightarrow H_2O + HIO$ (fast and reversible)
    $HIO + HI \rightarrow H_2O + I_2$ (slow)

    (a) Rate $= k[H_2O_2][HI]$
    (b) Rate $= k[HIO][HI]$
    (c) Rate $= k[H_2O_2]^2[HI]$
    (d) Rate $= k[H_2O_2][HI]^2$

11. The following reaction at 25°C in the gas phase is exothermic

    $$2\,SO(g) + O_2(g) \rightleftharpoons 2\,SO_3(g)$$

    What change occurs if the total volume of the system is increased at a constant 25 °C?

    (a) Pressure increases.   (b) The value of K changes.
    (c) More SO_3 forms.   (d) More O_2 forms.

12. The following system is at equilibrium at 25 °C.

    $$N_2(g) + O_2(g) \rightleftharpoons 2\,NO(g) \qquad\qquad K_c = 0.0112 \text{ at } 25\,°C$$

    What change in the position of equilibrium occurs if the concentrations of the gases are changed to N_2, 2.00 M, O_2, 1.00 M, and NO, 0.500 M?

    (a) More NO forms.   (b) More N_2 forms.
    (c) No change occurs.   (d) Temperature increases.

ANSWERS

1. (a) The reactions shown in Equations 2, 3, and 4 provide the information required to answer this question. All have equilibrium constants that are not large. Thus, in all cases, the reactants are favored compared to the products at equilibrium. This means that only a small portion of dissolved carbon dioxide forms carbonic acid and the subsequent step-wise ionization products of carbonic acid.

2. (a) According to Equation 6, adding hydrogen ions to solid calcium carbonate causes the formation of calcium ions. Making a solution more acidic increases the amount of hydrogen ions and therefore the amount of calcium ion. Adding more solid calcium carbonate does not affect the position of equilibrium in a saturated solution. Adding KCl does not change concentrations because neither of its ions reacts with calcium carbonate, calcium ion, or carbonate ion. Adding K_2CO_3, a source of carbonate ions, shifts the equilibrium to the left and reduces the amount of calcium ions at equilibrium.

3. (a) An equilibrium constant expression is a ratio of product concentrations to reactant concentrations. Pure solids and pure liquids do not appear in it. Calcium carbonate is a solid and thus it is not in the equilibrium constant expression.

4. (c) In the deep layer of water the amount of dissolved carbon dioxide gas in water is greater than at the surface because of the increased partial pressure of carbon dioxide. When this water rises to the surface the partial pressure of carbon dioxide above the solution decreases. According to Henry's law, this results in less dissolved carbon dioxide and the system responds by releasing carbon dioxide gas. Reducing the quantity of dissolved carbon dioxide gas also reduces the amount of carbonic acid and hydrogen ions.

5. (d) The data in Table 1 shows that the Henry's law constant for O_2 is greater than those for N_2 and He and is similar to that of Ar. The larger the value of Henry's law constant, the more soluble the gas is in water with a given partial pressure of gas. Thus, oxygen is about twice as soluble in water than nitrogen or helium. However, London dispersion forces should be somewhat equivalent for O_2 and N_2 given their similar molecular sizes and number of electrons. This normally results in similar solubilities and similar Henry's law constants. Helium is smaller in size, is less polarizable, and it should be somewhat less soluble. Ar is in the third period, and is a more polarizable species with its valence electrons in a higher n level; therefore it has a greater solubility. The reason for the greater solubility of oxygen in water compared to nitrogen is its paramagnetic property. Note: A paramagnetic O_2 molecule is more polarizable than a diamagnetic N_2 molecule, and this leads to oxygen's greater solubility.

6. (b) You are told in the descriptive passage that the reaction is first order in hydrogen ion and also in calcium carbonate. The rate expression in **(b)** shows the concentrations of both hydrogen ion and calcium carbonate with exponents of one. A component with a first order has a concentration raised to the first power. In chemical kinetics a solid may appear in a rate expression; however, a pure solid is not shown in equilibrium constant expressions.

7. (b) If the concentration of hydrogen ion is doubled and all other factors are held constant, then the rate will double: rate $= k[2 \times$ conc. $H^+][$constant calcium carbonate quantity]. This becomes rate $= k[2][$conc. $H^+][$constant calcium carbonate quantity]. The later equation shows the rate doubles.

8. (c) A negatively charged ion is repelled by the negative region of a dipole and is attracted to the positive region. The opposite occurs for a positively charged ion. A chloride ion is attracted to the positive region of water's dipole. Breaking of NaCl bonds requires energy. Sodium and chloride ions are hydrated as ions and do not move as independent species.

9. (a) The increase in boiling point is directly proportional to the molality of all solute species. $CaCl_2$ is a strong electrolyte and dissociates to form three ions per formula unit. Thus, the concentration of ions used in calculating boiling point elevation is $3 \times 0.10\ m$ or $0.30\ m$. The other substances form fewer ions per formula unit when they dissociate compared to calcium chloride. This condition along with the molalities of the substances results in smaller total molalities of ions in solution leading to smaller boiling point elevations. Ammonia is a weak electrolyte and only slightly dissociates.

10. (d) The rate depends on the slowest step, the second step of the mechanism. We can write Rate $= k[\text{HIO}][\text{HI}]$ for that step. However, HIO is an intermediate because it is formed in the first step and disappears in the second step. Intermediates do not appear in rate expressions. Answer **(a)** is not valid because it is a rate expression for a fast step. However, an equilibrium exists for the first step, thus $K = [\text{H}_2\text{O}][\text{HIO}]/[\text{H}_2\text{O}_2][\text{HI}]$. Water is in a large excess as the solvent and its concentration can be included with K to form a new equilibrium constant, K'. Therefore, we can write: $K' = [\text{HIO}]/[\text{H}_2\text{O}_2][\text{HI}]$. We can rearrange the equation to obtain the concentration of HIO: $[\text{HIO}] = K'[\text{H}_2\text{O}_2][\text{HI}]$. If this is substituted in the rate expression for the slow step we obtain: Rate $= kK'[\text{HI}][\text{H}_2\text{O}_2][\text{HI}] = kK'[\text{H}_2\text{O}_2][\text{HI}]^2$. This is equivalent to the rate expression for answer **(d)** because kK' is also a constant.

11. (d) Increasing the total volume of the system decreases the total pressure from the equilibrium total pressure because P is proportional to $1/V$. According to LeChatelier's principle the reaction responds to restore the state of equilibrium by increasing the total pressure of the system. The total pressure increases if the total number of moles of gas increases. We see from the stoichiometry of the reaction that the total number of moles of gases increases when products are converted to reactants ($\Delta n_g = +1$ towards the reactants). The system shifts in this direction and more $\text{O}_2(g)$ forms. The value of K changes only if the temperature is changed.

12. (b) At the new concentration conditions the ratio $\dfrac{[\text{NO}]^2}{[\text{N}_2][\text{O}_2]}$ becomes $\dfrac{[0.500]^2}{[2.00][1.00]} = 0.125$. This is larger than the value of K_c. The system reacts to restore equilibrium by forming more reactants and less product until the value of the ratio equals K_c. N_2 is one of the reactants and choice **(b)** is the correct response.

Acid–Base Equilibria

Chapter 16

OVERVIEW OF THE CHAPTER

Review: Acids and bases (4.3); hydrogen bond (11.2).

Learning Goals: You should be able to:

1. List the general properties that characterize acidic and basic solutions and identify the ions responsible for these properties.
2. Define the terms Brønsted-Lowry acid and base, and conjugate acid and base.
3. Identify the conjugate base associated with a given Brønsted-Lowry acid and the conjugate acid associated with a given Brønsted-Lowry base.

Review: Concept of equilibrium and equilibrium constant (15.1, 15.2).

Learning Goals: You should be able to:

1. Explain what is meant by the autoionization of water and write the ion-product constant expression for the process.
2. Define pH; calculate pH from a knowledge of $[H^+]$ or $[OH^-]$, and perform the reverse operation.

Review: Concentration units (13.4); strong electrolytes (4.1).

Learning Goal: You should be able to identify the common strong acids and bases and calculate the pHs of their aqueous solutions given their concentrations.

Review: Use of quadratic formula (see Appendix in the text); weak electrolytes (4.2); equilibrium calculations (15.6).

Learning Goals: You should be able to:

1. Calculate the pH for a weak acid solution in water, given the acid concentration and K_a; calculate K_a given the acid concentration and pH.
2. Calculate the pH for a weak base solution in water, given the base concentration and K_b; calculate K_b given the base concentration and pH.
3. Calculate the percent ionization for an acid or base, knowing its concentration in solution and the value of K_a or K_b.

16.9 SALT SOLUTIONS: HYDROLYSIS

Review: Ions (2.7); hydration (13.1).

Learning Goals: You should be able to:

1. Determine the relationship between the strength of an acid and that of its conjugate base; calculate K_b from a knowledge of K_a, and vice versa.
2. Predict whether a particular salt solution will be acidic, basic, or neutral.

16.10 FACTORS AFFECTING ACID–BASE STRENGTH

Review: Bond strength (8.8); electronegativity (8.4); polarization (8.4); Lewis structures (8.5, 8.6).

Learning Goals: You should be able to:

1. Explain how acid strength relates to the polarity and strength of the H—X bond.
2. Predict the relative acid strengths of oxyacids.

16.11 THE LEWIS MODEL OF ACIDS AND BASES

Review: Hydration (13.2); Lewis structures (8.5, 8.6).

Learning Goals: You should be able to:

1. Define an acid or base in terms of the Lewis concept.
2. Predict the relative acidities of solutions of metal salts from a knowledge of metal-ion charges and ionic radii.

TOPIC SUMMARIES AND EXERCISES

ACIDS AND BASES: BRØNSTED-LOWRY DEFINITION

In Section 4.3 of the text acid–base properties of substances in water are characterized.

- Acidic solutions are characterized by the presence of hydrated protons, $H^+(aq)$ in greater quantity than $OH^-(aq)$. An acid transfers a proton to water to form a hydrogen-bonded species. This species, sometimes called the **hydronium ion**, is often written as H_3O^+.
- Basic solutions are characterized by the presence of $OH^-(aq)$ in greater quantity than $H^+(aq)$.

J. Brønsted and T. M. Lowry proposed a model for characterizing the reactions of acids and bases in terms of proton (H^+) transfer.

- An **acid** is defined as any substance able to donate a proton in a chemical reaction, and a **base** is defined as any substance capable of accepting a proton in a chemical reaction.
- For example, look at the following reaction:

$$\underset{\text{Base}}{H_2O(l)} + \underset{\text{Acid}}{\textcircled{H}Cl(aq)} \longrightarrow H_3O^+(aq) + Cl^-(aq)$$

Proton transfer

Water is a base in this reaction because it accepts a proton from the acid HCl to form a hydrated proton, $H_3O^+(aq)$. For the remainder of this Student's

Guide, the hydrated proton will be represented as $H^+(aq)$. The above reaction becomes

$$HCl(aq) \longrightarrow H^+(aq) + Cl^-(aq)$$

- Acids such as HCl, HNO_3, or H_2SO_4 ionize in water to form a hydrated proton and a species called the **conjugate base** of the acid:

$$\underset{\text{Acid}}{HCl(aq)} \longrightarrow H^+(aq) + \underset{\substack{\uparrow \\ \text{Conjugate} \\ \text{base of HCl}}}{Cl^-(aq)}$$

Note that the conjugate base of HCl results from the *loss of one proton*.

- Bases such as NH_3 react with water to form a hydrated hydroxide ion and a species called the **conjugate acid** of the base.

$$\underset{\text{Base}}{NH_3(aq)} + H_2O(l) \rightleftharpoons \underset{\substack{\uparrow \\ \text{Conjugate} \\ \text{acid of } NH_3}}{NH_4^+(aq)} + OH^-(aq)$$

Note that the conjugate acid of NH_3 results from the *gain of one proton*.

- A **conjugate acid–base pair** is therefore an acid and a base, such as HF and F^-, that differ only in the presence or absence of *one* proton.
- When water reacts with ammonia it acts as an acid. When water reacts with HCl it acts as a base. A substance capable of acting both as an acid and a base is called **amphoteric**.

The strength of an acid is measured by its tendency to donate a proton, and that of a base by its tendency to form hydroxide ions or accept a proton.

- **Strong acids** react completely with water to form $H^+(aq)$ and their conjugate bases. Strong acids are said to undergo 100 percent dissociation in terms of proton donation to water.
- **Strong bases** ionize completely in water—that is, all ionizable hydroxide units in the base come off and are hydrated.
- The ionization reactions of strong acids and bases with water are written with a single arrow between products and reactants. For example, the ionization reactions of HBr and KOH are represented as follows:

$$\text{Strong acid: } HBr(aq) \longrightarrow H^+(aq) + Br^-(aq)$$

$$\text{Strong base: } KOH(aq) \longrightarrow K^+(aq) + OH^-(aq)$$

- **Weak acids** and **bases** only partly ionize in water. The reactions of weak acids and bases with water are written with double arrows between products and reactants. We do this to show that an equilibrium condition exists. For example, the ionization of the weak acid HF is represented as follows:

$$HF(aq) \rightleftharpoons H^+(aq) + F^-(aq)$$

- *It is important to realize that the stronger the acid, the weaker its conjugate base and the stronger the base, the weaker its conjugate acid.* For example, the weak acid HF has a conjugate base F^- that has a strong tendency to attract a proton to form HF. You need to understand that the conjugate base of a weak acid is a weak base, not a strong base in water. Similarly, the conjugate acid of a weak base is a weak acid, not a strong acid.

EXERCISE 1   Writing reactions of acids and bases with water and identifying acid/base properties of ions

(a) Write a chemical equation describing the ionization or dissociation in water of the following acids and bases: $HCl(aq)$, strong acid; $HBr(aq)$, strong acid; $HNO_2(aq)$, weak acid; $NH_3(aq)$, weak base; $NaOH(s)$, strong base. (b) Which of the following would you expect to have a significant capacity to react with a hydrated proton: Cl^-, Br^-, NO_2^-, NH_4^+?

SOLUTION: *Analyze*: (a) We are given several acids and bases and asked to write a chemical equation describing their ionization or dissociation in water. (b) We are given three anions and one cation and asked which are basic.

Plan: (a) We need to determine which acids are weak and strong and repeat for the bases. In this exercise the information is given; however, if it were missing we would have to find it in other sources or know it. Molecular acids and bases undergo ionization and metallic bases undergo dissociation in water. (b) Anions that are weak or strong bases react with hydrogen ions. Anions derived from weak acids or the hydroxide ion meet this requirement. We need to determine which ions are weak bases in this group of ions.

Solve:(a) Remember that one arrow between products and reactants is used with a strong acid or base and double arrows are used with weak acids and bases:

$$HCl(aq) \longrightarrow H^+(aq) + Cl^-(aq)$$

$$HBr(aq) \longrightarrow H^+(aq) + Br^-(aq)$$

$$HNO_2(aq) \rightleftharpoons H^+(aq) + NO_2^-(aq)$$

$$NH_3(aq) + H_2O(l) \rightleftharpoons NH_4^+(aq) + OH^-(aq)$$

$$NaOH(s) \xrightarrow{H_2O} Na^+(aq) + OH^-(aq)$$

(b) Only the following anion is derived from a weak acid: NO_2^-. HNO_2 is a weak acid. The ammonium ion is a weak acid.

Comment: It is important to know the common weak acids and bases. We will be using them often and it is easier to do problems if we know them. In problem (b) note that the ammonium ion, which has a positive charge, will not react with a hydrated hydrogen ion because it also has a positive charge.

EXERCISE 2   Identifying Brønsted-Lowry acids and bases in chemical reactions

Indicate which reactant is the acid and which is the base in each of the following reactions:

(a) $CH_3COOH(aq) + NaOH(aq) \longrightarrow CH_3COONa(aq) + H_2O(l)$
(b) $ZnS(s) + 2\,HCl(aq) \longrightarrow H_2S(g) + ZnCl_2(aq)$
(c) $KCN(aq) + H_2O(l) \longrightarrow KOH(aq) + HCN(aq)$

SOLUTION: *Analyze*: We are given three reactions and asked which reactants are acids and which are bases.

Plan: We need to recognize that an acid donates a proton and a base accepts the proton in a chemical reaction. Therefore we need to look at each reaction and determine how a proton is transferred.

Solve: (**a**) CH_3COOH is the acid, and NaOH is the base. (The OH^- ion accepts the proton.) (**b**) HCl is the acid and ZnS is the base. (The S^{2-} ion accepts the protons to form H_2S.) (**c**) H_2O is the acid and KCN is the base. (The CN^- ion accepts the proton.)

Check: If we know the common acids and bases we can confirm our analysis. For example, acetic acid is a weak acid and sodium hydroxide is a strong base and this agrees with our conclusion in (**a**). In (**b**) we should know that HCl is a strong acid and therefore zinc (II) sulfide must be the base. In (**c**) water is amphoteric and we have to determine if potassium cyanide is a base. If we remember that the cyanide ion is the conjugate base of the weak acid HCN then we know it is a weak base.

EXERCISE 3   Writing conjugate acids

Write the conjugate acid for each of the following species: (**a**) H_2O; (**b**) OH^-; (**c**) CO_3^{2-}; (**d**) NH_3.

SOLUTION: *The conjugate acid is formed by the addition of **one** proton:* (**a**) $H_3O^+(aq)$ or $H^+(aq)$; (**b**) H_2O; (**c**) HCO_3^-; (**d**) NH_4^+.

EXERCISE 4   Writing conjugate bases

Write the conjugate base for each of the following species: (**a**) H_2O; (**b**) OH^-; (**c**) H_2SO_4; (**d**) NH_3.

SOLUTION: *The conjugate base is formed by the removal of **one** proton:* (**a**) OH^-; (**b**) O^{2-}; (**c**) HSO_4^-; (**d**) NH_2^-.

EXERCISE 5   Identifying the stronger Brønsted-Lowry acid

CN^- reacts with water as follows:

$$CN^-(aq) + H_2O(l) \rightleftharpoons HCN(aq) + OH^-(aq)$$

A Cl^- ion does *not* react with water to form HCl. Which acid is stronger, HCl(*aq*) or HCN(*aq*)? Why?

SOLUTION: *Analyze*: We are given a chemical equation describing how CN^- reacts with water to form OH^- and are also told that a chloride ion does not react with water to form an acid. From this information we are asked to determine which acid is stronger, hydrochloric acid or hydrocyanic acid, and to explain our conclusion.

Plan: We are told in the summary section that the stronger a base is, the weaker its conjugate acid, and *vice versa*. We can use this information to answer the question.

Solve: The cyanide ion is basic because in its reaction with water it forms the hydroxide ion (a strong base); its conjugate acid is HCN. We also know that the chloride ion is not basic because it does not react with water to form a hydroxide ion; its conjugate acid is HCl. Therefore HCl is a stronger acid than HCN because a chloride ion is a weaker base than a cyanide ion.

ACIDS AND BASES IN WATER: BRØNSTED-LOWRY THEORY

Water undergoes **autoionization**, as shown in the chemical equation

$$H_2O(l) \rightleftharpoons H^+(aq) + OH^-(aq)$$

- The equilibrium-constant expression for the above is called the **ion-product constant** for water, K_w. At 25 °C,*

$$K_w = [H^+][OH^-] = 1.0 \times 10^{-14}$$

- In pure water at 25 °C, $[H^+] = [OH^-] = 10^{-7}M$.
- The autoionization of water forms H^+ and OH^- ions in any aqueous solution. We can calculate $[H^+]$ or $[OH^-]$ if we know the value of the other quantity:

$$[OH^-] = \frac{1.0 \times 10^{-14}}{[H^+]} \quad \text{or} \quad [H^+] = \frac{1.0 \times 10^{-14}}{[OH^-]}$$

For convenience, the hydrogen-ion concentration of a solution is commonly expressed in terms of **pH**, where pH is defined as:

$$pH = -\log[H^+] = \log\left(\frac{1}{[H^+]}\right)$$

- When we are given pH and asked to solve for $[H^+]$, we can use the 10^x (where $x = -pH$) function of a scientific calculator as follows:

$$[H^+] = 10^{-pH}$$

- You can describe acidity and basicity in terms of either $[H^+]$ or pH (see Table 16.1).
- Most solutions have a pH range of 0–14, but pH values greater than 14 and less than zero are possible.

TABLE 16.1 Relationship of $[H^+]$ and pH to Acidity and Basicity

Kind of solution	$[H^+]$, $[OH^-]$ relationship	pH requirement
Neutral	$[H^+] = [OH^-]$	pH = 7
Acidic	$[H^+] > [OH^-]$	pH < 7
Basic	$[OH^-] > [H^+]$	pH > 7

Other useful relationships in doing problems are:

$$pOH = -\log[OH^-]$$

$$pK_w = -\log K_w = 14$$

$$pK_w = pH + pOH = 14$$

---

*You should note that the value for the concentration of pure water, $[H_2O]$, does not appear in the K_w expression. The true equilibrium expression is $K_c = [H^+][OH^-]/[H_2O]^2$.
The value of $[H_2O]$ is essentially constant in solutions of dilute acids and bases, 55.5 M.

EXERCISE 6   Calculating [OH⁻] and pH

A solution has a hydroxide-ion concentration of 1.5×10^{-5} M. **(a)** What is the concentration of hydronium ions in this solution? **(b)** What is the pH of this solution? **(c)** Is the solution acidic, basic, or neutral?

SOLUTION: *Analyze*: We are given $[OH^-]$ and are asked to determine $[H^+]$ and the pH of the solution and to conclude whether the solution is acidic, basic, or neutral.

Plan: We can use the relation $K_w = [H^+][OH^-] = 1 \times 10^{-14}$ to calculate $[H^+]$.

The pH of the solution is calculated by using $pH = -\log[H^+]$; the nature of the solution is determined from the pH.

Solve:**(a)** Rearranging the equation for K_w to solve for $[H^+]$ gives:

$$[H^+] = \frac{K_w}{[OH^-]} = \frac{1.0 \times 10^{-14}}{1.5 \times 10^{-5}} = 6.7 \times 10^{-10} \ M$$

(b) Substituting the value of $[H^+]$ into $pH = -\log[H^+]$ gives:

$$pH = -\log[H^+] = -\log(6.7 \times 10^{-10}) = -(-9.17) = 9.17$$

With a scientific calculator this calculation is easily made by using the $\log_{10}$ function (*not* $\ln_e$).

(c) The solution is basic because pH is greater than 7, which is a definition of a basic solution. Alternatively you can recognize that $[OH^-] > [H^+]$.

EXERCISE 7   Calculating [H⁺] from pH

A solution has a pH of 5.60. Calculate the hydrogen-ion concentration.

SOLUTION: *Analyze*: We are given a pH of 5.60 and asked to calculate $[H^+]$.

Plan: We can use $pH = -\log[H^+]$ to solve for $[H^+]$.

Solve: We need to rearrange the previous equation so that $\log[H^+]$ is on the left-hand side of the equation.

$$-\log[H^+] = pH$$

or

$$\log[H^+] = -pH$$

Substituting for the value of pH gives

$$\log[H^+] = -5.60$$

The number -5.60 is a sum of a logarithmic characteristic and mantissa. Using a calculator we can find the value of 10^{-pH}, which equals $[H^+]$.

$$[H^+] = 10^{-5.60} = 2.5 \times 10^{-6} \ M$$

The number of significant figures in the final answer is determined by the number of digits to the right of the decimal point in -5.60, two.

Check: We can check that the value of $[H^+]$ is correct by calculating pH from it.

$pH = -\log[H^+] = -\log[2.5 \times 10^{-6}] = 5.60.$

EXERCISE 8   Calculating [H⁺] from pOH

Calculate $[H^+]$ for a solution with a pOH of 4.75.

SOLUTION: *Analyze*: We are given a pOH of 4.75 and asked to calculate $[H^+]$.

Plan: This problem is similar to Exercise 7 except we are given pOH instead of pH. We can use the relationship pOH + pH = 14 to solve for pH and then $[H^+]$.

Solve: pOH is defined as $-\log[OH^-]$. The pH of a solution with a pOH of 4.75 is

$$pH = 14 - pOH = 14 - 4.75 = 9.25$$

Using the procedure applied in Exercise 7 to calculate $[H^+]$ gives

$$\log[H^+] = -pH = -9.25 \qquad [H^+] = 5.6 \times 10^{-10}\,M$$

STRONG ACID AND BASE SOLUTIONS: pH CALCULATIONS

We should know these common strong acids: HCl, HBr, HI, HNO_3, $HClO_4$, and H_2SO_4 and strong bases: $LiOH$, $NaOH$, KOH, $RbOH$, $CsOH$, $Mg(OH)_2$, $Ca(OH)_2$, $Ba(OH)_2$, and $Sr(OH)_2$.

- All dissolved strong acids and bases completely transfer their protons or hydroxide units to water.
- In the case of H_2SO_4, only one proton is completely donated to water, forming HSO_4^-. HSO_4^- is a weak acid.
- The hydrogen-ion concentration of a strong acid solution is calculated by assuming that the protons are completely transferred to water. A similar assumption is made for hydroxide units in strong bases.

EXERCISE 9   Calculating the pH of a strong acid solution

Calculate the pH of a 0.10 M HNO_3 solution.

SOLUTION: *Analyze*: We are asked to calculate the pH of a 0.10 M nitric acid solution.

Plan: We need $[H^+]$ to calculate pH. First we write the chemical equation describing the ionization of nitric acid in water. Then we construct a concentration table to help us determine changes in concentrations of important species in solution, particularly $[H^+]$. Finally we use $[H^+]$ to determine pH.

Solve: The chemical equation describing the reaction of nitric acid with water is:

$$HNO_3(aq) \longrightarrow H^+(aq) + NO_3^-(aq)$$

The second step is to write the initial concentrations and then the concentrations of all species after HNO_3 completely transfers its proton to water. It is useful to tabulate the concentrations in solving problems of this type:

$HNO_3 \xrightarrow{100\%}$	H^+	$+$ NO_3^-
Initial (M)	0.10	0	0
Change (M):	−0.10	+0.10	+0.10
Final (M)	0	0.10	0.10

We use the final concentration of H^+, 0.10 M, to calculate pH.

$$pH = -\log[H^+] = -\log(0.10) = -(-1.00) = 1.00$$

EXERCISE 10 Calculating the pH and pOH of a strong base solution

Calculate the pOH and pH of a 0.02 M KOH solution.

SOLUTION: *Analyze*: We are asked to calculate both pOH and pH of a 0.020 M KOH solution.

Plan: We will use a similar plan as in Exercise 9. KOH is a strong base and dissociates 100 percent. We will determine $[OH^-]$ and then pOH and finally pH. Remember that $pOH = -\log[OH^-]$.

Solve: Write the dissociation reaction of KOH in water and develop a concentration table:

	KOH $\longrightarrow$	K$^+$	+	OH$^-$
Initial (M):	0.02	0		0
Change (M):	-0.02	$+0.02$		$+0.02$
Final (M)	0	0.02		0.02

We determine pOH and pH from the final concentration of hydroxide ion.

$$pOH = -\log[OH^-] = -\log(0.02) = 1.7$$
$$pH = pK_a - pOH = 14 - 1.7 = 12.3$$

The equilibrium constants associated with weak acids and bases are called the **acid-dissociation constant**, K_a, and the **base-dissociation constant**, K_b.

Examples of dissociation constants are shown for the following reactions of HF and NH$_3$ with water:

$$HF(aq) \rightleftharpoons H^+(aq) + F^-(aq) \qquad K_a = \frac{[H^+][F^-]}{[HF]}$$

$$NH_3(aq) + H_2O(l) \rightleftharpoons NH_4^+(aq) + OH^-(aq) \quad K_b = \frac{[NH_4^+][OH^-]}{[NH_3]}$$

- In all K_a and K_b expressions the term $[H_2O]$ is not shown.
- You should know common weak acids (for example, NH$_4^+$, HCN, HF, CH$_3$COOH) and weak bases (NH$_3$ and amines and basic anions).

H$_2$CO$_3$, H$_2$S, and H$_3$PO$_4$ are examples of **weak polyprotic acids**; they have more than one ionizable proton.

- When describing the reactions of weak polyprotic acids with water, you can treat the ionization of each proton in a stepwise manner. Each ionization equilibrium produces one hydrated proton, and each ionization has an associated K_a value.
- For example, the stepwise ionization reactions of H$_2$S are described as follows:

$$H_2S(aq) \rightleftharpoons H^+(aq) + HS^-(aq) \qquad K_{a1} = \frac{[H^+][HS^-]}{[H_2S]}$$

$$HS^-(aq) \rightleftharpoons H^+(aq) + S^{2-}(aq) \qquad K_{a2} = \frac{[H^+][S^{2-}]}{[HS^-]}$$

WEAK ACID AND BASE SOLUTIONS: pH CALCULATIONS

Weak bases are generally of two kinds.

- The first includes ammonia (NH_3) and compounds related to NH_3 that are called amines. Amines are formed when one or more $N-H$ bonds in ammonia are converted to an $N-C$ bond. Hydroxylamine, NH_2OH, contains an $N-OH$ bond instead of an $N-C$ bond.
- The second includes anions of weak acids, such as CH_3COO^-, F^-, CN^-, and S^{2-}. Conjugate bases of weak acids react with water in a manner that is illustrated by the reaction of F^- with H_2O:

$$F^-(aq) + H_2O(l) \rightleftharpoons HF(aq) + OH^-(aq) \quad K_b = \frac{[HF][OH^-]}{[F^-]}$$

Note that the OH^- ion is produced in all reactions of a weak base with water.

- Solutions that are basic are typically formed from salts containing a basic anion and a group 1A or group 2A cation. These cations have no significant acidic properties. Examples of basic salts are NaF, $Ca(CN)_2$, and $KC_2H_3O_2$.

Problem-solving techniques used for calculating pH and percent ionization of weak acid and base solutions are shown in the following exercises and in the next topic section.

EXERCISE 11   Calculating percent ionization of a weak base

Calculate the percentage of hydroxylamine ionized in a 0.020 M hydroxylamine solution. K_b is 1.1×10^{-8} for hydroxylamine (NH_2OH).

SOLUTION: *Analyze*: We are asked to determine the percentage of hydroxylamine ionized in a 0.020 M hydroxylamine solution.

Plan: The definition of percent ionized is

$$\text{Percent ionization} = \frac{\text{amount ionized}}{\text{original concentration}} \times 100$$

Thus, we need to solve for the amount of hydroxylamine that ionizes. We need to write the chemical equation describing the reaction of hydroxylamine with water. Is it an acid or base? Look at the other piece of information given in the question, K_b. The subscript b tells us that it is an equilibrium constant for a weak base. Alternatively we may recognize that the OH in the chemical formula suggests a base. We can use a concentration table to help us calculate the amount of hydroxylamine ionized along with the equilibrium-constant expression in a manner shown in Sample Exercises 16.10 in the text.

Solve: The chemical equation describing the equilibrium of hydroxylamine with water is:

$$H_2NOH(aq) + H_2O(l) \rightleftharpoons H_3NOH^+(aq) + OH^-(aq)$$

$$K_b = \frac{[H_3NOH^+][OH^-]}{[H_2NOH]} = 1.1 \times 10^{-8}$$

Tabulate the initial and equilibrium concentrations of all species involved in the equilibrium-constant expression. Let x equal the molarity of $H_3NOH^+(aq)$ ions formed.

	$H_2NOH \rightleftharpoons$	H_3NOH^+ +	OH^-
Initial (M):	0.020	0	0
Change (M):	$-x$	$+x$	$+x$
Equilibrium (M):	$(0.020 - x)$	x	x

Note that for every molecule of H_2NOH reacting with water, one $OH^-(aq)$ ion and one $H_3NOH^+(aq)$ ion form. Thus, if x moles per liter of $H_3NOH^+(aq)$ are formed at equilibrium, then x moles per liter of OH^- must also have formed, and x moles per liter of H_2NOH must have reacted with water. Substitute the equilibrium concentrations into the equilibrium-constant expression:

$$K_b = \frac{[H_3NOH^+][OH^-]}{[H_2NOH]} = \frac{(x)(x)}{0.020 - x} = 1.1 \times 10^{-8}$$

The small size of K_b tells you that very little hydroxylamine reacts with water; thus the value of x is much smaller than 0.020 M. Making the approximation that the value of x is much smaller than 0.020 M gives:

$$K_b = \frac{(x)(x)}{0.020 - x} \simeq \frac{(x)(x)}{0.020} = 1.1 \times 10^{-8}$$

Solving for x:

$$x^2 = (0.020)(1.1 \times 10^{-8}) = 2.2 \times 10^{-10}$$
$$x = [H_3NOH^+] = [OH^-] = \sqrt{2.2 \times 10^{-10}} = 1.5 \times 10^{-5} \, M$$

The approximation used in solving for x is considered valid if x is no more than 5 percent of the quantity it is subtracted from. In this case, x is $(1.5 \times 10^{-5}/0.020)(100) = 7.5 \times 10^{-2}\%$ of the quantity it is subtracted from; thus the assumption is valid. The percentage of hydroxylamine ionized is as follows:

$$\% \text{ ionized} = \left(\frac{\text{amount of species ionized}}{\text{original concentration}} \right)(100)$$

$$\% \text{ ionized} = \left(\frac{x}{0.020} \right)(100) = \left(\frac{1.5 \times 10^{-5}}{0.020} \right)(100) = 0.075\%$$

Comment: The small value for percent ionized is consistent with our knowledge that hydroxylamine is a weak acid (only partially ionizes); also, the value of K_b is very small, indicating very little ionization.

EXERCISE 12   Calculating the pH of a weak acid solution

Calculate the pH of a 0.015 M acetic acid solution. The value of K_a is 1.8×10^{-5} for acetic acid.

SOLUTION: *Analyze*: We are asked to calculate the pH of a 0.015 M acetic acid solution, given K_a for acetic acid.

Plan: We will use a problem-solving procedure similar to the one used in Exercise 11. We need to write a chemical equation describing the equilibrium of acetic acid with water, construct a concentration table showing $[H^+]$ and other species whose concentrations change in the equilibrium, and use the equilibrium-constant expression to solve for $[H^+]$ and pH.

Solve: First, we write the equilibrium reaction and the K_b expression:

$$CH_3COOH \rightleftharpoons H^+(aq) + C_2H_3O_2^-(aq)$$
$$K_a = \frac{[H^+][CH_3COO^-]}{[CH_3COOH]} = 1.8 \times 10^{-5}$$

Next, develop a concentration table, letting x equal the molarity of H^+ ions formed at equilibrium. Note that x must also equal the equilibrium concentration of CH_3COO^- because CH_3COO^- ions are formed in the same molar amount as H^+ ions.

$$CH_3COOH \rightleftharpoons H^+ + CH_3COO^-$$

	CH_3COOH	H^+	CH_3COO^-
Initial (M):	0.01	0	0
Change (M):	$-x$	$+x$	$+x$
Equilibrium (M):	$(0.015 - x)$	x	x

Again, assume that x is smaller in value than the number from which it is subtracted: $0.015 - x \simeq 0.015$.

$$K_a = \frac{(x)(x)}{0.015} = \frac{x^2}{0.015} = 1.8 \times 10^{-5}$$

Solving for x we have

$$x^2 = (0.015)(1.8 \times 10^{-5}) = 2.7 \times 10^{-7}$$

$$x = \sqrt{2.7 \times 10^{-7}} = 5.2 \times 10^{-4} = [H^+]$$

and

$$pH = -\log[H^+] = -\log(5.2 \times 10^{-4}) = 3.29$$

EXERCISE 13   Explaining the trend in K_as for a weak polyprotic acid

Explain for a triprotic acid why the following trend in equilibrium constants is observed:

$$K_{a1} > K_{a2} > K_{a3}$$

SOLUTION: The first ionization step involves dissociation of one hydrogen ion from a neutral molecule. This requires less energy than the second ionization step, which involves dissociation of a hydrogen ion from a species with a 1− charge. The hydrogen ion and the species with a 1− charge experience coulombic electrostatic attraction for each one. Similarly, the third ionization step involves dissociation of a hydrogen ion from a species with a 2− charge, which requires even more energy for ion separation than the previous two ionizations. The greater the energy required to remove a proton from a weak acid, the smaller the value of K_a.

EXERCISE 14   Calculating the pH of a weak polyprotic acid

(a) Explain why the hydrogen-ion concentration for a 0.10 M H_3PO_4 solution can be determined from its first ionization step. The successive equilibrium constants are: $K_{a1} = 7.5 \times 10^{-3}$, $K_{a2} = 6.2 \times 10^{-8}$, and $K_{a3} = 4.8 \times 10^{-13}$. **(b)** Calculate the pH of this solution.

SOLUTION: *Analyze*: We are asked to explain why $[H^+]$ can be calculated from the first ionization step for H_3PO_4, phosphoric acid, and to calculate the pH of the solution.

Plan: Phosphoric acid is a weak acid and can lose three hydrogen ions. (a) We need to write the three equilibria associated with the loss of each hydrogen ion and the K_a associated with each equilibrium. Qualitatively we can deduce the relative amount of hydrogen ion produced by each equilibrium and determine which is the most important in determining $[H^+]$. (b) We solve for $[H^+]$ using a problem-solving procedure similar to that used in Exercise 11.

Solve: **(a)** The three equilibria involved are:

$$H_3PO_4(aq) \rightleftharpoons H^+(aq) + H_2PO_4^-(aq) \quad K_{a1} = 7.5 \times 10^{-3}$$

$$H_2PO_4^-(aq) \rightleftharpoons H^+(aq) + HPO_4^{2-}(aq) \quad K_{a2} = 6.2 \times 10^{-8}$$

$$HPO_4^{2-}(aq) \rightleftharpoons H^+(aq) + PO_4^{3-}(aq) \quad K_{a3} = 4.2 \times 10^{-13}$$

The amounts of hydrogen ion produced by the second and third ionizations are not appreciable compared to the amount produced by the first ionization: the values of K_{a2} and K_{a3} are considerably smaller than the value of K_{a1}—and the concentrations of $H_2PO_4^-$ and HPO_4^{2-} are much lower than the concentration of H_3PO_4. Therefore, we can assume that the amount of hydrogen ions produced approximately equals that formed from the first ionization. The total amount of $H_2PO_4^-(aq)$ produced from the first ionization is also effectively the same as the hydrogen-ion concentration because the amount of $H_2PO_4^-$ dissociated is small, as indicated by the small value of K_{a2}. **(b)** Using the first equilibrium to solve for the pH, we can construct the following:

	H_3PO_4	$\rightleftharpoons$ H^+	+ $H_2PO_4^-$
Initial (M):	0.10	0	0
Change (M):	$-x$	$+x$	$+x$
Equilibrium (M):	$(0.10 - x)$	x	x

Since the value of K_{a1}, 7.5×10^{-3}, is moderately large, we will not be able to assume that 0.10 is much larger than the value of x:

$$K_{a1} = \frac{[H^+][H_2PO_4^-]}{[H_3PO_4]} = \frac{(x)^2}{0.10 - x} = 7.3 \times 10^{-3}$$

Rearranging the above into a quadratic form gives

$$x^2 + 7.3 \times 10^{-3}x - 7.3 \times 10^{-4} = 0$$

Solving for x using the quadratic formula yields

$$x = [H^+] = \frac{-(7.3 \times 10^{-3}) \pm \sqrt{(7.3 \times 10^{-3})^2 - 4(1)(-7.3 \times 10^{-4})}}{2}$$

$$= 0.024\ M$$

Thus,

$$pH = -\log[H^+] = -\log(0.024) = 1.62$$

SALT SOLUTIONS

0.1 M solutions of NH_4NO_3, $Fe(NO_3)_2$, or $Al(NO_3)_3$ are acidic. Examples of basic salt solutions are 0.1 M solutions of CH_3COONa, KCN, and CaF_2. The basicity or acidity of these solutions is caused by the reaction of an ion with water to produce $H^+(aq)$ or $OH^-(aq)$. Such a reaction is often referred to as **hydrolysis**.

- *Anions that are conjugate bases of weak acids undergo hydrolysis to form basic solutions.* An illustration of this is the reaction of a weak base, the acetate ion, with water:

$$CH_3COO^-(aq) + H_2O(l) \rightleftharpoons CH_3COOH(aq) + OH^-(aq)$$

- *Anions of strong acids, such as NO_3^-, Cl^-, Br^-, and ClO_4^-, do **not** undergo hydrolysis.*
- *Some common cations that react with water to form acidic solutions are NH_4^+, Fe^{3+}, Fe^{2+}, Cu^{2+}, Al^{3+}, Zn^{2+}, and Cr^{3+}.* The chemical equations describing the reactions of cations other than NH_4^+ with water will be discussed in the Lewis acid–base theory section.
- Alkali metal ions (Na^+, Li^+, Cs^+, and Rb^+) and heavier alkaline earth metal ions (Mg^{2+}, Ca^{2+}, Ba^{2+}, and Sr^{2+}) do not form acidic solutions.

Sometimes K_b values for basic anions or K_a values for acidic cations are not available in tables.

- Fortunately, there is a relationship (see derivation in Section 16.8 of the text) that enables you to calculate one of these values if the other is known:

$$K_a \text{ (of acid) } \times K_b \text{ (conjugate base of acid) } = K_w$$

- Therefore,

$$K_a = \frac{K_w}{K_b} \quad \text{or} \quad K_b = \frac{K_w}{K_a}$$

EXERCISE 15   Identifying salt solutions as neutral, acidic, or basic

Classify the following aqueous solutions as neutral, basic, or acidic: (a) KCl; (b) NH_4NO_3; (c) Na_2CO_3; (d) RbF.

SOLUTION: *Analyze*: We are given four salt solutions and are asked to determine if each is acidic, basic, or neutral.

Plan: The nature of each solution is determined by the acidic or basic character of the cation or anion in the salt. We need to assess whether the cation is acidic or the anion is basic using the principles in this section.

Solve: (a) Neutral. K^+ is a group 1A metal and therefore is not acidic. Cl^- is the conjugate base of the strong acid HCl and thus is not basic. (b) Acidic. NH_4^+ is the conjugate acid of the weak base NH_3, and therefore is acidic. NO_3^- is the conjugate base of the strong acid HNO_3 and thus is not basic. (c) Basic. Na^+, like K^+, is not acidic. CO_3^{2-} is the conjugate base of the weak acid HCO_3^- and is therefore basic. (d) Basic. Rb^+, like Na^+ and K^+, is not acidic. F^- is derived from the weak acid HF and is therefore a basic ion.

EXERCISE 16   Calculating the pH of a salt solution containing a basic anion

Calculate the pH of a 0.25 M sodium formate ($NaCHO_2$) solution, given that the value of K_a of formic acid ($HCHO_2$) is 1.8×10^{-4}.

SOLUTION: *Analyze*: We are given a 0.25 M $NaCHO_2$ solution and are asked to determine its pH.

Plan: Sodium formate is a salt (contains a metal ion and an anion). We need to determine if the cation is a weak acid or the anion is a weak base. Then we should proceed as in previous exercises by writing the equation describing the chemical equilibrium that occurs, developing a concentration table and using the equilibrium-constant expression to solve for $[H^+]$ or $[OH^-]$. Alkali metal ions are not acidic; thus, we focus our attention on the formate ion. Formic acid is a weak acid (it has a K_a value associated with it); therefore, its conjugate base, the formate ion, is a weak base. We are not given K_b but it can be calculated using the relationship:

$$K_w = K_a \times K_b.$$

Solve: We first write the dissociation reaction for sodium formate and the chemical equilibrium for the formate ion. Sodium ion is a spectator ion and therefore we can ignore it. Since CHO_2^- is derived from a weak acid, $HCHO_2$, it will react with water to form a *basic* solution and its conjugate acid, $HCHO_2$:

$$NaCHO_2(aq) \xrightarrow{\;100\%\;} Na^+(aq) + CHO_2^-(aq)$$

$$CHO_2^-(aq) + H_2O(l) \rightleftharpoons HCHO_2(aq) + OH^-(aq)$$

We need the value of K_b for the equilibrium. (Note that a base, OH^-, is formed. This tells us that K_b is used.) Using $K_w = K_a \times K_b$ and the value of K_a given in the exercise we can solve for K_b:

$$K_b = \frac{[HCHO_2][OH^-]}{[CHO_2^-]} = \frac{K_w}{K_a} = \frac{1.0 \times 10^{-14}}{1.8 \times 10^{-4}} = 5.6 \times 10^{-11}$$

Next we construct a concentration table and solve for the concentration of $[OH^-]$. The initial concentration of formate ion is the concentration of the sodium formate solution because for every mole of sodium formate that dissociates one mole of formate ion forms.

$$CHO_2^- + H_2O \rightleftharpoons HCHO_2 + OH^-$$

	CHO_2^-	$HCHO_2$	OH^-
Initial (M):	0.25	0	0
Change (M):	$-x$	$+x$	$+x$
Equilibrium (M):	$(0.25 - x)$	x	x

The second column shows the formate ion concentration. Since the value of K_b is small, we can assume $x \ll 0.25$, resulting in $0.25 - x \simeq 0.25$. Substituting equilibrium concentration values into the K_b expression gives

$$K_b = \frac{(x)(x)}{0.25} = 5.6 \times 10^{-11}$$

Solving for x, pOH, and pH yields

$$x = \sqrt{(0.25)(5.6 \times 10^{-11})} = 3.7 \times 10^{-6}\ M$$
$$pOH = -\log[OH^-] = -\log(x) = 5.43$$
$$pH = 14 - pOH = 8.57$$

FACTORS AFFECTING ACID–BASE STRENGTH

The acids discussed in Chapter 16 of the text are primarily of two types: (a) binary acids, such as HCl, HF, and H_2S; and (b) oxyacids, such as H_3PO_4, H_2SO_4, and CH_3COOH. Binary acids of the type H_nX show acidic properties when the strength of the H—X bond is low, and X is a highly electronegative atom (for example, F, Br, or Cl), or the X^- ion is very stable.

- For a series of H_nX in the same family of nonmetallic elements, the acidity tends to increase with increasing atomic number of X.
- In the case of the hydrogen halides, hydrofluoric acid is a weak acid whereas HCl, HBr, and HI are strong acids. The H—F bond is extremely strong and thus H—F does not easily ionize.
- Metallic hydrides are either basic or show very limited acid–base behavior.

Oxyacids contain a nonmetal central atom that is bonded to one or more oxygen atoms and to one or more hydroxide groups. For example, the structural formulas of H_2SO_4 and CH_3COOH are

sulfuric acid   acetic acid

A general formula representation of an oxyacid is $O_xY(OH)_y$, where Y is the central element.

- The relative acidity of oxyacids, $O_xY(OH)_y$, that have the same number of oxygen atoms and hydroxide units (that is, the same values of x and y) but differing central atoms, increases with increasing electronegativity of the central atom.
- The relative acidity of oxyacids that have different numbers of oxygen atoms and hydroxide units (that is, different values of x and y), but with the same central atom, increases with increasing oxidation number of the central atom. The oxidation number of the central atom will increase with an increasing number of oxygen atoms attached to it.
- The carboxyl group is found in a group of organic acids known as carboxylic acids. The general formula of a carboxylic acid is RCOOH, where R represents all groups attached to the —COOH carboxyl group.

Bases generally involve compounds with ionizable hydroxide groups, or with a nitrogen atom that has an available electron pair.

- Base strengths of amines (R—NH_2, for example) decrease with increasing electronegativity of attached R groups.
- Ionizable hydroxide units are found with the more electropositive elements, such as Na^+, K^+, Ca^{2+}, and Ba^{2+}.

EXERCISE 17   Explaining why Y—OH substances can have different acid/base properties

Although NaOH and HOCl have the same formula type, Y—OH, they have different acid–base properties. Explain why NaOH is a base and HOCl is an acid.

SOLUTION: *Analyze*: We are asked to explain why NaOH is a base and HOCl is an acid even though they both have a Y—OH formula type, where Y is Na or Cl.

Plan: We need first to determine whether each is ionic or covalent. If covalent, we assess the effect of Y on the strength of the O—H bond. If ionic then the OH group becomes a hydroxide ion in water and the substance is a base.

Solve: H—O—Cl is a polar covalent compound. Chlorine is a highly electronegative atom; it draws the bonding electrons in the polar covalent O—H bond toward the polar covalent O—Cl bond. This further polarizes the O—H bond and weakens it sufficiently so that the proton can ionize in water to form $H^+(aq)$. Sodium is an electropositive atom and forms ionic compounds. Thus, NaOH is an ionic solid with discrete Na^+ and OH^- ions, and it dissolves in water to give $Na^+(aq)$ and $OH^-(aq)$ ions.

EXERCISE 18   Determining relative acid strengths of oxyacids from Lewis structures

Determine which acid is stronger in water, sulfurous acid, whose Lewis structure is

$$H - \ddot{O} - \ddot{S} - \ddot{O} - H$$
$$|$$
$$:\ddot{O}:$$

or sulfuric acid, whose Lewis structure is

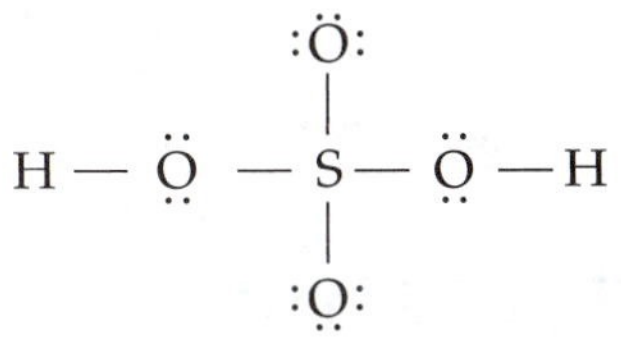

$$
\begin{array}{c}
:\ddot{O}: \\
| \\
H - \ddot{O} - S - \ddot{O} - H \\
| \\
:\ddot{O}:
\end{array}
$$

SOLUTION: *Analyze*: We are given the structures of sulfurous acid and sulfuric acid and are asked to conclude which is a stronger acid in water.

Plan: Both acids are oxyacids with differing numbers of oxygen atoms and hydroxide groups, but with the same central atom. The information in the summary section tells us to look at the oxidation number of the central atom in such cases and determine how it affects acidity.

Solve: The sulfur atom in sulfuric acid has a +6 oxidation number; in sulfurous acid, the sulfur atom has a +4 oxidation number. Sulfuric acid is a stronger acid than sulfurous acid because the sulfur atom, with a +6 oxidation number, strongly polarizes the bonding electrons in the $O - H$ bond toward the oxygen atom so that the proton is more easily removed by a base. In sulfurous acid, the sulfur atom has a smaller oxidation number of +4, resulting in a less polarized $O - H$ bond and thus a lesser acidity.

The Brønsted-Lowry definition of an acid and a base is limited to substances that are able to donate or accept protons. A more general concept of acids and bases was proposed by G. N. Lewis.

THE LEWIS MODEL OF ACIDS AND BASES

- A **Lewis acid** is any substance that accepts an electron pair.
- A **Lewis base** is any substance that donates an electron pair.
- For example, BCl_3 is a Lewis acid because boron has an empty orbital that can accept an electron pair from a Lewis base like NH_3, as shown in the following reaction:

$$
BCl_3 + :NH_3 \longrightarrow Cl_3B:NH_3
$$

- Note that all Brønsted-Lowry acids and bases are Lewis acids and bases, but the opposite is not necessarily true. BCl_3 is a Lewis acid but not a Brønsted-Lowry acid because it has no ionizable protons.

The Lewis model enables you to understand the mechanisms of many acid–base reactions. Of particular interest in Section 16.11 of the text is the hydration of metal ions and the hydrolysis of metal ions.

- Dissolved metal ions are hydrated in water. For example, the aluminum ion occurs as $Al(H_2O)_6^{3+}$. It forms when the Lewis acid Al^{3+} reacts with the Lewis base water as follows:

$$
Al^{3+} + 6 :\ddot{O}H_2 \longrightarrow Al(:\ddot{O}H_2)_6^{3+}
$$

This is an example of a hydration reaction.

- Protons of hydrated water molecules bonded to certain metal ions may dissociate to form $H^+(aq)$. For example, a proton in one of the water molecules bonded to Al^{3+} is easily lost to the bulk solvent:

$$
Al(H_2O)_6^{3+} \rightleftharpoons Al(H_2O)_5(OH)^{2+} + H^+(aq)
$$

This is another example of a hydrolysis reaction.

• Metal ions with a high charge/ionic radius ratio, such as Fe^{3+}, Al^{3+}, Cu^{2+}, Zn^{2+}, and Cr^{3+}, hydrolyze to the greatest extent, and thus form acidic solutions. Large ions with a $+1$ charge, such as Na^+, K^+, Cs^+, exhibit no acidic behavior.

EXERCISE 19   Explaining why water is both a Brønsted-Lowry acid and a Lewis acid

Using as an example the reaction of H^+ with OH^-, show that the Lewis definition of an acid is consistent with the Brønsted-Lowry definition.

SOLUTION: *Analyze:* We are asked to show that the reaction between H^+ and OH^- can be viewed both as a Lewis acid–base reaction and a Brønsted-Lowry acid–base reaction.

Plan: We need to look at the neutralization reaction between H^+ and OH^- and at the characteristics that relate to the two definitions of acids and bases. That is, a Lewis acid is an electron-pair acceptor and a Lewis base is an electron-pair donor. A Brønsted-Lowry acid is a proton donor and a Brønsted-Lowry base is a proton acceptor.

Solve: H^+ is a Lewis acid because it acts as an electron-pair acceptor from electron-donor bases such as OH^-. This is shown by the following Lewis diagram:

$$H^+ + :\ddot{O}-H^- \longrightarrow [H^+ + \overset{\frown}{O}\ddot{O}-H^-] \longrightarrow H-\ddot{O}-H$$

Therefore, any proton-donating substance (a Brønsted acid) is also a Lewis acid.

EXERCISE 20   Identifying Lewis acids and bases in a chemical reaction

Which reactant molecule or ion in the reaction

$$Ag^+(aq) + 2\,NH_3(aq) \longrightarrow Ag(NH_3)_2^+(aq)$$

is the Lewis acid? Which is the Lewis base?

SOLUTION: *Analyze:* We are asked to determine which reactant is a Lewis acid and which is a Lewis base.

Plan: We apply the definitions of Lewis acid and base to the reaction. This may require writing Lewis structures for each reactant to determine which can accept an electron pair and which can donate an electron pair.

Solve: Metal ions often act as Lewis acids. Also, from our reading of the text we may have learned that ammonia is a weak base in water. In this reaction the Lewis acid is $Ag^+(aq)$ because it accepts electron density from the Lewis base NH_3 to form the Lewis acid–base compound $Ag(NH_3)_2^+$ as follows:

$$H_3N: \longrightarrow Ag^+ \longleftarrow :NH_3$$

EXERCISE 21   Writing a hydrolysis reaction

Write the hydrolysis reaction for $Zn(H_2O)_4^{2+}$.

SOLUTION: *Analyze:* We are asked to write the hydrolysis reaction for the $Zn(H_2O)_4^{2+}$ ion.

Plan: A hydrolysis reaction involves a reaction of a species with water to form hydrogen ions or hydroxide ions. $Zn(H_2O)_4^{2+}$ is a cation and its reaction with water should form hydrogen ions. The solvent water acts as a base and accepts a proton from one of the hydrated water molecules.

Solve: The reaction is:

$$Zn(H_2O)_4^{2+} \overset{H_2O}{\rightleftharpoons} Zn(H_2O)_3(OH)^+ + H^+(aq)$$

SELF-TEST QUESTIONS

Key Terms

Having reviewed key terms in Chapter 16, match key terms with phrases and identify statements as true or false. If a statement is false, indicate why it is incorrect.

Match each phrase with the best term:

16.1 The reaction of an ion with water to produce an acidic or basic solution.

16.2 For a binary weak acid it has the form: $\dfrac{[H^+][X^-]}{[HX]}$.

16.3 For a weak base it has the form: $\dfrac{[OH^-][BH^+]}{[B]}$.

16.4 It has the form: $[H^+][OH^-]$.

16.5 Its value for a 0.01 M NaOH solution is 12.

16.6 $SO_3(g)$ can be characterized by this term when it reacts as follows:

$$SO_3(g) + H_2O(l) \longrightarrow H_2SO_4(aq)$$

16.7 $BF_3(g)$ can be characterized by this term when it reacts as follows:

$$BF_3(g) + N(CH_3)_3(g) \longrightarrow F_3B{:}N(CH_3)_3(s)$$

16.8 H_2SO_4 can be characterized by this term but not HCl.

16.9 A substance that donates a proton in a chemical reaction.

16.10 A substance that accepts a proton in a chemical reaction.

16.11 The relationship of HSO_4^- to H_2SO_4.

Terms:

(a) acid-dissociation constant
(b) base-dissociation constant
(c) Brønsted-Lowry acid
(d) Brønsted-Lowry base
(e) conjugate base
(f) hydrolysis
(g) ion-product expression
(h) Lewis acid
(i) Lewis base
(j) oxy acid
(k) pH

True-False Statements:

16.12 A *conjugate acid* is a substance formed by the addition of one proton to a Brønsted base.

16.13 CH_4 in water is an example of a *polyprotic acid*.

16.14 In the *autoionization* of water, equal molar quantities of hydrogen and hydroxide ions are formed.

16.15 The *hydronium ion* is a hydrated hydrogen ion and carries a positive charge.

16.16 H_2O and O^{2-} form *a conjugate acid–base pair*.

16.17 The hydroxide ion cannot act as an *amphoteric* substance.

16.18 For a substance to exist as an *amine* it must have a nitrogen atom with a lone pair of electrons.

16.19 A *carboxylic acid* contains the $-COOH$ group, and this group contains a hydrogen attached to a carbon atom.

16.20 A *hydronium ion* can hydrogen bond to other water molecules and form larger clusters of hydrated hydrogen ions such as $H_9O_4^+$.

16.21 NH_4OH is an oxyacid.

Problems and Short-Answer Questions

16.22 Identify in the following diagram the Brønsted-Lowry acids and bases. Explain your answer.

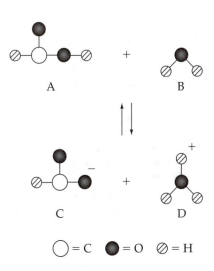

16.23 Identify which figure represents an aqueous solution of a binary weak acid, HX, for which pH = pK_a. Water molecules and hydrated protons are not shown for convenience. Explain your answer.

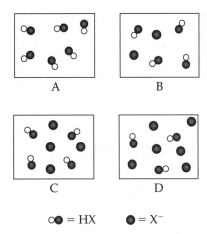

A   B

C   D

$\circ\!\bullet$ = HX $\bullet$ = X$^-$

16.24 Identify the following substances as strong or weak acids or bases in water:

(a) HBr   (d) NH$_3$
(b) HF   (e) H$_2$CO$_3$
(c) KOH   (f) H$_2$SO$_4$

16.25 Give the conjugate acid of each of the following bases:

(a) O^{2-} (c) H$_2$O
(b) HSO$_4^-$ (d) F$^-$

16.26 Give the conjugate base of each of the following acids:

(a) H$_2$S   (c) HCN
(b) H$_2$O   (d) HCO$_3^-$

16.27 Identify each reactant either as an acid or base by its behavior in the reaction:

(a) HCO$_3^-$(aq) + H$_2$O(l) $\rightleftharpoons$ H$_2$CO$_3$(aq) + OH$^-$(aq)

(b) HCN(aq) + H$_2$O(l) $\rightleftharpoons$ H$^+$(aq) + CN$^-$(aq)

(c) NH$_3$(aq) + H$_2$O(l) $\rightleftharpoons$ NH$_4^+$(aq) + OH$^-$(aq)

16.28 You are given the following information about three substances (HA, HB, and HC) in solution. The pH of 0.1M solutions of NaA, NaB, and NaC are 9.0, 10.5, and 7.0, respectively. Rank HA, HB, and HC in increasing order of acid strength.

16.29 Are the following aqueous solutions acidic, basic, or neutral? Write a net ionic equation to explain any acidity or basicity.

(a) Ba(NO$_3$)$_2$ (c) Al(NO$_3$)$_3$
(b) NaF   (d) Ca(NO$_2$)$_2$

16.30 CH$_4$, methane, has hydrogen atoms yet when it is placed in contact with water it does not act as an acid.

(a) Suggest a reason for this observation.
(b) How would you characterize the acid or base properties of the CH$_3^-$ ion?

16.31 Rank in increasing order of base strength the following ions: ClO$^-$, BrO$^-$, and IO$^-$. Justify your ranking.

16.32 Rank in increasing order of acid strength the following: H$_2$O, H$_2$S, H$_2$Se, and H$_2$Te. Justify your ranking.

16.33 Rank in order of increasing acid strength the following: Mn(OH)$_2$, Mn(OH)$_3$, and OMn(OH)$_3$. Justify your ranking.

16.34 Explain how the following reaction can be characterized as a Lewis acid–base reaction:

$$Ag^+(aq) + 2\,NH_3(aq) \longrightarrow Ag(NH_3)_2^+$$

16.35 Ethylamine (C$_2$H$_5$NH$_2$) has a K_b value that is 4.0 × 10^{-6}. A solution of ethylamine has a pH of 8.52. Is there sufficient information to determine the initial concentration of ethylamine in the solution? If there is insufficient information, what additional information is needed and why?

16.36 10.0 mL of an aqueous solution of a weak acid, HX, has a pH of 4.85. Is there sufficient information to determine K_a for HX? If there is insufficient information, what additional information is needed and why?

16.37 Calculate the pH of the following solutions:

(a)   1.20 M HCl   (c)   0.020 M NaOH
(b)   0.0030 M HNO$_3$ (d)   0.0010 M Ba(OH)$_2$

16.38 The pH of a 0.10 M nitrous acid solution is 2.17. Calculate K_a for a nitrous acid using this information.

16.39 What is the molarity of an acetic acid solution if its pH is 3.50? K_a = 1.8 × 10^{-5} for acetic acid.

16.40 What is the pH of a 0.20 M HClO solution? K_a for HClO is 3.2 × 10^{-8}.

16.41 What is the pH of a 0.35 M solution of NH$_4$NO$_3$ solution? K_b = 1.8 × 10^{-5} for NH$_3$. Would you expect a 0.35 M NH$_4$F solution to have the same pH? Explain.

16.42 Calculate the pH of a solution labeled as 0.055 M NaCN. K_a = 4.9 × 10^{-10} for HCN.

16.43 Predict whether the salt KHCO$_3$ will form an acidic or basic solution in water, given the following data:

$$HCO_3^-(aq) \rightleftharpoons H^+(aq) + CO_3^{2-}(aq) \quad K_{a2} = 5.6 \times 10^{-11}$$

$$H_2CO_3(aq) \rightleftharpoons H^+(aq) + HCO_3^-(aq) \quad K_{a1} = 4.3 \times 10^{-7}$$

16.44 Which member of the following pairs would you expect to produce the more acidic solution:

(a)   NaCl or Na$_2$S;   (c) Cr(NO$_3$)$_3$ or Ca(NO$_3$)$_2$;
(b)   FeCl$_2$ or FeCl$_3$;   (d) NaH or HBr?

16.45 Ascorbic acid, H$_2$C$_6$H$_6$O$_6$, is a polyprotic acid containing two ionizable hydrogens. K_{a1} is 8.0 × 10^{-5} and K_{a2} is 1.6 × 10^{-12}. Calculate the pH of a 2.0 × 10^{-3} M solution of ascorbic acid.

Integrative Exercises

16.46 Sulfuric acid is a strong acid and students often write the ionization reaction as

$$H_2SO_4(aq) \longrightarrow 2\,H^+(aq) + SO_4^{2-}(aq)$$

(a) Given that the second ionization constant K_{a2} is 1.1×10^{-2}, should the ionization reaction be written in the form above? Explain.

(b) What are $[H^+]$, $[HSO_4^-]$, and $[SO_4^{2-}]$ in a 0.10 M H_2SO_4 solution?

16.47 In many equilibrium problems involving aqueous solutions you have used the value of K_w as 1.0×10^{-14} at 25° C. At 50° C the value of K_w is 5.5×10^{-14}.

(a) Compare the pH of a 0.010 M KOH solution at the two temperatures.

(b) What is the pH of a neutral solution at 50° C?

(c) Is the autoionization reaction of water exothermic or endothermic? Explain.

16.48 (a) Calculate the number of hydroxide ions in 1.0 mL of pure water at 25° C and 1 atm pressure.

(b) If 0.50 mL of 0.10 M NaOH is added to 1.0 mL of pure water at 25° C, what is the hydroxide ion concentration?

(c) If the pressure above one liter of pure water at STP is increased to 3 atm, will there be a significant change in the concentration of hydroxide ions? Explain.

Multiple-Choice Questions

16.49 Which of the following salts should form a basic solution in water: (1) NH_4Cl; (2) AlI_3; (3) $KC_2H_3O_2$; (4) NaF; (5) Cs_2HPO_4?

(a) (1) and (2)
(b) (3) and (4)
(c) (2), (3), and (5)
(d) (3), (4), and (5)
(e) all

16.50 Which of the following salts would form an acidic solution in water: (1) NH_4Cl; (2) $Fe(NO_3)_3$; (3) KCl; (4) NaF; (5) RbS?

(a) (1) and (2)
(b) (3) and (4)
(c) (4) and (5)
(d) all of them
(e) none of them

16.51 What is the pH of a solution containing 2.5 g of NaOH dissolved in 100 mL of water?

(a) 0.2
(b) 13.8
(c) 1.2
(d) 12.8
(e) 3.2

16.52 Which of the following 0.1 M solutions has a pH closest to 7?

(a) KF
(b) Na_2CO_3
(c) NH_4I
(d) CsBr
(e) HNO_2

16.53 Which 0.10 M solution has the smallest pH?

(a) NH_4Cl
(b) $NaNO_3$
(c) BaI_2
(d) NH_3
(e) Na_2CO_3

16.54 The pOH of a solution of NaOH is 11.30. What is the $[H^+]$ for this solution?

(a) 2.0×10^{-3}
(b) 2.5×10^{-3}
(c) 5.0×10^{-12}
(d) 4.0×10^{-12}
(e) 6.2×10^{-8}

16.55 Which one of the following binary compounds would you expect to be the most acidic?

(a) NaH
(b) CH_4
(c) SnH_4
(d) H_2O
(e) H_2S

16.56 Which one the following would you expect to be the most acidic?

(a) HClO
(b) HBrO
(c) HIO
(d) acetic acid
(e) all are basic, not acidic

16.57 Which one of the following is a Lewis acid, but *not* a Brønsted acid?

(a) HCl
(b) BBr_3
(c) NH_3
(d) KOH
(e) CH_4

16.58 Which ion should show the greatest acidic property in water?

(a) S^{2-}
(b) Cu^{2+}
(c) Ca^{2+}
(d) Br^-
(e) Fe^{3+}

16.59 Which of the following binary hydrides is the most basic?

(a) HCl
(b) H_2S
(c) PH_3
(d) SiH_4
(e) NaH

16.60 What is the percent ionization of a 0.15 M HCN solution? K_a (HCN) $= 4.9 \times 10^{-10}$

(a) 7.35×10^{-11}
(b) 8.57×10^{-3}
(c) 8.57×10^{-6}
(d) 5.71×10^{-3}
(e) 5.71×10^{-5}

16.61 How many moles of CH_3COOH in a 200 mL solution are required to produce a solution with pH = 2.90? K_a (CH_3COOH) $= 1.8 \times 10^{-5}$

(a) 1.79×10^{-2}
(b) 8.82×10^{-2}
(c) 8.95×10^{-2}
(d) 1.62×10^{-3}
(e) 1.79×10^{-3}

16.62 What is the pH of a 0.030 M $Ba(OH)_2$ solution?

(a) 1.52
(b) 12.48
(c) 1.22
(d) 12.78
(e) 0.03

16.63 What is the pH of a 0.25 M weak acid that is 2.2% ionized?

(a) 0.60
(b) 1.00
(c) 1.60
(d) 2.00
(e) 2.26

16.64 The pH of a 0.15 M weak base B is 9.25. What is K_b for B?

(a) 1.38×10^{-10} (d) 2.39×10^{-7}
(b) 2.11×10^{-9} (e) 6.92×10^{-3}
(c) 6.25×10^{-8}

SELF-TEST SOLUTIONS

16.1 (f).   **16.2** (a).   **16.3** (b). **16.4** (g). **16.5** (k). **16.6** (h). **16.7** (i). **16.8** (j). **16.9** (c). **16.10** (d). **16.11** (e). **16.12** True. **16.13** False. Although CH_4 contains hydrogen atoms, the hydrogen atoms do not ionize in water; in fact, CH_4 is insoluble in water. Carbon is not sufficiently electronegative to cause the hydrogens to become acidic. **16.14** True. **16.15** True. **16.16** False. The substances forming a conjugate acid–base pair differ only by the presence or absence of a single proton. Thus, H_2O and OH^- form a conjugate acid–base pair. **16.17** False. An amphoteric substance can act both as an acid and a base. The hydroxide ion can gain a proton to form water, and lose a proton to form the oxide ion. These occur in different types of reactions. In water, the more common reaction is as a base. **16.18** True. **16.19** False. The hydrogen atom is attached to an oxygen atom of a OH group in —COOH. This is the acidic hydrogen in carboxylic acids. **16.20** True. **16.21** False. An oxyacid is an acid in which OH groups are attached to another atom and the O—H bond is acidic. Ammonium hydroxide contains an OH group but it is not acidic. Ammonium hydroxide is the name given to an aqueous solution of ammonia in which small quantities of ammonium ion and hydroxide ion are produced.

16.22 Substances A and D are the Brønsted-Lowry acids because each has an ionizable hydrogen atom which is donated as a proton. Substances B and C are Brønsted-Lowry bases because each accepts a proton.

16.23 Figure C represents a solution for which pH = pK_a or $[H^+] = K_a$. This occurs for a binary weak acid HX when $[X^-] = [HX]$ at equilibrium and thus these concentrations cancel in the expression: $\dfrac{[H^+][X^-]}{[HX]} = K_a$. In figure C the number of particles of X^- and HX are the same and therefore their concentrations are equivalent.

16.24 (a) strong acid;
(b) weak acid; of the HX acids in the halogen series, HF is the only weak acid.
(c) strong base; metallic hydroxides are strong bases.
(d) weak base;
(e) weak polyprotic acid;
(f) strong polyprotic acid.

16.25 The conjugate acid of a base is formed by adding *one* H^+.

(a) OH^-;   (c) H_3O^+;
(b) H_2SO_4;   d) HF.

16.26 The conjugate base of an acid is formed by removing *one* H^+.

(a) HS^-;   (c) CN^-;
(b) OH^-;   (d) CO_3^{2-}.

16.27 An acid donates a proton to a base. Thus, you should determine which substance is losing a proton and which one is gaining a proton.

(a) HCO_3^- is a base because it accepts a proton and forms carbonic acid. Water is an acid because it donates an H^+ and forms the hydroxide ion.

(b) HCN is an acid because it donates a proton to water to form a hydrated proton, also known as the hydronium ion. Water is the base because it accepts the proton.

(c) NH_3, ammonia, is a base because it accepts a proton from water.

16.28 Based on the pHs of the salt solutions, we can rank the base strengths of the anions as follows: $B^- > A^- > C^-$. The greater the pH of the solution the greater the relative base strength. The sodium ion is a spectator ion. Based on the principle that the stronger the acid the weaker its conjugate base, we can rank the acids: HC > HA > HB.

16.29 All of the salts listed are soluble, strong electrolytes, and dissociate to form their ions.

(a) Neutral. Ba^{2+} is a metal ion from the group 2A family and is a neutral ion. The nitrate ion is the conjugate base of the strong acid, nitric acid, and thus is not basic.

(b) Basic. Na^+ is a metal ion from the group 1A family and thus is a neutral cation. The fluoride ion is the conjugate base of the weak acid hydrofluoric acid and thus is a weak base. $F^-(aq) + H_2O(l) \longrightarrow HF(aq) + OH^-(aq)$. The nitrate ion is the conjugate base of the strong acid, nitric acid, and thus is not basic.

(c) Acidic. Ions with a large positive charge and a small ion radius will be acidic: $Al^{3+}(aq) + H_2O(l) \longrightarrow Al(OH)^{2+}(aq) + H^+(aq)$.

(d) Basic. The calcium ion is from the group 2A family and is a neutral ion. The nitrite ion is the conjugate base of the weak acid nitrous acid and thus is a weak base: $NO_2^-(aq) + H_2O(l) \longrightarrow HNO_2 + OH^-(aq)$.

16.30 (a) The C—H bond is essentially nonpolar and thus the electrons in the bond are not drawn toward the carbon atom. For the hydrogen atom to become acidic the electrons in the C—H bond would have to be polarized toward the carbon atom, which would make the hydrogen atom electropositive and attracted to a base. This does not happen.

(b) CH_3^- should be a very strong base. A proton should be easily attracted to it and form the stable CH_4. If it were a weak base, CH_4 would be acidic, which is not true.

16.31 Oxyacids of the type H—O—Y show increasing acid strength with increasing electronegativity of Y. With increasing electronegativity of Y, the H—O bond is more polarized, and the release of a proton to a base becomes more easily accomplished. Thus, HOCl is a stronger acid than HOBr, which is a stronger acid than HOI. Using the principle the stronger the acid the weaker the conjugate base, we predict the following trend in base strengths: $ClO^- < BrO^- < IO^-$.

16.32 The substances in the problem are all binary acids of the form H_2X. In general the primary determinant for acidity is the strength of the H—X bond, although other factors do operate. For a series of binary acids of the type H_nX, the acidity increases down a family because increasing atomic size results in a weaker H—X bond: $H_2O < H_2S < H_2Se < H_2Te$.

16.33 In a series of oxyacids with the same central atom the acidity increases with increasing oxidation number of the atom. Another way of stating this is that the acidity increases with increasing number of oxygen atoms. This results in the following trend in acidity: $Mn(OH)_2 < Mn(OH)_3 < OMn(OH)_3$. As the oxidation number of the central atom increases, polarization of the O—H bond increases and this assists in the release of the proton to a base.

16.34 To answer this question you need to know the definition of a Lewis acid and a Lewis base. A Lewis acid accepts electron density in a chemical reaction and a Lewis base donates electron density in a chemical reaction. If you write the Lewis structure of ammonia, you will observe it has a lone pair of electrons on the nitrogen atom:

$$H - \overset{\displaystyle ..}{\underset{\displaystyle |}{N}} - H$$
$$H$$

The electron domain on nitrogen is not tightly held and can interact with the empty orbitals of the silver ion to form a Ag—N bond in the product. Thus, ammonia is the electron donor and is the Lewis base; Ag^+ is the electron acceptor and is the Lewis acid.

16.35 There is sufficient information. From the pH the concentrations of H^+ and OH^- are calculated. The equation describing the equilibrium of ethylamine with water is:

$$C_2H_5NH_2 + H_2O \rightleftharpoons C_2H_5NH_3^+ + OH^-.$$

The concentration of $C_2H_5NH_3^+$ is equivalent to the concentration of OH^- and also represents the amount of ethylamine that reacts because there is a 1:1 relationship in the stoichiometric coefficients for ethylamine and its conjugate acid.

$$K_b = \frac{[C_2H_5NH_3^+][OH^-]}{[C_2H_5NH_2]}.$$

We can solve for the concentration of ethylamine in the denominator because all other values are known. The concentration of $C_2H_5NH_2$ in the denominator represents the equilibrium concentration. The initial concentration is the sum of the equilibrium concentration of $C_2H_5NH_2$ and the concentration of $C_2H_5NH_3^+$.

16.36 There is insufficient information. The equation describing the reaction of HX with water is: $HX(aq) + H_2O(aq) \rightleftharpoons H^+(aq) + X^-(aq)$. From the pH we can calculate $[H^+]$; the concentration of X^- is the same. To calculate K_a we also need the equilibrium concentration of HX. This could be calculated from the relationship: Initial concentration of HX = Equilibrium concentration of HX + equilibrium concentration of X^- (represents the amount of HX that has reacted). However, we do not know the initial concentration of HX; if it was given in the problem we could answer the question.

16.37 pH is calculated using the relation: $pH = -\log[H^+]$. In this problem all the examples are of the strong acid/base type.

(a) It is a strong acid; the hydrogen ionizes completely. Thus, the ionizations forms $1.20\ M\ H^+$. $pH = -\log[H^+] = -\log[1.20] = -0.0792$. Note that the pH is negative. This occurs because the concentration of the acid is greater than one molar.

(b) Nitric acid is also a strong acid. $pH = -\log[H^+] = -\log[0.0030] = 2.52$.

(c) Sodium hydroxide is a strong base and forms hydroxide ion in solution. We can first calculate $pOH = -\log[0.020] = 1.70$. To calculate pH, we can use the relationship: $pH + pOH = 14$. $pH = 14 - pOH = 14 - 1.70 = 12.30$.

(d) When $Ba(OH)_2$ dissolves and ionizes it forms two hydroxide ions per barium hydroxide empirical formula: $Ba(OH)_2 \longrightarrow Ba^{2+} + 2\ OH^-$. This means the concentration of hydroxide ion is twice that of the concentration of barium hydroxide. $pOH = -\log[2 \times 0.0010] = 2.70$. $pH = 14 - pOH = 11.30$.

16.38 $HNO_2(aq) \rightleftharpoons H^+(aq) + NO_2^-(aq)$. The equilibrium concentration of $H^+(aq)$ is calculated from the given pH: $\log[H^+] = -2.17$; $[H^+] = 6.76 \times 10^{-3}$. $[NO_2^-] = \times 10^{-3}$ because at equilibrium equal quantities of NO_2^- and H^+ are formed. The equilibrium concentration of HNO_2 is its initial concentration minus the amount ionized: $0.10\ M - 6.76 \times 10^{-3}\ M = 9.3 \times 10^{-2}\ M$. Substituting these values into the K_a expression yields $K_a = [H^+][NO_2^-]/[HNO_2] = (6.76 \times 10^{-3})(6.76 \times 10^{-3})/9.3 \times 10^{-2} = 4.9 \times 10^{-4}$.

16.39 We want to calculate $[CH_3COOH]$ given K_a and the pH. From $pH = 3.50$, we can calculate $[H^+]$ as $3.16 \times 10^{-4}\ M$. As explained in the solution to exercise 12, the concentration of CH_3COO^- is also $3.16 \times 10^{-4}\ M$, the same as $[H^+]$. Thus, $K_a = [H^+][CH_3COO^-]/[CH_3COOH] = (3.16 \times 10^{-4})(3.16 \times 10^{-4})/[CH_3COOH] = 1.8 \times 10^{-5}$; $[CH_3COOH] = (3.16 \times 10^{-4})^2/1.8 \times 10^{-5} = 5.6 \times 10^{-3}$. Solving for

[CH$_3$COOH] gives (CH$_3$COOH] = 5.6×10^{-3} M. The initial molarity of acetic acid is the sum of the ionized and unionized portions: 5.6×10^{-3} M + 3.16×10^{-4} M = 5.9×10^{-3} M.

16.40 Using a procedure analogous to the one in the solution to Exercise 12, HClO(aq) $\rightleftharpoons$ H$^+$(aq) + ClO$^-$(aq); K_a = [H$^+$][ClO$^-$]/[HClO] = 3.2×10^{-8}.

	HClO	$\rightleftharpoons$ H$^+$(aq) +	ClO$^-$(aq)
Initial (M):	0.20	0	0
Equilibrium (M):	(0.20 − x)	x	x

Assume x is much smaller than 0.20. K_a = $(x)(x)/0.20$ = 3.2×10^{-8} or x = [H$^+$] = 8.0×10^{-5} M. The small size of x as compared to 0.02 justifies the assumption. The pH is $-\log(8.0 \times 10^{-5})$ = 4.10.

16.41 Ammonium nitrate is a soluble, strong electrolyte and dissociates to form NH$_4^+$(aq) and NO$_3^-$(aq). The nitrate ion is a spectator ion and the ammonium ion is a weak acid because it is the conjugate acid of a weak base. NH$_4^+$(aq) + H$_2$O(l) $\rightleftharpoons$ NH$_3$(aq) + H$^+$(aq).

	NH$_4^+$	$\rightleftharpoons$ NH$_3$ +	H$^+$
Initial (M):	0.35	0	0
Equilibrium (M):	(0.35 − x)	x	x

Assume that x is significantly less than 0.35; therefore $0.35 - x \simeq 0.35$. K_a = [NH$_3$][H$^+$]/[NH$_4^+$] = K_w/K_b(NH$_3$) = $1.0 \times 10^{-14}/1.8 \times 10^{-5}$ = 5.6×10^{-10}; K_a = $(x)(x)/0.35$ = 5.6×10^{-10}; x=[H$^+$]=1.4×10^{-5}M; pH=4.85. 0.35 M NH$_4$F will not have the same pH. The fluoride ion is the conjugate base of the weak acid HF. Thus, it has basic properties and will cause the pH to be greater.

16.42 NaCN(aq) $\xrightarrow{100\%}$ Na$^+$(q) + CN$^-$(aq); CN$^-$(aq) + H$_2$O(l) $\rightleftharpoons$ HCN(aq) + OH$^-$(aq).

	CN$^-$	$\xrightarrow{H_2O}$ HCN +	OH$^+$
Initial (M):	0.055	0	0
Equilibrium (M):	(0.055 − x)	x	x

Assume that x is significantly less than 0.055; therefore $0.055 - x \simeq 0.055$. K_b = [HCN][OH$^-$]/[CN$^-$] = K_w/K_a(HCN) = $1.0 \times 10^{-14}/4.9 \times 10^{-10}$ = 2.0×10^{-5}; K_b = (x) K_b = $(x)(x)/0.055$ = 2.0×10^{-5}; x = [OH$^-$] = 1.0×10^{-3}; pH = 14 − pOH = 14 − 3.0 = 11.0.

16.43 The two possible reactions of HCO$_3^-$ in water are HCO$_3^-$(aq) $\rightleftharpoons$ H$^+$(aq) + CO$_3^{2-}$(aq), with K_{a2} = 5.6×10^{-11}, and HCO$_3^-$(aq) + H$_2$O $\rightleftharpoons$ H$_2$CO$_3$(aq) + OH$^-$(aq), with an unknown K_b. The solution will be acidic or basic depending on which equilibrium constant has the larger value. The value of K_b can be calculated from the relation K_a(H$_2$CO$_3$) × K_b(HCO$_3^-$) = K_w. Thus, K_b = $1 \times 10^{-14}/(4.3 \times 10^{-7})$ = 2.3×10^{-8}. Since K_b is larger than K_a for HCO$_3^-$, the reaction to form the OH$^-$ ion predominates, producing a basic solution.

16.44 (a) A solution of NaCl will be more acidic because the other salt, Na$_2$S, contains a basic anion, S^{2-}.

(b) The solution of FeCl$_3$ is more acidic because it contains iron with a higher charge.

(c) The solution of Cr(NO$_3$)$_3$ is more acidic because chromium has a 3+ charge where as calcium has a 2+ charge. In general, the higher the charge of the metal ion, the more acidic the solution.

(d) HBr is more acidic; it is a strong acid. NaH contains the basic ion H$^-$.

16.45 (a) Ascorbic acid has two ionizable hydrogen atoms. You can calculate the [H$^+$] of the solution by using the first ionization only: If K_a values differ by a factor of 10^3 or more, you can obtain a reasonable estimate of the pH by using only K_{a1}.

	H$_2$C$_6$H$_6$O$_6$	H$^+$	C$_6$H$_6$O$_6^-$
Initial (M):	2.0×10^{-3}	0	0
Change (M):	−x	+x	+x
Equil (M):	2.0×10^{-3} − x	x	x

You can assume that $x \ll 2.0 \times 10^{-3}$ because of the small value of K_{a1}. Substituting the equilibrium values into the equilibrium-constant expression gives

$$K_{a1} = \frac{(x)(x)}{2.0 \times 10^{-3}} = 8.0 \times 10^{-5}$$

Solving for the hydrogen ion concentration, x, gives

$$x = 4.0 \times 10^{-4} \text{ M and pH} = 3.40$$

16.46 (a) Sulfuric acid is a strong acid in its first ionization:

$$H_2SO_4(aq) \xrightarrow{100\%} H^+(aq) + HSO_4^-(aq)$$

However, since it has a second ionization constant, HSO$_4^-$ is a weak acid in water:

$$HSO_4^-(aq) \rightleftharpoons H^+(aq) + SO_4^{2-}(aq)$$

Therefore, the reaction originally written in the question is not a proper description of the ionization of sulfuric acid. It implies that it readily loses two protons, which is not correct.

(b) The concentrations are calculated by analyzing each ionization step sequentially. The first ionization results in a H$^+$ concentration of 0.10 M and a HSO$_4^-$ concentration of 0.10 M because there is complete ionization of sulfuric acid. The second ionization produces sulfate ion and more hydrogen ion. A concentration table for this equilibrium is constructed:

	HSO$_4^-$	H$^+$	SO$_4^{2-}$
Initial (M):	0.10	0.10	0
Change (M):	−x	+x	+x
Equilibrium (M):	0.10 − x	0.10 + x	x

The large value of K_{a2} means that you cannot make any assumptions.

$$K_{a2} = \frac{(x)(0.10 + x)}{0.10 - x} = 1.1 \times 10^{-2}$$
$$x = 0.0087 \text{ M}$$

Thus, the equilibrium concentrations are:

$$[H_2SO_4] \approx 0$$
$$[H^+] = 0.11 \, M$$
$$[HSO_4^-] = 0.09 \, M$$
$$[SO_4^{2-}] = 0.01 \, M$$

16.47 (a) The hydroxide concentration in a $0.010 \, M$ KOH solution is $0.010 \, M$ because KOH is a strong base. At 25 °C, $[H^+]$ is

$$[H^+] = K_w/[OH^-] = 1.0 \times 10^{-14}/0.010$$
$$= 1.0 \times 10^{-12} \, M \text{ or pH} = 12$$

At 50 °C, $[H^+]$ is

$$[H^+] = K_w/[OH^-] = 5.5 \times 10^{-14}/0.010$$
$$= 5.5 \times 10^{-12} \, M \text{ or pH} = 11.26$$

(b) In a neutral solution $[H^+] = [OH^-]$. Thus, at 50 °C,

$$[H^+][OH^-] = 5.5 \times 10^{-14}$$
$$[H^+]^2 = 5.5 \times 10^{-14}$$
$$[H^+] = 2.34 \times 10^{-7}$$
$$\text{pH} = 6.63$$

Note that the pH of a solution also depends on the temperature.

(c) The autoionization reaction is:

$$H_2O(l) \rightleftharpoons H^+(aq) + OH^-(aq)$$

The value of K_w increases with temperature as indicated by the given data. This occurs if the reaction is endothermic. If heat is added to an endothermic system, the added energy drives the equilibrium to the right, thereby increasing K_w.

16.48 (a) In pure water at 25 °C the concentration of hydroxide ion is 1.0×10^{-7} mol/L or 1.0×10^{-7} mol/1000 mL or 1.0×10^{-10} mol/1 mL. The number of ions is $(1.0 \times 10^{-10} \text{ mol}/1 \text{ mL})(6.02 \times 10^{23} \text{ ions/mol}) = 6.0 \times 10^{13}$ ions/1 mL.

(b) The concentration of hydroxide ion provided by the NaOH after adding it to the pure water is $M = (0.50 \text{ mL})(0.10 \, M)/1.5 \text{ mL} = 0.033 \, M$. The amount of hydroxide ion provided by water is negligible compared to that provided by the NaOH and can be neglected.

(c) No. Most liquids are relatively incompressible; thus the effect of increasing the pressure from 1 atm to 3 atm is to only very slightly decrease the volume of water.
16.49 (d). 16.50   (a). 16.51 (b).

$$M_{NaOH} = \frac{(2.5 \text{ g})\left(\dfrac{1 \text{ mol}}{40.01 \text{ g}}\right)}{(100 \text{ mL})\left(\dfrac{1 \text{ L}}{1000 \text{ mL}}\right)} = 0.62 \, M$$

$$\text{pOH} = -\log(0.62) = 0.21; \text{pH} = 14 - 0.21$$
$$= 13.79.$$

16.52 (d). 16.53 (a). 16.54 (a).

16.55 (e) H_2S has the weakest H—X bond.

16.56 (a)

16.57 (b) The boron atom accepts a pair of electrons; yet BBr_3 has no ionizable protons.

16.58 (e) A positive ion with the largest charge and smallest ion radius should show the greatest acidic properties.

16.59 (e) Alkali and alkaline-earth binary hydrides contain the H$^-$ ion, which is a basic ion.

16.60 (d)

$$\frac{[H^+]}{C_0(HCN)} \times 100 = \frac{8.57 \times 10^{-6}}{0.15} \times 100 = 5.7 \times 10^{-3}\%$$

HCN is an extremely weak acid as shown by the percent dissociation.

16.61 (a)

$$\frac{[H^+][CH_3COO^-]}{[CH_3COOH]} = 1.8 \times 10^{-5}$$
$$[CH_3COOH] = (1.26 \times 10^{-3})^2/1.8 \times 10^{-5}$$
$$= 8.82 \times 10^{-2}$$

The initial concentration of acetic acid, C_0, is calculated by adding to the equilibrium concentration of acetic acid the amount that ionized, which is equal to the hydrogen ion concentration:

$$C_0 = [HC_2H_3O] + [H^+] = 8.82 \times 10^{-2} + 1.26 \times 10^{-3}$$
$$= 8.95 \times 10^{-2} \, M$$

Moles of acetic acid $= 8.95 \times 10^{-2}$ mol/liter $\times 0.200$ liter

$$= 1.79 \times 10^{-2} \text{ mol.}$$

16.62 (d) $Ba(OH)_2$ produces two OH$^-$ ions per $Ba(OH)_2$ formula.   Thus, $[OH^-] = 2 \times 0.030M = 0.060$ M. pOH $= 1.22$; pH $= 12.78$.

16.63 (e) $HX \rightleftharpoons H^+ + X^-$. If 2.2 percent of HX is ionized, then $(0.022)(0.25 \, M) = 5.5 \times 10^{-3} \, M$ of hydrogen ion is formed; pH $= 2.26$.

16.64 (b) $B + H_2O \rightleftharpoons BH^+ + OH^-$. From pH, pOH is obtained: pOH $= 4.75$. Therefore $[OH^-] = [BH^+]$ $= 1.78 \times 10^{-5}$, and $[B] = 0.15 - 1.78 \times 10^{-5} \simeq 0.15 \, M$.

$$K_b = \frac{[BH][OH^-]}{[B]} = \frac{(1.78 \times 10^{-5})}{0.15}$$

$$= 2.11 \times 10^{-9}$$

Additional Aspects of Aqueous Equilibria

OVERVIEW OF THE CHAPTER

17.1 COMMON-ION EFFECT

Review: Net ionic equations (4.2); LeChatelier's principle (15.7).

Learning Goals: You should be able to predict qualitatively and calculate quantitatively the effect of an added common ion on the pH of an aqueous solution of a weak acid or base.

17.2 BUFFERS

Review: pH (16.4); weak acids and bases (16.6, 16.7).

Learning Goals: You should be able to:

1. Calculate the concentrations of each species present in a solution formed by mixing an acid and a base.
2. Describe how a buffer solution of a particular pH is made and how it operates to control pH.
3. Calculate the change in pH of a simple buffer solution of known composition caused by adding a small amount of strong acid or base.

17.3 TITRATIONS OF ACIDS AND BASES: TITRATION CURVES

Review: Neutralization (4.3); titrations and indicators (4.6); pH calculations involving strong and weak acids and bases (16.5, 16.6, 16.7).

Learning Goals: You should be able to:

1. Describe the form of the titration curves for titration of a strong acid by a strong base, a weak acid by a strong base, or a strong acid by a weak base.
2. Calculate the pH at any point, including the equivalence point, in acid–base titrations.

17.4 SOLUBILITY EQUILIBRIA

Review: Heterogeneous equilibria (15.2); hydrolysis of ions (16.9); LeChatelier's principle (15.7); solubility principles (13.1, 13.3, 15.3).

Learning Goals: You should be able to:

1. Set up the expression for the solubility-product constant for a salt.
2. Calculate K_{sp} from solubility data and solubility from the value for K_{sp}.
3. Calculate the effect of an added common ion on the solubility of a slightly soluble salt.

Review: Hydration (13.1); Lewis acid–base concepts (16.11); concept of reaction quotient (15.5).

Learning Goals: You should be able to:

1. Predict whether a precipitate will form when two solutions are mixed, given appropriate K_{sp} values.
2. Explain the effect of pH on a solubility equilibrium involving a basic or acidic ion.
3. Formulate the equilibrium between a metal ion and a Lewis base to form a complex ion of a metal.
4. Describe how complex formation can affect the solubility of a slightly soluble salt.
5. Calculate the concentration of a metal ion in equilibrium with a ligand with which it forms a soluble complex ion, from a knowledge of initial concentrations and K_f.
6. Explain the origin of amphoteric behavior and write equations describing the dissolution of an amphoteric metal hydroxide in either an acidic or basic medium.

TOPIC SUMMARIES AND EXERCISES

A 0.10 M CH$_3$COOH solution has a pH equal to 2.9, whereas a solution containing 0.10 M CH$_3$COOH and 0.10 M CH$_3$COONa has a pH equal to 4.7. Why has the addition of CH$_3$COONa to a 0.10 M CH$_3$COOH solution caused a decrease in hydrogen ion concentration (increase in pH)? Section 17.1 in the text provides answers to this question.

- When CH$_3$COO$^-$ (from the strong electrolyte CH$_3$COONa) is added to an acetic acid solution that is at equilibrium, the equilibrium condition is displaced. According to LeChatelier's principle, the system restores itself to an equilibrium state by removing the added stress, that is, some of the added acetate ion.

$$CH_3COOH(aq) \rightleftharpoons H^+(aq) + CH_3COO^-(aq)$$

Equilibrium   Stress:
CH$_3$COO$^-$ is added

shifts to left
to remove some of
the added CH$_3$COO$^-$

This simultaneously removes H$^+$ ions from solution with the formation of CH$_3$COOH and thereby increases the pH.

- The above change in equilibrium is an example of the **common-ion effect:** *addition of one of the ions involved in the equilibria represses dissociation of the weak electrolyte.*

A solution containing a significant quantity of an ion common to the weak electrolyte is often produced by one of two methods:

1. Addition of a salt containing an ion common to the weak acid or base. Examples are the addition of NaF to an HF solution, KNO$_2$ to a HNO$_2$ solution, and NH$_4$Cl to an NH$_3$ solution.

**17.5, 17.6, 17.7
CRITERIA FOR
PRECIPITATING OR
DISSOLVING
SLIGHTLY
SOLUBLE SALTS**

**COMMON-ION
EFFECT**

2. Addition of a strong base to partially neutralize a weak acid (HX) and thereby form its conjugate base (X^-):

$$HX(aq) + OH^-(aq) \xrightarrow{(100\%)} H_2O(l) + X^-(aq)$$

or addition of a strong acid (B) to partially neutralize a weak base to form its conjugate acid (BH^+):

$$B(aq) + H^+(aq) \xrightarrow{(100\%)} BH^+(aq)$$

EXERCISE 1   Identifying the principal components in a mixture and their acid/base properties

(a) An aqueous solution contains K_2CO_3 and $KHCO_3$. What are the principal species in the mixture other than water? (b) K_a for the bicarbonate ion is smaller than its K_b value and K_b for the carbonate ion is greater than K_b for the bicarbonate ion. Write the equilibrium that primarily determines the pH of the solution. (c) Explain what happens to the pH of the solution when HCl is added.

SOLUTION: *Analyze*: We are asked to identify the principal species in a solution and their acid/base properties. From this information we are to write an equilibrium which primarily determines the pH of the solution. We are also asked how the pH changes when HCl is added.

Plan: (a) Both salts are soluble and strong electrolytes; therefore the principal species are the cations and anions of the salts. (b) To write an equilibrium which is primarily responsible for the pH of the solution we must determine which species has the largest equilibrium constant in water. From this information we can write an equilibrium. (c) We have to assess the effect on adding HCl on the hydrogen ion concentration in the presence of the two salts.

Solve: (a) The salts dissociate to form their respective cations and anions: K^+, CO_3^{2-}, and HCO_3^-. The potassium ion is a spectator ion (group 1A cations act as spectator ions) and it does not affect the pH. (b) The bicarbonate ion, HCO_3^-, is amphoteric as shown by its two K values. It is basic overall because K_a has a smaller value than K_b. The carbonate ion, CO_3^{2-}, is also basic and it has the larger K_b value. Therefore, the pH of the solution is primarily determined by the equilibrium involving the carbonate ion reacting with water to form a basic solution.

$$CO_3^{2-}(aq) + H_2O(l) \rightleftharpoons HCO_3^-(aq) + OH^-(aq)$$

(c) Adding a strong acid to a solution will increase the hydrogen ion concentration. In this solution it will react with hydroxide ion and this will reduce the concentration of hydroxide ion, thereby decreasing pH.

EXERCISE 2   Predicting changes in pH when a salt, an acid, and a base are added to a solution of ammonia

Predict the changes in pH and condition of equilibrium when the following are added to a solution of NH_3: (a) NH_4Cl; (b) KOH; (c) HCl.

SOLUTION: *Analyze*: We are asked what is the effect on pH and the equilibrium of an aqueous solution of ammonia when three substances are added.

Plan: We first write the equilibrium of ammonia with water to determine what species form:

$$NH_3(aq) + H_2O(l) \rightleftharpoons NH_4^+(aq) + OH^-(aq)$$

We then apply LeChatelier's principle to predict in which direction the equilibrium is shifted when each substance is added.

Solve: (a) Addition of NH_4^+ from the strong electrolyte NH_4Cl causes the following equilibrium to shift to the left:

$$NH_3(aq) + H_2O(l) \rightleftharpoons NH_4^+(aq) + OH^-(aq)$$

shift left ← | added

The concentration hydroxide of ion decreases; pH decreases. (b) KOH is a strong base forming $K^+(aq)$ and $OH^-(aq)$. The addition of OH^- ions also drives the above equilibrium to the left, reducing the concentration of NH_4^+. However, the pH increases because not all of the added OH^- ion from KOH is removed. (c) Hydrogen ion from the strong acid HCl reacts with the weak base NH_3 to form additional NH_4^+.

$$NH_3(aq) + H^+(aq) \longrightarrow NH^+_4(aq)$$

The increased quantity of ammonium ion is partially reduced by the equilibrium shifting to the left. This decreases the concentration of hydroxide ion. Therefore the pH decreases.

EXERCISE 3 Calculating the pH of a solution containing a weak acid and a salt containing the conjugate base of the acid

Calculate the pH of a 0.15 M formic acid ($HCHO_2$) solution that also contains 0.050 M sodium formate ($NaCHO_2$). $K_a = 1.8 \times 10^{-4}$ for formic acid.

SOLUTION: *Analyze*: We are asked to calculate the pH of an aqueous solution containing both formic acid and sodium formate. We should recognize this as a common ion solution because it contains formic acid and a salt containing its conjugate base, the formate ion.

Plan: We first write any equilibria that occur in solution and the chemical reaction describing the dissociation of the salt, sodium formate. Formic acid is a weak acid and forms the following equilibrium:

$$HCHO_2(aq) \rightleftharpoons H^+(aq) + CHO_2^-(aq)$$

Sodium formate is a strong electrolyte and completely dissociates:

$$NaCHO_2(aq) \xrightarrow{100\%} Na^+(aq) + CHO_2^-(aq)$$

Note that the dissociation of $NaCHO_2$ forms CHO_2^-, an ion common to the equilibria involving $HCHO_2$. Therefore, both $HCHO_2$ and CHO_2^- are initially present in the solution.

Solve: As we did in Chapter 16, we can develop a concentration table of species involved in the formic acid equilibria.

	HCH_2O	$\rightleftharpoons$	H^+	+	CHO_2^-	
Initial (M):	0.15		0		0.050	from $NaCHO_2$
Reaction (M):	$-x$		$+x$		$+x$	
Equilibrium (M):	$0.15 - x$		x		$0.050 + x$	

Assume that $x \ll 0.15$ and 0.050 because K_a is reasonably small. Solve for x by using the equilibrium constant expression.

We now use K_a and x to solve for the hydrogen ion concentration:

$$K_a = \frac{[H^+][CHO_2^-]}{[HCHO_2]} = \frac{(x)(0.05)}{0.15}$$

$$x = \frac{K_a(0.15)}{(0.050)} = \frac{(1.8 \times 10^{-4})(0.15)}{(0.050)}$$

$$x = [H^+] = 5.40 \times 10^{-4}\ M$$

$$pH = -\log[H^+] = 3.27$$

Comment: A 0.15 M $HCHO_2$ solution *without* 0.050 M sodium formate present has a pH = 2.28. The effect of adding sodium formate is to reduce $[H^+]$ and increase pH.

BUFFER SOLUTIONS

A solution containing 0.1 M CH_3COOH and 0.1 M CH_3COONa resists *large* changes in pH when *small* amounts of acid or base are added. Similarly, a solution of 0.1 M NH_3 and 0.1 M NH_4Cl resists large changes in pH under similar conditions. Solutions that resist large changes in pH when small amounts of acid or base are added are called **buffer solutions.** We will encounter two types of buffers:

1. *Acid buffer:* A solution containing a weak acid and the conjugate base of the weak acid (for example, CH_3COOH and CH_3COONa);
2. *Base buffer:* A solution containing a weak base and the conjugate acid of the weak base (for example, NH_3 and NH_4Cl).

A buffer solution must have an acidic component that can neutralize added base and a basic component that can neutralize added acid.

Buffer capacity refers to the effectiveness of a buffer system in resisting changes in pH.

- Buffer solutions are the most effective when the concentrations of the conjugate acid–base pair are about equal, that is, when the pH of the buffer solution is close to pK_a.
- Buffer capacity also depends on the concentrations of the conjugate acid–base pair. For example, a 0.1 M $NaNO_2$ and 0.1 M HNO_2 solution has a larger buffer capacity than a solution containing 0.001 M $NaNO_2$ and 0.001 M HNO_2. The first solution has more acid and base particles that can neutralize added acid or base than the second solution.

EXERCISE 4   Writing net ionic equations for a buffer solution when an acid, a base, or salt is added to it

Write the net ionic equations describing the reactions that occur in a CH_3COOH and CH_3COONa buffer system upon addition of (**a**) HCl; (**b**) NaOH; (**c**) NaCl.

SOLUTION: *Analyze*: We are given a buffer system containing acetic acid and sodium acetate and are asked to write chemical reactions that describe what happens when three substances are added. We should recognize that it is a buffer system because the solution contains a weak acid and its conjugate base.

Plan: First identify the acid and base components of the buffer. Then identify the acid and/or base component of the substance added to the buffer. The net ionic equation is written by recognizing that the acid (base) component of the buffer reacts with the base (acid) component of the substance added.

Solve: (a) The base component of the buffer, CH_3COO^-, reacts with the added acid (HCl):

$$CH_3COO^-(aq) + H^+(aq) \longrightarrow CH_3COOH(aq)$$

(b) The acid component of the buffer, CH_3COOH, reacts with the added base (NaOH):

$$CH_3COOH(aq) + OH^-(aq) \longrightarrow H_2O + CH_3COO^-(aq)$$

(c) There is no reaction because $Na^+(aq)$ is not acidic and $Cl^-(aq)$ is not basic.

EXERCISE 5   Calculating the pH of a buffer solution before and after addition of a strong base

A 1.00-L buffer solution contains 0.100 M NH_4Cl and 0.0200 M NH_3. K_b for NH_3 is 1.8×10^{-5}. (a) What is the pH of the solution? (b) What is the pH of the solution if 0.0010 mol of NaOH is added?

SOLUTION: *Analyze*: We are given a buffer solution containing ammonia and its conjugate acid, the ammonium ion. We are asked to calculate the pH of the buffer solution and the pH after sodium hydroxide is added. We are also given K_b, which is the equilibrium-constant for ammonia.

Plan: (a) As we have done in previous exercises involving weak acids and bases we write the appropriate equilibria and develop a concentration table that shows how concentrations change either when the equilibrium forms. (b) When acids or bases are added to an existing equilibrium we first should write a chemical equation describing how the added acid or base reacts with a species involved in the buffer equilibrium. Then we develop a concentration table that shows how the concentrations of the species involved in the buffer equilibrium are changed by the addition of the acid or base. Finally, we develop a second concentration table that shows how these concentrations are changed when equilibrium is once again achieved and use K_a or K_b to solve for pH.

Solve: (a) The equation describing the chemical equilibrium of ammonia and the corresponding equilibrium-constant expression are written first:

$$NH_3(aq) + H_2O(l) \rightleftharpoons NH_4^+(aq) + OH^-(aq)$$

$$K_b = \frac{[NH_4^+][OH^-]}{[NH_3]} = 1.8 \times 10^{-5}$$

If we let x equal the molarity of the amount of NH_3 that ionizes, we can construct the following concentration table:

	NH_3	$\underset{\rightleftharpoons}{H_2O}$	NH_4^+	+	OH^-
Initial:	0.0200 M		0.100 M		0 M
Ionization effect:	$-x$ M		$+x$ M		$+x$ M
Equilibrium:	$(0.0200 - x)$ M		$(0.100 + x)$ M		x M
Equilibrium*:	0.0200 M		0.100 M		x M

Solving for x by appropriate substitutions into the K_b expression yields

$$\frac{[NH_4^+][OH^-]}{[NH_3]} = 1.8 \times 10^{-5}$$

---

*The value of x is small because the presence of NH_4^+ from NH_4Cl represses the ionization of NH_3; so you can assume that the equilibrium concentrations are 0.0200 M for NH_3, and 0.100 M for NH_3.

$$\frac{(0.100\ M)(x)}{0.0200\ M} = 1.8 \times 10^{-5}$$

$$x = \left(\frac{0.0200}{0.100}\right)(1.8 \times 10^{-5}) = 3.6 \times 10^{-6}\ M = [\text{OH}^-]$$

$$\text{pOH} = -\log[\text{OH}^-] = 5.44$$

The pH is calculated using the relation

$$\text{pH} + \text{pOH} = 14.00$$

$$\text{pH} = 14 - \text{pOH} = 14.00 - 5.44 = 8.56$$

(b)   The added NaOH reacts with the acid component of the buffer as shown below:

$$\text{OH}^-(aq) + \text{NH}_4^+(aq) \longrightarrow \text{H}_2\text{O}(l) + \text{NH}_3(aq)$$

This reaction proceeds in the forward direction to a very large extent. We assume that the added $\text{OH}^-(aq)$ completely reacts with $\text{NH}_4^+(aq)$ to form an equivalent amount of $\text{NH}_3(aq)$. Thus, we must change the initial concentrations of NH_4^+ and NH_3 to account for this reaction before we consider the equilibrium condition.

	OH^-	+	NH_4^+	$\longrightarrow$ H_2O +	NH_3
Initial:	0.0010 M		0.100 M		0.0200 M
Reaction (100%):	−0.0010 M		−0.0010 M		+0.0010 M
New Conc.:	0		0.099 M		0.0210 M

Note: The hydroxide ion initially is 0.0010 M and comes from the concentration of the NaOH added. It is a limiting reactant situation. The amount of hydroxide ion is less than the concentration of ammonium ion.

In terms of problem-solving strategies we now need to consider what type of solution exists after addition of the hydroxide ion. To do this, refer to the previous table and the last line. What species have concentrations other than zero? An equilibrium will exist if one of them is a weak electrolyte. (If one of the remaining species is OH^- or H^+ then it is likely that the residual H^+ or OH^- concentration controls the pH and not the weak electrolyte, depending on the magnitude of the concentrations.) The table shows that ammonium ion and ammonia still exist and thus they should be involved in an equilibrium. We can write the following concentration table with the associated equilibrium, which is the same as in part **(a)**.

	NH_3	+ H_2O $\longrightarrow$	NH_4^+	+	OH^-
Initial:	0.0210 M		0.099 M		0 M
Ionization effect:	−x M		+x M		+x M
Equilibrium:	(0.0210 − x) M		(0.099 + x) M		x
Assume that:	x ≪ 0.0210 and 0.099				
Equilibrium:	0.0210 M		0.099 M		x

The initial concentrations of ammonia and ammonium ion are from the last line of the concentration in part **(b)**. Solving for x using the K_b expression yields

$$\frac{[\text{NH}_4^+][\text{OH}^-]}{[\text{NH}_3]} = 1.8 \times 10^{-5}$$

$$\frac{(0.099)(x)}{0.0210} = 1.8 \times 10^{-5}$$

$$x = \left(\frac{0.0210}{0.099}\right)(1.8 \times 10^{-5})\ M = 3.8 \times 10^{-6}\ M = [\text{OH}^-]$$

$$\text{pOH} = -\log[\text{OH}^-] = 5.42$$

The pH is 8.58. The effect of the added OH$^-$ is to change the pH by only 0.02 pH units, which is a very small change.

EXERCISE 6 Determining which buffer solution has the greater buffer capacity

Which buffer solution has the larger buffer capacity toward added acid, 0.010 M CH$_3$COOH and 0.010 M CH$_3$COONa or 0.010 M CH$_3$COOH and 0.0020 M CH$_3$COONa?

SOLUTION: The buffer capacity of a solution is defined as its ability to resist changes in pH when an acid or a base is added. The smaller the change in pH upon addition of an acid or a base, the greater the buffer capacity. The buffer solution containing 0.010 M CH$_3$COOH and 0.010 M CH$_3$COONa has the larger buffer capacity toward added acid because it contains more base particles to neutralize added acid.

EXERCISE 7 Preparing a buffer solution of given pH

What concentrations of acetic acid and sodium acetate are required to prepare a buffer solution with a pH = 4.60? $K_a = 1.8 \times 10^{-5}$ for acetic acid.

SOLUTION: *Analyze*: We are asked to determine the initial concentrations of acetic acid and sodium acetate that result in a buffer solution of 4.60. This is a different type of situation compared to previous problems in that we know the pH and now we must work "backwards" to find the initial concentrations that give this pH.

Plan: As in previous exercises involving equilibrium conditions, we write the appropriate equilibria involving acetic acid and acetate ion and the chemical reaction describing the dissociation of the salt. Sodium acetate is a strong electrolyte and completely dissociates to form sodium ions and acetate ions.

$$CH_3COONa(aq) \xrightarrow{100\%} Na^+(aq) + CH_3COO^-(aq)$$

Acetic acid is a weak acid,

$$CH_3COOH(aq) \rightleftharpoons H^+(aq) + CH_3COO^-(aq)$$

It is this equilibrium that determines pH. In this problem, we are given the pH of the buffer solution. From the pH, the equilibrium concentration of H$^+$ at equilibrium is determined:

$$[H^+] = 10^{-4.60} = 2.51 \times 10^{-5} \, M$$

The required concentrations of acetic acid and acetate ion to produce a pH = 4.60 are calculated by first developing a concentration table and then using K_a. Since we do not know the initial concentrations of acetic acid and acetate ion, we can let x = *initial* molarity of CH$_3$COOH and y = *initial* concentration of acetate ion, CH$_3$COO$^-$. As in prior problems we can use a concentration table to help us solve the problem. The amount of acetic acid that dissociated is simply the concentration of hydrogen ion determined from the pH of the solution.

	CH$_3$COOH(aq)	$\rightleftharpoons$	H$^+$(aq)	+	CH$_3$COO$^-$(aq)
Initial (M)	x		0		y
Reaction (M)	-2.51×10^{-5}		$+2.51 \times 10^{-5}$		$+2.51 \times 10^{-5}$
Equilibrium (M)	$x - 2.51 \times 10^{-5}$		2.51×10^{-5}		$y + 2.51 \times 10^{-5}$
Choose x and y so that $x, y \gg 2.51 \times 10^{-5}$. Therefore,					
Equilibrium (M)	x		2.51×10^{-5}		y

Solve: We can use the equilibrium-constant expression for acetic acid to determine the ratio of the concentrations of acetic acid and acetate ion:

$$K_a = \frac{[H^+][CH_3COO^-]}{[CH_3COOH]} = \frac{(2.51 \times 10^{-5})(y)}{x} = 1.8 \times 10^{-5}$$

$$\frac{y}{x} = \frac{K_a}{[H^+]} = \frac{1.8 \times 10^{-5}}{2.51 \times 10^{-5}} = 0.72$$

Note that we have one equation with two unknowns. We cannot solve for x and y unless we assign a value to one of them. We can design a buffer by choosing arbitrarily $[CH_3COOH] = x = 0.10\ M$ and then solving for $[CH_3COO^-](= y)$.

$$\frac{y}{0.10\ M} = 0.72 \text{ or } y = (0.10\ M)(0.72) \text{ or } y = 0.072\ M$$

The acetate ion concentration (and therefore the sodium acetate concentration) must be 0.072 M in a 0.10 M CH_3COOH buffer solution to produce a pH = 4.60.

TITRATION OF ACIDS AND BASES: TITRATION CURVES

In Section 4.7 of the text, titrations were briefly described. The use of acid–base indicators to signal the presence of the equivalence point of an acid–base reaction was also discussed. In Section 17.3 of the text, you learn how to determine the appropriate acid–base indicator for a titration by examining titration curves. A **titration curve** is a plot of pH against the volume of added titrant.

Figure 17.6 in the text shows the general shape of a titration curve for a strong acid–strong base titration.

- Note that the equivalence-point pH occurs at 7.
- The titration curve shows a large pH change for a small change in the volume of added titrant near the equivalence point.
- To detect the equivalence point in a strong acid–base titration, you need to choose an indicator whose color changes in the pH range of this rapidly rising section of the titration curve. Since this pH range is large, many indicators are available, and they do not need to change color precisely at pH = 7.
- You should be able to calculate pH at various points along the titration curve. There are four types of calculations:

  1. Initial pH of the strong acid or strong base solution before the titrant is added.
  2. pH of the strong acid or strong base solution that exists after *partial* neutralization of initial acid or base by the added titrant.
  3. Equivalence-point pH, which in this case is 7, because water and a neutral salt are produced.
  4. pH of the solution formed by addition of excess titrant beyond the equivalence point.

- See Sample Exercise 17.6 in the text for examples of these calculations.

Titration curves for weak acid–strong base and weak base–strong acid titrations are shown in Figures 17.9 and 17.12 in the text.

- Note the characteristics of these curves compared to those in Figure 17.6. Observe in these weak acid–base titration curves that a smaller pH change occurs near the equivalence points. A suitable indicator to detect the equivalence point in these titrations must show a color change very near the equivalence-point pH.

- Also note that the equivalence-point pH for the titration of a weak acid with a strong base occurs at a pH that is greater than 7, and that for the titration of a weak base with a strong acid it occurs at a pH that is less than 7.

The calculations of pH changes during the titration of a weak acid with a strong base or the titration of a weak base with a strong acid are more complex than the calculations of pH changes during the titration of a strong acid or base. The dissociation of the weak acid or base must be considered. There are four distinct types of calculations required:

1. At the start of the titration, the pH of the solution is determined by the initial concentration of the weak acid or base being titrated and its dissociation constant.
2. The pH values up to, but not including the equivalence point, are calculated from the equilibrium concentrations of the salt produced and the residual weak acid or base after addition of titrant. (See Sample Exercises 17.7 and 17.8 in the text for an example.)
3. At the equivalence point, the solution contains the conjugate base or acid of the titrated weak acid or base; the pH is calculated from the concentration of the conjugate base or acid and the value of the respective K_b or K_a.
4. Beyond the equivalence point, the pH is primarily controlled by the equilibrium concentration of excess titrating agent.

EXERCISE 8   Interpreting a titration curve

In Figure 17.1 a graph of the pH of solutions of two weak bases, A and B, as a function of added 0.1 M HCl is shown. Both bases have the same initial concentrations. (a) Which weak base, A or B, has associated with it the larger K_b value? (b) Is segment 1 on the graph a buffer region? (c) Why is segment 2 on the graph the same for both bases?

SOLUTION: (a) For a series of weak bases with the same initial concentrations, the larger the K_b value of a weak base, the greater is the extent of its reaction with H_2O to produce OH^-, and the larger is the initial pH of the solution. Weak base A has associated with it the larger K_b value because its initial pH is larger. (b) Yes, because changes in pH are small as a function of added HCl. (c) After the equivalence point has been reached, the pH is determined by the concentration of excess titrant, HCl, which is the same for the titrations of both weak acids.

EXERCISE 9   Calculating pH during the titration of a weak base

Calculate the pH in the titration of 50.00 mL of 0.200 M ammonia by 0.200 M hydrochloric acid: (a) at the start of the titration; (b) after 20.00 ml of HCl has been added; and (c) at the equivalence point. K_b for ammonia is 1.8×10^{-5}.

SOLUTION: *Analyze*: We are given 50.00 mL of a 0.200 M ammonia solution and asked to calculate the pH at three different points when it is titrated with 0.200 M HCl.

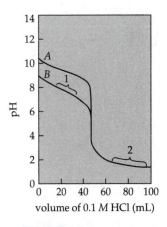

▲ **FIGURE 17.1** The pH of solutions of two weak bases, A and B, as a function of added HCl.

Plan: (**a**) At the start of the titration, the solution contains only aqueous ammonia; thus, the pH is determined from the initial concentration of ammonia and its equilibrium constant.

(**b**) The addition of 20.00 mL of HCl partially neutralizes the ammonia. Also, because the addition of HCl changes the volume of the final solution, you need to account for the dilution when doing calculations. In this problem, first calculate the molarities of HCl and ammonia in the mixture before they react. The total volume of the solution of HCl and ammonia is 20.00 mL + 50.00 mL = 70.00 mL. After we have calculated the molarities of the components after mixing, we need to consider what happens to their concentrations after the HCl reacts with the base in the solution, NH_3. This is a neutralization reaction. We should note that the concentration of HCl is less than that of NH_3. This means that HCl will be a limiting reactant and will be entirely consumed. After considering the concentration changes resulting from neutralization we use the equilibrium in part (**a**) to solve for the concentration of hydrogen ion and pH.

(**c**) At the equivalence point the ammonia is completely neutralized. We need to analyze the acid/base properties of the products formed to determine if any will dissociate or react with water to form hydrogen or hydroxide ions. We use this reaction to determine the pH of the solution.

Solve: (**a**)

$$NH_3(aq) \rightleftharpoons NH_4^+(aq) + OH^-(aq)$$

	NH_3	NH_4^+	OH^-
Initial (*M*):	0.200	0	0
Reaction (*M*):	$-x$	$+x$	$+x$
Equilibrium (*M*):	$0.200 - x$	x	x

We can solve for *x* by using the K_b expression for NH_3 and by making the assumption that $x \ll 0.200\ M$.

$$K_b = \frac{[NH_4^+][OH^-]}{[NH_3]} = \frac{(x)(x)}{0.200} = 18 \times 10^{-5}$$

Solving for the value of *x* gives:

$$x^2 = 3.60 \times 10^{-6} \text{ or } x = [OH^-] = 1.90 \times 10^{-3}\ M$$
$$pOH = 2.72 \text{ or } pH = 11.28$$

(**b**) To calculate the molarities in the mixture, we can use the following relationship:
$$M_iV_i = M_fV_f$$

where *i* refers to the components before mixing and *f* refers to the final mixture

$$M_{HCl}(\text{mixture}) = \frac{M_{HCl}V_{HCl}}{V_{\text{mixture}}} = \frac{(0.200\ M)(20.00\ mL)}{70.00\ mL} = 0.0571\ M$$

$$M_{NH_3}(\text{mixture}) = \frac{M_{NH_3}V_{NH_3}}{V_{\text{mixture}}} = \frac{(0.200\ M)(50.00\ mL)}{70.00\ mL} = 0.143\ M$$

Write the neutralization reaction and underneath write the initial concentrations (the concentrations after dilution). Assume the reaction goes to completion and calculate the molarities after the reaction occurs:

$$H^+(aq) + NH_3(aq) \xrightarrow{100\%} NH_4^+(aq)$$

	H^+		NH_3		NH_4^+
Initial (*M*):	0.0571		0.143		0
Reaction (*M*):	−0.0571		−0.0571		+0.0571
Final (*M*):	0		0.086		0.0571

Note that HCl (H^+) is indeed the limiting reactant in the above reaction. The resulting solution contains 0.086 *M* ammonia and 0.0571 *M* ammonium ion, which is a common ion solution. We can now solve for pH using the equilibrium used in (**a**):

$$NH_3(aq) \quad \rightleftharpoons \quad NH_4^+(aq) \quad + \quad OH^-(aq)$$

Initial (M):	0.086		0.0571		0
Reaction (M):	$-x$		$+x$		$+x$
Equilibrium (M):	$0.086 - x$		$0.0571 + x$		x

We can solve for x by using the K_b expression and assuming that $x \ll 0.086$ M or 0.0571 M:

$$K_b = \frac{(x)(0.0571)}{0.086} = 1.8 \times 10^{-5}$$

Solving for x gives: $x = [OH^-] = 2.7 \times 10^{-5}$ M. Thus pOH = 4.57 and pH = 9.43. Note that the pH decreased upon addition of HCl. This is to be expected because acid is added to the original solution. If the pH had increased, we would have known that an error had been made.

(c) We need to calculate the volume of HCl required for complete neutralization of 50.00 mL of 0.200 M NH_3 ammonia. We can use the following relationship to calculate the volume of HCl required because HCl and ammonia react in a 1:1 mole ratio:

$$M_{HCl}V_{HCl} = M_{NH_3}V_{NH_3} \quad [\text{Remember: moles} = M \times V]$$

$$V_{HCl} = \frac{M_{NH_3}V_{NH_3}}{V_{HCl}} = \frac{(0.200\ M)(50.00\ mL)}{0.200\ M} = 50.00\ mL$$

The solution formed at the equivalence point has a volume of 50.00 mL + 50.00 mL = 100.00 mL. We can now calculate the new molarities of HCl and ammonia upon mixing as we did in (b); however, we should observe that the final solution is simply twice the volume of either starting solution. This means that the molarities after dilution are simply one-half of their original molarities.

$$M_{HCl} = \frac{1}{2}(0.200\ M) = 0.100\ M$$

$$M_{NH_3} = \frac{1}{2}(0.200\ M) = 0.100\ M$$

Next, calculate the concentration of ammonium ion that forms at the equivalence point after all of the ammonia is neutralized.

	$H^+(aq)$	+	$NH_3(aq)$	$\xrightarrow{100\%}$	$NH_4^+(aq)$
Initial (M):	0.100		0.100		0
Reaction (M):	−0.100		−0.100		+0.100
Final (M):	0		0		0.100

The solution now only contains ammonium ion, a weak acid. The new equilibrium that is responsible for the pH of the solution is:

	$NH_4^+(aq)$	$\rightleftharpoons$	$NH_3(aq)$	+	$H^+(aq)$
Initial (M):	0.100		0		0
Reaction (M):	$-x$		$+x$		$+x$
Equilibrium (M):	$0.100 - x$		x		x

Solve for x by using K_a for ammonium ion and assuming that $x \ll 0.100$. K_a is calculated from the relationship $K_a \times K_b = K_w$.

$$K_a = \frac{[H^+[NH_3]}{[NH_4^+]} = \frac{K_w}{K_b(NH_3)} = \frac{1.0 \times 10^{-10}}{1.8 \times 10^{-5}} = 5.6 \times 10^{-10}$$

$$5.6 \times 10^{-10} = \frac{(x)(x)}{0.100}$$

Solving for x gives:

$$x^2 = 5.6 \times 10^{-11} \quad \text{or} \quad x = [H^+] = 7.48 \times 10^{-6}\,M \quad \text{or} \quad pH = 5.13$$

Note that the pH at the equivalence point of the titration of a weak base is in the acid region.

SOLUBILITY EQUILIBRIA

When a saturated solution of the slightly soluble salt $PbCl_2$ is in contact with solid $PbCl_2$, there is an equilibrium between the solid $PbCl_2$ and the dissolved Pb^{2+} and Cl^- ions. This equilibrium is formatted as

$$PbCl_2(s) \rightleftharpoons Pb^{2+}(aq) + 2\,Cl^-(aq)$$

- The concentration of ions in a saturated solution are determined by the **solubility-product constant**, K_{sp}, which for $PbCl_2$ is given by

$$K_{sp} = [Pb^{2+}][Cl^-]^2 = 1.7 \times 10^{-5}$$

- The solubility-product expression for the generalized equilibrium

$$M_aX_b(s) \rightleftharpoons aM^{m+}(aq) + bX^{n-}(aq)$$

is

$$K_{sp} = [M^{m+}]^a[X^{n-}]^b$$

- The concentration units for dissolved particles are moles per liter and the concentration of the solid is included in the value of K_{sp}.

Solubility and solubility product constant are not the same. Solubility refers to the amount of substance that dissolves in a given quantity of solvent. The solubility of a slightly soluble salt is affected by the presence of an ion common to its equilibrium.

- The common-ion effect, which represses the dissociation of $M_aX_b(s)$, causes the solubility of $M_aX_b(s)$ to *decrease*.
- For example, the solubility of $CaSO_4$ is reduced in a Na_2SO_4 solution. The presence of SO_4^{2-} ion from Na_2SO_4 shifts the equilibrium

$$CaSO_4(s) \rightleftharpoons Ca^{2+}(aq) + SO_4^{2-}(aq)$$

to the left, thereby decreasing the amount of calcium sulfate that dissolves.
- Values of K_{sp} can be used to calculate solubilities of salts. However, experimental studies show that these calculations are generally only valid when the charges of the ions are low and the ions do not react with water to form new species. Thus, calculating the solubility of AgCl from its K_{sp} value should give reasonable agreement with the experimental

solubility, whereas doing the same calculation with $La(IO_3)_3$ will give poor agreement with the experimental value.

Exercise 10   Using K_{sp} to calculate solubility and concentrations I

(a) Calculate the solubility of AgI in water in moles per liter given its K_{sp} value of 8.3×10^{-17}. (b) A saturated solution of AgI also containing NaI is found to have an iodide ion concentration of 0.020 *M*. What is the concentration of silver ion?

SOLUTION: *Analyze*: We are given the solubility-product constant for AgI and asked to calculate the solubility of AgI in water. We are also asked to calculate the concentration of silver ion in a saturated AgI solution also containing NaI. The latter is a common-ion solution. Note that we are given the *equilibrium* concentration of iodide ion, not just that from NaI.

Plan: For both problems we write the solubility equilibrium for AgI and the associated solubility-product constant. (a) We consider the concentrations of silver and iodide ions in a saturated system and how they are related to one another based on the equation describing the chemical equilibrium. We use the solubility-product expression to solve for the unknown quantity. (b) We are given the equilibrium concentration of iodide ion (0.020 *M*) and thus only the concentration of silver ion is unknown.

Solve: (a) The solubility equilibrium is formulated as

$$AgI(s) \rightleftharpoons Ag^+(aq) + I^-(aq)$$

and the corresponding solubility-product expression is

$$K_{sp} = [Ag^+][I^-] = 8.3 \times 10^{-17}$$

If we let *x* equal the molar solubility of AgI, then *x* moles/liter of AgI dissolves and completely dissociates. This means that *x* moles/liter of Ag^+ and *x* moles/liter of I^- form.

$$[Ag^+] = [I^-] = x$$

Substituting these terms into the K_{sp} expression for AgI gives

$$K_{sp} = [Ag^+][I^-] = (x)(x) = 8.3 \times 10^{-17}$$
$$x^2 = 8.3 \times 10^{-17}$$
$$x = \sqrt{8.3 \times 10^{-17}} = 9.1 \times 10^{-9} \, M$$

Thus, the molar solubility of AgI is 9.1×10^{-9} mol/L. (b) The given concentration of iodide ion is substituted in the K_{sp} expression

$$K_{sp} = [Ag^+][I^-] = [Ag^+][0.020] = 8.3 \times 10^{-17}$$

Solving for the silver ion concentration gives

$$[Ag^+] = \frac{8.3 \times 10^{-17}}{0.020} = 4.2 \times 10^{-15}$$

Thus, the concentration of silver ion is 4.2×10^{-15} mol/L in the presence of 0.020 *M* I^- ions.

EXERCISE 11   Using K_{sp} to calculate solubility and concentrations II

(a) What is the solubility of $BaSO_4$ in moles per liter in a solution containing 0.020 M Na_2SO_4? K_{sp} is 1.1×10^{-10} for $BaSO_4$. (b) Does this solution illustrate the common-ion effect?

SOLUTION: *Analyze and Plan*: We are asked to calculate the solubility of barium sulfate in a 0.020 M solution of sodium sulfate. This problem is similar to the one in Exercise 10 except the concentration of sodium sulfate is given, not the equilibrium concentration of sulfate ion. Thus, we have to account for the concentration of sulfate ion from both barium sulfate and sodium sulfate. We use the solubility-product expression to solve for the unknown quantity.

Solve: (a) The solubility of $BaSO_4$ is controlled by its K_{sp} value, even if common ions are present in the solution. In this problem, there are two sources of the SO_4^{2-} ion, $BaSO_4$ and Na_2SO_4. If we let x be the solubility of $BaSO_4$, then x mol/L of Ba^{2+} and x mol/L of SO_4^{2-} form from the dissolution of $BaSO_4$. However, the equilibrium concentration of SO_4^{2-} is x plus the amount of SO_4^{2-} from Na_2SO_4— that is, $x + 0.020$ M. The equilibrium is formulated as

$$BaSO_4(s) \rightleftharpoons Ba^{2+}(aq) + SO_4^{2-}(aq)$$

$$K_{sp} = [Ba^{2+}][SO_4^{2-}] = 1.1 \times 10^{-10}$$

The equilibrium concentrations are

$$[Ba^{2+}] = x \; M \qquad \text{and} \qquad [SO_4^{2-}] = (x + 0.020) \; M$$

The value of x is very small and can be assumed negligible compared to 0.020 M. Therefore, we can say

$$(x + 0.020) \; M \approx 0.020 \; M$$

Solving for the value of x by appropriate substitutions into the K_{sp} expression for $BaSO_4$ yields

$$K_{sp} = [Ba^{2+}][SO_4^{2-}] = (x)(0.020) = 1.1 \times 10^{-10}$$

$$x = 5.5 \times 10^{-9} \; M$$

The solubility of $BaSO_4$ in a 0.020 M Na_2SO_4 solution is 5.5×10^{-9} mol/L. Note that 5.5×10^{-9} is indeed small compared to 0.020. (b) Yes. The common-ion effect refers to a solution in which there is present an ion common to a slightly soluble salt equilibrium. In this case the common ion is SO_4^{2-}, from Na_2SO_4. The common ion supresses the ionization of $BaSO_4$ and thereby reduces its solubility.

CRITERIA FOR PRECIPITATING OR DISSOLVING SLIGHTLY SOLUBLE SALTS

A precipitate of a slightly soluble salt can be prepared by mixing appropriate salt solutions. One solution should contain a soluble salt containing the cation of the slightly soluble salt, and the other solution should contain a soluble salt containing the anion.

• A precipitate forms after the solutions are mixed if the **ion product**, Q, for the slightly soluble salt is greater than the K_{sp} value for the slightly soluble salt.

- The ion product for the insoluble salt upon mixing of the two solutions has the same form as the K_{sp} expression but it is evaluated using the initial salt-solution concentrations. Summarized in Table 17.1 are the possible relationships between the values of Q and K_{sp}.

TABLE 17.1 Relationships Between Q and K_{sp}

Relationship	Solution condition
$Q > K_{sp}$	Precipitation occurs until $Q = K_{sp}$
$Q = K_{sp}$	Equilibrium and a saturated solution
$Q < K_{sp}$	Solution is not saturated, and no precipitate forms

Sometimes chemists need to dissolve slightly soluble salts other than by the procedure of sufficiently diluting a saturated solution. In Section 17.5 of the text, you explore different methods for dissolving such salts. These are summarized below.

1. *If a slightly soluble salt contains a basic anion, such as $BaCO_3$, the addition of a strong acid causes its solubility to increase.* A strong acid (H^+) reacts with a basic anion and removes it from the salt equilibrium, thereby causing more of the slightly soluble salt to dissolve. For example, the addition of HNO_3 to a saturated solution of $BaCO_3$ causes solid $BaCO_3$ to dissolve along with evolution of $CO_2(g)$. The net ionic equation describing the process is

$$BaCO_3(s) + 2\,H^+(aq) \longrightarrow H_2O(l) + CO_2(g) + Ba^{2+}(aq)$$

2. *Insoluble metal hydroxides and oxides that are amphoteric dissolve in both acids and bases.* The concept of amphoterism was introduced in Section 16.2 of the text, and you should review this section. Examples are amphoteric hydroxides and oxides containing the metal ions Al^{3+}, Cr^{3+}, Zn^{2+}, and Sn^{2+}. You should know these examples.

3. *A slightly soluble salt containing a metal ion that reacts with Lewis bases to form complex ions can have its solubility increased by addition of the appropriate Lewis base.* Metal ions that are Lewis acids will often react with Lewis bases to form complex ions. The term **complex ion** refers in general to an ion that contains several different atoms. In this section, it refers to an ion that contains a metal ion bonded to Lewis bases. Some examples of complex ions that you should know are $Ag(NH_3)_2^+$, $Zn(OH)_4^{2-}$, HgS_2^{2-}, and $Cu(CN)_2^-$. For example, the net ionic equation describing the dissolving of $AgCl$ using $Na_2S_2O_3$ is as follows:

$$AgCl(s) + 2\,S_2O_3^{2-}(aq) \longrightarrow Ag(S_2O_3)_2^{3-}(aq) + Cl^-(aq)$$

The stability of a complex metal ion is measured by its **formation constant**, K_f. A formation constant for a complex metal ion is associated with the reaction of the metal ion with Lewis bases to form the complex metal ion. For the formation of $Ag(S_2O_3)_2^{3-}$, the reaction is written as

$$Ag^+(aq) + 2\,S_2O_3^{2-}(aq) \rightleftharpoons Ag(S_2O_3)_2^{3-}(aq)$$

$$K_f = \frac{[Ag(S_2O_3)_2^{3-}]}{[Ag^+][S_2O_3^{2-}]^2}$$

EXERCISE 12   Predicting how the solubility of $Mg(OH)_2$ changes when an acid, a base, or a salt is added

Predict the change in solubility of $Mg(OH)_2$ when the following substances are added to a saturated solution of $Mg(OH)_2$: (a) NaOH; (b) HCl; (c) $MgCl_2$.

SOLUTION: *Analyze and Plan:* We are asked to predict the change in solubility of magnesium hydroxide when three different substances are added to a saturated solution. We can use LeChatelier's principle to make predictions.

Solve: The equilibrium of the slightly soluble salt $Mg(OH)_2$ is described by

$$Mg(OH)_2(s) \rightleftharpoons Mg^{2+}(aq) + 2\,OH^-(aq)$$

(a) Adding NaOH, which contains an ion common to $Mg(OH)_2$, drives the equilibrium to the left, thus decreasing the solubility of $Mg(OH)_2$. (b) Hydrogen ions from HCl react with OH^- ions to form water. This removes OH^- ions and drives the equilibrium to the right, thus increasing the solubility of $Mg(OH)_2$. (c) $MgCl_2$ is a soluble salt and a strong electrolyte. The additional magnesium ions from $MgCl_2$ shift the equilibrium to the left and thus decrease the solubility of $Mg(OH)_2$.

EXERCISE 13   Writing the equilibrium for the formation of a complex ion of aluminum

Which of the following is a more informative description of the formation of $Al(OH)_4^-$ from an aluminum ion and hydroxide ions?

(a) $Al(H_2O)_6^{3+}(aq) + 4\,OH^-(aq) \rightleftharpoons Al(OH)_4(H_2O)_2^-(aq) + 4\,H_2O(l)$
(b) $Al^{3+}(aq) + 4\,OH^-(aq) \rightleftharpoons Al(OH)_4^-(aq)$

SOLUTION: Reaction (a) is a better description because Al^{3+} exists in water as the hydrated ion, not as a free ion. The reaction clearly shows that four hydroxide ions are displacing water molecules; or it can be viewed as hydroxide ions removing protons from bound water molecules and thus forming OH^- ions bonded to Al^{3+}.

EXERCISE 14   Writing chemical reactions of $Zn(OH)_2$ with an acid and base

(a) Write chemical equations that describe the addition of two hydroxide ions to the slightly soluble metal hydroxide $Zn(OH)_2(H_2O)_2$ and the addition of one proton to this same solution. (b) What is the term used to describe the behavior of $Zn(OH)_2(H_2O)_2$?

SOLUTION: *Analyze and Plan:* We are asked to write the chemical reaction describing the reaction of $Zn(OH)_2(H_2O)_2$ with hydroxide and hydrogen ions. The question implies it is an amphoteric metal hydroxide. The reaction in each case should be an acid–base reaction and a complex ion may form.

Solve: (a) The reaction of $Zn(OH)_2(H_2O)_2$ with two hydroxide ions is formulated as

$$Zn(OH)_2(H_2O)_2(s) + 2\,OH^-(aq) \rightleftharpoons Zn(OH)_4^{2-}(aq) + 2\,H_2O(l)$$

and its reaction with one proton is formulated as

$$Zn(OH)_2(H_2O)_2(s) + H^+(aq) \rightleftharpoons Zn(OH)(H_2O)_3^+(aq)$$

Note that the total number of groups attached to zinc does not change in both cases. (b) Amphoterism. $Zn(OH)_2(H_2O)_2$ is an amphoteric hydroxide.

EXERCISE 15   Writing a chemical reaction and K_f for the formation of a complex ion of silver

AgCl dissolves in water when CN^- is added because the very stable complex ion $Ag(CN)_2^-$ is formed. (a) Write the overall reaction equation for formation of $Ag(CN)_2^-$ from CN^- and AgCl. (b) Write two stepwise reaction equations that, when summed, yield the reaction equation you wrote in (a). (c) Write the equilibrium-constant expression associated with each stepwise reaction in (b).

SOLUTION: *Analyze*: We are asked to write the chemical reaction that describes the formation of $Ag(CN)_2^-$ when cyanide ion is added to the slightly soluble salt AgCl. We are also asked to write two chemical reactions that when added together form the first reaction. Finally we are asked to write equilibrium-constant expressions for each reaction in (b).

Plan: (a) The text discusses formation reactions involving metal complexes and we should review it. Formation reactions of metal complexes involve a metal ion reacting with a base to form a metal complex. (b) It is likely that one is a solubility equilibrium and the other is a metal complex reaction because $AgCl(s)$ and $Ag(CN)_2^-$ appear in the overall reaction. After we write each reaction and ensure that they add to the overall one, we can write equilibrium-constant expressions.

Solve:(a) The overall reaction is

$$AgCl(s) + 2\,CN^-(aq) \rightleftharpoons Ag(CN)_2^-(aq) + Cl^-(aq)$$

Note that Cl^- is displaced from AgCl.

(b)   The two equilibria involved when AgCl dissolves in an aqueous cyanide solution are

$$AgCl(s) \rightleftharpoons Ag^+(aq) + Cl^-(aq) \qquad [\text{solubility equilibrium}]$$

$$Ag^+(aq) + 2\,CN^-(aq) \rightleftharpoons Ag(CN)_2^-(aq) \quad [\text{complexation}]$$

When these two reaction equations are added together, the result is the overall reaction equation

$$AgCl(s) + 2\,CN^-(aq) \rightleftharpoons Ag(CN_2)^-(aq) + Cl^-(aq)$$

(c) The equilibrium-constant expressions for the first reaction in **(b)** is

$$K_{sp} = [Ag^+][Cl^-]$$

and for the second reaction.

$$K_f = \frac{[Ag(CN)_2^-]}{[Ag^+][CN^-]^2}$$

EXERCISE 16   Determining if a precipitate forms when two salt solutions are mixed

Will $AgIO_3$ precipitate when 100.00 mL of 0.0100 M $AgNO_3$ is mixed with 50.00 mL of 0.0100 M KIO_3? K_{sp} of $AgIO_3$ is 3.0×10^{-8}.

SOLUTION: *Analyze*: We are asked to predict whether $AgIO_3$ will precipitate when solutions of silver nitrate and potassium iodate are mixed. This involves an increase in volume; therefore, the concentrations of the salts are smaller in the final solution. We are given K_{sp} of $AgIO_3$ suggesting we need it to solve the problem.

Plan: KIO_3 and $AgNO_3$ undergo a reaction to form $AgIO_3$, which may precipitate. To determine whether $AgIO_3$ precipitates we must calculate the ion product, Q, for $AgIO_3$ and compare it with K_{sp}. The ion product is evaluated using the initial concentrations of silver ion and iodate ion: $Q = C_{Ag^+}C_{IO_3^-}$, where C means initial concentration. We cannot use the given concentrations because the solutions are mixed; thus, we must first calculate the molarities of the ions after mixing. The new volume is 150.00 mL.

Solve: Using $M_f = \dfrac{M_i V_i}{V_f}$ we can calculate the molarities of the salts after dilution:

$$M_f(AgNO_3) = \frac{(0.0100\ M)(100.00\ mL)}{150.00\ mL} = 0.00667\ M$$

$$M_f(KIO_3) = \frac{(0.0100\ M)(50.00\ mL)}{150.00\ mL} = 0.00333\ M$$

Therefore, the starting concentrations of the ions in the mixture are $C_{Ag^+} = 0.00667\ M$ and $C_{IO_3^-} = 0.00333\ M$. We can now solve for Q,

$$Q = C_{Ag^+}C_{IO_3^-} = (0.00667)(0.00333) = 2.22 \times 10^{-5}$$

Because $Q > K_{sp}$, $AgIO_3$ precipitates.

EXERCISE 17   Determining if a precipitate of a metal hydroxide occurs at a specified pH

A sodium hydroxide solution with a pH of 6.00 has sufficient cerium nitrate added to make the total cerium ion concentration 0.00300 M. Will $Ce(OH)_3$ precipitate? K_{sp} of $Ce(OH)_3$ is 1.5×10^{-20}. Assume the addition of cerium nitrate does not change the pH.

SOLUTION: *Analyze*: We are given a cerium nitrate solution that is 0.00300 M in cerium ion with a pH of 6.00 and K_{sp} of cerium hydroxide. Using this information we are asked to determine if cerium hydroxide will precipitate under these conditions. This problem will require us to compare K_{sp} to the ion-product when the solution is first formed.

Plan: The equilibrium of cerium ion with hydroxide ion is

$$Ce(OH)_3(s) \rightleftharpoons Ce^{3+}(aq) + 3\,OH^-(aq)$$

To determine whether cerium hydroxide precipitates, we must solve for the value of the ion product, Q. We are given the concentration of cerium ion and the pH of the solution. From the pH of the solution we can determine the hydroxide ion concentration:

$$pOH = 14 - pH = 14 - 6.00 = 8.00 \quad or \quad [OH^-] = 1.00 \times 10^{-8}\ M$$

Q can now be calculated:

$$Q = C_{Ce^{3+}}C_{OH^-}^3 = (0.00300)(1.00 \times 10^{-8})^3 = 3.00 \times 10^{-27}$$

Because $Q < K_{sp}$, cerium hydroxide does not precipitate.

SELF-TEST QUESTIONS

Key Terms

Having reviewed key terms in Chapter 17, match key terms with phrases and identify statements as true or false. If a statement is false, indicate why it is incorrect.

Match each phrase with the best term:

17.1 Term given to the effect when NaCl is added to a saturated solution of AgCl.

17.2 The species formed when Ag^+ reacts with an excess of ammonia.

17.3 A method for identifying which ions are in a solution.

17.4 A solution that resists large changes in pH when small amounts of acid or base are added.

17.5 The type of equilibrium expression for barium sulfate in equilibrium with its aqueous ions.

Terms:

(a) buffer solution   (d) qualitative-analysis
(b) common-ion effect   (e) solubility-product
(c) complex ion

True-False:

17.6 The *formation-constant* for $Ni(NH_3)_6^{2+}$ is 2×10^8. From this information we know that the following equilibrium favors the products:

$$Ni(NH_3)_6^{2+}(aq) \rightleftharpoons Ni^{2+}(aq) + 6\,NH_3(aq)$$

17.7 The *titration curve* for a weak acid, with a K_a of 3×10^{-9}, shows a steeply rising curve near the equivalence point.

17.8 A buffer solution containing 0.00100 M acetic acid and 0.00100 M sodium acetate will have a high *buffer capacity*.

17.9 The *Henderson-Hasselbalch* equation shows for a buffer solution that changing the total volume of the solution does not change the pH.

17.10 *Quantitative analysis* requires the use of an accurate balance, or an accurate volume dispensing device.

Problems and Short-Answer Questions

17.11 The figure below shows a saturated solution of $FeCO_3$ in water. The partially shaded spheres represent Fe^{2+} ions, the unshaded spheres represent CO_3^{2-} ions and the fully shaded region represents solid $FeCO_3$. Draw a figure representing the condition when additional solid $FeCO_3$ is added to the solution. Explain.

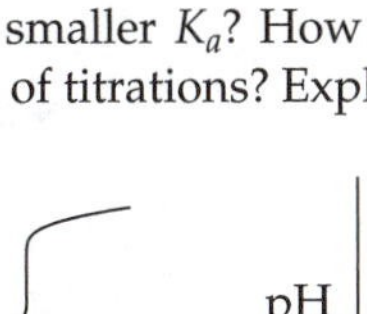

17.12 Two different weak acids with the same concentrations are titrated with the same strong base. The pH titration curves (A and B) for the two acids are shown below. Which one has the smaller K_a? How does this affect the precision of a series of titrations? Explain.

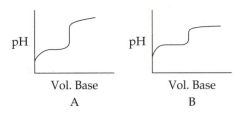

pH   Vol. Base   pH   Vol. Base
               A   B

17.13 What is a buffer solution? How can adding some HCl(*aq*) to a $NH_3(aq)$ solution produce a buffer solution?

17.14 A buffer solution contains 0.110 M $NaHCO_2$ (sodium formate) and 0.100 M HCO_2H (formic acid) and has a pH of 3.80. If a small amount of HCl (*aq*) is added, predict qualitatively what changes occur in the concentrations of sodium formate and formic acid and in the pH. Explain.

17.15 Conceptualizing how to do a chemistry problem without doing the math or exact solution is an important skill. In written words only describe the key aspects of the following problem and how you would solve it. A solution that is approximately 0.1 M HCl is to be used as a standard solution in a titration. Another solution containing 0.100 g of 99.95% Na_2CO_3 is prepared and it takes 21.50 mL of the HCl solution to completely titrate the sodium carbonate solution. What is the molarity of the HCl solution?

17.16 Critique the following statement: HgI_2 is completely insoluble because its K_{sp} is less than 10^{-25}.

17.17 When KOH is added to a solution of $NiCl_2(aq)$ a pale-green precipitate forms. When NH(*aq*) is added to a solution of $NiCl_2(aq)$ a pale-green precipitate initially forms but when more ammonia is added, a deep-blue solution forms. Explain these observations and write chemical equations to describe the observed reactions.

17.18 Which of the following solutions would make the best buffer? Explain.

(a) 0.1 M HCl and 0.1 M NaCl
(b) 0.1 M NH_4Cl and 0.1 M NaCl
(c) 0.1 M NH_4Cl and 0.1 M CH_3COOH
(d) 0.1 M NH_3 and 0.1 M NH_4Cl

17.19 A 0.20 M solution of Na_2CO_3 is titrated to the first equivalence point with 25.0 mL of 0.20 M HCl.

(a) Write the net ionic equation that describes the reaction leading to the first equivalence point.
(b) Given that $K_{a1} = 4.5 \times 10^{-7}$ and $K_{a2} = 4.7 \times 10^{-11}$ for H_2CO_3, qualitatively predict the pH region at the first equivalence point.

17.20 (a) Write the solubility-product expressions for $HgBr_2$ and $Ni_3(PO_4)_2$.
(b) What is wrong with the following solubility-product expression for $FeCO_3$? Write the correct form.

$$K_{sp} = \frac{[Fe^{2+}][CO_3^{2-}]}{[FeCO_3]}$$

17.21 The concentration of an acid is determined by titrating it with 25.0 mL of 0.100 M KOH. Is there sufficient information to calculate the concentration? Explain.

17.22 Explain why the pH at the equivalence point for the titration of nitrous acid with sodium hydroxide is in the basic region and that of nitric acid titrated with sodium hydroxide is seven.

17.23 In Exercise 4 you calculated the pH of the buffer solution after addition of NaOH.

(a) Calculate the pH after addition of 0.0015 mol of HCl.

(b) Would the solution still be a buffer if 0.0200 mol of HCl were added? In both problems, assume that no volume dilution occurs because of addition of HCl.

17.24 A solution initially contains $2.0 \times 10^{-5}\ M$ AgNO$_3$ and $3.0 \times 10^{-2}\ M$ NaCl. Given that $K_{sp} = 1.8 \times 10^{-10}$ for AgCl, is there a precipitate of AgCl in this solution?

17.25 Which conjugate acid–base pair should be used to prepare a buffer with a pH of 3.5? Explain.

Acid	Conjugate base	K_a
CH$_3$COOH	CH$_3$COO$^-$	1.8×10^{-5}
HCHO$_2$	CHO$_2^-$	1.7×10^{-4}
H$_2$CO$_3$	HCO$_3^-$	4.2×10^{-7}
H$_3$PO$_4$	H$_2$PO$_4^-$	7.5×10^{-3}

17.26 What is the solubility of AgCl in moles per liter in a $0.020\ M$ MgCl$_2$ solution?

$$K_{sp} = 1.8 \times 10^{-10} \text{ for AgCl.}$$

17.27 Calculate the pH in the titration of 50.00 mL of $0.1000\ M$ NH$_3$ by $0.2000\ M$ HCl:

(a) after addition of 10.00 mL HCl;

(b) at the equivalence point;

(c) and 10.00 mL of HCl beyond the equivalence point. K_b for NH$_3$ is 1.8×10^{-5}.

17.28 Calculate the concentration of cobalt ion present in a solution containing $0.10\ M$ Co(NO$_3$)$_2$ and $6.0\ M$ NH$_3$. $K_f = 1.1 \times 10^5$ for Co(NH$_3$)$_6^{2+}$.

17.29 Calculate the pH of a solution that is prepared by dissolving 3.25 g of CH$_3$COONa in 100 mL of $0.150\ M$ HCl. $K_a = 1.8 \times 10^{-5}$ for acetic acid. Assume that no volume change occurs when CH$_3$COONa is added.

17.30 A solution of an unknown weak acid has a pH of 4.54. Is there sufficient information to determine K_a for the weak acid? If not, what additional information is needed and why?

17.31 A solution contains $0.010\ M$ KOH and $0.0050\ M$ HCl. Is there sufficient information to determine if this is a buffer solution? If not, what additional information is needed and why?

17.32 A saturated solution of AgCl has 10.0 mL of $1.0\ M$ KCl added to it and solid AgCl is still visible. K_{sp} for silver chloride is given. Without doing a calculation, is there sufficient information to determine the concentration of chloride ion in the mixture? If not, what additional information is needed and why?

17.33 A chemical used in protein chemistry to prepare a buffer is TRIS:

$$\begin{array}{c} \text{OH} \\ | \\ \text{CH}_2 \\ | \\ \text{HO}-\text{CH}_2-\underset{\underset{\text{CH}_2}{\overset{|}{\underset{|}{}}}}{\overset{|}{\text{C}}}-\text{NH}_2 \\ | \\ \text{OH} \end{array} \quad \text{or} \quad (\text{HOCH}_2)_3\text{CNH}_2$$

(a) Given that TRIS is a weak base, which is the basic site, the group —OH or the —NH$_2$ group? Explain.

(b) A TRIS buffer is prepared by adding $12\ M$ HCl to a TRIS solution until the desired pH is achieved. Write a chemical reaction showing why the p changes.

(c) Write the net ionic equation showing how a base reacts with a TRIS buffer.

(d) What is the pH of a TRIS buffer prepared by adding 10.0 mL of $12\ M$ HCl to $1.00\ L$ of $0.50\ M$ TRIS. The pK_b of TRIS is 5.7.

17.34 (a) Ca(OH)$_2$ is a basic salt in water but it has a limited ability to form ions. What is the pH of a saturated solution if its K_{sp} is 6.5×10^{-6}?

(b) If K$_2$CO$_3$ is added to a saturated calcium hydroxide solution, what equilibrium occurs and what is K for it? K_{sp} of CaCO$_3$ is 4.5×10^{-9}. What happens to the pH compared to the solution in (a)?

17.35 (a) Compare the pH of a $0.10\ M$ acetic acid and $0.10\ M$ sodium acetate buffer solution to that of a $0.010\ M$ acetic acid and $0.010\ M$ sodium acetate buffer solution.

(b) Does the acetic acid in the more dilute buffer solution have a greater percent ionization than in the more concentrated buffer solution? Explain.

Multiple-Choice Questions

17.36 Which occur(s) when HCl is added to an aqueous buffer solution containing ammonia and ammonium chloride?

(a) pH of the solution decreases.

(b) More NH$_4^+$ forms.

(c) Less NH$_3$ exists.

(d) Less OH$^-$ exists.

(e) All of the preceding are correct.

17.37 What happens when K$_2$S is added to an aqueous solution of Na$_2$S?

(a) The concentration of sulfide ion decreases.

(b) The pH of the solution increases.

(c) More hydrogen ion forms.

(d) The pH of the buffer solution does not change.

(e) No change occurs in the physical properties of the solution.

17.38 Which of the following solutions should have the greatest buffering capacity toward added base?

- (a) 0.1 M CH_3COOH and 0.01 M CH_3COONa
- (b) 0.1 M CH_3COOH and 0.1 M CH_3COONa
- (c) 1 M CH_3COOH and 0.1 M CH_3COONa
- (d) 1 M HCl and 1 M NaOH
- (e) None of the above solutions can function as a buffer.

17.39 The net ionic equation describing the reaction of added NaOH to a buffer solution containing HCO_3^- and CO_3^{2-} is

(a) $OH^-(aq) + CO_3^{2-}(aq) \longrightarrow HCO_4^{3-}(aq)$

(b) $OH^-(aq) + HCO_3^-(aq) \longrightarrow H_2O(l) + CO_3^{2-}(aq)$

(c) $2\,Na^+(aq) + CO_3^{2-}(aq) \longrightarrow Na_2CO_3(aq)$

(d) $Na^+(aq) + HCO_3^-(aq) \longrightarrow NaHCO_3(aq)$

(e) $Na^+(aq) + H_2O(l) \longrightarrow H^+(aq) + NaOH(aq)$

17.40 What is the pH of a 0.150 M HCl solution at the start of its titration with 0.150 M NaOH?

- (a) 0.150
- (b) −0.150
- (c) 0.82
- (d) −0.82
- (e) 13.2

17.41 What is the equivalence-point pH of the solution formed by the titration of 50.00 mL of 0.150 M HCl using 50.00 mL of 0.150 M NaOH?

- (a) 1
- (b) 4
- (c) 7
- (d) 9
- (e) 10

17.42 What is the equivalence-point pH of the solution formed by the titration of 50.00 mL of 0.150 M CH_3COOH using 25.00 mL of 0.300 M NaOH?

- (a) 3.22
- (b) 4.53
- (c) 7.00
- (d) 8.26
- (e) 8.88

17.43 Which of the following has a solubility-product value in an aqueous environment?

- (a) $FeCl_3$
- (b) $Mg(NO_3)_3$
- (c) SO_3
- (d) HCl
- (e) $Ca_3(PO_4)_2$

17.44 If x equals the solubility of Ag_2CO_3 in moles per liter, then how is the value of K_{sp} for Ag_2CO_3 related to the value of x?

- (a) $K_{sp} = 4x^3$
- (b) $K_{sp} = x^3$
- (c) $K_{sp} = 2x^3$
- (d) $K_{sp} = x^2$
- (e) $K_{sp} = 2x^2$

17.45 The solubility of $BaCO_3$ in water is increased by which of the following: (1) addition of NaOH; (2) addition of HNO_3; (3) increasing pH of solution; (4) decreasing pH of solution?

- (a) (1)
- (b) (2)
- (c) (1) and (3)
- (d) (2) and (4)
- (e) (3)

17.46 What is the formation-constant expression for $Fe(CN)_6^{4-}$?

- (a) $K_f = [Fe^{2+}][CN^-]$
- (b) $K_f = [Fe^{2+}][CN^-]^6$
- (c) $K_f = [Fe^{2+}][CN^-]^6/[Fe(CN)_6^{4-}]$
- (d) $K_f = [Fe(CN)_6^{4-}]/[Fe^{2+}][CN^-]^6$
- (e) $K_f = [Fe(CN)_6^{4-}]/[Fe^{2+}][CN^-]$

17.47 Which of the following metal ions form amphoteric hydroxides: (1) Fe^{3+}; (2) Cr^{3+}; (3) Zn^{2+}; (4) Ca^{2+}; (5) Na^+?

- (a) (1) and (2)
- (b) (2) and (3)
- (c) (3) and (4)
- (d) (4) and (5)
- (e) (2) and (4)

17.48 The pH of a standard solution of H_2S is too low to precipitate ZnS. Which of the following could be added to the solution to cause ZnS to precipitate?

- (a) HCl
- (b) HNO_3
- (c) H_2S
- (d) NaCl
- (e) NaOH

17.49 Which of the following is the primary form of aluminum in a strongly basic solution?

- (a) $Al(H_2O)_6^{3+}$
- (b) $Al(H_2O)_5(OH)^{2+}$
- (c) $Al(H_2O)_4(OH)^{2+}$
- (d) $Al(H_2O)_3(OH)_3(s)$
- (e) $Al(H_2O)_2(OH)_4^-$

SELF-TEST SOLUTIONS

17.1 (b). **17.2** (c). **17.3** (d). **17.4** (a) **17.5** (e). **17.6** False. The reaction shown is for the dissociation of $Ni(NH_3)_6^{2+}$, not its formation. Thus for the given reaction, $K = 1/K_f = 1/(2 \times 10^8) = 5 \times 10^{-9}$, $Ni(NH_3)_6^{2+}$ is a very stable metal-complex ion, and the reaction $Ni(NH_3)_6^{2+}(aq) \rightleftharpoons Ni^{2+}(aq) + 6\,NH_3(aq)$ lies far to the left. **17.7** False. Weak acids with very small K_as show only a small rise near the equivalence point. **17.8** False. The concentrations are small, and thus the ability of the two components to neutralize added acid or base is limited.

17.9 True. **17.10** True.

17.11 Adding a pure solid to a saturated solution of a solid in equilibrium with its ion does not change the position of equilibrium. The solubility-product expression does not include the solid. Thus, the figure representing the solution shows the same number and type of ions but with more solid present:

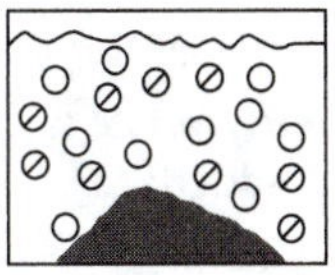

17.12 The weak acid associated with titration curve B has a K_a which is smaller. As the value of K_a becomes smaller the ΔpH near the equivalence point becomes smaller. The smaller ΔpH near the equivalence point makes it more difficult to reproduce good titration data.

17.13 A buffer solution is one that resists large changes in pH when *small* amounts of acid or base are added to it or when diluted. When some HCl(*aq*) is added to NH$_3$(*aq*) partial neutralization occurs: H$^+$(*aq*) + NH$_3$(*aq*) → NH$_4^+$(*aq*). The resulting solution contains both NH$_3$ and NH$_4^+$, the conjugate acid of ammonia. The equilibrium between ammonia and the ammonium ion, if their concentration ratio is in the range of 10/1 to 1/10, results in a buffer solution. A buffer solution has both an acid and a base component that can neutralize the added acid or base and then the system can readjust concentrations via an equilibrium.

17.14 HCl(*aq*) is a strong acid and ionizes completely to form H$^+$(*aq*), which will react with the base component of the buffer solution: H$^+$ + HCO$_2^-$ → HCO$_2$H. This increases the concentration of formic acid, the acid component of the buffer, and decreases the concentration of the formate ion, the base component of the buffer. The pH should decrease with increasing concentration of formic acid. This is logical since a strong acid is added and this in itself should reduce the pH.

17.15 A standard solution is one whose concentration is accurately known and can be used to determine the concentration of other solutions. You are asked to determine the molarity of an unknown acid solution and you know the volume (21.50 mL) it took to react with 0.100 g of 99.95% sodium carbonate. Write the reaction to determine the stoichiometry. You will need this to develop mole–mole relationships between the hydrochloric acid and sodium carbonate. The carbonate ion has a 2− charge and thus can add two H$^+$ ions: 2 H$^+$ + CO$_3^{2-}$ → H$_2$CO$_3$. You also have to recognize that the sodium carbonate is not pure and the actual number of moles of carbonate ion in it is 99.95% of the number of moles of sodium carbonate. From the stoichiometry and the later information we can say this:

mol carbonate ion = mol sodium carbonate × 0.9995
mol H$^+$ that react with carbonate ion =
$$2(\text{mol carbonate ion})$$

$$\text{molarity HCl} = \frac{\text{mol H}^+}{(21.50 \text{ mL})\left(\frac{1 \text{ L}}{1000 \text{ mL}}\right)}$$

17.16 HgI$_2$ is very insoluble as indicated by its K_{sp} value. However, even with a K_{sp} that small there are ions of Hg^{2+} and I$^-$ in solution resulting from the equilibrium:
$$\text{HgI}_2(s) \rightleftharpoons \text{Hg}^{2-}(aq) + 2\,\text{I}^-(aq)$$

17.17 The pale-green precipitate is logically a hydroxide of nickel since KOH is added:
$$\text{Ni}^{2+}(aq) + 2\,\text{OH}^-(aq) \longrightarrow \text{Ni(OH)}_2(s)$$

Ammonia is a base and forms hydroxide ion in the following equilibrium:
$$\text{NH}_3(aq) + \text{H}_2\text{O}(l) \rightleftharpoons \text{NH}_4^+(aq) + \text{OH}^-(aq)$$
In a low ammonia concentration the nickel ion is reacting with the hydroxide ion and forms nickel(II) hydroxide as in the first reaction. However, with excess ammonia, the ammonia is able to displace the hydroxide ion and form a complex that is soluble:
$$\text{Ni(OH)}_2(s) + 6\,\text{NH}_3(aq) \longrightarrow [\text{Ni(NH}_3)_6]^{2+}(aq) + 2\,\text{OH}^-(aq)$$

The complex ion has the deep blue color.

17.18 A buffer normally consists of a weak acid in equilibrium with its conjugate base. The only one that meets this requirement is (**d**). It contains the weak acid NH$_4^+$ in equilibrium with its conjugate base NH$_3$. (**c**) contains two weak acids. (**b**) contains only a weak acid. (**a**) contains a strong acid.

17.19 (**a**) CO$_3^{2-}$(*aq*) + H$^+$(*aq*) → HCO$_3^-$(*aq*)

(**b**) At the first equivalence point the bicarbonate ion, HCO$_3^-$, is formed. You have to ask whether this ion reacts with water. Does it have acidic or basic properties? The bicarbonate ion can act both as an acid and as a base. The equilibria and their equilibrium constants follow:
$$\text{HCO}_3^-(aq) \rightleftharpoons \text{H}^+(aq) + \text{CO}_3^{2-}(aq)$$
$$K_{a2} = 4.7 \times 10^{-11}$$
$$\text{HCO}_3^-(aq) + \text{H}_2\text{O}(l) \rightleftharpoons \text{H}_2\text{CO}_3(aq) + \text{OH}^-(aq)$$
$$K_b = \frac{K_w}{K_{a1}} = \frac{1 \times 10^{-14}}{4.5 \times 10^{-7}} = 2.2 \times 10^{-8}$$

Comparison of the two K values shows that the bicarbonate ion is a better base than it is an acid. Thus, the pH at the equivalence point will be in the basic range.

17.20 (**a**) $K_{sp} = [\text{Hg}^{2+}][\text{Br}^-]^2$;
$K_{sp} = [\text{Ni}^{2+}]^3[\text{PO}_4^{3-}]^2$;

(**b**) A solubility-product expression does not show a solid in the denominator; the correct form is $K_{sp} = [\text{Fe}^{2+}][\text{CO}_3^{2-}]$.

17.21 Two key pieces of information are missing. The first is the type of acid: Is it monoprotic, diprotic, or triprotic? You need this information to develop the stoichiometric equivalences between the acid and base to determine the number of moles of acid titrated. You also need the initial volume of the acid. Volume is in the denominator of the molarity expression and thus is required along with the number of moles of acid to solve for concentration.

17.22 The net ionic equation for the titration of nitrous acid (a weak acid) with sodium hydroxide (a strong base) is HNO$_2$(*aq*) + OH$^-$(*aq*) → H$_2$O(*l*) + NO$_2^-$(*aq*). At the equivalence point the nitrite ion, NO$_2^-$, is formed along with water. The nitrite ion is the conjugate base of a weak acid and thus is weakly basic. It reacts with water to form a basic solution [OH$^-$ forms] at the equivalence point: H$_2$O(*l*) + NO$_2^-$(*aq*) → HNO$_2$(*aq*) + OH$^-$(*aq*). The net ionic equation for the titration of nitric acid is: $HNO_3(aq) + OH^-(aq) \rightarrow H_2O(l) + NO_3^-(aq)$. The nitrate ion is the conjugate base of a strong acid, nitric acid, and thus is not basic. It acts as a spectator ion. The solution contains it and water; the pH is that of pure water, seven.

17.23 (a) The added HCl reacts with the base component of the buffer: $H^+(aq) + NH_3(aq) \longrightarrow NH_4^+(aq)$. This reaction essentially goes to completion. Therefore, the additional amount of NH_4^+ formed equals the amount of acid reacting with NH_3. Using similar logic, the amount of NH_3 removed from solution also equals the amount of acid reacting with NH_3. The result of this effect is shown in the following concentration table for the equilibrium

$$NH_3(aq) + H_2O(l) \rightleftharpoons NH_4^+(aq) + OH^-(aq):$$

	NH_3	NH_4^+	OH^-
Initial (*M*)	0.020	0.100	0
Effect of Added H^+ (*M*):	$(0.0200 - 0.0015)$	$(0.100 + 0.0015)$	0
Equilibrium (*M*):	$(0.0185 - x)$	$(0.1015 + x)$	x

Assuming that the value of x is negligible compared to 0.0185 and 0.1015 and substituting appropriate values into the K_b expression yields

$$K_b = \frac{[NH_4^+][OH^-]}{[NH_3]} = \frac{(0.1015)(x)}{0.0185} = 1.8 \times 10^{-5}$$

$$x = [OH^-] = 3.3 \times 10^{-6}$$

$$pOH = 5.48$$
$$pH = 14 - 5.48 = 8.52$$

(b) Addition of 0.200 mol of HCl, which is equivalent in quantity to the amount of NH_3 in the buffer solution, would completely neutralize NH_3 to form NH_4^+. Thus the solution is no longer a buffer because it contains primarily NH_4^+ and only a very small quantity of NH_3 from the reaction of NH_4^+ with water.

17.24 There is a precipitate if the reaction quotient for AgCl, Q, is greater in value than the value of K_{sp}; $Q = (2.0 \times 10^{-5})(3.0 \times 10^{-2}) = 6.0 \times 10^{-7}$. A precipitate of AgCl forms because $Q > K_{sp}$.

17.25 The optimal buffer system is the one whose pK_a is the closest to the required pH. The $HCHO_2$—CHO_2^- pair has a pK_a equal to 3.8, which is the closest one to the required pH of 3.5.

17.26 Let x equal the molar solubility of AgCl. Then $[Ag^+] = x$, and $[Cl^-] = x + 2(0.02 M)$. The factor of $2(0.02 M)$ in the Cl^- expression occurs because 1 mol of $MgCl_2$ dissolves and releases 2 mol of chloride ion. The value of x is solved for as follows: $K_{sp} = [Ag^+][Cl^-] = (x)(x + 0.04) \simeq (x)(0.04) = 1.8 \times 10^{-10}$; $x = 4.5 \times 10^{-9}$ mol/L.

17.27 (a) 10.00 mL of HCl reacts with 50.00 mL of ammonia to form 60.00 mL of solution as follows:

$$H^+(aq) + NH_3(aq) \xrightarrow{100\%} NH_4^+(aq)$$

Initial molarity of $H^+ = \dfrac{(0.01000\ L)(0.2000\ M)}{(0.06000\ L)}$

$$= 0.03333\ M$$

Initial molarity of $NH_3 = \dfrac{(0.05000\ L)(0.01000\ M)}{(0.06000\ L)}$

$$= 0.08333\ M$$

Use these values for the initial concentrations for the reaction between hydrogen ion and ammonia.

	$H^+(aq)$	+	$NH_3(aq)$	$\xrightarrow{100\%}$	$NH_4^+(aq)$
Initial (*M*):	0.03333		0.08333		0
Reaction (*M*):	−0.03333		−0.03333		+0.03333
New conc. (*M*):	0		0.05000		0.03333

Next, calculate the pH of a solution composed of 0.05000 M ammonia and 0.03333 M ammonium ion.

	$NH_3(aq)$	$\overset{H_2O(l)}{\rightleftharpoons}$	$NH_4^+(aq)$	+	$OH^-(aq)$
Initial (*M*):	0.05000		0.03333		0
Reaction (*M*):	$-x$		$+x$		$+x$
Equil (*M*):	$0.05000 - x$		$0.03333 + x$		x

Assume x is << 0.05000 and 0.03333

$$K_b = \frac{[NH_4^+][OH^-]}{[NH_3]} = \frac{(0.0333)(x)}{(0.05000)} = 1.8 \times 10^{-5}$$

$$x = [OH^-] = \left(\frac{0.05000}{0.0333}\right)(1.8 \times 10^{-5}) = 2.7 \times 10^{-5}$$

$$pOH = 4.57 \qquad pH = 9.43$$

(b) To calculate the equivalence point pH, you first must calculate the volume of HCl required to neutralize the ammonia completely. At the equivalence point,

$$\text{Moles HCl} = \text{moles } NH_3$$
$$M_{HCl}V_{HCl} = M_{NH_3}V_{NH_3}$$
$$(0.2000\ M)(V_{HCl}) = (0.1000\ M)(0.5000\ L)$$
$$V_{HCl} = 0.02500\ L \text{ or } 25.00\ mL$$

At the equivalence point, both HCl and NH_3 are in equal molar quantities and completely react to form NH_4^+. The concentration of NH_4^+ is

$$M_{NH_4^+} = \frac{(0.02500\ L)(0.2000\ M)}{0.07500\ L} = 0.06666\ M$$

The numerator represents the number of moles of HCl or NH_3 that reacted to form NH_4^+. The denominator is the new volume of the solution, 50.00 mL + 25.00 mL. Because NH_4^+, a weak acid forms, the new equilibrium and concentration table are

	NH_4	$\rightleftharpoons$	NH_3	+	H^+
Initial (*M*)	0.0666 *M*		0		0
Reaction (*M*)	$-x$		$+x$		$+x$
Equilibrium (*M*)	$0.0666 - x \simeq 0.0666$		x		x

$$NH_4^+(aq) + H_2O(l) \rightleftharpoons NH_3(aq) + H^+(aq)$$

$$K_a = \frac{K_w}{K_b} = \frac{[NH_3][H^+]}{[NH_4^+]} = \frac{1 \times 10^{-14}}{1.8 \times 10^{-5}} = \frac{x^2}{0.06666}$$

$$x^2 = 0.06666 \,(5.6 \times 10^{-10}) = 3.7 \times 10^{-11}$$

$$x = [H^+] = 6.1 \times 10^{-6} \, M$$

$$pH = 5.21$$

(c) The pH when 10.00 mL of HCl is added beyond the equivalence point is controlled by the excess HCl. The molarity of excess HCl is

$$M_{HCl} = \frac{(0.01000 \text{ L})(0.2000 \, M)}{(0.07500 \text{ L} + 0.01000 \text{ L})} = 0.02353 \, M$$

$$pH = -\log(0.02353) = 1.63$$

17.28 Because K_f is very large and the concentration of NH_3 is much larger than that of Co^{2+}, you can initially assume that all the Co^{2+} is in the complex ion form.

	Co^{2+}	+	$6 \, NH_3$	$\rightleftharpoons$	$CO(NH_3)_6^{2+}$
*In (*M*)	0		6.0 − 6(0.10)		0.10
**Rxn (*M*)	+x		+6x		−x
Eq (*M*)	x		5.4 + 6x ≈ 5.4		0.10 − x ≈ 0.10

$NH_3 \, Co^{2+}$.

$$K_f = \frac{[Co(NH_3)_6^{2+}]}{[Co^{2+}][NH_3]^6} = \frac{(0.10)}{(x)(5.4)^6} = 1.1 \times 10^5$$

$$x = [Co^{2+}] = 3.7 \times 10^{-11} \, M$$

---

*The data for the first line is based on the assumption that all the cobalt ion is completely complexed. The concentration of ammonia is reduced by 6.0(0.10) M because six moles of ammonia react for every mole of cobalt(II) ion reacting.

**The data for the second line is developed by recognizing that if an equilibrium is to exist, some cobalt(II) ion must now reform. If cobalt(II) ion forms, then ammonia must also form.

17.29 The acetate ion from sodium acetate combines with the hydrogen ion from HCl to form acetic acid. If excess acetate ion remains after this reaction, then we have a common ion solution. First, determine the number of moles of CH_3COONa and HCl in the solution. From this information, we can determine which species, CH_3COO^- or H^+, is in excess after the following reaction occurs: $H^+(aq) + CH_3COO^-(aq) \longrightarrow CH_3COOH(aq)$. The number of moles of acetate ion initially present in solution is calculated as follows: $82.0 \text{ g } CH_3COONa = 1 \text{ mol } NaC_2H_3O; 1 \text{ mol } C_2H_3O_2 = 1 \text{ mol } CH_3COONa;$

moles $CH_3COO^- = (3.25 \text{ g } CH_3COONa)(1 \text{ mol } CH_3 COONa / 82.0 \text{ g } CH_3COONa) (1 \text{ mol } CH_3COO^-/1 \text{ mol } CH_3COONa) = 0.0396$ mol. The number of moles of hydrogen ion present is determined as follows:

0.150 mol HCl ≐ 1 L of HCl solution

(from molarity statement)

1 mol H^+ = 1 HCl (HCl is a strong acid in solution)

$$\text{Moles } H^+(aq) = (100 \text{ mL HCl})\left(\frac{1 \text{ L}}{1000 \text{ mL}}\right)$$

$$\times \left(\frac{0.150 \text{ mol HCl}}{1 \text{ L}}\right)\left(\frac{1 \text{ mol } H^+}{1 \text{ mol HCl}}\right)$$

$$= 0.0150 \text{ mol } H^+$$

Since mol CH_3COO^- is greater in magnitude than mol H^+, excess CH_3COO^- exists after the reaction between H^+ and CH_3COO^-. This excess number of acetate ions is calculated as follows: excess moles of $CH_3COO^- =$ moles $C_2H_3O_2 -$ moles $H^+ = 0.0396$ mol $- 0.0150$ mol $= 0.0246$ mol. The molarity of the remaining CH_3COO^- in solution is

$$M_{C_2H_3O_2} = \frac{0.0246 \text{ mol } CH_3COO^-}{(100 \text{ mL})\left(\frac{1 \text{ L}}{1000 \text{ mL}}\right)} = 0.246 \, M$$

Also note that the reaction between H^+ and CH_3COO^- forms CH_3COOH. The amount of CH_3COOH formed equals the amount of HCl consumed during the reaction because 1 mol ≐ 1 mol CH_3COOH from the stoichiometry of the reaction. Therefore the molarity of CH_3COOH in the solution is

$$M = \frac{(0.0150 \text{ mol HCl})\left(\frac{1 \text{ mol } CH_3COOH}{1 \text{ mol HCl}}\right)}{(100 \text{ mL})\left(\frac{1 \text{ L}}{1000 \text{ mL}}\right)} = 0.150 \, M$$

You now can solve for pH using procedures previously developed. First write the equilibrium and concentration table:

	CH_3COOH	$\rightleftharpoons$	H^+	+ CH_3COO^-
Initial (*M*):	0.150		0 *M*	0.246
Reaction (*M*):	−x		+x M	+x
Equilibrium (*M*)	(0.150 − x)		x M	(0.246 + x)

Substituting equilibrium concentrations (with the assumption that the value of x is negligible compared to 0.150 and 0.246) into the K_a expression yields.

$$K_a = \frac{[H^+][CH_3COO^-]}{[HC_2H_3O_2^-]} = \frac{(x)(0.246)}{0.150} = 1.8 \times 10^{-5}$$

$$x = [H^+] = 1.1 \times 10^{-5} \, M$$

$$pH = 4.96$$

17.30 There is insufficient information. The form of K_a for a weak acid is $[H^+][A^-]/[HA]$. From the pH, $[H^+]$ and $[A^-]$ can be calculated because they form in equal amounts. $[H^+]$ also represents the amount of HA that ionizes; however, we do not know either the original concentration of the weak acid or the equilibrium concentration. Therefore, we have a missing quantity. If we are given the initial concentration of HA we can then solve for K_a.

17.31 There is sufficient information. KOH is a strong base and HCl is a strong acid. When they react they form KCl and water, which is not a buffer solution. This is a salt solution and does not contain a weak acid and its conjugate base or a weak base and its conjugate acid.

17.32 There is insufficient information. When the solution of KCl is added to the saturated solution of AgCl it becomes diluted. Without knowing the volume of the saturated AgCl solution we cannot calculate the concentration of added KCl in the mixture [it will be less than 1.0 M]. If we are given the volume of the saturated AgCl solution we can calculate the concentration of added KCl in solution. This permits us to use $K_{sp} = [Ag^+][Cl^-]$ to solve for the chloride ion concentration.

17.33 (a) TRIS contains a basic amine group, $-\overset{|}{\underset{|}{C}}-\ddot{N}H_2$

The nitrogen atom is the basic site because of the lone pair on nitrogen.
Carbon has a moderate electronegativity value and therefore

the $-\overset{|}{\underset{|}{C}}-\ddot{O}-H$ unit does not have a significant

ability to release a OH^- ion as is the case for NaOH. The oxygen atom is less basic than a nitrogen atom in TRIS because it has a higher electronegativity value. (b)

$$(HOCH_2)_3NH_2 + H^+ \longrightarrow (HOCH_2)_3NH_3^+$$
$$\text{TRIS} \qquad\qquad\qquad \text{TRISH}^+$$

A weak acid forms and the pH is lowered.

(b) A TRIS buffer contains a weak base, TRIS, and a weak acid, TRISH$^+$. The added base reacts with the weak acid as follows:
$$(HOCH_2)_3NH_3^+ + OH^- \longrightarrow (HOCH_2)_3NH_2 + H_2O$$

(c) The strong acid, HCl, reacts with TRIS as in (b) above. This reduces the concentration of TRIS and increases the concentration of TRISH$^+$. The effect on concentrations is summarized by the following concentration table:

	TRIS	+	H$^+$	$\longrightarrow$	TRISH$^+$
Initial (M)	0.50 mol 1.01 L		0.12 mol 1.01 L		0
Change (M)	−0.12		−0.12		+0.12
End (M)	0.38		0		0.12

The initial concentrations reflect the fact that dilution occurs; however, the effect is minimal because of the relative amounts of TRIS and HCl added together. At equilibrium:

$$\text{TRIS} + H_2O \rightleftharpoons \text{TRISH}^+ + OH^-$$
$$pK_b = 5.7 \text{ or } K_b = 2.0 \times 10^{-6}$$

	TRIS	TRISH$^+$	OH$^-$
Initial (M)	0.38	0.12	0
Change (M)	−x	+x	+x
Equilibrium (M)	0.38 − x	0.12 + x	x

To calculate equilibrium concentrations using the prior equilibrium the following concentration table is developed: Assume that x ≪ 0.38 or 0.12

$$K_b = \frac{(x)(0.12)}{0.38} = 2.0 \times 10^{-6}$$

Solving for x gives $x = [OH^-] = 6.3 \times 10^{-6}$ M. Thus pOH = 5.20 and pH = 8.80.

17.34 (a) $Ca(OH)_2(s) \rightleftharpoons \underset{s}{Ca^{2+}(aq)} + \underset{2s}{2\,OH^-(aq)}$

s is the solubility of calcium hydroxide.
$$K_{sp} = [Ca^{2+}][OH^-]^2 = (s)(2s)^2 = 6.5 \times 10^{-6}$$
$$s = 0.012\ M$$
$$[OH^-] = 2s = 0.024\ M$$
$$pOH = 1.62$$
$$pH = 12.38$$

(b) The carbonate ion from potassium carbonate (a strong electrolyte) is able to displace hydroxide ion from calcium hydroxide if K ≫ 1:
$$Ca(OH)_2(s) + CO_3^{2-}(aq) \rightleftharpoons CaCO_3(s) + 2\,OH^-(aq)$$
$$K = ?$$

This equilibrium is the sum of two equilbria:
$$Ca(OH)_2(s) \rightleftharpoons Ca^{2+}(aq) + 2\,OH^-(aq) \quad K_{sp}(Ca(OH)_2)$$

$$Ca^{2+}(aq) + CO_3^{2-}(aq) \rightleftharpoons CaCO_3(s) \quad 1/K_{sp}(CaCO_3)$$

The overall equilibrium constant is the product of the equilibrium constants for the individual steps:
$$K = K_{sp}(Ca(OH)_2) \times \frac{1}{K_{sp}(CaCO_3)} = \frac{5.6 \times 10^{-6}}{4.5 \times 10^{-9}}$$
$$= 1.4 \times 10^3$$

Note that K for the equilibria is large and hydroxide ion is produced; thus, the solution will be very basic and the pH will increase dramatically.

17.35 (a) The pH can be calculated using the Henderson-Hasselbach equation since the concentrations are reasonably dilute and the value of the equilibrium constant is small.

$$pH = pK_a + \log\frac{[\text{base}]}{[\text{acid}]}$$

The amount of base and acid present in the buffer solution is usually approximated by their initial concentrations. In this problem their concentrations are identical and the ratio is one: log 1 = 0. Thus, in both cases

$pH = pK_a = -\log(1.8 \times 10^{-5}) = 4.74$. **(c)** Yes. In both solutions the molarity of the hydrogen ions is the same; however, the percent ionizations are different. The denominator in the equation for percent ionization is the initial concentration of the acid; this is different for the two buffer solutions. In a 0.10 M acetic acid and 0.10 M sodium acetate buffer solution the percent ionization is

$$\frac{1.8 \times 10^{-5}}{0.10} \times 100 = 0.018\%$$

The percent ionization in a 0.010 M acetic acid and 0.010 M sodium acetate solution is

$$\frac{1.8 \times 10^{-5}}{0.010} = 0.18\%$$

17.36 (**e**) **17.37**   (**b**)

17.38 (**c**) This solution contains the acid component in the greatest amount, thereby buffering most effectively the added base. The solution described by response (**d**) contains only H_2O and NaCl because NaOH and HCl react together.

17.39 (**b**)

17.40 (**c**) $pH = -\log[H^+] = -\log(0.150) = 0.82$.

17.41 (**c**) The equivalence-point pH of the titration of a strong acid and strong base is always 7.

17.42 (**e**) $CH_3COOH(aq) + OH^-(aq) \rightarrow H_2O(l) + CH_3COO^-(aq)$ (neutralization). After neutralization, you need to consider the reaction of CH_3COO^- with water:

$$CH_3COO^-(aq) + H_2O(l) \longrightarrow OH^-(aq) + CH_3COOH(aq)$$

$$M_{CH_3COO^-} = \frac{\text{moles } CH_3COO^-}{V}$$
$$= \frac{\text{moles } CH_3COOH \text{ that reacted}}{V}$$
$$= \frac{(M_{CH_3COOH})(V_{CH_3COOH})}{V}$$
$$= \frac{(0.150\ M)(0.05000\ L)}{(0.05000\ L + 0.02500\ L)}$$
$$= 0.100\ M$$

$$K_b = \frac{[OH^-][CH_3COO^-]}{[CH_3COOH]} = \frac{x^2}{0.100} = \frac{K_w}{K_a}$$

$$K_b = \frac{1.0 \times 10^{-14}}{1.8 \times 10^{-5}} = 5.6 \times 10^{-10}$$

$$x = [OH^-] = 7.5 \times 10^{-6}\ M$$
$$pOH = 5.12;\ pH = 14 - 5.12 = 8.88$$

17.43 (**e**) It is the only species that is a salt and insoluble in water.

17.44 (**a**) $Ag_2CO_3(s) \rightleftarrows 2\ Ag^+(aq) + CO_3^{2-}(aq)$; $K_{sp} = [Ag^+]^2[CO_3^{2-}] = (2x)^2(x) = 4x^3$.

17.45 (**d**) Acid addition (decreasing pH) causes the reaction $H^+(aq) + CO_3^{2-}(aq) \rightarrow HCO_3^-(aq)$ to occur, thereby shifting the following equilibrium to the right: $BaCO_3(s) \rightleftarrows Ba^{2+}(aq) + CO_3^{2-}(aq)$.

17.46 (**d**) **17.47** (**b**) **17.48** (**e**) **17.49** (**e**)

Chemistry of the Environment

Chapter

18

OVERVIEW OF THE CHAPTER

Review: Bond-dissociation energy (8.7); ionization (2.8, 6.3); radiant energy (6.1); nature of photons (6.2); quantum of light and $E = hv$ (6.1, 6.2).

Learning Goals: You should be able to:

1. Sketch the manner in which the atmospheric temperature varies with altitude and list the names of the various regions of the atmosphere and the boundaries between them.
2. Sketch the manner in which atmospheric pressure decreases with elevation and explain in general terms the reason for the decrease.
3. Describe the composition of the atmosphere with respect to its four most abundant components.
4. Explain what is meant by the term *photodissociation*, and calculate the maximum wavelength of a photon that is energetically capable of producing photodissociation, given the dissociation energy of the bond to be broken.
5. Explain what is meant by *photoionization*, and relate the energy requirement for photoionization to the ionization potential of the species undergoing ionization.

18.1, 18.2 THE EARTH'S ATMOSPHERE: REGIONS, COMPOSITION, AND REACTIONS

Review: Structure of ozone (8.7); bond enthalpies (8.8).

Learning Goals: You should be able to explain the presence of ozone in the mesosphere and stratosphere in terms of appropriate chemical reactions.

18.3 OZONE IN THE ATMOSPHERE

Review: Lewis structures (8.5, 8.6); nomenclature (2.8).

Learning Goals: You should be able to:

1. List the names and chemical formulas of the more important pollutant substances present in the troposphere and in urban atmospheres.
2. List the major sources of sulfur dioxide as an atmospheric pollutant.
3. List the more important reactions of nitrogen oxides and ozone that occur in smog formation.

18.4 CHEMISTRY OF ELEMENTS AND COMPOUNDS IN THE TROPOSHPERE

4. Explain why carbon monoxide constitutes a health hazard.
5. Explain why the concentration of carbon dioxide in the troposphere has an effect on the average temperature at the earth's surface.

18.5 THE WORLD OCEAN: COMPOSITION AND DESALINATION

Review: Concentration units (13.4).

Learning Goals: You should be able to:

1. List the more abundant ionic species present in seawater.
2. Explain the principles involved in desalination of water by distillation and by reverse osmosis.

18.6 FRESH WATER: IMPROVEMENT OF WATER QUALITY

Learning Goals: You should be able to:

1. List and explain the various stages of treatment that are used to treat freshwater sources and waste water.
2. Describe the chemical principles involved in the lime-soda process for reducing water hardness.

18.7 GREEN CHEMISTRY

Learning Goals: You should be able to:

1. Define the term green chemistry.
2. State the major goals governing green chemistry.
3. Give examples of solvents and reagents that are of concern in green chemistry and explain why there is concern.
4. Explain why trihalomethanes are of concern in public water systems.

TOPIC SUMMARIES AND EXERCISES

THE EARTH'S ATMOSPHERE: REGIONS, COMPOSITION, AND REACTIONS

The temperature of the atmosphere above the earth's surface changes in a complex manner as a function of altitude. Table 18.1 lists four regions in which temperature tends to rise or fall with increasing altitude. The direction of the temperature change, the approximate altitude ranges for these regions, and the average pressure exerted by the earth's atmosphere in each region are indicated in the table.

At the junctions of the temperature regions, temperature attains a minimum or maximum value. These junctions are, in order of increasing altitude:

*   The **tropopause**, at 12 km (temperature minimum)
*   The **stratopause**, at 50 km (temperature maximum)
*   The **mesopause**, at 85 km (temperature minimum)

TABLE 18.1 Regions of the Atmosphere

Region	Direction of temperature change with increasing altitude	Approximate altitude range (km)	Approximate average pressure (torr)
Thermosphere	Increasing	85–200	10^{-5}
Mesosphere	Decreasing	50–85	10^{-1}
Stratosphere	Increasing	10–50	10
Troposphere	Decreasing	0–12	100

The molecular and atomic composition of the atmosphere differs from region to region.

- Near the earth's surface, the four most abundant components, in order of decreasing concentration, are N_2, O_2, Ar, and CO_2. They make up about 99 percent of the entire atmosphere.
- With increasing altitude we observe a decreasing average molecular weight. Only the lightest atoms, molecules, or ions exist in the higher altitudes (for example, He, O, N_2, and N^+_2).
- In the upper atmosphere, electromagnetic radiation and high-energy particles from the sun bombard species such as O_2, N_2, and H_2O. Two types of reactions in the upper atmosphere are discussed in detail in Section 18.2 in the text:

Name of reaction	Description of reaction	Example
Photodissociation	Solar photons cause molecules to split into smaller molecules or atoms.	$O_2 + h\nu \longrightarrow 2\,O$
Photoionization	Solar photons cause atoms or molecules to form ions.	$N_2 + h\nu \longrightarrow N_2^+ + e^-$

EXERCISE 1 Applying Hess's law to photoionization reactions

Refer to the following data to answer the questions in this exercise:

$$NO + h\nu \longrightarrow NO^+ + e^- \qquad E = 890 \text{ kJ/mol}$$
$$N_2 + h\nu \longrightarrow N_2^+ + e^- \qquad E = 1495 \text{ kJ/mol}$$

(a) Of the two reactions

$$N_2^+ + NO \longrightarrow N_2 + NO^+$$
$$N_2 + NO^+ \longrightarrow N_2^+ + NO$$

which reaction is the one that occurs? (b) What is the energy change in the reaction that occurs?

SOLUTION: *Analyze and Plan:* We are given two reactions and asked to determine which one is likely to occur having been given two photoionization reactions and their energies. We are also asked to calculate the energy change. We will use the principles of Hess's law to determine which reaction is exothermic and thus more likely to occur.

Solve: (a) The exothermic reaction,

$$N_2^+ + NO \longrightarrow N_2 + NO^+$$

occurs. It involves the transfer of an electron from NO to N_2^+ because more energy is released when N_2^+ adds an electron than when NO^+ does. (b) Using Hess's law, we can write the reaction as the sum of two reactions:

$$N_2^+ + e^- \longrightarrow N_2 + h\nu \quad E = -1495 \text{ kJ/mol}$$
$$\underline{NO + h\nu \longrightarrow NO^+ + e^- \quad E = 890 \text{ kJ/mol}}$$
$$NO + N_2^+ \longrightarrow NO^+ + N_2 \quad E = -605 \text{ kJ/mol}$$

EXERCISE 2   Calculating the concentration of argon in air

After nitrogen and oxygen, argon is the most abundant element in the air near sea level. Calculate the concentration of argon in air in ppm (parts per million) given that its concentration in percent by volume is 0.93.

SOLUTION: *Analyze and Plan*: We are asked to calculate the concentration of Ar in ppm in air having been given that its concentration in percent by volume is 0.93. We can convert from percent by volume to parts per million by noting that percentage is also a ratio but in parts per hundred. Thus, the concentration of argon in percent by volume can be restated as:

$$0.93 \% \text{ means } \frac{0.93}{100}$$

Also, x parts per million can be stated as

$$\frac{x}{1,000,000}$$

We can relate the two to each other and solve for x.

Solve: You can use a ratio to calculate parts per million of argon

$$\frac{0.93}{100} = \frac{x}{1,000,000}$$

Solving for x gives

$$x = \frac{(0.93)(1,000,000)}{100} = (0.93)(1 \times 10^4) = 9300 \text{ ppm}$$

This example demonstrates that you can always convert from a percent to ppm by taking the original percent composition and multiplying by 10,000 (1×10^4).

Comment: For gases, one ppm means one part by volume in one million volume units of the whole. Also, mole fraction and volume fraction are the same; therefore, ppm = mole fraction $\times 10^6$.

OZONE IN THE ATMOSPHERE

The concentration of oxygen molecules in the upper atmosphere is low because of their photodissociation into oxygen atoms.

- In the mesosphere and stratosphere, atomic oxygen produced by the photodissociation of O_2 reacts with available O_2 to form ozone, O_3.
- Ozone absorbs harmful ultraviolet radiation with wavelengths in the range of 200 nm to 310 nm. This harmful radiation is not absorbed at lower altitudes. The layer of ozone in the mesosphere and stratosphere, which is called the "ozone shield," is thus very important for us.
- Depletion of ozone in the stratosphere may result from chlorofluorocarbons (CFCs) rising from the troposphere to the stratosphere and undergoing reaction with ozone. In the presence of light of wavelength 190–225 nm, CFCs lose chlorine atoms which then react with ozone to form ClO and O_2. The ClO can regenerate a Cl atom to once again react with ozone. Be sure you review reactions 18.6–18.10 in the text.

EXERCISE 3   Identifying regions in atmosphere containing allotropes of oxygen

In what regions of the atmosphere are molecular oxygen and ozone most abundant?

SOLUTION: *Analyze and Plan*: This is a question requiring factual information, the regions of the atmosphere in which O_2 and O_3 are most abundant. We can review Sections 18.2 and 18.3 in the text to find this information.

Solve: Molecular oxygen is found primarily in the troposphere and stratosphere, whereas ozone is found primarily in the stratosphere.

EXERCISE 4   Identifying the primary reactions of oxygen in the atmosphere

What are the primary reactions of O_2 above 90 km and below 90 km?

SOLUTION: *Analyze and Plan*: We are asked to write the primary reactions of O_2 above 90 km and below it. As in Exercise 3, this is a question requiring factual information and a similar approach is used.

Solve: Above 90 km, photodissociation of O_2 to form oxygen atoms dominates:

$$O_2 + hv \longrightarrow O + O \text{ (photodissociation)}$$

Below 90 km two reactions are important:

$$O + O_2 \longrightarrow O_3^* \text{ (ozone formation)}$$
$$O_3^* + M \longrightarrow M^* + O_3$$

In the latter reaction M represents a third atom or molecule that removes excess energy from O_3. (The asterisk tells us that a species has excess energy.) The presence of M is necessary in the reaction because if excess energy is not removed from O_3 it will dissociate back to O_2 and O.

EXERCISE 5   Explaining why temperature changes in the stratosphere

Why does temperature rise with altitude in the stratosphere?

SOLUTION: *Analyze and Plan*: We are asked to explain why temperature increases with altitude in the stratosphere region of the atmosphere. We can review Section 18.3 in the text to find this information.

Solve: The following reaction is exothermic:

$$O + O_2 + M \longrightarrow M^* + O_3$$

The rate of the reaction increases in the atmosphere from 10 km to 50 km and reaches its maximum rate at 50 km, the stratopause. As the rate of this reaction increases, so does the amount of heat liberated. This results in an increasing temperature with increasing altitude in the stratosphere. Above 50 km in the mesosphere, the rate of the reaction decreases with increasing altitude, which results in a decreasing temperature. The reaction of ozone with NO also contributes to the temperature profile.

Several minor constituents of the atmosphere can create health hazards if they are localized in the environment. This localization effect is called **air pollution**. The five substances that account for more than 90 percent of the worldwide air pollution are carbon monoxide (CO), nitrogen oxides (NO_x), hydrocarbons (compounds containing carbon and hydrogen), sulfur oxides (SO_x), and particulates:

CHEMISTRY OF ELEMENTS AND COMPOUNDS IN THE TROPOSPHERE

- **Sulfur oxides:** SO_2 is considered one of the most harmful of polluting gases. Coal mined in the eastern United States often has a high sulfur content, and when it is burned one of the products is SO_2. SO_2 is oxidized in the atmosphere to SO_3. Acid rain forms during a rainstorm when SO_3 is present in the air. Raindrops containing H_2SO_4 are a menace to human health and the environment.

• **Nitric oxide**, NO, is formed during the combustion of gasoline in an internal-combustion engine. In the presence of air, NO is slowly oxidized to nitrogen dioxide, NO_2. Nitric oxides, along with carbon monoxide and other products formed during the combustion of gasoline, yield a mixture of polluting gases that can produce photochemical smog in the presence of sunlight.

• **Carbon monoxide**, CO, is formed during the combustion of coal, gasoline, tobacco, and other carbon-containing compounds. It is a relatively unreactive species. However, it binds readily to hemoglobin in the human blood system. Hemoglobin is a very large iron-containing molecule that carries oxygen to muscle tissue. When CO is bound to hemoglobin, the ability of the hemoglobin to carry oxygen is diminished.

Water and CO_2 play a critical role in absorbing outgoing infrared radiation from the earth's surface.

• This absorption of infrared radiation holds heat at the earth's surface and thus keeps our nights from becoming extremely cold.

• Unfortunately, with excess CO_2 production now occurring as a product of industrial processes including coal burning, there is a danger of too much heat being held at the earth's surface. This will raise the average temperature of the troposphere and may cause significant changes in our climate.

Exercises 6–8 are questions requiring specific information that can be found in Section 18.4 of the text. The *Analyze and Plan* approach is to find the appropriate information in this section. In Exercise 7, we may find useful information in Sections 18.2 and 18.3.

EXERCISE 6   Explaining how sulfuric acid forms in the atmosphere

Pollution from sulfur oxides is severe near industries that have as one of their processes the reaction of a metal sulfide with oxygen at high temperatures. **(a)** Write balanced chemical reaction describing the conversion of PbS to PbO by oxidation with $O_2(g)$. **(b)** H_2SO_4 can be formed near industries emitting sulfur dioxide if water vapor is present. Explain how H_2SO_4 is formed under these conditions.

SOLUTION: (a) SO_2 is a product in the oxidation of metal sulfides:

$$2\,PbS + 3\,O_2 \longrightarrow 2\,PbO + 2\,SO_2$$

(b) SO_2 is converted to SO_3 by oxidative processes. The SO_3 reacts with H_2O to form H_2SO_4:

$$SO_3 + H_2O \longrightarrow H_2SO_4$$

EXERCISE 7   Writing chemical reactions showing cycling of NO and NO_2 in the atmosphere

Write the chemical reactions that describe the cycling of NO and NO_2 in the atmosphere to maintain their relative concentrations, assuming no hydrocarbons are present.

SOLUTION: The following reactions describe the cyclic process:

$$NO_2 + hv \longrightarrow NO + O \text{ (photodissociation of } NO_2)$$

$$O + O_2 + M \longrightarrow O_3 + M^* \text{ (ozone formation)}$$

$$NO + O_3 \longrightarrow NO_2 + O_2 \text{ (ozone reacts with NO to re-form } NO_2).$$

EXERCISE 8   Explaining why CO is dangerous to humans

CO is a relatively unreactive molecule, yet when present in the air in concentrations higher than 50 ppm it is dangerous to humans. Why?

SOLUTION: The CO in the air we breathe binds readily to hemoglobin, reducing the ability of hemoglobin to transfer O_2 from the lungs to other areas of the body.

Seawater consists of salts, gases, and organic molecules dissolved in water.

THE WORLD OCEAN: COMPOSITION AND DESALINATION

- There are 11 ionic species that make up 99.9 percent of the dissolved salts in seawater, with sodium and chloride ions as the dominant constituents.
- A measure of the amount of salts in seawater is **salinity**. It is defined as the mass in grams of dry salts present in one kilogram of seawater. The average value of salinity for the world ocean is 35. Because of mixing, the relative concentrations of ionic species are constant in the world ocean.

Desalination is the process of the removal of salts to form useable water.

- One approach to desalination is flash-distillation.
- **Reverse osmosis** is also used to desalinate seawater. Refer to Section 13.5 of the text for a discussion of osmosis which involves a solvent passing through a semipermeable membrane from a dilute solution into a more concentrated one. In reverse osmosis, the opposite process occurs. If a sufficient external pressure is applied, a solvent will move from a concentrated solution to a more dilute solution.

EXERCISE 9   Writing chemical reactions involving oyster shells and lime

Oyster shells in the oceans are a source of $CaCO_3$. By roasting oyster shells we can obtain lime, CaO, and CO_2. If lime is placed in seawater, a suspension of $Mg(OH)_2(s)$ is obtained. (Milk of magnesia is a suspension of $Mg(OH)_2$ in water.) Write reactions showing the roasting of oyster shells to produce lime and the subsequent reaction of lime with Mg^{2+} in seawater to form $Mg(OH)_2$.

SOLUTION: *Analyze and Plan*: We are asked to write two reactions that show how magnesium hydroxide is formed from oyster shells. From the information provided about the commercial process we can deduce the chemical reactions and write them.

Solve: The roasting of oyster shells causes the decomposition of $CaCO_3(s)$:

$$CaCO_3(s) \longrightarrow CaCO(s) + CO_2(g)$$

Seawater contains magnesium ions which react with lime as follows:

$$Mg^{2+}(aq) + CaO(s) + H_2O(l) \longrightarrow Mg(OH)_2(s) + Ca^{2+}(aq)$$

This reaction is possible because CaO is a basic oxide; it reacts with water to form OH^- ions.

EXERCISE 10   Calculating the concentration of magnesium in seawater

What is the concentration of magnesium, in grams per kilogram, in seawater with a salinity of 34 parts per million (ppm)? The concentration of magnesium is 19.35 g/kg in seawater with a salinity of 35 ppm.

SOLUTION: *Analyze and Plan*: We are asked to calculate the concentration of magnesium in seawater with a salinity of 34 ppm having been given its concentration in seawater with salinity of 35 ppm. From our reading of Section 18.5 we know that the salinity of seawater is equal to the mass in grams of dry salts of seawater dissolved in 1 kg of water. The concentration of an ion in seawater is usually reported with respect to seawater of standard salinity of 35 ppm. Since the relative concentrations of ionic species are constant throughout the world, the relative concentrations of magnesium in seawater of 34 and 35 ppm is the same. We can equate the magnesium concentrations at two different seawater salinities to calculate the concentration at seawater salinity of 34 ppm.

Solve: The ratio of magnesium concentration at different seawater salinities is

$$\frac{19.35 \text{ g kg}}{x \text{ g kg}} = \frac{35 \text{ ppm}}{34 \text{ ppm}}$$

Solving for x gives 18.79 g of magnesium in 1000 kg of seawater with a salinity of 34 ppm.

FRESH WATER: IMPROVEMENT OF WATER QUALITY

Water is normally thought of as an inexhaustible natural resource because it is recycled in the natural environment. However, because some water has a high mineral content or has been polluted, it is not always suitable for human use and must be purified. Sewage water must be treated to remove harmful organisms and a variety of other substances such as organic chemicals and metals before it is returned to the environment.

- Sewage water is usually treated by a primary treatment, in which solids are allowed to settle and some organic materials are removed, followed by a secondary treatment that involves aeration and decomposition of biodegradable organic matter by aerobic bacteria.
- Some water departments also provide a tertiary treatment to remove other material in the water, such as excess phosphorus or other substances potentially harmful to the environment.
- You should look at the reactions in A CLOSER LOOK: Water Softening in your text. Water may also be treated to reduce the concentrations of Ca^{2+} and Mg^{2+}, which are responsible for water hardness. The lime-soda process is used to treat water-hardness in large-scale municipal operations.

Exercises 11–13 are questions requiring specific information that can be found in Section 18.4 of the text. The *Analyze and Plan* approach is to review this section to find the appropriate information.

EXERCISE 11   Identifying compounds formed by anaerobic organisms with sewage water

What types of chemical compounds are usually formed by anaerobic organisms acting on sewage water? Why are these compounds objectionable?

SOLUTION: Anaerobic organisms produce compounds that do not contain oxygen. The two most common compounds formed during anaerobic decay are CH_4 and H_2S. Such decay produces sludges and noxious gases such as H_2S or flammable ones such as CH_4.

EXERCISE 12   Characterizing hard water

What is hard water and why is it objectionable?

SOLUTION: Hard water contains calcium, magnesium, and iron salts; the negative ions are usually chloride, sulfate, and hydrogen carbonate. The presence of the cations causes the formation of insoluble soaps that have no cleansing ability and leave a "ring" in tubs and sinks. Also, under appropriate conditions, such as very hot water, insoluble salts such as calcium sulfate and calcium carbonate form.

EXERCISE 13   Writing a chemical equation showing how ammonia reduces water hardness

An aqueous solution of ammonia may be used to remove hardness resulting from the presence of calcium and hydrogen carbonate ions. Write a net ionic equation that shows how ammonia removed this hardness.

SOLUTION: Ammonia makes the water basic, which increases the carbonate ion concentration so that calcium carbonate precipitates. The reaction is similar to the lime-soda process:

$$Ca^{2+}(aq) + 2\,HCO_3^-(aq) + 2\,NH_3(aq) \rightarrow CaCO_3(s) + 2\,NH_4^+(aq) + CO_3^{2-}(aq)$$

GREEN CHEMISTRY

Green chemistry is the design of chemical processes and materials that are environmentally benign and involves the reduction of hazardous wastes. Key themes involve prevention of chemical wastes, use and production of chemical substances at minimal levels and with minimal or no toxicity, low-energy chemical processes, use of renewable feedstocks for chemical processes, use of catalytic reagents, and minimization of solvents. The text focuses on the following areas of green chemistry:

- Reduction of volatile solvents and reagents. Supercritical fluids are now used as solvents that can be recovered. Dimethylcarbonate is a polar reagent and more environmentally friendly than other substances such as methyl halides and phosgene, $COCl_2$.
- Supercritical carbon dioxide is being used in the dry cleaning business to replace C_2Cl_4.
- Use of ozone to replace chlorine in water purification to prevent formation of trihalomethanes.

EXERCISE 14   Characterizing phosgene and writing its Lewis structure

Phosgene, $COCl_2$, has been used in a variety of organic reactions. It is highly toxic and often forms carbon tetrachloride, which is also a toxic compound. Its use is being eliminated through procedures involving green chemistry. What is its Lewis structure?

SOLUTION: *Analyze and Plan*: We are asked to draw the Lewis structures of phosgene. We can use the rules for drawing found in Section 9.2 of the text.

Solve: Phosgene is $COCl_2$. Carbon is the central atom and oxygen and chlorine atoms surround it. Twenty-four valence electrons in total are available to form bonds and nonbonding electrons. It has the following Lewis structure:

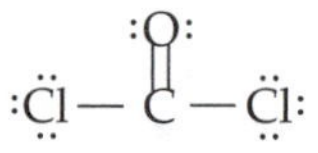

EXERCISE 15   Identifying undesirable substances formed in treating water with chlorine

What type of undesirable substances form when chlorine is used to treat water? Why are they of concern?

SOLUTION: *Analyze and Plan*: We are asked to identify the undesirable substances formed when chlorine is used to treat water. Section 18.7 in the text has information about the use of chlorine to treat water.

Solve: Chlorination of water produces byproducts known as trihalomethanes. CHX_3 (X = Cl or Br). They are suspected carcinogens (cancer-producing agents).

SELF-TEST QUESTIONS

Key Terms

Having reviewed key terms in Chapter 18, match key terms with phrases and identify statements as true or false. If a statement is false, indicate why it is incorrect.

Match each phrase with the best term:

18.1 A region of the atmosphere which is lowest in altitude.

18.2 A region of the atmosphere between 10 km and 50 km.

18.3 An iron-containing molecule that readily binds to oxygen.

18.4 A type of compound known to degrade ozone.

18.5 Rain that interacts with sulfur trioxide gas.

18.6 Solar photons cause the formation of smaller molecules.

18.7 Solar photons cause the formation of ions.

Terms:

(**a**) acid rain
(**b**) chlorofluoromethanes
(**c**) hemoglobin
(**d**) photodissociation
(**e**) photoionization
(**f**) stratosphere
(**g**) troposphere

True-False Statements:

18.8 Water with a dissolved salt concentration of 10 g/kg has a lower *salinity* than that of seawater.

18.9 In the *reverse osmosis* process of desalination, solute molecules pass through a semipermeable membrane, thereby removing salts from seawater.

18.10 In general, *desalination* processes involve small energy use.

18.11 In the *lime-soda process* for removing water hardness resulting from the presence of Ca^{2+} and Mg^{2+} ions, soda ash is added to remove Mg^{2+} ions by precipitating $MgCO_3$.

18.12 *Biodegradable* organic material in water provides a source of oxygen.

18.13 The photodissociation of $NO_2(g)$ initiates the reactions associated with *photochemical smog*.

18.14 The study of chemicals that are hazardous to humans can involve *green chemistry*.

Problems and Short-Answer Questions

18.15 Construct a diagram showing how molecules A, B, and C might be relatively distributed above the earth's surface in the troposphere and stratosphere given that the molar masses of the three are A≫B≫C. Explain.

18.16 What is a CFC? How do you think a HCFC and a HCF differ from a CFC? Explain. Draw a Lewis structure of a CFC, a HCFC, and a HCF.

18.17 Which portions of the atmosphere contain >99% of the mass of the atmosphere? Why?

18.18 Why do charged particles exist in the upper atmosphere? What are some typical examples?

18.19 N_2, O_2, and O absorb photons in the upper atmosphere. Why don't these processes completely protect us from UV radiation?

18.20 Rank the following constituents of the troposphere in increasing order of concentration: O_3, CO, CO_2, and CH_4. Which can form from decomposition of organic material?

18.21 Are volcanoes the primary source of SO_2 in the atmosphere? Explain.

18.22 The pH of most natural waters is acidic. Why?

18.23 K for the formation of $NO(g)$ from $N_2(g)$ and $O_2(g)$ at 300 K is about 10^{-15} and at 2400 K it is about 0.05. Based on this information, is the reaction exothermic or endothermic? Why?

18.24 Chemistry laboratories often require the availability of deionized water. Why have many companies and universities converted from a distillation system to a reverse osmosis system to provide deionized water?

18.25 Methane (CH_4) comprises 1.5 ppm of our atmosphere. What is this concentration in percent by volume?

18.26 Why does N_2 undergo very little solar photolysis in the upper atmosphere?

18.27 The troposphere region in the atmosphere is transparent to visible light but not to infrared radiation. What gases in the troposphere cause the lack of transparency in the infrared region of radiation?

18.28 Atmospheric reactions are too slow to account for the removal of CO from the atmosphere; in the lower atmosphere, these reactions only account for approximately 0.1 percent of CO removal. What other mechanisms are available to remove CO from the atmosphere?

18.29 In an urban atmosphere the level of NO increases rapidly from 6 to 8 A.M. but decreases rapidly after 8 A.M. Why?

18.30 Acute injury to plants by SO_2 results from sulfate salts forming at the tips or edges of leaves. These salts eventually cause leaf-dropping in plants. Suggest a mechanism for the formation of sulfate salts in plants.

18.31 Write equations to describe the following steps in the lime-soda process for softening water:

   (a) the addition of $Ca(OH)_2$ to remove carbonate (HCO_3^-) hardness;
   (b) the addition of excess $Ca(OH)_2$ to remove Mg^{2+};
   (c) the addition of Na_2CO_3 to remove the Ca^{2+} ion after much of the HCO_3^- ion is removed.

18.32 One of the gaseous impurities found in water is hydrogen sulfide. The process of aeration is used to remove most of the dissolved hydrogen sulfide. When water is aerated, it becomes saturated with oxygen. The dissolved oxygen reacts with hydrogen sulfide to form sulfur, thus removing hydrogen sulfide impurities. Write a chemical equation that describes the removal of hydrogen sulfide by the process of aeration.

18.33 The total ionic concentration of seawater is approximately 1.100×10^{-3} mol/L, and the total ionic concentration in the body fluids of bony saltwater fish is approximately 3.70×10^{-4} mol/L. Using the concept of osmotic pressure, explain why bony fish in the ocean swallow seawater and excrete salts through their gills.

18.34 Chlorine is used to sterilize water in water treatment plants. Explain why household bleach, which is a dilute solution of hypochlorous acid, can be used to sterilize dishes during their final rinse after they have been washed with soap.

18.35 During the chlorination of water, what chemical substance is primarily responsible for the formation of trihalomethanes? Why is ozone being considered as a substitute for chlorine and what problems are associated with it?

18.36 The concentration of argon in the atmosphere is 9340 ppm. Is this sufficient information to conclude that 9340 moles of argon are found in 1 million moles of gas? If not, what additional information is needed and why?

18.37 Oxygen undergoes photodissociation in the upper atmosphere to form oxygen atoms which then react with oxygen molecules to form ozone. If it is known that 550 kJ of energy will cause photodissociation, can we conclude that 450 kJ of energy will also cause it? If not, what additional information is needed and why?

Integrative Exercises

18.38 Can a photon with a wavelength of 230 nm cause the dissociation of an oxygen molecule in the thermosphere? The bond enthalpy of O_2 is 499 kJ/mol.

18.39 $CO_2(g)$ is a toxic gas in the atmosphere if it has too high of a concentration. It can be removed from contaminated air by passing it through a solution of limewater, a suspension of $Ca(OH)_2$ in water. When 1.5 L of an air sample is passed through limewater at 20 °C and at a pressure of 760 torr it is found that 0.12 g of $CaCO_3$ precipitates. Write the reaction of carbon dioxide with limewater. Assuming that the carbon dioxide gas was completely removed, what was the partial pressure of carbon dioxide in the air sample?

18.40 CCl_2F_2 has been used as a propellant in aerosol cans. Solar photolysis of it occurs at 213.9 nm at low atmospheric temperatures.

(a) What is the energy in kJ/mol required for photolysis?

(b) Write the photodissociation reaction for its photolysis.

(c) Why are scientists concerned about CCl_2F_2 in the atmosphere?

(d) Suggest a common gas that could replace it.

Multiple-Choice Questions

18.41 In which region of the atmosphere is the rate of production of O_3 at its maximum value?

(a) troposphere   (d) thermosphere
(b) stratosphere   (e) thermopause
(c) mesosphere

18.42 When environmentalists talk about acid rain in certain industrial regions, which of the following acids is the major species involved?

(a) H_2SO_4 (d) H_3PO_4
(b) HI   (e) $HClO_4$
(c) HNO_2

18.43 Which trace pollutant is normally present in the greatest concentration in a polluted urban atmosphere?

(a) SO_2 (d) NO_2
(b) C_2H_4 (e) CO
(c) NH_3

18.44 The concentration of one of the molecules in the atmosphere that prevents the loss of heat radiated from the earth's surface is appreciably affected by human activities. What is this molecule?

(a) PAN   (d) CO_2
(b) H_2 (e) CO
(c) CH_4

18.45 In the stratosphere, O_3 and ultraviolet radiation from the sun are converted into which of the following?

(a) $O_2 + O$ (d) N_2 + heat
(b) O_3 + heat   (e) NO + heat
(c) $3 O$ + heat

18.46 Which of the following is a region in the atmosphere in which temperature reaches a maximum value?

(a) stratosphere;   (d) mesopause;
(b) stratopause;   (e) none of the above.
(c) mesosphere;

18.47 A catalyst not only speeds up the rate of a reaction, but it also is usually regenerated during the process. Which of the following species acts as a catalyst in the destruction of ozone?

(a) O_2 (d) Cl
(b) O_3 (e) H_2O
(c) CO

18.48 The mole-fraction content of O_2 in dry air near sea level is 0.209. What is the partial pressure of O_2 in dry air when the total dry air pressure is 750 mm Hg?

(a) 750 mm Hg   (d) 3589 mm Hg
(b) 700 mm Hg   (e) 157 mm Hg
(c) 0.00637 mm Hg

18.49 At an altitude of 400 km, which form of oxygen is the most abundant?

(a) O_3 (d) O
(b) O_2 (e) O^{2-}
(c) O_2^+

18.50 What is the primary form of $CO_2(g)$ dissolved in seawater?

(a) $CO_2(aq)$ (d) $CO_3^{2-}(aq)$
(b) $H_2CO_3(aq)$ (e) $CaCO_3(s)$
(c) $HCO_3^-(aq)$

18.51 Anaerobic bacteria in polluted water consume organic material and thereby form gases. What is one of the gases formed?

(a) CH_3Cl (d) SO_2
(b) H_2S (e) HNO_3
(c) H_2O

18.52 Which of the following elements is removed in the tertiary treatment of municipal sewage water but not usually removed in the primary and secondary treatment stages?

(a) C   (d) S
(b) H   (e) P
(c) O

18.53 Most nations are moving towards what is called a "hydrogen-based fuel economy." Hydrogen is being used to replace fossil fuels. Which is one of the important reasons for this movement?

(a) Hydrogen gas is a commonly available free element in nature.

(b) It is a dense gas at room temperature.

(c) It is easily stored.

(d) Its combustion forms water.

(e) The carbon economy is dead.

18.54 Which technical problem do you think will be a major limiting factor in forming a hydrogen-based economy?

(a) Availability of hydrogen-containing compounds.

(b) Lack of catalysts to break apart hydrogen-containing compounds.

(c) Energy cost to form hydrogen gas.

(d) Political will of politicians.

(e) Manufacturers are not interested.

18.55 If hydrogen gas becomes readily produced, which of the following factors may create major obstacles?

    (**a**) Infrastructure to distribute hydrogen gas.

    (**b**) Its low density.

    (**c**) Cost to convert to a liquid.

    (**d**) Loss of hydrogen to power compressors in pipelines.

    (**e**) All of the above.

SELF-TEST SOLUTIONS

18.1 (g). **18.2** (f). **18.3** (c). **18.4** (b). **18.5** (a). **18.6** (d). **18.7** (e). **18.8** True. **18.9** False. Solvent molecules pass through the semipermeable membrane. **18.10** False. Both multistage flash distillation and reverse osmosis require energy expenditure; however, the latter process requires a lesser amount of energy. **18.11** False. Soda ash, Na_2CO_3, causes the precipitation of $CaCO_3$, not $MgCO_3$; Mg^{2+} precipitates as $Mg(OH)_2$. **18.12** False. Aerobic bacteria consume dissolved oxygen in water when they oxidize organic materials; therefore the amount of oxygen decreases. **18.13** True. **18.14** True. To properly find ways to minimize hazardous chemicals we must understand hazardous chemicals.

18.15 Gravitation forces will cause the more massive particles to be nearer the surface of the earth and the lighter particles will be distributed further away from the surface.

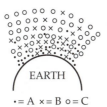

$$\bullet = A \quad \times = B \quad \circ = C$$

18.16 CFC is the abbreviation for the class of compounds known as chlorofluorocarbons. These contain C, F, and Cl atoms. The abbreviation HCFC suggests a hydrogen atom is now part of the molecule and replaces a fluorine or chlorine atom. The abbreviation HCF suggests hydrogen atoms have replaced chlorine atoms in a CFC or HCFC. Examples are given in the following figure:

Cl	F	H F
Cl—C—F	H—C—Cl	H—C—C—H
Cl	F	H F
CFC-11	HCFC-22	HFC-152

18.17 Gravitational forces are greater near the surface of the earth and keep most of the particles in the troposphere and stratosphere. Only the very light particles rise to the upper atmosphere.

18.18 Solar radiation in the upper atmosphere provides photons of sufficient energy to cause ionization of small particles. Examples of photoionization are N_2 to form N_2^+ and e^-, and O_2 to form O_2^- and e^-.

18.19 The three particles are effective at absorbing photons with wavelengths shorter than about 240 nm. However, not all of the radiation in the UV region is absorbed, particularly above 240 nm. Ozone in the stratosphere absorbs radiation above 240 nm and removes a large portion of the residual UV energy.

18.20 $O_3 < CH_4 < CO < CO_2$. The carbon-containing particles are produced from organic (carbon-containing) materials.

18.21 Although volcanoes can spew significant quantities of sulfur dioxide, these events are sporadic, whereas the burning of sulfur-containing coal occurs on a daily basis. Coal burning is the primary source of sulfur dioxide.

18.22 There are two primary reasons: The dissolving of carbonate-containing rocks slowly over time releases the carbonate ion, which reacts with water to form HCO_3^- and H_2CO_3; and the carbon dioxide in the air in contact with the surface of natural waters dissolves to a limited extent and forms the same species.

18.23 The reaction is endothermic. Note that the value of the equilibrium constant increases with increasing temperature; thus, the equilibrium is driven to the products as more heat is added. LeChatelier's principle tells you that adding heat to an endothermic reaction will drive the reaction to the right.

18.24 Energy costs are high with a distillation system. Reverse osmosis requires high pressure plumbing and some input of electrical energy but not in the number of joules as required for distillation.

18.25 The concentration unit parts per million (ppm) for a gas is based on volume measurements and represents the volume of a compound contained in 1 million volumes of air. The concentration of methane is 1.5 ppm or 1.5 volumes of methane per 1 million volumes of air: percentage composition is

$$CH_4 = [1.5/(1 \times 10^6)](100) = 1.5 \times 10^{-4} \%.$$

18.26 The N_2 bond is a triple bond: $:N{\equiv}N:$. High-energy photons are required to break it. Also, N_2 does not readily absorb photons.

18.27 Primarily H_2O causes the lack of transparency, but CO_2 also contributes.

18.28 The two primary mechanisms are (1) microorganisms in the soil that remove CO from air and (2) the ocean which dissolves some of the CO.

18.29 The level of NO increases rapidly from 6 to 8 A.M. because of NO production from cars driven by commuters. After 8 A.M. the rate of production decreases; the NO reacts rapidly with ozone to form NO_2 and O_2.

18.30 The plant absorbs SO_2, which then reacts with H_2O to form H_2SO_3. The plant metabolism then converts the H_2SO_3 to sulfate salts.

18.31 **(a)** The addition of $Ca(OH)_2$ to water produces hydroxide ions, which react with HCO_3^-:
$HCO^-(aq) + OH^-(aq) \rightarrow H_2O + CO_3^{2-}(aq)$.
The carbonate ion precipitates Ca^{2+} in the form of $CaCO_3$: $Ca^{2+}(aq) + CO_3^{2-}(aq) \rightarrow CaCO_3(s)$.

   (b) The excess $Ca(OH)_2$ increases the pH of the solution sufficiently to precipitate Mg^{2+}:
$Mg^{2+}(aq) + 2\,CO_3^{2-}(aq) + 2\,H_2O \rightarrow 2\,HCO_3^-(aq) + Mg(OH)_2(s)$.

   (c) If any Ca^{2+} ion remains, the addition of CO_3^{2-} from Na_2CO_3 precipitates Ca^{2+}: $Ca^{2+}(aq) + CO_3^{2-}(aq) \rightarrow CaCO_3(s)$.

18.32 Hydrogen sulfide and oxygen undergo a redox reaction in water as follows: $H_2S(aq) + \frac{1}{2}O_2(g) \rightarrow S(s) + H_2O(l)$.

18.33 Since the total ionic concentration in body fluids of bony fish is less than that of seawater, the osmotic pressure of body fluids in bony fish is less than that of seawater ($\pi \propto$ total ionic concentration). As a result of the difference in osmotic pressures, water leaves bony fish through their cellular tissues in order to equalize the osmotic pressures. To prevent dehydration and to maintain the low osmotic pressure of their body fluids, bony fish swallow seawater to replace water lost through osmosis. Since salts are contained in the swallowed seawater, bony fish excrete salts through their gills in order to maintain a constant total ionic concentration.

18.34 The sterilizing action of chlorine is believed to be due to hypochlorous acid, which is the same component as occurs in bleach. Hypochlorous acid is formed when chlorine reacts with water: $Cl_2(aq) + H_2O(l) \rightarrow HOCl(aq) + HCl(aq)$.

18.35 Chlorine reacts with water to form hypochlorous acid, HOCl, which is an oxidizing agent. It reacts with organic materials to form trihalomethanes. Ozone also sterilizes water but it produces a byproduct, hypobromous acid, HOBr, which is also capable of reacting with organic materials. Ozone, however, does produce fewer trihalomethanes.

18.36 There is sufficient information. When dealing with a solution of entirely gases, one part per million refers to one part by volume in one million volume units. Using the ideal gas law and Dalton's law of partial pressures, it can be shown that volume fraction is the same as mole fraction. Therefore, 1 ppm of Ar is the same as one mole of Ar in one million moles of gas.

18.37 There is insufficient information. We are told that 550 kJ of energy causes photodissociation but we are not told if this is the minimum energy necessary. We need to know the bond dissociation enthalpy of $O_2(g)$, the minimum energy required to break one mole of O—O bonds. The text reports that it is 495 kJ/mol; thus 450 kJ is insufficient to cause photodissociation.

18.38 We need to calculate the minimum energy required to dissociate one *molecule* of oxygen. From this energy we can then calculate the required wavelength.

$$E_{molecule} = \left(499 \times 10^3 \frac{J}{mol}\right)\left(\frac{1\ mol}{6.02 \times 10^{23}\ molecules}\right)$$

$$= 8.29 \times 10^{-19} \frac{J}{molecule}$$

$$E = h\upsilon = \frac{hc}{\lambda}$$

$$\lambda = \frac{hc}{E} = \frac{(6.63 \times 10^{-34}\,J - s)\left(3.00 \times 10^8 \frac{m}{s}\right)}{8.29 \times 10^{-19}\,J}$$

$$\lambda = 2.40 \times 10^{-7}\,m = 240\ nm$$

Any photon with a $\lambda \leq 240$ nm should be able to dissociate an oxygen molecule. A photon of 230 nm thus should be able to dissociate it.

18.39 The reaction is:

$$Ca(OH)_2(s) + CO_2(g) \longrightarrow CaCO_3(s) + H_2O(l)$$

The partial pressure of carbon dioxide can be calculated if the number of moles of it are known in the air sample. From the number of moles of calcium carbonate precipitated the number of moles of carbon dioxide gas can be determined. The stoichiometry of the reaction shows you that the number of moles of carbon dioxide gas reacting is the same as the number of moles of calcium carbonate precipitated.

The number of moles of calcium carbonate and therefore carbon dioxide is:

moles of $CaCO_3$ = moles of CO_2

$$= (0.12\ g)(1\ mol/100.09\ g) = 1.2 \times 10^{-3}\ mol$$

The partial pressure of carbon dioxide is calculated using the ideal-gas law:

$$P_{CO_2} = \frac{nRT}{V} = \frac{(1.2 \times 10^{-3}\ mol)\left(0.0821 \frac{L\text{-atm}}{mol\text{-}K}\right)(293\ K)}{1.5\ L}$$

$$P_{CO_2} = 0.019\ atm$$

$$P_{CO_2} = 14\ torr$$

Therefore, carbon dioxide contributes 14 torr to the total pressure of the air sample.

18.40 **(a)** $E = h\frac{c}{\lambda} = (6.625 \times 10^{-34}\,J - s)$

$$\times \left(\frac{3.00 \times 10^{10} \frac{cm}{s}}{213.9\ nm\left(\frac{10^{-7}\ cm}{1\ nm}\right)}\right) = 9.29 \times 10^{-19}\,J$$

This is the energy per photon. To calculate the energy per mole: E = (9.29 × 10^{-19} J/photon)(6.022 × 10^{23} photons/mol) = 5.60 × 10^5 J/mol = 560 kJ/mol.

(b) $CCl_2F_2 \xrightarrow{h\nu} CClF_2 + Cl$.

(c) The chlorine atom produced when CCl_2F_2 undergoes photolysis can react with ozone to form ClO and oxygen. The removal of ozone from the atmosphere would allow high-energy radiation to reach the surface of earth.

(d) Carbon dioxide gas is now used as a replacement propellant in many aerosol cans.

18.41 (b). **18.42** (a). **18.43** (e).

18.44 (d) A low concentration of water vapor in the atmosphere would allow infrared heat to escape, causing the earth's surface to be cold at night.

18.45 (a). **18.46** (b).

18.47 (d) $Cl + O_3 \longrightarrow ClO + O_2$ (occurs in

$$\frac{ClO + O \longrightarrow Cl + O_2 \text{ ozone layer})}{O_3 + O \longrightarrow 2\,O_2}$$

18.48 (e) See solution to problem 10.40 in the *Student's Guide* for the derivation of a similar expression: $P_{O_2} = (n_{O_2}/n_t)P_t$, where n_{O_2}/n_t is the mole fraction of O_2. Thus, $P_{O_2} = (0.209)(750 \text{ mm Hg}) = 157 \text{ mm Hg}$.

18.49 (d) At 400 km, 99% of elemental oxygen is in the form of O, not O_2, due to solar photolysis.

18.50 (c). **18.51** (b).

18.52 (e).

18.53 (d) A hydrogen economy will be challenging to create. Hydrogen is relatively rare as a free gas. It has the major advantage of being "green" in that it form harmless water instead of carbon dioxide.

18.54 (c) Politicians and manufacturers are very interested in using hydrogen to replace fossil fuels. We can produce hydrogen from water and natural gas. Unfortunately the cost is not yet competitive with producing a gallon of gas.

18.55 (e) There are major challenges ahead in converting to a hydrogen-based fuel economy.

Sectional MCAT and DAT Practice Questions VI

Passage I

A student places an excess of solid calcium carbonate in 500 mL of water and forms a slurry. The slurry is stirred and left to equilibrate for 24 hours at 25 °C. The equilibrium between the undissolved calcium carbonate and its ions is shown in Equation 1.

$$CaCO_3(s) \rightleftharpoons Ca^{2+}(aq) + CO_3^{2-}(aq)$$
Equation 1

The slurry is filtered and the filtrate is collected and is free of solid particles. The student pipets 25.0 mL of the filtrate into a flask. This solution is then buffered to a pH of 10. The calcium ions in the flask are titrated with 0.0100 M EDTA. The reaction of calcium ions with EDTA is given in Equation 2.

$$Ca^{2+}(aq) + Y^{4-}(aq) \rightarrow CaY^{2-}$$
Equation 2

EDTA is ethylenediaminetetracetic acid, abbreviated H_4Y, which is in the form of Y^{4-} at a pH of 10. After three titrations the average volume of EDTA required to react completely with calcium ions is 12.5 mL.

Table 1 lists K_{sp} values at 25 °C of several salts.

TABLE 1 K_{sp} values at 25 °C

Salt	K_{sp}
$BaSO_4$	1.1×10^{-10}
Hg_2Cl_2	1.2×10^{-18}
$Mg(OH)_2$	1.8×10^{-11}
$AgCl$	1.8×10^{-10}
$AgBr$	5.0×10^{-13}
Ag_2SO_4	1.5×10^{-5}
$PbSO_4$	6.3×10^{-7}

The reaction quotient for a salt, Q, has the same form as the salt's K_{sp} expression but is evaluated at any specified concentration conditions.

1. What is the concentration of calcium ions in the filtrate?
   (a) $5.00 \times 10^{-3}\ M$ (b) $1.25 \times 10^{-2}\ M$
   (c) $1.0 \times 10^{-2}\ M$ (d) $0.125\ M$

2. What is the value of K_{sp} for calcium carbonate based on the experimental data?
   (a) 1.6×10^{-2} (b) 1.0×10^{-4} (c) 1.5×10^{-4} (d) 2.5×10^{-5}

3. Which equation shows the correct K_{sp} for silver(I) sulfate if the molar solubility of silver(I) sulfate is given the symbol s.
   (a) $K_{sp} = s^2$ (b) $K_{sp} = 2s^2$ (c) $K_{sp} = 2s^3$ (d) $K_{sp} = 4s^3$

4. Which salt is the least soluble in 100 mL of water at 25 °C?
   (a) $BaSO_4$ (b) $PbSO_4$ (c) AgBr   (d) AgCl

5. Will AgCl precipitate if a student mixes 50.0 mL of $2 \times 10^{-5}\ M$ solution of $Ag^+(aq)$ with 50.0 mL of a $6 \times 10^{-5}\ M$ solution of $Cl^-(aq)$?
   (a) No, because the reaction quotient is larger than K_{sp}.
   (b) No, because the K_{sp} is larger than the reaction quotient.
   (c) Yes, because the reaction quotient is larger than K_{sp}.
   (d) Yes, because the K_{sp} is larger than the reaction quotient.

6. A saturated solution of $Mg(OH)_2$ with excess solid present at a pH of 10.5 is prepared (solution A) and then the pH is altered (solution B). Which experimental measurement for solution B shows that the solubility of $Mg(OH)_2$ decreased compared to its solubility in solution A?
   (a) pH = 12
   (b) pH = 7
   (c) Red litmus paper changes to blue when solution B is tested.
   (d) Red litmus paper does not change color when solution B is tested.

7. A 100 mL solution containing acetic acid and sodium acetate is buffered to a pH of 5. If 5 mL of 0.01 M perchloric acid is added to the buffer, which statement best describes what happens in the solution?
   (a) Sodium perchlorate precipitates.
   (b) pOH increases slightly.
   (c) The concentration of acetic acid does not change.
   (d) The concentration of acetate ions increases.

Passage II

MCAT questions based on analysis of a passage with descriptive chemical reactions and qualitative analysis and your general knowledge of the chemistry of ions and compounds have occurred in several past tests.

A class of students studies the properties of solutions of 0.1 M nitrate salts with the following cations: Fe^{3+}, Al^{3+}, Mn^{2+}, Co^{2+}, and Ni^{2+}.

The students react a small volume of each solution in a test tube with the following reagents: Na_2S under alkaline conditions, NaOH with slight warming, KSCN, potassium thiocyanate, in an alcoholic solution at a pH slightly above seven, and KSCN in an aqueous solution at a pH slightly less than seven. The reagents are slowly added drop wise and observations recorded. If a precipitate forms with Na_2S or NaOH, then another test is done to determine the solubility of the precipitate as shown in Table 2.

The students are told by their instructor that they should first separate the metal ions from one another before doing any confirmatory tests. Each individual test is done in the absence of the other metal ions. When a mixture of metal ions is studied the reactions or colors of solutions or precipitates may not be as expected compared to those observed in isolated tests. Interferences in confirmatory tests can occur with mixtures.

TABLE 2 Experimental results for testing ions with different reagents

Metal Ion	Na_2S (basic) Color of precipitate	Solubility of precipitate in 6 M HCl	NaOH Color of precipitate	Solubility of precipitate in excess NaOH	1M KSCN(aq) Color of solution	1M KSCN in alcohol Color of solution
Fe^{3+}	black	dissolves	red-brown	insoluble	red	red, depending on SCN^- concentration
Al^{3+}	white	dissolves	white	dissolves	—	—
Mn^{2+}	pink	dissolves	white which gradually darkens	insoluble	—	
Co^{2+}	black	insoluble	pink which very gradually turns dark brown	insoluble	—	blue
Ni^{2+}	black	slightly soluble	green	insoluble	—	—

8 A solution in a test tube has a basic solution of sodium sulfide added to it. A black precipitate forms. 6 M HCl is added and a black precipitate remains. Which of the following ions are possibly present in the unknown solution?
(a) Co^{2+}, Fe^{3+}, and Ni^{2+} (b) Co^{2+} and Fe^{3+}
(c) Co^{2+} and Ni^{2+} (d) Co^{2+}

9 A white precipitate forms when NaOH is slowly added to an aqueous solution of Al^{3+}. What species forms when an excess of NaOH is added to the solution?
(a) $Al(H_2O_6)_6^{3+}$ (b) $Al(OH)_3$
(c) $Al(OH)_4^-$ (d) Al_2O_3

10 Which ions appear to undergo a change in oxidation state at some point in the reactions with base?
(a) Fe^{3+}, Al^{3+}, and Mn^{2+} (b) Co^{2+} and Ni^{2+}
(c) Al^{3+} and Fe^{3+} (d) Mn^{2+} and Co^{2+}

11 Which reagent will most likely cause a precipitate to form when added to a test tube containing 2 mL of 0.5 M $Fe(NO_3)_3$?
(a) 3 M H_2SO_4 (b) 6 M NH_3
(c) 6 M $KClO_4$ (d) 3 M NH_4Cl

12 A student wants to determine if Fe^{3+} is present in an unknown aqueous solution possibly containing Fe^{3+} and Co^{2+}. Which experimental procedure will best enable the student to make this determination?

(a) Add an alcoholic solution of KSCN to the unknown solution in a test tube and adjust the pH per instructions.

(b) To a solution of the unknown in a test tube add an alkaline solution of sodium sulfide. Pour any mixture through a filter and collect any solid. To any solid collected add 6 M NaOH. To the resulting solution add an aqueous solution of KSCN and adjust the pH per instructions.

(c) To a solution of the unknown in a test tube add NaOH. Pour this mixture through a filter and collect any solid and the solution. To any solid add excess NaOH. To the solution add an aqueous solution of KSCN and adjust the pH per instructions.

(d) To a solution of the unknown in a test tube add an alkaline solution of sodium sulfide. Pour this mixture through a filter and collect ANY solid, and the solution. Add 6 M HCl to any solid and filter to separate any remaining solid from the solution. Add an aqueous solution of KSCN to the last solution and adjust the pH per instructions.

Questions 13 through 17 are **not** based on a descriptive passage.

13 Which characterizes an aqueous solution of K_2S?
(a) Acidic   (b) Basic
(c) Neutral   (d) Buffer

14 Which solution is a buffer with a pH of less than 7 and also has the best buffering capacity when small amounts of KOH are added?
(a) 0.1 M HCl and 0.1 M NaOH
(b) 0.05 M NH_4Cl and 0.1 M NH_3
(c) 0.01 M $HC_2H_3O_2$ and 0.05 M $NaC_2H_3O_2$
(d) 0.15 M KCl and 0.15 M $KC_2H_3O_2$

15 What is the pH of an aqueous solution containing 0.00050 M $Ba(OH)_2$?
(a) 3.00   (b) 3.30
(c) 10.7   (d) 11.00

16 What is the approximate pH at the equivalence point in the titration of 25.0 mL of 0.10 M NH_3 with 0.10 M HCl?
(a) 2   (b) 6
(c) 7   (d) 9

17 Which ion is matched with its correct name?
(a) PO_4^{3-}, phosphite   (b) CO_3^{2-}, carbonoate
(c) CN^-, cyanate   (d) SO_3^{2-}, sulfite

ANSWERS

1. (a) The filtrate is the solution in the filter flask after $CaCO_3(s)$ is removed. Equation 2 shows that the reaction between calcium ions and the anion of EDTA is in a 1:1 mole ratio. Thus, at the equivalence point in the titration, the number of moles of calcium ions equals the number of moles of the anion of EDTA, or

$$M_{Ca^{2+}}V_{Ca^{2+}} = M_{Y^{4-}}V_{Y^{4-}}. \ (25.0 \text{ mL})(M) = (0.01 \ M)(12.5 \text{ mL}). \ M = 5.00 \times 10^{-3}$$

2. (d) $K_{sp} = [Ca^{2+}][CO_3^{2-}] = [5.00 \times 10^{-3}][5.00 \times 10^{-3}] = 2.5 \times 10^{-5}$. The concentration of calcium ions and sulfate ions are the same in the filtrate, which is a saturated solution of calcium carbonate. The concentrations are from the answer to question one.

3. (d). The solubility equilibrium is: $Ag_2SO_4 \rightleftharpoons 2\,Ag^+ + SO_4^{2-}$. Thus, if s moles/L of silver sulfate dissolve, $2s$ moles per liter of silver ions form and s moles per liter of sulfate ions form. $K_{sp} = [Ag^+]^2[SO_4^{2-}] = (2s)^2(s) = 4s^3$.

4. (c) All of the salts have a 1:1 ratio of cation to anion. Therefore we can determine relative solubilities by comparing the values of K_{sp}. AgBr has the smallest K_{sp} therefore it is the least soluble. Note, if the ratio of cation to anion had not been 1:1 for all salts, we would have had to solve for the solubilities of each salt using their values of K_{sp}.

5. (c) The form of the reaction quotient is $Q = C_{Ag^+} \times C_{Cl^-}$. Precipitation occurs if initially in the solution $Q > K_{sp}$. After the two solutions are mixed, the concentrations of the silver and chloride ions are halved because the volumes of the two solutions added are equal. $Q = (1 \times 10^{-5})(3 \times 10^{-5}) = 3 \times 10^{-10}$. From the descriptive passage we are given K_{sp} of AgCl, 1.8×10^{-10}. AgCl precipitates because $Q > K_{sp}$.

6. (a) The solubility equilibrium is: $Mg(OH)_2(s) \rightleftharpoons Mg^{2+}(aq) + 2\,OH^-(aq)$. An increase in the pH of solution A to form solution B increases the hydroxide ion concentration. According to LeChatelier's principle this will cause the equilibrium to shift to the left and decrease the amount of $Mg(OH)_2$ dissolving. Answers (c) and (d) only tell us if a solution is acidic or basic and not the actual pH. For example, the pH could decrease from 10.5 to 8 and red litmus paper changes to blue; yet this change would increase the solubility of $Mg(OH)_2$.

7. (b) The original solution is an acidic buffer consisting of $HC_2H_3O_2$ and $C_2H_3O_2^-$. The equilibrium is: $HC_2H_3O_2 \rightleftharpoons H^+ + C_2H_3O_2^-$. Perchloric acid is a strong acid and forms H^+ ions. H^+ ions react with $C_2H_3O_2^-$ ions to form $HC_2H_3O_2$ and the system is no longer at equilibrium. However, the system adjusts to this change by partially reforming H^+ and $C_2H_3O_2^-$ ions to reestablish equilibrium. The result is that most of the H^+ ions from perchloric acid are consumed to form $HC_2H_3O_2$, but not all. The pH decreases slightly, which is characteristic of a buffer solution when a small amount of acid is added. The relationship pH + pOH = 14 shows that when pH decreases, pOH increases. Sodium perchlorate is a soluble salt based on solubility rules.

8. (a) Co^{2+}, Fe^{3+}, and Ni^{2+} form black solids when an alkaline solution of sodium sulfide is added. When acid is added to the solid, the precipitate with Fe^{3+} dissolves, the precipitate with Ni^{2+} slightly dissolves and the precipitate with Co^{2+} does not dissolve. Both sulfides of Co^{2+} and Ni^{2+} could remain after adding HCl. Fe^{3+} may also be present because it could be in the solution after HCI is added. Thus, all three ions are possible and further tests are required to further separate and identify them.

9. (c) When NaOH is slowly added a white precipitate forms. This is likely to be $Al(OH)_3$, an amphoteric hydroxide. When an excess of NaOH is added, the hydroxide ion reacts with $Al(OH)_3$ to form a complex ion, $Al(OH)_4^-$. If an acid were added instead of a base to $Al(OH)_3$ then we expect a reaction to form $Al(H_2O_6)_6^{3+}$. Al_2O_3 is a solid and does not form in a solution under these conditions. Thus, **(c)** is the best choice.

10. (d) Both precipitates of Mn^{2+} and Co^{2+} slowly change color in the presence of NaOH. This indicates that new substances are formed, which result from air oxidation of the metal ions to higher oxidation states. The other precipitates in the presence of NaOH do not change color and therefore the oxidation states of these metals are stable with respect to air oxidation.

11. (b) This question is answered using the data in Table 2 and knowing solubility rules for salts. Fe^{3+} precipitates in the presence of hydroxide ions. NH_3 is basic and forms hydroxide ion when it reacts with water: $NH_3(aq) + H_2O(l) \rightleftharpoons NH_4^+(aq) + OH^-(aq)$. Thus $Fe(OH)_3$ should precipitate in 6 M NH_3. $Fe^{3+}(aq)$ in the presence of SO_4^{2-}, ClO_4^-, or Cl^- does not form a precipitate.

12. (d) Adding an alkaline solution of Na_2S precipitates both Co^{2+} and Fe^{3+}. Adding HCl to the precipitate causes Fe^{3+} to be in solution and the Co^{2+} remains as a solid. When aqueous KSCN is added to the solution after the solid is separated, and if a red solution forms, this confirms the presence of Fe^{3+}. Procedure **(a)** by itself is not a conclusive test because Co^{2+} also forms a colored substance in alcoholic KSCN and this interferes with detecting Fe^{3+}. Procedure **(b)** involves a procedure not shown in Table 2. We do not know if the procedure will confirm the presence of Fe^{3+}. Procedure **(c)** does not work because both Co^{2+} and Fe^{3+} form precipitates in the presence of NaOH and both precipitates are insoluble in excess NaOH. Thus, the filtrate does not contain Fe^{3+} to react with SCN^-.

13. (b) K_2S is a strong electrolyte because it is a salt containing an alkali metal ion and the sulfide ion. The potassium ion does not change the pH of water but the sulfide ion is a basic anion.

14. (c) An acidic buffer contains a weak acid and its conjugate base in equilibrium. Answer **b** reflects a solution that is a basic buffer. Answer **a** has a strong acid and a strong base, and answer **d** has a neutral salt and a salt with a basic anion.

15. (d) In dilute solutions $Ba(OH)_2$ acts as a strong electrolyte. $Ba(OH)_2$ dissociates to form two hydroxide ions per barium hydroxide formula unit. Thus, the concentration of hydroxide ions is $2 \times 5.0 \times 10^{-4}\ M$ or $1.0 \times 10^{-3}\ M$. $pOH = -\log[OH^-] = -\log(1.0 \times 10^{-3}) = 3$. We can solve for pH by using $pH + pOH = 14$, or $pH = 14 - 3 = 11$.

16. (b) When NH_3 is neutralized by HCl at the equivalence point, NH_4Cl forms. NH_4^+ is a weak acid and it reacts with water to form a solution whose pH is slightly less than 7.

17. (d). The others should be: PO_4^{3-}, phosphate; CO_3^{2-}, carbonate; and CN^-, cyanide.

Chapter

19

Chemical Thermodynamics

OVERVIEW OF THE CHAPTER

19.1 SPONTANEOUS CHANGES

Review: Activation energy (14.5); chemical equilibrium (15.1); first law of thermodynamics (5.2); concept of rate of reaction (14.1).

Learning Goals: You should be able to:

1. Define the terms *spontaneous* and *spontaneity* and apply them in identifying spontaneous processes.
2. Define the terms *reversible* and *irreversible* and apply them to identifying reversible and irreversible processes.

19.2, 19.3, 19.4, 19.5 ENTROPY: ITS INTERPRETATION AND CALCULATION

Review: Properties of gases, liquids, and solids (10.1, 11.1); solution processes (13.1); thermodynamic state functions (5.2, 5.3).

Learning Goals: You should be able to:

1. Describe how entropy is related to the total number of microstates in a thermodynamic system. Describe how the number of microstates is related to the ideas of dispersion of energy or randomness or disorder in a system.
2. State the second law of thermodynamics.
3. Predict whether the entropy change in a given process is positive, negative, or near zero.
4. State the third law of thermodynamics.
5. Describe how and why the entropy of a substance changes with increasing temperature or when a phase change occurs, starting with the substance as a pure solid at 0 K.
6. Calculate $\Delta S°$ for any reaction from tabulated absolute entropy values, $S°$.

19.5, 19.6, 19.7 FREE ENERGY AND THE EQUILIBRIUM STATE

Review: Enthalpy (5.3); standard state and heat of formation (5.7); equilibrium constant (15.2); reaction quotient (15.5).

Learning Goals: You should be able to:

1. Define free energy in terms of enthalpy and entropy.
2. Explain the relationship between the sign of the free-energy change, ΔG, and whether a process is spontaneous in the forward direction.

3. Calculate the standard free-energy change at constant temperature and pressure, $\Delta G°$, for any process from tabulated values for the standard free energies of reactants and products.
4. List the usual conventions regarding standard states in setting the *values* for standard free energies.
5. Calculate $\Delta G°$ from K and perform the reverse operation.
6. Describe the relationship between ΔG and the maximum work that can be derived from a spontaneous process, or the minimum work required to accomplish a nonspontaneous process.
7. Calculate the free-energy change under nonstandard conditions, ΔG, given $\Delta G°$, temperature, and the data needed to calculate the reaction quotient.

Review: Melting and freezing processes (11.4).

Learning Goals: You should be able to:

1. Predict how ΔG will change with temperature, given the signs for ΔH and ΔS.
2. Estimate $\Delta G°$ at any temperature, given $\Delta S°$ and $\Delta H°$.

19.6 THE DEPENDENCE OF FREE ENERGY ON TEMPERATURE

TOPIC SUMMARIES AND EXERCISES

Thermodynamic principles can be used to answer the question "Does a particular chemical reaction spontaneously proceed in the forward direction?" By spontaneous we mean processes or reactions that can move forward without continuous intervention from external sources. This chapter focuses on how two important thermodynamic quantities, entropy and free energy, are used to answer the previous question. First, what do we already know from our prior studies about spontaneous processes?

SPONTANEOUS CHANGES

- Processes that occur spontaneously in a specific direction are not spontaneous in the opposite direction.
- Systems not at equilibrium tend to proceed toward a state of equilibrium.
- Most exothermic reactions at room temperature are spontaneous; however, there are endothermic processes that are also spontaneous, such as KNO_3 dissolving in water.

What can we conclude? An important driving force for spontaneous processes is their tendency to proceed toward a state of minimum energy, but it cannot be the only factor favoring spontaneity. In the next section, we will learn about another thermodynamic term, entropy, which increases in the universe when a process is spontaneous.

Spontaneous processes can in principle do work. Reversing a spontaneous process requires the application of work from an external source on the system. Key ideas that are presented in this section of the chapter include:

- A reversible process is one that in theory is always possible to go exactly between the initial and final states in both directions. An irreversible process is one which cannot be simply returned to the original state. An important example of a reversible process in the text is a phase change at constant temperature in which both phases are in equilibrium and the change occurs slowly.

• A spontaneous process cannot be reversible; otherwise it could easily return to its original state.

EXERCISE 1   Relating enthalpy to spontaneity

At 25 °C, ΔH is -572 kJ for the reaction

$$2 H_2(g) + O_2(g) \longrightarrow 2 H_2O(l)$$

Can you conclude from the sign of ΔH that the reaction is spontaneous?

SOLUTION: Most exothermic reactions ($\Delta H < 0$) are spontaneous. However, because there are exceptions to this statement, the prediction that the reaction is spontaneous because it is exothermic may be in error. In this case the reaction is spontaneous.

EXERCISE 2   Identifying spontaneous processes

Which of the following processes are spontaneous: (a) the dissolving of solid NaCl in water to form a solution; (b) the decomposition of carbon dioxide into its elements at 25 °C and 1 atm pressure; (c) a ball rolling up a hill; (d) the freezing of water at -3 °C and 1 atm pressure?

SOLUTION: (a) Spontaneous. Table salt (NaCl) readily dissolves in H_2O. (b) Not spontaneous. The decomposition of most compounds into their constituent elements is not spontaneous. (c) Not spontaneous. The rolling of a ball downhill under the influence of gravity is spontaneous. (d) Spontaneous. We know from our experiences that the freezing of water at 0 °C or lower at 1 atm pressure forms ice.

EXERCISE 3   Identifying processes which can provide useful work

Which of the following processes could, in principle, be used to obtain useful work? (a) A rapid phase change of liquid water to steam at the normal boiling point? (b) Combustion of gasoline at 350 K?

SOLUTION: *Analyze and Plan*: We are asked to identify which of two processes could be used to obtain useful work. A process that is spontaneous has the potential to do useful work because it is irreversible. We need to analyze each situation to determine if the process is spontaneous.

Solve: (a) A continuous input of energy is required to convert a liquid to a gas. This is not a spontaneous process and useful work is not available. (b) Combustion of gasoline is a spontaneous process and in principle can be made to produce useful work.

ENTROPY: ITS INTERPRETATION AND CALCULATION

To predict whether a process is spontaneous, we will need to know the change in both the enthalpy and the entropy of the system. **Entropy** is a thermodynamic state function represented by the symbol S.

• Entropy depends only on the initial and final states of a system: $\Delta S = S_{final} - S_{initial}$

• For an isothermal process (constant temperature): $\Delta S = \dfrac{q_{rev}}{T}$

• The units of entropy are J/K. For a change involving one mole of a substance the units are J/K-mol.

- Entropy is a measure of the total number of microstates (positions and energies of particles) for a particular thermodynamic state. We often use the ideas of an increase in the dispersion of energy or randomness in a system to describe an increase in the entropy of the system.

Boltzmann's equation relates the probability of the number of microstates, W, in a system to entropy: $S = k \ln W$. k is the Boltzmann's constant, 1.38×10^{-23} J/K.

In general, we can associate an increase in the entropy of a system with an increase in the number of particles, an increase in its temperature or an increase in its volume. For example,

- $S_{solid} < S_{liquid} \ll S_{gas}$
- $\Delta S_{melting} > 0$
- $\Delta S_{mixing} > 0$
- $\Delta S_{vaporization} > 0$

The second law of thermodynamics involves entropy:

- *In any spontaneous process the entropy of the universe always increases.* For all spontaneous processes:

$$\Delta S_{universe} = \Delta S_{system} + \Delta S_{surroundings} > 0.$$

- For a system that is at equilibrium, $\Delta S_{universe} = 0$.

Note: A process for which $\Delta S_{system} < 0$ can be spontaneous provided it causes even greater positive entropy changes in the surroundings.

Note: An isolated system (no exchange of energy or matter between the surroundings and system) always increases its entropy when it undergoes a spontaneous change. For such a change, $\Delta S_{surroundings} = 0$.

Entropy changes for chemical reactions can be calculated from standard entropy values.

- Standard entropies (symbol, $S°$) have been determined for many substances, and these values are based on the **third law of thermodynamics:** *The entropy of a perfect, pure crystalline substance at 0 K is zero.* As the temperature of a substance increases from *0 K*, its entropy also increases. Increasing temperature causes an increase in thermal motions, which results in a greater number of microstates.
- Entropy, like enthalpy, is a state function; therefore you can calculate the entropy change for a reaction as follows:

$$aA + bB + cC \rightleftharpoons pP + qQ + rR$$

$S°=$ (sum of standard entropies for products) $-$ (sum of standard entropies for reactants)

$S°= [pS°(P) + qS°(Q) + rS°(R)] - [aS°(A) + bS°(B) + cS°(C)]$

or

$$S° = \Sigma n S°_{products} - \Sigma m S°_{reactants}$$

where n and m are the coefficients in front of the species in the balanced chemical reaction.

EXERCISE 4   Identifying which substances have the greater entropy

For each of the following pairs, which has the greater molecular entropy if all substances are at 1 atm pressure: (a) $CO_2(s)$ or $CO_2(g)$; (b) $NH_3(l)$ or $NH_3(g)$; (c) a crystal of pure magnesium at 0 K or a crystal at 200 K?

SOLUTION: *Analyze and Plan*: We are given pairs of substances and asked to determine which has the greater molecular entropy. To do this we need to look at the various properties of each substance such as the state of the system, the temperature, and the complexity of the molecules within the system.

Solve: (a) $CO_2(s)$ molecules are held rigidly in a crystal and undergo only limited thermal motion. $CO_2(g)$ molecules are distributed throughout a much larger volume and are more free to move about in this volume than an equivalent amount of solid CO_2. This random movement of $CO_2(g)$ molecules compared to the rigid arrangement of $CO_2(s)$ molecules results in $CO_2(g)$ molecules being more randomly distributed and thus having a higher entropy than the $CO_2(s)$ molecules. (b) $NH_3(g)$ has a higher entropy than $NH_3(l)$ for reasons similar to those discussed in (a). $NH_3(l)$ molecules occupy a smaller volume than $NH_3(g)$ molecules. Therefore $NH_3(l)$ molecules move less randomly about the container in which they are enclosed; their motions are more restricted. Thus $NH_3(g)$ molecules are more randomly distributed and have a higher entropy. (c) Mg(s) at 0 K contains magnesium atoms in a rigid lattice with no vibrational motion. At 200 K, the atoms have vibrational motion, which results in a more random arrangement of atoms and thus a higher entropy.

EXERCISE 5   Applying the third law of thermodynamics to an impure substance

The standard entropy for a perfect, pure, crystalline substance is zero at 0 K. Why is the entropy for an impure substance greater than zero at 0 K?

SOLUTION: *Analyze and Plan*: We are asked to explain why the entropy of an impure substance has an entropy greater than zero at absolute zero. The third law of thermodynamics tells us that a pure crystalline substance at absolute zero has a standard entropy of zero. In other words, there is only one way energy is distributed (one microstate). How does an imperfect substance then differ in terms of distribution of energy states or particles?

Solve: All molecular motion has stopped for an impure substance at 0 K just as it has for a pure crystalline crystal; thus this is not the determining factor. In an imperfect crystal the impurities can be distributed in more than one way. Consequently, there are a variety of positional microstates possible, and the entropy is greater than zero.

EXERCISE 6   Predicting the sign of ΔS

Predict for each of the following changes whether ΔS is greater than zero or less than zero:

(a) sugar + water $\longrightarrow$ sugar dissolved in water

(b) $Na_2SO_4(s) \longrightarrow 2\,Na^+(aq) + SO_4^{2-}(aq)$

(c) $2\,H_2(g) + O_2(g) \longrightarrow 2\,H_2O(g)$

SOLUTION: *Analyze and Plan*: We are asked to determine the sign of ΔS for three systems. We need to compare the characteristics of the reactants and products, such as: complexity of molecules, ways in which the molecules move or are distributed, state of each substance or mixture, and the number of particles. We consider these characteristics to determine if the final state has a greater number of microstates in its thermodynamic state than the initial state. This condition will be reflected in a greater dispersal of energy or increasing randomness (disorder) in the system and thus in an increasing entropy.

Solve: (**a**) The entropy change in this process is greater than zero because the sugar solution contains a more random arrangement of sugar and water molecules. In the solid state, sugar molecules are more rigidly confined than in the solution state. (**b**) The entropy change for the dissolving of a salt to form solvated ions is greater than zero. In the liquid phase, the solvated ions Na^+ and SO_4^{2-} are free to move about the entire volume of the liquid, whereas in the solid crystal lattice the ions are confined. Therefore dissolved ions have more random movements than ions in a crystal. (**c**) Note that two mol of products are formed for every three mol of reactants. Therefore, the entropy change is less than zero because fewer moles of gas exist after the reaction than before.

EXERCISE 7   Calculating ΔS given tabulated standard entropies

Calculate the entropy change for the following reaction at 25 °C;

$$2\,SO_2(g) + O_2(g) \longrightarrow 2\,SO_3(g)$$

given the following standard entropy values at 25 °C: $SO_2(g)$, 248.5 J/K-mol; $O_2(g)$, 205.0 J/K-mol; and $SO_3(g)$, 256.2 J/K-mol.

SOLUTION: *Analyze and Plan*: We are given standard entropy values for reactants and products of a chemical reaction and asked to calculate the entropy change. To calculate the entropy change for the reaction we use the relation:

$$\Delta S^{\circ}_{rxn} = \Sigma n S^{\circ}\,(\text{products}) - \Sigma m S^{\circ}\,(\text{reactants})$$
$$= 2\,S^{\circ}(SO_3) - [2\,S^{\circ}(SO_2) + S^{\circ}(O_2)]$$

Solve: We substitute the given standard entropy values into the previous equation to calculate ΔS°_{rxn}:

$$\Delta S^{\circ}_{rxn} = \left[(2\,\text{mol})\left(256.2\,\frac{J}{K\text{-mol}} \right) \right]$$
$$- \left[(2\,\text{mol})\left(248.5\,\frac{J}{K\text{-mol}} \right) + (1\,\text{mol})\left(205.0\,\frac{J}{K\text{-mol}} \right) \right]$$

$$\Delta S^{\circ}_{rxn} = 512.4\,J/K - 702.0\,J/K = -189.6\,J/K$$

The negative sign of ΔS° tells us that this reaction involves a decrease in randomness. We can infer this result from the reaction equation because three moles of reactants form fewer moles of products, two.

EXERCISE 8   Calculating ΔS for a phase change

The enthalpy of vaporization for one mole of Ar gas at −185.7 °C is 6.52 kJ/mol. What is the entropy change for the phase change?

SOLUTION: *Analyze and Plan*: We are given the enthalpy of vaporization and the boiling point for argon gas and asked to calculate the entropy change for the phase change and determine if the sign of ΔS is positive. Phase changes that occur along a reversible pathway at constant temperature have a ΔS that is calculated as follows:

$$\Delta S = \frac{q_{rev}}{T} = \frac{\Delta H_{rev}}{T}$$

where T is in Kelvin. The unit for entropy for a substance is J/K-mol; thus, enthalpy should be expressed in the unit joule in this equation.

Solve: We can substitute the given enthalpy of vaporization in units of J/mol and the temperature in units of kelvin into the previous equation:

$$\Delta S = \frac{6520 \text{ J/mol}}{87.5 \text{ K}} = 74.5 \text{ J/K-mol}$$

FREE ENERGY AND THE EQUILIBRIUM STATE

We have learned that two factors must be considered in determining whether a process or reaction is spontaneous:

1. Exothermic processes ($\Delta H < 0$) are more likely to be spontaneous than endothermic processes ($\Delta H > 0$).
2. An increase in the entropy of a system is often associated with a spontaneous process.

Enthalpy and entropy are connected together by the thermodynamic state function **free energy** (symbol G, and sometimes called Gibb's free energy). The change in free energy for any process or reaction at *constant temperature and pressure* is given by the relation

$$\Delta G = \Delta H - T \Delta S$$

- If $\Delta G < 0$, a reaction spontaneously proceeds in the forward direction until equilibrium is reached.
- If $\Delta G = 0$, a system is at equilibrium.
- If $\Delta G > 0$, a reaction is not spontaneous as written, but the reverse direction is spontaneous.
- The free-energy change represents the maximum amount of useful work that can be obtained from a process that occurs in a reversible manner at constant temperature and pressure: $\Delta G = -w_{max}$. If a process is not spontaneous, the increase in free energy represents the minimum amount of work that must be done on the system to cause the process to occur.

Gibb's free energy and enthalpy are alike in that we cannot measure absolute values for either and both are state functions. Therefore we can define and tabulate standard free energies of formation.

- The concepts of standard states and of heats of formation have been discussed in Chapter 5 of the text. Just as ΔH_f°s for elements are set to zero, ΔG_f°s for elements are also set to zero.
- Standard free-energy changes for reactions are calculated using the relation:

ΔG_{rxn}° = (sum of standard free energies of formation of products)
     − (sum of standard free energies of formation of reactants)

$$\Delta G_{rxn}^\circ = \Sigma n \, \Delta G_f^\circ \text{ (products)} - \Sigma m \, \Delta G_f^\circ \text{ (reactants)}$$

- The free-energy change for a reaction at non-standard conditions (ΔG) is related to the free-energy change at standard conditions (ΔG°) as follows:

$$\Delta G = \Delta G^\circ + RT \ln Q$$

where $R = 8.314$ J/K-mol, T is in Kelvin, and Q is the reaction quotient (see Section 19.5 in the text).

$\Delta G°$ is related to the equilibrium constant at constant temperature and pressure as follows:

$$\Delta G° = -RT \ln K$$

- Relationships between $\Delta G°$ and K are:

$\Delta G° < 0 \quad K > 1$ Products are favored over reactants at equilibrium

$\Delta G° = 0 \quad K = 1$

$\Delta G° > 1 \quad K < 1$ Reactants are favored over products at equilibrium

EXERCISE 9 Differentiating between ΔG and $\Delta G°$

Differentiate between the functions ΔG and $\Delta G°$ for the generalized reaction occurring at 298 K

$$a\text{A} + b\text{B} \longrightarrow c\text{C} + d\text{D}$$

SOLUTION: $\Delta G°$ is the standard free-energy change for the reaction when the reactants A and B in their standard states at 298 K react to form products C and D in their standard states at 298 K.

ΔG is the free-energy change for any process in which one or more reactants or products are not in their standard states.

EXERCISE 10 Identifying standard states

At 298 K, are the following substances in their standard states: (a) Al(s); (b) Fe(s) with some oxide impurity; (c) $H_2(g)$ at 2 atm pressure; (d) 1 M solution of sugar in water?

SOLUTION: (a) Yes. The standard state for a solid or liquid is the pure substance in its stable form at 298 K. (b) No. The iron is not pure. (c) No. The standard-state concentration for a gas is 1 atm pressure. (d) Yes. The standard state for a solute is a 1 M concentration solution.

EXERCISE 11 Predicting if a phase change is spontaneous given its ΔG value

Given the ΔG value for each of the following phase changes at 1 atm, predict whether each change is spontaneous: (a) at 283 K, $\Delta G = -250$ J/mol for $H_2O(s) \rightarrow H_2O(l)$; (b) At 273 K, $\Delta G = 0$ J/mol for $H_2O(s) \rightleftarrows H_2O(l)$; (c) At 263 K, $\Delta G = 210$ J/mol for $H_2O(s) \rightarrow H_2O(l)$.

SOLUTION: (a) The reaction is spontaneous since $\Delta G < 0$. We expect this because ice spontaneously melts at 283 K (10 °C). (b) The system is at equilibrium because $\Delta G = 0$; there is no net reaction. Ice and water coexist at the melting point of ice, 273 K (0 °C). (c) The reaction is not spontaneous because $\Delta G > 0$. Ice does not spontaneously melt below its melting point, 273 K.

EXERCISE 12 Calculating $\Delta G°$ for a chemical reaction using $\Delta G_f°$ values

(a) Given that $\Delta G_f°$ for $C_6H_{12}O_6(s)$ equals -907.9 kJ/mol, $\Delta G_f°$ for $CO_2(g)$ equals -394.6 kJ/mol, and $\Delta G_f°$ for $H_2O(l)$ equals -237.2 kJ/mol at 298 K, calculate $\Delta G°$ for the oxidation of glucose:

$$C_6H_{12}O_6(s) + 6\,O_2(g) \longrightarrow 6\,CO_2(g) + 6\,H_2O(l)$$

(b) Is the oxidation of glucose spontaneous at standard-state conditions?

(c) Calculate the equilibrium constant for the reaction.

SOLUTION: *Analyze and Plan*: We are given free-energy data for the combustion of glucose and asked to calculate $\Delta G°$ and the equilibrium constant. We can calculate $\Delta G°$ by using the fact that it is a state function and its value depends only on the values for the initial and final states. The combustion is spontaneous if its sign is positive. We can calculate the value of the equilibrium constant from the relation: $\Delta G° = -RT \ln K$.

Solve: (a) The standard free-energy change for the reaction is calculated using the expression

$$\Delta G° = [6\,\Delta G_f°(CO_2) + 6\,\Delta G_f°(H_2O)] - [\Delta G_f°(C_6H_{12}O_6) + 6\,\Delta G_f°(O_2)]$$

The standard free energy of formation for an element in its stable state at 298 K is zero. Substituting into this equation the appropriate $\Delta G_f°$ values yields

$$\Delta G° = [6\,\text{mol}(-394.6\,\text{kJ/mol}) + 6\,\text{mol}(-237.2\,\text{kJ/mol})]$$
$$- [(1\,\text{mol})(-907.9\,\text{kJ/mol}) + (6\,\text{mol})(0\,\text{kJ/mol})] = -2882.9\,\text{kJ}$$

(b) Yes, because $\Delta G° < 0$. The equilibrium constant is calculated using the expression:

(c)
$$\Delta G° = -RT \ln K$$
$$\ln K = -\frac{\Delta G°}{RT} = \frac{-(-2882.9 \times 10^3\,\text{J/mol})}{(8.314\,\text{J/K-mol})(298\,\text{K})} = 1.16 \times 10^3$$

Taking the antiln of $\ln K$ yields $K = 1 \times 10^{505}$. (antiln $[1.16 \times 10^3] = e^{1.16 \times 10^3}$) The value of K is very large; the reaction essentially goes to completion. Note that the units of $\Delta G°$ are changed to joules from kilojoules because the units of R contain the unit joule.

EXERCISE 13   Calculating ΔG_{rxn} to determine if a chemical reaction is spontaneous at non-standard conditions

Is the following reaction spontaneous at 25 °C under the given conditions?

$$N_2(g) + O_2(g) \longrightarrow 2\,NO(g)$$

$P(\text{atm}) =$	1.0	1.0	4.0

$\Delta G°_{rxn} = +173.1\,\text{kJ}$

SOLUTION: *Analyze and Plan*: We are given the initial pressures of nitrogen, oxygen, and nitrogen (II) oxide and $\Delta G°_{rxn}$ and asked to determine if the reaction is spontaneous at the given conditions. We observe that the reaction is not at standard state conditions because the pressures are not all one atmosphere. Thus, we will have to calculate the reaction quotient, Q, and use the relation

$$\Delta G_{rxn} = \Delta G°_{rxn} + RT \ln Q$$

to solve for ΔG_{rxn} and interpret its sign.

Solve: First calculate the value of Q:

$$Q = \frac{P_{NO}^2}{P_{N_2}P_{O_2}} = \frac{(4.0)^2}{(1.0)(1.0)} = 16$$

Then use the following relation and substitute for the value of Q:

$$\Delta G_{rxn} = \Delta G^{\circ}_{rxn} + RT \ln Q$$

$$= 173.1 \text{ kJ} + \left(8.314 \frac{\text{J}}{\text{K}}\right)(298 \text{ K}) \ln 16$$

$$= 173.1 \text{ kJ} + 6.9 \text{ kJ} = 180.0 \text{ kJ}$$

The reaction is not spontaneous because ΔG_{rxn} is greater than zero.

The value of ΔG for a physical or chemical change depends on the values of ΔH and ΔS, and the temperature at which the change occurs: $\Delta G = \Delta H - T \Delta S$.

- As the temperature increases, the importance of the term $T \Delta S$ in the expression $\Delta G = \Delta H - T \Delta S$ increases (at very high temperatures, ΔG approaches the value of $-T \Delta S$).
- As the temperature decreases, the importance of ΔH increases (at very low temperature ΔG approaches the value of ΔH).
- Look at Table 19.4 in the text and note how the sign of ΔG depends on temperature and on the signs of ΔH and ΔS.

THE DEPENDENCE OF FREE ENERGY ON TEMPERATURE

EXERCISE 14   Calculating the temperature at which NaCl reversibly melts

What is the temperature at which sodium chloride reversibly melts, that is, when the melting occurs so slowly that the solid phase is always in equilibrium with the liquid phase? The enthalpy of melting is 30.3 kJ/mol, and the entropy change upon melting is 28.2 J/mol-K.

SOLUTION: *Analyze and Plan*: We are given the enthalpy of melting for NaCl and the entropy change and asked to calculate the melting point. Assuming that melting occurs reversibly, we can say that $\Delta G = 0$ and therefore $\Delta G = 0 = \Delta H - T\Delta S$. We can solve for temperature by rearranging the equation so that T is by itself.

Solve: Rearranging the equation when free-energy change is zero gives:

$$\Delta H = T\Delta S$$

$$T = \frac{\Delta H}{\Delta S} = \frac{30{,}300 \text{ J/mol}}{28.2 \text{ J/mol-K}} = 1070 \text{ K}$$

Sodium chloride reversibly melts at 1070 K.

EXERCISE 15   Determining how the signs of ΔS and ΔH affect the sign of ΔG

Is a reaction spontaneous if $T\Delta S > 0$ and $\Delta H < 0$?

SOLUTION: *Analyze and Plan*: The sign of ΔG for a reaction is determined by the relation: $\Delta G = \Delta H - T\Delta S$. We need to determine how the signs of ΔH and ΔS affect the sign of ΔG.

Solve: If $T\Delta S > 0$, then $-T\Delta S < 0$, and the sign of ΔG for this set of conditions is always negative because ΔH is also less than zero. The reaction is spontaneous at all temperatures.

EXERCISE 16   Using ΔG° and ΔH° to calculate ΔS° and to identify the primary driving force for a chemical reaction

At 298 K, $\Delta G^{\circ} = -190.5$ kJ and $\Delta H^{\circ} = -184.6$ kJ for the reaction

$$H_2(g) + Cl_2(g) \longrightarrow 2 \text{ HCl}(g)$$

(a) What is the standard entropy change for the reaction? (b) What is the primary driving force for the reaction at 298 K?

SOLUTION: *Analyze and Plan*: We are given the standard free energy and standard enthalpy for the reaction. We are asked to calculate the standard entropy change and to determine which thermodynamic term primarily determines the sign and magnitude of the standard free energy. We can determine the standard entropy change from the relation: $\Delta G° = \Delta H° - T\Delta S°$. Then we compare the signs and magnitudes of $\Delta H°$ and $-T\Delta S°$ to see which term dominates.

Solve: (a) $\Delta S°$ is calculated using the expression $\Delta G° = \Delta H° - T\Delta S°$. Rearranging the equation and solving for $\Delta S°$ yields

$$\Delta S° = \frac{\Delta H° - \Delta G°}{T}$$

Substituting the values of $\Delta H°$ and $\Delta G°$ in the equation yields

$$\Delta S° = \frac{-184.6 \text{ kJ} - (-190.5 \text{ kJ})}{298 \text{ K}} = \frac{5.9 \text{ kJ}}{298 \text{ K}}$$

$$\Delta S° = 2.0 \times 10^{-2} \text{ kJ/K} = 20 \text{ J/K}$$

(b) The primary driving force for the reaction at 298 K is the enthalpy change because the term $-T\Delta S$, which equals -5.9 kJ, is much smaller in magnitude than ΔH, -184.6 kJ.

SELF-TEST QUESTIONS

Key Terms

Having reviewed key terms in Chapter 19, match key terms with phrases and identify statements as true or false. If a statement is false, indicate why it is incorrect.

Match each phrase with the best term:

19.1 The value of this term for an isothermal system is $\frac{q_{rev}}{T}$.

19.2 A statement that the overall entropy change during a spontaneous process is greater than zero.

19.3 For a system at equilibrium at constant pressure and temperature we find that the change in the value of this term is zero.

19.4 It proceeds on its own without external assistance and occurs in a definite direction.

19.5 A term associated with a molecule moving in one direction.

19.6 Atoms in a molecule moving internally with respect to one another.

19.7 Molecules moving around an internal axis.

Terms:

(a) entropy
(b) Gibb's free-energy
(c) rotational motion
(d) second law of thermodynamics
(e) spontaneous process
(f) translational motion
(g) vibrational motion

True-False Statements:

19.8 According to the *third law of thermodynamics*, the entropy of any crystalline solid at absolute zero temperature is zero.

19.9 When heat is released in a chemical reaction in an *isolated system*, the entropy of the surroundings increases.

19.10 When a molecule moves in a linear direction it undergoes *translational motion*.

19.11 The *standard free energy of formation* of any pure substance at standard state conditions is zero.

19.12 The *standard molar entropy* for $N_2(g)$ at 290 K is zero as is the standard molar heat of formation.

19.13 A reaction that is spontaneous in the forward direction only is an *irreversible* reaction.

19.14 The pathway between reactants and products in a state of equilibrium is a *reversible* pathway.

Problems and Short-Answer Questions

19.15 Figures A and B show the initial and final states, respectively, of a change in a system. Is the entropy change positive or negative? What is the name given to this change? Explain.

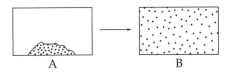

19.16 Figure 19.1 shows two different atoms, A (shaded) and B (unshaded), as gases in separate compartments connected by a valve which can be opened or closed. Suppose the condition in Figure 19.2 is formed. What can you say about the signs of ΔS and ΔG? Explain your answer.

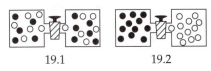

19.1   19.2

19.17 (a) When KNO_3 is cooled from 30 °C to 5 °C, what is the expected sign of the entropy change?

(b) Explain the sign in terms of changes at the molecular/ion level.

19.18 At 25 °C, $\Delta G° = -103.7$ kJ for the reaction

$$CCl_4(l) + H_2(g) \longrightarrow HCl(g) + CHCl_3(l)$$

(a) What are the pressures of the gases at standard conditions?

(b) If the partial pressures of the gases are changed from standard conditions to a pressure of 5.0 atm for hydrogen gas and 20.0 atm for hydrogen chloride gas, in which direction does the reaction proceed?

19.19 Why must the molar-free energies of two phases in equilibrium be the same?

19.20 Rank the following substances in order of increasing entropy: $Mg(s)$; $MgCl_2(s)$; $CuSO_4(s)$; and $C_2H_6(g)$. Justify your response.

19.21 Rank the following substances in order of increasing entropy: $HCOOH(g)$ and $HCOOH(aq)$. Justify your response.

19.22 Rank the following gases in order of increasing entropy: F_2 and Cl_2. Justify your response.

19.23 The standard free-energy change for the vaporization of liquid water at 298 K is a positive quantity. Given this information about the vaporization of water, why does water evaporate at the surface of the Earth?

19.24 Using your basic understanding of thermodynamic principles and properties of substances, indicate whether you think the following two reactions are spontaneous at room temperature and pressure. Justify your answer.

(a) $2 O(g) \rightleftharpoons O_2(g)$
(b) $N_2(g) + O_2(g) \rightleftharpoons 2 NO(g)$

19.25 (a) Calculate ΔG at 298 K for the following reaction if the reaction mixture consists of 1.0 atm of SO_2, 2.0 atm of O_2, and 2.0 atm of SO_3:

$$2 SO_2(g) + O_2(g) \rightleftharpoons 2 SO_3(g) \quad \Delta G°_{298} = -140.0 \text{ kJ}$$

(b) Does a larger "driving force" to produce SO_3 exist at these conditions compared to standard-state conditions?

19.26 (a) Calculate $\Delta S°$ for the reaction:

$$4 CuO(g) \longrightarrow 2 Cu_2O(s) + O_2(g)$$

at 298 K, given the following:

$$\Delta H°_f(CuO) = -157.3 \text{ kJ/mol}$$
$$\Delta H°_f(Cu_2O) = -168.6 \text{ kJ/mol}$$
$$\Delta G°_f(CuO) = -129.7 \text{ kJ/mol}$$
$$\Delta G°_f(Cu_2O) = -146.0 \text{ kJ/mol}$$

(b) Is this reaction feasible for the preparation of $Cu_2O(s)$ at standard state conditions?

(c) If the reaction is not feasible, how could it be made to occur?

19.27 (a) $\Delta G°$ equals $+ 20.00$ kJ/mol for the reaction

$$CuS(s) + H_2(g) \longrightarrow Cu(s) + H_2S(g)$$

at 298 K and 1 atm of pressure. Calculate the equilibrium constant for the reaction.

(b) At 298 K, does the equilibrium lie largely in the direction of products or of reactants?

19.28 Given the following data at 25 °C

	$S°$(J/mol-K)	$\Delta H°_f$(kJ/mol)
$NO_2(g)$	240.45	33.8
$N_2O_4(g)$	304.33	9.66

(a) Calculate the value of $\Delta G°$ for the following reaction at 25 °C:

$$2 NO_2(g) \rightleftharpoons N_2O_4(g)$$

(b) Is the formation of $N_2O_4(g)$ a spontaneous process at 25 °C and standard-state conditions?

(c) What is the value of the equilibrium constant for this reaction?

19.29 Calculate the equilibrium constant for the following reaction at 25 °C

$$2 H_2(g) + O_2(g) \rightleftharpoons 2 H_2O(l)$$

given the following standard free-energy: $H_2O(l)$, -237.2 kJ/mol at 25 °C.

19.30 ΔS for the reversible expansion of an ideal gas at constant temperature can be calculated from:

$$\Delta S = nR \ln\left(\frac{V_2}{V_1}\right)$$

Two moles of an ideal gas at a pressure of 3.0 atm, a volume of 2.00 L, and a temperature of 300 K are allowed to reversibly expand at constant temperature to a pressure of 1.00 atm. Is there sufficient information to determine the entropy for the expansion of the two moles of ideal gas? If not, explain what additional information is needed.

19.31 ΔS° at 25 °C is -176 J/K for the dimerization of $NO_2(g)$ to form $N_2O_4(g)$. Is there sufficient information given to determine whether the reaction is spontaneous? If not, explain what additional information is needed.

Integrative Exercises

19.32 Films of silicon nitride, Si_3N_4, are used as a high temperature structural material and in chemical-vapor deposition of thin films.

(a) Calculate the standard free-energy of formation of silicon nitride given that the standard free-energy of its reaction with oxygen gas to form silicon dioxide and nitrogen gas is -1927 kJ at 25 °C and $\Delta G^\circ_f(SiO_2(s))=-856.7$ kJ/mol.

(b) For the reaction you wrote for (a) is it thermodynamically favored? Is the formation of silicon nitride from its elements favored? Why?

(c) The reaction described in (a) only occurs when silicon nitride is strongly heated in air, otherwise in practice it is stable. Is this in agreement with your conclusion in (b)? Suggest an explanation if there is a difference.

19.33 The bond enthalpies for B-F and B-Cl are 645 kJ/mol and 445 kJ/mol, respectively.

(a) Would you expect the following reaction to occur based on its enthalpy of reaction?

$$BF_3(g) + BCl_3(g) \longrightarrow BF_2Cl(g) + BCl_2F(g)$$

(b) Given that the reaction in (a) does tend to occur, and given your answer in (a), explain this tendency.

19.34 Calculate the pH of water at 40 °C given the following information for the equilibrium at 25 °C:

$$H_2O(l) \rightleftharpoons H^-(aq) + OH^-(aq)$$

	$H_2O(l)$	$H^-(aq)$	$OH^-(aq)$
$(\Delta G^\circ{}_f)_{298}$ (kJ/mol)	-237.2	0	-157.3
$(\Delta H^\circ{}_f)_{298}$ (kJ/mol)	-285.8	0	-230.0

(Hint: Determine ΔG° and K at 40 °C to solve for the pH.)

19.35 Diamond and graphite are allotropic forms of carbon. The standard enthalpies of formation for diamond and graphite are 1.88 kJ/mol and 0 kJ/mol, respectively; the standard molar entropies are 2.43 J/mol-K and 5.69 J/mol-K, respectively.

(a) What does the term allotropic mean?

(b) Which allotropic form exists in a more "disordered" state?

(c) Can the transformation of diamond to graphite at 298 K and 1 atm pressure occur spontaneously? Does this agree with your knowledge of diamonds?

Multiple-Choice Questions

19.36 What are the expected signs of ΔH and ΔS° for the sublimation of dry ice, $CO_2(s)$?

	ΔH	ΔS°
(a)	+	+
(b)	−	−
(c)	+	−
(d)	−	+

(e) More information is needed

19.37 Under what conditions can we absolutely say a system is at equilibrium at constant pressure and temperature?

(a) $\Delta H = 0$ (d) $\Delta G > 0$
(b) $\Delta H > 0$ (e) $\Delta S > 0$
(c) $\Delta G = 0$

19.38 Which of the following substances would you expect to have the highest entropy at 1 atm of pressure?

(a) $H_2O(s)$(1 mol)
(b) $H_2O(l)$(1 mol)
(c) $H_2O(g)$(1 mol)
(d) 0.5 mol of ethanol in 0.5 mol of $H_2O(l)$
(e) 0.5 mol of $H_2O(g)$ and 0.5 mol of $CH_4(g)$

19.39 For any *isolated* spontaneous process that is endothermic, which of the following conditions are true:

(1) $\Delta G_{system} < 0$; (2) $\Delta G_{system} > 0$; (3) $\Delta S_{system} < 0$;
(4) $\Delta S_{system} > 0$; (5) $\Delta S_{universe} < 0$; (6) $\Delta S_{universe} > 0$?

(a) (1), (3), and (4)   (d) (1), (4), and (6)
(b) (2), (4), and (6)   (e) (2), (4), and (6)
(c) (1), (4), and (5)

19.40 For which of the following changes would you expect ΔS to be less than zero?

(a) $H_2O(l) \longrightarrow H_2O(g)$
(b) $CO(g) + \frac{1}{2} O_2(g) \longrightarrow CO_2(g)$
(c) $2 SO_3(g) \longrightarrow 2 SO_2(g) + O_2(g)$
(d) $CO_2(s) \longrightarrow CO_2(g)$
(e) $2 AgCl(s) \longrightarrow 2 Ag(s) + Cl_2(g)$

19.41 When KNO_3 dissolves in water at room temperature, ΔH is positive for the dissolution process. Given this information, what can you conclude?

(a) $\Delta G > 0$ for the dissolution process.
(b) $\Delta G = 0$ for the dissolution process.
(c) The dissolving of salts in water is always a spontaneous process.
(d) $\Delta S > 0$ for the dissolution process.
(e) $\Delta S < 0$ for the dissolution process.

19.42 What is K_p for the equilibrium

$H_2(g)+I_2(g) \rightleftharpoons 2 HI(g)$ if ΔG° is -24.2 kJ at 490 ° C?

(a) 1.7   (d) 46
(b) 2.6   (e) 6.6×10^3
(c) 38

19.43 Which statement is true about the reversible conversion of a liquid to its solid at its melting point?

    (**a**) It is spontaneous.

    (**b**) $\Delta H° = \Delta S°$

    (**c**) $\Delta G < 0$

    (**d**) The entropy of the system increases.

    (**e**) The enthalpy of the system increases.

19.44 Calculate $\Delta G°$ for the following reaction at 25 °C

$$4\,NO(g) + 6\,H_2O(g) \longrightarrow 5\,O_2(g) + 4\,NH_3(g)$$

given the following $\Delta G_f°$ values: NO, 86.71 kJ/mol; H_2O, −228.61 kJ/mol; O_2, 0.0 kJ/mol; NH_3, −16.66 kJ/mol.

    (**a**) 958.2 kJ   (**d**)   −1318.18 kJ

    (**b**) −2145.14 kJ   (**e**)   1318.18 kJ

    (**c**) 2145.14 kJ

19.45 Estimate the normal boiling point for HCl, given the following information:

$$\Delta H°_{vaporization} = 16.13 \text{ kJ/mol}$$

$$\Delta S°_{vaporization} = 85.77 \text{ J/mol-K.}$$

    (**a**) 188 °C   (**d**)   0.189 K

    (**b**) 188 K   (**e**)   100 °C

    (**c**) 0.189 °C

SELF-TEST SOLUTIONS

19.1 (a). **19.2** (d). **19.3** (b). **19.4** (e). **19.5** (f). **19.6** (g). **19.7** (c). **19.8** False. Only for pure crystalline solids does $S = 0$ at 0 K. The presence of impurities results in $S > 0$. **19.9** False. An isolated system cannot exchange heat with its surroundings. Therefore no heat is transferred to the surroundings and its entropy is not changed. **19.10** True. **19.11** False. It is zero only for elements in their standard states, not all substances. **19.12** False. Only at absolute zero is the standard molar entropy for a perfect, pure crystalline substance equal to zero. **19.13** True. **19.14** True.

19.15 The entropy change is positive because a solid is changed to a gas. The term for this process is sublimation. A gas consists of particles in a chaotic motion, with many accessible microstates. A solid contains particles in relatively fixed positions with far fewer microstates. An increase in available microstates is associated with an increasing entropy.

19.16 ΔS is negative and ΔG is positive. Figure 19.1 shows the two atoms in a random distribution in both compartments. Figure 19.2 shows the two atoms separated in each compartment; thus the change is one of a more ordered distribution, less dispersal of energy and fewer microstates leading to a decrease in entropy. Separating a mixture into its components in separated compartments is not a spontaneous event. Thus, ΔG is positive for the change.

19.17 (**a**) Negative. (**b**) As the temperature is lowered, the ions in the crystal structure have less kinetic energy; also

the vibrational energy decreases. This results in less random motion of the ions in the crystal, which means the final state is more ordered than the initial state and thus there is less entropy.

19.18 (**a**) One atm pressure each.

    (**b**) $\Delta G = \Delta G° + RT \ln Q$

$$\Delta G = -103{,}700 \text{ J} + (8.314 \text{ J/K})(298 \text{ K}) \ln \frac{20.0}{5.00}$$

$$= -100{,}300 \text{ J} = -103.3 \text{ kJ}$$

The value of ΔG is now less negative; however, the reaction is still spontaneous.

19.19 The value of ΔG for a phase change occurring under equilibrium conditions is zero. This occurs because the molar-free energies for the two phases must be the same. If the values were not equal then there would be a spontaneous change in favor of one of the phases and equilibrium could not exist.

19.20 Entropy tends to increase with increasing complexity of chemical structure as well as increasing molar mass. In addition entropy increases in the following order: solid < liquid < gas. The predicted order is: $Mg(s) < MgCl_2(s) < CuSO_4(s) < C_2H_6(g)$. The first three are solids and have ordered solid structures, whereas the last one is a gas with a high degree of disorder. The solids are ranked according to increasing complexity and molar mass.

19.21 The predicted trend is: $HCOOH(aq) < HCOOH(g)$. The aqueous state for a substance has fewer microstates than its gaseous state, primarily because of increased thermal motions and separation of particles in a gaseous state.

19.22 Flourine and chlorine are gases. Therefore, the greater number of electrons and molar mass for chlorine compared to fluorine primarily determines the trend in entropies: $F_2 < Cl_2$.

19.23 The sign of $\Delta G°$ is positive, which implies that the transformation of liquid water to vapor water is not spontaneous. This is contrary to our observations in the physical world. What we have to realize is the effect of concentrations on the value of ΔG. We learned in the text that $\Delta G = \Delta G° + RT \ln Q$. Q depends on the vapor pressure of water vapor only since liquid water does not appear in the equilibrium-constant expression. If $Q = 1$, the vapor pressure is 1 atm and standard conditions exist. When this condition exists, $\Delta G = \Delta G°$ because the value of $\ln(1) = 0$. In our world, the pressure of the atmosphere may be near one at the surface of the Earth but the partial pressure of water in the air is only a few torr. Thus, $Q<<1$; ΔG becomes less than one and the process is spontaneous.

19.24 (**a**) For a reaction to be spontaneous, the free-energy change must be less than zero. This is favored by an exothermic reaction and one in which entropy increases. In this reaction the enthalpy of reaction should be strongly exothermic, because a bond is formed and none is broken. Entropy decreases because there are two moles of reactants versus one mole of product and the product has organized chemical bonds. The reaction, however, should be spontaneous because of the strong exothermic character.

(b) The enthalpy of reaction should be endothermic. The breaking of an O—O bond and the forming of a N—O bond should result in a net difference that is not overly large, either exothermic or endothermic compared to the cost of breaking a N—N triple bond. Thus, one should expect the reaction to be endothermic (It is: 90 kJ/mol). The change in entropy should be a negative value. The reactants are a mixture of gases, whereas the product is one type of gas with chemical bonds that are equivalent. Thus, both factors work against the reaction being spontaneous.

19.25 Equation 19.19 in the text is used to solve for ΔG at nonstandard state conditions: $\Delta G = \Delta G° + RT \ln Q$. $Q = P_{SO_3}/P_{SO_2}{}^2 P_{O_2} = (2.0)^2/(1.0)^2(2.0) = 2.0$. From the given value of $\Delta G°_{298}$, we calculate ΔG_{298}: $\Delta G = (-140.0 \text{ kJ}) + (8.314 \text{ J/K})(298 \text{ K})(1 \text{ kJ}/10^3 \text{ J}) \ln(2.0) = -138.3 \text{ kJ}$ **(b)** The free-energy change becomes less negative, changing from -140.0 kJ to -138.3 kJ, $\Delta G°$). The smaller negative value for ΔG indicates a lesser, not a larger, driving force to produce SO_3.

19.26 **(a)** In order to calculate $\Delta S°$ using the relation $\Delta S° = (\Delta H° - \Delta G°)/T$, the values of $\Delta H°$ and $\Delta G°$ must first be calculated. $\Delta G° = [\Delta G_f°(O_2) + 2\Delta G_f°(Cu_2O)] - [4\Delta G_f°(CuO)] = [0 \text{ kJ} + (2 \text{ mol})(-146.0 \text{ kJ/mol})] - [(4 \text{ mol})(-129.7 \text{ kJ/mol})] = 226.8 \text{ kJ}$. And $\Delta H° = [\Delta H_f°(O_2) + 2\Delta H_f°(Cu_2O)] - [4\Delta H_f°(CuO)] + [0 \text{ kJ} + (2 \text{ mol})(-168.6 \text{ kJ/mol})] - [(4 \text{ mol})(-157.3 \text{ kJ/mol})] = 292.0 \text{ kJ}$ Solving for $\Delta S°$, $\Delta S° = (\Delta H° - \Delta G°)/T = (292.0 \text{ kJ} - 226.8 \text{ kJ})/298 \text{ K} = 0.219 \text{ kJ/K} = 219 \text{ J/K}$.

(b) At standard state conditions $\Delta G° = 226.8$ kJ; the reaction is not feasible because the sign of $\Delta G°$ is positive.

(c) 226.8 kJ is the *minimum* amount of energy needed to cause the reaction to occur. We could couple this reaction with a spontaneous one that has a $\Delta G°$ more negative than -226.8 kJ; the second reaction should not interfere with the decomposition of CuO(s).

19.27 **(a)** $\ln K = \dfrac{-\Delta G°}{RT}$

$$= \frac{-(20,000 \text{ J/mol})}{(8.314 \text{ J/K-mol})(298 \text{ K})} = -8.07$$

$$K = 3.10 \times 10^{-4}$$

(b) The equilibrium lies far on the side of reactants because $K < 1$.

19.28 **(a)** To calculate $\Delta G°$ for the reaction, first calculate $\Delta S°$ and $\Delta H°$: $\Delta S° = S°(N_2O_4) - 2 S°(NO_2) = (1 \text{ mol})(304.33 \text{ J/mol-K}) - (2 \text{ mol})(240.45 \text{ J/mol-K}) = -176.57 \text{ J/K}$. And $\Delta H° = \Delta H_f°(N_2O_4) - 2 \Delta H_f°(NO_2) = (1 \text{ mol})(9.66 \text{ kJ/mol}) - 2(33.8 \text{ kJ/mol}) = -57.94 \text{ kJ}$. Thus $\Delta G° = \Delta H° - T\Delta S° = -57,940 \text{ J} - (298 \text{ K})(-176.57 \text{ J/K}) = -5320 \text{ J} = -5.32 \text{ kJ}$.

(b) The reaction is spontaneous at standard-state condition at 25 °C because $\Delta G° < 0$.

(c) $\ln K = \dfrac{-\Delta G°}{RT}$

$$= \frac{-(-5320 \text{ J})}{(8.314 \text{ J/K-mol})(298 \text{ K})} = 2.15$$

$$K = 8.57$$

19.29 The equilibrium constant is calculated using the relationship: $\Delta G°_{rxn} = -RT \ln K$. To solve for K, you first need to calculate the standard free-energy change.

$\Delta G°_{rxn} = [2 \Delta G_f°(H_2O(l))] - [2 \Delta G_f°(H_2(g)) - \Delta G_f°(O_2(g))]$
$= [(2 \text{ mol})(-237.2 \text{ kJ/mol})] - [(2 \text{ mol})(0 \text{ kJ/mol}) + (1 \text{ mol})(0 \text{ kJ/mol})]$

$$= -474.4 \text{ kJ}$$

$\Delta G°_{rxn} = -RT \ln K$

$-474.4 \text{ kJ} \times \dfrac{1000 \text{ J}}{1 \text{ kJ}} = -(8.314 \text{ J/K-mol})(298 \text{ K}) \ln K$

$\ln K = 191.5$

$K = 1.4 \times 10^{83}$

19.30 There is sufficient information. The change in entropy for the isothermal (constant temperature) expansion of a gas along a reversible pathway is calculated from $\Delta S = nR \ln\left(\dfrac{V_2}{V_1}\right)$. n is the number of moles of gas, R is the gas law constant, V_2 is the final volume of the gas, and V_1 is the initial volume of the gas. We are given the initial volume of the gas, 2.00 L, n is two moles, P_1 is 3.0 atm, P_2 is 1.00 atm, and R is a known constant; thus V_2 is the only unknown. V_2 can be calculated using the relationship:

$$\frac{P_1 V_1}{T_1} = \frac{P_2 V_2}{T_2}$$

After calculating V_2 we can use the given equation to solve for ΔS.

19.31 There is insufficient information. To determine whether the reaction is spontaneous we need to calculate $\Delta G°$ which equals $\Delta H° - T\Delta S°$. The value of $\Delta H°$ is not given nor are standard enthalpies of formation which would permit us to calculate the enthalpy of the dimerization. We could look up the values in a table of standard enthalpies of formation.

19.32 **(a)** To answer this question we need to write the balanced chemical equation:

$$Si_3N_4(s) + 3 O_2(g) \longrightarrow 3 SiO_2(s) + 2 N_2(g)$$

$\Delta G°_{rxn} = \Sigma \Delta G_f°(\text{Products}) - \Sigma \Delta G_f°(\text{Reactants})$

$-1927 \text{ kJ} = (3 \text{ mol})(-856.7 \text{ kJ/mol}) + (0)$
$- (1 \text{ mol})(\Delta G_f°(Si_3N_4)) - (0)$

$\Delta G_f°(Si_3N_4) = -643 \text{ kJ/mol}$

(b) The free-energy of formation of silicon nitride is negative; thus, its formation from the elements is thermodynamically favored. Similarly, the negative free-energy of reaction tells us that the reaction of silicon nitride with oxygen is thermodynamically favored.

(c) The observation that the reaction of silicon nitride with oxygen does not readily occur does not agree with the thermodynamic conclusion. However, thermodynamics does not tell you anything about the kinetics of a reaction or other effects. Note that $SiO_2(s)$ is formed as a product. This is a very stable material as discussed in earlier chapters. Once it is made it forms a surface film that protects the silicon nitride from further reaction with oxygen.

19.33 (a) No. The enthalpy of reaction can be estimated by summing the bond enthalpies of the reactants and subtracting the bond enthalpies of the product. However, since all bonds in both reactants and products are either B—F or B—Cl, the number and types of bonds do not change. This should result in an enthalpy of reaction that is zero (or very close to zero since bond enthalpies are slightly different depending on the environment in the molecule). **(b)** The problem says that the reaction tends to occur, which means $\Delta G_{rxn} < 0$. Thus, the $T\Delta S$ term in $\Delta G = \Delta H - T\Delta S$ must be positive; therefore, ΔS must also be positive. Inspection of the reaction shows that the bonds in the products are arranged in a more random manner than those in the products. Thus, this supports the idea that ΔS is indeed greater than zero and therefore the entropy change drives the reaction.

19.34 Use the relation $\ln K = -\dfrac{\Delta G°}{RT}$ to solve for K at 40 °C. $\Delta H°$ and $\Delta S°$ are relatively constant for the temperature range of 298 K (25 °C) to 313 K (40 °C); thus you first calculate $\Delta G°$, $\Delta H°$, and $\Delta S°$ at 25 °C and then calculate $\Delta G°$ at 40 °C.

First calculate $\Delta G°$:

$$\Delta G° = (\Delta G_f°[H^+] + \Delta G_f°[OH^-]) - \Delta G_f°(H_2O) = 79.9 \text{ kJ}$$

In a similar manner $\Delta H°$ is calculated: $\Delta H° = 55.8$ kJ. $\Delta S°$ is calculated as at 25 °C:

$$\Delta S° = \frac{\Delta H° - \Delta G°}{T} = -0.0809 \text{ kJ/K} = -80.9 \text{ J/K}$$

We can solve for $\Delta G°$ at 40 °C (313 K):

$$\Delta G° = 55.8 \text{ kJ} - (313 \text{ K})(-0.0809 \text{ kJ/K}) = 81.1 \text{ kJ}$$

Solving for K:

$$\ln K = \frac{-81,100 \text{ J}}{(8.314 \text{ J/K-mol})(313 \text{ K})} = -31.2$$

$$K = e^{-31.2} = 3 \times 10^{-14} = [H^+][OH^-]$$

At equilibrium, $[H^+] = [OH^-]$ thus $[H^+]^2 = 3 \times 10^{-14}$

or $[H^+] = 2 \times 10^{-7}$ or pH = 6.7.

19.35 (a) Allotropes are different forms of the same element existing in the same physical state.

(b) Graphite should exist in a more disordered solid state form than diamond because it has a larger standard molar entropy value. Diamond is a covalently bonded giant molecule with each carbon atom bonded to four others in an extensive network. Graphite has carbon atoms bonded to three other carbon atoms to form a sheet of hexagonal units bonded together. The sheets are able to slide with respect to each other.

(c) To answer this question we calculate $\Delta G°$ for one mole of diamond converting to one mole of graphite.

$$\Delta G° = \Delta H° - T\Delta S°$$

$$= (\Delta H°_{products} - \Delta H°_{reactants}) - T(S°_{products} - S°_{reactants})$$

$$\Delta G° = (0 - 1.88 \text{ kJ}) - (298 \text{ K})(5.69 \text{ J/K} - 2.43 \text{ J/K})$$

$$= -1.88 \text{ kJ} - 0.971 \text{ kJ} = -2.85 \text{ kJ}$$

The conversion should be spontaneous as the sign of the free-energy change is negative. Diamond, however, is considered a durable gem. Although the application of thermodynamic principles tells us the transformation is spontaneous, our observations should convince us that the kinetics of the transformation are extremely slow.

19.36 (a). 19.37 (c).

19.38 (e) A mixture of gases shows the greatest random motion of particles.

19.39 (d).

19.40 (b) A mixture of gases with $1\frac{1}{2}$ particles reacting yields a single gas with fewer particles of gas, one particle.

19.41 (d). 19.42 (d). 19.43 (b). 19.44 (a). 19.45 (b).

Chapter

20

Electrochemistry

OVERVIEW OF THE CHAPTER

20.1, 20.2 OXIDATION-REDUCTION REACTIONS: BALANCING

Review: Balancing chemical equations (3.1); oxidation-reduction (4.4).

Learning Goals: You should be able to:

1. Identify the oxidizing agent (oxidant) and reducing agent (reductant) in an oxidation-reduction reaction.
2. Balance simple oxidation-reduction reactions by the method of half-reactions.

20.3, 20.4 VOLTAIC CELLS: STANDARD ELECTRODE AND CELL POTENTIALS

Review: Electrolyte solutions (4.2).

Learning Goals: You should be able to:

1. Diagram simple voltaic and electrolytic cells, labeling the anode, the cathode, the directions of ion and electron movement, and the signs of the electrodes.
2. Given appropriate electrode potentials, calculate the emf generated by a voltaic cell.
3. Use electrode potentials to predict whether a reaction will be spontaneous.

20.5, 20.6 RELATIONSHIP OF EMF TO CONCENTRATION, FREE-ENERGY CHANGES, AND THE EQUILIBRIUM CONSTANT

Review: Equilibrium constant (15.2); free-energy function (19.5); free energy and the equilibrium state (19.5, 19.6).

Learning Goals: You should be able to:

1. Interconvert $E°$, $\Delta G°$, and K for oxidation-reduction reactions.
2. Use the Nernst equation to calculate the concentration of an ion, given E, $E°$, and the concentrations of the remaining ions.
3. Use the Nernst equation to calculate an emf under nonstandard conditions.

20.7, 20.8 PRACTICAL EXAMPLES OF ELECTRO-CHEMISTRY

Learning Goals: You should be able to:

1. Describe the lead storage battery, the dry cell, and the nickel–cadmium cell.
2. Describe corrosion in terms of the electrochemistry involved, and explain the principles that underlie cathodic protection.

418

Learning Goals: You should be able to:

1. Diagram simple electrolytic cells, labeling the anode, the cathode, the directions of ion and electron movement, and the sign of the electrodes.
2. Given appropriate electrode potentials, calculate the minimum emf required to cause electrolysis.
3. Interrelate time, current, and the amount of substance produced or consumed in an electrolysis reaction; given two of the three quantities, you should be able to calculate the third.
4. Calculate the maximum electrical work performed by a voltaic cell, and the minimum electrical work required for an electrolytic process.

20.9 ELECTROLYSIS: FARADAY'S LAW

TOPIC SUMMARIES AND EXERCISES

The concepts of oxidation state and oxidation number were introduced in Section 4.4. In Section 20.1 we learn how changes in oxidation numbers of atoms are used to understand a class of reactions known as oxidation-reduction reactions. To illustrate the concepts, consider the following reaction, with the oxidation state of each atom shown above its elemental symbol.

OXIDATION-REDUCTION REACTIONS: BALANCING

$$\overset{+1\ +5\ -2}{3\ AgNO_3(aq)} + \overset{0}{Al(s)} \longrightarrow 3\ \overset{0}{Ag(s)} + \overset{+3\ +5\ -2}{Al(NO_3)_3(aq)}$$

- Silver undergoes **reduction** because its oxidation number is decreasing from +1 to 0 in the reaction.
- Aluminum undergoes **oxidation** because its oxidation number is increasing from 0 to +3.
- Silver is an **oxidizing agent** or **oxidant** because it causes aluminum to be oxidized.
- Aluminum is a **reducing agent** or **reductant** because it causes silver to be reduced.

   Balancing oxidation-reduction reactions is an important skill. The method discussed in the text requires us to write two half-reactions and balance them in acid or base solution.

- An oxidation-reduction reaction (redox) can be broken down into two separate reactions, called **half-reactions**. One half-reaction represents the oxidation process and the other half is the reduction process. The reaction between $AgNO_3$ and Al is broken down as follows:

   Reduction: $3\ AgNO_3(aq) + 3\ e^- \longrightarrow 3\ Ag(s) + 3\ NO_3^-(aq)$

   Oxidation: $Al(s) \longrightarrow Al^{3+}(aq) + 3\ e^-$

Notice (1) that the number of electrons gained in the oxidation half-reaction equals the number lost in the reduction half-reaction and (2) that the sums of atoms and charges must be the same on each side of the arrow in each half-reaction.

- In Section 20.2, rules are given for using half-reactions to balance redox reactions in both acid and base solutions. Exercise 4 illustrates how to apply these rules.

EXERCISE 1 Identifying atoms undergoing oxidation and reduction

Identify the atoms undergoing oxidation and reduction in the following reactions:

(a) $2\,AuCl_4^-(aq) + 3\,Cu(s) \longrightarrow 2\,Au(s) + 8\,Cl^-(aq) + 3\,Cu^{2+}(aq)$
(b) $Cl_2(g) + OH^-(aq) \longrightarrow Cl^-(aq) + ClO^-(aq) + H^+(aq)$

SOLUTION: *Analyze*: We are given two reactions and asked to identify the atoms in the reactants that undergo oxidation and reduction.

Plan: First we assign oxidation numbers to all atoms in each reaction. Then we use the definitions of oxidation and reduction to determine which atoms undergo oxidation and reduction. Remember that the sum of oxidation numbers in a substance equals the charge carried by the substance. As an example we have for $AuCl_4^-$: $+3 + 4(-1) = -1$.

Solve:
(a)

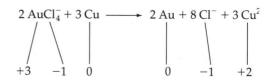

$$2\,AuCl_4^- + 3\,Cu \longrightarrow 2\,Au + 8\,Cl^- + 3\,Cu^2$$

Oxidation number: $+3 \quad -1 \quad 0 \qquad 0 \quad -1 \quad +2$

Cu has been oxidized from 0 to +2, and Au has been reduced from +3 to 0.

(b)

$$Cl_2 + OH^- \longrightarrow Cl^- + ClO^- + H^+$$

Oxidation number: $0 \quad -2 \ +1 \qquad -1 \quad +1 \ -2 \ +1$

Cl has been oxidized from 0 to +1, and Cl has been reduced from 0 to −1.

Comment: Note in (b) that chlorine is both oxidized and reduced simultaneously in the reaction. An atom can be both the source of electrons and the acceptor of electrons in an oxidation-reduction reaction.

EXERCISE 2 Identifying whether a substance is an oxidizing or reducing agent

In each of the following equations, is hydrogen peroxide acting as a reducing or oxidizing agent?

(a) $2\,MnO_4^-(aq) + 5\,H_2O_2(aq) + 6\,H^+(aq) \longrightarrow 2\,Mn^{2+}(aq) + 5\,O_2(g) + 8\,H_2O(l)$
(b) $PbS(s) + 4\,H_2O_2(aq) \longrightarrow PbSO_4(s) + 4\,H_2O(l)$

SOLUTION: *Analyze*: We are given two reactions and asked whether hydrogen peroxide, H_2O_2, acts as a reducing or oxidizing agent.

Plan: We first note that hydrogen gas is not produced in either reaction; this means that hydrogen ions and hydrogen atoms with an oxidation state of +1 are not being reduced. Thus, we can focus on oxygen. First assign oxidation numbers to oxygen in hydrogen peroxide and then to all products containing oxygen. Then we use the definitions of oxidation and reduction to determine whether oxygen undergoes oxidation or reduction.

Solve: (**a**) H_2O_2 acts as a reducing agent. Oxygen in H_2O_2 has a -1 oxidation number; it increases to 0 in elemental O_2. This is verified by observing that manganese in MnO_4^- is reduced; the oxidation number of manganese decreases from $+7$ to $+2$. (**b**) H_2O_2 acts as an oxidizing agent. Oxygen in H_2O_2 decreases in oxidation number from -1 to -2 in H_2O; this is verified by observing that the oxidation number of sulfur in PbS increases from -2 to $+6$ in $PbSO_4$.

EXERCISE 3   Writing half-reactions

Write two half-reactions that when added together form the following balanced redox reaction:

$$S_2O_3^{2-}(aq) + 4\,I_2(aq) + 10\,OH^-(aq) \longrightarrow 2\,SO_4^{2-}(aq) + 8\,I^-(aq) + 5\,H_2O(l)$$

SOLUTION: *Analyze*: We are given a balanced chemical reaction and asked to separate it into two half-reactions.

Plan: The two half-reactions when added must give the balanced chemical reaction. First assign oxidation numbers to all atoms in the overall chemical reaction. Next, determine which atoms undergo oxidation and reduction. Then, using the atom which undergoes oxidation, write a half-reaction that shows a loss of electrons. Similarly, using the atom which undergoes reduction, write a half-reaction that shows a gain of electrons. Water, $H^+(aq)$, and $OH^-(aq)$ may have to be included to balance each half-reaction.

Solve: The oxidation numbers are:

$$\overset{+2\ -2}{S_2O_3^{2-}}(aq) + 4\,\overset{0}{I_2}(aq) + 10\,\overset{-2\ +1}{OH^-}(aq) \longrightarrow 2\,\overset{+6\ -2}{SO_4^{2-}}(aq) + 8\,\overset{-1}{I^-}(aq) + 5\,\overset{+1\ -}{H_2O}$$

Next, determine which atoms are involved in oxidation and reduction. Sulfur in thiosulfate, $S_2O_3^{2-}$, is oxidized from $+2$ to $+6$ in sulfate, SO_4^{2-}. Iodine, I_2, is reduced from 0 to -1 in I^-. Write the oxidation half-reaction by including $S_2O_3^{2-}$ and other species in the overall reaction that are necessary for the number of atoms to be the same on both sides of the arrow:

$$S_2O_3^{2-}(aq) + 10\,OH^-(aq) \longrightarrow 2\,SO_4^{2-}(aq) + 5\,H_2O(l)$$

(OH^- and H_2O are necessary for the oxygen atoms in $S_2O_3^{2-}$ and SO_4^{2-} to be balanced.) Then add enough electrons so that the sum of charges is the same on both sides of the arrow. In this example, there is a $12-$ total charge $[(2-) + 10(1-)]$ on the left and a $4-$ total charge $[2(2-) + 5(0)]$. Eight negative charges are missing on the right. The oxidation half-reaction requires $8\,e^-$ on the right for it to be properly balanced.

$$S_2O_3^{2-}(aq) + 10\,OH^-(aq) \longrightarrow 2\,SO_4^{2-}(aq) + 5\,H_2O(l) + 8\,e^-$$

The reduction half-reaction is formed using the same procedure. Iodine is assigned a zero oxidation state because it is in its elemental form. The ion I^- is in the

-1 oxidation state, the same as its ion state. I_2, which contains two iodine atoms, must form two I^- ions. Thus the reaction describing the reduction of iodine is

$$I_2(aq) + 2\,e^- \longrightarrow 2\,I^-(aq)$$

The number of electrons lost in the oxidation half-reaction must equal the number of electrons gained in the reduction half-reaction. Eight electrons are lost in the oxidation half-reaction; therefore you must multiply the reduction half-reaction for iodine by four to make the number of electrons equal:

$$4\,I_2(aq) + 8\,e^- \longrightarrow 8\,I^-(aq)$$

Check: If we add the two half-reactions we will find that they add to the overall balanced chemical reaction given in the question.

EXERCISE 4   Balancing an oxidation-reduction chemical reaction

Use the procedures given in section 20.2 of the text to balance the following redox reactions:

(a) $Zn(s) + NO_3^-(aq) \longrightarrow Zn^{2+}(aq) + N_2(g)$ (acidic solution)
(b) $Br_2(aq) \longrightarrow BrO_3^-(aq) + Br^-(aq)$ (basic solution)

SOLUTION: *Analyze*: We are given two skeletal redox (reduction-oxidation) reactions and asked to balance them. Note that the first one is in an acid solution and the second is in a basic solution.

Plan: We will use the steps summarized in the text for balancing redox reactions. We will need to remove hydrogen ion in the second reaction and show instead hydroxide ion because it is a basic solution.

Solve: *First*, write the skeletal half-reactions for oxidation and reduction. To do this, you must be able to identify the atoms undergoing oxidation and reduction:

$$\text{Oxidation: } Zn(s) \longrightarrow Zn^{2+}(aq)$$
$$\text{Reduction: } NO_3^-(aq) \longrightarrow N_2(g)$$

Note that in the reduction half-reaction only species containing nitrogen are included and that the oxygen atoms are not balanced at this time.
  Second, balance all atoms *other than* oxygen and hydrogen in each half-reaction:

$$Zn \longrightarrow Zn^{2+} \qquad 2\,NO_3^- \longrightarrow N_2$$

Two NO_3^- ions are required to balance the two nitrogen atoms contained in N_2.
  Third, balance oxygen atoms in each reaction by adding an appropriate number of water molecules:

$$Zn \longrightarrow Zn^{2+} \qquad 2\,NO_3^- \longrightarrow N_2 + 6\,H_2O$$

Six H_2O molecules are added as products to balance the six oxygen atoms contained in two NO_3^- ions.
  Fourth, balance hydrogen atoms by adding an appropriate number of hydrogen ions:

$$Zn \longrightarrow Zn^{2+} \qquad 12\,H^+ + 2\,NO_3^- \longrightarrow N_2 + 6\,H_2O$$

Twelve H^+ ions are added as reactants to balance the 12 hydrogen atoms contained in six H_2O molecules.

Fifth, make the sum of charges on each side of the arrow equal by adding the required number of electrons to the appropriate side:

$$Zn \longrightarrow Zn^{2+} + 2\,e^- \qquad 10\,e^- + 12\,H^+ + 2\,NO_3^- \longrightarrow N_2 + 6\,H_2O$$

Sixth, make the number of electrons lost equal the number of electrons gained for each half-reaction by multiplying each half-reaction by the smallest integer that makes the number of electrons in each half-reaction the same:

$$5(Zn \longrightarrow Zn^{2+} + 2\,e^-) \qquad 1(10\,e^- + 12\,H^+ + 2\,NO_3^- \longrightarrow N_2 + 6\,H_2O)$$

Seventh, add the two half-reactions together and cancel equal amounts of electrons, H^+ ions, and H_2O molecules that appear on both sides of the arrow:

$$
\begin{aligned}
5\,Zn &\longrightarrow 5\,Zn^{2+} + 10\,e^- \\
\underline{10\,e^- + 12\,H^+ + 2\,NO_3^-} &\underline{\longrightarrow N_2 + 6\,H_2O} \\
5\,Zn + 12\,H^+ + 2\,NO_3^- &\longrightarrow 5\,Zn^{2+} + N_2 + 6\,H_2O
\end{aligned}
$$

Eighth, check the final equation to make sure that the sums of atoms and charges on each side of the arrow are the same.

(**b**) This reaction demonstrates the procedures for balancing redox reactions in basic solution. It also illustrates the phenomenon of a single substance acting both as an oxidant and reductant. Note that in this example Br_2 is reduced to Br^- and oxidized to BrO_3^-. Therefore Br_2 will be a reactant in both half-reactions.

First, when balancing a redox reaction in basic solution, follow all the steps in (**a**). This results in the following equations for this redox reaction:

$$
\begin{aligned}
5(Br_2 + 2\,e^- &\longrightarrow 2\,Br^-) \\
\underline{1(Br_2 + 6\,H_2O} &\underline{\longrightarrow 2\,BrO_3^- + 12\,H^+ + 10\,e^-)} \\
6\,Br_2 + 6\,H_2O &\longrightarrow 10\,Br^- + 2\,BrO_3^- + 12\,H^+
\end{aligned}
$$

Dividing by 2 yields reactants and products with the smallest integral coefficients possible:

$$3\,Br_2 + 3\,H_2O \longrightarrow 5\,Br^- + BrO_3^- + 6\,H^+$$

Second, add sufficient OH^- ions to both sides of the arrow so that all the H^+ ions are converted to H_2O by the reaction

$$H^+ + OH^- \longrightarrow H_2O$$

Cancel equal numbers of water molecules on both sides of the arrow. In this exercise add 6 OH^- to both sides of the arrow:

$$
\begin{aligned}
3\,Br_2 + 3\,H_2O &\longrightarrow 5\,Br^- + BrO_3^- + 6\,H^+ \\
\underline{+6\,OH^-} &\underline{\qquad\qquad\qquad + 6\,OH^-} \\
3\,Br_2 + 3\,H_2O + 6\,OH^- &\longrightarrow 5\,Br^- + BrO_3^- + 6\,H_2O
\end{aligned}
$$

Cancel three H_2O molecules from each side and reduce the equation to the smallest coefficients:

$$3\,Br_2(aq) + 6\,OH^-(aq) \longrightarrow 5\,Br^-(aq) + BrO_3^-(aq) + 3\,H_2O(l)$$

Check: The balanced chemical reaction is checked by counting atoms and charges. In (**a**) there are 5 Zn, 12 H, 2 N, and 6 O atoms and a net charge of 10- on both sides of the arrow. In (**b**) there are 6 Br, 6 O, and 6 H and a net charge of 6- on both sides of the arrow.

VOLTAIC CELLS: STANDARD ELECTRODE AND CELL POTENTIALS

A spontaneous oxidation-reduction reaction involves the transfer of electrons from oxidant to reductant. In principle, it is possible to design a device in which electrons are forced to move through an external circuit rather than transferring directly between the oxidant and reductant. Such a device is called a **voltaic** (or **galvanic**) **cell**.

The design of a voltaic cell prevents the oxidant and reductant from being in direct physical contact with one another. A voltaic cell which separates the two half-reactions is shown in Figure 20.5 in the text. The principal features of a voltaic cell are:

- One cell compartment contains the chemicals that involve the oxidation half-reaction; the other contains those needed for the reduction half-reaction.
- A solid electrical connector, called an **electrode**, is placed in each cell compartment to provide a means for electrons to be transferred in the external circuit. An electrode may participate in the chemical reaction.
- The electrode at which reduction occurs is termed the **cathode**; the one at which oxidation occurs is termed the **anode**.
- By convention, the anode is assigned a negative charge; the cathode is assigned a positive charge.
- Electrons flow from the anode to the cathode in the external circuit through a wire that connects them.
- The solutions are connected by a **salt bridge** or semi-permeable membrane that allows ions to flow into or out of each cell compartment. A salt bridge enables each cell compartment to maintain a net charge of zero without allowing the two solutions to mix.
- Positive ions move in solution away from the anode to the cathode; negative ions move in solution away from the cathode to the anode.

In a voltaic cell, electrons are pumped through the outer circuit with a force that varies depending on the nature of the species reacting.

- The driving force is called **electromotive force (emf)** and is measured in the unit of volts (symbol V); it is also referred to as the cell potential or voltage.

$$1 \text{ volt (V)} = \frac{1 \text{ joule (J)}}{1 \text{ coulomb (C)}}$$

- The emf generated by a cell is represented by the symbol E.
- A voltaic cell operating under standard-state conditions (all gases at 1 atm pressure and all solutes at 1 M concentration) generates a standard cell emf, $E°$. Unless otherwise stated, the temperature is assumed to be 25 °C.
- The superscript° indicates that standard-state conditions exist and therefore $E°_{cell}$ is the **standard emf** or **standard cell emf** or **standard cell potential**.

It is impossible to measure directly the electromotive force generated by an oxidation or reduction half-reaction, but we can easily measure the potential difference between a cathode and an anode using a voltaic cell.

- A cell emf is simply the potential difference between the cathode and anode.
- If we assign a zero potential to a reference, standard electrode reaction, we can then measure the potentials of a series of half-reactions with respect to this reference. The chosen reference electrode system is the **standard hydrogen electrode (SHE):**

$$2\,H^+(aq)(1\,M) + 2\,e^- \xrightarrow{\;25\,°C\;} H_2(g)(1\,\text{atm}) \quad E^°_{red} = 0\;V$$

- By convention, **standard reduction potentials** (or simply standard potential), $E^°_{red}$, are tabulated for reduction half-reactions and are used in calculations. *Note*: Whenever a potential is assigned to a half-reaction we write the standard reduction emf. This is done even if we are referring to an oxidation half-reaction.
- A cell potential under standard-state conditions is calculated from standard reduction potentials by the equation:

$$E^°_{cell} = E^°_{red}(\text{cathode}) - E^°_{red}(\text{anode})$$

This equation reflects the concept that a standard cell potential is a difference in potential energy per electrical charge between the cathode and anode.

When using standard reduction potentials of half-reactions to determine the standard potential of a spontaneous reaction, we need to remember the following:

1. *Always use standard reduction potentials when giving potentials for half-reactions.* Thus, for the reduction and oxidation of nickel you would write:

$$\text{Reduction: } Ni^{2+}(aq) + 2\,e^- \longrightarrow Ni(s) \quad E^°_{red} = -0.28\;V$$
$$\text{Oxidation: } Ni(s) \longrightarrow Ni^{2+}(aq) + 2\,e^- \quad E^°_{red} = -0.28\;V$$

2. *The more positive the standard reduction potential for a half-reaction, the greater the tendency for that reduction half-reaction to occur.* Thus, in the following series:

	$E^°_{red}(v)$
$\uparrow\; Ag^+ + e^- \longrightarrow Ag \;\downarrow$	0.80
Ease of reduction $\;|2\,H^+ + 2\,e^- \longrightarrow H_2|\;$ Ease of oxidation	0.00
$Fe^{2+} + 2\,e^- \longrightarrow Fe$	-0.44

Ag^+ is more easily reduced than H^+ or Fe^{2+}. The negative emf of Fe^{2+} means that Fe^{2+} is more difficult to reduce than H^+. Note that Fe is more easily oxidized than Ag or H_2.

3. *A voltaic cell (uses a spontaneous reaction) exhibits a positive emf; thus the difference in reduction potentials between the cathode and anode must yield a positive emf.*

4. *A standard reduction potential is not changed in value or sign if the reduction half-reaction is multiplied by a number.*

EXERCISE 5   Diagramming an electrochemical cell and using E_{red}°

A voltaic cell is to be designed at standard-state conditions using the following two half-reactions:

$$Mn^{2+}(aq) + 2\,e^- \longrightarrow Mn(s) \quad E_{red}^{\circ} = -1.18\ V$$

$$Cr^{3+}(aq) + 3\,e^- \longrightarrow Cr(s) \quad E_{red}^{\circ} = -0.74\ V$$

(a) Write the chemical equation describing the spontaneous cell reaction and calculate E_{cell}°. (b) Draw a figure representing the voltaic cell, using these reactions. Label all components and indicate the direction of electron flow.

SOLUTION: Analyze: (a) We are given two half-reactions with their standard reduction potentials and asked to write the spontaneous reaction that occurs when they are used to construct a spontaneous cell reaction (a voltaic cell). (b) We are asked to diagram a voltaic cell and label components for the cell described in (a).

Plan: (a) A spontaneous cell reaction has a positive cell potential. We must reverse one of the given half-reactions so that when they are added a positive standard cell potential results. (b) The two half-reactions must be separated in different cell compartments to prevent the reaction from occurring directly in solution; this is accomplished using a salt bridge. If an electrode is not required as one of the reactants we can use an inert electrode for the transfer of electrons. Refer to Sample Exercise 20.4 in the text for a similar approach.

Solve: (a) In order to produce a spontaneous cell reaction at standard-state conditions, one of the given reduction reactions must be reversed. The *reduction half-reaction with the more negative standard reduction potential is always the one reversed.* In this exercise, the reduction reaction of manganese is reversed:

$$3[Mn(s) \longrightarrow Mn^{2+}(aq) + 2\,e^-] \qquad E_{red}^{\circ} = -1.18\ V^* \ \text{Anode}$$

$$\underline{2[Cr^{3+}(aq) + 3\,e^- \longrightarrow Cr(s)]} \qquad E_{red}^{\circ} = -0.74\ V \ \text{Cathode}$$

$$3Mn(s) + 2\,Cr^{3+}(aq) \longrightarrow 3\,Mn^{2+}(aq) + 2\,Cr(s) \quad E_{cell}^{\circ} = E_{red}^{\circ}(\text{cathode}) - E_{red}^{\circ}(\text{anode})$$

$$E_{cell}^{\circ} = -0.74\ V - (-1.18\ V) = 0.44\ V$$

$^*\ E_{red}^{\circ}$ without a sign change is written even when the half-reaction is oxidation.

(b)

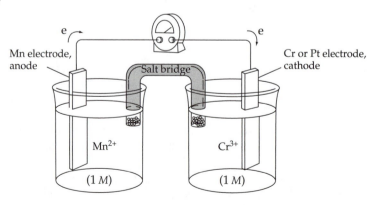

Mn is needed as an anodic electrode because Mn is a reactant in the oxidation half-reaction. It must be supplied. An inert electrode, such as Pt or a carbon rod, can be used as the cathodic electrode in this cell. In the reduction half-reaction you see that Cr is formed and deposits on an electrode. Since the electrode does not provide any reactant material, but only serves as an electron transfer surface, it does not have to be Cr. In such cases, an inert electrode is usually used.

Comment: Note in (**a**) that although the reduction half-reaction is multiplied by two and the oxidation half-reaction is multiplied by three so that the number of electrons lost and gained are equal, their standard reduction potentials are *not* multiplied by these factors.

EXERCISE 6   Using E°_{red} to determine whether a chemical reaction is spontaneous

Given the following E° values at 25 °C

$$Fe^{2+} + 2\,e^{-} \longrightarrow Fe \qquad E^{\circ}_{\text{red}} = -0.44 \text{ V}$$
$$Sn^{2+} + 2\,e^{-} \longrightarrow Sn \qquad E^{\circ}_{\text{red}} = -0.14 \text{ V}$$

determine which reaction is spontaneous at standard conditions

$$Fe^{2+} + Sn \longrightarrow Fe + Sn^{2+} \text{ or } Sn^{2+} + Fe \longrightarrow Fe^{2+}$$

Also identify the anode and cathode for the reaction in a voltaic cell.

SOLUTION: *Analyze*: We are given two half-reactions and their standard reduction potentials and asked to determine which reaction is spontaneous and to also identify the anode and cathode if the reactions were used in a voltaic cell.

Plan: We need to determine which reaction has a positive standard emf. We learned in the previous Exercise that we need to reverse the half-reaction with the more negative standard reduction potential. If we do this and add the half-reactions we determine the spontaneous reaction.

Solve: The half-reaction with the more negative reduction potential involves iron. We will reverse this one:

Anode: $Fe \longrightarrow Fe^{2+} + 2\,e \qquad\quad E^{\circ}_{\text{red}} = -0.44 \text{ V}$

Cathode: $\underline{Sn^{2+} + 2\,e^{-} \longrightarrow Sn \qquad\quad E^{\circ}_{\text{red}} = -0.14 \text{ V}}$

$\qquad\quad Sn^{2+} + Fe \longrightarrow Sn + Fe^{2+} \quad E^{\circ}_{\text{rxn}} = E^{\circ}_{\text{cathode}} - E^{\circ}_{\text{anode}}$

$$= -0.14 \text{ V} - (-0.44 \text{ V}) = 0.30 \text{ V}$$

The balanced chemical reaction has a positive standard emf, which means it is spontaneous at standard conditions.

EXERCISE 7   Using E°_{red} to determine the best reducing agent

Determine which one of the three metals Zn, Fe, or Na is the most active metal at standard conditions, given the following data:

$$Na^{+}(aq) + e^{-} \longrightarrow Na(s) \qquad E^{\circ}_{\text{red}} = -2.71 \text{ V}$$
$$Zn^{2+}(aq) + 2\,e^{-} \longrightarrow Zn(s) \quad E^{\circ}_{\text{red}} = -0.76 \text{ V}$$
$$Fe^{2+}(aq) + 2\,e^{-} \longrightarrow Fe(s) \quad E^{\circ}_{\text{red}} = -0.44 \text{ V}$$

SOLUTION: *Analyze*: We are given three metals with their standard reduction potentials and asked to determine which is the most active metal.

Plan: An active metal is one that is easily oxidized (that is, a good reducing agent). The more negative the standard reduction potential, the more active the metal. We thus look for the most negative standard reduction potential.

RELATIONSHIP OF EMF TO CONCENTRATION FREEENERGY CHANGES, AND THE EQUILIBRIUM CONSTANT

Solve: Sodium metal is the most active metal because it has the largest negative standard potential, $-2.71\ V$.

A spontaneous oxidation-reduction reaction has an accompanying negative free-energy change and a positive cell potential.

• The relationship between cell emf and the free-energy change for any redox process is given by

$$\Delta G = -nFE$$

where n is the number of moles of electrons transferred in the redox reaction and F is faraday's constant, which equals 96,500 J/V-mol e^-. A faraday represents the charge carried by one mole of electrons.

• If all reactants and products are in their standard states, the relation is $\Delta G° = -nFE°$.

• Note that a positive emf yields a negative ΔG.

Most reactions are not done at standard state conditions; thus cell emf is expressed as E not $E°$. The Nernst equation relates E to $E°$ and concentrations.

• At 298 K, the potential for a redox reaction at nonstandard conditions is related to $E°$ as given by the Nernst equation:

$$E = E° - \frac{0.0592\ V}{n} \log Q$$

where Q is the reaction quotient for the redox reaction and n is the number of moles of electrons involved in the redox reaction. The concentrations of pure liquids and solids and the solvent of a dilute mixture do not appear in the reaction quotient.

If a redox reaction has achieved an equilibrium state, then $\Delta G = 0$ and $E = 0$ V. At 298 K, the equilibrium constant for a redox reaction is related to $E°$ for the reaction by the relation

$$E° = \frac{0.0592\ V}{n} \log K$$

or

$$\log K = \frac{nE°}{0.0592\ V}$$

EXERCISE 8   Calculating $\Delta G°$, K, and E

Calculate $\Delta G°$, K, and E at 298 K for the reaction

$$2\ Br^-(aq) + Cl_2(g) \longrightarrow Br_2(l) + 2\ Cl^-(aq)$$

given $E° = 0.300$ V and the following concentrations: $[Br^-] = 0.10\ M$, $P_{Cl_2} = 0.50$ atm, and $[Cl^-] = 0.010\ M$.

SOLUTION: *Analyze*: We are given a chemical reaction, its standard emf, and a set of concentrations. Using this data we are asked to calculate $\Delta G°$, K, and E.

Plan: We can use $\Delta G° = -nFE°$ to determine $\Delta G°$ from the given standard potential for the reaction. K is determined from the standard potential using the relation:

$$\log K = \frac{nE°}{0.0592\ V}$$

E, the non-standard state potential, is determined using the Nernst equation. The value of n is determined by determining how many moles of electrons are transferred in the chemical reaction. This can be done by inspection or by writing half-reactions.

Solve: Inspection of the redox reaction shows that two moles of electrons are transferred. The standard free-energy change is calculated as follows:

$$\Delta G° = -nFE° = -(2 \text{ mol e}^-)\left(96{,}500\frac{\text{J}}{\text{V} - \text{mol e}^-}\right)(0.300 \text{ V})$$

$$= -5.97 \times 10^4 \text{ J} = -57.9 \text{ kJ}$$

The value of $\log K$ is obtained using the relation

$$\log K = \frac{nE°}{0.0592 \text{ V}} = \frac{(2)(0.300 \text{ V})}{0.0592 \text{ V}} = 10.2$$

Taking the antilog of both sides yields $K = 2 \times 10^{10}$. The value of E is calculated using the Nernst equation:

$$E = E° - \left(\frac{0.0592 \text{ V}}{n}\right)\log\frac{[\text{Cl}^-]^2}{[\text{Br}^-]^2 P_{\text{Cl}_2}}$$

The concentration of Br_2 is not shown because it is a pure liquid.

$$E = 0.300 \text{ V} - \left(\frac{0.0592 \text{ V}}{2}\right)\log\frac{(0.010)^2}{(0.10)^2(0.50)}$$

$$E = 0.300 \text{ V} - (0.0296 \text{ V})(-1.70) = 0.300 \text{ V} + 0.050 \text{ V} = 0.350 \text{ V}$$

Check: The sign of $\Delta G°$ is negative which is in agreement with the positive sign of $E°$ and both signs indicate a spontaneous process. An estimate of the value is $-2(100{,}000)(.3) = -60{,}000$ J, which is in close agreement with the calculated answer in the solution. K is a number greater than one, which is in agreement with the idea that a positive $E°$ tells us that the products are favored compared to the reactants. An estimate of the value is

$$\frac{2(0.3)}{0.06} = 10 = \log K$$

and $K = 10^{10}$ which is in close agreement with the calculated answer. E is larger than $E°$ which is consistent with the observation that the concentrations of the reactants are significantly greater than that of the product and this will tend to drive the reaction to the right. The value of Q is 0.02 which is less than K; this also tells you that the reaction is driven to the product side until $Q = K$.

EXERCISE 9 Calculating E for a concentration cell

Calculate the voltage generated by a hydrogen-concentration cell utilizing the following concentrations:

$$H_2(2 \text{ atm}) + 2 \text{ H}^+(0.1 \text{ M}) \rightleftharpoons H_2(0.5 \text{ atm}) + 2 \text{ H}^+(0.5 \text{ M})$$

A **concentration cell** involves a reaction having the same reactants and products, but with differing concentrations. In which direction is the reaction spontaneous?

SOLUTION: *Analyze*: We are given a concentration cell and asked to determine in which direction the cell is spontaneous.

Plan: The concentrations are not at standard-state conditions; therefore, we will use the Nernst equation to calculate the value of the cell emf.

Solve: The voltage of the concentration cell is calculated using the Nernst equation:

$$E = E° - \left(\frac{0.0592\ V}{n}\right) \log Q$$

Inspection of the redox equation shows $n = 2$. Substituting the value of $E°$ into the Nernst equation and remembering that $Q = [\text{products}]/[\text{reactants}]$, we have

$$E = 0.00\ V - \left(\frac{0.0592\ V}{2}\right) \log \frac{P_{H_2}[H^+]^2}{P_{H_2}[H^+]^2}$$

Substituting the given concentration values yields

$$E = 0.00\ V - \left(\frac{0.0592\ V}{2}\right) \log \frac{(0.5)(0.5)^2}{(2)(0.1)^2} = 0.00\ V - (0.0295\ V)(0.80)$$
$$= -0.02\ V$$

The concentration cell is spontaneous toward the reactant side of the reaction equation because $E < 0$ for the given reaction.

PRACTICAL EXAMPLES OF ELECTROCHEMISTRY

Several practical applications of electrochemical principles are discussed in Chapter 20. Of particular importance are the lead–acid storage battery, the rechargeable nickel–cadmium battery, fuel cells, and corrosion of metals. Exercises 10–14 are designed so that you can check your understanding of some of these applications.

EXERCISE 10   Determining if a lead storage battery needs charging

The density of the sulfuric acid solution in a lead storage battery is found to be 1.19 g/cm^3. Does the battery need recharging?

SOLUTION: Yes. During the discharge of a lead storage battery the sulfuric acid is consumed and the density of the solution in the battery decreases. If the density is below 1.20 g/cm^3, it needs recharging.

EXERCISE 11   Differentiating between a fuel cell and a battery

Differentiate between a fuel cell and a battery.

SOLUTION: Batteries store electrical energy, and their lifetime is limited by the amount of reactants initially present. Fuel cells do not store electrical energy because the reactants are continuously supplied to the electrodes; they are energy-conversion devices.

EXERCISE 12   Using a metal for cathodic protection

Why is magnesium often used in the cathodic protection of iron?

SOLUTION: The standard reduction potentials for iron and magnesium are

$$Mg^{2+} + 2\,e^- \longrightarrow Mg \qquad E°_{red} = -2.37\ V$$

$$Fe^{2+} + 2\,e^- \longrightarrow Fe \qquad E°_{red} = -0.44\ V$$

Because $E°_{red}(Mg)$ is more negative than $E°_{red}(Fe)$, Mg is more easily oxidized than Fe. Thus when iron and magnesium are connected, magnesium is preferentially oxidized, and iron becomes the site at which the following reduction reaction occurs:

$$O_2(g) + 4\,H^+(aq) + 4\,e^- \longrightarrow 2\,H_2O$$

Thus iron is acting as a cathodic electrode and is inert to oxidation until all the magnesium is used.

EXERCISE 13   Explaining why the corrosion of iron is pH dependent

Explain why the corrosion of iron becomes less favorable as the pH is increased.

SOLUTION: The initial reaction of iron in corrosion is

$$Fe(s) \longrightarrow Fe^{2+}(aq) + 2\,e^-$$

This reaction is coupled to the cathodic reaction

$$O_2(g) + 4\,H^+(aq) + 4\,e^- \longrightarrow 2\,H_2O(l)$$

Note that H^+ takes part in the reaction with oxygen. As the pH is increased, the amount of H^+ present in solution decreases (more OH^-). According to LeChatelier's principle, reducing the concentration of H^+ will cause the cathodic reaction to shift to the left, thereby making the reduction of O_2 less favorable. If the cathodic reaction becomes less favorable, then the anodic oxidation of iron also becomes less favorable.

EXERCISE 14   Using relative activities of metals

The kinds of metals used in constructing ships is extremely important. Suppose part of a ship was constructed using steel (mostly iron) and it was attached to a bracket consisting of aluminum. Why would this not be a good idea?

SOLUTION: If you refer to Table 20.1 in the text you see that aluminum is a more active metal than iron. You previously learned that magnesium also is a more active metal than iron and when it is attached to iron it acts as a cathodic protector; in the process it is consumed. Similarly, aluminum would act as a cathodic protector for the steel. It would be consumed and the aluminum bracket would undergo extreme corrosion near the steel.

ELECTROLYSIS: FARADAY'S LAW

A nonspontaneous oxidation-reduction reaction can be made to occur by the application of a direct current of electricity to the reactants. This process is called **electrolysis**, and the special apparatus in which it takes place is called an electrolytic cell.

The minimum applied emf that causes an electrolytic reaction to proceed is calculated using standard electrode potentials and the Nernst equation.

- The reduction half-reaction with the least negative E_{red} will occur first at the cathode.
- Similarly, at the anode, the oxidation half-reaction with the least positive E_{red} will occur first.
- If a half-reaction is kinetically slow then additional energy will be required, an overvoltage, to cause the reaction to occur.

Faraday's law, a result of the work of Michael Faraday, states that the quantity of a substance undergoing a change during electrolysis is directly proportional to the quantity of electricity that passes through the cell.

- The quantity of electricity passing through a cell is expressed in number of faradays (F). One faraday is equal to 96,500 coulombs or one mole of electrons. Therefore,

$$\text{Number of faradays} = (\text{number of coulombs})\left(\frac{1\,F}{96{,}500\,C}\right)$$

Number of moles of electrons

$$= (\text{number of coulombs})\left(\frac{1\text{ mol electrons}}{96{,}500\,C}\right)$$

- The number of coulombs (abbreviated as C) is equal to the product of current passing through the cell (in amperes) and the amount of time (in seconds) that the current is passed through the cell:

$$\text{Coulombs} = \text{amperes} \times \text{seconds} = it \qquad 1\,C = 1\text{ amp-sec}$$

- A useful relationship when using Faraday's law is that one mole of electrons equals one faraday. We can relate the number of moles of electrons in a half-reaction to the number of moles of reactants or products. The maximum amount of work that a voltaic cell can do is:

$$w_{\max} = -nFE$$

A nonspontaneous electrochemical cell has a cell potential that is negative. To force the nonspontaneous reaction to occur we must do work on the electrolytic system that meets the following requirement:

$$w_{\min} > -nFE_{\text{cell}}$$

where E_{cell} is the cell potential for the nonspontaneous reaction, a negative number.

- Remember that work done *by* a system *on* its surroundings is indicated by a *negative* sign. A *positive* sign means that the surroundings is doing work on the system.
- Electrical work is more commonly expressed as watts $\times$ time, where a watt is the rate of energy expenditure:

$$1\text{ watt} = \frac{1\,J}{\sec}$$

Note: $3.6 \times 10^6\,J = 1$ kilowatt-hour and $1\,J = 1$ watt-s

EXERCISE 15   Identifying the requirements for an electrolytic cell

What are the minimum requirements for an electrolytic cell?

SOLUTION: A battery (voltaic cell) or the rectified current from an electric generator is required as a direct-current source of electrons. The electrons are forced onto one electrode and removed from another. Electrodes are composed of substances that are good conductors; they are in contact with a liquid or solution containing the reactants. Reduction occurs at the surface of the electrode to which electrons are pumped, and oxidation occurs at the electrode from which electrons are removed.

EXERCISE 16   Identifying reactions at the anode and cathode in an electrolytic cell

The electrodes in an electrolytic cell are given the names anode and cathode. What processes occur at the surface of these electrodes in the electrolytic solution?

SOLUTION: Oxidation occurs at the anode (the positive electrode in an electrolytic cell), and reduction occurs at the cathode (the negative electrode in an electrolytic cell). Notice that the signs of the electrodes are opposite to those found in a voltaic cell.

EXERCISE 17   Calculating the mass of a metal deposited in electrolysis I

What mass of zinc metal is produced at the cathode in an electrolysis cell when a constant current of 10.00 amp is passed for 1.00 hr?

SOLUTION: *Analyze*: We are asked to calculate the mass of Zn produced in an electrolysis cell when a constant current of 10.00 amp is passed for one hour.

Plan: Figure 20.0 in the text provides a road map for determining the mass of material formed in an electrolysis cell. We can determine the number of moles of electrons that are passed through the electrolysis cell by using the relationship:

$$\text{Number of moles of electrons} = (\text{number of coulombs})\left(\frac{1 \text{ mol electrons}}{96,500 \text{ C}}\right)$$

where the number of coulombs is equal to: $it = (\text{amp})(\text{seconds})$. One hour is also 3600 s. We then look at the half reaction for Zn, $Zn^{2+} \longrightarrow Zn + 2\,e^-$, and determine the number of moles of zinc that react for every mole of electrons transferred. Using stoichiometric equivalences we can determine the number of moles of Zn that form and thus the grams of Zn.

Solve: First calculate how many moles of electrons are passed through the cell (remember that 1 hr = 3600 sec):

$$\text{Number of moles of electrons} = (\text{number of coulombs})\left(\frac{1 \text{ mol electrons}}{96,500 \text{ C}}\right)$$

$$= (\text{amperes})(\text{seconds})\left(\frac{1 \text{ mol electrons}}{96,500 \text{ C}}\right)$$

$$= (10.00 \text{ amps})(3600 \text{ s})\left(\frac{1 \text{ mol electrons}}{96,500 \text{ C}}\right)$$

$$= (3.600 \times 10^4 \text{ C})\left(\frac{1 \text{ mol electrons}}{96,500 \text{ C}}\right) = 0.373 \text{ mol electrons}$$

The half-reaction for zinc tells us that two moles of electrons passing through a cell containing zinc ions causes one mole of zinc metal to form. Calculate the number of moles of zinc metal deposited from this information:

$$\text{Number of moles of Zn} = (0.373 \text{ mol electrons})\left(\frac{1 \text{ mol Zn}}{2 \text{ mol electrons}}\right) = 0.187 \text{ mol Zn}$$

Thus the mass of zinc metal produced is

$$\text{Mass of Zn} = (0.187 \text{ mol Zn})\left(\frac{65.4 \text{ g Zn}}{1 \text{ mol Zn}}\right) = 12.2 \text{ g}$$

EXERCISE 18   Predicting the expected products in an electrolysis cell

(a) An aqueous solution of $CuSO_4$ is electrolyzed between two copper electrodes. What are the expected electrolysis products given the following data?

$$E^\circ_{red}(V)$$

$$2\,H_2O(l) + 2\,e^- \longrightarrow H_2(g) \longrightarrow 2\,OH^-(aq) \qquad -0.83$$

$$Cu^{2+}(aq) + 2\,e^- \longrightarrow Cu(s) \qquad +0.337$$

$$4\,H^+(aq) + O_2(g) + 4\,e^- \longrightarrow 2\,H_2O(l) \qquad +1.23$$

$$S_2O_8^{-2}(aq) + 2\,e^- \longrightarrow 2\,SO_4^{2-}(aq) \qquad +2.01$$

(b) Could this process be used to refine impure copper?

SOLUTION: *Analyze*: We are asked to determine what electrolysis products form when copper (II) sulfate is electrolyzed and whether this process can be used to refine impure copper. We are also given a set of standard reduction potentials to help us make our decision.

Plan: (a) To answer this question we need the standard reduction potentials given in the problem. If they were not provided we would have to examine a table of standard reduction potentials in the text or other reference. Initially we have to determine what species in solution can be reduced. We look at the standard reduction potentials to determine which species in the electrolytic cell show a possibility for reduction; in other words, do any of the species in solution appear to the left of the arrows in the given half-reactions? Similarly, we do this analysis for oxidation (do any of the species appear to the right of the arrows?). We then use the standard reduction potentials to determine which half-reactions are more likely to occur as the external potential is slowly applied. (b) We can look at the predicted products in (a) and determine if the production of copper metal occurs.

Solve: (a) Only $H_2O(l)$ and $Cu^{2+}(aq)$ can possibly be reduced because only H_2O, Cu^{2+}, and SO_4^{2-} are present in the solution. The reduction emf of $Cu^{2+}(aq)$ is more positive than that of $H_2O(l)$; thus Cu^{2+} will be electrolyzed preferentially at the copper cathode:

$$Cu^{2+}(aq) + 2\,e^- \longrightarrow Cu(s)$$

At the anode there are three possible reactions:

$$E^\circ_{red}(V)$$

$$Cu(s) \longrightarrow Cu^{2+}(aq) + 2\,e^- \qquad +0.337$$

$$2\,H_2O(l) \longrightarrow 4\,H^+(aq) + O_2(g) + 4\,e^- \qquad +1.23$$

$$SO_4^{2-}(aq) \longrightarrow S_2O_8^{2-}(aq) + 2\,e^- \qquad +2.01$$

Note that although the half-reactions are written as oxidation half-reactions the standard potentials for the half-reactions are given for the reduction half-reactions, by convention. This requires us to interpret the standard potentials for oxidation. As the standard potential increases the tendency for the oxidation process decreases. At the anode the oxidation half-reaction with the least positive standard potential of the three will occur first in electrolysis. The oxidation of $Cu(s)$ has the least positive standard potential and therefore requires the least amount of applied potential during electrolysis; therefore, at the anode the reduction of the copper electrode occurs. (b) Yes. Impure copper could be used as the anode, and a solution of $CuSO_4$ electrolyzed. The copper in the anode electrode would dissolve and pure copper would plate out at the cathode.

EXERCISE 19 Calculating the mass of a metal deposited in electrolysis II

Calculate the mass of Mg produced in an electrolytic cell if 2.67×10^2 kwh of electricity is passed through a solution of $MgCl_2$ at 4.20 volts.

SOLUTION: *Analyze*: We are asked to calculate the mass of Mg produced in an electrolysis cell when 2.67×10^2 kwh of electricity is passed through a solution of magnesium ion at a constant voltage of 4.20 V.

Plan: This question is similar to Exercise 17 except we are given different types of information. We need to develop a plan that uses the given data to calculate the number of coulombs passed through the cell. We can do this by referring to a table of unit relationships: 1 kwh $= 3.6 \times 10^6$ J. From the definitions of joules and volts you can calculate the number of coulombs that pass through the cell: joule $=$ coulomb $\times$ volts. Then we can solve the problem as we did in Exercise 17.

Solve: The number of joules of work required is:

$$\text{Joules} = (2.67 \times 10^2 \text{ kwh})\left(\frac{3.6 \times 10^6 \text{ J}}{1 \text{ kwh}}\right) = 9.6 \times 10^8 \text{ J}$$

From the joules of work and volts you can then calculate the number of coulombs passed through the cell from the relationship

$$\text{joule} = \text{coul} \times \text{volt}$$

or

$$\text{Coul} = \frac{\text{J}}{\text{volt}} = \left(\frac{9.6 \times 10^8 \text{ J}}{4.20 \text{ V}}\right)\left(\frac{1 \text{ C} - \text{V}}{1 \text{ J}}\right) = 2.3 \times 10^8 \text{ C}$$

This corresponds to the following number of Faradays

$$\text{Faradays} = (2.3 \times 10^8 \text{ C})\left(\frac{1 \text{ F}}{96{,}500 \text{ C}}\right) = 2.4 \times 10^3 \text{ F}$$

This also corresponds to the number of moles of electrons passed through the cell because $1 \text{ F} = 1$ mol electrons. Therefore, from the half-reaction

$$Mg^{2+}(aq) + 2 \text{ e}^- \longrightarrow Mg(s)$$

we can state $2 \text{ F} \simeq 1$ mol Mg.
The grams of Mg(s) produced are thus

$$\text{Mass Mg} = (2.4 \times 10^3 \text{ F})\left(\frac{1 \text{ mol Mg}}{2 \text{ F}}\right)\left(\frac{24.3 \text{ g Mg}}{1 \text{ mol Mg}}\right)$$

$$= 2.9 \times 10^4 \text{ g Mg} = 29 \text{ kg Mg}$$

SELF-TEST QUESTIONS

Key Terms

Having reviewed key terms in Chapter 20, match key terms with phrases and identify statements as true or false. If a statement is false, indicate why it is incorrect.

Match each phrase with the best term:

20.1 The quantity of charge required when one mole of silver(I) ion is reduced.

20.2 The attachment of Mg to certain metals gives this type of protection.

20.3 An electrochemical process causing the unwanted reactions of metals.

20.4   A cell potential when temperature is 25 °C, partial pressures are one atmosphere, and concentrations are one molar.

20.5   Oxidation occurs at this electrochemical site.

20.6   Reduction occurs at this electrochemical site.

20.7   A type of cell that requires energy to operate.

20.8   The potential difference between two electrodes of a voltaic cell.

20.9   A relationship between emf at standard state conditions and emf at non-standard state conditions.

20.10 A type of voltaic cell which is designed to feed reactants continuously.

Terms:

   (**a**) anode
   (**b**) cathode
   (**c**) cathodic protection
   (**d**) corrosion
   (**e**) electrolytic
   (**f**) electromotive force
   (**g**) faraday
   (**h**) fuel cell
   (**i**) Nernst equation
   (**j**) standard electrode potential

True-False Statements:

20.11 The anode in a *voltaic cell* is assigned a positive sign because electrons are produced at the cathode.

20.12 Electrochemistry studies the production of electricity using oxidation-reduction reactions.

20.13 A process involving an oxidation-reduction reaction that requires an outside source of electrical energy is termed *electrolysis*.

20.14 The following *half-reaction* is an example of oxidation

$$Zn^{2+}(aq) + 2\,e^- \longrightarrow Zn(s)$$

20.15 The following is an example of an *oxidation-reduction reaction*

$$CuCl_2(aq) + 4\,NH_3(aq) \longrightarrow [Cu(NH_3)_4]Cl_2(aq)$$

20.16 Oxygen is the *oxidant*, or *oxidizing agent*, in the reaction

$$4\,NH_3(g) + 3\,O_2(g) \longrightarrow 2\,N_2(g) + 6\,H_2O(l)$$

20.17 A *reductant*, or *reducing agent*, is always reduced in a chemical reaction.

20.18 The term *standard cell potential*, or *standard emf*, means that all reagents are at standard state conditions at the given temperature.

20.19 A *cell potential* represents the difference between the potential energy per electrical charge between two electrodes in a voltaic cell.

20.20 The standard potential of a *concentration cell* is zero.

Problems and Short-Answer Questions

20.21 Can the following be balanced to form an oxidation-reduction reaction? Explain.

$$H^+(aq) + Cr_2O_7{}^{2-}(aq) \longrightarrow Cr^{2+}(aq) + H_2O(l)$$

20.22 Construct a diagram of a voltaic cell for the following generalized spontaneous reaction:

$$A^+(aq) + B(s) \longrightarrow A(s) + B^+(aq)$$

Show the direction that the ions and electrons move in the voltaic cell. Do electrons move in the solution? Explain.

20.23 Referring to Table 20.1 in the text, which element can most easily oxidize zinc metal at standard state conditions? Explain.

20.24 Referring to Table 20.1 in the text, which element is the strongest reducing agent? Why?

20.25 A student writes the following statement: A positive value for both E and ΔG indicates that a reaction is spontaneous. Critique this statement.

20.26 The anodic half-reaction in a simple alkaline battery is

$$Zn(s) + 2\,OH^-(aq) \longrightarrow Zn(OH)_2(s) + 2\,e^-$$

If a small quantity of an additional base is accidentally added, would it increase or decrease the emf of the half-reaction? Explain.

20.27 Why does metallic aluminum undergo very slow corrosion rather than rapid corrosion as expected by its standard potential? Why does corrosion of aluminum accelerate in an acidic aqueous environment?

20.28 The electrolysis of an aqueous NaF solution produces oxygen at the anode from the oxidation of water. In contrast, the electrolysis of an aqueous NaCl solution produces chlorine gas at the anode. Given the following half-reactions and their standard potentials, explain why there is a difference:

$2\,H_2O(l) \rightarrow O_2(g) + 4\,H^+(aq) + 4\,e^-$	1.23 V
$2\,F^-(aq) \rightarrow F_2(g) + 2\,e^-$	2.87 V
$2\,Cl^-(aq) \rightarrow Cl_2(g) + 2\,e^-$	1.36 V

20.29 Complete and balance the following redox reactions by the half-reaction method:

   (**a**) $Cr_2O_7^{2-} + I^- \longrightarrow Cr^{3+} + I_2$ (acid solution)

   (**b**) $MnO_4^- + C_2O_4^{2-} \longrightarrow$
   $$Mn^{2+} + CO_2 \text{ (acid solution)}$$

   (**c**) $HSO_3^- + NO_3^- \longrightarrow$
   $$HSO_4^- + NO \text{ (acid solution)}$$

   (**d**) $H_2O_2 + ClO_2 \longrightarrow$
   $$ClO_2^- + O_2 \text{ (base solution)}$$

(e) $Br_2 + AsO_2^- \longrightarrow$

$$Br^- + AsO_4^{3-} \text{ (base solution)}$$

(f) $Bi(OH)_3 + SnO_2^- \longrightarrow$

$$Bi + SnO_3^{2-} \text{ (base solution)}$$

20.30 How many grams of aluminum are formed when its ions are reduced to metallic aluminum in an electrolytic cell and a constant current of 0.20 amp is passed for 30.00 sec?

20.31 Given the following data, explain why metallic copper can be dissolved by nitric acid but not by hydrochloric acid at standard-state conditions:

$$Cu^{2+} + 2\,e^- \longrightarrow Cu \qquad\qquad E_{red}^\circ = 0.34 \text{ V}$$

$$NO_3^- + 4\,H^+ + 3\,e^- \longrightarrow NO + H_2O \quad E_{red}^\circ = 0.96 \text{ V}$$

$$Cl_2 + 2\,e^- \longrightarrow 2\,Cl^- \qquad\qquad E_{red}^\circ = 1.36 \text{ V}$$

20.32 Given the following data:

$$Fe^{2+} + 2\,e^- \longrightarrow Fe \qquad E_{red}^\circ = -0.44 \text{ V}$$

$$Ag^+ + e^- \longrightarrow Ag \qquad E_{red}^\circ = 0.80 \text{ V}$$

Answer the following questions with respect to the reaction

$$Fe^{2+}(aq) + 2\,Ag(s) \longrightarrow Fe(s) + 2\,Ag^+(aq)$$

(a) What is E° for the reaction?
(b) Is the reaction spontaneous at standard-state conditions?
(c) What is the value of E at equilibrium?
(d) What is the value of the equilibrium constant at 25 °C?
(e) If $[Fe^{2+}] = 0.100\,M$ and $[Ag^+] = 0.0100\,M$, what is the magnitude of E at 25 °C?
(f) For the reaction that is spontaneous, what is the maximum amount of work that can be performed?

20.33 How is the corrosion of iron affected by pH? A steel wool pad does not rust as rapidly in a soap solution as in pure water. What does this suggest about the pH of soap?

20.34 Calculate E° for the following unbalanced cell reactions. Indicate if the reactions are spontaneous as written.

(a) $Al^{3+}(aq) + Mg(s) \longrightarrow Al(s) + Mg^{2+}(aq)$

(b) $ClO_4^-(aq) + Mn^{2+}(aq) + H_2O(l) \longrightarrow$

$$MnO_4^-(aq) + H^+(aq) + ClO_3^-(aq)$$

Given data:

$$Al^{3+} + 3\,e^- \longrightarrow Al \qquad E_{red}^\circ = -1.66 \text{ V}$$

$$Mg^{2+} + 2\,e^- \longrightarrow Mg \qquad E_{red}^\circ = -2.37 \text{ V}$$

$$MnO_4^- + 8\,H^+ + 5\,e^- \longrightarrow Mn^{2+} + 4\,H_2O$$
$$E_{red}^\circ = 1.51 \text{ V}$$
$$ClO_4^- + 2\,H^+ + 2\,e^- \longrightarrow ClO_3^- + H_2O$$
$$E_{red}^\circ = 1.19 \text{ V}$$

20.35 Given the following reaction,

$$2\,Al(s) + 3\,Mn^{2+}(aq) \longrightarrow 2\,Al^{3+}(aq) + 3\,Mn(s)$$

what is the concentration of Al^{3+} if $E_{cell}^\circ = 0.48$ V, $E_{cell} = 0.47$ V, and the concentration of Mn^{2+} is 0.49 M at 25 °C?

20.36 A voltaic cell is constructed using two silver electrodes and different concentrations of silver(I) nitrate solutions. The reaction for the concentration cell is described by

$$Ag(s) + Ag^+(1\,M, aq) \longrightarrow Ag(s) + Ag^+(0.001\,M, aq)$$

Is there sufficient information to predict the standard emf and direction of the reaction from the given information? If not, explain what additional information is needed.

20.37 You are given the following chemical equation and initial concentration in parentheses,

$$Cu^{2+}(aq), (2.0\,M) + H_2(g) \longrightarrow Cu(s) + 2\,H^+(aq), (1.0\,M).$$

Is there sufficient information to determine the emf for the reaction? If not, explain what additional information is needed.

Integrative Questions

20.38 Given the following data

$$Fe^{2+}(aq) + 2\,e^- \longrightarrow Fe(s) \qquad E_{red}^\circ = -0.440 \text{ V}$$

$$Cr^{3+}(aq) + e^- \longrightarrow Cr^{2+}(aq) \qquad E_{red}^\circ = -0.407 \text{ V}$$

answer the following questions:

(a) What is the spontaneous reaction when a voltaic cell is constructed using these two half-reactions? What is E° for the reaction?
(b) What is K_{eq} and ΔG°?
(c) If the concentrations of Cr^{3+} and Cr^{2+} are each 0.10 M and the cell potential is 0.100 V, what is the concentration of Fe^{2+}?
(d) What is E for the reaction at equilibrium?

20.39 Corrosion of many metals occurs in aerobic conditions. However, it is possible to have corrosion in anaerobic conditions. Certain microorganisms may increase the rate of corrosion under these conditions; one important type is sulfate-reducing bacteria. These bacteria are responsible for the corrosion of iron pipes in soil when anaerobic conditions exist.

(a) What are aerobic and anaerobic conditions?
(b) One view of the anodic reaction of buried iron in anaerobic conditions is that the metal reacts with HS^- in a basic environment to form ferrous hydroxide and ferrous sulfate. Write the half-reaction for this anodic reaction.
(c) At the cathodic site sulfate ion is reduced to the HS^- ion in a basic medium. Write the half-reaction

for this cathodic reaction.

(d) Write the overall reaction.

(e) What observable condition of corroded iron pipe is an indicator that anaerobic bacteria were involved?

20.40 Given the following standard potentials at 25 °C:

$$Cd^{2+}(aq) + 2\,e^- \longrightarrow Cd(s) \qquad E^{\circ}_{red} = -0.403\ V$$

$$CdS(s) + 2\,e^- \longrightarrow Cd(s) + 2\,S^{2-}(aq) \quad E^{\circ}_{red} = -1.21\ V$$

(a) Calculate K_{sp} for CdS.

(b) If the concentration of Cd^{2+} is increased in a saturated CdS(s) solution and the system is permitted to return to a state of equilibrium, what impact does this have on the value of ΔG°? Explain.

(c) Qualitatively what impact would the change in ion concentration in (b) have on the value of E?

20.41 KMnO₄ is used to titrate ferrous ion to ferric ion in acid solution with manganese (II) ion also produced.

(a) Write the balanced chemical reaction in a net ionic form.

(b) What is the molarity of a ferrous nitrate solution if 100.0 mL of it is titrated with 50.0 mL of 1.0 M KMnO₄?

Multiple-Choice Questions

Use the following data at 25 °C for questions 20.42 to 20.45.

$$Ni^{2+}(aq) + 2\,e^- \longrightarrow Ni(s) \qquad E^{\circ}_{red} = -0.28\ V$$

$$Mg^{2+}(aq) + 2\,e^- \longrightarrow Mg(s) \qquad E^{\circ}_{red} = -2.37\ V$$

20.42 What is the standard cell potential for a voltaic cell constructed using the two half-reactions?

(a) −2.65 V   (d) 2.09 V
(b) −2.09 V   (e) 2.65 V
(c) 0

20.43 What is the cell potential if $[Mg^{2+}] = 0.50\ M$ and $[Ni^{2+}] = 1.0\ M$?

(a) 1.95 V   (d) 2.10 V
(b) 2.00 V   (e) 2.16 V
(c) 2.08 V

20.44 If NiCl₂ is added to the compartment containing the nickel electrode and solution at standard-state conditions how would the cell emf change?

(a) no change   (c) decrease
(b) increase

20.45 What is K for the equilibrium at 25 °C?

(a) 4×10^{-70} (d) 4×10^{70}
(b) 1×10^{-35} (e) 1×10^{170}
(c) 70

20.46 Chromium metal can be electrolytically plated out from an acidic solution containing CrO₃. Assuming that all of the CrO₃ is in a soluble form, how many coulombs are required to cause 3.68 g of Cr to be deposited on the cathodic electrode?

(a) 20,500 coul   (d) 61,500 coul
(b) 41,000 coul   (e) 96,500 coul
(c) 10,250 coul

20.47 Which of the following is the half-cell potential for

$$H_2(2\ atm) \longrightarrow 2\,H^+(0.001\ M) + 2\,e^-$$

(a) 0.19 V   (d) −0.04 V
(b) 0.25 V   (e) −0.10 V
(c) 0.01 V

20.48 Given the following data:

$$Ca^{2+}(aq) + 2\,e^- \longrightarrow Ca(s) \qquad E^{\circ}_{red} = -2.87\ V$$

$$Zn^{2+}(aq) + 2\,e^- \longrightarrow Zn(s) \qquad E^{\circ}_{red} = -0.76\ V$$

$$Co^{2+}(aq) + 2\,e^- \longrightarrow Co(s) \qquad E^{\circ}_{red} = -0.28\ V$$

$$Sn^{2+}(aq) + 2\,e^- \longrightarrow Sn(s) \qquad E^{\circ}_{red} = -0.14\ V$$

$$Pb^{2+}(aq) + 2\,e^- \longrightarrow Pb(s) \qquad E^{\circ}_{red} = -0.13\ V$$

which of the following correctly describes the ease of oxidation of the substances listed under standard state conditions?

(a) $Ca^{2+} > Zn^{2+} > Co^{2+} > Sn^{2+} > Pb^{2+}$
(b) $Pb^{2+} > Sn^{2+} > Co^{2+} > Zn^{2+} > Ca^{2+}$
(c) $Ca > Zn > Co > Sn > Pb$
(d) $Pb > Sn > Co > Zn > Ca$
(e) $Pb > Ca^{2+} > Zn > Sn^{2+} > Pb$

20.49 Given that $K_{eq} = 3.76 \times 10^{14}$ for the reaction

$$2\,Fe^{3+}(aq) + Cu(s) \longrightarrow 2\,Fe^{2+}(aq) + Cu^{2+}(aq)$$

at 25 °C, which of the following corresponds to E° for the reaction?

(a) −0.43 V   (d) −0.22 V
(b) 0.43 V   (e) 0.00 V
(c) 0.22 V

20.50 Which of the following metals could provide cathodic protection for Mn? (Refer to Appendix E in the text for emf data.)

(a) Zn   (d) Cu
(b) Co   (e) Mg
(c) Sn

20.51 Which of the following is formed during the discharging of a lead–acid battery?

(a) Pb(s)   (d) $H^+(aq)$
(b) $PbO_2(s)$ (e) $HSO_4^-(aq)$
(c) $PbSO_4(s)$

20.52 Which species disappears at the anode in Figure 20.1? Cu(II) and Cr(III) ions are in appropriate cell compartments and nitrate ions are in both; all ions are initially at standard state conditions.

(a) Cr   (c) Cu
(b) Pt   (d) Cu^{2+}

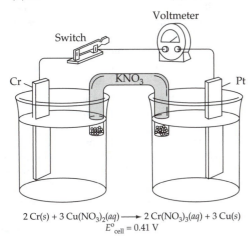

$$2\,Cr(s) + 3\,Cu(NO_3)_2(aq) \longrightarrow 2\,Cr(NO_3)_3(aq) + 3\,Cu(s)$$
$$E^{\circ}_{cell} = 0.41\ V$$

▲ FIGURE 20.1

20.53 Which electrode is the anode in Figure 20.1?
(a) Cr   (b) Pt

20.54 At which electrode does reduction occur in Figure 20.1?
(a) Cr   (b) Pt

20.55 Which cation initially exists in the anodic compartment in Figure 20.1?
(a) Cu^{2+} (b) Cr^{3+}

20.56 In which compartment does the reducing agent exist in Figure 20.1?
(a) anodic compartment
(b) cathodic compartment

20.57 Which statement about the salt bridge in Figure 20.1 is true?
(a) The concentration of the salt is dilute.
(b) The ions in the salt bridge undergo oxidation or reduction.
(c) Its presence creates an irreversible cell.
(d) A gel is often used to store the salt.
(e) A molecular species could also be used in the salt bridge.

20.58 In Figure 20.1, toward which electrode do Cr^{3+} ions migrate?
(a) Cr   (b) Pt

20.59 Into which compartment do NO_3^- ions in the salt bridge in Figure 20.1 migrate?
(a) anodic compartment
(b) cathodic compartment

20.60 In Figure 20.1, what is the polarity of the anode?
(a) +   (b) −

20.61 For the unbalanced redox reaction

$$Ag_2S(s) + NO_3^-(aq) \longrightarrow Ag^+(aq) + S(s) + NO(g)$$

in acidic solution, what is the balanced reduction half-reaction?

(a) $Ag_2S(s) \longrightarrow 2\,Ag^+(s) + S(s) + 2\,e^-$
(b) $S^{2-} \longrightarrow S(s) + 2\,e^-$
(c) $NO_3^-(aq) + 4\,H^+(aq) + 3\,e^- \longrightarrow NO(g) + 2\,H_2O$
(d) $2\,H^+(aq) + 2\,e^- \longrightarrow H_2(g)$
(e) $NO_3^-(aq) + 2\,H^+(aq) + e^- \longrightarrow NO(g) + 2\,OH^-(aq)$

20.62 For the unbalanced redox reaction

$$Cl_2(g) \longrightarrow ClO_4^-(aq) + Cl^-(aq)$$

in basic solution, what is the balanced oxidation half-reaction?

(a) $Cl_2 + 2\,e^- \longrightarrow Cl^-$
(b) $Cl_2 + 4\,O_2 + 2\,e^- \longrightarrow 2\,ClO_4^-$
(c) $Cl_2 + 8\,H_2O \longrightarrow 2\,ClO_4^- + 16\,H^+ + 14\,e^-$
(d) $Cl_2 + 16\,OH^- \longrightarrow 2\,ClO_4^- + 8\,H_2O + 14\,e^-$
(e) $2\,Cl_2 + 8\,OH^- \longrightarrow 2\,ClO_4^- + Cl_2 + 6\,e^-$

20.63 What is the coefficient in front of the underlined substance in the balanced form of the following redox reaction?

$$Pb(NO_3)_2(s) \longrightarrow PbO(s) + \underline{NO_2(g)} + O_2(g)$$

(a) 1   (d) 4
(b) 2   (e) 5
(c) 3

20.64 What is the coefficient in front of the underlined substance in the balanced form of the following redox reaction?

$$\underline{MnO_4^-(aq)} + SO_2(g) + H_2O(l) \longrightarrow Mn^{2+} + SO_4^{2-} + H^+$$

(a) 1   (d) 4
(b) 2   (e) 5
(c) 3

20.65 Which of the following is capable of acting both as an oxidizing and reducing agent?
(a) $KMnO_4$ (d) NaI
(b) Cl_2 (e) $SnCl_2$
(c) H_2O_2

20.66 Which of the following is the oxidizing agent in the reaction

$$SO_3^{2-}(aq) + Br_2(l) + H_2O(l) \longrightarrow SO_4^{2-}(aq) + 2\,Br^-(aq) + 2\,H^+(aq)$$

(a) SO_3^{2-} (d) SO_4^{2-}
(b) Br_2 (e) H^+
(c) H_2O

SELF-TEST SOLUTIONS

20.1 (g). **20.2** (c). **20.3** (d). **20.4** (j). **20.5** (a). **20.6** (b). **20.7** (e). **20.8** (f). **20.9** (i). **20.10** (h). **20.11** False. The anode is assigned a negative sign because it is the source of electrons in the outer circuit. **20.12** True. **20.13** True. **20.14** False. The reaction shows reduction, the gain of electrons. **20.15** False. An oxidation-reduction reaction involves atoms changing oxidation numbers. The reaction shown is a complex ion reaction, with all atoms retaining the same oxidation state. **20.16** True. An oxidizing agent is reduced; oxygen is reduced. **20.17** False. A reducing agent causes another species to gain electrons; thus it is oxidized. **20.18** True. **20.19** True. **20.20** True.

20.21 The given reaction will be an oxidation-reduction reaction if an atom is oxidized and the same atom or another is reduced. Assign oxidation numbers to the atoms to see if these criteria are met: Hydrogen has a +1 oxidation number as an ion and in water, thus there is no change; chromium has an oxidation number of +6 in $Cr_2O_7^{2-}$ and +2 in Cr^{2+}, thus there is a reduction in the oxidation number. Oxygen has a −2 oxidation number in $Cr_2O_7^{2-}$ and in water, thus there is no change. Chromium is reduced but no atom undergoes oxidation; therefore we cannot write a balanced oxidation-reduction reaction using this reaction alone.

20.22 The electrodes are constructed from the metallic forms of A and B. The electrons move only in the outer circuit and the metal ions only move in the solution. The movement of the ions in solution and the electrons in the circuit complete the electrical circuit.

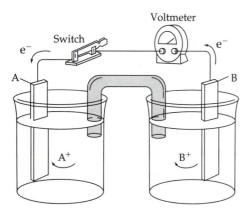

20.23 $F_2(g)$. The ability to oxidize $Zn(s)$ increases with increasing standard reduction potential above Zn^{2+} ion in the table.

20.24 $Li(s)$. A reducing agent is oxidized. Thus, in a standard reduction potential table the half-reaction with the most negative standard reduction potential has a substance on the product side that is the strongest reducing agent.

20.25 The statement is false. The relation between the two is $\Delta G = -nFE$. An increasing positive potential is accompanied by a more negative value of ΔG.

20.26 A basic material in water produces additional hydroxide ion. According to LeChatelier's principle this should drive the reaction to the right, thereby increasing the emf of the half-reaction.

20.27 Aluminum metal is reactive as shown by its standard potential. Normally we should expect corrosion to be rapid as it reacts with oxygen. However, it quickly forms an oxide coating of $Al_2O_3(s)$ and this provides a coating that prevents oxygen from reaching the residual aluminum metal. If an acidic environment is present, $H^+(aq)$ reacts with aluminum oxide and dissolves it; this then exposes the aluminum metal to oxygen and further corrosion occurs.

20.28 The half-reactions shown are for oxidation but the standard potential refers to a reduction half-reaction. The more positive the standard potential for an oxidation half-reaction, the greater the applied energy it takes to cause the reaction to occur in an electrolysis cell. Although water should be the most easily oxidized because it has the smallest standard potential, it has a significant overvoltage and it takes around 1.5 V to cause it to be oxidized. Thus, in aqueous NaCl the chloride ion is preferentially oxidized and chlorine gas forms. The standard potential for fluorine is very large and is greater than the minimum potential required to oxidize water. Thus, in aqueous NaF the fluoride ion is not oxidized but instead water is oxidized and oxygen gas forms.

20.29

(a)
$$3(2\,I^- \longrightarrow I_2 + 2\,e^-)$$
$$\underline{(Cr_2O_7^{2-} + 14\,H^+ + 6\,e^- \longrightarrow 2\,Cr^{3+} + 7\,H_2O)}$$
$$6\,I^- + Cr_2O_7^{2-} + 14\,H^+ \longrightarrow 3\,I_2 + 2\,Cr^{3+} + 7\,H_2O$$

(b)
$$2(MnO_4^- + 8\,H^+ + 5\,e^- \longrightarrow Mn^{2+} + 4\,H_2O)$$
$$\underline{5(C_2O_4^{2-} \longrightarrow 2\,CO_2 + 2\,e^-)}$$
$$2\,MnO_4^- + 5\,C_2O_4^{2-} + 16\,H^+ \longrightarrow 2\,Mn^{2+} + 10\,CO_2 + 8\,H_2O$$

(c)
$$3(H_2O + HSO_3^- \longrightarrow HSO_4^- + 2\,H^+ + 2\,e^-)$$
$$\underline{2(3\,e^- + 4\,H^+ + NO_3^- \longrightarrow NO + 2\,H_2O)}$$
$$3\,HSO_3^- + 2\,NO_3^- + 2\,H^+ \longrightarrow 3\,HSO_4^- + 2\,NO + H_2O$$

(d)
$$1(H_2O_2 \longrightarrow O_2 + 2\,H^+ + 2\,e^-)$$
$$\underline{2(ClO_2 + e^- \longrightarrow ClO_2^-)}$$
$$H_2O_2 + 2\,ClO_2 \longrightarrow O_2 + 2\,ClO_2^- + 2\,H^+$$
$$\underline{+2\,OH^- \qquad\qquad\qquad + 2\,OH^-}$$
$$H_2O_2 + 2\,ClO_2 + 2\,OH^- \longrightarrow O_2 + 2\,ClO_2^- + 2\,H_2O$$

(e)
$$(Br_2 + 2\,e^- \longrightarrow 2\,Br^-)$$
$$\underline{(2\,H_2O + AsO_2^- \longrightarrow AsO_4^{3-} + 4\,H^+ + 2\,e^-)}$$
$$2\,H_2O + Br_2 + AsO_2^- \longrightarrow 2\,Br^- + AsO_4^{3-} + 4\,H^+$$
$$\underline{+ 4\,OH^- \qquad\qquad\qquad\qquad + 4\,OH^-}$$
$$Br_2 + AsO_2^- + 4\,OH^- \longrightarrow 2\,Br^- + AsO_4^{3-} + 2\,H_2O$$

(f)
$$3\,e^- + 3\,H^+ + Bi(OH)_3 \longrightarrow Bi + 3\,H_2O$$
$$\underline{3(H_2O + SnO_2^- \longrightarrow SnO_3^{2-} + 2\,H^+ + e^-)}$$
$$Bi(OH)_3 + 3\,SnO_2^- \longrightarrow Bi + 3\,SnO_3^{2-} + 3\,H^+$$
$$\underline{+ 3\,OH^- \qquad\qquad + 3\,OH^-}$$
$$Bi(OH)_3 + 3\,SnO_2^- + 3\,OH^- \longrightarrow Bi + 3\,SnO_3^{2-} + 3\,H_2O$$

20.30 We write the half-reaction describing the reduction process: $Al^{3+}(aq) + 3 e^- \rightarrow Al(s)$.

One mole of aluminum metal is formed for every mole of aluminum ions reduced. This requires three faradays of charge since three moles of electrons are required for the reduction of aluminum ions. The number of coulombs used is:

$$\text{Coulombs} = (0.20 \text{ amp})(30.00 \text{ s})\left(\frac{1 \text{ C}}{1 \text{ amp-s}}\right) = 6.0 \text{ C}$$

The number of moles of electrons that passes in the cell is:

$$\text{Moles } e^- = (6.0 \text{ C})\left(\frac{1 \text{ mol } e^-}{96,500 \text{ C}}\right) = 6.2 \times 10^{-5} \text{ mol } e^-$$

The number of moles of aluminum formed is therefore:

$$\text{Grams Al} = (6.2 \times 10^{-5} \text{ mol } e^-)$$
$$\left(\frac{1 \text{ mol Al}}{3 \text{ mol } e^-}\right)\left(\frac{27.0 \text{ g Al}}{1 \text{ mol Al}}\right) = 5.6 \times 10^{-4} \text{ g}$$

20.31 The NO_3^- ion can oxidize metallic Cu at standard-state conditions with $E° = 0.96 \text{ V} - 0.34 \text{ V} = 0.62 \text{ V}$. However, the Cl^- ion cannot oxidize Cu because Cl^- cannot be reduced further.

20.32 (a)

$Fe^{2+} + 2e^- \longrightarrow Fe$	$E°_{red} = -0.44 \text{ V}$
$2(Ag \longrightarrow Ag^+ + e^-)$	$E°_{red} = 0.80 \text{ V}$
$Fe^{2+} + 2 Ag \longrightarrow 2 Ag^+ + Fe$	$E° = -1.24 \text{ V}$

(b) The reaction is not spontaneous because $E° < 0$.
(c) $E = 0$ for any reaction at equilibrium.
(d) You can calculate $\log K$ by using the equation $\log K = nE°/0.0592 \text{ V} = (2)(-1.24 \text{ V})/0.0592 \text{ V} = -42.0$.
Taking the antilog of both sides yields $K = 1.0 \times 10^{-42}$.
(e) E for the reaction is calculated using the Nernst equation:

$$E = E° - \left(\frac{0.0592 \text{ V}}{2}\right)\log\frac{[Ag^+]^2}{[Fe^{2+}]}$$
$$= -1.24 \text{ V} - (0.0296 \text{ V})\log\frac{(0.0100)^2}{(0.100)}$$
$$= -1.24 \text{ V} - (-0.887 \text{ V}) = -1.15 \text{ V}$$

(f) The reaction that is spontaneous is

$$2 Ag^+ + Fe \longrightarrow Fe^{2+} + 2 Ag \text{ with } E = 1.24 \text{ V}.$$
$$w_{max} = -nFE = -(2 \text{ mol } e^-)$$
$$\times \left(\frac{96.500 \text{ C}}{1 \text{ mol } e^-}\right)(1.24 \text{ V})\left(\frac{1 \text{ J}}{1 \text{ C-V}}\right)$$
$$= -2.39 \times 10^5 \text{ J}$$

or in kwh,

$$(2.39 \times 10^5 \text{ J})\left(\frac{1 \text{ kwh}}{3.6 \times 10^6 \text{ J}}\right) = 6.6 \times 10^{-2} \text{ kwh}$$

20.33 When iron corrodes it is oxidized: $Fe(s) \rightarrow Fe(aq) + 2 e^-$. The cathodic reaction is $O_2(g) + 2 H^+(aq) + 4 e^- \rightarrow 2 H_2O(l)$. This half-reaction is pH dependent because hydrogen ion is a reactant. As the concentration of hydrogen ions increases (a decreasing pH), the potential of the half-reaction increases and this results in a larger cell potential. This drives the corrosion of iron to iron(II). The pH of soap must be basic since it inhibits the corrosion of iron; it reduces the concentration of hydrogen ions which drives the cell reaction to the left.

20.34 (a) Note the half-reaction for Mg must be reversed:

$(2 Al^{3+} + 3 e^- \longrightarrow Al)$	$E°_{red} = -1.66 \text{ V}$
$3(Mg \longrightarrow Mg^{2+} + 2 e^-)$	$E°_{red} = -2.37 \text{ V}$

$$2 Al^{3+} + 3 Mg \longrightarrow 2 Al + 3 Mg^{2+}$$
$$E°_{cell} = -1.66 \text{ V} - (-2.37 \text{ V}) = 0.71 \text{ V}$$

The reaction is spontaneous as $E°_{cell} > 0$. (a) The MnO_4^- half-reaction must be reversed.

$$5(ClO_4^- + 2 H^+ + 2 e^- \longrightarrow ClO_3^- + H_2O)$$
$$2(Mn^{2+} + 4 H_2O \longrightarrow 2 MnO_4^- + 8 H^+ + 5 e^-)$$
$$2 Mn^{2+} + 3 H_2O + 5 ClO_4^- \longrightarrow 2 MnO_4^- + 6 H^+ + 5 ClO_3^-$$
$$E°_{cell} = 1.19 \text{ V} - (1.51 \text{ V}) = -0.32 \text{ V}$$

The reaction is *not* spontaneous as $E°_{cell}$ is <0.

20.35 The Nernst equation is used to solve for the unknown concentration

$$E = E° - \frac{0.0592}{n}\log Q$$

$n = 6$ because the half-reactions are

$$2 Al \longrightarrow 2 Al^{3+} + 6 e^-$$
$$3 Mn^{2+} + 6 e^- \longrightarrow 3 Mn$$

$$0.47 \text{ V} = 0.48 \text{ V} - \frac{0.0592}{6}\log Q$$

$$Q = \frac{[Al^{2+}]^2}{[Mn^{2+}]^3} = \frac{[Al^{3+}]^2}{(0.49)^3}$$

$$\log Q = \frac{6(-0.01)}{-0.0592} = 1.02$$

$$Q = 10$$

$$\frac{[Al^{2+}]^2}{(0.49)^3} = 10$$

$$[Al^{3+}]^2 = (0.118)(10) = 1.2 \text{ } M^2$$

$$[Al^{2+}] = \sqrt{1.2 \text{ } M^2} = 1.1 \text{ } M$$

20.36 Yes, there is sufficient information. This is a concentration cell and because the reactants and products are the same: $E^\circ_{cell} = E^\circ_{red}(\text{cathode}) - E^\circ_{red}(\text{anode}) = 0. E^\circ_{red}$ is the same for the cathode and anode. The reaction will proceed in a direction until an equilibrium state is reached and the cell emf is zero, that is E. The concentration of the reactant is greater than that of the product; thus, $Q < K$ ($K = 1$ for a concentration cell), and the reaction proceeds to the right until equilibrium is achieved.

20.37 There is insufficient information. The emf for the reaction is calculated using the Nernst equation:

$$E = E^\circ - \frac{0.0592\ V}{n}\log Q$$

E° can be determined from Table 20.1 in the text. However, the form of Q is

$$\frac{[H^+]}{[Cu^{2+}]P_{H_2}}$$

and calculation of its value requires the partial pressure of hydrogen gas which is not given.

20.38 (a) The half-reaction with the more negative E°_{red} is the one reversed:

$$2\,Cr^{3+}(aq) + Fe(s) \longrightarrow 2\,Cr^{2+}(aq) + Fe^{2+}(aq)$$
$$E^\circ = E^\circ_{red}(\text{cathode}) - E^\circ_{red}(\text{anode}) =$$
$$-0.407\ V - (-0.440\ V) = 0.033\ V$$

(b) $\log K = \dfrac{nE^\circ}{0.0592\ V} = \dfrac{2(0.033\ V)}{0.0592\ V} = 1.11$

$$K = 13$$

$$\Delta G^\circ = -nFE^\circ = -(2\ \text{mol})$$

$$\times \left(96{,}500\frac{J}{V\text{-mol}}\right)(0.033\ V) = -6400\ J$$

$$\Delta G^\circ = -6.4\ kJ$$

(c) The concentration of Fe^{2+} can be calculated using the Nernst equation:

$$E = E^\circ - \frac{0.0592\ V}{n}\log Q$$

$$0.100\ V = 0.033\ V - \frac{0.0592\ V}{2}\log\frac{(C^2_{Cr^{2+}})(C_{Fe^{2+}})}{(C_{Cr^{3+}})^2}$$

$$0.100\ V = 0.033\ V - \frac{0.0592\ V}{2}\log\frac{(0.10)^2 C_{Fe^{2+}}}{(0.10)^2}$$

$$= 0.033\ V - \frac{0.0592\ V}{2}\log C_{Fe^{2+}}$$

Solving for the concentration of Fe^{2+} gives $C_{Fe^{2+}} = 5.4 \times 10^{-3}\ M$.

(d) At equilibrium, there is no further net change in concentrations; therefore $E = 0\ V$

20.39 (a) Aerobic conditions are those with oxygen present and anaerobic conditions are those without oxygen.

(b) $4\,Fe(s) + HS^-(aq) + 7\,OH^-(aq) \longrightarrow$

$$3\,Fe(OH)_2(s) + FeS(s) + H_2O(l) + 8\,e^-$$

(c) $SO_4^{2-}(aq) + 5\,H_2O(l) + 8\,e^- \longrightarrow$

$$HS^-(aq) + 9\,OH^-(aq)$$

(d) $4\,Fe(s) + SO_4^{2-}(aq) + 4\,H_2O(l) \longrightarrow$

$$3\,Fe(OH)_2(s) + FeS(s) + 2\,OH_2(aq)$$

(e) The formation of FeS, which is black, gives rise to black regions on the pipe surface. If these black regions are removed, the anodic site will show a pit and metallic iron.

20.40 (a) The two half-reactions can be arranged into an oxidation and reduction half-reaction which when added together form the solubility equilibrium:

$$Cd(s) \longrightarrow Cd^{2+}(aq) + 2\,e^- \qquad E^\circ = -0.403\ V$$
$$\underline{CdS(s) + 2\,e^- \longrightarrow Cd(s) + 2\,S^{2-}(aq)}\ E^\circ = -1.21\ V$$

$$CdS(s) \rightleftharpoons Cd^{2+}(aq) + S^{2-}(aq)$$
$$E^\circ = -1.21\ V - (-0.403\ V) = -0.81\ V$$

K_{sp} is the equilibrium constant and it is related to the standard potential as follows:

$$\log K_{sp} = \frac{nE^\circ}{RT} = \frac{nE^\circ}{0.0592\ V} = \frac{2(-0.81\ V)}{0.0592\ V} = -27;$$

taking the antilog gives $K_{sp} = 1 \times 10^{-27}$

(b) $\Delta G^\circ = -nFE^\circ$. The value of E° is a constant at the given temperature and at standard-state conditions. The addition of additional cadmium ion does not change its value; thus the value of ΔG° also does not change.

(c) When the system is at equilibrium the value of E (not E°) is zero. The addition of additional cadmium ion would drive the reaction to the left, reducing the concentration of sulfide ion and increasing the amount of cadmium sulfide; this reduces the value of E (in this case it would temporarily make it more negative). However, when the system is restored to equilibrium the value of E would return to zero.

20.41 (a) The balanced chemical reaction is

$$5\,Fe^{2+} + MnO_4^- + 8\,H^+ \longrightarrow Mn^{2+} + 4\,H_2O + 5\,Fe^{3+}$$

(b) The number of moles of MnO_4^- used in the titration is calculated using:

$$M \times V = 1.00\ \frac{mol}{L} \times 0.0500\ L = 0.0500\ mol$$

Use the stoichiometry of the chemical reaction to calculate the number of moles of ferrous ion that reacted:

$$(0.0500\ \text{mol}\ MnO_4^-)\left(\frac{5\ \text{mol}\ Fe^{2+}}{1\ \text{mol}\ Mn^{2+}}\right)$$

$$= 0.2500\ \text{mol}\ Fe^{2+}$$

Finally calculate the molarity of ferrous ion as follows:

$$M = \frac{mol}{liters} = \frac{0.2500\ \text{mol}\ Fe^{2+}}{0.100\ L} = 2.50\frac{mol}{liters}$$

20.42 (**d**) The half-reaction with the most negative standard potential becomes the oxidation half-reaction. $E° = -0.28$ V $- (-0.237$ V$) = 2.09$ V. The overall reaction is $Mg(s) + Ni^{2+}(aq) \rightarrow Mg^{2+}(aq) + Ni(s)$.

20.43 (**d**) The Nernst equation is used to solve for the cell potential. Note that two moles of electrons are transferred in the reaction.

$$E = 2.09 \text{ V} - \frac{0.5092 \text{ V}}{2} \log \frac{0.50}{1.0} = 2.10 \text{ V}$$

20.44 (**b**) The addition of $Ni^{2+}(aq)$ to the nickel cathodic compartment will increase the cell emf. Application of LeChatelier's principle tells us that the system will respond by removing some of the added nickel ion, which drives the reaction to the right and increasing cell emf.

20.45 (**d**) $\log K = \dfrac{nE°}{0.0592} = \dfrac{2(2.09)}{0.0592} = 70.6$

$$K = 10^{70.6} = 4 \times 10^{70}$$

20.46 (**b**) $CrO_3(aq) + 6 H^+(aq) + 6 e^- \longrightarrow$
$$Cr(s) + 3 H_2O(l)$$

$$\text{Coulombs} = (3.68 \text{ g Cr})\left(\frac{1 \text{ mol Cr}}{52.0 \text{ g Cr}}\right)\left(\frac{6 F}{1 \text{ mol Cr}}\right)$$
$$\times \left(\frac{96,500 \text{ C}}{1 F}\right) = 41,000 \text{ coul}$$

20.47 (**a**) $E = 0.00$ V $- (0.0592$ V$/2) \log([H^+]^2/P_{H_2})$
$$= 0.19 \text{ V}.$$

20.48 (**c**)

20.49 (**b**) $E° = (0.0592$ V$) \log K/n = (0.0592$ V$)$
$$\log(3.76 \times 10^{14}) = 0.43 \text{ V}.$$

20.50 (**e**) Of those metals listed, only Mg has a more positive oxidation emf than the oxidation emf of Mn (1.81 V), and thus it is preferentially oxidized.

20.51 (**c**). **20.52** (**a**). **20.53** (**a**). **20.54** (**b**). **20.55** (**b**). **20.56** (**a**). **20.57** (**d**).

20.58 (**b**). **20.59** (**a**). **20.60** (**b**). **20.61** (**c**). **20.62** (**d**) **20.63** (**d**) **20.64** (**b**) **20.65** (**c**).

20.66 (**b**). An oxidizing agent causes an element to be oxidized; thus an oxidizing agent is reduced: $2 e^- + Br_2 \rightarrow 2 Br^-$.

Chapter

21

Nuclear Chemistry

OVERVIEW OF THE CHAPTER

Review: Atoms (2.1, 2.2); atomic number (2.4); lanthanides and actinides (2.5); mass number (2.2); subatomic particles (2.1, 2.2).

Learning Goals: You should be able to:

1. Write the nuclear symbols for protons, neutrons, electrons, alpha particles, and positrons.
2. Complete and balance nuclear equations, having been given all but one of the particles involved.
3. Write the shorthand notation for a nuclear reaction or given the shorthand notation, write the nuclear reaction.
4. Determine the effect of different types of decay on the proton-neutron ratio and predict the type of decay that a nucleus will undergo based on its composition relative to the belt of stability.
5. Calculate the binding energies of nuclei, having been given their masses and the masses of protons, electrons, and neutrons.

Review: Early radioactivity experiments (2.2); energy changes (5.1, 5.2); first-order reactions and concept of half-life (14.3); fuel values (5.8).

Learning Goals: You should be able to:

1. Use the half-life of a substance to predict the amount of radioisotope present after a given period of time.
2. Calculate half-life, age of an object, or the remaining amount of radioisotope, having been given any two of these pieces of information.
3. Use Einstein's relation, $\Delta E = c^2 \Delta m$, to calculate the energy change or the mass change of a reaction, having been given one of these quantities.
4. Explain how radioisotopes can be used in dating objects and as radiotracers.
5. Explain how radioactivity is detected, including a simplified description of the basic design of a Geiger counter.
6. Explain the roles played by the chemical behavior of an isotope and its mode of radioactivity in determining its ability to damage biological systems.

7. Define the units used to describe the level of radioactivity (*curie*) and to measure the effects of radiation on biological systems (*rem* and *rad*).
8. Describe various sources of radiation to which the general population is exposed, and indicate the relative contributions of each.

Review: Energy use (5.8).

Learning Goals: You should be able to:

1. Define fission and fusion, and state which types of nuclei produce energy when undergoing these processes.
2. Describe the design of a nuclear power plant, including an explanation of the role of fuel elements, control rods, moderator, and cooling fluid.

21.7, 21.8 NUCLEAR POWER: FISSION AND FUSION

TOPIC SUMMARIES AND EXERCISES

A change in the structure of a nucleus may occur spontaneously, or it may be brought about artificially.

- The spontaneous emission of particles and electromagnetic radiation from unstable nuclei is referred to as **radioactivity**. An unstable nucleus that spontaneously transmutes to the nucleus of another element is said to undergo a process of **radioactive decay.**
- The types of radioactive decay discussed in Section 21.1 of the text are summarized in Table 21.1.

RADIOACTIVITY, NUCLEAR STABILITY, AND NUCLEAR TRANSMUTATIONS

TABLE 21.1 Types of Radioactive Decay

Type	Example	Comments
Alpha-particle emission ($^4_2$He Nuclei)	$^{280}_{90}\text{Th} \longrightarrow {}^{226}_{88}\text{Ra} + {}^4_2\text{He}$	New isotope has a mass number that is four less and an atomic number that is two less than decaying nucleus
Beta-particle emission (high-speed electrons, $^0_{-1}$e)	$^{231}_{90}\text{Th} \longrightarrow {}^{231}_{91}\text{Pa} + {}^0_{-1}\text{e}$	New isotope has the same mass number as the decaying nucleus, but an atomic number one greater
Gamma-ray emission (high-energy electromagnetic radiation)	Gamma emission often accompanies alpha and beta emission.	Involves energy released during organization of nucleus Symbol for gamma ray, $^0_0\gamma$, is usually not shown in a nuclear equation
Positron emission (a particle of same mass as electron, but with opposite charge, $^0_1$e)	$^{11}_{6}\text{C} \longrightarrow {}^{11}_{5}\text{B} + {}^0_1\text{e}$	New isotope has the same mass number as the decaying nucleus, but an atomic number that is one less Emission of positron particles involves conversion of proton to neutron: $^1_1\text{p} \rightarrow {}^1_0\text{n} + {}^0_1\text{e}$
Electron capture	$^{81}_{37}\text{Rb} + {}^0_{-1}\text{e} \longrightarrow {}^{81}_{36}\text{Kr}$	New isotope has the same mass number as the decaying nucleus, but with an atomic number that is one less Decaying nucleus captures electron from its surrounding electron cloud: $^1_1\text{p} + {}^0_{-1}\text{e} \longrightarrow {}^1_0\text{n}$

Many radioactive nuclei have been prepared by bombarding existing nuclei with subatomic particles such as neutrons, electrons, protons, alpha particles, and deuterons ($^2_1$H) or with high-energy radiation, such as gamma rays.

• These types of reactions are referred to as induced nuclear reactions or **nuclear transmutations**. The newly formed nucleus may undergo radioactive decay and initiate a nuclear-disintegration series.
• For electrically charged subatomic particles to combine with a target nucleus so as to effect a nuclear transmutation, they first must be accelerated to high speeds using a particle accelerator, such as a cyclotron or synchrotron. Acceleration of these particles provides them with sufficient kinetic energy to overcome electrostatic repulsions between them and the target nucleus.

Neutrons and particles such as $^{12}_6$C, $^{14}_7$N, and $^{10}_5$B are used to induce the formation of some **transuranium elements**, the elements occurring after uranium in the periodic table. Radioactive transuranium elements have short half-lives and consequently are difficult to isolate and identify.

An induced nuclear reaction, such as

$$^6_3\text{Li} + ^1_0\text{n} \longrightarrow ^3_1\text{H} + ^4_2\text{He}$$

is often abbreviated using a shorthand notation system. The shorthand notation for the given reaction is $^6_3\text{Li}(\text{n}, \alpha)^3_1\text{H}$.

• In the shorthand notation the target nucleus is the first item listed, followed by the symbols for the bombarding particle (or radiation) and the ejected particle (or radiation), both enclosed in parentheses, with the product nucleus listed last:
Target nucleus (bombarding particle, ejected particle) product nucleus.

Look at Figure 21.3 in your text and note that as the number of protons increases, the number of neutrons necessary to create a stable nucleus rapidly increases. Note that all nuclei with 84 or more protons are radioactive.

• We use the belt of stability in Figure 21.3 to determine whether a nucleus will naturally decay. A nucleus with a neutron/proton ratio outside the belt of stability is unstable and radioactive.
• The stability of a nucleus depends on the forces holding nuclear particles together. For a heavy nucleus (with an atomic number greater than 20) to be stable it must have more neutrons than protons. The presence of excess neutrons in a stable nucleus counteracts the increased proton–proton coulombic repulsions that normally would cause the nucleus to disintegrate.

Exercise 1   Using an isotopic symbol to determine number of neutrons, protons, and electrons

Indicate the number of protons, neutrons, electrons, and nucleons possessed by each of the following: (a) $^{14}_8$O; (b) ^{238}U; (c) bismuth-214.

SOLUTION: *Analyze and Plan:* We are asked to state the number of protons, neutrons, electrons, and nucleons possessed by three isotopes. The sum of the number of protons and neutrons equals the number of nucleons. The superscript before each isotopic symbol is the mass number of the element (sum of protons and neutrons). The atomic number, the number of protons, is the subscript. A "-Number" after the name of an element tells you the mass number of the element.

Solve: (a) $^{14}_{8}O$ has 8 protons, 8 electrons, 6 neutrons (the mass number minus the atomic number), and 14 nucleons. (b) ^{238}U has 238 nucleons. Its atomic number, 92, can be determined from the periodic table. Therefore ^{238}U has 92 protons, 92 electrons, and 146 neutrons. (c) In bismuth-214 the number 214 is the mass number of bismuth. Its atomic number, determined from the periodic table, is 83. Thus, bismuth has 83 protons, 83 electrons, 131 neutrons, and 214 nucleons.

EXERCISE 2   Completing and balancing nuclear equations

Complete and balance the following nuclear equations by supplying the missing particle or energy ray. Identify the type of radioactive decay for each reaction.

(a) $^{224}_{88}Ra \longrightarrow {}^{220}_{86}Rn +$ ____

(b) $^{226}_{88}Ra \longrightarrow {}^{222}_{86}Rn + {}^{4}_{2}He +$ ____

(c) $^{232}_{90}Th \longrightarrow {}^{232}_{91}Pa +$ ____

(d) $^{13}_{7}N \longrightarrow {}^{13}_{6}C +$ ____

SOLUTION: *Analyze and Plan*: We are asked to complete and balance four nuclear equations by supplying missing particles or types of energy emitted and to identify the type of radioactivity decay. We can complete the nuclear reactions by analyzing the change in mass numbers and atomic numbers. The sum of atomic numbers and the sum of mass numbers for the reactants must equal the same sums for the products. The missing particle must have an atomic number and a mass number fulfilling this requirement.

Solve: From the nuclear equation provided, we can write the following:

$$\text{Mass numbers:}\quad 224 = 220 + x \qquad x = 224 - 220 = 4$$

$$\text{Atomic numbers:}\quad 88 = 86 + y \qquad y = 88 - 86 = 2$$

The missing particle has a mass number of 4 and an atomic number of 2. A helium nucleus (alpha particle) possesses these properties. The balanced equation is thus

$$^{224}_{88}Ra \longrightarrow {}^{220}_{86}Rn + {}^{4}_{2}He$$

Alpha emission involves forming an isotope that has a mass number four less and an atomic number two less than the isotope undergoing radioactive decay. (b) Proceeding in a similar fashion as in (a),

$$\text{Mass numbers:}\quad 226 = 222 + 4 + x \qquad x = 0$$

$$\text{Atomic numbers:}\quad 88 = 86 + 2 + y \qquad y = 0$$

The nuclear equation is already balanced; therefore if there is a missing species it probably is high-energy gamma radiation.

$$^{226}_{86}Ra \longrightarrow {}^{222}_{86}Rn + {}^{4}_{2}He + {}^{0}_{0}\gamma$$

Gamma emission often accompanies alpha or beta emission. (c) Proceeding in a similar fashion as in (a),

$$\text{Mass numbers:}\quad 232 = 232 + x \qquad x = 0$$

$$\text{Atomic numbers:}\quad 90 = 91 + y \qquad y = 90 - 91 = -1$$

An electron, $^{0}_{-1}e$, has a mass number of 0 and atomic number of -1. Therefore the balanced nuclear reaction is

$$^{232}_{90}Th \longrightarrow {}^{232}_{91}Pa + {}^{0}_{-1}e$$

Beta emission involves forming an isotope that has the same mass number and an atomic number one greater than those of the isotope undergoing radioactive decay. (**d**) Proceeding in a similar fashion as in (**a**),

$$\text{Mass numbers:} \quad 13 = 13 + x \qquad x = 0$$
$$\text{Atomic numbers:} \; 7 = 6 + y \qquad y = 7 - 6 = 1$$

A positron, $_{1}^{0}e$, has a mass number of 0 and an atomic number of 1. Therefore the balanced nuclear equation is

$$_{7}^{13}N \longrightarrow \, _{6}^{13}C + \, _{1}^{0}e$$

Positron emission involves forming an isotope that has the same mass number and an atomic number one less than those of the isotope undergoing radioactive decay.

EXERCISE 3   Determining a missing particle in a nuclear equation

The synthesis of transuranium element 104 was followed by a controversy. Both a University of California group and a Russian group claimed to have been the first to synthesize the element, and each suggested a name for the new element. Supply the missing particle in each of the following two reactions, which describe the reported syntheses:

$$(\mathbf{a}) \; _{94}^{242}Pu + ? \longrightarrow \, _{104}^{260}X + 4\,_{0}^{1}n$$

$$(\mathbf{b}) \; _{98}^{249}Cf + ? \longrightarrow \, _{104}^{257}X + 4\,_{0}^{1}n$$

SOLUTION: *Analyze and Plan*: This exercise is similar to Exercise 2. We are asked to find a missing reactant instead of a product. We will use the same procedure as in Exercise 2.

Solve: (**a**) The sums of mass numbers and atomic numbers on the right are 264 and 104, respectively. The missing particle must have a mass number of 22 and an atomic number of 10 so that the sums on the left are 264 and 104, respectively. By referring to a periodic table, we find that the nuclide with an atomic number of 10 (that is, containing 10 protons) is neon. Consequently it must be the missing particle:

$$_{94}^{242}Pu + \, _{10}^{22}Ne \longrightarrow \, _{104}^{260}X + 4\,_{0}^{1}n$$

(**b**) Using the same approach as in (**a**), we can obtain the following nuclear equation:

$$_{98}^{249}Cf + \, _{6}^{12}C \longrightarrow \, _{104}^{257}X + 4\,_{0}^{1}n$$

Note: Element X is now called rutherfordium, Rf.

EXERCISE 4   Writing a nuclear equation given its shorthand notation or vice versa

Write a balanced nuclear equation for each of the following processes, or write the shorthand notation for the given reaction:

(**a**) $_{5}^{10}B(\alpha, p)_{6}^{13}C$

(**b**) $_{47}^{107}Ag(n, 2n)_{47}^{106}Ag$

(**c**) $_{11}^{23}Na + \, _{0}^{1}n \longrightarrow \, _{11}^{24}Na + \gamma$

(**d**) $_{92}^{238}U + \, _{6}^{12}C \longrightarrow \, _{98}^{246}Cf + 4\,_{0}^{1}n$

SOLUTION: *Analyze and Plan*: We are asked to either write a balanced nuclear reaction given its shorthand notation or *vice versa*. In the summary section the structure of the shorthand notation is described. We will use this structure to answer each question.

Solve: (**a**) The symbol α is the abbreviation for an alpha particle, and p is the abbreviation for a proton. The nuclear reaction is

$$^{10}_{5}B + ^{4}_{2}He \longrightarrow ^{13}_{6}C + ^{1}_{1}p$$

(**b**) The letter n is the abbreviation for a neutron. There are two neutrons ejected as a product. The nuclear equation is

$$^{107}_{47}Ag + ^{1}_{0}n \longrightarrow ^{106}_{47}Ag + 2^{1}_{0}n$$

(**c**) The notation is $^{23}_{11}Na$ $(n, \gamma)^{24}_{11}Na$, where γ is the symbol for gamma radiation.

(**d**) The notation is $^{238}_{92}U$ $(^{12}_{6}C, 4n)^{246}_{98}Cf$.

EXERCISE 5 Determining which nuclides are likely to be radioactive

Using Figure 21.3 in the text, predict which of the following three nuclides are likely to be radioactive: (**a**) $^{64}_{30}Zn$; (**b**) $^{90}_{35}Br$; (**c**) $^{103}_{47}Ag$. Briefly justify your choice.

SOLUTION: *Analyze and Plan*: We are asked to use Figure 21.3 in the text to predict whether three nuclides are likely to be radioactive. Figure 21.3 in the text shows the belt of nuclear stability for nuclides. We first calculate the neutron-to-proton ratio for each nuclide. Those nuclides whose neutron-to-proton ratios are not within the stability region are likely to be radioactive.

Solve: The neutron/proton ratios are: (**a**) $34/30 = 1.1$ for $^{64}_{30}Zn$; (**b**) $55/35 = 1.6$ for $^{90}_{35}Br$; and (**c**) $56/47 = 1.2$ for $^{103}_{47}Ag$. Inspection of Figure 21.3 shows that the neutron/proton ratio for $^{64}_{30}Zn$ is in the region of stability and the neutron/proton ratios for $^{90}_{35}Br$ and $^{103}_{47}Ag$ lie above and below the region of stability, respectively. $^{90}_{35}Br$ and $^{103}_{47}Ag$ are therefore likely to be radioactive because their neutron/proton ratios do not lie in the belt of stability.

RADIOACTIVITY: RATES OF DECAY, DETECTION, ENERGY CHANGES, AND BIOLOGICAL EFFECTS

$^{235}_{92}U$ undergoes radioactive decay very slowly in nature and thus can be isolated and stored for long periods of time. By contrast, $^{132}_{53}I$ undergoes substantial decay over a period of several hours.

- The rate of radioactive decay of a nucleus is characterized by its half-life.
- Since radioactive decay is a first-order process, we can use the concept of half-life developed earlier in Section 14.3 of the text to characterize the stability of a radioisotope. *The half-life of a decaying substance which obeys a first-order rate law is independent of the amount of substance reacting; it depends only on the rate constant:*

$$t_{1/2} = \frac{0.693}{k} \qquad [21.1]$$

where k is the rate constant and $t_{1/2}$ is the half-life.

- An important relationship that relates the time, t, required for an initial number, N_0, of nuclei of a particular isotope to decay to N_t number of particles is:

$$\ln\frac{N_t}{N_0} = -kt \qquad [21.2]$$

Remember: $N_t < N_0$

Carbon-14 is radioactive and is used to determine the age of carbon-based materials. Carbon-14 dating is based on the assumption that the ratio of $^{14}_{6}C$ to $^{12}_{6}C$ in the atmosphere has remained constant for at least 50,000 years. We assume that the ratio of $^{14}_{6}C$ to $^{12}_{6}C$ in an air-breathing organism or plant at its death is the same as if it were to die today. After its death $^{14}_{6}C$ from the atmosphere is no longer incorporated and the quantity of $^{14}_{6}C$ begins to decrease. The half-life of $^{14}_{6}C$ is 5715 years; thus the ratio of $^{14}_{6}C$ to $^{12}_{6}C$ will decrease by 50% in 5715 years.

Nuclear binding energy is the energy required to break apart a nucleus into its component nucleons. To calculate the binding energy for a nucleus, use the following procedure:

1. Calculate the mass change, Δm, when a nucleus is separated into its constituent protons and neutrons. Note: **Mass defect** is the mass loss that occurs when a nucleus is formed from its constituent protons and neutrons.
2. Use Einstein's relationship $\Delta E = c^2 \Delta m$ to calculate binding energy, which is a positive quantity.

Energy changes for nuclear reactions are calculated in a manner similar to above.

- Δm = nuclear mass of products − nuclear mass of reactants.
- Unless the total number of electrons for the products and reactants differs, you can use isotopic masses instead of nuclear masses. If they are not equal, you must subtract the mass of the electrons from the isotopic masses.

In Section 21.5 of the text, three methods for detecting emissions from radioactive substances are described: photographic plates, Geiger counters, and scintillation counters. You should note how each detects radioactive emissions.

The operation of a Geiger counter to detect radioactivity is based on the ability of radiation to ionize matter. Ionization of cellular matter is of special concern to scientists.

- Biological damage to cells occurs because of the ability of radiation to ionize and fragment molecules. Such processes in human tissue can lead to various forms of cancer: Leukemia is the most common form observed.
- *The relative ability of radiation to penetrate human tissue is gamma > beta > alpha.* Although alpha rays do not deeply penetrate into human tissue, they are excellent ionizing agents. Therefore it is important to prevent ingestion of alpha emitters. Once they are within the body extensive biological damage occurs.
- Ionizing radiation passing through living tissue reacts with water to form a neutral OH molecule. This molecule possesses an unpaired electron. It is called a free radical, a substance with one or more unpaired electrons. Free radicals readily attack cells and tissues.

As indicated by the previous discussion of the characteristics of alpha rays, biological effects of radiation depend on several factors.

- A measure of the energy of radiation received by tissue is a **rad**, the amount of radiation that deposits 1×10^{-2} J of energy per kilogram of tissue. The gray (Gy) is the amount of radiation that deposits 1 J of energy per kilogram of tissue.

- The relative biological damage caused by a form of radiation is given by its **RBE factor**. If the rad value for a particular form of radiation is multiplied by its RBE factor, we obtain a radiation dose measurement which has the unit of rem:

$$\text{Number of rems} = (\text{rads})(\text{RBE})$$

- The SI unit for nuclear radioactivity is the **becquerel**, defined as one nuclear disintegration per second. An older, but more commonly used measure of nuclear activity is the **curie** (Ci), the number of disintegrations that 1 g of radium undergoes in 1 sec; this number is 3.7×10^{10} disintegrations/s.

EXERCISE 6 Deriving a relationship between t and $t_{1/2}$

Use Equations [21.1] and [21.2] in the Student's Guide to show that the relationship between t and $t_{1/2}$ is

$$t = 1.44\, t_{1/2} \ln N_0/N_t$$

where t is the time after a radioactive decay process begins. *This equation is very useful when you are solving problems that involve radioactive dating and half-lives.*

SOLUTION: *Analyze and Plan*: We are asked to use equations 21.1 and 21.2 given in the summary to develop a new relationship between t and $t_{1/2}$. If we look at the desired equation we see that k is not in it. Thus, we want to eliminate k from equation 21.2. We will start with equation 21.1 and rearrange it so that k is on the left hand side of the equation. Then we will substitute it into equation 21.2 and rearrange this equation to obtain the desired one.

Solve: Use the expression for k in Equation [21.1] in the Student's Guide

$$k = \frac{0.693}{t_{1/2}}$$

to replace k in Equation [21.2]

$$\ln(N_t/N_0) = -kt = -(0.693/t_{1/2})t$$

Rearrange the previous equation to solve for t:

$$t = -(t_{1/2}/0.693)\ln(N_t/N_0) = -1.44 t_{1/2}\ln(N_t/N_0)$$

Use the relationship $-\ln x = \ln(1/x)$ to eliminate the minus sign in the above expression:

$$t = 1.44 t_{1/2}\ln(N_0/N_t) \qquad\qquad [21.3]$$

EXERCISE 7 Determining the age of a carbon-based sample using $^{14}_{6}\text{C}$ dating

What is the approximate age of Crater Lake in Oregon if the ratio of $^{14}_{6}\text{C}$ to $^{12}_{6}\text{C}$ in a charcoal sample formed at the same time as the lake was formed is 48% of the same ratio in a freshly cut tree sample in the year 2008?

SOLUTION: *Analyze and Plan*: We are given information that today the ratio of $^{14}_{6}\text{C}$ to $^{12}_{6}\text{C}$ in the charcoal sample is 48% of the original value at the formation of the charcoal and we are asked to determine the approximate age of Crater Lake. We will use Equation [21.3] to answer the question.

Solve: Equation [21.3] is:

$$t = 1.44 t_{1/2}\ln(N_0/N_t)$$

where t is the time since the original sample started undergoing $^{14}_{6}C$ radioactive decay, $t_{1/2}$ is the half-life of $^{14}_{6}C$, N_0 is the amount of $^{14}_{6}C$ in the sample at its death, and N_t is the amount of $^{14}_{6}C$ in the sample today (t years have passed). The ratio N_t/N_0 is equivalent to the percentage of $^{14}_{6}C$ to $^{12}_{6}C$ remaining in the sample because the amount of carbon-12 does not change. Note that this ratio is the inverse of N_0/N_t. Thus the value of N_0/N_t is $1/0.48$ or 0.73. We can now solve for the age of the charcoal sample:

$$t = 1.44(5715 \text{ yr})(\ln(0.73))$$
$$t = 6.0 \times 10^3 \text{ yr}$$

It took approximately 6.0×10^3 years for the amount of $^{14}_{6}C$ in the original charcoal sample to decay to 48% of this original value. Therefore the age of Crater Lake is the difference of 2008 A.D. -6000 yr or approximately 4000 B.C.

EXERCISE 8   Using the half-life of carbon-14 to determine the original amount of a carbon-based substance

What is the original mass of ^{14}C in a sample if 10.00 mg of it remains after 20,000 years? The half-life of ^{14}C is 5715 years.

SOLUTION: *Analyze and Plan*: We are asked to determine the original mass of a carbon-14 sample if 10.00 mg of it remains after 20,000 years. We know: t, 20,000 yr; $t_{1/2}$, 5715 yr; and N_t, 10.00 mg. We can substitute this data into Equation [21.3].

Solve: The initial mass of ^{14}C can be calculated using the equation

$$t = 1.44 \, t_{1/2} \ln \frac{N_0}{N_t}$$

Substituting the given data into the equation yields

$$20,000 \text{ yr} = (1.44)(5715 \text{ yr}) \ln \frac{N_0}{10.00 \text{ mg}}$$

$$\ln \frac{N_0}{10.00 \text{ mg}} = \frac{20,000 \text{ yr}}{(1.44)(5715 \text{ yr})} = 2.43$$

Taking the antiln of both sides of the equation yields

$$\frac{N_0}{10.00 \text{ mg}} = 11.4$$

Solving for N_0 gives

$$N_0 = (11.4)(10.00 \text{ mg}) = 114 \text{ mg}$$

The original mass of ^{14}C in the sample was 114 mg.

EXERCISE 9   Determining the relationship between one amu and joule

Calculate the energy of 1 amu in joules using the relation $\Delta E = c^2 \, \Delta m$. 1 J is equivalent to 1 kg-m^2/sec^2 and 1 g $= 6.022 \times 10^{23}$ amu.

SOLUTION: *Analyze and Plan*: We are asked to calculate the energy of one amu in joules using Einstein's equation. This equation requires mass in kilograms. Thus, we will have to convert amu to grams to kilograms using the equivalence given in the question. c is the speed of light, which is a constant given in tables as $3.00 \times 10^8 \frac{\text{m}}{\text{s}}$.

Solve: Substitute the appropriate values into Einstein's equation:

$$\Delta E = c^2 \, \Delta m = \left(3.00 \times 10^8 \frac{m^2}{s}\right)(1 \text{ amu})\left(\frac{1 \text{ g}}{6.022 \times 10^{23} \text{ amu}}\right)\left(\frac{1 \text{ kg}}{10^3 \text{ g}}\right)$$

$$\Delta E = 1.49 \times 10^{-10} \text{ kg-m}^2/\text{s}^2 = 1.49 \times 10^{-10} \text{ J}$$

The equivalence 1 amu $= 1.49 \times 10^{-10}$ J is a useful equivalence for solving binding-energy problems. Sometimes binding energies are expressed in units of MeV (million electron volts): 1 amu $= 931$ MeV.

EXERCISE 10 Calculating the amount of energy change in a nuclear reaction

For the nuclear reaction

$$^{14}_{7}\text{N} + ^{4}_{2}\text{He} \longrightarrow ^{17}_{8}\text{O} + ^{1}_{1}\text{H}$$

calculate the energy in joules associated with the reaction of one nuclide of $^{14}_{7}\text{N}$ with one of $^{4}_{2}\text{He}$, given that the isotopic masses (amu) are: $^{14}_{7}\text{N}$, 14.00307; $^{4}_{2}\text{He}$, 4.00260; $^{17}_{8}\text{O}$, 16.9991; and $^{1}_{1}\text{H}$, 1.007825.

SOLUTION: *Analyze*: We are given a nuclear equation and asked to calculate the energy change in joules under the stated conditions. Note the problem states that there is one nuclide of each reactant, not one atom. Isotopic masses in amu are given, but we are not given masses of the nuclides. We need to consider how to do the calculation using isotopic masses instead of masses of nuclides. That is, how do we account for the masses of electrons using isotopic masses?

Plan: We first calculate Δm, the difference in nuclear masses between the products and reactants, and then convert this quantity to energy in joules. The listed masses are the isotopic masses for one atom of each element. These values are not the nuclear masses that are used to calculate Δm; an isotopic mass can be converted to a nuclear mass by subtracting the mass of the electrons in the atom. We can use isotopic masses in the calculation because the sum of the number of electrons in the reactants equals the same number in the products. Their masses cancel in the subtraction of isotopic masses.

Solve:

$$\Delta m = \text{mass } ^{17}_{8}\text{O} + \text{mass } ^{1}_{1}\text{H} - \text{mass } ^{14}_{7}\text{N} - \text{mass } ^{4}_{2}\text{He}$$

$$= 16.9991 \text{ amu} + 1.007825 \text{ amu} - 14.00307 \text{ amu} - 4.00260 \text{ amu}$$

$$= 0.0013 \text{ amu}$$

The positive value for Δm indicates that energy is converted to mass during the reaction. From Exercise 9, we know that 1 amu $= 1.49 \times 10^{-10}$ J. Therefore we can convert mass to energy:

$$\Delta m = (0.0013 \text{ amu})\left(1.49 \times 10^{-10} \frac{\text{J}}{\text{amu}}\right) = 1.9 \times 10^{-13} \text{ J}$$

EXERCISE 11 Using gamma rays for irradiating foods

Cobalt-60 and cesium-137 are used as sources of radiation by the food industry to preserve food. Suggest a reason for irradiating potatoes with gamma rays before they are shipped to markets.

SOLUTION: The eyes on the surface of a potato will form buds if not irradiated with gamma rays. Gamma rays destroy the reproductive abilities of potatoes, causing them to be sterile and not form buds (eyes).

EXERCISE 12   Explaining the role of strontium-90 in somatic damage

Linus Pauling led a group of scientists in a relatively successful endeavor to stop nuclear bomb tests. One of their concerns was biological damage caused by radiation from ^{90}Sr. Sr is one of the group 2A elements and replaces an element of this group that is found extensively in animals. What is the element that Sr replaces?

SOLUTION: Radiation affects an organism during its lifetime. ^{90}Sr readily replaces a common element found in bones, calcium. ^{90}Sr emits beta rays that cause biological damage. As a result, the production of blood cells, which form in the marrow of bones, is diminished.

EXERCISE 13   Using rad dose and RBE factor to determine relative radiation damage

Slow and fast neutrons have RBE factors of 5 and 10, respectively, for producing cataracts in the eyes. What is the relative radiation damage caused by a 0.4-rad dose of slow neutrons as compared with a 0.8-rad dose of fast neutrons?

SOLUTION: The biological effect is measured by number of rems, which equals number of rads times RBE.

$$\text{Rems for slow neutrons} = (0.4 \text{ rad})(5 \text{ RBE}) = 2.0 \text{ rems}$$
$$\text{Rems for fast neutrons} = (0.8 \text{ rad})(10 \text{ RBE}) = 8.0 \text{ rems}$$

The 0.8-rad dose of fast neutrons would cause damage to the eyes four times greater than that caused by a 0.4-rad dose of slow neutrons.

NUCLEAR POWER: FISSION AND FUSION

A **fission reaction** is a reaction in which a nuclide breaks apart into other nuclides of smaller mass number. The process is exothermic and the energy released can be used for production of electrical power. In Section 21.7 of the text, the nuclear fission reaction of $^{235}_{92}$U, which is initiated by bombardment with neutrons, is described.

- Note in Equations 21.24 and 21.25 in the text that more neutrons per nuclear fission reaction are formed as products than are needed to initiate the fission reaction of $^{235}_{92}$U. Therefore the reaction can be self-sustaining if other fissionable materials capture the emitted neutrons.
- If a sufficient nuclear mass of $^{235}_{92}$U is present, a large number of other nuclear fission reactions occur, with a violent explosion resulting from the large amount of energy released. Such a series of multiplying nuclear fission reactions is called a **branching chain reaction**.
- A minimum mass of fissionable material is required, called **critical mass**. A mass in excess of this is called **supercritical mass.**

A **fusion reaction** is one in which lightweight nuclei combine to form heavier nuclei and sometimes other subatomic particles.

- Temperatures of around 1×10^8 °C are required to give the combining atoms sufficient kinetic energy to overcome the coulombic repulsion forces that exist between nuclei.
- Once nuclei combine, a large amount of energy is evolved. Therefore, considerable interest is focused upon developing techniques to control fusion reactions in order to harness this energy.

EXERCISE 14 Explaining how rates of fission reactions are controlled in nuclear reactors

How are the rates of fission reactions controlled in nuclear reactors?

SOLUTION: Control rods made of cadmium, tantalum, or boron carbide readily capture neutrons. The control rods are placed into the fissionable material to regulate the number of neutrons that are available to initiate fission reactions.

EXERCISE 15 Explaining how fast neutrons can be slowed

How are fast neutrons, released by a fission process, slowed down so that they may be captured by fissionable material?

SOLUTION: A moderator, such as H_2O, D_2O, or graphite, is used to slow down neutrons. A neutron collides with, but is not captured by, the moderator and transfers some of its kinetic energy to the moderator. With a decreased kinetic energy, the velocity of a neutron decreases.

SELF-TEST QUESTIONS

Key Terms

Having reviewed key terms in Chapter 21, match key terms with phrases and identify statements as true or false. If a statement is false, indicate why it is incorrect.

Match each phrase with the best term:

21.1 A term describing the following nuclear reaction: $^1_1p + ^0_{-1}e \rightarrow ^1_0n$.

21.2 Radiation with a high value of this quantity is dangerous to human health.

21.3 It is a measure of the amount of energy that deposits on a kilogram of tissue.

21.4 The nuclear mass of $^{59}_{27}Co$ is 58.91837 amu. The sum of masses of nucleons forming this isotope is larger than this number by 0.55563 amu. What term describes this loss of mass when an isotope forms from its nucleons?

21.5 When this particle is emitted it causes an isotope to lose two protons and two neutrons.

21.6 When this particle is emitted by an isotope undergoing radioactive decay, a new isotope forms with one more proton and no change in mass number.

21.7 The name given to 3.7×10^{10} disintegrations per second.

21.8 The term given to the following nuclear process: $^1_0n + ^{235}_{92}U \rightarrow ^{103}_{42}Mo + ^{131}_{80}Sn + 2^1_0n$.

21.9 The term given to the following nuclear process: $3^4_2He \longrightarrow ^{12}_6C + $ energy.

21.10 A form of radiation that travels at the speed of light with a frequency above 10^{20} Hz.

21.11 A particle with no protons or neutrons which when emitted causes a decrease in the number of protons.

21.12 A term given to this type of nuclear reaction: $^{98}_{42}Mo + ^2_1H \rightarrow ^{99}_{43}Tc + ^1_0n$.

21.13 These types of isotopes undergo first-order nuclear decay.

21.14 A sustained reaction that occurs in a nuclear reactor to produce energy.

21.15 The name given to 16.28 kg, the amount of $^{239}_{94}Pu$ required to maintain a chain reaction involving $^{239}_{94}Pu$ reacting with neutrons.

21.16 The name given to 18 kg of $^{239}_{94}Pu$ in reference to the situation in **21.15**.

Terms:

(a) alpha particle	(i) gamma radiation
(b) beta particle	(j) mass defect
(c) chain reaction	(k) nuclear transmutation
(d) critical mass	(l) positron
(e) curie	(m) rad
(f) electron capture	(n) radioisotope
(g) fission	(o) rem
(h) fusion	(p) supercritical mass

21.17 A nucleus containing a *magic number* of nucleons, such as 18 protons, is likely to be radioactive.

21.18 The *radioactive series* for ^{238}U involves 14 decay steps leading to a stable ^{206}Pb.

21.19 There are many *nuclear disintegration* series that occur in nature.

21.20 *Particle accelerators* involve shooting particles into a vacuum in the presence of alternating electrical or magnetic fields.

21.21 The *transuranium elements* immediately follow uranium in the periodic table.

21.22 In a typical *Geiger counter*, gamma rays enter a tube through a thin window and ionize a gas.

21.23 A *scintillation counter* is an instrument used to detect radiation by fluorescence.

21.24 A *radiotracer* used to study reactions of carbon-based compounds is carbon-12.

21.25 Low temperatures can be used to induce *thermonuclear reactions*.

21.26 *Genetic damage* is short term in nature.

21.27 The SI unit of radioactivity is the *becquerel*, which is equivalent to one curie.

21.28 If a nucleus emits radiation only with the input of energy, it is said to be *radioactive*.

21.29 The *half-life* of strontium-90 is 29 years. This means that in three half-lives 10.0 g of it decays to 3.3 grams (1/3 of original mass).

21.30 A *radionuclide* is a nucleus that is radioactive and had a specified number of protons and neutrons.

21.31 The neutral OH molecule is unstable and is an example of a *free radical* because it possesses one unpaired electron.

Problems and Short-Answer Questions

21.32 The following diagram represents two different nuclear decay processes. Identify the decay process labeled **A** from among the following selections and explain your reasoning:

(a) electron capture;
(b) gamma emission;
(c) beta emission;
(d) alpha emission.

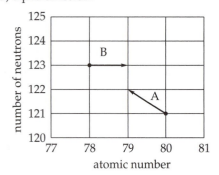

21.33 Identify the decay process labeled **B** in the previous figure from the selections given in 21.32.

21.34 What information can be derived from the name uranium-233? What is the isotopic symbol?

21.35 What are the differences among α, β, and γ radiations in terms of charge and mass?

21.36 What are the symbols for alpha and beta particles. Explain the symbols and their components.

21.37 What particle is similar to an electron but has a different charge? What is its symbol? If this particle is emitted, what effect does it have on the nucleus?

21.38 If protons repel one another, why is it possible for nuclei of more than one proton to be stable?

21.39 Refer to Figure 21.3 in the text and answer the following questions:

(a) Why does the plot end at element 83, bismuth?
(b) Why does the belt of stability in most cases lie above the 1:1 neutron-to-proton ratio?
(c) What generally happens to nuclei with atomic numbers greater than 83?

21.40 How can the concept of magic numbers help you determine relative nuclear stability?

21.41 Critique the following statement: In chemical reactions that are exothermic there is no change in mass.

21.42 What is the origin of mass defect?

21.43 The storage of nuclear wastes from nuclear power plants is a very controversial issue in the United States. Many states do not want to become a storehouse of these wastes. From a half-life viewpoint for radioactive nuclei, why is the public concerned?

21.44 Water is the primary component of living tissue in humans. Water can absorb radiation and prevent it from destroying tissue. Which types of radiation discussed in the text interact with water and ionize it? Why is this a problem? What species is initially responsible for a cascading sequence of events that leads to disruption of cell operations?

21.45 Radon-222 is continually generated in rocks and soil samples that contain uranium-238. Its half-life is only 3.82 days. Why is there such concern about its presence in the environment if the half-life is so short?

21.46 Write balanced nuclear equations for the following nuclear transformations:

(a) niobium-99 undergoes beta decay
(b) chromium-48 undergoes electron capture
(c) calcium-39 undergoes positron emission
(d) carbon-16 undergoes neutron emission
(e) gadolinium-148 undergoes alpha emission

21.47 Supply the missing subatomic particle, energy form, or nuclide in the following induced nuclear reactions:

(a) ____ $(^3_1\text{H}, \text{n})^4_2\text{He}$
(b) $^{253}_{99}\text{Es}(\underline{\quad}, \text{n})^{256}_{101}\text{Mv}$
(c) $^{81}_{35}\text{Br}(\gamma, \text{n})$ ____

21.48 One theory of stellar evolution has $^{16}_{8}\text{O}$ forming by the reaction of helium-4 and carbon-12 in the core of a sun at very high temperatures.

(a) Write the nuclear equation describing this nuclear reaction.
(b) Is this a fission or fusion reaction?

21.49 Calculate the energy associated with the reaction you wrote for problem 21.48. Does the result of your energy calculation support this particular stellar-evolution theory? Required atomic masses (amu) are: $^4_2$He, 4.00260; $^{12}_6$C, 12.00000; $^{16}_8$O, 15.99491.

21.50 In 2008, you and your friends barbecue hamburgers using charcoal that was made from freshly cut wood. Later you bury some unburned charcoal in the ground. Suppose that many years later an archeologist digs up your charcoal and by experimental methods finds that it has a ^{14}C:^{12}C ratio that is only 30 percent of the ^{14}C:^{12}C ratio in charcoal made from a recently cut tree. Assuming that the ^{14}C:^{12}C ratio does not change in the atmosphere, approximately how far in the future will the archeologist have found the buried charcoal? The half-life of ^{14}C is 5715 years.

21.51 How much radioactivity, in curies, does 1 μg of Ra-226 produce? What is this in units of disintegrations per second?

21.52 Technetium was first prepared by the following nuclear reaction:
$$^{97}_{42}Mo + ^2_1H \longrightarrow ^{93}_{43}Tc$$
Is this sufficient information to describe the nuclear synthesis? If not, what additional information is needed and why?

21.53 A particular form of radiation is found to be shielded by 3 mm of aluminum metal. Is this sufficient information to conclude that the radiation is relatively non-penetrating and thus the radiation must be alpha particles? If not, what additional information is needed and why?

21.54 Isotopes that lie above the belt of stability in Figure 21.3 in the text tend to disintegrate because they have too high of a neutron to proton ratio. Is this sufficient information to determine how a particular neutron-rich isotope disintegrates, such as $^{87}_{36}$Kr? If not, what additional information is needed and why?

Integrative Exercises

21.55 Carbon-14 is produced in the atmosphere by the action of cosmic rays on nitrogen-14. Cosmic rays from the sun have a high concentration of protons and other nuclei. In the upper atmosphere there are collisions of atoms with cosmic rays that produce neutrons and these are responsible for the transmutation of nitrogen-14.

(a) Write the reaction that describes the interaction of nitrogen-14 with neutrons to form carbon-14.
(b) What is the primary chemical form of carbon-14 in the atmosphere and why?
(c) The specific activity of carbon-14 in all living animals and plants is 0.227 Bq/g-C. What is a Bq? How is it related to the curie?
(d) The half-life of carbon-14 is 5715 yr. What is the age of an old bone found in a tomb if it has a specific activity of 0.142 Bq/g-C?

21.56 In the 1980s a worker at a nuclear power plant in Pennsylvania set off radiation alarms yet there was no logical reason for the incidence based on conditions in the area where the worker labored. The problem was traced to his home where radiation levels were about 2700 pCi/L. An average level in a home is about 1.35 pCi/L. The primary culprit was the presence of radioactive radon-222 in the home which was traced to the ground under the house. The worker carried the radon-222 in his clothes and this triggered alarms.

(a) What was the radiation level in the home in units of Becquerel?
(b) Rn-222 has a half-life of four days and decays to Po-218. Write the expected nuclear reaction for this decay.
(c) If the half-life is four days, why is there concern about Rn-222 in air? Consider the properties of Rn.

21.57 The half-life of $^{131}_{53}$I is 8.07 days. A saturated solution of Pb($^{131}_{53}$I$^-$)$_2$ is prepared and the excess solid is removed. Given that the K_{sp} of PbI$_2$ is 7.1 × 10^{-9}, how many grams of $^{131}_{53}$I$^-$ remain in solution after two weeks? The isotopic mass of $^{131}_{53}$I is 130.906 amu.

Multiple-Choice Questions

21.58 How many protons, neutrons, and electrons does a $^7_3$Li atom contain?

(a) 7p, 3n, 4e   (d) 4p, 3n, 3e
(b) 4p, 3n, 7e   (e) 3p, 4n, 3e
(c) 3p, 3n, 4e

21.59 Which of the following particles cannot be accelerated in a cyclotron?

(a) alpha;   (d) neutron;
(b) beta;   (e) positron.
(c) proton;

21.60 How is the presence of radioactivity detected in a scintillation counter?

(a) by fluorescence of ZnS upon interaction with radiation
(b) by interaction of the radiation with a photographic plate
(c) by ionization of a gas
(d) by precipitation of a radioactive substance
(e) by gas chromatography techniques

21.61 What is the missing particle or energy in the reaction $^{209}_{83}$Bi + $^2_1$H $\longrightarrow$ $^{210}_{84}$Po + _____

(a) $^0_{-1}$e   (d) α
(b) $^0_1$e   (e) $^4_2$He
(c) $^1_0$n

21.62 What is the binding energy of $^{35}_{17}$Cl, given that the atomic mass equals 34.95953 amu?

(a) 4.77232 × 10^{-11} J   (d) 5.27231 × 10^{-12} J
(b) 4.77232 × 10^{-10} J   (e) 5.27231 × 10^{-13} J
(c) 5.27231 × 10^{-11} J

21.63 Which of the following reactions illustrates a nuclear process that occurs within a control rod in a nuclear reactor?

(a) $^1_1H + ^1_1H \longrightarrow ^2_1H + ^0_1e$

(b) $^9_4Be + ^2_1H \longrightarrow ^{10}_5B + ^1_0N$

(c) $^0_{-1}e + ^{106}_{47}Ag \longrightarrow ^{106}_{46}Pd$

(d) $^9_4Be + ^4_2He \longrightarrow ^{12}_6C + ^1_0N$

(e) $^{113}_{48}Cd + ^1_0n \longrightarrow ^{114}_{48}Cd + \gamma$

21.64 What is the major type of biological damage in humans following sub-lethal exposure to radiation?

(a) leukemia;   (d) loss of sight;

(b) burns;   (e) loss of a human

(c) skin cancer;   appendage.

21.65 Which statement is true?

(a) A rem involves a relatively large amount of energy.

(b) The heat equivalent of a rem is responsible for major damage to tissue.

(c) The ionization of water yields a hydroxyl free radical.

(d) Free radicals contain an even number of electrons.

(e) Free radicals are stable in biochemical tissues.

21.66 Fluorine-20

(a) is expected to be a stable isotope because its neutron/proton ratio is 11/9.

(b) is expected to undergo alpha decay to increase the neutron/proton ratio.

(c) is expected to undergo beta decay to increase the neutron/proton ratio.

(d) lies in the band of stability in Figure 21.2.

(e) probably has a half-life too short to enable it to be detected.

21.67 A piece of wood from an ancient artifact has a carbon-14 activity of 11.7 disintegrations per minute per gram of carbon. Current carbon-14 activity in fresh samples is 15.3 disintegrations per minute per gram of carbon. The half-life of carbon is 5715 yr. Which statement is true?

(a) The age of the sample is 1270 years.

(b) The rate constant for the decay is 1.21×10^4 yr.

(c) Insufficient data is available to calculate the age of the artifact.

(d) The carbon-14 activity in the freshly cut sample is presumed to be different than in carbon dioxide in the air.

(e) The artifact could have been from the age of Ptolemy V in Egypt during the period 203 to 181 BC.

SELF-TEST SOLUTIONS

21.1 (f). **21.2** (o). **21.3** (m). **21.4** (j). **21.5** (a). **21.6** (b). **21.7** (e). **21.8** (g). **21.9** (h). **21.10** (l). **21.11** (l). **21.12** (k). **21.13** (n). **21.14** (c). **21.15** (d). **21.16** (p). **21.17** False.

Magic numbers refer to nucleons with a closed shell set of protons or neutrons; these nuclei are more stable. **21.18** True. **21.19** False. There are only three in nature. **21.20** True. **21.21** True. **21.22** False. Alpha or beta particles, not gamma rays. **21.23** True. **21.24** False. Carbon-12 is not radioactive; thus it cannot be used as a radiotracer. Carbon-14 is commonly used. **21.25** False. High temperatures must be used to overcome repulsions between nuclei. **21.26** False. It involves damage to chromosomes and genes, thus having an effect upon offspring. **21.27** False. The first phrase is true; however, it is equivalent to significantly less than one curie. One Bq is one disintegration per second. **21.28** False. A nucleus is radioactive if it spontaneously emits radiation. **21.29** False. For each half-life, strontium-90 loses one-half of its mass. Thus, in three half-lives, it loses $\frac{1}{2} \times \frac{1}{2} \times \frac{1}{2}$ of its original mass, or $\frac{1}{8} \times 10.0 \text{ g} = 1.25 \text{ g}$. **21.30** True. **21.31** True.

21.32 Decay process **A** involves an increase in the number of neutrons by one and a loss of one proton. When a proton captures an electron a neutron is formed and the atom has an atomic number of one less. The decay process is (a), electron capture.

21.33 Decay process **B** involves an increase in the number of protons by one and no change in the number of neutrons. Beta emission, (c), shows this type of change.

21.34 The number 233 in uranium-233 tells you the mass number, the sum of protons and neutrons. Uranium has 92 protons, therefore, it has $233 - 92 = 140$ neutrons. The isotopic symbol is $^{233}_{92}U$.

21.35 An alpha particle, α, has a 2+ charge; a beta particle, β, has a 1− charge; and gamma radiation, γ, has no charge. Both the alpha and beta particles have exceedingly small masses and the gamma radiation has no mass.

21.36 An alpha particle, α, is the equivalent of a helium-4 nucleus, ^4_2He. The superscript is the mass number and the subscript is the atomic number of helium. A beta particle, β, is a high-speed electron and is represented by $^0_{-1}e$. The superscript is zero because there are no neutrons and the subscript is −1 to imply a particle that has negative charge instead of a positive charge associated with a proton.

21.37 A positron has the same mass as an electron but carries a positive charge, ^0_1e. Positron emission causes the atomic number of a nucleus to decrease by one unit but the mass number remains the same.

21.38 Over very close distances, there are very strong nuclear forces involving nucleons. Neutrons are part of the reason why nuclei are stable. Although they carry no charge, their presence is critical for the existence of stable nuclei. As more protons occupy the same small region, an increasing number of neutrons are necessary to stabilize the nucleus.

21.39 (a) All elements above bismuth are radioactive and not stable.

(b) As discussed in problem 21.38, an excess of neutrons to protons is necessary to "glue" protons together as they increase in number in a nuclei. Thus, there will be an excess of neutrons to protons for most elements, particularly the heavier ones.

(c) These are radioactive elements and will spontaneously undergo radioactive decay. Alpha decay is very common.

21.40 The magic numbers for protons are 2, 8, 28, 50, and 82; for neutrons they are 2, 8, 20, 50, 82, or 126. Nuclei containing one of these numbers of protons or neutrons tend to be more stable than those without these numbers.

21.41 According to the Einstein equation $E = mc^2$, a mass change is associated with an energy change. If a system loses mass, it loses energy (exothermic). In normal chemical reactions the mass change associated with an energy change is exceedingly small and difficult to detect; we say the law of conservation of mass applies to chemical reactions. The mass changes in nuclear reactions are much larger.

21.42 Mass defect is the difference in mass between a nucleus and its constituent nucleons. It arises because it takes energy to separate the nucleons contained within a nucleus. The energy added to a nucleus to separate the nucleons is converted to mass and thus the sum of masses of the nucleons is greater than that of the nucleus.

21.43 Nuclear reactors produce fission products that are radioactive. The fuel rods need to be replaced or reprocessed at periodic intervals. It was the plan to separate the fission products from the nuclear fuel and recover usable nuclear fuel. However, this process has not been effective and fuel rods are being stored. It has been estimated that it will take at least 20 half-lives for the radioactive materials to reach radioactivity levels acceptable to biological species—depending on the fission product or fuel it could take 600 to more than 24,000 years. How can these be stored safely for so many years? How can they be transported safely? These are concerns of the public.

21.44 Alpha, beta, and gamma radiation along with X rays possess sufficient energies (greater than 1216 kJ/mol) to ionize water. Electrons are removed from water molecules and form H_2O^+ ions. This ion can react with water to form the OH radical, which has an unpaired electron. This free radical can attack cell tissues to form other free radicals and these are responsible for disrupting normal operations of cells.

21.45 The decay of radon-222 leads to the formation of alpha particles and polonium-218. Alpha particles are ionizing sources or radiation that lead to the formation of disruptive free radicals in the human body; cancer can be one result of significant exposure. Furthermore, polonium-218 has an even shorter half-life and it also produces alpha particles.

21.46 (a) $^{99}_{41}Nb \rightarrow ^{99}_{42}Mo + ^{0}_{-1}e$.

(b) $^{48}_{24}Cr + ^{0}_{-1}e \rightarrow ^{48}_{23}V$.

(c) $^{39}_{20}Ca \rightarrow ^{39}_{19}K + ^{0}_{1}e$.

(d) $^{16}_{6}C \rightarrow ^{15}_{6}C + ^{1}_{0}n$.

(e) $^{148}_{64}Gd \rightarrow ^{144}_{62}Sm + ^{4}_{2}He$.

21.47 (a) $^{2}_{1}H$ ($^{3}_{1}H$, n)$^{4}_{2}He$: $^{2}_{1}H + ^{3}_{1}H \rightarrow ^{4}_{2}He + ^{1}_{0}n$.

(b) $^{253}_{99}Es(\alpha, n)^{256}_{101}Mv$: $^{253}_{99}Es + ^{4}_{2}He \rightarrow ^{256}_{101}Mv + ^{1}_{0}n$.

(c) $^{81}_{35}Br(\gamma, n)^{80}_{35}Br$: $^{81}_{35}Br + \gamma \rightarrow ^{80}_{35}Br + ^{1}_{0}n$.

21.48 (a) $^{4}_{2}He + ^{12}_{6}C \rightarrow ^{16}_{8}O$.

(b) This is a fusion reaction because lightweight nuclides combine to form a heavier nuclide. The fact that the reaction occurs at a high temperature classifies it as a thermonuclear reaction.

21.49 The mass change taking place during the reaction is Δm = mass products − mass reactants = 15.99491 amu − 4.00260 amu − 12.00000 amu = −0.00769 amu. The energy change is $\Delta E = (\Delta m)$ (conversion factor that converts mass to J) $= \Delta m(1.49 \times 10^{-10} J/amu) = (-0.00769 amu)(1.49 \times 10^{-10} J/amu) = -1.15 \times 10^{-12} J$. The fact that energy is released during the reaction shows than an oxygen-16 nucleus is more stable than the separate helium-4 and carbon-12 nuclei. This gives support to the theory that oxygen-16 can be made from a fusion of carbon-12 and helium-4 in a thermonuclear reaction.

21.50 Assuming that the $^{14}C:^{12}C$ ratio in charcoal made in year x is the same as in charcoal made in 2008, you can calculate the value of the ratio $N_0:N_t$ where N_0 is the concentration of ^{14}C in recently made charcoal and N_t is the concentration in year x. In this problem, the ratio is 1:0.30 because only 30 percent of the original value of the $^{14}C:^{12}C$ ratio in the charcoal made in 2008 remains in year x. The concentration of ^{12}C nuclides in the charcoal sample remains constant because ^{12}C is a stable isotope. The time required for the decay of ^{14}C nuclides from 2008 to year x is calculated as follows: $t = 1.44 t_{1/2} \ln(N_0/N_t) = 1.44 t_{1/2} \ln(1/0.30) = (1.44)(5715 yr) \ln(1/0.30) = 9900 yr$. The year that the buried charcoal will be discovered is 9900 + 2008 = 11,908.

21.51 1 μCi. One curie (Ci) is equivalent of 1 g of Ra-226 producing 3.7×10^{10} disintegrations per second. Thus, 1 μg of Ra-226 in disintegrations per second is equivalent to

$$(10^{-6} g \text{ Ra-226})\left(3.7 \times 10^{10} \frac{\text{disintegrations/s}}{1 g \text{ Ra-226}}\right)$$

$$= 3.7 \times 10^4 \text{ disintegrations/s}$$

21.52 There is insufficient information. Formation of new species by bombarding them with high energy elementary particles, such as deuterium, involves energy changes. In such reactions other particles or energy are released and need to be identified or information provided to indicate additional particles or energy are not involved. In this particular case neutrons are ejected.

21.53 There is insufficient information. Beta particles are also relatively low penetrating, although more penetrating than alpha particles. A thin sheet of aluminum metal is sufficient to provide shielding from beta particles. A piece of thick paper could be used for additional testing. It will shield alpha radiation but not beta radiation.

21.54 There is insufficient information for a complete picture. If an isotope is neutron-rich it will decay to a different isotope with a smaller neutron to proton ratio. This could happen either by β decay which increases the number of protons or by spontaneous neutron emission. The type of particle emitted must also be stated.

21.55 (a) Neutrons formed in cosmic rays react with nitrogen-14:

$$^1_0n + ^{14}_7N \longrightarrow ^{14}_6C + ^1_1p$$

(b) The primary chemical form is carbon dioxide gas which is formed by oxidation of carbon-14 atoms in the presence of oxygen.

(c) A Bq(Becquerel) is the SI unit of activity and is one nuclear disintegration per second. The curie (Ci) is the older unit of activity and one curie is 3.7×10^{10} Bq.

(d) Activity is a measure of concentration; thus [A] in the equation $\ln\frac{[A]_0}{[A]_t} = kt$ can represent activity. This equation permits you to calculate the age (t) of the bone because you know the specific activity of carbon-14 today and that of the old bone. k is calculated from the equation $t_{1/2} = \frac{0.693}{k}$. Substituting the value of $t_{1/2}$ into this equation gives

$$k = 1.21 \times 10^{-4}/\text{yr}.$$

Substituting the known values into the first equation gives:

$$\ln\frac{[0.227\ \text{Bq/g-C}]}{[0.142\ \text{Bq/g-C}]} = 0.469 = (1.21 \times 10^{-4}/\text{yr})t$$

$$t = \frac{1}{1.21 \times 10^{-4}/\text{yr}}(0.469) = 3880\ \text{yr}$$

The age of the bone is 3880 yr or about 39 centuries.

21.56 (a) The activity in units of Becquerel is: $(2700\ \text{pCi/L})$ $(10^{-12}\ \text{C/1 pCi})(3.7 \times 10^{10}\ \text{Bq/1C}) = 100\ \text{Bq/L}$

(b) When Rn-222 decays to Po-218 it loses a particle with four mass numbers, which should be the alpha particle. Along with alpha particles gamma rays are often given off:

$$^{222}_{86}\text{Rn} \longrightarrow ^{218}_{84}\text{Po} + ^4_2\text{He} + \gamma$$

(c) Rn-222 is a noble gas and is relatively unreactive. As a gas it diffuses readily through rocks, soil, and into buildings. When it decays it emits alpha particles and gamma rays which are dangerous to humans. If Rn-222 is breathed some of it will stay in the lungs and the emission of alpha particles and gamma rays will cause lung cancer.

21.57 In the saturated solution $K_{sp} = [\text{Pb}^{2-}][^{131}_{53}\text{I}^-]^2$. From this relationship we can calculate the quantity of $^{131}_{53}\text{I}^-$ that is initially present. $7.1 \times 10^{-9} = (s)(2s)^2 = 4s^3$. Assuming ideal conditions, $s = 1.2 \times 10^{-3}\ \text{mol/L}$. This is the quantity of lead iodide that is soluble; it is also the amount of lead ion dissolved and twice the quantity is the concentration of $^{131}_{53}\text{I}^-$ in solution, $3.4 \times 10^{-3}\ \text{mol/L}$. To determine the mass of $^{131}_{53}\text{I}^-$ present we need to first

determine the number of moles present. However, the initial volume of the solution is 500.0 mL, not one liter. Thus in 500.0 mL there are 1.2×10^{-3} moles of $^{131}_{53}\text{I}^-$. The mass of 1.2×10^{-3} mol of $^{131}_{53}\text{I}^-$ is

$$\text{Mass} = (1.2 \times 10^{-3}\ \text{mol})\left(\frac{130.906\ \text{g}}{1\ \text{mol}\ ^{131}_{53}I}\right) = 0.16\ \text{g}$$

We can use the half-life of $^{131}_{53}\text{I}$ and the relation from Exercise 6, $t = 1.44\ t_{1/2}\ \ln\frac{N_0}{N_t}$ to solve for the quantity of $^{131}_{53}\text{I}^-$ remaining after two weeks. Two weeks is 14 days and the half-life is 8.07 days. We can rearrange the previous equation to solve for N_t.

$$\ln\frac{N_0}{N_t} = \frac{t}{1.44\ t_{1/2}} = \frac{14.0\ \text{days}}{(1.44)(8.07\ \text{days})} = 1.20$$

$$\ln\frac{0.16\ \text{g}}{N_t} = 1.20$$

solving, $N_t = 0.048$ g, the amount of $^{131}_{53}\text{I}^-$ remaining after two weeks.

21.58 (e). **21.59 (d)**. Only charged particles are accelerated in a cyclotron; a neutron has no charge.

21.60 (a) 21.61 (c)

21.62 (a)   Binding energy = (mass of nucleons − mass of nucleus)$(1.49 \times 10^{-10}\ \text{J/amu})$ = $[(17)(1.00728\ \text{amu}) + 18 \times (1.00867\ \text{amu}) - 34.95953\ \text{amu}](1.49 \times 10^{-10}\ \text{J/amu})$ = $(0.32029\ \text{amu})(1.49 \times 10^{-10}\ \text{J/amu})$ = 4.77232×10^{-11} J.

21.63 (e) 21.64 (a)

21.65 (c) A rem is a relatively small amount of energy, but it is sufficient to cause ionization of cellular material, such as water. When water ionizes it forms a OH free radical which has one unpaired electron.

21.66 (c) The ideal neutron/proton ratio for the lighter elements is 1:1. Thus, fluorine-20 can achieve this ratio by beta decay, which converts a neutron to a proton and an electron.

21.67 (e) The age of Ptolemy V was about 2208 years ago (add about 200 years to 2008, the date of this writing, because the age of Ptolemy V was B.C.). The rate constant for the decay of carbon-14 is 1.21×10^{-4} yr, which is calculated from the relationship:

$$k = \frac{0.693}{t_{1/2}} = \frac{0.693}{5715\ \text{yr}} = 1.21 \times 10^{-4}\ \text{yr}^{-1}$$

The age of the sample can be calculated from the relation $t = 1.44\ t_{1/2}\ \ln(N_0/N_t) = 1.44(5715\ \text{yr})(\ln(15.3\ \text{min}\ /11.7\ \text{min})) = 2200\ \text{yr}$. 2200 yr is close to 2208 years ago.

Sectional MCAT and DAT Practice Questions VII

Several oxides of nitrogen found in the environment are also used in industrial processes. N_2O, NO_2, and NO comprise the three most common stable oxides of nitrogen. All are gases at room temperature. Nitrous oxide is also known as laughing gas, nitric oxide is a toxic air pollutant produced by automobiles and power plants, and nitrogen dioxide is an orange-brown gas and is also an air pollutant.

Nitric oxide is prepared commercially by the following reaction:

$$4\,NH_3(g) + 5\,O_2(g) \xrightarrow{\text{Pt, 85 °C}} 4\,NO(g) + 6\,H_2O(g)$$
Equation 1

Nitrogen dioxide is readily formed by oxidation of $NO(g)$ with $O_2(g)$:

$$2\,NO(g) + O_2(g) \rightarrow 2\,NO_2(g)$$
Equation 2

Nitrogen dioxide also exists in equilibrium with its dimer:

$$2\,NO_2(g) \rightleftarrows N_2O_4(g)$$
Equation 3

Table 1 has thermochemical data at 298 K for substances in the previous reactions.

TABLE 1 Thermochemical Data, 298 K

Property	$NH_3(g)$	O_2	$NO(g)$	$H_2O(g)$	$NO_2(g)$	$N_2O_4(g)$
ΔH_f^0 (kJ mol^{-1})	−80.3	0	90.4	−241.8	33.8	9.66
ΔG_f^0 (kJ mol^{-1})	−26.5	0	86.7	−228.6	51.8	98.3
S^0 (J K^{-1} mol^{-1})	111.3	237.6	210.6	188.3	240.5	304.3

The equilibrium constant for a reaction is related to ΔG^0 by the relationship:
$$\Delta G^0 = -RT \ln K$$

R = 8.314 J/K-mol.

1.   Which Lewis structure best represents N_2O_4?

(a)

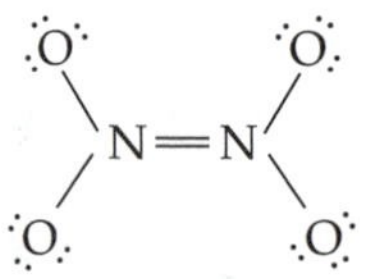

(b)

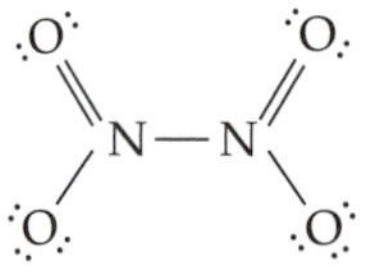

(c)

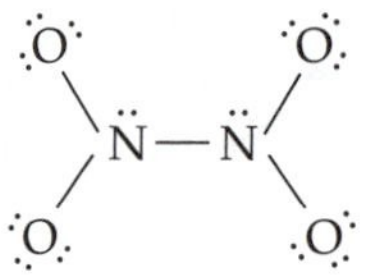

(d)

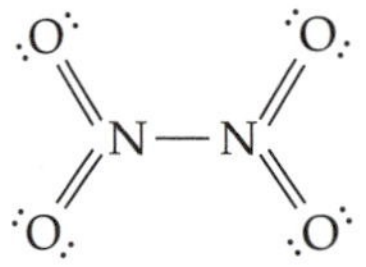

2.   Which statement is true concerning the formation reaction for one mole of nitrogen dioxide gas from its elements at 298 K and 1 atm pressure?
    (a)   Reaction is spontaneous because ΔG^0 for the reaction is < 0.
    (b)   Reaction is spontaneous because ΔS^0 for the reaction is < 0.
    (c)   Reaction is not spontaneous because ΔG^0 for the reaction is > 0.
    (d)   Reaction is not spontaneous because ΔS^0 for the reaction is > 0.

3.   Which is the free-energy change in kJ for the reaction shown by Equation 1?
    (a)   4(86.7)
    (b)   4(86.7) + 6(−228.6)
    (c)   [4(86.7) + 6(−228.6)] − [4(−80.3) + 5(0)]
    (d)   [4(86.7) + 6(−228.6)] − [4(−26.5) + 5(0)]

4.   Why does Pt appear in Equation 1?
    (a)   To change ΔG^0 for the reaction to be less than zero.
    (b)   To lower E_a.
    (c)   To provide a means to raise the temperature to 85 °C.
    (d)   To change ΔH^0 for the reaction to be exothermic.

5.   A reason for the sign of ΔS^0 for Equation 2 is:
    (a)   The value of ΔH_f^0(kJ mol^{-1}) for NO(g) is very positive.
    (b)   The number of moles of product is less than the number of moles of reactants.
    (c)   The free-energy of formation for $NO_2(g)$ favors the reaction compared to that for NO(g).
    (d)   All substances are in the gas phase.

6. What can be concluded about the value of K for the reaction shown in Equation 3?
   (a) It equals one
   (b) It is greater than one
   (c) It is less than one
   (d) It is evaluated from the relation $K = \dfrac{P_{N_2O_4}}{P_{NO_2}}$

7. How many grams of $NO(g)$ (Molar mass = 30 g mol^{-1}) are produced by the reaction in Equation 1 if one mole of ammonia and one mole of oxygen react and the reaction goes to completion?
   (a) 37.5 g　　(b) 30 g　　(c) 24 g　　(d) 18 g

| Questions 8 through 12 are **not** based on a descriptive passage. |

8. Which substance has the lowest entropy at 20 °C and at 1 atm pressure?
   (a) 1 mole of $H_2O(l)$　　　(b) 1 mole of $NaCl(s)$
   (c) 1 mole of $O_2(g)$　　　(d) 1 mole of $CH_3OH(l)$

9. What is the coefficient in front of $Fe(OH)_2$ when the following oxidation-reduction reaction is balanced in basic solution?

$$Fe(OH)_2 + O_2 \rightarrow Fe(OH)_3$$

   (a) 2　　　(b) 4　　　(c) 5　　　(d) 6

10. Given the following data determine which element or ion reduces Cu^{2+} to $Cu(s)$ and produces the largest E^0 for the reaction.

Half-reaction	E^0 (V)
$F_2 + 2e^- \rightarrow 2F^-$	+2.87
$Ce^{4+} + e^- \rightarrow Ce^{3+}$	+1.61
$Cu^{2+} + 2e^- \rightarrow Cu$	+0.337
$Sn^{2+} + 2e^- \rightarrow Sn$	−0.136
$Cd^{2+} + 2e^- \rightarrow Cd$	−0.403

   (a) Ce^{4+}　　(b) F^-　　(c) Sn　　(d) Cd

11. What is the half-cell reaction in the cathodic compartment of a voltaic cell using the following spontaneous reaction?

$$3\,Zn(s) + Cr_2O_7^{2-}(aq) + 14\,H^+(aq) \rightarrow 3\,Zn^{2+} + 2\,Cr^{3+}(aq) + 7\,H_2O(l)$$

   (a) $Zn(s) \rightarrow Zn^{2+}(aq) + 2e^-$

   (b) $Zn^{2+}(aq) + 2e^- \rightarrow Zn(s)$

   (c) $6e^- + Cr_2O_7^{2-}(aq) + 14\,H^+(aq) \rightarrow 2\,Cr^{3+}(aq) + 7\,H_2O(l)$

   (d) $2\,Cr^{3+}(aq) + 7\,H_2O(l) \rightarrow 6e^- + Cr_2O_7^{2-}(aq) + 14\,H^+(aq)$

12. What is E^0 for the following reaction

$$5\,AgI(s) + Mn^{2+}(aq) + 4\,H_2O(l) \rightarrow 5\,Ag(s) + 5\,I^-(aq) + MnO_4^-(aq) + 8\,H^+(aq)$$

given the following data:

Half-reaction	E^0 (V)
$AgI + e^- \rightarrow Ag + I^-$	−0.152
$Ag^+ + e^- \rightarrow Ag$	+0.800
$MnO_4^- + 8\,H^+ + 5e^- \rightarrow Mn^{2+} + 4\,H_2O$	+1.507
$Mn^{2+} + 2e^- \rightarrow Mn$	−1.185

(a) 0.707 V   (b) 1.355 V   (c) −1.659 V   (d) −2.267 V

ANSWERS

1. (b) N_2O_4 has 34 valence electrons. Structure (b) is the only one that has 34 valence electrons with all atoms showing an octet of electrons.

2. (c) A reaction is spontaneous if $\Delta G_{rxn} < 0$. The formation reaction as described in the question, $\frac{1}{2}N_2(g) + O_2(g) \rightarrow NO_2(g)$, is at standard state conditions. Thus, we can write $\Delta G_{rxn}^0 = \Delta G_f^0(NO_2)$. Table I shows a positive value for $NO_2(g)(+51.8\ \text{kJ/mol})$ and therefore (c) is the correct answer.

3. (d) $\Delta G_{rxn}^0 = \sum \Delta G_f^0(P) - \sum \Delta G_f^0(R)$. Using the data in Table 1 we have:
$$[4\ \text{mol}\ (86.7\ \text{kJ/mol}) + 6\ \text{mol}\ (-228.6\ \text{kJ/mol})] - [4\ \text{mol}\ (-26.5\ \text{kJ/mol}) + 5\ \text{mol}\ (0\ \text{kJ/mol})]$$

4. (b) Pt is a catalyst. It increases the rate of the chemical reaction by providing a new reaction pathway which has a lower E_a (activation energy).

5. (b) $\Delta S_{rxn}^0 = \sum S^0(P) - \sum S^0(R) =$

$(2\ \text{mol})(240.5\ \text{J/K-mol}) - (2\ \text{mol})(210.6\ \text{J/K-mol}) - (1\ \text{mol})((237.6\ \text{J/K-mol}) = -177.8\ \text{J/K}$. The sign is negative and is interpreted as follows: the reaction changes to a system of less dispersion of energy, or more "order." This is reflected by a decrease in the net number of moles of gas from three to two in the reaction and a change from a mixture of gases in the reactants to one type of gas in the product.

6. (b) Using $\Delta G^0 = -RT \ln K$ we can relate the value of K to the sign and value of ΔG_{rxn}^0 for the reaction. If $\Delta G_{rxn}^0 < 0$ then $K > 1$, $\Delta G_{rxn}^0 > 0$ then $K < 1$, and $\Delta G_{rxn}^0 = 0$ then $K = 1$. $\Delta G_{rxn}^0 = \sum \Delta G_f^0(P) - \sum \Delta G_f^0(R) =$ $(1\ \text{mol})(98.3\ \text{kJ/mol}) - (2\ \text{mol})(51.8\ \text{kJ/mol}) = -5.3\ \text{kJ}$. K is greater than one. Answer (d) is wrong because the form of K for the equilibrium is incorrect. The term in the denominator should be squared.

7. (c) First the limiting reactant, if one exists, must be determined. From the stoichiometry of the reaction we see that four moles of NH_3 react with five moles of O_2. If we divide the reaction by 4, we then see that one mole of NH_3 requires 5/4 moles (1.25 moles) of O_2. Oxygen is initially present in the quantity of one mole, not 1.25 moles, thus it is the limiting reactant. The

amount of NO produced based on the limiting reactant is
$(1 \text{ mol } O_2)(4 \text{ mol NO/5 mol } O_2)(30 \text{ g NO/1 mol NO}) = 24 \text{ g NO}.$

8. (b) The substance with the lowest entropy has particles arranged in the most organized manner and energy is dispersed in the fewest ways possible. A solid has these characteristics compared to a liquid or gas.

9. (b) $4 \text{ Fe(OH)}_2 + O_2 + 2 \text{ H}_2\text{O} \rightarrow 4 \text{ Fe(OH)}_3$

10. (d) An oxidation half-reaction is needed to combine with the half-reaction for the reduction of Cu^{2+} to Cu and produce the largest positive E^0: $E^0_{rxn} = E^0_{reduction} - E^0_{oxidation}$. E^0's are standard reduction potentials and are used for $E^0_{oxidation}$ without a change in sign. If we use the half-reaction, $Cd \rightarrow Cd^{2+} + 2 e^-$, the largest E^0_{rxn} is obtained, $0.337 \text{ V} - (-0.403 \text{ V}) = +0.740 \text{ V}$. Sn can also be used but it results in a smaller value of E^0_{rxn}. F^- cannot be used because it results in a negative value of E^0_{rxn}. Answer **(a)** is not possible because Ce^{4+} is reduced in the half-reaction as shown in the table.

11. (c) Reduction occurs at the cathode. Reduction involves the gain of electrons in a half-reaction. The reactants in **(c)** correspond to reactants in the overall reaction. In answer **(d)** we see that this is reversed and thus is not correct.

12. (c) The two half-reactions and associated E^0 values are:

Cathode:	$5(\text{AgI} + e^- \rightarrow \text{Ag} + I^-)$	$E^0 = -0.152 \text{ V}$
Anode:	$Mn^{2+} + 4 \text{ H}_2\text{O} \rightarrow MnO_4^- + 8 \text{ H}^+ + 5 e^-$	$E^0 = +1.507 \text{ V}$

$E^0_{rxn} = E^0_{reduction} - E^0_{oxidation} = -0.512 \text{ V} - (1.507 \text{ V}) = -1.659 \text{ V}$

Standard reduction potentials are not changed in sign when oxidation half-reactions are written. When a half-reaction is multiplied by a number, such as five in the first half-reaction above, the standard reduction potential is **not** multiplied by the number.

Chapter

Chemistry of the Nonmetals

OVERVIEW OF THE CHAPTER

22.1 PERIODIC TRENDS: METALS AND NONMETALS

Review: Atomic radii (7.2); ionic radii (8.3); electronegativity (8.5); periodic properties and electron configurations (6.9); sigma and pi bonds (9.5, 9.6).

Learning Goals: You should be able to:

1. Identify an element as a metal, semimetal, or nonmetal on the basis of its position in the periodic table or its properties.
2. Give examples of how the first member in each family of nonmetallic elements differs from the other elements of the same family, and account for these differences.
3. Predict the relative electronegativities and metallic character of any two members of a periodic family or a horizontal row of the periodic table.
4. Predict the products of chemical reactions involving oxidation or combustion involving oxygen or proton-transfer reactions.

22.2 HYDROGEN

Review: Ionization energy (7.3).

Learning Goals: You should be able to:

1. Cite the most common occurrences of hydrogen and how it is obtained in its elemental form.
2. Cite at least two uses for hydrogen.
3. Describe and name the three isotopes of hydrogen.
4. Distinguish among ionic, metallic, and molecular hydrides.

Review: Ionization energy (7.3); prediction of the geometric shape of covalent molecules from Lewis structures (9.1, 9.2).

Learning Goals: You should be able to:

1. Cite the most common occurrence of the noble gases.
2. Write the formulas of the known fluorides, oxyfluorides, and oxides of xenon, give the oxidation state of Xe in each, and describe the relative stabilities of the oxides as compared with the fluorides.

22.3 GROUP 8A: NOBLE GASES

3. Account for the fact that xenon forms several compounds with fluorine and oxygen, krypton forms only KrF_2, and no chemical reactivity is known for the lighter noble-gas elements.

4. Describe the electron-domain and molecular geometries of the known compounds of xenon.

Review: Chlorofluorocarbons in the atmosphere (18.3); emf calculations (20.4); factors affecting oxyacid strengths (16.10).

22.4 GROUP 7A: HALOGENS

Learning Goals: You should be able to:

1. Predict the maximum and minimum oxidation state of each halogen discussed in this chapter. Give an example of a compound containing the element in each of these oxidation states.

2. Cite the most common occurrences of each halogen discussed in this chapter.

3. Write balanced chemical equations describing the preparation of each of the hydrogen halides.

4. Describe at least one important use of each halogen element.

5. Write a balanced chemical equation describing the preparation of each of the hydrogen halides.

6. Give examples of diatomic and higher interhalogen compounds, and describe their electron-pair and molecular geometries.

7. Name and give the formulas of the oxyacids and oxyanions of the halogens.

8. Describe the variation in acid strength and oxidizing strength of the oxyacids of chlorine.

Review: Allotropes (7.7); VSEPR model (9.2).

22.5 OXYGEN

Learning Goals: You should be able to:

1. Cite the most common occurrences of oxygen and how it is obtained in its elemental form.

2. Cite at least two uses for oxygen.

3. Describe the allotropes of oxygen.

4. Describe the chemical and physical properties of hydrogen peroxide and its method of preparation.

Review: Catalysts (15.6); nomenclature of salts (2.6); polyprotic acids (16.6); use of hydrogen sulfide in qualitative analysis (17.7).

22.6 GROUP 6A: OXYGEN FAMILY

Learning Goals: You should be able to:

1. Predict the maximum and minimum oxidation state of any group 6A element discussed in the chapter. Give an example of a compound containing the element in each of those oxidation states.

2. Cite the most common occurrences of each group 6A element discussed in this chapter.

3. Cite the most common form of each group 6A element discussed in this chapter.

4. Indicate the formulas of the common oxides of sulfur and the properties of their aqueous solutions.
5. Write balanced chemical equations for formation of sulfuric acid from sulfur and describe the important properties of the acid.
6. Compare the chemical behaviors of selenium and tellurium with that of sulfur, with respect to common oxidation states and formulas of oxides and oxyacids.

22.7 NITROGEN

Review: Haber process (15.1).

Learning Goals: You should be able to:

1. Cite the most common occurrences of nitrogen and how it is obtained in its elemental form.
2. Cite at least two uses for nitrogen.
3. Write balanced chemical equations for formation of nitric acid via the Ostwald process starting from NH_3.
4. Cite examples of nitrogen compounds in which nitrogen is in the oxidation states from -3 to $+5$ and be able to name them.

22.8 GROUP 5A: NITROGEN FAMILY

Review: Haber process (15.1); polyprotic acids (16.6).

Learning Goals: You should be able to:

1. Predict the maximum and minimum oxidation state of each group 5A element discussed in the chapter. Give an example of a compound containing the element in each of those oxidation states.
2. Cite the most common occurrences of each group 5A element discussed in this chapter.
3. Cite the most common form of each group 5A element discussed in the chapter.
4. Describe the preparation of elemental phosphorus from its ores, using balanced chemical equations.
5. Describe the formulas and structures of the stable halides and oxides of phosphorus.
6. Write balanced chemical equations for the reactions of the halides and oxides of phosphorus with water.
7. Describe a condensation reaction and give examples involving compounds of phosphorus.

22.9 CARBON

Learning Goals: You should be able to:

1. Cite the most common occurrences of carbon and how it is obtained in its elemental form.
2. Cite at least two uses for carbon.
3. Describe the allotropes of carbon.
4. Distinguish among ionic, interstitial, and covalent carbides.

22.10 GROUP 4: CARBON FAMILY

Review: Exceptions to octet rule (8.8).

Learning Goals: You should be able to:

1. Cite the most common occurrences of silicon and the most common form of elemental silicon.
2. Cite the most common occurrences of carbon.

3. Describe the structures possible for silicates and their empirical formulas (for example, silicate tetrahedra can combine through bridging oxygens to form a single-string silicate chain whose empirical formula is SiO_3^{2-}).
4. Correlate the physical properties of certain silicate minerals, such as asbestos, with their structures.
5. Explain the changes in composition and properties that accompany substitution of Al^{3+} for Si^{4+} in a silicate.
6. Describe what is meant by a clay mineral, and explain the role of clay minerals in soil fertility.
7. Describe the composition and manufacture of soda-lime glass.

Learning Goals: You should be able to:

| **22.11 BORON**

1. Cite the most common occurrence of boron.
2. Describe the structure of diborane, and explain its unusual feature.
3. Describe a condensation reaction involving compounds of boron.

TOPIC SUMMARIES AND EXERCISES

Elements are grouped into three classes: metals, metalloids, and nonmetals.

| **PERIODIC TRENDS: METALS AND NONMETALS**

- **Nonmetals** are located in the right-hand portion of the periodic table.
- **Metalloids** are shown in Figure 22.1 of the text and occur between the metals and nonmetals.
- **Metals** lie to the left of metalloids.

Metals have lower electronegativity values than nonmetals.

- Electronegativity decreases down a given family and increases from left to right in a period (horizontal row of periodic table).
- Substances containing metals and nonmetals with large differences in electronegativities tend to be ionic.
- Nonmetals combine to form covalent substances.

There is one general periodic trend that metallic and nonmetallic elements have in common: *The first member of a metallic or nonmetallic family often has marked physical and chemical differences from subsequent members of the family.*

- For example, beryllium forms covalent compounds, whereas magnesium forms ionic ones.
- Another example is the fact that both carbon and nitrogen can more readily form multiple bonds to themselves (for example, $:N{\equiv}N:$ and $H{-}C{\equiv}C{-}H$), than other members of their families.

A large number of chemical reactions of nonmetals are presented in Chapter 22. Some principles of chemical reactions you should remember are:

- Water is produced when hydrogen-containing compounds are oxidized by $O_2(g)$. CO_2 is also produced when the compound contains carbon (in a limited amount of oxygen, CO or C can form). N_2 is usually formed when the compound also contains nitrogen. See reactions [22.1] and [22.2] in the text for examples.
- H^-, O^{2-}, NH_2^-, N^{3-}, CH_3^-, C^{4-}, and C_2^{2-} are all very strong bases toward protons. See reactions [22.2] and [22.3] in the text for examples.

EXERCISE 1   Classifying elements as metals, nonmetals, or metalloids

Carefully look at a periodic table and try to remember the relative locations of the more common elements. Without referring to the periodic table, try to classify the following elements as metals, nonmetals, or metalloids: Li; C; F; Ar; Sn; Sb; S; I.

SOLUTION: Metals: Li, Fe, Ba, Sn
                 Metalloid: Sb
                 Nonmetals: C, F, Ar, S, I

If you had trouble identifying the elements as to type, review the periodic table again. This will help you later with your studies of the properties of elements.

EXERCISE 2   Predicting trends in physical properties of elements

Write the elements in order of increasing value of the stated property.
Electronegativity: S, Na, Al, Cs
Metallic character: Ca, Cl, Al
Tendency to form pi bonds: C, Si, N
Atomic radius: Sr, Ca, Ba
Electrical conductivity: Ge, Na, S
Ionic character: H_2S, KBr, F_2

SOLUTION: *Analyze*: We are given a list of physical properties of elements and for each asked to write the elements given in increasing order of the indicated physical property.

Plan: Section 22.1 in the text reviews many of the physical properties and their trends. Review this section before determining the trends for each physical property.

Solve: Electronegativity: Cs < Na < Al < S (increases from left to right in a period and decreases down a family)

      Metallic character: Cl < Al < Ca (decreases from left to right across the periodic table)

      Tendency to form pi bonds: Si < N < C (first member of a nonmetal family more readily forms pi bonds)

      Atomic radius: Ca < Sr < Ba (increases down a family)

      Electrical conductivity: S < Ge < Na (increases with increasing metallic character)

      Ionic character: F_2 < H_2S < KBr (increases with increasing difference in electronegativity values between the bonded atoms)

EXERCISE 3   Explaining why tin can bond to six atoms but carbon does not

Explain why tin can form the ion $SnCl_6^{2-}$, but carbon cannot form CCl_6^{2-}.

SOLUTION: *Analyze*: We are asked why tin can form a complex ion with six chloride ions but carbon cannot.

Plan: We need to consider the different physical properties and electronic configuration of tin compared to carbon. We are looking for some aspect of silicon that would permit it to form six bonds to the chloride ion but prevents carbon from also forming the same type of complex ion. It likely deals with electronic structures and orbitals available for bonding.

Solve: Tin has low-energy, empty $4d$ orbitals in addition to its valence $5s$ and $5p$ orbitals which are able to form bonds with orbitals of other atoms. Thus tin can form more than four bonds; in the case of $SnCl_6^{2-}$, tin forms six bonds to chlorine atoms. Carbon does not have accessible low-energy, empty *d* orbitals that can bond with orbitals of other atoms. As a result, carbon can form only four bonds, utilizing its 2*s* and 2*p* valence orbitals. The larger size of tin compared to carbon also permits it to accommodate six bonded atoms.

HYDROGEN

Hydrogen occurs mainly in a combined state, primarily in water and in organic compounds (C,H-containing compounds).

- Hydrogen gas (H_2) is not very reactive as an element; however it will react spontaneously with the most electropositive elements (for example, Na, K) and the most electronegative elements (for example, F_2). The low reactivity of dihydrogen is attributable to the large energy required to break H—H bonds (436 kJ/mol).
- One of the strongest covalent bonds is H—O, with a bond enthalpy of 463 kJ/mol; thus reactions of H_2 (to form O—H-containing compounds) occur readily with many oxygen-containing compounds.

Summarized below is other important information contained in Section 22.3.

Abundance: 11% of the earth's crust by mass; 70% of the universe contains hydrogen.

Uses: In the manufacture of methanol (CH_3OH).

Preparation:

1. Reaction of active metals with mineral acid

$$Zn(s) + 2\,H^+(aq) \longrightarrow Zn^{2+}(aq) + H_2(g).$$

2. As a byproduct in the production of methane (CH_4).

$$CH_4(g) + H_2O(g) \xrightarrow[\text{1200 K, 30 atm}]{\text{Ni catalyst}} \underbrace{CO(g) + 3\,H_2(g)}_{\text{water gas}}$$

Isotopes:

1. ^1H: (protium) 99.984% abundance.
2. ^2H (deuterium, D); 0.016% abundance.
3. ^3H (tritium, T); rare; prepared by nuclear transformation, radioactive.

Deuterium and tritium undergo reactions at slower rates than ^1H. Both are used to "label" reactions by replacing ^1H in compounds and thus enabling one to follow how hydrogen changes locations in reactions.

Binary Compounds:

1. **Ionic hydrides:** H^- forms with the most electropositive metals (group 1A and all of group 2A except Be and Mg). The H^- ion is a strong base: $KH(s) + H_2O(l) \longrightarrow H_2(g) + KOH(aq)$.
2. **Molecular hydrides:** Hydrogen binds covalently to nonmetals or metalloids; it may exist in the +1 or −1 oxidation state in these compounds. The thermal stability of molecular hydrides decreases down a family.

3. **Metallic hydrides:** These are compounds of transition metals and hydrogen, which possess metallic properties. Many are nonstoichiometric hydrides, called interstitial hydrides (M_yH_x, where the ratio of y to x is not a ratio of whole numbers). Some metallic hydrides are solutions of hydrogen atoms in a transition metal matrix.

EXERCISE 4   Writing chemical reactions involving hydrogen

Write equations to describe the following: (**a**) Al reacts slowly with steam to form hydrogen gas. (**b**) Ni reacts with a strong acid to form hydrogen gas. (**c**) Ba reacts with cold water to form hydrogen gas. (**d**) Oxidation of metallic Cu by NO_3^- in acid solution forms $NO_2(g)$.

SOLUTION: *Analyze*: We are given descriptions of four chemical reactions and are asked to write the chemical reactions from the information.

Plan: We first need to convert the chemical names to chemical formulas and place reactants to the left of a reaction arrow. Then, from the given information about products, write the products. In some cases we may have to determine the other products from the one product given. Our familiarity with chemical compounds will help with this.

Solve: (**a**) Metals such as Al, Mn, Zn, Cr, Cd, and Fe react slowly with steam. The reaction for Al is $2\,Al(s) + 6\,H_2O(g) \rightarrow 2\,Al(OH)_3(s) + 3\,H_2(g)$. (**b**) Most metals, except Cu, Bi, Hg, and Ag, react with strong acids to form H_2. The reaction of Ni is $Ni(s) + 2\,H^+(aq) \rightarrow Ni^{2+}(aq) + H_2(g)$. (**c**) Active metals, such as Cs, K, Na, Ba, Sr, and Ca react with cold H_2O to form $H_2(g)$. The reaction of barium is

$$Ba(s) + 2\,H_2O(l) \longrightarrow Ba(OH)_2(s) + H_2(g).$$

(**d**) $Cu(s) + 2\,NO_3^-(aq) + 4\,H_3O^+(aq) \longrightarrow Cu^{2+}(aq) + 2\,NO_2(g) + 6\,H_2O(l).$

EXERCISE 5   Identifying hydrides as ionic, metallic, or molecular

Identify the following hydrides as ionic, metallic, or molecular: NaH; SiH_4; HCl; PdH_2; CaH_2; SbH_3.

SOLUTION: *Analyze*: We are asked to identify six compounds as ionic, metallic, or molecular.

Plan: Ionic hydrides are a class of hydrides with Group 1A and Group 2A elements. Metallic hydrides are a class of hydrides with transition metals. Molecular hydrides are a class of hydrides that contain nonmetals. By looking at the nonhydride element in each compound we should be able to identify the class of hydride.

Solve: Ionic hydrides: NaH and CaH_2. Metallic hydrides: PdH_2. Nonmetallic hydrides: SiH_4, SbH_3, and HCl.

GROUP 8A: NOBLE GASES

The noble gas family consists of helium, neon, argon, krypton, xenon, and radon. Because of their low concentration in the earth's crust and atmosphere, they are sometimes referred to as rare gases.

- Isotopes of radon are radioactive and are formed from the decay of other radioisotopes.
- Helium is found in natural gas wells under the surface of the earth. It boils at 4.2 K under one atmosphere pressure.

- Neon, krypton, argon, and xenon can be obtained by fractional distillation of liquid air.

Helium has a valence-shell electron configuration of $1s^2$; all other noble gases have the general valence-shell electron configuration of ns^2np^6.

- Note that all noble gases have "closed" outer valence shells. Thus noble gases do not readily combine with other atoms.
- The first ionization energies for xenon and krypton are similar to those for nitrogen, bromine, and iodine. Therefore, it is not unrealistic to expect xenon to exist in positive oxidation states and to form bonds with highly electronegative elements. Examples of xenon compounds are XeF_2, XeF_4, XeO_3, XeO_4, and XeO_2F_2. Only one binary compound of krypton is known, KrF_2; it decomposes at $-10\ °C$.

EXERCISE 6 Predicting the trend in molar heats of vaporization of the noble gases

Predict the trend in molar heats of vaporization of the noble gases.

SOLUTION: *Analyze and Plan*: We are asked to predict the molar heats of vaporization of the noble gases. This process involves $X(l) \rightarrow X(g)$, which requires breaking intermolecular forces between noble gas atoms in the liquid state. We need to analyze the types of intermolecular forces and how they affect heats of vaporization.

Solve: The molar heat of vaporization (energy required to vaporize a liquid at its normal boiling temperature) is an indicator of the magnitude of intermolecular attractive forces between atoms in the liquid state. We learned in Section 11.2 of the text that London dispersion forces between atoms increase with increasing mass of the atoms. Since noble gases exist as monatomic elements and are nonpolar, the only forces holding them together in the liquid state are London dispersion forces. Therefore the molar heats of vaporization of noble gas elements should increase with increasing atomic mass. The predicted trend in molar heats of vaporization is thus Xe > Kr > Ar > Ne > He.

EXERCISE 7 Predicting the geometry of the electron domains about Xe in XeF_2 and the molecular shape of XeF_2

Describe the electron-domain and geometrical shape of XeF_2.

SOLUTION: *Analyze and Plan:* We are asked to describe the electron-domain geometry and geometrical shape of XeF_2. In Chapter 9 we learned to use the VSEPR model to answer this type of question. We will apply the principles of this model to the question.

Solve: First write the Lewis structure of XeF_2, using procedures found in Section 8.5 of the text:

Xe:	8 valence-shell electrons
2 F:	7 valence-shell electrons each, or a total of 14 electrons
XeF_2:	22 valence-shell electrons

To construct the Lewis structure of XeF_2, we will have to expand the number of electrons about xenon beyond an octet:

$$:\ddot{F} - .\ddot{Xe}. - \ddot{F}:$$

We see xenon has five electron domains about it. Using procedures given in Sections 9.1 and 9.2 of the text, we predict that the electron domains of xenon are in a trigonal bipyramidal arrangement. To minimize lone pair–lone pair and lone pair–bonding pair electron-domain repulsions, the fluorine atoms are placed in the axial positions in a trigonal bipyramidal structure. Therefore we expect XeF_2 to be linear in shape. Experimental evidence verifies this prediction.

GROUP 7A; HALOGENS

The halogen family consists of the elements fluorine, chlorine, bromine, iodine, and astatine. Since all isotopes of astatine are radioactive and have short half-lives, the discussion of the chemistry of halogens in Section 22.4 of the text concentrates on fluorine, chlorine, bromine, and iodine. Some of the properties of these elements are summarized in Table 22.1.

The halogens are highly reactive. The relatively low bond enthalpy of F_2 causes elemental fluorine to be extremely reactive.

- Halogens have high electronegativities and thus tend to act as oxidizing agents. A given halogen is able to oxidize the anions of halogens below it in the family. For example, Br_2 will oxidize I^- but not Cl^- and F^-.
- Elemental fluorine is prepared from the electrolysis of molten KHF_2 to form elemental fluorine and hydrogen along with potassium fluoride. It is too reactive to be prepared in an aqueous medium.
- Elemental chlorine is produced from the electrolysis of aqueous sodium chloride or molten sodium chloride.
- Both elemental chlorine and bromine can be prepared from brines containing the halide ions. Chlorine gas is used to oxidize the halide ions to the elemental halogens.

Hydrogen halides, HX, are covalent gases at room temperature and pressure.

- All hydrogen halides dissolve in water to form acidic solutions.
- All but HF are strong acids in water.
- The weak acid characteristics of HF(aq) are partly attributable to the fact that it forms hydrogen bonds to other HF molecules and water.

TABLE 22.1 **Properties of Elements in the Halogen Family**

Element	Elemental form	Highest and lowest oxidation state, with example	Common source	Example of use
Fluorine	$F_2(g)$	-1, HF 0, F_2	CaF_2	Teflon
Chlorine	$Cl_2(g)$	-1, HCl $+7$, Cl_2O_7	NaCl in seawater	Bleaching powder, Ca(ClO)Cl, NaClO
Bromine	$Br_2(l)$	-1, HBr $+7$, $HBrO_4$	NaBr in seawater	Organic bromine compounds
Iodine	$I_2(s)$	-1, HI $+7$, IF_7	NaI in seawater	Tincture of iodine (iodine in ethyl alcohol)

- Several synthetic methods are used to prepare hydrogen halides, and you should learn at least one method for preparing each as given in Section 22.4 of the text.

Oxyacids are another group of halogen compounds that act as acids in water.

- A discussion of the relationships between their structures and acid properties occurs in Section 16.10 and 16.11 of the text.
- Examples of oxyacids are HFO, HClO, HBrO, HIO, $HClO_2$, $HClO_3$, $HBrO_3$, HIO_3, $HClO_4$, and H_5IO_6.
- We should review the nomenclature rules for oxyanions (Section 2.8 of the text).
- Both oxyacids and their oxyanions are excellent oxidizing agents, and several are thermally unstable.
- Reduction potentials of oxyanions decrease with increasing oxidation number of the halogen. Thus ClO_4^- is a weaker oxidizing agent than ClO^-.

Since halogens are similar in many ways, we find many binary compounds in which different halogens are bonded to one another.

- Examples are ClF, BrF_3, ClF_3, BrF_5, IF_5, and IF_7.
- For interhalogens of type XX'_n ($n = 3, 5,$ or 7) where $X = $ Cl, Br, or I, X' is nearly always fluorine.
- The structures of interhalogen compounds can be prediced using procedures found in Sections 9.1 and 9.4 of the text. These structures are consistent with bonding models utilizing d-orbital participation and hybridization of valence-shell orbitals of the central halogen in a positive oxidation state.

EXERCISE 8   Providing examples of compounds of halogens in different oxidation states

Review the oxidation states of halogen compounds. Try to remember examples for each case. Give three examples of halogen compounds in which a halogen atom is in each of the following oxidation states: +1, +3, +5, +7.

SOLUTION: The positive oxidation states for the halogens and examples for each are as follows:

+1	+3	+5	+7
ClF,	ClF_3,	BrF_5, IF_5, I_2O_5,	IF_7, $HClO_4$,
Cl_2O,	BrF_3,	$HClO_3$	Cl_2O_7
HBrO	$HClO_2$		

EXERCISE 9   Providing examples of compounds of halogens in which the halogen atoms have different characteristics

Give examples of halogen compounds in which: (**a**) the central halogen uses d orbitals to expand its octet of electrons (**b**) the halogen occurs as a halide ion (X^-) and forms an ionic bond; (**c**) the halogen forms polar and nonpolar bonds with other like or differing elements; (**d**) the halogen acts as electron donor by reacting as a halide ion with a Lewis acid; (**e**) a halogen forms a hydrogen bond.

SOLUTION: *Analyze and Plan*: We are asked to give examples of halogen compounds existing in six different types of conditions. We need to use the information in Section 22.4 of the text and the summary in the *Student's Guide*, and our knowledge of electronic configurations, polarity, and characteristics of different types of bonding to develop responses.

Solve: (**a**) All but fluorine can use d orbitals to expand their octet of electrons. Examples are ClF_2^-, ICl_4^-, and ClF_3. You should draw the Lewis structures for these ions and compounds to check that their octets are in fact expanded. (**b**) All the halogens form ionic salts with group 1A and 2B cations. Examples are NaF, KI, $CaBr_2$, and $CaCl_2$. (**c**) Halogens form nonpolar covalent bonds in homonuclear diatomic molecules such as F_2, Cl_2, and Br_2. Polar covalent bonds result when a halogen bonds to an element with a different electronegativity. Examples are ICl, CH_3F, and OF_2. (**d**) Halide ions act as donor atoms when reacting with compounds that are good Lewis acids. Two examples are the following:

$$BF_3 + F^- \longrightarrow BF_4^-$$

$$SnCl_4 + 2\,Cl^- \longrightarrow SnCl_6^{2-}$$

(**e**) Fluorine bonded to hydrogen in HF is the primary example of a halogen compound that forms hydrogen bonds.

EXERCISE 10   Naming oxyacids and listing oxyacids in order of acid strength

(**a**) Name the following oxychlorine acids and determine the oxidation number of chlorine in each: HClO; $HClO_2$; $HClO_3$; $HClO_4$. (**b**) List these acids in order of increasing acid strength. Briefly justify your order.

SOLUTION: *Analyze*: We are asked to name the oxyacids of chlorine and determine the oxidation state of chlorine in each. In addition we are asked to list the oxyacids in order of their increasing acid strength and to draw the Lewis structure of ClO_3^-.

Plan: In Section 2.8 we learned how to name oxyacids and we can apply these principles. The oxidation number of the central halogen atom can be determined by assigning each oxygen atom an oxidation number of -2 and each hydrogen atom an oxidation number of $+1$ and remembering that the sum of oxidation numbers equals the charge of the substance, which is zero for compounds.

Solve: (**a**) The names and oxidation states are as follows:

Oxyacid	Name	Halogen oxidation number
HClO	Hypochlorous	+1
$HClO_2$	Chlorous	+3
$HClO_3$	Chloric	+5
$HClO_4$	Perchloric	+7

(**b**) The order of acid strengths is $HClO < HClO_2 < HClO_3 < HClO_4$. As the oxidation number of the central atom increases in a series of oxyacids containing the same central atom, the acid strengths increase.

OXYGEN

Summarized below is important information contained in Section 22.5.

Abundance: Oxygen occurs as O_2 in the atmosphere (21% by volume, 23% by weight). Air, earth, and seas contain about 50% by weight.

Allotropes: Two forms of elemental oxygen exist: O_2 (oxygen) and O_3 (ozone). Ozone is less stable than O_2 at room temperature and pressure:

$$3\,O_2(g) \longrightarrow 2\,O_3(g) \qquad [\Delta H^\circ = +285\text{ kJ}].$$

Uses: O_2 is used as an oxidizing agent, particularly in combustion reactions, such as oxyacetylene welding. Ozone is used as a powerful oxidizing agent, with O_2 often formed as one of the products.

Preparation: O_2: fractional distillation of liquid air is the industrially important method. Thermal decomposition of oxides may be used in the laboratory:

$$2\,KClO_3(s) \xrightarrow[\Delta]{MnO_2} 2\,KCl(s) + 3\,O_2(g).$$

O_3 (ozone): Pass O_2 through an electrical discharge:

$$3\,O_2(g) \xrightarrow[\text{discharge}]{} 2\,O_3(g).$$

Compounds:

1. **Oxides:** Compounds containing oxygen in the -2 oxidation state. Oxides of metallic and nonmetallic elements exhibit differing chemical behaviors. Metal oxides are called **basic oxides/basic anhydrides.** Soluble metal oxides, such as CaO, react with water to form basic solutions:

$$CaO(s) + H_2O(l) \longrightarrow Ca(OH)_2(s)$$

Insoluble metal oxides, such as Al_2O_3, react with acids to give a salt and water:

$$Al_2O_3(s) + 6\,HNO_3(aq) \longrightarrow 2\,Al(NO_3)_3(aq) + 3\,H_2O(l)$$

Soluble nonmetallic oxides, such as SO_3, react with water to form acidic solutions:

$$SO_3(g) + H_2O(l) \longrightarrow H_2SO_4(aq)$$

Insoluble nonmetallic oxides react with bases. **Acid anhydride** or **acidic oxide** are other terms for nonmetallic oxides.

2. **Peroxides:** Compounds containing O—O bonds with each oxygen in an oxidation state of -1. Ionic peroxides are Na_2O_2, CaO_2, SrO_2, and BaO_2. Hydrogen peroxide, H_2O_2, contains a covalently bonded O—O group. A 3%-by-weight aqueous H_2O_2 solution is commonly used as a bleaching agent. H_2O_2 is formed by reacting water with the persulfate ion, $S_2O_8^{2-}(aq)$. Concentrated H_2O_2 is dangerous because it can decompose to $H_2O(l)$ and $O_2(g)$ with explosive violence.

3. **Superoxides:** Compounds containing the O—O$^-$ ion with each oxygen in a $-\frac{1}{2}$ oxidation state: examples are CsO_2, RbO_2, KO_2. The superoxide ion only occurs with the most active metals. KO_2 is used as a source of $O_2(g)$ because it readily reacts with H_2O to form $O_2(g)$, KOH(aq), and $H_2O_2(aq)$.

EXERCISE 11 Providing examples of compounds of oxygen in which the oxygen atoms have different characteristics

Give an example of an oxygen-containing compound in which oxygen: (**a**) forms a π bond; (**b**) is in a positive oxidation state; and (**c**) expands its octet.

SOLUTION: *Analyze and Plan:* We are asked to give an example of an oxygen-containing compound in four different conditions. We can use the principles used in Exercise 9, the information in Section 22.5, and the summary in the *Student's Guide* to determine the examples.

Solve: (a) Oxygen can form π bonds using its p orbitals such as in CO. (b) Oxygen occurs in a positive oxidation state only with fluorine: O_2F (Oxygen is in a $+\frac{1}{2}$ oxidation state). (c) Oxygen cannot expand its octet of electrons because it has no empty d orbitals of low energy; thus there are no examples.

EXERCISE 12   Identifying oxides as acidic or basic and writing their reactions with water

Identify each of the following as a basic or acidic oxide and write its reaction with water: (a) $N_2O_5(g)$; (b) $Na_2O(s)$; (c) $SO_2(g)$; (d) $MgO(s)$.

SOLUTION: *Analyze and Plan:* We are asked to determine for four different compounds if they are basic or acidic oxides and to write their chemical reactions with water. We can determine the type of oxide by inspection: A basic oxide contains a metal and oxygen atoms whereas an acidic oxide contains a nonmetal with oxygen atoms. The oxides have to be soluble in water or react with acid or base, depending on the type.

Solve: (a) An acidic oxide because nitrogen is a nonmetal:

$$N_2O_5(g) + H_2O(l) \longrightarrow 2\,HNO_3(aq)$$

(b) A basic oxide because sodium is a metal. The oxide ion reacts with water to form hydroxide ion—that is, $O^{2-}(aq)$ combines with H_2O to form $2\,OH^-(aq)$:

$$Na_2O(s) + H_2O(l) \longrightarrow 2\,NaOH(aq)$$

(c) An acidic oxide because sulfur is a nonmetal:

$$SO_2(g) + H_2O(l) \longrightarrow H_2SO_3(aq)$$

(d) A basic oxide because magnesium is a metal:

$$MgO(s) + H_2O(l) \longrightarrow Mg(OH)_2(s)$$

EXERCISE 13   Identifying a superoxide ion and writing its chemical reaction with carbon dioxide

Space-science chemists are interested in the use of the superoxide ion to remove CO_2 from the atmosphere to form the carbonate ion. There is another product of this reaction. What is it? Write the reaction.

SOLUTION: *Analyze and Plan:* We are asked how a superoxide ion reacts with carbon dioxide to form a carbonate ion. We need to recognize that this will be an oxidation-reduction reaction and apply the principles of oxidation-reduction, including changes in oxidation numbers, to predict the products of the chemical reaction and to write the chemical reaction.

Solve: If O_2^- reacts with CO_2 to form CO_3^{2-}, oxygen in O_2^- must be both oxidized and reduced—the oxidation states of C and O in CO_2 and CO_3^{2-} are the same. In forming CO_3^{2-}, the change in oxidation state of oxygen in O_2^- is $-\frac{1}{2}$ to -2 (reduction). The other part of the reaction must then involve oxidation—the increase in the oxidation number of oxygen. Thus O_2 (zero oxidation state) is the likely other product:

$$4\,O_2^- + 2\,CO_2 \longrightarrow 2\,CO_3^{2-} + 3\,O_2.$$

TABLE 22.2 Properties of Elements of the Oxygen Family

Element	Common elemental form	Highest and lowest oxidation state, with example	Common source	Example of use
Oxygen	O_2 (O_3 is an allotropic form)	-2, MgO $+2$, OF_2	Air	A combustion agent as O_2
Sulfur	Rhombic sulfur, S_8 (polymeric)	-2, H_2S $+6$, SF_6	CuS	H_2SO_4, used in industrial processes
Selenium	Se (polymeric solid)	-2, H_2Se $+6$, SeO_3	Cu_2Se	Used in copying machines as conductor
Tellurium	Te (polymeric solid)	$+2$, H_2Te $+6$, TeO_3	Cu_2Te	Used in inorganic syntheses, but not commonly

GROUP 6A: THE OXYGEN FAMILY

The oxygen family consists of the elements oxygen, sulfur, selenuim, tellurium, and polonium. Since polonium is radioactive, its chemistry is not discussed. Some of the properties of these elements and their oxidation states are summarized in Table 22.2.

- Elements of the oxygen family have the general valence-shell electron configuration ns^2np^4.
- Several oxidation states are known for oxygen, but it usually occurs in the -2 oxidation state.
- Oxygen is highly electronegative and readily gains two electrons to form the oxide ion, O^{2-}.
- Sulfur commonly occurs in the -2, $+4$, and $+6$ oxidation states.

The chemistry of sulfur is the primary emphasis in this section.

- Sulfur has several allotropic forms; the more familiar form is yellow rhombic sulfur, which consists of puckered S_8 rings. Elemental sulfur occurs naturally in rocks deep below the surface of the earth and is brought to the surface and extracted by the Frasch process.
- Combustion of sulfur in air produces SO_2 and small amounts of SO_3. Both oxides dissolve in water to form acidic solutions. Gaseous SO_2 dissolves to form $SO_2(aq)$, often written as $H_2SO_3(aq)$. $SO_3(g)$ dissolves slowly to form $H_2SO_4(aq)$. Sulfuric acid, H_2SO_4, is a strong acid, extensively used in industrial and manufacturing processes.
- An ion related to the sulfate ion, SO_4^{2-}, is the thiosulfate ion, $S_2O_3^{2-}$. The prefix *thio-* tells us a sulfur atom has been substituted for an oxygen atom. $Na_2S_2O_3 \cdot 5\,H_2O$ is referred to as hypo; it is used in photography as a source of $S_2O_3^{2-}$ for complexing silver ions. An important quantitative application of $S_2O_3^{2-}$ is its use as an oxidizing agent for I_2.

E X E R C I S E 14 Providing examples of compounds of oxygen and sulfur with similar characteristics

Give an example of an oxygen-containing compound and a sulfur-containing compound in which the oxygen and sulfur atoms: (**a**) form π bonds; (**b**) expand their octet of electrons and (**c**) are in a positive oxidation state.

SOLUTION: *Analyze and Plan*: We are asked to give examples of oxygen- and sulfur-containing compounds in four different conditions. We can use the principles given in Exercise 13, the information in Sections 22.5 and 22.6 of the text and the summary section to provide examples.

Solve: (a) Oxygen can form π bonds using its p orbitals, such as in

$$\ddot{O}=C=\ddot{O}$$

Sulfur forms π bonds in compounds such as

$$\ddot{S}=C=\ddot{S}$$

(b) Oxygen cannot expand its octet of electrons because it has no empty d orbitals of low energy: thus there are no examples. Sulfur can expand its octet of electrons because it has empty $3d$ orbitals of sufficiently low energy; SF_6 is an example.
(c) Oxygen occurs in the $+\frac{1}{2}$ oxidation state in O_2F. Oxygen is assigned a positive oxidation number because it has a lower electronegativity than fluorine. Sulfur occurs in the $+4$ oxidation state in SF_4.

EXERCISE 15   Identifying the acidic or basic properties of $SO_2(g)$

(a) What type of acid or base is $SO_2(g)$? (b) Write its reaction with water. (c) Write a balanced net ionic equation for the reaction of $SO_2(g)$ with aqueous NaOH.

SOLUTION: *Analyze and Plan*: We are asked to determine what type of acid $SO_2(g)$ is, to write its reaction with water, and to write the net ionic equation for its reaction with NaOH. It is a nonmetallic oxide, thus it should be an acidic oxide; when it reacts with water it should form the hydronium ion. When it reacts with NaOH it should form water.

Solve: (a) Group 6A dioxides dissolve in water to form slightly acidic solutions. $SO_2(g)$ acts thus as a Lewis acid (it possesses no ionizable protons) when it dissolves in water. (b) $SO_2(g) + H_2O(l) \rightarrow H^+(aq) + HSO_3^-(aq)$. (c) Note that the weakly acidic ion HSO_3^- is formed in water. Therefore the reaction of $SO_2(g)$ with a NaOH(aq) can be viewed as the reaction of $HSO_3^-(aq)$ with $OH^-(aq)$ to form $SO_3^{2-}(aq)$ and $H_2O(l)$:

$$HSO_3^-(aq) + 2\,OH^-(aq) \longrightarrow SO_3^{2-}(aq) + H_2O(l)$$

EXERCISE 16   Characterizing the term "thio"

(a) What does the term "thio" mean? (b) Write the reaction for the reduction of iodine in the presence of aqueous thiosulfate ion.

SOLUTION: *Analyze and Plan*: We are asked to identify what the term "thio" means. We are also asked to write a chemical reaction between iodine and thiosulfate ion. We need to review Section 22.6 and the summary section to determine the meaning of the terms. We are told that the chemical reaction we need to write is an oxidation-reduction type: Iodine is reduced, that is, it gains electrons to form the iodide ion. This helps us to complete the reaction.

Solve: (a) The term "thio" means that a sulfur has been substituted for an oxygen in a compound. When an oxygen atom in the sulfate ion, SO_4^{2-}, is substituted by a sulfur atom, the thiosulfate ion is formed. (b) The thiosulfate ion is used in quantitative analysis as a reducing agent for iodine.

$$2\,S_2O_3^{2-}(aq) + I_2(s) \longrightarrow 2\,I^-(aq) + S_4O_6^{2-}(aq)$$

Nitrogen is a gaseous diatomic element (N_2). | **NITROGEN**

- The high $N\equiv N$ bond enthalpy of +944 kJ/mol causes an extremely stable molecule.
- Because nitrogen's valence electronic structure is $2s^2 2p^3$, nitrogen tends principally to form the oxidation states of +3 and +5; however, nitrogen exhibits all integral oxidation numbers from +5 to −3.
- The ability of nitrogen to acquire three electrons to form N^{3-} reflects its high electronegativity value. This high electronegativity partially accounts for nitrogen's ability to be involved in hydrogen bonding.

Summarized below is other important information contained in Section 22.7.

Abundance: The greatest source of N_2 is the atmosphere, which contains 78% nitrogen by volume. It is also found as a constituent of living matter, particularly in proteins.

Uses: Nitrogen is used extensively in fertilizers such as $(NH_4)_3PO_4$. It is also used in nitrogen fixation, the formation of nitrogen compounds in certain plants.

Preparation: Obtained commercially by fractional distillation of liquid air.

Compounds:

1. Important hydrogen-containing compounds of nitrogen are ammonia (NH_3), hydrazine (N_2H_4), hydroxylamine (NH_2OH), and hydrogen azide (HN_3). Ammonia is the most stable of these; the others are highly reactive.
2. Nitrogen forms a number of oxides; the three most common are N_2O, NO, and NO_2.
3. Nitric acid (HNO_3), a commercially important strong acid, is formed by the catalytic conversion of NH_3 to NO, followed by the reaction of NO with O_2 to form NO_2, and the final step being the reaction of NO_2 with water to form nitric acid. The three-step process for the formation of nitric acid is known as the Ostwald process.
4. Nitrous acid (HNO_2) is less stable than HNO_3 and tends to disproportionate (nitrogen is both oxidized and reduced) to NO and HNO_3.

EXERCISE 17   Naming the oxides of nitrogen

Name the following oxides of nitrogen and determine the oxidation number of nitrogen in each: N_2O, NO, N_2O_3, NO_2, and N_2O_5.

SOLUTION: *Analyze and Plan*: We are asked to name five oxides of nitrogen and to determine the oxidation number of nitrogen in each. We should review the principles of naming nonmetallic compounds in Section 2.8 of the text. The oxidation number of nitrogen is determined as we did in Exercise 12.

Solve: These five oxides of nitrogen are examples of nitrogen-containing compounds that show that nitrogen can have all integral positive oxidation numbers from +1 to +5.

Oxide	Name	Oxidation number of nitrogen
N_2O	Nitrogen(I) oxide or nitrous oxide	+1
NO	Nitrogen(II) oxide or nitric oxide	+2
N_2O_3	Nitrogen(III) oxide or dinitrogen trioxide	+3
NO_2	Nitrogen(IV) oxide or nitrogen dioxide	+4
N_2O_5	Nitrogen(V) oxide or dinitrogen pentoxide	+5

EXERCISE 18   Writing Lewis structures for dinitrogen oxide

Dinitrogen oxide possesses a linear, unsymmetrical structure. Write all reasonable Lewis structures for dinitrogen oxide.

SOLUTION: *Analyze and Plan*: We are asked to write the Lewis structures for dinitrogen oxide and are told information about its structure. The information about the structure helps us identify the arrangement of the atoms in the compound; it has to be nonsymmetrical. We can use the principles in Section 8.5 to write Lewis structures.

Solve: N_2O possesses 16 valence electrons that are used in forming the Lewis structure. Because the structure is unsymmetrical, the skeletal arrangement of atoms must be N—N—O. Two reasonable resonance forms exist using this structure:

$$:\ddot{N} = N = \ddot{O}: \longleftrightarrow :N \equiv N - \ddot{\underset{..}{O}}:$$

GROUP 5A: NITROGEN FAMILY

The nitrogen family consists of the elements nitrogen, phosphorus, arsenic, antimony, and bismuth. Some of the properties of these elements and their oxidation states are summarized in Table 22.3 and below.

- Elements of the nitrogen family have the general valence-shell electron configuration ns^2np^3.
- All elements exhibit the +5 oxidation state in compounds, but only nitrogen forms a 3− ion (nitride ion) with the more active metals.

TABLE 22.3 Properties of Elements of the Nitrogen Family

Element	Common elemental form	Highest and lowest oxidation state, with example	Common source	Example of use
Nitrogen fertilizers	$N_2(g)$	-3, NaN_3 $+5$, HNO_3	Air	HNO_3, nitrogen-containing
Phosphorus	White phosphorus, $P_4(s)$; red phosphorus, $P_x(s)$	-3, Be_3P_2 $+5$, P_4O_{10}	$Ca_3(PO_4)_2$	Phosphate-containing fertilizers
Arsenic $As_4(g)$	$As(s)$ As_2S_3	-3, Mg_3As_2 $+5$, AsF_5	As_4S_4	Pesticides, poisons
Antimony	$Sb(s)$ $Sb_4(g)$	0, $Sb(s)$ $+5$, SbF_5	Sb_2S_3	Fe-Sb alloy in pewter
Bismuth	$Bi(s)$	0, $Bi(s)$ $+5$, $BiCl_5$	Bi_2S_3	Used in emetics

Phosphorus exists in several allotropic forms: white, red, and black.

- White phosphorus consists of P_4 tetrahedra. This form of the element is highly reactive.
- Red phosphorus is the most stable allotropic form; it consists of chains of phosphorus atoms.
- Phosphorus is prepared by the reduction of phosphate in $Ca_3(PO_4)_2(s)$ with coke (a special form of carbon) and SiO_2. See equation (22.51) in the text.

All of the elements form gaseous hydrides with the general formula MH_3.

- The stability of the hydrides decreases with increasing atomic mass in the family, with SbH_3 and BiH_3 being thermally unstable.
- NH_3 differs significantly in its physical properties from the other group 6A hydrides because it associates through hydrogen bonding.

Halide compounds are known for all group 6A elements.

- Both MX_3 and MX_5 halides exist for all group 6A elements except nitrogen. Nitrogen does not have low-energy, empty d orbitals to expand its octet; therefore, it forms only MX_3 halides.
- Phosphorus halides are extensively discussed in Section 22.8 of the text. You should notice how they are prepared, what their structures are, and how they react with water.

Oxides of nitrogen and phosphorus form an important class of compounds.

- Nitrogen is represented among the nitrogen oxides in all integral oxidation numbers ranging from +1 to +5. These compounds are all covalent. Nitrogen oxides are strong oxidizing agents.
- Phosphorus forms two important oxides, P_4O_6 and P_4O_{10}. Phosphorus(III) oxide is the anhydride of phosphorous acid (H_3PO_3), and phosphorus(V) oxide is the anhydride of phosphoric acid (H_3PO_4). Phosphoric acid undergoes a condensation reaction when heated to form pyrophosphoric acid ($H_4P_2O_7$):

$$2\,H_3PO_4 \xrightarrow{\Delta} H_4P_2O_7 + H_2O$$

EXERCISE 19   Explaining why phosphoric acid is triprotic and phosphorous acid is diprotic

A hydrogen atom attached to phosphorus (P—H) is less easily replaced than a hydrogen atom attached to an oxygen atom in a P—O—H linkage in oxyacids of phosphorus. Suggest a reason for the observation that phosphoric acid is triprotic (3 ionizable hydrogens) and phosphorus acid is diprotic (2 ionizable hydrogens) even though both contain three hydrogen atoms.

SOLUTION: *Analyze and Plan*: We are asked to explain why phosphoric acid is triprotic and phosphorous acid is diprotic. We are given information about the stability of a P—H bond compared to a O—H bond when it is in the form P—O—H. The information tells us that the acidity is due to the presence of a P—O—H bond and not a P—H bond. We can write skeletal structures based on this analysis.

Solve: The formulas for phosphoric and phosphorus acids are H_3PO_4 and H_3PO_3, respectively. Hydrogen atoms attached to oxygen atoms that are bonded to phosphorus atoms in oxyacids of phosphorus are expected to be replaceable, and therefore acidic compared to hydrogen atoms attached to phosphorus. Since H_3PO_4 has three ionizable hydrogen atoms, all three must be attached to oxygen atoms. All of the hydrogen atoms cannot be attached to oxygen atoms in H_3PO_3 because only two hydrogen atoms ionize. The third hydrogen atom in H_3PO_3 must be attached to an atom other than oxygen, which only leaves the phosphorus atom. The skeletal structures for the acids are

$$
\begin{array}{cc}
\text{OH} & \text{H} \\
| & | \\
\text{O}-\text{P}-\text{OH} \qquad & \text{O}-\text{P}-\text{OH} \\
| & | \\
\text{OH} & \text{OH}
\end{array}
$$

Phosphoric acid   Phosphorous acid

EXERCISE 20   Writing the combination reactions of P_4 with oxygen, a halogen, and sulfur and naming the products

Molecular weight studies of white phosphorus indicate that it exists as tetrahedral P_4 molecules. Write equations illustrating the combination reactions of P_4 with oxygen, a halogen, and sulfur. Name the products formed.

SOLUTION: *Analyze and Plan*: We are asked to write the chemical reaction of P_4 with oxygen, a halogen, and sulfur. All reactions described are direct combinations. We need to review Section 22.8 in the text and the summary to help us determine the nature of the product in each reaction.

Solve: All of the reactions are direct combinations:

$$P_4(s) + 5\,O_2(g) \longrightarrow 2\,P_2O_5(s) \qquad \text{(or } P_4O_{10} \text{ its actual form)}$$
$$\text{diphosphorus pentaoxide}$$

$$P_4(s) + 10\,Cl_2(g) \longrightarrow 4\,PCl_5(g) \qquad \text{phosphorus pentachloride}$$

$$P_4(s) + \tfrac{5}{4}\,S_8(s) \longrightarrow 2\,P_2S_5 \qquad \text{diphosphorus pentasulfide}$$

EXERCISE 21   Identifying the main source of phosphorus and explaining how phosphorus can form "superphosphate"

(a) What is the main source of phosphorus? (b) Why isn't this source used as a phosphatic fertilizer? (c) How can it be converted into the fertilizer "superphosphate"?

SOLUTION: *Analyze and Plan*: We are asked to give the main source of phosphorus, to explain why this source isn't used to make phosphatic fertilizers and how this source can be used to make "superphosphate." We need to review Section 22.8 in the text and the summary to help us answer these questions.

Solve: (a) The main source of phosphorus is rock phosphate, primarily $Ca_3(PO_4)(s)$. (b) The solubility of calcium phosphate in water is too low for it to be used as a source of phosphorus in fertilizer. (c) "Superphosphate" is formed by treating $Ca_3(PO_4)_2$ with 70 percent sulfuric acid.

$$Ca_3(PO_4)_2(s) + 2\,H_2SO_4(aq) \longrightarrow Ca(H_2PO_4)_2(s) + 2\,CaSO_4(s)$$
$$\text{"superphosphate"}$$

Although carbon is not widely abundant as an element, it plays an important role in living organisms and in the petroleum industry. C,H-containing compounds form a large area of chemistry known as organic chemistry. Section 22.9 focuses on some of the inorganic forms of carbon.

CARBON

Abundance: Over half of carbon occurs in carbonate compounds such as $CaCO_3$. It is also found in coal and petroleum deposits.

Uses: The use of carbon is widespread: polymers, diamonds, graphite, natural gas, and carbonates, such as baking soda ($NaHCO_3$), are just a few examples.

Allotropes: Three forms exist: diamond, graphite, and fullerenes. At room temperature graphite is slightly more thermodynamically stable than diamond, but the conversion between them is kinetically negligibly slow. Diamond consists of a three-dimensional network of single carbon–carbon bonds whereas graphite consists of sheets of carbon atoms connected by alternating carbon–carbon double bonds in hexagonal arrays. Buckminsterfullerene consists of C_{60} molecules resembling the surface of soccer balls.

Preparation: Carbon black is formed by heating hydrocarbons in a small amount of oxygen: $CH_4(g) + O_2(g) \rightarrow C(s) + 2\,H_2O(g)$. The burning of wood forms charcoal, a form of carbon.

Compounds:

1. Carbon monoxide (CO) is a colorless gas that acts as a weak Lewis base using the electron pair on carbon. It is used in metallurgy to reduce oxides of metals to the elemental form of the metal.
2. Carbon dioxide (CO_2) is a colorless gas formed in the combustion of organic compounds or by the action of acids on carbonates (CO_3^{2-}). When CO_2 is dissolved in water, H_2CO_3 (a weak, diprotic acid) forms. The anions of carbonic acid in water, HCO_3^-, and CO_3^{2-}, are basic.
3. Carbonate-containing compounds are plentiful. The principal form is calcite, $CaCO_3$. Carbonates dissolve in slightly acidic water.
4. Binary compounds with elements other than H are known as carbides. Some important examples are ionic acetylides containing the ion $:C\equiv C:^{2-}$ (for example, CaC_2), interstitial carbides formed by many transition elements (WC is extremely hard), and covalent carbides formed by boron and silicon (SiC is known as carborundum).

EXERCISE 22   Writing the Lewis structure of phosgene and determining the oxidation number of carbon

Carbon monoxide reacts with chlorine gas in the presence of heat and a platinum catalyst to form phosgene, $COCl_2$, a toxic gas. What is the oxidation number of carbon in $COCl_2$? Draw the Lewis structure of $COCl_2$.

SOLUTION: *Analyze and Plan*: We are asked to draw the Lewis structure of phosgene and to determine the oxidation number of carbon in it. We can use the principles in Exercise 8 and Exercise 12 to do this exercise.

Solve: By assigning oxidation numbers of -2 to oxygen and -1 to chlorine in $COCl_2$, we find that the oxidation number of carbon is $+4$: $(+4) + (-2) + 2(-1) = 0$. To draw the Lewis structure of $COCl_2$, we first count the total number of valence-shell electrons.

C: 4 valence-shell electrons
O: 6 valence-shell electrons
2 Cl: 2 × (7 valence-shell electrons) = 14

COCl$_2$: 24 valence-shell electrons or 12 pairs

The Lewis structure of COCl$_2$ is

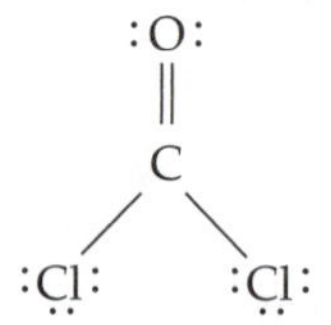

EXERCISE 23   Comparing bonding forces between carbon atoms and comparing hybridization of carbon atoms in diamond and graphite

What types of forces hold the carbon atoms together in diamond and graphite? What is the hybridization of carbon in each?

SOLUTION: *Analyze and Plan*: We are asked to state what types of forces hold carbon atoms together in diamond and graphite and the type of hybridization used by each carbon atom. Use Figures 11.41(b) and 11.41(a) in the text to help you determine the types of forces holding the carbon atoms together. Also the information in Section 22.9 provides additional information. The type of hybridization can be determined using the principles found in Section 9.5 in the text.

Solve: Each carbon atom in diamond is bonded to four other carbon atoms via single covalent bonds. Each carbon uses sp^3 hybrids. The hardness of diamond results from this three-dimensional structure. In graphite, each carbon atom is bonded covalently to three other carbon atoms, all arranged in hexagonal, planar arrays to form layers of stacked sheets of hexagons. Each carbon atom is sp^2 hybridized. Van der Waals forces hold the sheets of carbon layers together; these sheets may move with respect to one another.

EXERCISE 24   Writing chemical reactions of carbon and carbon-containing compounds

Write balanced equations for the following reactions:

(a) C(s) + small amount of O$_2$(g)
(b) Al$_2$O$_3$(s) + CO(g)
(c) MgCO$_3$(s) + HCl(aq)
(d) MgC$_2$(s) + H$_2$O(l)

SOLUTION: *Analyze and Plan*: We are given reactants for five chemical reactions and are asked to complete the chemical reactions. We can do this by analogy to known chemical reactions in Section 22.9 in the text. The equation number for a chemical reaction similar to that given in the problem is shown in parentheses after the answer for each part.

Solve:
(a) $2\,C(s) + O_2(g) \longrightarrow 2\,CO(g)$ (see 22.60 in text)
(b) $Al_2O_3(s) + 3\,CO(g) \longrightarrow 2\,Al(s) + 3\,CO_2(g)$ (see 22.62 in text)
(c) $MgCO_3(s) + 2\,HCl(aq) \longrightarrow MgCl_2(aq) + CO_2(g) + H_2O(l)$ (see 22.69 in text)
(d) $MgC_2(s) + 2\,H_2O(l) \longrightarrow Mg(OH)_2(s) + C_2H_2(g)$ (see 22.74 in text)

TABLE 22.4 Properties of Carbon and Silicon

Element	Elemental form	Common oxidation states, with examples	Common source	Example of use
Carbon	$C(s)$, diamond structure or graphite	$0, C(s)$ +4, CF_4	Carbonates and CO_2 in air	Graphite is a lubricant
Silicon	$Si(s)$	$0, Si(s)$ +4, $SiCl_4$, SiO_2	$SiO_2(s)$ silicates	Solar cells and transistors

Some of the properties of the group 4A elements carbon and silicon are summarized in Table 22.4.

GROUP 4A: CARBON FAMILY

- Note that with increasing atomic number in group 4A elements, there is a trend from nonmetallic (C) to metalloid (Si,Ge) to metallic (Sn,Pb).
- Another contrasting feature is the ability of carbon to form multiple bonds to itself and other nonmetals (N, O, and S) and to undergo catenation (to form compounds containing a chain of $-C-C-$ bonds). Silicon and germanium undergo catenation, but to a far lesser extent.

The two important oxides of carbon are CO and CO_2.

- Oxides of silicon have different characteristics than oxides of carbon. CO and CO_2 are molecular substances, whereas silicon oxides are solids with a network of $-Si-O-Si-$ linkages.

Silicate minerals have as their basic structural unit a silicon atom bound to four oxygen atoms that form a tetrahedron about silicon.

- The simplest structural unit is the orthosilicate ion, SiO_4^{4-}; it is found only in a few simple mineral structures.
- More complex silicate structural units are formed by a sharing of oxygens between silicate tetrahedra, yielding a network of $Si-O-Si$ bonds. For example, the dimer of SiO_4^{4-} is $Si_2O_7^{2-}$. Varying arrangements of SiO_4^{4-} tetrahedra sharing oxygen atoms yield chains, sheets, or three-dimensional arrays. See Figures 22.45 and 22.46 in the text.

EXERCISE 25   Determining the empirical formula of the mineral orthoclase

The mineral orthoclase is a feldspar mineral formed by replacing one quarter of the silicon atoms in quartz, SiO_2, with aluminum ions and by maintaining charge balance with additional potassium ions. What is the empirical formula for this mineral?

SOLUTION: *Analyze and Plan:* We are asked to determine the empirical formula of orthoclase given that it is a feldspar mineral with one quarter of the silicon atoms replaced by aluminum ions with the presence of potassium ions. An empirical formula shows the simplest whole number ratio of atoms in a substance. We need to replace a silicon atom, which is viewed to be in a +4 oxidation state, with an Al^{3+} ion and maintain charge balance using potassium ions.

Solve: If we replace a Si atom in SiO_2 with Al^{3+}, the neutral silicate group SiO_2 becomes AlO_2^-. Orthoclase is formed by replacing one AlO_2^- group for one SiO_2

group in every four SiO_2 groups. Therefore every four SiO_2 groups in quartz are converted to the $AlO_2 \cdot 3\, SiO_2^-$ ion in orthoclase. In order to maintain charge balance in the mineral orthoclase, one K^+ ion is required. Thus the empirical formula of orthoclase is $KAlO_2(SiO_2)_3$, or $KAlSi_3O_8$.

EXERCISE 26   Explaining why calcium ions in certain silicate minerals are leached by acidic rainwater

The mineral plagioclase is a mixture of $NaAlSi_5O_8$ and $CaAl_2Si_2O_8$. Explain why rainwater containing CO_2 causes Ca^{2+} to be leached when the rainwater comes in contact with plagioclase.

SOLUTION: *Analyze:* We are asked to explain why calcium in the mineral plagioclase is leached by rainwater containing carbon dioxide.

Plan: We will need to use some of the information in Section 22.10 to develop an explanation. One of the key issues is why rainwater containing dissolved carbon dioxide leaches calcium ions from minerals. If we remember that carbon dioxide reacts with water to form an acidic solution, then we have the start of our explanation.

Solve: $CO_2(g)$ dissolves in rainwater to form carbonic acid:

$$CO_2(g) + H_2O \rightleftharpoons H_2CO_3(aq)$$

H_2CO_3 is a weak acid and dissociates to form H^+ and HCO_3^- ions:

$$H_2CO_3(aq) + H_2O \rightleftharpoons H^+(aq) + HCO_3^-(aq)$$

CO_3^{2-} ions in a small concentration are also formed:

$$HCO_3^-(aq) + H_2O \rightleftharpoons H^+(aq) + CO_3^{2-}(aq)$$

Ca^{2+} in $CaAl_2Si_2O_8$ reacts with carbonic acid in water to release hydrogen ions which form hydroxide units in plagioclase. Thus, Ca^{2+} is slowly leached from plagioclase as described by the following overall reaction:

$$CaAl_2Si_2O_8(s) + H_2O + CO_2(aq) \rightleftharpoons$$
$$Ca^{2+}(aq) + 2\, HCO_3^-(aq) + Al_2Si_2O_6(OH)_2(s)$$

BORON | Boron is the only element of group 3A that is nonmetallic.

- Boron has a valence configuration of $2s^2 2p^1$ and exhibits a valency of 3 in all of its compounds; however, it does not form the B^{3+} ion. Instead, its valence electrons are shared to form covalent bonds.
- Note that the Lewis structures of BX_3 compounds show boron to have six valence electrons, and thus boron does not always obey the octet rule.
- Boron oxide, B_2O_3, is the anhydride of boric acid, H_3BO_3. Boric acid readily undergoes condensation reactions to form metaboric acid, $(HBO_2)_x$, and other condensed forms of boric acid.
- An important anionic compound containing boron is the borohydride ion, BH_4^-. The hydrogen atoms in BH_4^- carry a partial negative charge and are said to be "hydridic." Borohydrides are good reducing agents.

- Boranes are compounds of boron and hydrogen, B_xH_y, such as B_2H_6 which is formed from the combination of two BH_3 units. They are very reactive compared to the oxides of boron.

EXERCISE 27 Identifying the acidic or basic properties of a hydroxide of boron

(a) What is the formula of the hydroxide of boron? (b) Why is it acidic rather than basic as the name implies? (c) Write its reaction with water.

SOLUTION: *Analyze and Plan*: We are asked for the formula of a hydroxide of boron and to state whether it is an acidic or basic oxide and to write its reaction with water.

Plan: We can write the chemical formula if we use the common oxidation state of boron, +3, with the hydroxide ion to generate an empirical formula. Boron hydroxide can be viewed as a nonmetallic oxide. The chemical reaction of boron hydroxide with water should reflect its acidic or basic property.

Solve: (a) Boron has an oxidation number of +3; therefore, the formula of its hydroxide is $B(OH)_3$. (b) Because boron is nonmetallic, its hydroxide is acidic. Instead of writing $B(OH)_3$, we usually write H_3BO_3 (boric acid). (c) The reaction is

$$B(OH)_3(s) + H_2O \longrightarrow B(OH)_4^-(aq) + H^+(aq)$$

SELF-TEST QUESTIONS

Key Terms

Having reviewed key terms in Chapter 22, match key terms with phrases and identify statements as true or false. If a statement is false, indicate why it is incorrect.

Match each phrase with the best term:

22.1 A compound which forms two molecules of an acid when it reacts with water.

22.2 A soluble metal oxide which reacts with water to form a base.

22.3 An industrial process in which one of the steps is the oxidation of NO to NO_2.

22.4 A chemical reaction in which a single substance undergoes both oxidation and reduction.

22.5 A mixture containing only $H_2(g)$ and $CO(g)$.

22.6 The most common isotope of the element hydrogen.

22.7 An isotope of hydrogen that contains one neutron.

22.8 An isotope of hydrogen that is radioactive.

22.9 NaH is an example of this type of compound.

22.10 A compound which can be viewed as hydrogen atoms dissolved in a transition metal.

Key terms:

(a) acid anhydride
(b) basic anhydride
(c) deuterium
(d) disproportionation
(e) interstitial anhydride
(f) ionic anhydride
(g) Ostwald process
(h) protium
(i) tritium
(j) water gas

True-False Statements:

22.11 Ammonia is an example of a *molecular hydride*.

22.12 *Acidic oxides* typically contain a metal as the cation.

22.13 Most *basic oxides* are ionic oxides.

22.14 *Charcoal*, an amorphous form of carbon, is very dense.

22.15 When carbon is heated strongly in the absence of air, *coke* is formed.

22.17 The simplest structural unit found in *silicate minerals* is the $Si_2O_7^{2-}$ ion.

22.18 The addition of B_2O_3 to soda-lime *glass* results in a glass with a lower melting point.

22.19 The simplest *borane* is B_2O_3.

Problems and Short-Answer Questions

22.20 Use the following diagrams of first ionization energies and electronegativities to help you determine which family of elements are listed for elements A, B, C, D, and E. Explain how you reached your decision.

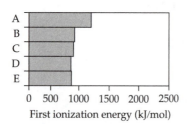

First ionization energy (kJ/mol)

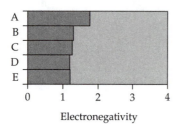

Electronegativity

22.21 In the following figure, which shows two molecular oxyacids, the solid spheres represent an element, the partially shaded spheres represent a second element, and the unshaded spheres represent a third element, which are the same elements in both structures. From the two diagrams, and given that the fully shaded atom is from the second period, determine which oxyacids they represent. Explain your reasoning.

22.22 Chapter 22 focuses on the chemistry of nonmetals. Compare in general the following physical properties of elements for a metal such as sodium to a nonmetal such as chlorine: ionization energy, atomic radius, and electronegativity. How are these reflected in the way they bond to other elements?

22.23 Carbon exhibits three allotropic forms. What are these forms, and which ones exhibit carbon-carbon pi bonds? Does silicon exhibit similar allotropic behavior?

22.24 Why do we observe the formation of water, carbon dioxide, or elemental nitrogen in many combustion reactions? Which substance would you expect to have the largest bond enthalpy?

22.25 Compare the hydride ion to a halide ion. If the resemblances are minimal or close, explain.

22.26 For many years it was thought that noble gases were unreactive, but in 1962 Neil Bartlett prepared a crystalline compound of xenon. He also prepared three fluorides of xenon. What property of xenon gives an indication that xenon had the potential to form xenon fluorides?

22.27 The melting points of XeF_2 and XeF_4 are 129 °C and 117 °C, respectively. Suggest a reason for the very similar melting points.

22.28 Compare the oxidizing abilities of Cl_2, Br_2, and I_2.

22.29 Name the following acids: HIO, HIO_2, HIO_3, and HIO_4.

22.30 Oxides of phosphorus can form when white phosphorus is burned in oxygen. Depending on the quantity of oxygen present, two forms are obtained. What are these forms? In what ways do their molecular structures resemble and not resemble silicates?

22.31 What nitrogen ion is analogous to the conjugate base of OH^-? Write reactions to demonstrate the analogy.

22.32 Write the overall redox equation and the half-reactions for the oxidation of Fe^{2+} to Fe^{3+} by hydrogen peroxide in acid solution.

22.33 Complete and balance the following reactions:

(a) $Li(s) + O_2(g) \longrightarrow$

(b) $Na(s) + O_2(g) \longrightarrow$

(c) $Mg(s) + O_2(g) \longrightarrow$

(d) $CO_2(g) + H_2O(l) \longrightarrow$

(e) $CO(g) + Cl_2(g) \longrightarrow$

(f) $BaO(s) + H_2O(l) \longrightarrow$

(g) $NaN_3 \xrightarrow{\Delta}$

(h) $NO(g) + O_2(g) \longrightarrow$

22.34 N_2O_5 is the anhydride of what acid? Draw the Lewis structure of N_2O_5 given that it contains a $N-O-N$ bond.

22.35 Using MO theory, determine the bond order of O_2^-.

22.36 What oxide of carbon is isoelectronic with N_2? How do their reactivities compare?

22.37 Write balanced chemical equations that exemplify the following changes in oxidation states of nitrogen.

(a) 0 to −3

(b) +2 to +4

(c) +5 to +2

22.38 The free energy of formation of ammonia is −16 kJ/mol and for phosphine it is +13 kJ/mol. What is the chemical formula of phospine and is it in the same class of compounds as ammonia? Could you prepare phosphine and ammonia from their elements?

22.39 Why is elemental white phosphorus highly reactive, while elemental nitrogen is relatively inert?

22.40 Write chemical formulas for the following compounds:

(a) potassuim chlorate

(b) nitric oxide

(c) thiourea

(**d**) orthophosphoric acid

(**e**) magnesium nitride

22.41 Write names for the following compounds:

(**a**) $HClO_3$ (**e**) XeO_3

(**b**) $NaClO_4$ (**f**) ICl_3

(**c**) $HBrO$ (**g**) H_2Te

(**d**) PF_3

22.42 Xenon forms compounds in which it exhibits positive oxidation states of 2, 4, 6, and 8. For each oxidation state, give an example of a xenon compound with Xe possessing that oxidation state, draw its Lewis structure, and name it.

22.43 Write balanced reaction equations describing or illustrating the following:

(**a**) reaction of sulfur with hot, concentrated nitric acid to form sulfuric acid nitrogen dioxide, and water

(**b**) reaction of H_2S with Fe^{3+} to form sulfur and iron (II)

(**c**) reaction of CaF_2 with sulfuric acid to form calcium sulfate and HF

(**d**) reaction of white P_4 with excess oxygen

(**e**) formation of a condensed phosphate from phosphoric acid

22.44 What substance is B_2O_3 the anhydride of? What is the expected Lewis acid–base chemistry of B_2O_3? Why?

22.45 Phosphorus forms two fluorides, PF_3 and PF_5. Given that nitrogen is a member of the same family as nitrogen, is this sufficient information to conclude that nitrogen also forms NF_3 and NF_5? If not, what additional information is needed and why?

22.46 Sulfuric acid is classified as a strong acid. Is this sufficient information to conclude that it forms a significant quantity of sulfate ion when it is in an aqueous solution? If not, what additional information is needed and why?

22.47 Molecular nitrogen contains a triple bond, has a bond enthalpy of 941 kJ/mol and it normally does not react with molecular oxygen. Is this sufficient information to conclude that it does not react with an active metal such as magnesium? If not, what additional information is needed and why?

Integrative Exercises

22.48 BF_3 is a stronger Lewis acid than $B(CH_3)_3$ but a weaker one than BCl_3.

(**a**) What is a Lewis acid? Draw the Lewis structure for BF_3. What is its molecular structure? What is the Lewis acid site in BF_3?

(**b**) Based on the electronegativity difference between boron and the atom bonded to it in each species, is the trend in Lewis acidity expected? Explain.

(**c**) What does this data suggest about the B—F bond in terms of the electron density on boron?

22.49 Iodine forms a number of ions with fluorine, for example: IF_6^+, IF_4^+, and IF_2^+.

(**a**) Draw the Lewis structures of these ions. Describe the geometrical structures.

(**b**) In which cases must the valence shell of the central atom be expanded? Explain.

(**c**) Predict the approximate bond angles in each ion.

(**d**) Would you expect ClF_6^+ to form? Explain.

22.50 NH_4NO_3 is thermally explosive when detonated.

(**a**) What are the likely products?

(**b**) Write the chemical reaction for the thermal detonation.

(**c**) When 1.00 g of ammonium nitrate is exploded, what volume of gas is collected at 1.00 atm and 830 °C?

22.51 Ozone is bubbled through 50.0 mL of an aqueous potassium iodide solution to form KOH, iodine, and dioxygen. The iodine liberated is titrated with thiosulfate ion $(S_2O_3^{2-})$ to form NaI and $Na_2S_4O_6$. Calculate the number of moles of O_3 passed through the solution if 60.0 mL of 0.100 M $Na_2S_2O_3$ is used in the titration. Write balanced equations describing the reactions that take place.

Multiple-Choice Questions

22.52 Which of the following are acidic anhydrides: (1) BaO; (2) SO_3; (3) Cl_2; (4) Na_2O_2; (5) P_4O_{10}?

(**a**) (3) only   (**d**)   (1), (4), and (5)

(**b**) (1) and (2)   (**e**)   (2), (4), and (5)

(**c**) (2) and (5)

22.53 Which of the following contains a peroxide ion?

(**a**) H_2O_2 (**d**) Na_2O_2

(**b**) CO_2 (**e**)   CaO

(**c**) Li_2O

22.54 Which of the following statements is *not* true about hydrogen gas?

(**a**) Forces between H_2 molecules are weak.

(**b**) It is a colorless gas at room temperature and pressure.

(**c**) It is an effective reducing agent for many metal oxides.

(**d**) The H—H bond is weak.

(**e**) Igniting of H_2 in air produces H_2O.

22.55 Which hydride forms hydrogen gas when it is dissolved in water?

(**a**) H_2O_2 (**d**)   HI

(**b**) NaH   (**e**) CH_4

(**c**) NH_3

22.56 PH_3

(**a**) is an ionic hydride

(**b**) is a base

(**c**) is relatively non-reactive in water

(**d**) contains hydrogen in a zero oxidation state

(**e**) forms an acid in water

22.57 The most thermally stable substance listed below is:

(a) NH_3 (d) SbH_3
(b) PH_3 (e) H_2Te
(c) AsH_3

22.58 A compound containing an oxygen atom in a positive oxidation state is

(a) H_2O_2 (d) O_2
(b) CO_2 (e) OF_2
(c) Li_2O

22.59 Which of the following is a basic anhydride?

(a) SO_2 (d) $Ca(OH)_2$
(b) H_2O (e) BaO
(c) OF_2

22.60 Which of the following elements form superoxide compounds: (1) Li; (2) Na; (3) K; (4) Rb; (5) Cs?

(a) (5) only   (d) (1) and (2)
(b) (4) and (5)   (e) all
(c) (3), (4), and (5)

22.61 Which of the following reacts with water to form ammonia?

(a) NO_2 (d) Mg_3N_2
(b) Mg   (e) N_2
(c) NO_3^-

22.62 HCN can be produced in the laboratory by reaction of

(a) CH_4 with N_2 (d) NaCN with HNO_3
(b) H_2 with NaCN   (e) none of the above
(c) NaCN with H_2O

22.63 Which of the following elements has the capacity to form more than four covalent bonds with other elements?

(a) N   (d) F
(b) Na   (e) S
(c) Ca

22.64 In terms of commercial use, the most important halogen is

(a) F   (d) I
(b) Cl   (e) At
(c) Br

22.65 Which of the following is the active ingredient in many household bleaches?

(a) $NaClO_4$ (d) NaClO
(b) $NaClO_3$ (e) NaCl
(c) $NaClO_2$

22.66 Which hydrogen halide must be stored in a wax or plastic bottle because it reacts with SiO_2 in glass bottles?

(a) HF   (d) HI
(b) HCl   (e) all of the above
(c) HBr

22.67 Of the oxyacids listed below, which one possesses the greatest acid strength in H_2O?

(a) $HClO_4$ (d) HClO
(b) H_2CO_3 (e) HBrO
(c) H_3BO_3

22.68 Of the compounds listed below, which one has the following Lewis structure:

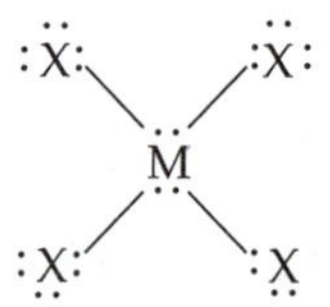

(a) CH_4 (d) NH_4^+
(b) BF_4^- (e) ClO_4^-
(c) ICl_4^-

22.69 Which of the following substances readily act as reducing agents: (1) BaO_2; (2) $NaBH_4$; (3) Cl_2; (4) O_2; (5) HNO_3?

(a) (1), (3), and (5) only
(b) (2) only
(c) (5) only
(d) (2) and (4) only
(e) all five substances

22.70 Whenever SiO_4^{2-} units share oxygen atoms, they do so only at the

(a) edges of tetrahedra
(b) faces of tetrahedra
(c) corners of tetrahedra
(d) faces of octahedra
(e) corners of octahedra

22.71 Whenever SiO_4^{2-} ions condense in infinite chains, what is the empirical formula of the resulting chain?

(a) Si_4O^{2-} (d) $S_3O_8^{8-}$
(b) SiO_3^{2-} (e) $S_{10}O_{12}^{12-}$
(c) $S_2O_8^{4-}$

22.72 What silicate structure is represented by the simplest formula $Si_4O_{11}^{6-}$?

(a) single tetrahedra
(b) double tetrahedra
(c) single chains
(d) double chains
(e) sheets

22.73 What silicate ion is represented by the following structure?

(a) $Si_2O_6^{4-}$ (d) $Si_2O_6^{4-}$
(b) $Si_4O_{11}^{6-}$ (e) $Si_2O_7^{6-}$
(c) $Si_4O_{10}^{4-}$

22.74 Glasses are essentially mixtures of

    (**a**) sodium oxides

    (**b**) calcium carbonates

    (**c**) lead oxides

    (**d**) boron oxides

    (**e**) silicates

22.75 Which statement is true?

    (**a**) Boron is a metallic element.

    (**b**) Boron has a higher melting point than carbon.

    (**c**) The hydrogen atoms in BH_4^- are "hydridic."

    (**d**) Boron in diborane, B_2H_6, is surrounded by three hydrogen atoms.

22.76 Limestone is

(**a**) $CaCO_3$		(**c**)	Na_2CO_3
(**b**) Na_2O		(**d**)	KCl

22.77 Silicones

    (**a**) consist of Si—Si—Si chains.

    (**b**) contain only silicon and oxygen atoms.

    (**c**) are rubberlike materials if crosslinking occurs.

    (**d**) are toxic.

SELF-TEST SOLUTIONS

22.1 (a). **22.2** (b). **22.3** (g). **22.4** (d). **22.5** (j). **22.6** (h). **22.7** (c). **22.8** (i). **22.9** (f). **22.10** (e). **22.11** True. **22.12** False. They are typically composed of two nonmetals. **22.13** True.

22.14 False. It has an open structure, with a large surface area. **22.15** True. **22.16** False. Bromine is not sufficiently large to accommodate seven fluorine atoms, whereas iodine is. **22.17** False. The simplest structural unit is orthosilicate, SiO_4^{4-}. **22.18** False. It raises the melting point.

22.19 False. Boranes are compounds containing only boron and hydrogen. The simplest borane is BH_3, which readily reacts with itself to form B_2H_6.

22.20 The family of elements belongs to the carbon group: C, Si, Ge, Sn, Pb. The range of electronegativities is approximately 1.8 – 2.5. Elements in the families 1A, 2A, and transition elements do not show this range, particularly with the lower limit of 1.8. A look at a table of electronegativities also shows the elements in families 5A, 6A, and 7A each have a greater upper limit. The first ionization energies are in the range of 715–1100 kJ/mol, which are also typical of a nonmetal family. A look at a table of first ionization energies confirms this conclusion.

22.21 Oxyacids are acids in which OH groups and possibly additional oxygen atoms are bound to the central atom. A general representation of an oxyacid formula is $(OH)_aX(O)_b$. If we look at the two figures we can see that there is only one possible O—H group represented by an unshaded sphere attached to a partially shaded sphere. Thus, the partially shaded sphere must be H and the unshaded sphere must be O. We now have the two

formulas: $OHXO_2$ and $OHXO$. The shaded sphere is the central atom, X, and is seen with four bonds to oxygen atoms and three bonds to oxygen atoms. Thus, in the second figure, the central atom must have one pair of nonbonded electrons. Given that it is a molecular oxyacid, the central atom must be a nonmetal and one of the following: B, C, N, or F. From our readings in this chapter we should know that boron forms the oxyacid, $B(OH)_3$, which does not fit the chemical formula; carbon forms H_2CO_3, which does not fit the chemical formula; and fluorine does not form a stable oxyacid because fluorine would have a positive oxidation state. Thus, the only element that fits the two given formulas is nitrogen: Nitric acid, HNO_3, and nitrous acid, HNO_2.

22.22 Metals have lower ionization energies than nonmetals, larger atomic sizes than a nonmetal in the same period, and smaller electronegativities than nonmetals. This is best characterized by the fact that metals tend to lose electrons and form ionic compounds, whereas nonmetals either gain electrons and form ionic compounds or share electrons with other nonmetals.

22.23 Carbon exists in the following forms: diamond, graphite, and fullerenes. In the latter two allotropes carbon-carbon pi bonds exist. Elemental silicon exhibits a diamondlike covalent-network structure with sigma bonds only. Si—Si pi bonds are not observed.

22.24 One of the driving forces for a combustion reaction is the formation of one or more of these compounds. They are all thermodynamically very stable. Energy is released when they from from their elements. Elemental nitrogen, N_2, contains the N≡N triple bond and has the largest bond enthalpy, 941 kJ/mol.

22.25 The hydride ion is H^- and has the electron configuration $1s^2$, which is the same as helium. A halide ion has the electron configuration ns^2np^6. Both will form ionic compounds, although the hydride ion is generally found with the more active metals. A hydride ion is relatively small compared to the halides. A hydride ion will readily react with water to form hydrogen gas, whereas a halide ion will not.

22.26 If you look at a table of ionization energies, you will observe that the ionization energy of xenon is almost equal to that of oxygen. Neil Bartlett knew that oxygen could undergo a reaction with PtF_6 and he thought xenon might. It did. From that discovery, several xenon fluorides were quickly synthesized.

22.27 XeF_2 is a linear molecule. Figure 22.8 in the text shows the structure of XeF_4 as square planar. Both molecules are nonpolar and the primary intermolecular forces are London forces. In both cases molecules of each can align themselves along their "flat" surface in the crystal state and similar London force interactions can occur. It is interesting to note that although xenon tetrafluoride is a more massive molecule, its melting point is slightly lower, contrary to what we might expect based on what we learned earlier in the chapter on intermolecular forces. This is an example where forces within crystals are difficult to predict since the orientations of molecules may significantly differ, leading to different intermolecular attractions.

22.28 The oxidizing abilities of the halogens decrease down the family. Thus, Cl_2 can oxidize Br^- and I^-, Br_2 can oxidize I^-, but I_2 cannot oxidize Cl^- or Br^-.

22.29 HIO, hypoiodous acid; HIO_2, iodous acid; HIO_3, iodic acid; and HIO_4, periodic acid. Notice the trend in naming the acids. The two acids with iodine in the lower two oxidation states end in *-ous* and the two acids with iodine in the upper two oxidation states end in *-ic*. The acid with iodine in the lowest oxidation state has the prefix *-hypo* and the one in the highest oxidation state has the prefix *-per*.

22.30 In a limited supply of oxygen, phosphorus(III) oxide is obtained: P_4O_6. In excess oxygen, phosphorus(V) oxide forms: P_4O_{10}. If you look at Figure 22.34 in the text, you should see that the oxygen atoms bridge the phosphorus atoms as they bridge silicon atoms in silicates. However, silicates form an extended network covalent structure, whereas the units in the two oxides are more like independent P_4 units.

22.31 The nitride ion (N^{3-}) is analogous to O^{2-}, the conjugate base of OH^-. The O^{2-} ion is formed by removing the proton from OH^-. An oxide ion in water acts as a base by accepting a proton from water to form the OH^- ion, as illustrated by the following chemical reaction: $CaO(s) + H_2O \rightarrow Ca(OH)_2(s)$. Similarly, the nitride ion in water acts as a base by accepting protons from water to form NH_3 and OH^- ions as illustrated by the following chemical reaction: $Mg_3N_2(s) + 6 H_2O \rightarrow 2 NH_3(aq) + 3 Mg(OH)_2(s)$.

22.32
$$2(Fe^{2+} \longrightarrow Fe^{3+} + e^-)$$
$$\frac{2 H^+ + H_2O_2 + 2 e^- \longrightarrow 2 H_2O}{2 Fe^{2+} + 2 H^+ + H_2O_2 \longrightarrow 2 H_2O + 2 Fe^{3+}}$$

22.33 (a) $4 Li(s) + O_2(g) \rightarrow 2 Li_2O(s)$

(b) $2 Na(s) + O_2(g) \rightarrow Na_2O_2(s)$

(c) $2 Mg(s) + O_2(g) \rightarrow 2 MgO(s)$

(d) $CO_2(g) + H_2O(l) \rightarrow H_2CO_3(aq)$

(e) $CO(g) + Cl_2(g) \rightarrow COCl_2(g)$

(f) $BaO(s) + H_2O(l) \rightarrow Ba(OH)_2(s)$

(g) $2 NaN_3 \rightarrow 2 Na(s) + 3 N_2(g)$

(h) $2 NO(g) + O_2(g) \rightarrow N_2O_4(g)$.

22.34 N_2O_5 is the anhydride of HNO_3 (Two HNO_3 molecules less a water molecule).

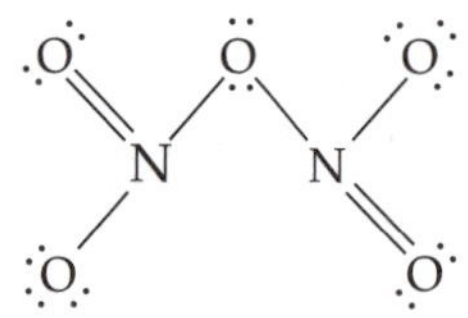

22.35 $(\sigma_{1s})^2(\sigma_{1s}^*)^2(\sigma_{2s})^2(\sigma_{2s}^*)^2(\pi_{2p})^4(\sigma_{2p})^2(\pi_{2p}^*)^3$

$$\text{Bond order} = \frac{10 - 7}{2} = 1\frac{1}{2}$$

22.36 N_2 has 14 electrons as does CO. Both contain a triple bond $:N\equiv N:$, $:C\equiv O:$, which is broken only with great difficulty. However, the carbon atom's lone pair of valence electrons is not as tightly held as those of nitrogen's because of carbon's lower electronegativity. Thus the carbon atom in CO acts as a Lewis base site toward some metals.

22.37 (a) $3 Mg(s) + N_2(g) \longrightarrow Mg_3N_2(s)$;

(b) $2 NO(g) + O_2(g) \longrightarrow 2 NO_2(g)$

(c) $3 Cu(s) + 8 HNO_3(aq) \longrightarrow$
$$3 Cu(NO_3)(aq) + 2 NO(g) + 4 H_2O(l).$$

22.38 Phosphine is PH_3 and ammonia is NH_3. They both are covalent hydrides with the same type of formula, MH_3. Ammonia could be made from its elements (the Haber process) because it has a negative free-energy of formation, but phosphine could not because it has a positive free energy of formation

22.39 Elemental white phosphorus, P_4, contains P—P single bonds in a tetrahedron that is sterically strained. These bonds are easily broken. The N—N bond in N_2, which is a triple bond ($:N\equiv N:$), is very difficult to rupture.

22.40 (a) $KClO_3$; (d) H_3PO_4;

(b) NO; (e) Mg_3N_2.

(c) H_2NCSNH_2;

(a) chloric acid;

(b) sodium perchlorate;

(c) hypobromous acid;

(d) phosphorus trifluoride;

(e) xenon trioxide;

(f) iodine trichloride;

(g) telluric acid.

22.42

Compound	Oxidation state	Lewis structure	Name
XeF_2	+2		Xenon difluoride
XeF_4	+4	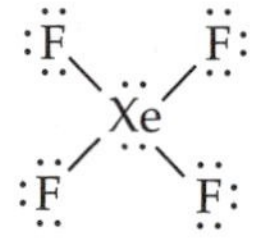	Xenon tetrafluoride
XeO_3	+6	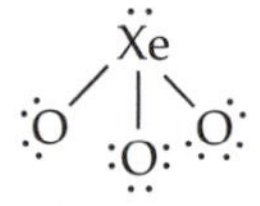	Xenon trioxide
XeO_4	+8	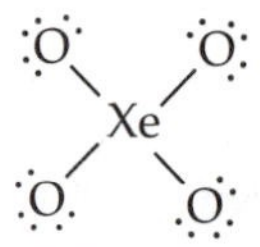	Xenon tetraoxide

22.43 (a) $(S(s) + 6\,HNO_3(aq) \rightarrow$
$$H_2SO_4(aq) + 6\,NO_2(g) + 2\,H_2O(l).$$

(b) $2\,Fe^{3+}(aq) + H_2S(g) \rightarrow$
$$2\,Fe^{2+}(aq) + 2\,H^+(aq) + S(s).$$

(c) $CaF_2(s) + H_2SO_4(l) \rightarrow$
$$CaSO_4(s) + 2\,HF(g).$$

(d) $P_4(s) + 5\,O_2(g) \rightarrow P_4O_{10}(s).$

(e) $2\,H_3PO_4 \rightarrow H_4P_2O_7 + H_2O.$

22.44 B_2O_3 is the anhydride of H_3BO_3 (two H_3BO_3 units minus three H_2O units). B_2O_3, like H_3BO_3, is a Lewis acid because boron is nonmetallic.

22.45 This is insufficient information without considering the types of orbitals available on both phosphorus and nitrogen available for bonding. Although both elements are in the same family, phosphorus is in the third period whereas nitrogen is in the second period. Nonmetallic elements in the third and higher periods have empty, low-energy d orbitals that permit them to accommodate more than eight valence electrons. Thus, phosphorus forms not only PF_3 (with an octet of valence electrons about phosphorus in three bond and a lone pair) but also PF_5 (with ten valence electrons about phosphorus in five bonds). Nitrogen, a second row element, does not have low-energy $3d$ orbitals available for additional bonding and it is limited to eight valence electrons when it bonds to other atoms.

22.46 There is insufficient information. A diprotic acid, such as sulfuric acid, may be a strong acid with respect to the loss of one hydrogen ion but not with respect to the second one. To determine if sulfuric acid forms a significant quantity of sulfate ion we need to know for each ionization step the value of the equilibrium constant associated with it. The ionization of HSO_4^- to form H^+ and SO_4^{2-} has a $K_a = 1.1 \times 10^{-2}$. This tells us that the bisulfate ion is a weak acid and therefore the amount of sulfate ion forming is not as significant as expected for a strong acid.

22.47 There is insufficient information. Although nitrogen has a bond enthalpy of 941 kJ/mol, suggesting it is highly unreactive, it does react with some active metals such as magnesium. One has to look at the nature of potential products formed when nitrogen reacts to determine if their stability is sufficient to overcome the bond enthalpy of the N—N triple bond. In the case of magnesium, we find in the text that it reacts with nitrogen to form $Mg_3N_2(s)$. The crystal lattice energy of Mg_3N_2 is expected to be quite large given that the charge of magnesium is 2+ and nitrogen is 3−. There is sufficient energy released to permit breaking the triple bond.

22.48 (a) A Lewis acid is a substance that accepts electron density in a chemical reaction. An atom in the substance must be capable of accepting additional electron density. Boron is electron deficient and is the Lewis acid site. The Lewis structure for BF_3 is

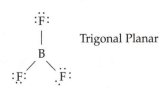

Trigonal Planar

(b) The difference in electronegativity values for B—F is 2, for B—C it is 0.5 and for B—Cl it is 1. Thus, the boron atom in BF_3 should be the most electropositive since the fluorine atom should most strongly attract bonding electrons to itself. Thus it should be a better Lewis acid site than in either of the other two species. However, the information provided agrees with this prediction only for $B(CH_3)_3$. BCl_3 is a stronger Lewis acid than BF_3, contrary to our prediction.

(c) The fact that BCl_3 is a stronger Lewis acid suggests that the boron atom in BF_3 is not as electropositive as expected. In some way electron density must be kept on boron or transferred from the fluorine atom. One model that has been suggested is that pi bonding exists between boron and fluorine in which a fluorine p orbital donates electron density to the empty boron p orbital. However, this model has several problems, one of which is that the resonance forms for this model have fluorine atoms with positive formal charges.

22.49 (a)

[Bent, F—I—F bond angle is smaller than the tetrahedral angle of 109° because of lone pair-bonding pair repulsions]

[Distorted "see-saw"]

[Regular octahedron with bond angles of 90°]

(b) IF_6^+ and IF_4^+ have iodine atoms with an expanded octet; thus the iodine atom is using additional higher energy orbitals for bond formation.

(c) See Lewis structures above.

(d) No. A chlorine atom is too small to accommodate six fluorine atoms about it.

22.50 (a) The likely products are $N_2(g)$, $O_2(g)$, and $H_2O(g)$. All will be gases at the given temperature.

(b) $2 NH_4NO_3(s) \rightarrow 2 N_2(g) + O_2(g) + 4 H_2O(g)$.

(c) Note that two moles of ammonium nitrate when detonated yield seven moles of gas $(2 + 1 + 4)$. The number of moles of ammonium nitrate is

$$(1.00 \text{ g})\left(\frac{1 \text{ mol}}{80.05 \text{ g}}\right) = 0.0125 \text{ mol.}$$

The number of moles of gas produced is

$$(0.00125 \text{ mol})\left(\frac{7 \text{ mol gas}}{2 \text{ mol solid}}\right) = 0.0438 \text{ mol gas.}$$

The volume of gas produced is calculated using the ideal-gas law: $PV = nRT$.

$$V = \frac{nRT}{P} = \frac{(0.0438 \text{ mol})\left(0.081 \dfrac{\text{L-atm}}{2 \text{ mol-K}}\right)(1103 \text{ K})}{1.00 \text{ atm}}$$

$$= 3.92 \text{ L}$$

Note that a small 1.00 g sample of solid produces a significant volume of gas that will act as a detonation wave.

22.51

$$2 KI + O_3 + H_2O \rightarrow 2 KOH + I_2 + 2 O_2;$$

$$I_2 + 2 Na_2S_2O_3 \rightarrow 2 NaI + Na_2S_4O_6$$

$$\text{Moles } O_3 = (\text{moles } S_2O_3^{2-})\left[\frac{1 \text{ mol } I_2}{2 \text{ mol } S_2O_3^{2-}}\right]\left[\frac{1 \text{ mol } O_3}{1 \text{ mol } I_2}\right]$$

$$= (0.0600 \text{ L } S_2O_3^{2-})(0.100 \text{ M } S_2O_3^{2-})\left[\frac{1 \text{ mol } O_3}{2 \text{ mol } S_2O_3^{2-}}\right]$$

$$= 0.00300 \text{ mol } O_3$$

22.52 (c).

22.53 (d) O_2^{2-} is the peroxide ion. Hydrogen peroxide is a covalent substance.

22.54 (d).

22.55 (b) An ionic hydride with a H^- ion can react with water to form hydrogen gas and a metallic hydroxide.

22.56 (a).

(b) It is basic just as NH_3 is a base.

22.57 (a).

22.58 (e) Fluorine is the most electronegative atom and this results in oxygen possessing a positive oxidation state.

22.59 (e). **22.60** (c). **22.61** (d). **22.62** (d).

22.63 (e) For example, sulfur forms SF_6. It is the only element listed that has low-energy d orbitals available for bonding.

22.64 (b) Chlorine is extensively used in forming the strong acid HCl.

22.65 (d). **22.66** (a). **22.67** (a).

22.68 (c) All other compounds listed show a tetrahedral arrangement of atoms about the central element.

22.69 (b). **22.70** (c). **22.71** (b). **22.72** (d). **22.73** (c). **22.74** (e). **22.75** (c). **22.76** (a). **22.77** (c).

Metals and Metallurgy

Chapter

23

OVERVIEW OF THE CHAPTER

Review: Metals (2.4, 22.1); periodic table (2.4).

Learning Goals: You should be able to:

1. Describe what is meant by the term *mineral*, and provide a few examples of common minerals.
2. Define various terms employed in discussions of metallurgy, notably *calcination, roasting, slag, smelting, refining*, and *leaching*.

Review: Electrode potentials and electrochemical cells (20.3, 20.4); free energy (19.5).

Learning Goals: You should be able to:

1. Distinguish among pyrometallurgy, hydrometallurgy, and electrometallurgy, and provide examples of each type of metallurgical process.
2. Describe the nature of slag, and indicate the means by which it is formed in pyrometallurgical operations.
3. Describe the pyrometallurgy of iron. You should know which ores are employed; the general design of a blast furnace; the ingredients in the blast furnace; the ingredients in the blast furnace reactions; and the chemical reactions of major importance. You should know how a converter is used to refine the crude pig iron that is the product of the blast furnace operation.
4. Describe the hydrometallurgy of gold, including the chemical reactions of major importance.
5. Describe the Bayer process for purification of bauxite, including balanced chemical equations.
6. Describe the electrometallurgical purification of copper, including balanced chemical equations for electrode processes.
7. Describe the process by which sodium metal is obtained from NaCl, including balanced chemical equations for electrode processes.
8. Describe the Hall process for obtaining aluminum, including balanced chemical equations for electrode processes.

23.1 METALS: SOURCES AND PRETREATMENT

23.2, 23.3, 23.4 METALS: ENRICHMENT PROCESS

23.5 METALS: PHYSICAL PROPERTIES AND ELECTRONIC STRUCTURES

Review: Electron configuration of elements (6.9); molecular orbital theory (9.7, 9.8); delocalization of electrons (8.7).

Learning Goals: You should be able to:

1. Describe how the extent of metallic bonding among the first transition elements varies with the number of valence electrons.
2. Discuss the simple electron-sea model for metals, and indicate how it accounts for certain important properties of metals.
3. Describe the molecular orbital model for metals, including the idea of bands of allowed energy levels. You should also be able to distinguish between metals and insulators in terms of this model.

23.6 ALLOYS

Review: Solutions and mixtures (1.2, 4.1).

Learning Goals: You should be able to:

1. Name the important types of alloys, and distinguish between them.
2. Describe the nature of intermetallic compounds, and provide examples of such compounds and their uses.

23.7 TRANSITION METALS: PHYSICAL PROPERTIES

Review: Atomic radii (7.3); *d* orbitals (7.6); electron configurations (7.8, 7.9); ionization energy (8.4); paramagnetism (9.7); metal complexes (16.11); standard reduction potential (20.4).

Learning Goals: You should be able to:

1. Identify the transition elements in a periodic table.
2. Write the electron configurations for the transition metals.
3. Describe the manner in which atomic properties such as ionization energy and atomic radius vary within each transition element series, and within each vertical group of transition elements.
4. Describe the lanthanide contraction and explain its origin.
5. Describe the general manner in which the maximum observed oxidation state varies as a function of group number among the transition elements, and account for the observed variation.
6. Name and describe the types of magnetic behavior discussed in this chapter: diamagnetism, paramagnetism, ferromagnetism, antiferromagnetism and ferrimagnetism.
7. Provide an explanation of each type of magnetic behavior in terms of the arrangements in the lattice of metal ions with unpaired spins, and their interactions with one another.

23.8 SELECTED CHEMICAL PROPERTIES: Cr, Fe, Cu

Review: Acids and bases (4.3, 16.2).

Learning Goals: You should be able to write a balanced chemical equation corresponding to the simple aqueous solution chemistry of chromium, iron, nickel, and copper, and account for the variations in chemical properties observed among these elements in terms of the characteristics of the elements themselves.

TOPIC SUMMARIES AND EXERCISES

METALS: SOURCES AND PRETREATMENT

Metals are essential to a modern society. In this chapter of the text you explore the sources of metals, their extraction and purification, their physical properties, and the properties of mixtures of metals (alloys). A large number of new terms are introduced, and you need to pay careful attention to their definitions.

Metals are mined from the solid earth that we stand upon, the **lithosphere**.

- Active metals are found in a combined state in nature. Less active metals may occur in the uncombined native state.
- Whether in a combined state or a free state, metals can be found in deposits containing several compounds that can be economically worked to extract the desired metals. These deposits are called **ores**.
- Ores contain **minerals**, which are solid substances occurring in nature with identifiable composition and well-defined crystal structures. They can be classified into one of three groups: native elements, silicates, and nonsilicates.
- A solid collection of minerals is called a rock. In general, silicates and aluminosilicates are the primary components of most rocks. They also make up the bulk of the earth's crust and mantle.

An ore consists of valuable pure minerals along with impurities.

- After an ore is mined, the desired mineral is concentrated, and the impurities are removed. Techniques for separating impurities from minerals include separation of minerals by density differences, magnetic differences, and flotation techniques.
- Once an ore has been enriched in the desired metal, chemical processes are used to gain further enrichment of the ore in the desired metal.

> *Note:* Many of the Exercises in this chapter require you to recall facts and information about the physical and chemical properties of elements and compounds or refining technologies discussed in the chapter. For these Exercises you can use the following as the first part of the Solution to each Exercise, *Analyze and Plan*: You are asked for specific information about the elements or substances or refining technologies in the question. Unless you remember this information from your readings you will have review material in the text or find a different reference which has the required information. Exercises which do not provide an *Analyze and Solve* fit this type of question and only the *Solve* portion is provided.

EXERCISE 1 Providing examples of metals in combined states

Give some examples of metals that occur in combined states with the anions oxide and sulfide.

SOLUTION: If you can't remember some examples, look at Table 23.1 in your text. Some examples are

Anion	Example
Oxide	Al_2O_3 (corundum)
	Fe_3O_4 (magnetite)
	Fe_2O_3 (hematite)
Sulfide	PbS (galena)
	HgS (cinnabar)

EXERCISE 2   Determining order of abundance of elements by mole fraction

After silicon and oxygen, the most abundant elements in the earth's crust are aluminum, iron, and calcium, if percent by weight is the basis used for comparing their relative abundances. What is the order of the abundance of these three elements if they are measured by mole fraction instead of percent by weight? The percent-by-weight abundances of aluminum, iron, and calcium are 7.5, 4.7, and 3.4, respectively.

SOLUTION: *Analyze and Plan*: We are asked to list the relative abundances of Al, Fe, and Ca based on mole fraction rather than percent by weight, which is given data. We are not given a sample size; thus, we can assume an arbitrary sample size and calculate the mass of each element in the sample from the percent by weight data. From these masses the relative order of abundance based on mole fraction can be determined.

Solve: If we assume that we have a 100-g sample of the earth's crust, the weights of these elements in the sample are

$$\text{Al: } (100 \text{ g})(0.075) = 7.5 \text{ g}$$
$$\text{Fe: } (100 \text{ g})(0.047) = 4.7 \text{ g}$$
$$\text{Ca: } (100 \text{ g})(0.034) = 3.4 \text{ g}$$

The mole fraction of a component of a mixture is defined as follows:

$$\text{Mole fraction} = \frac{\text{number of moles of component}}{\text{total number of moles of all components in mixture}}$$

Since the denominator of the mole fraction is a constant for a given mixture, the order of abundances of Al, Fe, and Ca in the earth's crust based on mole fraction is simply the order of the number of moles of each in the 100-g sample:

$$\text{Al: } (7.5 \text{ g})\left(\frac{1 \text{ mol}}{26.98 \text{ g}}\right) = 0.28$$

$$\text{Fe: } (4.7 \text{ g})\left(\frac{1 \text{ mol}}{55.85 \text{ g}}\right) = 0.084$$

$$\text{Ca: } (3.4 \text{ g})\left(\frac{1 \text{ mol}}{40.08 \text{ g}}\right) = 0.085$$

The order of abundances based on mole fraction is therefore Al > Ca $\simeq$ Fe.

METALS: ENRICHMENT PROCESSES

Metallurgy involves three principal operations: Concentration, reduction, and refining. The last two processes involve reducing a metal to its free form or to another compound and then final purification and isolation of the free metal. The text broadly classifies these two operations into three areas:

Pyrometallurgy: Heat is used to convert a mineral into another form or into the free metal. Three such processes are discussed:

- *Calcination:* Heating an ore to form an oxide, sulfate, or free metal. Carbonates readily decompose to the oxide of the metal and $CO_2(g)$.
- *Roasting:* Heating an ore below its melting point in the presence of oxygen (or CO) to produce the free metal or another compound. If oxygen is the atmosphere in the furnace, oxides or sulfates are produced. The use of CO, a reducing gas, can produce the free metal.

- *Smelting:* An ore is heated above its melting point to form a molten solution containing at least two layers. One of the layers is slag. The slag contains impurities that still remains after ore concentration—primarily silica and silicate impurities. $CaCO_3$ is commonly added to the melt to form a slag with silica—$CaSiO_3(l)$. The slag is removed from the molten mixture; thus the ore becomes more concentrated in the desired metal.
- The pyrometallurgy of iron is extensively discussed. Note the design of the blast furnace, and the role of coke (85 percent carbon) as a fuel and as the source of reducing gases, CO and H_2. The iron produced is referred to as pig iron.

Hydrometallurgy: This technique involves separating minerals from the rest of the ore by an aqueous chemical process.

- *Leaching,* the selective dissolving of the desired metal, is the most important hydrometallurgical process. The text discusses the treatment of gold ores with CN^- and oxygen to form the complex ion $Au(CN)_2^-$. Zinc dust is used to precipitate Au.
- The metallurgy of aluminum involves treating bauxite, $Al_2O_3 \cdot xH_2O$, with a concentrated NaOH solution using the *Bayer process.* The Al_2O_3 dissolves to form $Al(H_2O)_2(OH)_4^-$ a soluble complex ion. The majority of impurities, particularly silica and iron, settle out of the NaOH solution.

Electrometallurgy: Active metals such as sodium, aluminum, and potassium are refined by electrolysis methods. Other less active metals, such as copper, are further refined by similar electrometallurgical techniques.

- Sodium, magnesium, and aluminum must be produced from a molten salt solution. Sodium is electrolyzed in a Downs cell. Molten NaCl is the electrolytic medium, with $CaCl_2$ added to lower the melting point of the cell medium. Na(*l*) and $Cl_2(g)$ are produced.
- Aluminum is commercially produced by the Hall process. The cell medium contains purified Al_2O_3 (from the Bayer process) and molten cryolite, Na_3AlF_6. Graphite (C) rods are used as anodes and are consumed during electrolysis.
- Copper can be further purified in an electrolytic cell. Copper sulfate is used in an acidic medium as an electrolyte. Impure copper is the anode, and pure copper is the cathode. Copper dissolves at the anode to form Cu^{2+}, which is reduced to pure Cu at the cathode.

EXERCISE 3   Identifying the major methods of reduction in metallurgy

What are the major methods of reduction used in metallurgy?

SOLUTION: The two general methods are chemical reduction and electrolytic reduction. Electrolytic reduction of minerals is used to form the most electropositive metals, such as K, Al, Na, and Ca. Chemical reduction usually involves reacting carbon, in the form of coke or CO, with a mineral to form the desired metal or one of its compounds.

EXERCISE 4   Writing chemical reactions for the reduction of the ores of Hg, Fe, and Al

(a) What are the ores used to produce the metals Hg, Fe, and Al? (b) For these metals, write chemical equations showing a specific reduction process.

SOLUTION: *Analyze and Plan*: We are asked to identify the ores used to produce Hg, Fe, and Al and write the chemical equations describing a reduction process. We can find this information in Sections 23.1, 23.2, and 23.3 of the text. By reviewing these sections we can also find the appropriate chemical reactions. (a) Mercury is produced from cinnabar, HgS; iron from hematite, Fe_2O_3, and magnetite, Fe_3O_4; and aluminum from bauxite, $Al_2O_3 \cdot xH_2O$. (b) The reduction processes are

Hg: Roasting of HgS in air:

$$HgS(s) + O_2(g) \longrightarrow Hg(g) + SO_2(g)$$

Fe: Reduced in the presence of coke and oxygen in a blast furnace:

$$2\,C(coke) + O_2(g) \longrightarrow 2\,CO(g)$$
$$Fe_2O_3(s) + 3\,CO(g) \longrightarrow 2\,Fe(s) + 3\,CO_2(g)$$

Al: Al_2O_3 is electrolyzed in molten cryolite, Na_3AlF_6:

$$2\,Al_2O_3(l) \longrightarrow 4\,Al(l) + 3\,O_2(g)$$

EXERCISE 5   Providing information about calcination, slag formation, and hydrometallurgy of gold

(a) Why is the formation of a sulfate in calcination sometimes advantageous? (b) Why does CaO react with SiO_2 to form a slag? Write their reaction. (c) Why must the solution of Au and CN^- in the hydrometallurgy of gold be kept basic?

SOLUTION: *Analyze and Plan*: We are asked three questions relating to calcination, slag, and the conditions for the hydrometallurgy of gold. We can find appropriate information in Sections 23.2 and 23.3 of the text. By reviewing these sections we can answer the questions.

Solve: (a) Sulfates are usually water soluble. The desired metal can be leached away from insoluble impurity. (b) CaO is a basic oxide, whereas SiO_2 is an acidic oxide. Their reaction is similar to that of an acid and base reacting to form a salt:

$$CaO(l) + SiO_2(l) \longrightarrow CaSiO_3(l)$$

(c) If the solution were to become acidic, CN^- would react to form HCN(*aq*). HCN is only moderately soluble in water, and some would escape as a toxic gas.

METALS: PHYSICAL PROPERTIES AND ELECTRONIC STRUCTURES

The physical and chemical properties of metals are varied, but most show the following characteristics.

Gray or silver luster	Electropositive
High electrical conductivity	Reducing agents
High thermal conductivity	Form cations
Malleable (hammered into thin sheets)	Low ionization energies
Ductible (drawn into wire)	Low electron affinities

The text discusses the nature of metallic bonding.

- Boiling points, melting points, heats of fusion, heats of vaporization, hardness, and densities of metals tend to increase as the strengths of the bonds between metal atoms in a metallic crystal increase.

- The strengths of metal–metal bonds are partly determined by the number and kinds of valence electrons that a metal possesses. Metals that have six outer *s* and *d* valence electrons per atom exhibit physical properties characteristic of very strong bonds between metal atoms.

A simple model for metallic bonding, the **electron-sea model**, pictures metallic bonding involving:

- a lattice of metal ions with valence electrons free to move completely about the lattice;
- electrostatic attractions between the positive metal ions and the electrons;
- electrons that are free to move under the influence of an electric field or with heating of the solid. The latter property gives rise to electrical and thermal conduction of metals.

In Sections 9.6 and 9.7 of the text, we saw that electrons can be delocalized over several nuclear centers. We also learned that several atomic orbitals could overlap to form delocalized molecular orbitals. A similar situation occurs in metals.

- The interaction of metal valence atomic orbitals throughout a metallic crystal yields a large number of molecular orbitals closely spaced in energy. These molecular orbitals form a continuous band of allowed energy levels.
- A metal will be stable when more valence electrons occupy bonding molecular orbitals than antibonding molecular orbitals.
- The incomplete filling of these energy bands gives rise to characteristic metallic properties. For example, the availability of higher-energy empty orbitals in an energy band allow electrons to move throughout the metallic crystal.

EXERCISE 6   Explaining why calcium has a higher melting point than potassium

The *ns* and *np* orbitals of the alkali and alkaline earth metals can overlap to form a band of allowed energy states. Given this information, explain why Ca has a higher melting point (850°C) than K (64°C).

SOLUTION: *Analyze*: We are asked why calcium has a higher melting point than potassium using information relating to the formation of energy bands from the overlap of *ns* and *np* orbitals.

Plan: The melting point of an alkali or alkaline-earth metal is related to the number of occupied bonding molecular orbitals in the molecular-orbital structure for the solid. We need to consider what types of molecular orbitals can form from the overlap of the *ns* and *np* atomic orbitals of potassium and calcium and determine which one has more bonding molecular orbitals occupied.

Solve: From the information provided, we know that the 4*s* and 4*p* atomic orbitals of Ca and K are involved in metallic bonding. Each metal atom contributes one 4*s* and three 4*p* atomic orbitals to the molecular orbital structure of the metal. If we assume that the molecular orbitals formed by the 4*s* and 4*p* atomic orbitals all overlap, then a continuous band of allowed energy states occurs. Because each Ca atom contributes two valence electrons ($4s^2$) to the valence band—compared to one valence electron ($4s^1$) contributed by K in its valence band—more molecular orbitals are occupied in the Ca valence band than in the K valence band. Since the melting point of a metal is directly related to the number of occupied bonding molecular orbitals in the molecular-orbital structure of the metal, Ca has a higher melting point than K.

EXERCISE 7   Explaining why beryllium is a metallic conductor

Beryllium has the electron configuration $1s^2 2s^2$. We might expect beryllium *not* to be a conductor of electricity because the $2s$ valence band appears to be full. Yet beryllium is a metallic conductor. Provide an explanation for this observation.

SOLUTION: *Analyze*: We are asked to explain why Be is a metallic conductor even though its $2s$ subshell is completely filled.

Plan: Be is a metal and we can consider the molecular orbital theory of metallic bonding to answer the question. If the highest energy bonding molecular orbital of Be is close in energy to the lowest energy empty molecular orbital then electrons can move from one to the other. This results in conduction and suggests that Be may have this inherent ability.

Solve: As the problem states, if the highest filled valence band were composed *only* of the $2s$ orbitals of beryllium, the energy band would be filled. To account for the metallic conduction property of beryllium, Be must "expand" the number of allowed energy levels in the valence conduction band, so that some of the higher energy orbitals are empty. If both the $2s$ and $2p$ orbitals of beryllium form energy bands that overlap, then the band resulting is only one-quarter filled. Thus electrons can move from the lower filled energy levels to the higher empty orbitals, and this results in beryllium having the capacity to conduct current.

ALLOYS

Alloys are mixtures of elements with characteristic properties of metals. Several types of alloys are discussed:

- **Solution alloys** are homogeneous mixtures with the components randomly dispersed.
- **Substitutional alloys** are homogeneous mixtures with atoms of the solute occupying positions normally occupied by the solvent atoms.
- **Interstitial alloys** are homogeneous mixtures with the solute occupying interstitial positions between the solvent atoms.
- **Heterogeneous alloys** are heterogeneous mixtures.
- **Intermetallic compounds** are homogeneous alloys with specific properties and composition, such as duralumin, $CuAl_2$.

Steels are interstitial alloys containing carbon and iron.

- As the percentage of carbon increases from less than 0.2% to 1.5%, the durability and toughness of the steel increases: mild steel (0.2%); medium steels (0.2–0.6%); high-carbon steel (0.6–1.5%).
- If an element other than carbon such as chromium is added, an *alloy steel* is formed. Stainless steel is an alloy of 0.4% carbon, 18% chromium, 1% nickel, and the rest iron.

Intermetallic compounds have an important role in modern technology. Examples discussed in the text include:

- Ni_3Al as a major component of jet aircraft engines.
- Cr_3Pt as a coating for the new, very hard razor blades.
- Co_5Sm as a permanent magnet.

EXERCISE 8   Distinguishing between high-carbon steel and alloy steel

How could you determine whether a knife blade is made from high-carbon steel or an alloy steel?

SOLUTION: A knife blade made from high-carbon steel is easily sharpened but quickly loses its sharpness after being used, whereas one made from an alloy steel has the opposite properties. An alloy steel blade is more wear-resistant and harder than one of the high-carbon steel.

EXERCISE 9   Explaining why an interstitial element in an interstitial alloy is often a nonmetal

Why is the interstitial element in an interstitial alloy typically a nonmetal that can participate in bonding to the metal?

SOLUTION: For the interstitial element to occupy the interstitial space between the solvent atoms, it must have a smaller radius than the solvent particles. As we learned in Section 7.2 of the text, atomic radius tends to decrease from left to right across a period for the representative elements. Thus nonmetals have sufficiently small covalent radii to allow them to occupy interstitial positions.

The term **transition elements** refers to those metals whose outer d orbitals are being filled. Inserted between lanthanum and hafnium are the **inner transition elements**, the lanthanides. For these metals, the f orbitals are being filled.

Many of the physical properties of transition elements depend upon their electronic structure.

TRANSITION ELEMENTS: PHYSICAL PROPERTIES

- Review Section 6.7 to recall the order of orbital energies when electrons are added to nuclei to form atoms.
- In particular, we should note that the $(n - 1)d$ and ns orbitals are very close in energy; therefore it should not be unexpected that in a few transition elements an electron may occupy a $(n - 1)d$ orbital instead of a valence ns orbital as in Cr ($3d^54s^1$ instead of $3d^44s^2$).
- Remember that the valence ns electrons are lost before the $(n - 1)d$ electrons when transition metals form ions.
- The first ionization energies of the first-row transition elements increase from left to right across the period; atomic radii decrease in the same direction. Both trends can be understood by recalling (Section 6.7) that the effective nuclear charge experienced by the outer electrons of an atom increases from left to right across a period and that this increase outweighs the greater repulsions among the d electrons.

The atomic radii of the transition elements of the fifth period are greater than those of the fourth period, as expected.

- However, the transition elements of the sixth period have almost the same radii as the elements immediately above them in the fifth period. This effect is referred to as the **lanthanide contraction**.
- The 14 lanthanide elements, with their poorly shielding f electrons, occur before hafnium. Therefore, the outer electrons of hafnium and the elements immediately following it experience a higher effective

nuclear charge than would normally be anticipated; this results in smaller atomic radii than expected.

Transition elements exhibit a variety of oxidation states.

- We should look at Figure 23.22 in the text and note the most common oxidation states of the first-row transition elements.
- Also we should recognize that the most common oxidation states of elements in the solid state are not always the same as those found in solution. For example, Mn^{4+} is not common in solution, but it is found in the very stable MnO_2.
- Some observations and trends in the oxidation states of the first-row transition elements are:

  1. The maximum oxidation state for Sc through Mn equals the total number of outer s and d electrons.
  2. In general, the highest oxidation states are found when the most electronegative elements (O, F, and sometimes Cl) are combined with transition elements.

In Section 9.7 you learned that the presence of an unpaired electron in the valence orbitals of an atom causes the atom to possess a magnetic moment.

- **Paramagnetic substances** possess unpaired electrons and are drawn into a magnetic field.
- When all electrons of an atom are paired, it exhibits diamagnetic behavior. **Diamagnetic substances** are repelled by a magnetic field.
- Metals of the iron triad (Fe, Co, Ni) exhibit **ferromagnetism** in the uncombined state. In the presence of a magnetic field, the magnetic moments of individual metal atoms become permanently magnetized—the electron spins of the unpaired electrons permanently orient themselves. The magnitude of ferromagnetism can be as much as a million times greater than that of simple paramagnetism.
- Figure 23.23 in the text shows two additional forms of magnetism: antiferromagnetism and ferrimagnetism. In the first case, the spins of unpaired electrons on a given atom are aligned in the opposite direction as the spins on adjacent atoms. The spins cancel one another. In ferrimagnetism the alignment of spins of unpaired electrons is the same as in ferromagnetism but there is a net magnetic moment. This occurs when adjacent atoms have differing number of unpaired electrons.

EXERCISE 10   Stating the most common oxidation state of three transition elements and writing their valence-electron configurations

For each of the following elements state the most common oxidation state(s) and the valence-electron configuration for the element in each oxidation state: V, Mn, and Co.

SOLUTION: *Analyze and Plan*: We are asked to state the most common oxidation states for five transition elements and to write the valence-electron configuration for each element in these oxidation states. We can read Section 23.7 to review the most common oxidation states. After we write each element with its common oxidation states we can use principles in Section 6.9 of the text to write valence-electron configurations.

Solve: V^{5+}: All of vanadium's $3d^3 4s^2$ valence electrons are lost to form the [Ar] electron configuration
V^{4+}: $3d^1$
Mn^{7+}: All of manganese's $3d^5 4s^2$ valence electrons are lost to form the [Ar] electron configuration

Mn^{4+}: $3d^3$ $\quad\quad\quad\quad\quad\quad$ Mn^{2+}: $3d^5$
Co^{3+}: $3d^6$ $\quad\quad\quad\quad\quad\quad$ Co^{2+}: $3d^7$

EXERCISE 11 Determining whether Co^{2+} or Co^{3+} is the better oxidizing agent

Which should be a better oxidizing agent in H_2O, Co^{2+} or Co^{3+}?

SOLUTION: *Analyze and Plan*: We are asked to determine which of three species is a better oxidizing agent. An oxidizing agent is reduced in a redox reaction. We can look up standard reduction potentials to answer this type of question: The larger the standard reduction potential of a species the more easily it is reduced.

Also, we generally find the higher the oxidation state of an element being reduced in a compound the better it is as an oxidizing agent.

Solve: Co^{3+} should be the better oxidizing agent in H_2O. Standard reduction-potentials for Co^{2+} (-0.28 V) and Co^{3+} (1.8 V) in acid solution show that Co^{3+} is more easily reduced (that is, a better oxidizing agent) than Co^{2+}.

EXERCISE 12 Understanding the spatial relationships of atoms in a ferromagnetic substance

Would you expect the atoms in a ferromagnetic substance to be extremely close to one another?

SOLUTION: No. If the atoms were extremely close to one another, then the unpaired electrons of adjacent atoms could pair with one another. The atoms of a ferromagnetic substance must be sufficiently far apart so unpaired electrons do not pair, but they also must not be so far apart that the unpaired electrons on adjacent atoms cannot interact cooperatively.

EXERCISE 13 Determining whether an ion is diamagnetic or paramagnetic

Would you expect the following isolated ions to be diamagnetic or paramagnetic in the presence of a magnetic field: Ti^{3+}, Co^{2+}, and Zn^{2+}?

SOLUTION: Write the valence electron configuration of each ion and determine if any of the electrons are unpaired or if they are all paired. Ti^{3+}: $3d^1$.

Paramagnetic—one unpaired electron. Co^{2+}: $3d^7$. ⊞

Paramagnetic—three unpaired electrons. Zn^{2+}: $3d^{10}$. Diamagnetic—all of the d electrons are paired.

SELECTED CHEMICAL PROPERTIES: Cr, Fe, Cu

Chromium

Source: Chromite, $FeCr_2O_4$ (a mixed oxide of formula $FeO \cdot Cr_2O_3$)
Use: Chromium-based alloys (~ 60 percent) and 20 percent into other uses.
Reaction in dilute H_2SO_4:

$$Cr(s) + 2\,H^+(aq) \longrightarrow Cr^{2+}(aq)[\text{blue}] + H_2(g)$$
$$4\,Cr^{2+}(aq) + O_2(g) + 4\,H^+(aq) \longrightarrow 4\,Cr^{3+}(aq) + 2\,H_2O(l) \quad [\text{rapid}]$$

In HCl, the green complex $Cr(H_2O)_4Cl_2^+$ forms.
Frequent form in aqueous state: +6 oxidation state.
Basic solution: yellow chromate ion, CrO_4^{2-}
Acidic solution: orange dichromate ion, $Cr_2O_7^{2-}$

Iron

Source: hematite or magnetite
Uses: steels and alloys
Reaction in warm dilute oxidizing acid (HNO_3):

$$Fe(s) + NO_3^-(aq) + 4\,H^+(aq) \longrightarrow Fe^{3+}(aq) + NO(g) + 2\,H_2O(l)$$

Frequent forms in aqueous state: Either as +2 or +3 ion. Air oxidation of +2 ion to +3 ion readily occurs:

$$4\,Fe^{2+}(aq) + O_2(g) + 4\,H^+(aq) \longrightarrow 4\,Fe^{3+}(aq) + 2\,H_2O$$

A basic solution of $Fe^{3+}(aq)$ yields a gelatinous red-brown precipitate, $Fe_2O_3 \cdot nH_2O$, although it is commonly written as $Fe(OH)_3$.

Copper

Source: chalcopyrite, $CuFeS_2$
Use: alloys (bronze, Cu and Sn; brass, Cu and Zn)
Reaction in dilute oxidizing acid:

$$3\,Cu(s) + 8\,HNO_3(aq) \longrightarrow 3\,Cu(NO_3)_2(aq) + 2\,NO(g) + 4\,H_2O(l)$$

Frequent forms in aqueous state: Primarily as the +2 ion, but Cu also exists in the +1 ion state. Salts of Cu^+ are usually insoluble in water and mostly white. Also Cu^+ readily disproportionates to Cu^{2+} and $Cu(s)$; thus Cu^{2+} is more common. Salts of Cu^{2+} are often water-soluble. $CuSO_4 \cdot 5\,H_2O$ (copper sulfate pentahydrate, or blue vitrol) contains four H_2O molecules strongly bound to Cu^{2+}; one is weakly attracted. $Cu(OH)_2$ [blue] forms in basic solution. It readily loses water on heating to form black $CuO(s)$. CuS [black] is one of the least soluble Cu(II) compounds. It will dissolve only in a strong oxidizing acid such as HNO_3.

$$3\,CuS(s) + 2\,NO_3^-(aq) + 8\,H^+(aq) \longrightarrow$$

$$3\,Cu^{2+}(aq) + 3\,S(s) + 2\,NO(g) + 4\,H_2O(l)$$

EXERCISE 14 Stating properties of chromium under different conditions

(a) What color of solution occurs when an acidified solution of $CrCl_2$ is in contact with air for a day or two? (b) What is the most common form of chromium in a basic solution? What ion forms when a basic solution of chromium is acidified?

SOLUTION: (a) Chromium (II) is readily oxidized to the violet-colored chromium (III) ion in the presence of air. (b) In basic solution the chromate ion, CrO_4^{2-}, is the most stable ion form; in acid it becomes the dichromate ion, $Cr_2O_7^{2-}$.

EXERCISE 15 Writing the chemical reaction of iron (II) in a nitric acid solution, identifying the gas formed, and stating what happens when sodium hydroxide is added

A solution containing Fe^{2+} is first treated with nitric acid and then evaporated. A colorless gas evolves. The residue is dissolved in water and treated with 6 M NaOH. What is the species formed upon treatment with NaOH? What is the effect of adding HNO_3? What is the colorless gas?

SOLUTION: Addition of nitric acid causes Fe^{2+} to be oxidized to Fe^{3+}:

$$3\,Fe^{2+}(aq) + 4\,H^+(aq) + NO_3^-(aq) \longrightarrow 3\,Fe^{3+}(aq) + NO(g) + 2\,H_2O(l)$$

As the equation shows, NO should evolve upon evaporation; therefore, NO must be the colorless gas. The Fe^{3+} residue dissolves in water and forms $Fe(OH)_3(s)$ upon addition of 6 M NaOH.

EXERCISE 16 Writing the chemical reaction of copper (II) sulfide in a nitric acid solution

CuS dissolves in nitric acid to form sulfur and $NO(g)$. Write a chemical equation that shows this reaction.

SOLUTION: The information in the question tells us that the NO_3^- ion is reduced to $NO(g)$ and that S^{2+} in CuS is oxidized to sulfur. A *skeleton* equation would be

$$CuS + H^+ + NO_3^- \longrightarrow Cu^{2+} + S + NO + H_2O$$

Using techniques for balancing redox equations in Sections 20.1 and 20.2, we write the balanced redox equation is

$$3\,CuS(s) + 8\,H^+(aq) + 2\,NO_3^-(aq) \longrightarrow 3\,Cu^{2+}(aq) + 3\,S(s) + 2\,NO(g) + 4\,H_2O(l)$$

SELF-TEST QUESTIONS

Key Terms

Having reviewed key terms in Chapter 23, match key terms with phrases and identify statements as true or false. If a statement is false, indicate why it is incorrect.

Match each phrase with the best term:

23.1 The part of our earth that contains ores.

23.2 An example of this class of materials is sterling silver.

23.3 The term given to heating $PbCO_3$ to form PbO_2 and CO_2.

23.4 The term given to heating an ore such as HgS in oxygen to produce the metal.

23.5 Material consisting of mostly molten silicate materials.

23.6 The melting of materials to form at least two molten layers.

23.7 A process to improve the purity of an impure metal product.

23.8 Using heat to change the composition of an ore.

23.9 The general term giving to any process using reactions in water to extract metals from minerals.

23.10 A term other than dissolving given to removing a metal from an ore in an aqueous environment.

Terms:

(a) alloy
(b) calcination
(c) hydrometallurgy
(d) leaching
(e) lithosphere
(f) pyrometallurgy
(g) refining
(h) roasting
(i) slag
(j) smelting

True-False Statements:

23.11 Copper is recovered from ores by the *Bayer process.*

23.12 The more active metals are commonly recovered from their compounds by *electrometallurgy.*

23.13 In the *Downs cell*, $CaCl_2$ is added to lower the melting point of the cell medium.

23.14 The *Hall process* uses cryolite to recover pure aluminum.

23.15 An *intermetallic compound* is a heterogeneous mixture of metals.

23.16 Because of the *lanthanide contraction*, Zr and Hf have similar atomic radii.

23.17 Typically *ores* contain desired metals in high concentrations.

23.18 A *mineral* is a solid compound containing a metal and is found in nature.

23.19 *Metallurgy* is the science and technology of refining metals.

23.20 Steel is a *solution alloy* because it has carbon dispersed uniformly in its matrix.

23.21 Rapidly cooling a liquid alloy can result in the formation of a *heterogeneous alloy.*

Problems and Short-Answer Questions

23.22 (a) Which atoms in their ground state have the following valence-electron configurations?

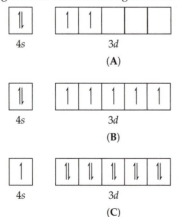

(A)

(B)

(C)

(b) Which atoms are diamagnetic or paramagnetic?

23.23 Look up the densities of the *d*-block elements of the sixth period. Plot the data using density for the y axis and the symbol of the element for the x axis.

23.24 Write chemical equations that show how gold is leached from ore by cyanide ions. The thermodynamically stable form of gold in solution is Au^{3+}.

23.25 What is the most common oxidation state of chromium? When the dichromate ion is reduced in an aqueous system, what chromium ion forms? When the pH of a chromate ion solution is decreased, the color of the solution changes from yellow to orange. Why?

23.26 What three anions are most commonly found in minerals?

23.27 When coke is burned in air in the lower portion of a blast furnace, heat is produced by the reaction. Another reaction is used to control the temperature. What are the two reactions and how does the second help control the temperature?

23.28 Many electrometallurgical processes must be done in a molten-salt mixture rather than in an aqueous environment. Why?

23.29 What is bauxite? Why cannot Al be readily recovered from Al_2O_3 in a molten form using electrometallurgical processes? In the Hall process, what is the purpose of cryolite?

23.30 What property of Ge permits it to be a semiconductor whereas carbon is not?

23.31 A necklace is made of 14-karat gold and another one is made of 10-karat gold. Does this mean the 14-karat gold necklace has 40% more gold than the 10-karat gold necklace?

23.32 Which would you expect to have a higher first ionization energy: Cr or Mn? Why?

23.33 Iron oxides are quite insoluble in water, yet natural water in a region containing iron minerals is found to contain ferric and ferrous ions. Explain.

23.34 Zinc is commonly found in nature in the form of ZnS. If ZnS is heated in the presence of oxygen, zinc oxide and sulfur dioxide form. Zinc metal is produced in a blast-furnace operation similar to the one used to reduce iron oxides to iron. Write chemical reactions to describe the formation of zinc oxide from zinc sulfide and the reduction of zinc oxide in the presence of coke.

23.35 Why does the electrical conductivity of a metal decrease when its temperature is increased?

23.36 Magnesium is primarily mined from the sea, which contains 0.13 percent magnesium by weight. Magnesium hydroxide is first precipitated by adding slaked lime, $Ca(OH)_2$, to seawater. The slaked lime is obtained by roasting crushed oyster shells ($CaCO_3$) and then adding a small amount of water. The $Mg(OH)_2$ is neutralized with HCl; from the solution $MgCl_2$ is recovered. Write chemical equations showing how $MgCl_2$ is recovered. How could $MgCl_2$ be further refined to produce $Mg(s)$?

23.37 Why does the roasting of HgS form $Hg(g)$, whereas roasting NiS produces $NiO(s)$ instead of $Ni(g)$?

23.38 Write the valence electron configurations for

(a) Fe (c) Ru^{3+}
(b) Ag^+ (d) Pt

23.39 Why are the physical properties of the second and third series of transition elements more similar to one another than they are to the first series of transition elements?

23.40 What is the most stable oxidation state for each of the following elements; (a) Sc; (b) Ti; (c) Cr; (d) Fe; (e) Ni; (f) Ag?

23.41 Which ion should exhibit a larger magnetic moment: Co^{2+} or Co^{3+}? Explain

23.42 Write reaction equations for (a) the reaction of $MnO_2(s)$ in a strongly basic medium to form MnO_3^{2-}; (b) the reaction of Ag(s) with dilute nitric acid; (c) the decomposition of $Cu_2SO_4(s)$ upon exposure to moisture.

23.43 What are the maximum oxidation states given in the text for

(a) Sc (d) Co
(b) V (e) Cu
(c) Mn (f) Zn

23.44 We have discovered a new ore that contains an unknown metal carbonate. We decide to calcine it and we find carbon dioxide is released. Do we have sufficient information to conclude that we now have the free metal remaining? If not, what additional information is needed?

23.45 We are mining low-grade ores of gold and concentrate the gold by using a solution of NaCN (see equations 23.12 and 23.13 in the text). We find we have no more zinc metal to reduce $Au(CN)_2^-$ but we have some copper metal. Cu is also a member of the transition elements and we believe it should also act in a manner similar to zinc. We decide to use it in place of zinc. Do we have sufficient information to conclude that the substitution of copper for zinc will be successful? If not, what additional information is needed?

23.46 We have a metal oxide that we think has a metal atom with unpaired electrons. When we place the metal oxide in a magnetic field we find it is very difficult to remove the substance. Furthermore, we notice that the metal oxide now acts as a permanent magnet. Do we have sufficient information to conclude that the metal oxide has unpaired electrons and to identify the type of magnetism?

23.47 Copper (I) salts are not common in nature. They are only slightly soluble in water and most are colorless.

(a) Given the following data

$$Cu^+(aq) + e^- \longrightarrow Cu(s) \quad E°_{red} = +0.52\ V$$

$$Cu^{2+}(aq) + 2\,e^- \longrightarrow Cu^+(aq) \quad E°_{red} = +0.16\ V$$

show that copper (I) spontaneously undergoes a disproportionation reaction (both oxidized and reduced) at standard conditions.

(b) What is K for the reaction at 25°C and what does this suggest about the extent of the disproportionation?

(c) CuI is insoluble in water. When iodide ion is added to an aqueous copper (I) solution the extent of disproportionation is extensively reduced. Explain.

23.48 $Cr(OH)_3(s)$ is amphoteric in water.

(a) What is amphoterism?
(b) Write net ionic equations describing the reactions of $Cr(OH)_3(s)$ with added acid and base.
(c) Do you expect chromium (II) to form from $Cr(OH)_3(s)$ in the presence of air?
(d) $Cr(OH)_2(s)$ is a typical hydroxide and is not amphoteric. Explain why it is not amphoteric.

23.49 Manganese(IV) oxide is a dark brown or black substance and is used in a variety of dry cell batteries. X-ray crystallography studies show that the formula is approximately $MnO_{1.9}$, not MnO_2. Several percent of oxygen atoms are missing and this results in holes occurring in the structure.

(a) Provide an explanation for the observation that manganese(IV) oxide can conduct current (which is one reason it is used in batteries).
(b) Manganese(IV) oxide is a nonstoichiometric compound. What does this mean and how can the charge balance in the crystalline substance be rationalized so that overall it carries a zero net charge?

Multiple-Choice Questions

23.50 Which pairing of metal and mineral source is correct?

(a) Chalcocite, CuF (d) Cinnabar, HgS
(b) Hematite, NiO (e) Bauxite, B_2O_3
(c) Rutile, $TiCl_2$

23.51 The flotation process in metallurgy involves which of the following?

(a) mining of an ore;
(b) concentration of an ore;
(c) reduction of a metal ion;
(d) smelting of iron;
(e) leaching of a component of an ore.

23.52 Which of the following is *not* a reasonable reaction in an industrial process for obtaining a metal by reduction?

(a) $FeO(s) + CO(g) \longrightarrow Fe(g) + CO_2(g)$
(b) $Zn^{2+}(aq) + 2\,e^- \xrightarrow{electrolysis} Zn(s)$
(c) $TiCl_4(s) + 4\,Na(s) \longrightarrow 4\,NaCl(s) + Ti(s)$
(d) $WO_3(s) + 3\,O_2(g) \longrightarrow W(s) + 3\,O_3(g)$
(e) $MoO_3(s) + 3\,H_2(g) \longrightarrow Mo(s) + 3\,H_2O(g)$

23.53 Molten NaCl is electrolyzed in a Downs cell to produce metallic sodium. Why is calcium chloride also added?

(a) To prevent the formation of chlorine gas
(b) To lower the melting point of molten NaCl
(c) To increase the rate of the electrolysis reaction
(d) To increase the concentration of chloride ions
(e) To raise the temperature of the system during electrolysis

23.54 Which of the following fourth-period elements should have the highest heat of fusion?

(a) K (d) Cr
(b) Ca (e) Fe
(c) Ti

23.55 Which of the following elements is likely to occur in its native state?

(a) Cu   (d) Al
(b) Ca   (e) Fe
(c) Mg

23.56 Which of the following elements should have the highest melting point at standard conditions?

(a) Cs   (d) W
(b) Ba   (e) Os
(c) Ta

23.57 Which of the following types of alloys is stainless steel?

(a) solution   (d) heterogeneous
(b) substitutional   (e) intermetallic
(c) interstitial

23.58 Which of the following has the greatest electrical conductivity in the solid state?

(a) I_2 (d) C
(b) Ni   (e) O_2
(c) Sr

23.59 Which of the following exhibits the greatest number of oxidation states?

(a) Zr   (d) Ni
(b) Ti   (e) Mn
(c) V

23.60 Which element has the greatest first ionization energy?

(a) Au   (d) Ta
(b) Os   (e) Hf
(c) Re

23.61 Which of the following pairs of atoms has the most similar atomic radii?

(a) Ti, V   (d) Mo, W
(b) Ti, Zr   (e) Ni, Pd
(c) Cr, Mo

23.62 Which of the following ions has all of its electrons paired?

(a) Ti^{2+} (d) Ni^{2+}
(b) Cr^{3+} (e) Cu^+
(c) Mn^{2+}

23.63 Which of the following exhibits the property of ferromagnetism?

(a) Ti   (d) Zn
(b) Mn   (e) Hg
(c) Co

23.64 Which of the following is *not* a characteristic of transition elements?

(a) colorless ions
(b) multiple oxidation states
(c) paramagnetism occurs
(d) *d* orbitals being filled
(e) groups 5B and 6B show maximum number of metallic bonds

23.65 Carbon in the form of diamond is not a conductor yet metallic silicon is a conductor. This is best explained by the fact that

(a) silicon is larger than carbon.
(b) silicon has more electrons than carbon.
(c) silicon has a lower ionization energy than carbon.
(d) silicon has a lower electronegativity than carbon.
(e) silicon has valence and outer orbitals closer in energy than that for carbon.

SELF-TEST SOLUTIONS

23.1 (e). **23.2** (a). **23.3** (b). **23.4** (h). **23.5** (i). **23.6** (j). **23.7** (g). **23.8** (f). **23.9** (c). **23.10** (d). **23.11** False. Al is recovered. **23.12** True. **23.13** True. **23.14** True. **23.15** False. They are homogeneous mixtures. **23.16** True. **23.17** False. Low concentrations. **23.18** True. **23.19** False. Metallurgy involves more than just refining metals. Also involved is mining, extraction, reduction, concentrating, and refining. **23.20** True. **23.21** True. **23.22(a)** (A) Ti; (B) Mn; (C) Cu; (b) Diamagnetic: None; Paramagnetic: All. All of the examples have unpaired electrons and thus are paramagnetic.

23.23

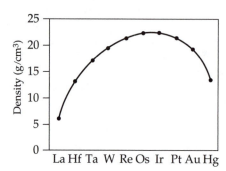

23.24 $4\,Au(s) + 8\,CN^-(aq) + O_2(aq) + 2\,H_2O(l) \longrightarrow$
$$4\,Au(CN)_2^-(aq) + 4\,OH^-(aq)$$

23.25 Cr (III) is the most common chromium ion. Cr (III) is produced when $Cr_2O_7^{2-}$ is reduced. The chromate ion exists in an acidic solution in an equilibrium with the dichromate ion as follows:

$$2\,CrO_4^{2-}(\text{yellow}) + 2\,H^+ \rightleftharpoons Cr_2O_7^{2-}(\text{deep orange}) + H_2O$$

When the pH is decreased (hydrogen ion concentration increases) the equilibrium shifts to the right and the orange dichromate ion forms in a greater concentration.

23.26 Oxide, sulfide, and carbonate.

23.27 The burning of coke is $2\,C(s) + O_2(g) \rightarrow 2\,CO(g)$ with the release of 221 kJ/mol of energy. Coke also reacts with water vapor with the absorption of 131 kJ/mol of energy: $C(s) + H_2O(g) \rightarrow CO(g) + H_2(g)$. By controlling the water vapor, the amount of energy consumed can be varied and this can ameliorate the heat generated by the first reaction.

23.28 The active metals such as Mg, Na, and Al cannot be recovered from an aqueous system because water is more easily reduced than the metal ions. See equations 23.15 and 23.16 in the text. In molten-salt systems, the active metal ion is the most easily reduced species.

23.29 Bauxite is a key mineral source for aluminum. It contains a hydrated aluminum oxide and several major impurities such as SiO_2 and Fe_2O_3. Al_2O_3 is formed from bauxite and if it is used to form metallic aluminum in a molten state, a temperature greater than 2000 °C is required. This is not an economical process. Thus, cryolite, Na_3AlF_6, is added to aluminum oxide to create a mixture that has a lower melting point, 1012 °C. The lower temperature makes feasible the economics of elec-trometallurgical recovery of aluminum.

23.30 In Ge the energy gap between the valence band MO and the empty conduction band is 67 kJ/mol, whereas for C the gap is 502 kJ/mol. In Ge the valence electrons can move from the valence MO to the conduction MO, leaving electron holes in the valance band. The presence of the electron holes enable electrons to move through the solid. In C the energy gap is too large and the valence electrons remain in the valence band MO.

23.31 Pure gold is defined as 24 karat. A 14-karat gold necklace has $14/24 \times 100 = 58\%$ gold and the 10-karat gold necklace has $10/24 \times 100 = 42\%$ gold. The increase in gold content is an additional 16%.

23.32 The valence electron configurations are Cr, $3d^54s^1$, and Mn, $3d^54s^2$. Removing an electron from the $4s$ orbital in Mn will require more energy because it has a filled subshell. This is not the case for Cr. Thus, Mn has the higher ionization energy.

23.33 Another type of iron mineral found in nature is $FeCO_3$ and it has a K_{sp} of 3.2×10^{-11}. Natural water contains dissolved carbon dioxide, which causes slight acidity. This permits the iron(II) carbonate to dissolve as follows:

$$FeCO_3(s) + CO_2(aq) + H_2O(l) \longrightarrow$$
$$Fe^{2+}(aq) + 2\,HCO_3^{-}(aq)$$

The ferrous ion can be oxidized by oxygen in air to form ferric ion.

23.34
$$2\,ZnS(s) + 3\,O_2(g) \longrightarrow 2\,ZnO(s) + 2\,SO_2(g)$$
$$2\,C(s) + O_2(g) \longrightarrow 2\,CO(g)$$
$$ZnO(s) + CO(g) \longrightarrow Zn(g) + CO_2(g)$$

23.35 With increasing temperature the metal atoms in the solid lattice vibrate more. This vibration disrupts the overlap of atomic orbitals between atoms. There are thus fewer vacant energy bands that electrons can move into under the influence of an electrical field.

23.36 $CaCO_3(s) \xrightarrow{\Delta} CaO(s) + CO_2(g)$
$$CaO(s) + H_2O(l) \longrightarrow Ca(OH)_2(s)$$
$$Ca(OH)_2(s) + Mg^{2+}(aq) \longrightarrow Mg(OH)_2(aq) + Ca^{2+}(aq)$$
$$Mg(OH)_2(aq) + 2\,HCl(aq) \longrightarrow MgCl_2(aq) + 2\,H_2O(l)$$

Evaporate off water from the $MgCl_2$ solution to form $MgCl_2$. Then using an electrolytic cell and molten $MgCl_2$, electrolyze Mg^{2+} to $Mg(l)$.

23.37 Hg is low on the activity series of metals; thus in the presence of $O_2(g)$ it only very slowly forms HgO. Nickel is higher on the activity series. If native Ni were formed in roasting, it would immediately react with oxygen to form $NiO(s)$.

23.38 (a) $3d^64s^2$;   (c) $4d^5$;
     (b) $4d^{10}$;   (d) $5d^96s^1$.

23.39 The phenomenon of the Lanthanide contraction results in the atomic radii of the second and third series of transition elements being almost the same within each group. There is a significant difference in atomic radii between the first and second series of transition elements; thus their physical properties are more different.

23.40 (a) +3;   (d) +3;
     (b) +4;   (e) +2;
     (c) +6;   (f) +1.

23.41 The valence-electron configurations are $3d^7$ for Co^{2+} and $3d^6$ for Co^{3+}. Co^{2+} possesses three unpaired electrons and Co^{3+} possesses four unpaired electron. The ion with the larger number of unpaired electrons will have a larger magnetic moment: Co^{3+}. Constructing an electronic "box" diagram may help you see the differences in electron population.

23.42 (a) $MnO_2(s) + 2\,OH^{-}(aq) \longrightarrow$
$$MnO_3^{2-}(aq) + H_2O(l)$$

(b) The reaction is similar to the one for Cu:
$$3\,Ag(s) + 4\,HNO_3(aq) \longrightarrow$$
$$3\,AgNO_3(aq) + NO(g) + 2\,H_2O(l)$$

(c) Cu in Cu_2SO_4 is in the +1 oxidation state, which can disproportionate.
$$Cu_2SO_4(s) \longrightarrow Cu(s) + CuSO_4(s)$$

23.43 (a) Sc^{3+};   (d) Co^{3+};
     (b) V^{5+};   (e) Cu^{3+};
     (c) Mn^{7+};   (f) Zn^{2+}.

23.44 There is insufficient information. Calcination brings about the decomposition of the metal carbonate with the elimination of carbon dioxide gas. Typically the metal oxide remains, not the free metal. You have to take the product remaining after calcination and do further analysis to determine, for example, if it is an oxide. You could react it with carbon monoxide and see if a new product, more likely the free metal, is produced.

23.45 There is insufficient information. We know that zinc is reduced in the reduction process and we have to compare how easily copper is reduced. We should look up in a table of standard reduction potentials the relative reducing ability of zinc versus iron. The standard reduction potential for zinc is -0.76 V and for copper it is $+0.34$ V. This data shows that copper is not easily reduced compared to zinc.

23.46 There is sufficient information. A substance with unpaired electrons is drawn into a magnetic field. The metal oxide has this property. Furthermore, the fact that it is strongly attracted to the magnetic field and then becomes a permanent magnet strongly suggests that it is ferromagnetic.

23.47 (**a**) The two half-reactions are:

$$2\,Cu^+(aq) + 2\,e^- \longrightarrow 2\,Cu(s) \quad E^\circ_{red} = +0.52\ V$$

$$Cu(s) + 2\,e^- \longrightarrow Cu^{2+}(aq) \quad E^\circ_{red} = +0.16\ V$$

The overall reaction and potential are:

$$2\,Cu^+(aq) \longrightarrow Cu(s) + Cu^{2+}(aq)$$

$$E^\circ = E^\circ_{red}(\text{cathode}) - E^\circ_{red}(\text{anode})$$
$$= 0.52\ V - (+0.16\ V) = 0.36\ V$$

Since the potential is greater than zero, the disproportionation reaction is spontaneous.

$$(\mathbf{b})\ \log K = \frac{nE^\circ}{0.0592\ V} = \frac{(2)(0.36\ V)}{0.0592\ V} = 12.2$$
$$K = 1.6 \times 10^{12}$$

The large value of K means that the equilibrium lies far to the right for the disproportionation reaction and that the concentration of copper (I) ion is small.

(**c**) If iodide ion is added to an aqueous solution of copper (I) the stable CuI(s) forms and removes copper (I) from solution. This drive the equilibrium to the left and therefore reduces the extent of disproportionation.

23.48 (**a**) The term amphoteric refers to the ability of a substance to act both as an acid and a base.

(**b**) Acts as an acid:

$$Cr(OH)_3(s) + OH^-(aq) \longrightarrow Cr(OH)_4^-(aq)$$

Acts as a base:

$$Cr(OH)_3(s) + 3\,H^+(aq) \longrightarrow Cr^{3+}(aq) + 3\,H_2O(l)$$

(**c**) No. Chromium (II) reacts readily with oxygen to form chromium (III), not the reverse.

(**d**) A metal hydroxide with a M—O—H unit can also be acidic in water if the O—H bond is sufficiently polarized. This requires that the metal has a large positive charge. Chromium (II) is less electropositive than chromium (III) and therefore does not sufficiently polarize the O—H bond to release H$^+$ ions to water.

23.49 (**a**) Conductivity in a solid can occur when electrons or ions move through the crystalline structure. In the case of manganese(IV) oxide, a deficiency of oxide ions exists, resulting in holes in the structure. An oxide ion can therefore move into one of the vacant holes and create a new hole. Another oxide ion can migrate to this newly created hole and by this migration the substance is able to conduct current.

(**b**) The charge of the substance is zero and the observed chemical formula is $MnO_{1.9}$ (approximately). If each manganese ion is in the +4 ion state there should be two oxide ions per manganese ion to balance the charge. A deficiency of oxide ions suggests that some of the manganese ions must be in a lower charge state; as the oxidation number is reduced the number of oxide ions required is also lowered. Apparently, some of the manganese ions are in the +2 ion state and the sum of positive charges requires *an average* 1.9 oxide ions per manganese ion. A nonstoichiometric compound is one that has a defect in the structure such that the ratio of elements is not in a simple whole-number ratio.

23.50 (**d**). See Table 23.1 in the text. **23.51** (**b**). **23.52** (**d**). **23.53** (**b**). **23.54** (**d**); Cr possesses six s and d valence electrons: $3d^54s^1$. **23.55** (**a**). **23.56** (**d**), a Group 6B element. **23.57** (**c**). **23.58** (**b**). **23.59** (**e**). **23.60** (**a**). **23.61** (**d**); effect of Lanthanide contraction. **23.62** (**e**), $3d^{10}$. **23.63** (**c**). **23.64** (**a**). **23.65** (**e**).

Chapter

Chemistry of Coordination Compounds

24

OVERVIEW OF THE CHAPTER

Review: Complex ions (16.11); Lewis bases (16.11); Lewis structures (8.5); oxidation numbers (4.4).

Learning Goals: You should be able to determine either the charge of a complex ion, having been given the oxidation state of the metal, or the oxidation state, having been given the charge of the complex. (You will need to recognize the common ligands and their charges.)

Review: Nomenclature (2.8).

Learning Goals: You should be able to name coordination compounds, having been given their formulas, or write their formulas, having been given their names.

Review: Molecular shapes (9.1).

Learning Goals: You should be able to:

1. Describe, with the aid of drawings, the common geometries of complexes. (You will need to recognize whether the common ligands are functioning as monodentate or polydentate ligands.)
2. Describe the common types of isomerism and distinguish between structural and stereoisomerism.
3. Determine the possible number of stereoisomers for a complex, having been given its composition.

Review: Planck's relationship (6.2).

Learning Goals: You should be able to:

1. Distinguish between inert and labile complexes.
2. Explain how the conductivity, precipitation reactions, and isomerism of complexes are used to infer their structures.
3. Explain how the magnetic properties of a compound can be measured and used to infer the number of unpaired electrons.
4. Explain how the colors of substances are related to their absorption and reflection of incident light.

24.6 BONDING IN TRANSITION METAL COMPLEXES: CRYSTAL-FIELD THEORY

Review:　　　　Nature of salts (2.8); shapes of $3d$ orbitals (6.6).

Learning Goals:　You should be able to:

1. Explain how the electrostatic interaction between ligands and metal d orbitals in an octahedral complex results in a splitting of energy levels.
2. Explain the significance of the spectrochemical series.
3. Account for the tendency of electrons to pair in strong-field, low-spin complexes.
4. Sketch representations of the d orbital energy levels in octahedral and tetrahedral complexes and explain the reason for a smaller crystal-field splitting in tetrahedral complexes as compared to octahedral complexes.
5. Sketch the d orbital energy levels in a square-planar complex.

TOPIC SUMMARIES AND EXERCISES

COORDINATION COMPOUNDS: TERMINOLOGY

The term metal **complexes** (or simply, complex) refers to any substance containing a central metal atom that is bonded to other atoms, ions, or molecules. Most complexes have a charge, but some do not.

- Groups bonded to the central metal of a metal complex ion are Lewis bases: these groups are known as **ligands**.
- For example, the salt $K_3[Fe(CN)_6]$ contains the complex ion $Fe(CN)_6^{3-}$. Six cyanide ions (ligands) surround and bind to the Fe^{3+} ion to form $Fe(CN)_6^{3-}$. The salt $K_3[Fe(CN)_6]$ is called a **coordination compound** because it contains a metal complex ion. The neutral complex, $Ni(CO)_4$, is a coordination compound and a metal complex.
- The central atom and surrounding ligands in a complex form the **coordination sphere**. In $K_3[Fe(CN)_6]$, the Fe^{3+} ion and the six cyanide ions constitute the coordination sphere. In the formula $K_3[Fe(CN)_6]$, the coordination sphere is identified by placing brackets about $Fe(CN)_6^{3-}$.
- Atoms of a ligand that bind to the central atom of a complex are referred to as **donor atoms**. In $[Fe(CN)_6]^{3-}$, the donor atoms are the six carbon atoms from the six cyanide groups.
- The *total number of donor atoms* bonded to a central metal atom in a complex is known as the **coordination number** of the metal atom. In $[Fe(CN)_6]^{3-}$, iron has a coordination number of six.

Ligands are classified according to the manner in which they bind to the central atom in a complex.

- Ligands such as F^-, Cl^-, Br^-, and NH_3 are said to be **monodentate ligands** because only one donor atom participates in the bond to the central atom.
- Other ligands, such as

$$H_2\ddot{N}CH_2CH_2\ddot{N}H_2 \qquad \underset{\substack{\| \quad | \\ O \quad O^-}}{CH_3CCHCCH_3} \qquad H_2\ddot{N}CH_2CH_2\ddot{N}HCH_2CH_2\ddot{N}H_2$$

(ethylenediamine)　　(anion of acetylacetone)　　(diethylenetriamine) attach themselves to the central metal atom of a complex using two or more of their donor atoms; these are referred to as **polydentate ligands**.

- The ligands ethylenediamine, diethylenetriamine, and the anion of acetylacetone are also referred to as **chelating agents** because their donor atoms bind simultaneously to the same central atom:

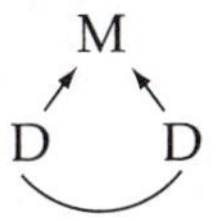

$$
\begin{array}{c}
M \\
\diagup \quad \diagdown \\
D \qquad D \\
\smile
\end{array}
$$

(D = donor atom)

- Note that the metal atom, the donor atoms, and all atoms bonded to one another between the donor atoms in the ligand form a ring system known as a **chelate ring**. We experimentally find that ligands with donor atoms separated by two or three other atoms make the best chelating ligands and form the most stable complexes. Therefore, chelate rings containing a total of five or six atoms tend to be most stable.

EXERCISE 1 Determining oxidation number of a metal atom in a complex

Determine the oxidation number of the metal atom in each of the following:
(a) $[Cr(NH_3)_6](NO_3)_3$; (b) $K_3[Fe(CO)(CN)_5]$; (c) $[Zn(NH_3)_2Cl_2]$.

SOLUTION: *Analyze*: We are asked to determine the oxidation number of the metal atom in each of three coordination compounds.

Plan: The oxidation number of the central metal atom in a complex is calculated from the relationship:

charge of complex = oxidation number of metal atom +

sum of charges of ligands

The first step, therefore, is to determine the charge of the metal complex, which is the species in brackets.

(a) *Solve:* The complex, $[Cr(NH_3)_6]^{n+}$ must have a positive charge (cation) because the counter ion is NO_3^-, the anion of the compound. The complex ion must have a 3+ charge because each NO_3^- group has a 1− charge: $[Cr(NH_3)_6]^{3+}$. Next, determine the charge of all ligands; in this case, the only ligand is ammonia, and it is a neutral molecule, of charge zero. Let n represent the oxidation number of chromium:

$$
[Cr(NH_3)_6]^{3+}
$$
$$
1(n) + 6(0) = +3
$$
$$
n + 0 = +3
$$
$$
n = +3
$$

The oxidation number (and ion charge) of chromium is therefore +3. (**b**) Since each potassium ion in $K_3[Fe(CN)_5(CO)]$ possesses a 1+ charge, the complex $[Fe(CN)_5(CO)]^{3-}$ must possess a 3− charge to maintain charge balance. Each cyanide ion has a 1− charge, and the CO group possesses a zero charge. You can calculate the oxidation number of iron as follows:

$$
[Fe(CN)_5(CO)]^{3-}
$$
$$
(1)(n) + 5(-1) + 1(0) = -3
$$
$$
n - 5 + 0 = -3
$$
$$
n = +2
$$

Therefore the oxidation state of iron is $+2$. (**c**) Proceeding in a manner similar to that used in (**a**) and (**b**), you can calculate the oxidation number of zinc as follows:

$$[Zn(NH_3)_2Cl_2]$$

$$(1)(n) + 2(0) + 2(-1) = 0$$
$$n + 0 - 2 = 0$$
$$n = +2$$

Therefore, zinc has an oxidation number of $+2$.

EXERCISE 2   Drawing Lewis structures of ligands and identifying donor atoms

Draw the Lewis structures of the following ligands and indicate the donor atoms in each: (**a**) NH_3; (**b**) CO; (**c**) oxalate ion, $C_2O_4^{2-}$; (**d**) ethylenediamine; Also identify each ligand in (**a**)–(**d**) as a monodentate ligand or a chelating agent.

SOLUTION: *Analyze and Plan*: We are asked to write Lewis structures for four ligands and to identify the donor atoms in each. The question essentially provides the plan: Write the Lewis structures. By looking at the Lewis structure we can identify atoms which have nonbonding electrons capable of being shared with another atom, that is, acting as a Lewis base. Chelating agents are ligands with two or more donor sites and a flexible ligand which permits the donor atoms to bind simultaneously to the same central metal atom.

Solve: (**a**) The Lewis structure for NH_3 is

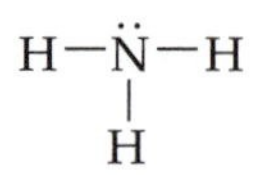

$$H-\overset{..}{N}-H$$
$$|$$
$$H$$

The nitrogen atom is the donor atom because it possesses a nonbonding electron pair capable of donating electron density to another atom. Since there is only one donor atom, it is a monodentate ligand. (**b**) The Lewis structure for CO is

$$:C\equiv O:$$

Both carbon and oxygen have nonbonding electron pairs, and therefore both can act as donor atoms. However, only one atom is capable of donating to the same metal atom; it is impossible for the other atom to bend and attach itself to the same metal atom. Carbon monoxide usually acts as a monodentate ligand, with carbon behaving as the donor atom.

(**c**) The Lewis structure for the oxalate ion is

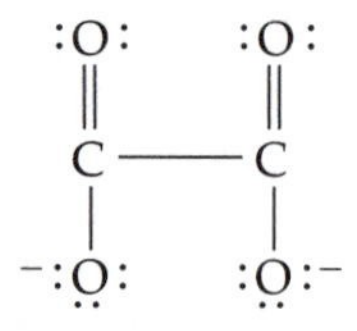

The oxalate ion acts as a chelating agent:

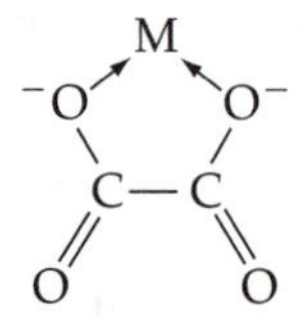

(**d**) The Lewis structure for ethylenediamine is

$$H-\overset{\overset{\displaystyle H}{|}}{\underset{\underset{\displaystyle H}{|}}{\ddot{N}}}-\overset{\overset{\displaystyle H}{|}}{\underset{\underset{\displaystyle H}{|}}{C}}-\overset{\overset{\displaystyle H}{|}}{\underset{\underset{\displaystyle H}{|}}{C}}-\overset{\overset{\displaystyle H}{|}}{\underset{\underset{\displaystyle H}{|}}{\ddot{N}}}-H$$

Both nitrogen atoms possess nonbonded electron pairs and therefore can behave as Lewis bases. Ethylenediamine is a chelating agent:

$$\begin{array}{c} {}_{\nearrow}M_{\searrow} \\ H_2N \qquad\qquad NH_2 \\ \diagdown\qquad\qquad\diagup \\ CH_2-CH_2 \end{array}$$

EXERCISE 3   Determining coordination number

What is the coordination number of the central metal atom in each of the following complexes: (a) $[Zn(NH_3)_2Cl_2]$; (b) $[Ni(en)_3]^{2+}$, where "en" is an abbreviation for ethylenediamine; (c) $[Cr(H_2O)_2Cl_2]^+$?

SOLUTION: *Analyze*: We are asked to determine the coordination number of the metal atom in each of three complexes. We are also told that "en" is an abbreviation for ethylenediamine, which we should know is a chelate.

Plan: The coordination number of a metal complex is the total number of donor atoms bonded to the central metal atom. Thus, we need to know for each ligand how many donor atoms bind to a metal. This may require writing a Lewis structure if we are not familiar with the ligand. Remember that the number of ligands is not necessarily the coordination number.

Solve: (a) Four, because each ligand has only one donor atom. (b) Six. Each ethylenediamine has two donor atoms, and there are three ethylenediamine ligands. (c) Four. Each ligand has only one donor atom.

The rules for naming complex ions and compounds are given in Section 24.3. In the next four exercises we are asked to name ligands, metal complexes and coordination compounds. This requires us to determine the oxidation state of the metal and the name of the ligands in each metal complex. Each exercise highlights particular cases and provides additional information.

NAMING OF COORDINATION COMPOUNDS

EXERCISE 4   Naming ligands

Before you can name complexes, you must be able to name ligands. Name the following ligands: (a) F^-; (b) NO_3^-; (c) O^{2-}; (d) $C_2O_4^{2-}$; (e) CO_3^{2-}; (f) SO_4^{2-}; (g) C_5H_5N (pyridine).

SOLUTION: Names of anionic ligands are derived from the commonly used name of the anion by dropping an *-ide* ending or the *-e* of an *-ate* or *-ite* ending and adding *-o*. Neutral ligands are named as the neutral molecule, except for H_2O, aqua, NH_3, ammine and CO, carbonyl. (a) Fluoro (derived from *fluor- + o*); (b) nitrato (derived from *nitrat- + o*); (c) oxo (derived from *ox- + o*); (d) oxalato (derived from *oxalat- + o*); (e) carbonato (derived from *carbonat- + o*); (f) sulfato (derived from *sulfat- + o*); (g) pyridine (a neutral molecule, and not a special case).

EXERCISE 5   Naming complexes: Cationic and neutral

Name the following complex cations and neutral coordination compounds
(a) $[Ag(NH_3)_2]^+$; (b) $[Co(en)_3]^{3+}$; (c) $[Ni(CO)_4]$; (d) $[Cr(H_2O)_4Cl_2]^+$.

SOLUTION: Complex cations and neutral complexes are named by first listing the ligands, in alphabetical order and with appropriate prefixes to indicate their number, and then naming the central metal, using its name as given
in the periodic table, followed by its oxidation number in parentheses.
(a) Diamminesilver(I) ion. *Note:* NH_3 is the only ligand containing an "ammine" with two *m*'s; all others have one *m*. (b) tris(ethylenediamine)cobalt(III)
ion (because ethylenediamine contains the prefix di-, its name is enclosed in
parentheses and the alternate prefix tris- is used); (c) tetracarbonylnickel(0);
(d) tetraaquadichlorochromium(III) ion (note that neither the terminal "a" of
tetra nor the "a" of aqua is dropped).

EXERCISE 6   Naming anionic complexes

Name the following anionic complexes: (a) $[FeCO(CN)_5]^{3-}$; (b) $[Ag(CN)_2]^-$;
(c) $[PtCl_6]^{2-}$.

SOLUTION: The rules for naming an anion complex are the same as those discussed in Exercise 5, except that the name of the central metal ends in -ate and in
certain cases the Latin stem names of some metals are used. Examples of these
are: Pt (platinate), Fe (ferrate), Ag (argentate), Sn (stannate), Pb (plumbate), Cu
(cuprate), and Au (aurate). (a) Carbonylpentacyanoferrate(II) ion; (b) dicyanoar-
gentate(I) ion. (c) hexachloroplatinate(IV) ion.

EXERCISE 7   Naming coordination compounds

Name the following coordination compounds: (a) $K[Ag(CN)_2]$; (b) $[Co(en)_3]Cl_3$;
(c) $Na[Cr(en)_2(SO_4)_2]$(en = ethylenediamine).

SOLUTION: Remember that cations of salts are named before the anions.
(a) Potassium dicyanoargentate(I); (b) tris(ethylenediamine)cobalt(III) chloride:
(c) sodium bis(ethylenediamine)disulfatochromate(III).

THE STRUCTURAL CHEMISTRY OF COORDINATION COMPOUNDS: ISOMERISM

Although complexes containing metal atoms with integral coordination
numbers from two to six and greater are observed, the great majority of
metal atoms have coordination numbers of four or six, with six the most
common one.

- Two structural configurations are commonly observed for four-coordinate
  complexes: tetrahedral and square planar. Tetrahedral transition metal
  complexes occur more commonly than square-planar ones. Square-planar
  transition metal complexes primarily occur with a metal atom possessing
  eight valence-shell electrons.
- The geometry of six-coordinate complexes usually corresponds to six
  coordinated atoms at the center of an octahedron or a distorted octahedron.
  In the idealized octahedral case, all atoms are an equivalent distance
  from the central atom.

In Section 24.4 several classes of isomerism are discussed. These are briefly summarized in the flowchart shown below. Stereoisomers form the most important class of isomers.

- An important form of geometrical isomerism is *cis-trans* isomerism. In a *cis* isomer like groups are in adjacent positions; in a *trans* isomer the like groups are across from one another in a diagonal relationship. Octahedral and square planar complexes may form geometrical isomers when two or more different ligands are present, but *cis-trans* geometrical isomerism is not observed in tetrahedral complexes.
- Optical isomerism exists when mirror images of a compound cannot be superimposed on each other. Nonsuperimposable mirror images are referred to as chiral. Exercise 10 shows you how to draw a pair of mirror images and to determine if they are superimposable.
- A solution of a chiral compound will rotate the plane of polarized light: Dextrorotatory optical isomers rotate the plane to the right, and levorotatory rotate it to the left. Equal amounts of a pair of optical isomers form a racemic mixture; the racemic mixture does not rotate the plane of polarized light.

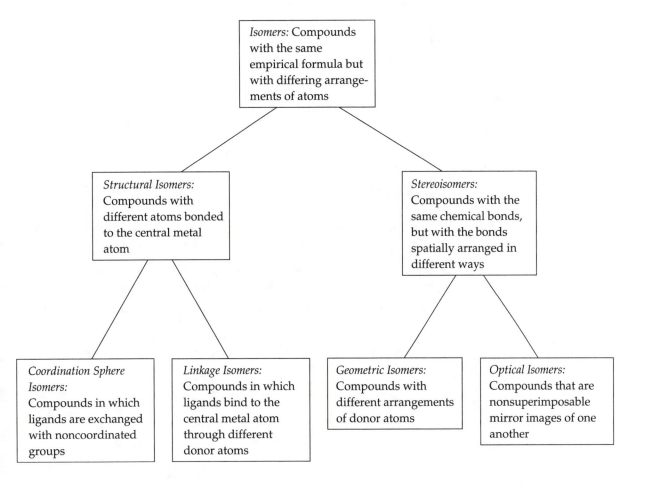

EXERCISE 8   Drawing geometric isomers of Pt(II) complexes

Pt(II) complexes are typically square-planar. (**a**) Sketch the geometrical isomers of $[Pt(NH_3)_2Cl_2]$. (**b**) An important type of geometrical isomerism is *cis-trans* isomerism. What is *cis-trans* isomerism? Identify the structures drawn for (**a**) as *cis* or *trans*.

SOLUTION: *Analyze*: We are told that platinum(II) complexes are typically square planar and to sketch the geometrical isomers for $[Pt(NH_3)_2Cl_2]$. Also we are asked to describe *cis-trans* isomerism and to label our sketches if this type of isomerism exists.

Plan: To sketch geometrical isomers we need to know the basic shapes of different geometries. Figures 24.3 and 24.4 in the text provide us this information. Note in Figure 24.3 that a square planar geometry has a square outline of donor atoms and the metal atom is at the center. We then consider what are the possible arrangements of the ligands around the square. We must be careful to recognize duplicate structures. If two structures can be rotated to superimpose upon each other, then they are not different. *Cis-trans* isomerism may occur in square planar or octahedral complexes. In a *cis* configuration involving monodentate ligands, two equivalent monodentate ligands are adjacent to one another. In a *trans* complex they are opposite one another, normally on a straight line through the central atom.

Solve: (**a**) Arrange the four ligands about platinum in a square-planar array. Only two distinct structures are possible:

$$\begin{bmatrix} H_3N & \diagdown & Cl \\ & Pt & \\ H_3N & \diagup & Cl \end{bmatrix} \quad \begin{bmatrix} H_3N & \diagdown & Cl \\ & Pt & \\ Cl & \diagup & NH_3 \end{bmatrix}$$
cis-form   *trans*-form

(**b**) The plan describes *cis-trans* isomerism and the isomers are labeled.

EXERCISE 9   Drawing geometric isomers of octahedral $[Ma_2b_4]$ complexes

Repeat questions (**a**) and (**b**) in Exercise 8 for the generalized octahedral complex $[Ma_2b_4]$ where a and b are monodentate ligands.

SOLUTION: *Analyze and Plan*: The question is the same as for exercise 8 except an octahedral complex is involved. We will follow the same approach.

Solve: (**a**) Arrange the ligands a and b at the corners of an octahedron with M at the center. Only two distinct structures are possible. (**b**) In an octahedral $[Ma_2b_4]$ complex, the terms *cis*- and *trans*- refer to the spatial configuration of the two equivalent donor atoms; in this example the two equivalent donor atoms are **a**.

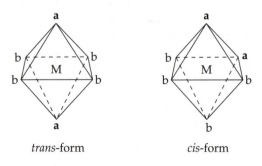

trans-form   *cis*-form

EXERCISE 10   Determining whether *cis*-[Co(en)₂(NO₂)₂]⁺ is chiral

Optical isomerism exists for a complex if the structure of the complex cannot be superimposed on its mirror image. Molecules that are not superimposable mirror images of one another are said to be chiral. Show that *cis*-[Co(en)₂(NO₂)₂]⁺ is chiral.

SOLUTION: *Analyze*: We are asked to show that a *cis* complex of cobalt is chiral.

Plan: In the question we are reminded that if a metal complex has a non-super-imposable mirror image the two mirror images (isomers) are chiral and optically active. First sketch the structure of the given cobalt complex and then draw its mirror image. In this problem we are told that the cobalt metal complex is optically active; thus, the mirror image should not be superimposable. Mentally we have to locate the centers of the two complexes so they coincide and then rotate one of the structures to see if they superimpose. This can be challenging and sometimes it is necessary to construct three-dimensional models.

Solve: First sketch one of the optical forms (**a**). Then draw the dotted line (**b**) to represent a mirror plan. Finally, sketch the mirror image (**c**) of structure (**a**) by reflecting all atoms through the mirror plan. Rotate structure (**c**) and superimpose on structure (**a**). If you cannot superimpose structure (**c**) on (**a**), the structures (**a**) and (**c**) are chiral. The curved lines represent ethylenediamine ligands. Note that if we rotate the molecules so that the −NO₂ groups of both complexes are superimposed, the curved lines of the ethylenediamine ligands are not; thus chiral forms of *cis*-[CO(en)₂(NO₂)₂]⁺ exist.

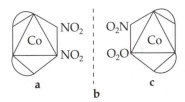

EXERCISE 11   Determining the coordination sphere of a complex and its geometric forms

(a) Co(NH₃)₅(NO₂)Cl₂ dissolves in water to form three moles of ions per mole of complex. One mole of the complex reacts with two moles of AgNO₃ to yield two moles of AgCl(*s*). Co(III) typically has a coordination number of six. Write the formula of the complex, bracketing the coordination sphere. (**b**) Does the complex exhibit *cis-trans* isomerism?

SOLUTION: *Analyze*: We are given a cobalt complex and told that it dissolves in water to form three moles of ions per mole of complex, reacts with silver nitrate to form two moles of silver chloride, and has a coordination number of six. Given this information we are asked to write the formula of the complex and to determine if it exhibits *cis-trans* isomerism.

Plan: (**a**) We analyze each piece of information to determine a specific characteristic of the cobalt complex: The number of ions comprising the metal complex, the number of ionizable chloride ions and the total number of donor atoms attached to the metal (which suggests the geometry). (**b**) We can use an approach such as in Exercise 8 to solve this question.

Solve: (**a**) The coordination sphere must contain cobalt and six donor atoms because cobalt has a coordination number of six. Because two moles of AgCl(*s*) form per one mole of the complex, the two chlorine atoms cannot be tightly bound to the cobalt and consequently must be outside of the coordination sphere. The formula $[Co(NH_3)_5(NO_2)]Cl_2$ fits the experimental observation that the substance dissolves in water to form three ions: $[CO(NH_3)_5NO_2]^{2+}$ and $2\ Cl^-$. The predicted formula also shows Co with a coordination number of six. (**b**) The complex $[Co(NH_3)_5(NO_2)]^{2+}$ has six ligands surrounding Co(III); thus it is an octahedral complex of type Ma_5b. *Cis-trans* isomerism cannot occur because there is no pair of *b* atoms.

EXERCISE 12   Determining the number of geometric isomers for a complex

How many geometric isomers are possible for $[Co(NH_3)_4(NO_2)_2]$?

SOLUTION: *Analyze and Plan*: We are asked to determine the number of geometric isomers possible for a cobalt complex. We can draw geometrical structures to answer this question. This requires us to determine the coordination number of the complex and then draw geometrical structures consistent with this coordination number.

Solve: Assume an octahedral arrangement of ligands exists about cobalt because the coordination number of cobalt is six. We see that there are two nitrite ligands which could be in *cis* or *trans* positions. In order to determine the number of isomers, we have to draw geometrical structures. Sometimes you may draw two structures that do not look identical but are identical if you can superimpose them. In this case there are two geometric isomers:

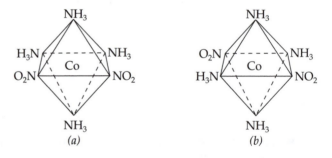

(a)   (b)

MAGNETISM AND COLORS OF TRANSITION METAL COMPLEXES

A physical property of complexes studied by chemists is their magnetic behavior.

- A substance with unpaired electrons is *strongly attracted* by a magnetic field; it is said to be **paramagnetic**.
- A substance possessing no unpaired electrons is *weakly repelled* by a magnetic field and is said to be **diamagnetic**.
- The magnetic properties of a transition metal complex depend on the metal ion, on how many *d* electrons it possesses, and on the number and kinds of ligands bonded to the metal.

The colors of transition metal complexes also depend on the metal ion and the nature of surrounding ligands.

- For a metal in a particular oxidation state, the color of the complex depends on the number and kinds of ligands attached.
- A complex exhibits color if it absorbs energy in the region of visible light. The color observed is not the color of the visible light absorbed; rather it

is the complementary color of the light absorbed. If several different energies in the visible spectrum are absorbed, the analysis is not straightforward. We should know the colors of the visible spectrum and their complementary colors.

In Section 24.6, a bonding model for transition metal complexes is presented that will further aid us in understanding magnetic properties and colors of complexes.

EXERCISE 13   Determining the number of unpaired 3*d* electrons possessed by cobalt(III)

Cobalt often exists as a 3+ ion. How many 3*d* electrons does a cobalt 3+ ion possess? As a free ion is it paramagnetic or diamagnetic? Why?

SOLUTION: *Analyze*: We are asked to determine how many 3*d* electrons exist in cobalt(III) and if it is diamagnetic or paramagnetic as a free ion and to explain our answers.

Plan: We first write the complete electron configuration of cobalt(III) to determine how many 3*d* electrons and unpaired electrons exist. The cobalt(III) ion is paramagnetic if it possesses one or more unpaired electrons.

Solve: Cobalt is a transition element and possesses 27 electrons. Therefore its electron configuration is $1s^2 2s^2 2p^6 3s^2 3p^6 3d^7 4s^2$. The valence electrons are the 3*d* and 4*s* electrons. The 4*s* electrons are of higher energy and leave first when a positive ion is formed. Therefore, the valence electrons of a cobalt 3+ ion are $3d^6$. There are five 3*d* orbitals and each can hold a maximum of two electrons. The first five electrons enter the orbitals singly, remaining unpaired. The sixth electron has to pair with one of the other five 3*d* electrons because there are only five 3*d* orbitals. This results in four of the electrons remaining unpaired. As a free ion Co^{3+} is paramagnetic because it possesses unpaired electrons.

EXERCISE 14   Predicting the color region and wavelength of absorbed light by a complex

The complex prussian blue, $[KFe(CN)_6 Fe]_x$, has a deep blue color, as indicated by its name. Using Figure 24.24 in the text, predict the color region and wavelength of visible light in which the complex primarily absorbs light.

SOLUTION: From Figure 24.24 in the text, you see that the complementary color of blue is orange, centered at a wavelength of about 600 nm. Thus, the complex absorbs light primarily in the orange region of the visible spectrum.

Crystal-field theory treats the bonding between ligands and the central metal in complexes as electrostatic. Key features of the crystal-field model are:

- The central metal atom with a positive charge is attracted electrostatically to the negative charge of a ligand or to the negative region of a polar ligand.
- The electrostatic field created by the surrounding ligands in a transition metal complex raises the energies of the *d* electrons in the central transition element compared to their energies in an isolated metal atom. This repulsive interaction between ligands and the outermost electrons of a transition metal is called the **crystal field**.
- Furthermore, the energies of the *d* orbitals of a transition metal are no longer equal in the presence of a crystal field. Depending on the

BONDING IN TRANSITION METAL COMPLEXES: CRYSTAL-FIELD THEORY

geometrical arrangement of ligands about the central transition metal atom, different energy groupings of d orbitals occur.

- For octahedral, tetrahedral, and square-planar complexes, the groupings of d energy levels are

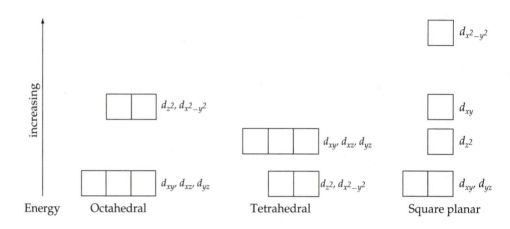

- The energy separation between the set of $d_{z^2}, d_{x^2-y^2}$ orbitals and the set of d_{xy}, d_{xz}, d_{yz} orbitals in the octahedral and tetrahedral diagrams are represented by Δ, delta. Delta is often called the crystal-field splitting energy.

You should learn how application of Hund's rule enables you to explain the existence of high-spin and low-spin complexes and other physical properties of transition metal complexes.

- The **low-spin (strong field)** case occurs when the six ligands surrounding the metal atom create a large crystal field (large Δ); in this case, less energy is required to first pair electrons in the lower energy set of d orbitals before placing them in the higher energy set.
- For a **high-spin (weak field)** case to occur, the value of Δ must be smaller than the energy required to pair electrons (spin-pairing energy); in this situation, a lower energy state results when electrons enter the five d orbitals singularly and with the same spins before pairing of electrons occurs.
- For example, high- and low-spin complexes exist for transition metals with four d electrons:

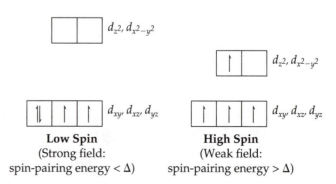

- For tetrahedral complexes, we observe only the weak-field case because Δ for tetrahedral crystal fields is always smaller than the energy required to pair electrons.

The values of Δ for transition metal complexes are often of the same order of magnitude as the energy per photon of light in the visible spectrum.

- A transition metal complex exhibits color if an electron in a lower-energy d orbital is excited into a higher-energy d orbital and if Δ is in the range of visible light.
- The magnitude of Δ for a given transition metal atom ion depends on the number and kinds of surrounding ligands and the charge of the metal ion.
- The **spectrochemical series** is a list of ligands arranged in order of their ability to increase Δ in a series of complexes of type ML_x in which M and X are constant:

$$\overset{\Delta \text{ increasing}}{\underset{\longrightarrow}{}}$$
$$Cl^- < F^- < H_2O < NH_3 < en < NO_2^- < CN^-$$

Ligands to the left of this series are said to be low in the spectrochemical series, and those to the right are said to be high. Thus in many complexes the ligand CN^- causes Δ to be sufficiently large so that a low-spin complex forms (electrons pair first in lower energy orbitals) whereas Cl^- causes a high-spin complex (maximum number of unpaired electrons).

EXERCISE 15   Applying crystal-field theory to octahedral complexes of cobalt(II)

Explain using crystal-field theory why octahedral cobalt(II) complexes exhibit different magnetic properties.

SOLUTION: *Analyze*: We are asked to explain using the principles of crystal-field theory why octahedral cobalt(II) complexes can have differing magnetic properties.

Plan: To apply the principles of crystal-field theory we need to determine the number of valence $3d$ electrons possessed by cobalt(II) and then construct crystal field diagrams for an octahedral complex with these $3d$ electrons. When we place the $3d$ electrons in the crystal field diagram we need to consider the possibility of high-spin and low-spin crystal field environments. Figure 24.33 in the text provides an example of this situation.

Solve: (1) The electron configuration of cobalt is $[Ar]3d^7 4s^2$. Since the $4s$ electrons leave first, the electron configuration of cobalt(II) is $[Ar]3d^7$. (2) Write the appropriate crystal field diagram for the given geometry. (3) Place electrons into the crystal field diagram. Depending on the ligand, two possible $3d$ electron populations of cobalt(II) in an octahedral environment can occur:

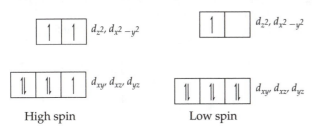

Octahedral crystal field

The high-spin case occurs when the value of Δ is smaller than pairing energy. Electrons enter the $3d$ orbitals one at a time and with parallel spins; the electrons remain unpaired until all d orbitals have one electron. The low-spin case occurs when the value of Δ is significantly larger than the energy required to pair electrons in $3d$ orbitals; electrons enter the low-energy set of $3d$ orbitals until all are filled before they occupy the higher-energy set of $3d$ orbitals.

EXERCISE 16   Applying crystal-field theory and drawing crystal-field energy-level diagrams

Draw the crystal-field energy-level diagram and show the placement of electrons for: (**a**) $CoCl_4^{2-}$ (tetrahedral); (**b**) $[MnF_6]^{4-}$ (high-spin); and (**c**) $[Cr(en)_3]^{2+}$ (four unpaired electrons).

SOLUTION: *Analyze*: We are asked to draw the crystal-field energy-level diagrams for three metal complexes and show the placement of $3d$ electrons in each.

Plan: For each case we (1) determine the number of $3d$ electrons possessed by the metal; (2) determine the geometry of the metal complex if it is not given; (3) draw the crystal-field diagram for the geometry; and (4) place the $3d$ electrons in the crystal-field diagram. The last step may require an assessment whether the ligands create a high-spin or low-spin environment using the spectrochemical series.

Solve: (**a**) The oxidation number of cobalt in $CoCl_4^{2-}$ is +2. Co(II) has an electron configuration of $[Ar]3d^7$. All tetrahedral complexes are high spin, even if there is the possibility of low-spin electron configuration. Place the electrons in the d orbitals singularly until all orbitals have one electron. Then begin to pair electrons in the lower energy set of orbitals first. The crystal-field energy-level diagram is

$\uparrow$ $\uparrow$ $\uparrow$ d_{xy}, d_{xz}, d_{yz}

$\uparrow\downarrow$ $\uparrow\downarrow$ $d_{x^2-y^2}, d_{z^2}$

Tetrahedral crystal field

(**b**) Mn(II) has an electron configuration of $[Ar]3d^5$. The complex is expected to be octahedral because the coordination number is six. The F^- ion typically forms high-spin complexes; it is low in the spectrochemical series.

$\uparrow$ $\uparrow$ $d_{x^2-y^2}, d_{z^2}$

$\uparrow$ $\uparrow$ $\uparrow$ d_{xy}, d_{xz}, d_{yz}

Octahedral crystal field

(**c**) Cr(II) has an electron configuration of $[Ar]3d^4$. $[Cr(en)_3]^{2+}$ is expected to be octahedral because the coordination number of the complex is six. Four electrons are placed in a high-spin environment to agree with the given information

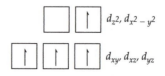

$\square$ $\uparrow$ $d_{z^2}, d_{x^2-y^2}$

$\uparrow$ $\uparrow$ $\uparrow$ d_{xy}, d_{xz}, d_{yz}

Octahedral crystal field

SELF-TEST QUESTIONS

Key Terms

Having reviewed key terms in Chapter 24, match key terms with phrases and identify statements as true or false. If a statement is false, indicate why it is incorrect.

Match each phrase with the best term:

24.1 The term given to the environment within the brackets in $[Ni(NH_3)_6]Cl_2$.

24.2 The number of donor atoms bonded to cobalt in $[Co(NH_3)_4Cl_2]^+$.

24.3 Acts as a Lewis base in metal coordination compounds.

24.4 A ligand which acts like a "claw" when bonded to a metal atom.

24.5 Possesses two or more active donor atoms.

24.6 Possesses only two active donor atoms.

24.7 Possesses only one active donor atom.

24.8 Compounds which have the same chemical formula but with different arrangement of atoms.

24.9 A type of isomerism in which the same chemical bonds are present but they are arranged differently in space.

24.10 A type of isomerism in which the same chemical bonds are present but like ligands are adjacent to each other.

24.11 A type of isomerism in octahedral complexes in which two identical ligands occupy opposite vertices and the other vertices have different, but identical, ligands.

24.12 A molecule or ion that does not have a superimposable mirror image is called this term.

24.13 An isomer that rotates the plane of polarized light to the right is labeled this term.

24.14 An isomer that rotates the plane of polarized light to the left is labeled this term.

24.15 All chiral molecules have this property.

24.16 A general term given to isomers which have different chemical bonds.

24.17 A general term given to isomers which have the same chemical bonds but the ligands occupy the space around the metal atom in different ways.

Terms:

(a) bidentate
(b) chiral
(c) chelating
(d) *cis*
(e) coordination number
(f) coordination sphere
(g) dextrorotary
(h) geometric
(i) isomers
(j) levorotary
(k) ligand
(l) monodentate
(m) optically active
(n) polydentate
(o) stereoisomers
(p) structural isomers
(q) *trans*

True-False:

24.18 A metal ion with a d^8 valence-electron configuration in an octahedral field of ligands can form *low-spin* complexes, depending on the nature of the ligands.

24.19 CN^- is high in the *spectrochemical series* because it creates a large crystal field.

24.20 Using the spectrochemical series and the principles of crystal-field theory, we predict $Co(CN)_6^{3-}$ to be a *high-spin* complex.

24.21 In the compound $[Au(CN)_2]Cl$, both ions are *complex ions*.

24.22 The *complex* in the compound in 24.21 is $[Au(CN)_2]^-$.

24.23 The ligand ethylenediamine has two *donor atoms*: nitrogen and oxygen.

24.24 $[Ni(en)_3]^{2+}$ has a larger formation constant than that of $[Ni(NH_3)_6]^{2+}$ because of the *chelate effect*.

24.25 Myoglobin is a globular protein containing a *porphyrin*, a large nitrogen-containing chelate.

24.26 The ligand NH_3 exhibits *linkage isomerism* because it can coordinate through either nitrogen or hydrogen.

24.27 *Coordination-sphere isomerism* occurs when two compounds exhibit the following behavior: $[MX_4Y_2]Z$ and $[MX_4YZ]Y$.

24.28 *Optical isomerism* is a form of stereoisomerism of the following type: A pair of mirror images that are superimposeable on each other.

24.29 When equal amounts of a pair of *d*- and *l*-isomers are present, a *racemic* mixture exists.

24.30 Green is the *complementary color* of red.

24.31 When a complex absorbs the color red, it will have an *absorption spectrum* in the visible range of light.

24.32 According to *crystal-field theory*, the interaction of a *d* electron with a negatively charged ligand will lower the energy of all *d* electrons.

24.33 Energy is released when electrons pair, and it is referred to as *spin-pairing energy*.

Problems and Short-Answer Questions

24.34 Given the following diagram for the geometry of a metal complex, what can you conclude about the type of geometry and isomerism? Draw another figure that is identical but has a different view.

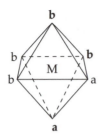

24.35 From the following crystal-field energy-level diagram for a metal complex, what can you say about its structure, its magnetism, and whether it is a weak field or strong field situation?

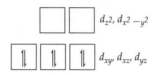

24.36 Should you expect the chemistry of $Ag^+(aq)$ to be similar to that of $[Ag(CN)_2]^-(aq)$? Explain.

24.37 Critique the following statement: The coordination number of $[ZnBr_2(en)]$ is three because there are three ligands attached.

24.38 If an oxalate ion is added to a solution containing $[NiCl_6]^{4-}(aq)$ what do you expect to happen and why?

24.39 If a chelating ligand is used to displace nonchelating ligand with the same donor atoms as in the chelate what enthalpy and entropy changes might you expect?

24.40 How do isomers differ in terms of number of atoms in their chemical formulas, solubilities, or structures?

24.41 What common property do enantiomers and chiral molecules possess?

24.42 What accounts for the differences between a high-spin metal complex and a low-spin metal complex? Which d electron populations in an octahedral metal complex can exhibit these two types?

24.43 In the spectrochemical series is there any relationship between the charge of a ligand and the relative impact of the ligand on the energy gap Δ? Is your conclusion in agreement with your expectation of the impact?

24.44 Name the following compounds:

   (a) $[Co(NH_3)_6]Cl_3$ (d) $Ca_2[Fe(CN)_6]$
   (b) $[Fe(CO)_5]$ (e) $[Fe(H_2O)_6]SO_4$
   (c) $K_2[ZnCl_4]$ (f) $[Ir(NH_3)_2(en)_2]Cl_3$.

24.45 Write the formula for each of the following compounds or ions:

   (a) potassium tetrachloroplatinate(II);
   (b) hexaammineplatinum(II) chloride;
   (c) ammonium diamminetetracyanochromate(III);

   (d) hexachlorostannate(IV) ion;
   (e) tetrachloroethylenediamineplatinum(IV)

24.46 Give the coordination number of the central metal ion in each of the complexes given in problem 24.44.

24.47 $[Ir(CO)(PR_3)_2Cl]$ has a square-planar arrangement of ligands about Ir. Draw and name all of its geometrical isomers. The name of the neutral ligand PR_3 is triphenylphosphine, where R is C_6H_5.

24.48 Sketch all isomeric structures for

   (a) $[Cr(H_2O)_5Cl]^+$
   (b) $[Fe(C_2O_4)_3]^{3-}$

24.49 List the ligands A, B, and C in order of increasing Δ, given that the following octahedral complexes absorb energy equivalent to Δ at the stated wavelengths: MA_6, 350 nm; MB_6, 400 nm; and MC_6, 300 nm.

24.50 The complex ion $[FeBr_4]^-$ has a tetrahedral structure. Using the crystal-field model, predict how many unpaired d electrons the complex possesses.

24.51 Using water and Cl^- as ligands,

   (a) give the formula of a six-coordinate Cr(III) complex that would be a nonelectrolyte in solution;
   (b) give the formula of a six-coordinate Cr(III) complex that has about the same electrical conductivity in solution as $MgCl_2$ in water;
   (c) give the formula of an octahedral complex of Cr(III) with a charge of 1−.

24.52 Predict the magnetic properties for $[V(NH_3)_6]^{3+}$ and $[MnF_6]^{3-}$.

24.53 A new complex has the following formula: $[NiX_2(NH_3)_2]$. Is this sufficient information for us to determine the charge of the ligand X? If not, what additional information is needed and why?

24.54 The name of a metal complex is bis(oxalato)dichloro-cobaltate(III) ion. Is this sufficient information for us to determine the charge of the complex and if it is a *trans-* complex?

24.55 $[Fe(CN)_6]^{4-}$ is diamagnetic. Is this sufficient information for us to conclude that it is a low-spin metal complex? If not, what additional information is needed and why?

Integrative Exercises

24.56 Vitamin B_{12s} is a metal complex containing cobalt and a large ligand; it is biologically important. Each molecule of Vitamin B_{12s} contains one cobalt atom and the percent of cobalt by mass is 4.43%. What is the molar mass of Vitamin B_{12s}?

24.57 When *trans*-$[Cr(en)_2(NCS)_2]SCN(s)$ is heated it is converted to $[Cr(en)_2(NCS)_2][Cren(NCS)_4](s)$ and en(g).

   (a) What is "en"?
   (b) Why is the ligand NCS^- written with the nitrogen first in the complex but when outside the complex it is written in the reverse order, SCN^-?

(c) Write a balanced chemical equation for the reaction.

(d) Is the reaction a redox type?

24.58 The addition of a ligand to a metal can be considered in a step-wise manner with each addition having an associated enthalpy and equilibrium constant. Given the following enthalpies in kJ/mol for adding ethylenediamine to nickel(II) and copper(II) in a step-wise manner answer the questions that follow:

	$-\Delta H_1$	$-\Delta H_2$	$-\Delta H_3$
Ni(II)	9.0	9.18	9.71
Cu(II)	13.0	12.4	Very small

(a) For each metal ion explain why the sequential enthalpy values are similar, except for the third step for copper(II) ion?

(b) What can you deduce about the relative thermodynamic and kinetic stability of the two ions with respect to the addition of ethylenediamine?

(c) What can you deduce about the structures of the final forms of the two metal complexes?

Multiple Choice Questions

24.59 Which metal ion possesses six $3d$ electrons?

(a) Fe^{2+} (d) Al^{3+}
(b) Cr^{3+} (e) Ca^{2+}
(c) Co^{3+}

24.60 What is the coordination number of platinum in $[PtCl_2(en)]$?

(a) 2   (d) 5
(b) 3   (e) 6
(c) 4

24.61 Which of the following complexes has a nonsuperimposable mirror image?

(a) $[Zn(H_2O)_4]^{2+}$ (tetrahedral)

(b) *trans*-$[Cr(H_2O)_4Cl_2]^+$

(c) $[Pt(en)_2]^{2+}$ (square planar)

(d) $[Cr(H_2O)_6]^{3+}$

(e) *cis*-$[Co(en)_2(Cl)(NO_2)]^+$

24.62 Which of the following metal ions could form both high- and low-spin complexes in an octahedral array of ligands?

(a) Cr^{3+} (d) Zn^{2+}
(b) Mn^{3+} (e) Ti^{3+}
(c) Cu^+

24.63 Which metal complex ion should have the largest crystal-field splitting energy?

(a) $[Co(CN)_6]^{3-}$ (d) $[Co(H_2O)_6]^{3+}$
(b) $[Co(CN)_6]^{4-}$ (e) $[Zn(OH)_4]^{2-}$
(c) $[Co(CN)_5(H_2O)]^{2-}$

24.64 Which of the following is the correct name for $[Co(en)_2Br_2]_2SO_4$?

(a) bromoethylenediaminecobalt(III) sulfate
(b) ethylenediaminebromocobalt(III) sulfate
(c) dibromodiethylenediaminecobalt(III) sulfate
(d) diethylenediaminedibromocobalt(III) sulfate
(e) dibromobis(ethylenediamine)cobalt(III) sulfate

24.65 Which of the following formulas is associated with the name sodium tetrafluorooxochromate(IV)?

(a) $Na_2[CrOF_4]$ (d) $[CrOF_4Na_2]$
(b) $Na[CrOF_4]$ (e) $Na_2[CrF_4]$
(c) $[NaCrOF_4]$

24.66 Which of the following statements are true about the complexes $Fe(H_2O)_6^{2+}$ and $Fe(CN)_6^{4-}$: (1) Both contain Fe^{2+}; (2) both are low-spin complexes; (3) $Fe(H_2O)_6^{2+}$ is paramagnetic and $Fe(CN)_6^{4-}$ is diamagnetic; (4) $Fe(H_2O)_6^{2+}$ absorbs light at a lower energy than $Fe(CN)_6^{4-}$; (5) the names of the complexes are hexaaquairon(II) ion and hexacyanoiron(II) ion, respectively?

(a) (1) and (2)
(b) (1) and (3)
(c) (3), (4), and (5)
(d) (1), (3), and (4)
(e) (1), (3), (4), and (5)

24.67 Which of the following is the correct name for $[Cu(en)_2]SO_4$?

(a) copperbis(ethylenediamine)(II) sulfate
(b) copper(II)bis(ethylenediamine) sulfate
(c) bis(ethylenediamine)copper(II) sulfato
(d) bis(ethylenediamine)copper(II) sulfate
(e) bis(ethylenediamine)copper(I) sulfate

24.68 Which of the following can exist as structure(s) of tetraamminebromochlorocobalt (III) ion?

(a) *Cis*-isomer;
(b) *trans*-isomer;
(c) optical isomers;
(d) (a) and (b);
(e) (a), (b), and (c).

24.69 How many unpaired electrons does $[CoI_6]^{3-}$ possess?

(a) 0   (d) 3
(b) 1   (e) 4
(c) 2

24.70 How many unpaired electrons does $[Sc(H_2O)_3Cl_3]$ possess?

(a) 0   (d) 3
(b) 1   (e) 4
(c) 2

24.71 What is the coordination number of Ni in $[Ni(C_2O_4)_3]^{4-}$?

(a) 2   (d) 5
(b) 3   (e) 6
(c) 4

24.72 Which of the following could exhibit coordination-sphere isomerism?

(a) $Li[Al(CN)_4]$

(b) $[Pt(H_2O)_4Cl_2]$

(c) $[Co(NH_3)_4Cl_2]Br$

(d) $K_3[Fe(CN)_6]$

(e) None of the above

24.73 What is the observed color for a complex that primarily absorbs radiation around 700 nm?

(a) yellow   (d) violet

(b) green   (e) red

(c) blue

24.74 Which of the following complexes has two unpaired d electrons in the set of d_{z^2} and $d_{x^2-y^2}$ orbitals?

(a) $[NiF_6]^{4-}$ (c) $[IrCl_6]^{3-}$

(b) $[V(H_2O)_6]^{2+}$ (d) $[ZrCl_6]^{3-}$

24.75 The $Ni(H_2O_6)^{2+}$ ion is green whereas the $Ni(NH_3)_6^{2+}$ ion is purple. Which statement is correct?

(a) The complementary color of green is yellow.

(b) The complementary color of purple is red.

(c) $Ni(H_2O)_6^{2+}$ absorbs light with a shorter wavelength than $Ni(NH_3)_6^{2+}$.

(d) $Ni(NH_3)_6^{2+}$ absorbs light with a shorter wavelength than $Ni(H_2O)_6^{2+}$.

24.76 Which complex is diamagnetic?

(a) $Co(CN)_6^{3-}$ (c) $Ti(H_2O)_6^{3+}$

(b) $Co(CN)_6^{4-}$ (d) $Fe(en)_3^{3+}$

SELF-TEST SOLUTIONS

24.1 (f). **24.2** (e). **24.3** (k). **24.4** (c). **24.5** (n). **24.6** (a). **24.7** (l). **24.8** (i). **24.9** (h). **24.10** (d). **24.11** (q). **24.12** (b). **24.13** (g). **24.14** (j). **24.15** (m). **24.16** (o). **24.17** (p). **24.18** False. A d^8 transition metal ion forms only one d electron population in an octahedral field of ligands:

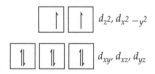

24.19 True. **24.20** False. Since CN^- is a strong-field ligand, d electrons preferentially pair in lower-energy orbitals before entering higher-energy orbitals. Therefore a low-spin complex is formed. **24.21** False. A complex ion contains two or more different elements and carries a charge. The complex ion is bracketed by []. Cl^- is an ion, but of one element, chlorine. **24.22** True. The term complex is another term for complex ion. **24.23** False. It has two nitrogen donor atoms. This ligand does not contain any oxygen atoms. **24.24** True. **24.25** True. **24.26** False. Ammonia can bind only through the nitrogen atom. Linkage isomerism occurs when a ligand can bind to a metal in two different ways. The ligand, NO_2^-, exhibits linkage isomerism by using either a nitrogen or oxygen atom when binding to a metal. **24.27** True. **24.28** False. It exists when a pair of mirror images can*not* be superimposed. **24.29** True. **24.30** True. **24.31** True. **24.32** False. The interaction of two negatively charged particles will raise the energy of the d-electrons. **24.33** False. Energy is required to pair electrons in the same orbital.

24.34 The structure has a central metal atom with six ligands about it, which is characteristic of an octahedral complex. Two of the same ligands are adjacent to one another and therefore it is a *cis*- geometric form. Another view of an identical form of the structure is:

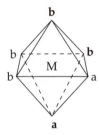

24.35 The structure is octahedral because the crystal-field energy-level diagram has two energy levels, with the lower one having three d orbitals and the higher one having two d orbitals. It is diamagnetic because all electrons are paired. It is a strong field case because all the electrons are in the lower energy level. If it was a weak field case the figure would have two electrons in the upper energy level and four electrons in the lower energy level.

24.36 In general the chemistry of complex ions is significantly different than that of the metal ion in a different environment. Look at equations 24.2 and 24.3 in your text for specific data involving silver ion. Different ligands can have a significant impact on structure, electronic properties, stability, etc.

24.37 The number of donor atoms attached to the central atom is referred to as the coordination number. Note it is not the number of ligands attached. The complex $[ZnBr_2(en)]$ has en, ethylenediamine, which is a bidentate ligand. This means it has two active nitrogen atoms that bind to the central zinc ion. Thus, the coordination number is four, not three.

24.38 The oxalate ion is a chelating ligand and will form a more stable six-coordinate complex than the six-coordinate $[NiCl_6]^{4-}(aq)$ complex. Thus, the oxalate ion should displace the chloroligands to form the complex $[Ni(ox)_3]^{4-}(aq)$.

24.39 *A Closer Look:* **Entropy and the Chelate Effect** in Section 24.2 in the text discusses this particular question.

If the donor atoms are the same in the chelate ligand and nonchelating ligand, then the enthalpy change should not be large because the nature of the metal-donor atom bond is not significantly different. However, there is a significant entropy change in a reaction of the following type

$$Ma_6 + 3bb \longrightarrow M(bb)_3 + 6a$$

where a is monodentate and bb is a bidentate chelate. Three bb ligands displace six a ligands. This results in an increase in the number of microstates and therefore an increase in entropy.

24.40 Isomers have the exact same chemical formula; thus, the number of atoms in each chemical formula is the same. They will have differing physical properties such as solubility. They may have the same basic geometrical structure (for example, octahedral) but within that structure there will be a difference in the arrangement of bonds or donor atoms or some other change in location of atoms.

24.41 Enantiomers are nonsuperimposable mirror images of one another and thus are optically active. A molecule that has a nonsuperimposable mirror image is said to be chiral. Thus, the common property is that enantiomers and chiral molecules are optically active.

24.42 High-spin metal complexes have the maximum number possible of unpaired electrons, whereas low-spin metal complexes have the minimum number. For octahedral complexes both can be observed for the following electron populations: $d^4, d^5, d^6,$ and d^7.

24.43 There is no specific relationship between a negative charge and the impact on the energy gap. Cl^- lies at the far left of the spectrochemical series and CN^- lies to the far right. Furthermore, neutral ligands are in the middle of the series. One might expect negatively charged ligands to produce a greater energy gap than neutral ligands. The energy gap results from d electron-ligand electron repulsions. If a ligand has a negative charge we might expect it to be drawn more closely to the nucleus than if it had no charge. This should cause the d electrons to interact more strongly with the ligand electrons and to increase the energy gap. Obviously, this picture is too simple and this model is not correct; other factors are operating including the transfer of electron density from ligand to the metal atom.

24.44 (a) hexaamminecobalt(III) chloride;
(b) pentacarbonyliron(0);
(c) potassium tetrachlorozincate(II);
(d) calcium hexacyanoferrate(II);
(e) hexaaquairon(II) sulfate;
(f) diamminebis(ethylenediamine)iridium(III) chloride.

24.45 (a) $K_2[PtCl_4]$;
(b) $[Pt(NH_3)_6]Cl_2$;
(c) $NH_4[Cr(NH_3)_2(CN)_4]$;
(d) $[SnCl_6]^{2-}$;
(e) $[Pt(en)Cl_4]$.

24.46 (a) six;
(b) five;

(c) four;
(d) six;
(e) six;
(f) six (ethylenediamine has two donor atoms).

24.47 There are two square-planar geometrical isomers:

(a)   (b)

where R_3P represents triphenylphosphine. The name of isomers are

24.47 (a) *cis*-carbonylchlorobis(triphenylphosphine)iridium (I)

    (b) *trans*-carbonylchlorobis(triphenylphosphine)iridium(I).

24.48 (a) No other forms exist.

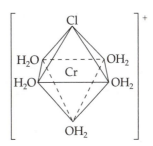

(b) Two optical isomers exist:

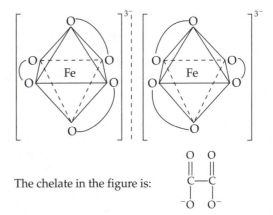

The chelate in the figure is:

$$\begin{matrix} O & & O \\ \| & & \| \\ C & - & C \\ | & & | \\ {}^-O & & O^- \end{matrix}$$

They are nonsuperimposable mirror images.

24.49 The energy of absorbed radiation is inversely proportional to the wavelength: $E = hc/\lambda$. The complex that absorbs energy at the lowest wavelength has associated with it the largest Δ. The order of increasing energies absorbed by the complexes is thus $MB_6 < MA_6 < MC_6$. Since the metal is a constant factor, the differences among the ligands, a, b, and c, cause the variances in the wavelengths of absorbed radiation among the complexes. The order of increasing Δ for the ligands is thus B < A < C.

24.50 (a) The Fe^{3+} ion has the electronic configuration $[Ar]3d^5$. Experimentally we observe that the crystal-field-splitting energy for tetrahedral complexes is small because only four ligands with their negative parts interact with the d electrons, rather than six as in an octahedral case. Consequently all tetrahedral complexes are high-spin complexes. The population of $Fe^{3+}3d$ electrons in a tetrahedral environment of ligands is

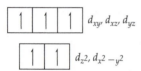

24.51 (a) $[Cr(H_2O)_3Cl_3]$;

  (b) $[Cr(H_2O)_5Cl]Cl_2$;

  (c) $[Cr(H_2O)_2Cl_4]^-$.

24.52

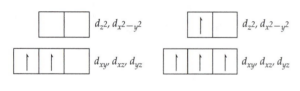

Paramagnetic
$[V(NH_3)_6]^{3+}$

Paramagnetic
$[MnF_6]^{3-}$

(Note: F^- is a weak-field ligand.)

24.53 There is insufficient information. It is likely that the charge of ligand X is 1− because ammonia is a neutral ligand, the charge of the overall complex is zero and nickel typically is a 2+ ion. However, without knowing the specific charge or oxidation state of nickel we cannot definitely make this conclusion. Nickel is known to have oxidation states of zero and +3, for example.

24.54 There is insufficient information to determine if it is a *trans-* complex but it is sufficient to determine charge of the complex. The formula is $[Co(C_2O_4)_2Cl_2]^{3-}$: The cobalt ion is 3+, the oxalate ion is 2−, and the chloride ion is 1−. The net charge for the complex is $(3+) + 2(2-) + 2(1-) = 3-$. The complex is octahedral because it has a coordination number of six [remember that the oxalate ion is a bidentate chelate]. Thus, the chloride ions could be adjacent to one another (*cis-*) or across from one another (*trans-*). Without knowing the specific arrangement of ligands in the octahedral structure we cannot conclude it is a *trans-* structure.

24.55 There is sufficient information. The Fe^{2+} ion in the complex has a $3d^6$ electron configuration. We are asked if we can conclude that it is low-spin metal complex. The

term "low-spin" is used in applications of crystal-field theory. If we construct an energy-level diagram using six d electrons, a diamagnetic electron structure results when all six electrons occupy and entirely fill the lower energy level. The upper energy level of two d orbitals is vacant. If we construct a high-spin case for Fe^{2+}, the lower level would have four d electrons and the upper level two. This results in a paramagnetic ion, which is not consistent with the given information.

24.56 Assume a sample of Vitamin B_{12s} has a mass of 1.000 g. This means that the mass of cobalt in the sample is 0.0443 g. Thus, the number of moles of cobalt in the sample is: $(0.0443 \text{ g})(1 \text{ mol}/58.93 \text{ g}) = 7.52 \times 10^{-4} \text{ mol}$ Co. The question tells us that there is one atom of cobalt per one molecule of Vitamin B_{12s}; thus, there is one mole of cobalt atoms per one mole of Vitamin B_{12s} molecules. This means that the 1.000 g sample contains 7.52×10^{-4} moles of cobalt atoms and the same number of Vitamin B_{12s} molecules. The molar mass of Vitamin B_{12s} is:

$$\frac{\text{mass of vitamin sample}}{\text{number of moles of sample}} = \frac{1.000 \text{ g}}{7.52 \times 10^{-4} \text{ moles}}$$

$$\text{Molar mass} = 1330 \frac{\text{g}}{\text{mol}}$$

24.57 (a) "En" is an abbreviation for the bidentate ligand ethylenediamine, $H_2NCH_2CH_2NH_2$. It is a chelating ligand using the nitrogen atoms as donor atoms.

  (b) The ligand NCS^- is able to coordinate through either the nitrogen atom or sulfur atom. When it coordinates through the nitrogen atom the N is written first. When it is outside the brackets it is not functioning as a ligand; it is a negatively charged ion that counterbalances the charge of the metal complex ion and in this state it is commonly written as SCN^-.

  (c) The balanced chemical equation is

$$2 \, trans\text{-}[Cr(en)_2(NCS)_2]SCN(s) \xrightarrow{\Delta}$$

$$[Cr(en)_2(NCS)_2][Cren(NCS)_4](s) + en(g)$$

  (d) To determine if an oxidation-reduction reaction has occurred, assign oxidation numbers to the metals in the three complexes and determine if there is a change. Inspection of the ligands shows that they have not undergone any changes in formula or charges so they can be ignored. Remember that en is a neutral ligand. The common oxidation states of chromium are +2, +3, and +6. In the first metal complex chromium has a +3 oxidation state. The metal complex has a 1+ charge and since the charge of NCS is 1− the chromium metal must have a +3 oxidation number $[(3+) + 0 + 2(1-) = 1+]$. The product containing two chromium metal atoms is more challenging. We do know that the charge of the cationic metal complex must equal the charge of the anionic metal complex since they are in a 1:1 ratio. By varying the oxidation number of the chromium metal among +2, +3, and +6, you will find that the only combination that works is if both chromium metals possess a +3 oxidation number. The cationic complex will have a 1+ charge and the anionic complex will have a 1− charge. This is not an oxidation-reduction reaction because there is no change in oxidation numbers NCS⁻.

24.58 (a) For each metal ion the addition of a ethylenediamine molecule in three steps gives: M^{2+} + en $\rightarrow M(en)^{2+}$, $M(en)^{2+}$ + en $\rightarrow M(en)_2^{2+}$, and $M(en)_2^{2+}$ + en $\rightarrow M(en)_3^{2+}$. Each step involves the formation of two M—N bonds and the displacement of solvent molecules from the metal ion and en molecule. For each step the interactions and the strengths of the M—N bonds formed should be similar. For the third step of copper(II) ion, the small value tells us that it is very difficult to add a third ethylenediamine ligand; essentially no reaction occurs.

(b) The larger exothermic values for copper tell us that when copper(II) ion adds ethylenediamine a relatively stronger M—N bond forms compared to nickel(II); thus the copper(II) complex should be thermodynamically more stable through the addition of two ethylenediamine ligands. The information tells you nothing about the kinetic stability of the complexes. You need rate data to deduce information about kinetic stability.

(c) The nickel(II) complex must be octahedral since it adds three ethylenediamine ligands. Each ethylenediamine ligand contributes two donor atoms. The copper(II) complex must have a coordination number of four since only two ethylenediamine ligands add. Thus, $[Cu(en)_2]^{2+}$ is either square planar or tetrahedral but we do not have sufficient information to know which structure exists.

24.59 (c) Remember that s electrons are removed before d electrons when forming transition metal ions.

24.60 (c) En, ethylenediamine is a bidentate chelate. Thus, there are four donor atoms.

24.61 (e)

24.62 (b). Mn^{3+} is a d^4 ion; high- and low-spin complexes can exist, depending on the ligand.

24.63 (a) For a complex with the same metal and the same ligands, the larger the charge of the metal ion the larger the crystal-field splitting energy. Also, the spectrochemical series of ligands shows the cyanide ion as a ligand which creates a larger crystal-field splitting energy than water. The replacement of a cyanide ligand with water reduces the crystal-field splitting energy. The four-coordinate zinc complex has a tetrahedral environment of ligands. The crystal-field splitting energy of tetrahedral complexes is smaller than for octahedral complexes.

24.64 (e). **24.65 (a)**.

24.66 (d). (5) is not correct because the name of $[Fe(CN)_6]^{4-}$ is hexacyanoferrate(II) ion.

24.67 (d). **24.68 (d)**. **24.69 (e)**. **24.70 (a)**. **24.71 (e)**.

24.72 (c). **24.73 (b)**. **24.74 (a)**. **24.75 (d)**. **24.76 (a)**. Co^{3+} with a strong field ligand.

The Chemistry of Life: Organic and Biological Chemistry

OVERVIEW OF THE CHAPTER

25.1, 25.2, 25.3 OVERVIEW OF HYDROCARBONS AND ALKANES

Review: Catalysis (14.6); electronegativity (8.4); Lewis structures (8.5).

Learning Goals: You should be able to:

1. List four groups of hydrocarbons and draw the structural formula from each group—in this case the alkanes.
2. Write the formulas and names of the first 10 members of the alkane series.
3. Write the structural formula of an alkane given its systematic (IUPAC) name.
4. Name an alkane, given its structural formula.
5. Give an example of structural isomerism in alkanes.

25.3 ALKENES AND ALKYNES

Review: See prior review.

Learning Goals: You should be able to:

1. List the four groups of hydrocarbons and draw the structural formula of an example from each group—in this case as applied to alkenes and alkynes.
2. Write the structural formula of an alkene or alkyne, given its systematic (IUPAC) name.
3. Name an alkene or alkyne, given its structural formula.
4. Give an example of structural and geometrical isomerism in alkenes and alkynes.
5. Give examples of addition reactions of alkenes and alkynes, showing the structural formulas of reactants and products.

25.3 AROMATIC HYDROCARBONS

Review: Delocalization of pi electrons (8.6).

Learning Goals: You should be able to:

1. Explain why aromatic hydrocarbons do not readily undergo addition reactions.
2. Give two or three examples of substitution reactions of aromatic hydrocarbons.

Learning Goals: You should be able to:

1. Identify the groups or arrangement of atoms in a molecule that correspond to the functional groups found in Table 25.4 in the text.
2. Give examples of the condensation reactions of alcohols to form ethers, of alcohols and carboxylic acids to form esters, and of amines and carboxylic acids to form amides.

25.4 FUNCTIONAL GROUPS

Learning Goals: You should be able to:

1. List the several functions of proteins in living systems.
2. Write the reaction for formation of a peptide bond between two amino acids.
3. Explain the structures of proteins in terms of primary, secondary, and tertiary structure.
4. Define the terms chiral, enantiomer, and racemic mixture and draw the enantiomer of a given chiral molecule.

25.5, 25.6, 25.7 PROTEINS; CHIRALITY

Learning Goals: You should be able to:

1. Describe two distinct reasons why energy is required by living organisms.
2. Describe the formation of cyclic structures for sugars from their open-chain forms, and distinguish between the α and β forms of the cyclic structures of glucose.
3. Describe the manner in which monosaccharides are joined together to form polysaccharides.
4. Enumerate the major groups of polysaccharides and indicate their sources and general functions.
5. Describe the structures of fats and oils and list the sources of these substances.
6. Describe the basic chemical structure of phospholipids.

25.8, 25.9 CARBOHY-DRATES AND LIPIDS

Learning Goals: You should be able to:

1. Draw the structures of any of the nucleotides that make up the polynucleotide DNA.
2. Describe the nature of the polymeric unit of polynucleotides.
3. Describe the double-stranded structure of DNA and explain the principle that determines the relationship between bases in the two strands.

25.10 NUCLEIC ACIDS: RNA AND DNA

TOPIC SUMMARIES AND EXERCISES

Hydrocarbons are compounds consisting of only the elements carbon and hydrogen. Except for methane, CH_4, hydrocarbons contain stable carbon–carbon bonds. Hydrocarbons containing extended chains containing single, double, or triple carbon–carbon bonds are well known. These compounds and related ones are also referred to as organic compounds. In this chapter we explore several classes of hydrocarbons; these are summarized in Table 25.1 on the next page. Note carefully the structural and carbon–carbon bonding differences among the different classes. They are important to understanding the material in this chapter.

TABLE 25.1 Characteristics of Several Classes of Hydrocarbons

Class	Simplest general formula	Example	Characteristics
Alkane (saturated hydrocarbon)	C_nH_{2n+2}	$\begin{array}{cc} H & H \\ \mid & \mid \\ H-C-C-H \\ \mid & \mid \\ H & H \end{array}$ Ethane	Contains an open chain of carbon–carbon bonds; all carbon atoms in compound are attached to four other atoms *via* single σ bonds
Cycloalkanes	C_nH_{2n}	$\begin{array}{c} CH_2 \\ \diagup \diagdown \\ H_2C-CH_2 \end{array}$	An alkane with one or more carbon rings; all bonds to carbon atoms are single bonds
Alkene (unsaturated hydrocarbon)	C_nH_{2n}	$\begin{array}{c} H \qquad\quad H \\ \diagdown \qquad \diagup \\ C=C \\ \diagup \qquad \diagdown \\ H \qquad\quad H \end{array}$ Ethylene	A hydrocarbon containing at least one carbon–carbon double bond $(>C=C<)$
Alkyne (unsaturated hydrocarbon)	C_nH_{2n-2}	$H-C\equiv C-H$ Ethyne	A hydrocarbon containing one or more carbon–carbon triple bonds $(-C\equiv C-)$
Aromatic hydrocarbon		Benzene	Planar, cyclic arrangement of carbon atoms bonded to one another by both σ and π bonds (the π electrons are delocalized over the carbon atoms and are indicated by the circle in the middle)

Alkanes contain all carbon–carbon single bonds and they are also called *saturated hydrocarbons*. The term saturated means that all carbon–carbon bonds are single.

- Note that hydrocarbon molecules are relatively nonpolar, dissolve in nonpolar solvents, and that their boiling points increase with increasing molar mass.
- If an alkane contains more than three carbon atoms in a continuous chain, the carbon atoms can be arranged in more than one way: a straight chain of carbon atoms or a branched chain. This is an example of **structural isomerism**: Compounds with the same molecular formula, but with different bonding arrangements.
- A chain of carbon atoms can also be cyclic, forming **cycloalkanes**. If a chain contains fewer than six carbon atoms, it will be strained, and thus more reactive. Note that the general formula for cycloalkanes, C_nH_{2n}, is different from that for straight-chain alkanes.

- Naming organic compounds such as alkanes is an important skill. Review Section 25.3 for the rules of naming alkanes. Exercises 1 and 2 help you with learning these rules.

Alkanes are generally unreactive because C—C single bonds and C—H bonds are relatively strong.

- Alkanes undergo combustion in air, which forms the products carbon dioxide (gas) and water (gas).
- Alkanes undergo substitution reactions with elemental fluorine, chlorine, and bromine in which one or more hydrogen atoms are replaced by a halogen.

EXERCISE 1 Naming alkanes and writing structural formulas

Name or write structural formulas for the following:

(a) $CH_3-CH-CH_2-CH-CH_3$
with CH_3 below the second carbon, and CH_2 below the fourth carbon, with CH_3 below that CH_2.

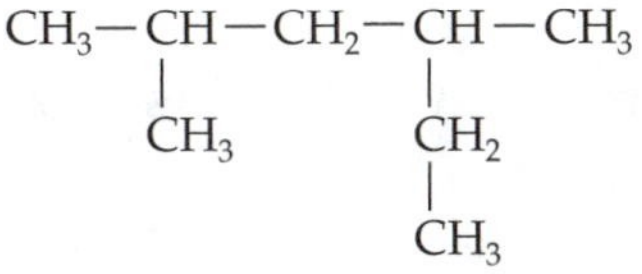

(b)
CH_3 at top, $CH_3-C-CH_2-CH-CH_3$ with CH_3 below the C, and CH_2 below the CH, and CH_3 below that CH_2.

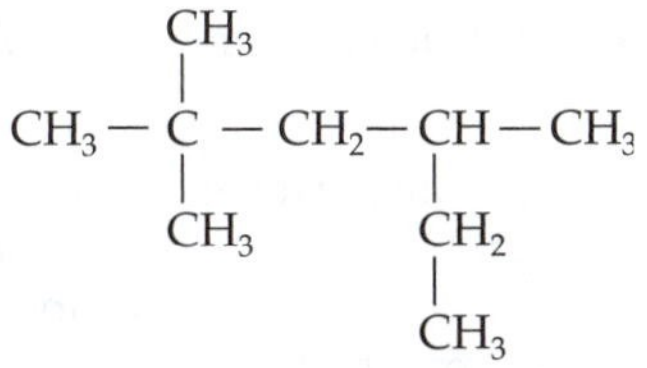

(c) 1,3-Diethylcyclohexane

SOLUTION: *Analyze*: In (a) and (b) we are given structural formulas of hydrocarbons and asked to name them. In (c) we are given a name of a hydrocarbon and asked to sketch its structural formula. We should recognize that they are alkanes: All C—C bonds are single.

Plan: For (a) and (b) we need to identify the longest continuous chain of carbon atoms and the positions and names of branched alkyl groups. We will use the procedures given in the text to identify the appropriate branched alkyl groups and base name of the alkanes. For (c) we need to identify how many carbon atoms are in the longest continuous chain and the number of carbon atoms in branched alkyl groups and their locations. The prefix cyclo- before the base, hexane, tells us the base of carbon atom forms a cyclic, closed chain. Then we sketch the formula based on the longest continuous chain and the locations of branched alkyl groups.

Solve: (a) The longest carbon chain contains six carbon atoms, as shown by this skeleton formula:

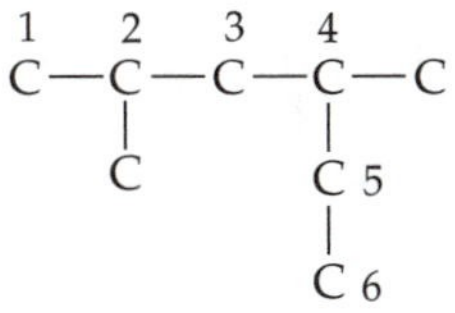

$$\overset{1}{C}-\overset{2}{C}-\overset{3}{C}-\overset{4}{C}-C$$
with C below carbon 2, and C 5 below carbon 4, and C 6 below C 5.

The six-carbon chain is named hexane. The CH_3 (methyl) groups are at the 2 and 4 carbon positions; consequently, they are named 2,4-dimethyl. The name of the compound is 2,4-dimethylhexane. (**b**) As in (**a**), the longest chain contains six carbon atoms and is named hexane. However, in this example, there are two methyl groups at the 2 position, and one methyl group at the 4 position. The location and number of methyl groups are indicated by 2,2,4-trimethyl, with the number 2 repeated once to indicate two methyl groups at that position. The name of the compound is 2,2,4-trimethylhexane. (**c**) The longest carbon chain is a cyclic six-carbon atom chain—the end name "cyclohexane" means six carbon atoms (hexane) arranged cyclically (cyclo- prefix). Ethyl groups occur at the carbon atoms numbered 1 and 3. The structural formula is shown below:

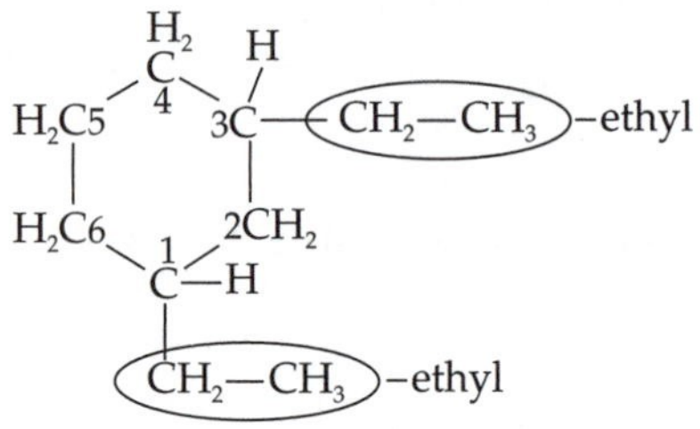

EXERCISE 2   Drawing and naming structural isomers of alkanes

Draw all structural isomers for a five-membered alkane and name them.

SOLUTION: *Analyze*: We are asked to draw all structural isomers for a five-membered alkane and name them.

Plan: When constructing a member of the alkane family we follow the rule of one bond per hydrogen atom and four bonds per carbon atom. A carbon atom may have zero to four hydrogen atoms attached and one to four carbon atoms attached; however, there may not be more than a total of four bonds. To draw all the isomers we must arrange the five carbon atoms in various ways. Structures will be different if no amount of moving, twisting, or rotating about the carbon–carbon atoms causes the structures to be the same as any others we have drawn.

Solve: First start with a straight chain of five carbon atoms in the base chain:

This structure is called n-pentane. The n- prefix means it is a straight chain struc-

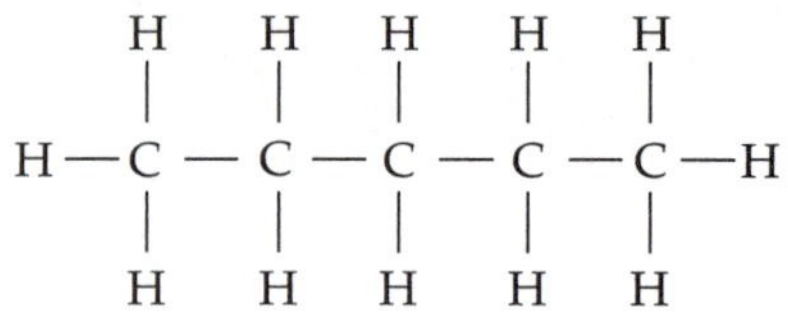

ture. Now, move an end carbon group to a different carbon atom in the middle of the chain:

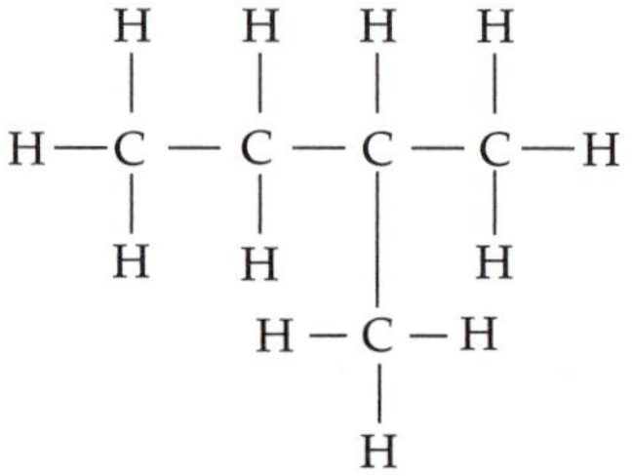

This structure is 2-methylbutane (sometimes it is called isopentane). Finally, move the far left carbon atom to the same carbon atom that has two carbon atoms attached to it:

$$
\begin{array}{c}
\quad\quad\quad\quad H \\
\quad\quad\quad\quad | \\
H \quad\quad H-C-H \quad H \\
| \quad\quad\quad\quad | \quad\quad\quad | \\
H-C \text{———} C \text{———} C-H \\
| \quad\quad\quad\quad | \quad\quad\quad | \\
H \quad\quad H-C-H \quad H \\
\quad\quad\quad\quad | \\
\quad\quad\quad\quad H
\end{array}
$$

The name of this structure is 2,2-dimethylpropane.

ALKENES AND ALKYNES

Alkenes are hydrocarbons that have one or more C=C bonds and simple alkenes have the general formula C_nH_{2n}. **Alkynes** have the general formula C_nH_{2n-2} and contain one or more carbon–carbon triple bonds. Both are also termed *unsaturated hydrocarbons* because the carbon atoms involved in double or triple bonds do not possess the maximum number of single bonds.

- Naming alkenes and alkynes is discussed in Section 25.3. Exercise 3 helps you with learning these rules.

Alkenes often exhibit geometrical isomerism. **Geometrical isomers** consist of compounds that have the same molecular formula and the same atoms bonded to one another but they differ in the spatial arrangement of groups of atoms.

- Compounds with double bonds may be in the *cis* form (same groups are spatially on the same side of the double bond) or in the *trans* form (same groups are spatially on opposite sides of the double bond, along a diagonal). *Cis*- and *trans*- geometrical isomers exist because of restricted rotation about a C–C double bond.

$$
\begin{array}{cc}
\text{Br} \qquad \text{Br} & \text{H} \qquad \text{Br} \\
\diagdown \qquad \diagup & \diagdown \qquad \diagup \\
\text{C}=\text{C} & \text{C}=\text{C} \\
\diagup \qquad \diagdown & \diagup \qquad \diagdown \\
\text{H} \qquad \text{H} & \text{Br} \qquad \text{H}
\end{array}
$$

cis-dibromoethylene   *trans*-dibromoethylene

- Restricted rotation about a C=C bond exists because the bond consists of two types: sigma (σ) and pi (π). Rotation about a carbon–carbon double bond requires the breaking of a pi bond, a process that requires a significant quantity of energy.
- Alkynes do not exhibit *cis-trans* isomerism.

Alkenes and alkynes are more reactive than alkanes because of the presence of carbon–carbon double or triple bonds. A pi bond is weaker than a single bond and thus pi bonds are more readily broken in chemical reactions.

- Alkenes and alkynes undergo oxidation and substitution reactions.
- Alkenes and alkynes also undergo addition reactions, which involve the addition of other atoms directly to carbon atoms participating in

double or triple bonds. The pi bonds are uncoupled and the electrons are available to combine with other atoms.

• The addition of hydrogen to an alkene to form an alkane is shown in equation (25.2) in the text; this reaction is termed hydrogenation. A catalyst is necessary.

EXERCISE 3   Naming alkenes and alkynes and writing structural formulas

Name or write structural formulas for the following:

(a) $CH_3 - C = CH - CH - CH_3$
           $\underset{|}{}$
           $CH_3 \qquad CH_3$

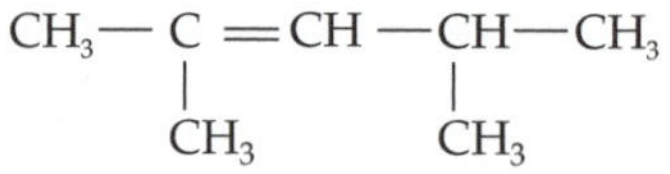

(b) $CH_3 - C \equiv C - CH - CH_3$
                   $\underset{|}{}$
                   CH_3

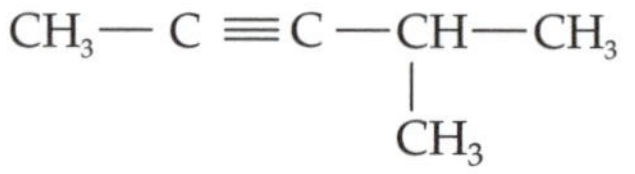

(c) 4-Methylcyclohexene

SOLUTION: *Analyze and Plan:* This problem is similar to the one in Exercise 1 except the structures contain unsaturated carbon atoms. We will follow the rules given in the text to answer each part of the question. The new feature is identification and naming of the location of the unsaturation in the longest continuous carbon chain.

Solve: (a) The longest carbon chain containing the carbon–carbon double bond has five carbon atoms; it is named 2-pentene. The carbon atoms are numbered as follows:

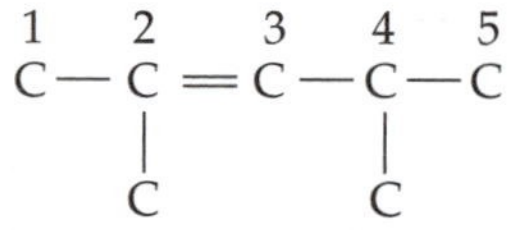

$$
\begin{array}{ccccc}
1 & 2 & 3 & 4 & 5 \\
C - & C = & C - & C - & C \\
 & | & & | & \\
 & C & & C &
\end{array}
$$

Note that the smallest number possible is assigned to the first atom of the carbon–carbon double bond; in this situation it is assigned the number 2. Since there are two methyl groups, one each at the 2 and 4 positions of the carbon chain, the name of the compound is 2,4-dimethyl-2-pentene. (b) As in part (a), the longest carbon chain contains five carbon atoms, but instead of containing a double bond it contains a triple bond. A triple bond is characteristic of an alkyne, whose name ends in -yne. Since one methyl group occurs at the 4 position of the carbon chain, the name of the compound is 4-methyl-2-pentyne. (c) The parent carbon chain is cyclohexene, which contains a six-member carbon ring with one carbon–carbon double bond. The prefix 4 before the methyl group in the name of the compound tells us that the methyl group occurs at the carbon atom numbered 4, with the first carbon atom of the carbon–carbon double bond assigned the number one position. Therefore the structural formula is

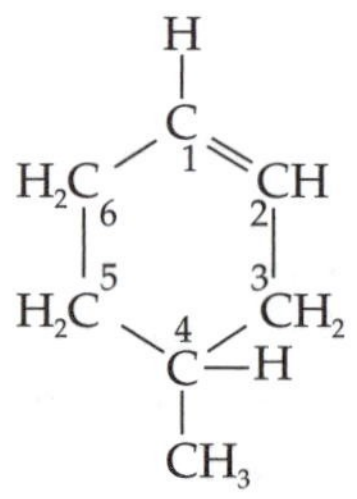

EXERCISE 4 Drawing *cis-* and *trans-* structural isomers

Draw the structural formulas for *cis*-2-pentene and *trans*-2-pentene.

SOLUTION: *Analyze*: We are given the names of the *cis-* and *trans-* isomers of pentene and asked to sketch their structural formulas. We should recognize that pentene has five carbon atoms and is an alkene with unsaturation.

Plan: The *cis-* prefix indicates that the two non-hydrogen groups, each attached to a different carbon atom, are located on the same side of the double bond; in the *trans-* form they are located on opposite sides. Using this information we can orient the groups correctly about the carbon–carbon double bond.

Solve:

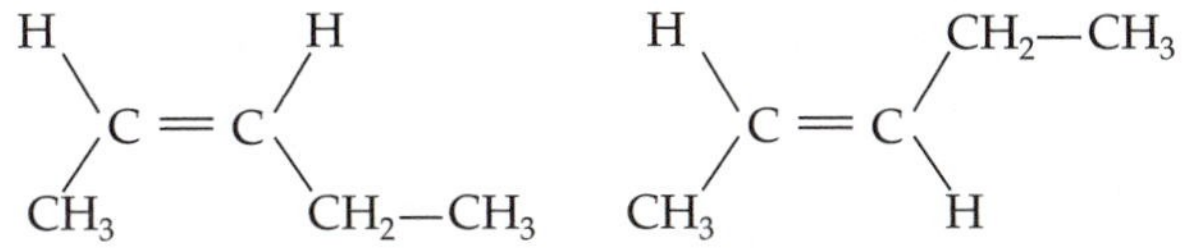

EXERCISE 5 Writing reactions of alkenes

Write the structural formulas of the organic substances formed when the following react: (**a**) 1-butene and hydrogen in the presence of catalyst; (**b**) One mole each of 1-butene and chlorine gas.

SOLUTION: *Analyze and Plan*: We are given two sets of reactants and asked to write the structural formulas of the products formed. To do this we need to review the section on addition reactions in Section 25.3 of the text. Then by using pattern recognition we can complete the two reactions and sketch the structural formulas.

Solve: All are addition reactions:

(**a**) $CH_2=CH-CH_2-CH_3 + H_2 \xrightarrow{Pt} CH_3-CH_2-CH_2-CH_3$

(**b**)

$$CH_2=CH-CH_2-CH_3 + Cl_2 \xrightarrow{uv} CH_2-CH-CH_2-CH_3$$
with Cl and Cl attached below the first two carbons.

Aromatic hydrocarbons are cyclic hydrocarbons containing p electrons delocalized over several carbon atoms. The simplest example of an aromatic hydrocarbon is benzene, C_6H_6, which is discussed in Section 9.6 of the text.

AROMATIC HYDROCARBONS

• The simplest way of showing an aromatic benzene ring is by the picture

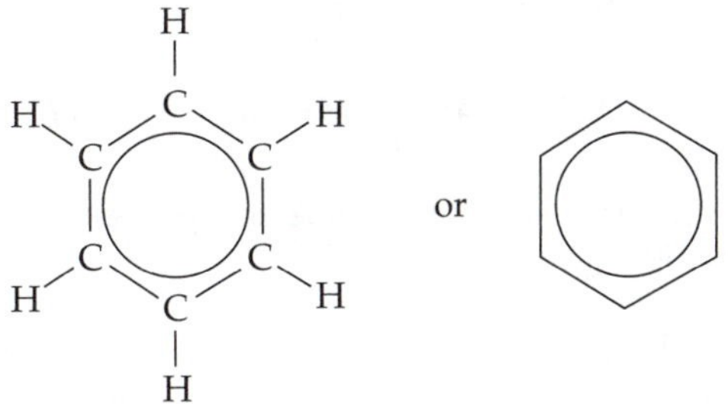

or

This picture is a shorthand description for the following arrangement of carbon and hydrogen atoms:

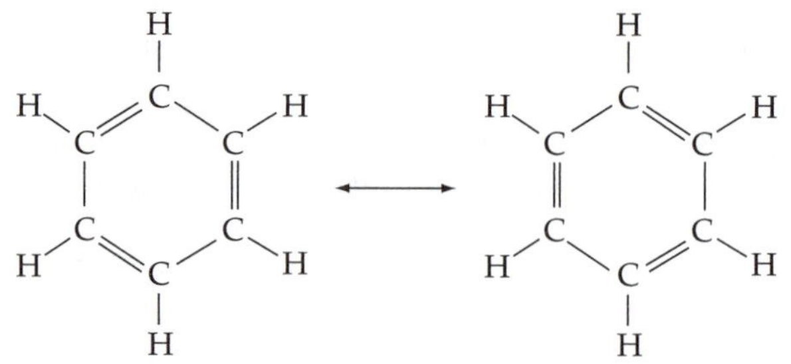

The circle in the middle of the ring indicates that the three pi bonds are delocalized over the entire ring structure.

- Several benzene rings can be fused together to form more complex structures. Look at Figure 25.13 in the text to see examples and the numbering system used to identify carbon atoms.
- There are three possible isomers of benzene when two groups are attached: ortho- (*o*-), meta- (*m*-), and para- (*p*). Examples are

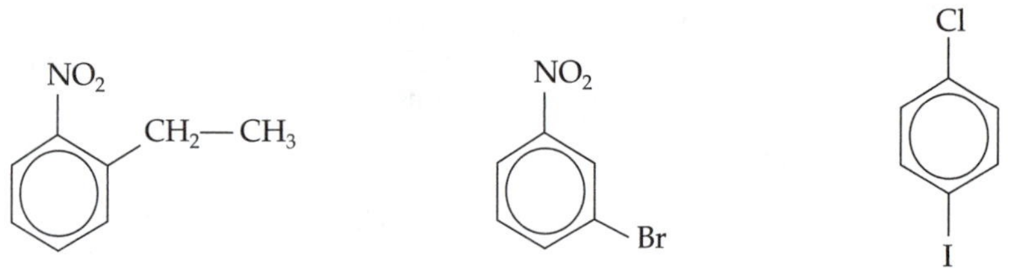

o-Ethylnitrobenzene   *m*-Bromonitrobenzene   *p*-Chloroiodobenzene

- Aromatic hydrocarbons do not readily undergo addition reactions; the most common type of reaction is substitution.
- Hydrogen atoms can be replaced by other atoms or groups such as $-NO_2$, $-Br$, $-CH_3$, and $-OH$.
- An important type of substitution reaction involving a catalyst are *Friedel-Crafts reactions*. Positively charged species are created by substances such as H_2SO_4, $FeCl_3$, and $AlCl_3$. The charged reactant attacks the aromatic ring and a hydrogen ion is eventually lost from the ring.

EXERCISE 6   Comparing reactions of cyclohexene and benzene

What are the expected products if cyclohexene and benzene are reacted with Br_2 in the solvent CCl_4?

SOLUTION: *Analyze:* We are asked to predict the products of bromine reacting with cyclohexene and benzene in carbon tetrachloride. We should recognize that cyclohexene is an alkene whereas benzene is an aromatic hydrocarbon and thus they are likely to react differently.

Plan: Alkenes undergo addition reactions with halogens but aromatic hydrocarbons do not. We can complete the reactions and write structural formulas based on this information.

Solve: Cyclohexene contains one double bond that should undergo rapid addition with bromine as follows:

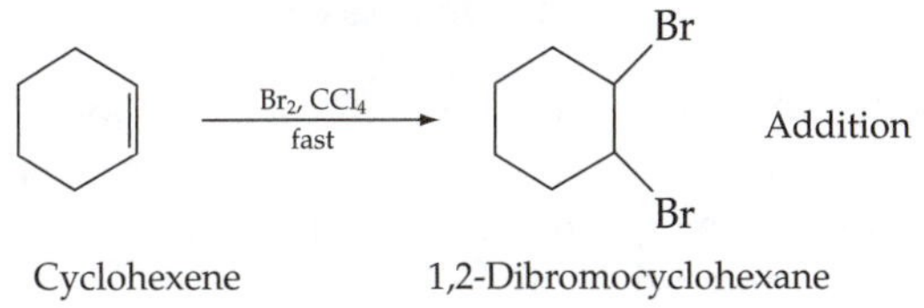

| Cyclohexene | 1,2-Dibromocyclohexane |

A similar reaction with benzene does not occur. Remember that aromatic hydrocarbons do not readily undergo addition reactions. Br_2 reacts with benzene in the presence of $FeBr_3$, a catalyst, to form bromobenzene.

$$C_6H_6 + Br_2 \xrightarrow{FeBr_3} C_6H_5{-}Br + H{-}Br$$

Benzene   Bromobenzene

EXERCISE 7 Writing structural formulas of substituted benzenes

Write structural formulas for the following: (**a**) ethylbenzene; (**b**) toluene (methylbenzene); (**c**) *o*-dichlorobenzene.

SOLUTION: *Analyze and Plan*: We are asked to write the structural formulas for three substituted benzene compounds. We can sketch the structural formulas by recognizing that they are monosubstituted chloro- compounds of benzene. The "*o*-" in front of dichlorobenzene tells us the location of the two chlorine atoms (ortho: attached to adjacent carbon atoms).

Solve: (**a**) Most monosubstituted compounds of benzene are named by placing the name of the substituent in front of the word benzene. In this case the substituent is ethyl, C_2H_5.

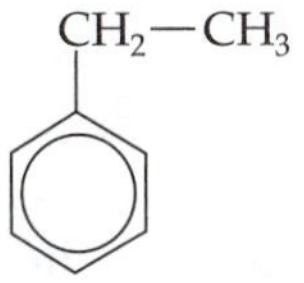

Ethylbenzene

(**b**) Toluene is the name given to benzene with a methyl group attached

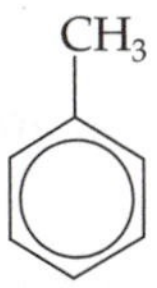

Toluene
(methylbenzene)

(**c**) *o*-Dichlorobenzene is a disubstituted benzene with one of the chloro- groups in the ortho position

o-Dichlorobenzene
(*ortho* isomer)

FUNCTIONAL GROUPS

The reactivity of hydrocarbons depends on what atoms or groups are attached to carbon atoms, or on the types of carbon–carbon bonds. These reactive sites are called **functional groups**. Important reactive sites (functional groups) discussed in the text include

$$\text{C} = \text{C} \quad \text{(alkene)} \qquad -\text{C} \equiv \text{C} - \quad \text{(alkyne)}$$

$-\text{Cl}$ (chloro) $-\text{OH}$ (hydroxo or alcohol)

$-\text{C} = \text{O}$ (carbonyl) $-\underset{\text{OH}}{\text{C}} = \text{O}$ (carboxylic acid)

$-\text{NH}_2$ (amine) $-\text{C}-\text{O}-\text{C}-$ (ether)

The functional group $-\text{OH}$ occurs in compounds called **alcohols** ($\text{R}-\text{O}-\text{H}$, where R is an alkyl group) and **phenols** ($\text{Ar}-\text{OH}$, where Ar is an aromatic group).

- The names of alcohols end in -ol, as in propanol ($\text{CH}_3\text{CH}_2\text{CH}_2\text{OH}$). Alcohols are formed by the reaction between water and an alkene in the presence of H_2SO_4.

Ethers are structurally related to alcohols. In ethers an oxygen atom holds two alkyl or aromatic groups, or one alkyl and one aromatic group (R-O-R, Ar-O-Ar, or R-O-Ar).

- Ethers are formed by alcohols in the presence of strong acids, such as sulfuric and phosphoric:

$$\text{ROH} + \text{HOR}' \xrightarrow{\text{H}_2\text{SO}_4} \text{ROR}' + \text{H}_2\text{O}$$

- The previous reaction is called a **condensation reaction**. Two reagents are combined with the elimination of water.

Alcohols can be oxidized to form aldehydes and ketones; both contain the **carbonyl** functional group,

$$\text{C} = \text{O}$$

- In a **ketone**, two organic groups are bonded to the carbon atom of the carbonyl group

$$\text{R}-\underset{\overset{\|}{\text{O}}}{\text{C}}-\text{R}'$$

- In an **aldehyde** at least one hydrogen atom is bonded to the carbon atom of the carbonyl group,

$$R\!-\!\underset{\underset{O}{\|}}{C}\!-\!H$$

Carboxylic acids contain the carboxyl functional group,

$$-\underset{\underset{OH}{|}}{C}\!=\!O$$

- This structure is sometimes abbreviated $-COOH$ or $-CO_2H$.
- The carboxyl functional group has distinctive chemical properties different from those of the carbonyl functional group because the OH and $C\!=\!O$ groups interact with each other.
- Oxidation of aldehydes produces carboxylic acids.

Esters form from the condensation reaction between an alcohol and a carboxylic acid

$$R'\!-\!\underset{\underset{}{\overset{O}{\|}}}{C}\!-\!OH + ROH \longrightarrow R'\!-\!\underset{\underset{}{\overset{O}{\|}}}{C}\!-\!OR + H_2O$$
$$\text{Ester}$$

A condensation reaction between an **amine** (a basic compound containing a nitrogen with a lone pair of electrons, RNH_2, for example) and a carboxylic acid produces **amides**:

$$R'\!-\!\underset{\underset{}{\overset{O}{\|}}}{C}\!-\!OH + RNH_2 \longrightarrow R'\!-\!\underset{\underset{}{\overset{O}{\|}}}{C}\!-\!\underset{\overset{H}{|}}{N}R + H_2O$$
$$\text{Amide}$$

EXERCISE 8 Writing structural formulas and reactions of alcohols

For each of the following alcohols, write the structural formula and write the structural formula of the ketone or aldehyde formed (if any) when the alcohol is oxidized: **(a)** 2-butanol; **(b)** 2-methyl-2-butanol; **(c)** ethanol.

SOLUTION: *Analyze*: We are given three alcohols and asked to sketch their structural formulas. We are also asked to sketch the structural formulas of the ketone or aldehyde formed when each alcohol is oxidized.

Plan: We will sketch the structural formulas using the principles found in Exercise 1 and also adding the feature that an alcohol ends in *-ol* and the prefix before the base name tells us the location(s) of the OH group(s) along the chain. If an alcohol group is oxidized a $-C\!=\!O$ group forms. An alcohol cannot be oxidized if the OH group is attached to a carbon atom that has three other carbon atoms bonded to it.

Solve: The *-ol* ending tells you that a compound is an alcohol, and the prefix number before the name containing the *-ol* ending tells you the location of the OH group.
(a) The structural formula is

$$CH_3\!-\!CH_2\!-\!\underset{\underset{OH}{|}}{CH}\!-\!CH_3$$

Since the carbon atom with the OH group has two other carbon atoms directly bonded to it, it can be oxidized. When oxidized, it forms a ketone; in this problem, it is methyl ethyl ketone,

$$CH_3\!-\!CH_2\!-\!\underset{\underset{O}{\|}}{C}\!-\!CH_3$$

(b) The structural formula is

$$CH_3-\underset{\underset{OH}{|}}{\overset{\overset{CH_3}{|}}{C}}-CH_2-CH_3$$

Since there are three carbon atoms bonded to the carbon atom with the OH group, it can*not* be oxidized to $>C=O$. (c) The structural formula is

$$CH_3-CH_2-OH$$

Since there is only one carbon atom attached to the carbon atom with the OH group, it can be oxidized. Oxidation forms an aldehyde; in this case, acetaldehyde,

$$CH_3-\underset{\underset{O}{\parallel}}{C}-H$$

EXERCISE 9   Predicting solubilities of alcohols in water

Which is more soluble in water, 1-butanol or 1-octanol? Explain.

SOLUTION: *Analyze and Plan*: We are asked to predict which is more soluble in water, 1-butanol or 1-octanol. To be soluble in water a substance must be sufficiently polar so that water molecules are attracted by dipole–dipole forces and surround it. We can use this concept to answer the question. From the names of the two alcohols we should recognize one has more carbon atoms than the other.

Solve: 1-butanol is more soluble in water than 1-octanol. The attraction of the polar OH group in 1-octanol toward polar water molecules is not sufficient to "pull" the nonpolar, eight-carbon-atom chain of 1-octanol into solution. The nonpolar, four-carbon atom chain in 1-butanol is sufficiently small that it is surrounded by water molecules attracted to the polar OH group. This hydration of the polar and nonpolar components of 1-butanol enables it to be soluble in water.

EXERCISE 10   Writing structural formulas: ester, carboxylic acid, and amide

Write structural formulas for each of the following compounds: (a) phenyl acetate; (b) benzoic acid; (c) acetamide.

SOLUTION: *Analyze*: We are asked to sketch the structural formulas of three substances that contain functional groups.

Plan: For each we need to identify the functional group present. The ending of the base group normally helps you identify the type of functional group, for example, an ester or alcohol or carboxylic acid. We can review examples of substances in Section 25.4 in the text and how they were named to help us with this question. An -ate ending tells us the substance is an ester; the -oic ending tells us it is a carboxylic acid; and the -amide ending tells us it is an amide.

Solve: (a) The "ate" ending of the name phenyl acetate tells us that the carboxyl group of acetic acid is in the form

$$CH_3-\underset{\underset{O}{\parallel}}{C}-O-$$

The phenyl prefix tells us that a phenyl group has replaced the hydrogen atom of the hydroxyl unit in the carboxyl group. The formula is

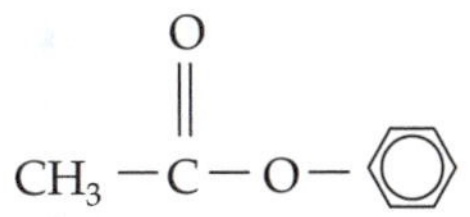

$$CH_3-\overset{\displaystyle O}{\overset{\|}{C}}-O-\bigcirc$$

(b) The structural formula of benzoic acid is

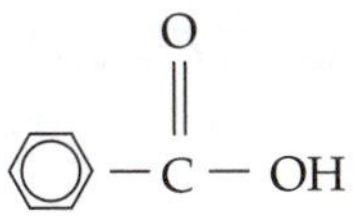

$$\bigcirc-\overset{\displaystyle O}{\overset{\|}{C}}-OH$$

(c) The "amide" ending of acetamide tells us that the $-NH_2$ group has replaced the $-OH$ group of acetic acid:

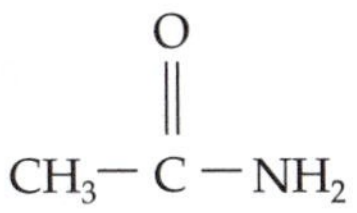

$$CH_3-\overset{\displaystyle O}{\overset{\|}{C}}-NH_2$$

Analysis of organisms shows that more than 90 percent of their dry weight is comprised of macromolecules, polymers of high-molecular weight. In Chapter 25, we study four broad classes of biopolymers: Proteins, carbohydrates, lipids, and nucleic acids. Each of these groups of biopolymers is discussed and reviewed.

PROTEINS; CHIRALITY

Amino acids, the building blocks of proteins, are carboxylic acids with an amine group at the carbon atom next to the carboxylic acid group (the α-carbon):

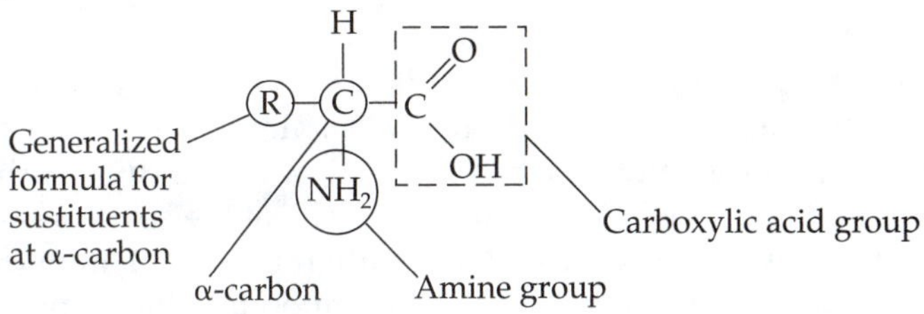

Proteins are the fundamental constituent of cells and tissues in the human body. They are large molecules with molecular weights varying from 10,000 to more than 50 million amu.

- The simplest proteins are composed entirely of amino acids; conjugated proteins consist of simple proteins bound to other kinds of biochemical structures.
- The characteristic bonding linkage between amino acids in proteins is the amide linkage (see Section 25.7 of the text):

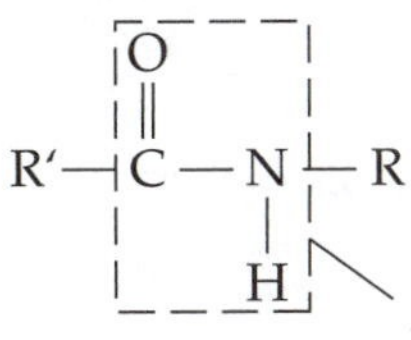

Amide linkage, referred to as **peptide linkage** in proteins

- **Polypeptides** are proteins that have molecular weights smaller than about 10,000 amu. Characteristically, polypeptides consist of about 50 or fewer amino acids bonded together via peptide linkages.

Proteins have a specific "shape" or conformation that enables them to perform specific biological functions. Section 25.7 describes three structures of proteins that together form the conformation.

- The specific sequence of amino acids in a protein is the *primary structure*. The primary structure contains the information needed by the protein to form the other two structures.
- The orientation in space of the primary structure is the *secondary structure*. Secondary structures arise from the formation of hydrogen bonds between the carbonyl group of one amino acid and an amino group of another. The alpha and triple helix structures are part of the secondary structure.
- A secondary structure undergoes twisting and folding to form a three-dimensional shape that is the *tertiary structure*. This folding leads to globular structures.
- Most biological catalysts, called **enzymes**, have a protein as a principal structural component.

Amino acids contain both acidic (carboxylic acid) and basic (amine) functional groups. Near a neutral pH there is a transfer of a hydrogen ion from the acid functional group to the nitrogen atom of the basic functional group. The doubly ionized form is called a *zwitterion*. For example, glycine (gly) at a neutral pH is in the form

$$H_3N^+-\overset{\overset{\displaystyle H}{\displaystyle |}}{\underset{\underset{\displaystyle H}{\displaystyle |}}{C}}-\overset{\overset{\displaystyle O}{\displaystyle ||}}{C}-O^-$$

Proteins and other organic compounds may have optically active sites; that is, the molecules are chiral. If a carbon-containing compound has a carbon atom with four different attached groups it will be a chiral carbon atom and thus the compound will be chiral. If you look at the structures of amino acids in your text you will see that the α-carbon is a chiral atom except for glycine. The enantiomeric forms (nonsuperimposable mirror images) of a chiral molecule have identical physical properties except for their ability to rotate the plane of polarized light. Also their reactions with nonchiral reagents are the same. Generally when a chiral molecule is synthesized the two optically active species form in equal quantities and the mixture does not rotate the plane of polarized light; it is called a *racemic* mixture.

EXERCISE 11   Writing and naming structural formulas of peptides

Write the structural representations of the two peptides that form by condensation reactions between alanine and serine.

$$\underset{\text{Alanine}}{HO-CH_2-\overset{\overset{\displaystyle }{\displaystyle |}}{\underset{\underset{\displaystyle NH_2}{\displaystyle |}}{CH}}-COOH} \qquad \underset{\text{Serine}}{CH_3-\overset{\overset{\displaystyle }{\displaystyle |}}{\underset{\underset{\displaystyle NH_2}{\displaystyle |}}{CH}}-COOH}$$

Name each dipeptide using the nomenclature rules found in Section 25.7 of the text.

SOLUTION: *Analyze:* We are given two amino acids and asked to write the two peptides that could form when a condensation reaction occurs. A condensation reaction involves two substances combining with the elimination of a smaller molecule, often water.

Plan: A carboxylic acid functional group can react with an amine group to form an amide linkage with water being eliminated. Two dipeptides are possible because two different condensation reactions can occur. One condensation reaction is between the $-NH_2$ group of alanine and the $-COOH$ group of serine (structure *a*) and the other is between the $-NH_2$ group of serine and the $-COOH$ group of alanine (structure *b*).

Solve: The two different peptides are shown below.

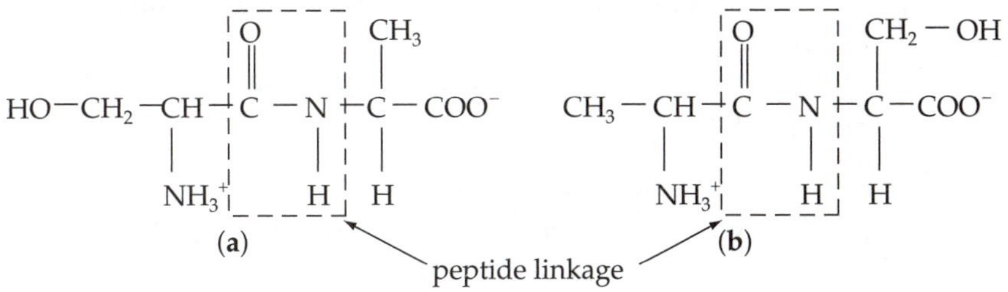

In naming simple amino acids, the amino acid that retains its carboxylic functional group is named last and retains its name as a free amino acid. The other amino acids in the polypeptide are named by adding *-yl* to the stem of the name of an amino acid. Therefore, structure **a** is named serylalanine and structure **b** is named alanylserine.

EXERCISE 12   Explaining why the folded structure of a protein may be altered in solvents

When a protein is placed in water, the α-helical structure is stabilized in part by hydrophilic interactions of polar side chains of the amino acids with water. Hydrophobic groups are directed primarily towards the interior of the helix, whereas hydrophilic groups are directed primarily toward the aqueous surroundings. Why can the folded structures of proteins be altered in weakly polar organic solvents?

SOLUTION: *Analyze*: We are given information about the α-helical structure of proteins and asked to explain why the folded structure of a protein may be different in water compared to when it is dissolved in a weakly polar solvent.

Plan: We are given information about the relative positions of hydrophobic (not attracted to water) groups and hydrophilic (attracted to water) groups in a protein when dissolved in water. We can use this information to help us predict structural changes when a protein is placed in a weakly polar solvent.

Solve: The hydrophobic groups of a protein tend to be in the interior of its structure to avoid interacting with polar water molecules. When a protein is placed in a weakly polar solvent the hydrophobic groups in the α-helical structure are no longer restricted to the interior. The hydrophobic groups can interact with the weakly polar solvent and this causes an unfolding of the α-helix. This changes the structure of the protein and the process is given the name denaturation.

One of the steps in the process of energy production in organisms involves the breaking down of complex carbohydrates into their precursory components: sugars.

CARBOHYDRATES AND LIPIDS

- An important simple carbohydrate is **glucose** ($C_6H_{12}O_6$), a sugar. In solution, glucose forms a hemiacetal structure: A cyclic arrangement of five carbon atoms and one oxygen atom with appropriate groups attached.

• Other simple sugars also exist. One example is **fructose**, which most commonly has a cyclic arrangement of four carbon atoms and one oxygen atom.

In general, **carbohydrates** are polyhydroxy aldehydes or ketones with empirical formulas $C_y(H_2O)_x$ and polymers derived from the polyhydroxy aldehydes and ketones.

• **Monosaccharides** are monomeric sugars that cannot be further broken apart by hydrolysis.
• **Polysaccharides** are polymers formed from monosaccharides combining.
• Starch, glycogen, and cellulose are all polysaccharides formed from the glucose monomer.

Lipids are a family of compounds that are insoluble in water and include fats, waxes, and steroids. *Fatty acids* are the simplest lipids and consist of a long hydrocarbon chain ending with a carboxylic acid functional group. A *saturated* fatty acid contains only single C—C bonds. An *unsaturated* fatty acid contains at least one C=C bonds. *Trans* fatty acids have the two hydrogen atoms on the opposite sides (diagonal) of the C=C bond and a *cis* fatty acid has the two hydrogen atoms on the same sides (adjacent) of the C=C bond. Fatty acids react with glycerol to form triglycerides.

EXERCISE 13   Identifying the structures of glucose and fructose

Identify the following structures as glucose or fructose.

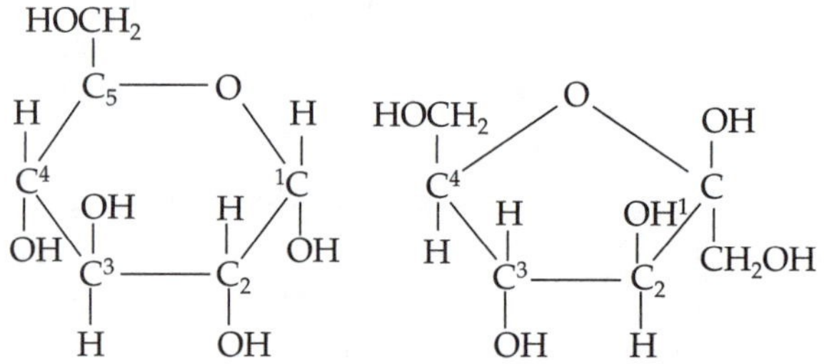

SOLUTION: *Analyze and Plan*: We are asked to identify which structure of two shown is glucose and which is fructose. This requires us to know the differences between glucose and fructose. In Section 25.8 we can find pictures and descriptions of the two. We can compare these to the ones shown in the problem and identify the key features in each structure that help us label them appropriately.

Solve: Structure (**a**) represents glucose. The glucose form of a sugar contains a cyclic arrangement of *five* carbon atoms and one oxygen atom. Structure (**b**) represents fructose. The fructose form of a sugar contains a cyclic arrangement of *four* carbon atoms and one oxygen atom.

EXERCISE 14   Writing the structural formula of sucrose

In this chapter you learned that a disaccharide is formed by linking two monosaccharide molecules and eliminating a water molecule. How is the disaccharide sucrose formed? Draw its structures. Indicate whether it has alpha or beta linkages.

SOLUTION: Sucrose is formed by the combination of glucose and fructose. The sugar units are also joined by an alpha linkage:

Sucrose

EXERCISE 15   Solubility of fatty acids in water

Fatty acids consist of a hydrocarbon chain with a polar carboxylic acid functional group at the end. Explain why a lipid, which has a polar group, is insoluble in water.

SOLUTION: Although water molecules are attracted to the polar carboxylic acid, they are not attracted to the nonpolar hydrocarbon chain. The hydrocarbon chain is long and the attraction of the carboxylic acid group to water molecules is not sufficient to overcome the lack of attraction between the hydrocarbon groups and water. If the hydrocarbon chain is short, for example, one to three carbon atoms, then the carboxylic acid has some solubility in water.

Chromosomes, which contain genes, consist of macromolecular substances called nucleic acids. **Nucleic acids** are bipolymers formed from repeating units called **nucleotides**.

NUCLEIC ACIDS: DNA AND RNA

- Look at the appropriate figure in Section 25.10 in the text and note the three parts of nucleotide units: (**a**) a phosphoric acid unit, (**b**) a sugar in the furanose (five-membered ring) form (ribose type), and (**c**) a nitrogen-containing organic base such as adenine.
- Polynucleotides form by condensation reactions between an —OH unit of phosphoric acid with an —OH unit of a sugar unit; an ester-like linkage is formed.

Nucleic acids are classified into two groups: **deoxyribonucleic acids** (DNA) and **ribonucleic acids** (RNA).

- DNA molecules are found almost exclusively in the nuclei of cells and are biopolymers of very high molecular weights. Nucleotide units within DNA contain a sugar unit known as deoxyribose. DNA molecules exist as double-stranded polymers wound in the form of a double helix.
- RNA molecules are found primarily in the cytoplasm (the cellular medium in which a nucleus is embedded) and are biopolymers of considerably lower molecular weights than DNA molecules. Nucleotide units within RNA molecules contain the sugar ribose. Ribose contains an —OH unit substituted for a hydrogen atom in deoxyribose.

EXERCISE 16   Identifying ribose and deoxyribose rings

Identify the following rings as either ribose or deoxyribose:

(a)   (b)

SOLUTION: *Analyze and Plan*: We are asked to identify which ring structure is ribose and which is deoxyribose. This requires us to know the difference between ribose and deoxyribose. In Section 25.10 we can find pictures and descriptions of the two. We can compare these to the ones shown in the problem and identify the key features in each structure that help us label them appropriately.

Solve: Both ribose and deoxyribose rings contain four carbon atoms and one oxygen atom in a cyclic arrangement. A key difference will be in the number of —OH groups and their locations. Figure (**b**) represents deoxyribose, and Figure (**a**) represents ribose. Note that at carbon atom 3 in ribose there is an —OH group, whereas at the same carbon atom in deoxyribose there is a hydrogen atom.

EXERCISE 17   Explaining the effect of heating on the structure of DNA in solution

What effect might heating a DNA solution have upon the double-helix structure of DNA?

SOLUTION: *Analyze and Plan*: We are asked to analyze what might happen to the double-helix structure of a DNA solution if it is heated. Adding heat increases the amount of energy in the solution and therefore increases the vibrational and kinetic energies. We should consider how this might affect the double-helix structure.

Solve: The strands of a DNA double helix are held together primarily by hydrogen bonds. When a solution of DNA is heated, the increased kinetic energy of the polymer causes the hydrogen bonds to break and the strands to separate. Thus, the double-helix structure is destroyed, or denatured.

SELF-TEST QUESTIONS

Key Terms

Having reviewed key terms in Chapter 25, match key terms with phrases and identify statements as true or false. If a statement is false, indicate why it is incorrect.

Match each phrase with the best term:

25.1 This class of hydrocarbons contains a cyclic arrangement of carbon atoms with alternating pi bonds.

25.2 This class of hydrocarbons is relatively unreactive.

25.3 The class of hydrocarbons containing a triple carbon–carbon bond.

25.4 The name given to atoms and groups when they replace a hydrogen atom in a hydrocarbon.

25.5 The functional group characterized by

$$-\overset{\textstyle |}{\underset{\textstyle |}{C}}-O-H$$

25.6 The name given to the reaction of bromine with ethene in the presence of UV light.

25.7 A hydrocarbon in which the carbon–carbon bonds are all single and have the formula C_nH_{2n}.

25.8 A functional group that consists of a carbon atom bonded to an oxygen atom *via* a double bond.

25.9 A type of isomerism exhibited by 2-dichloroethene.

25.10 The term given CH_3CH_2- when it is bonded to a carbon atom in a hydrocarbon chain.

25.11 A functional group that consists of C—O—C.

25.12 A functional group that consists of a carbonyl group as the ending carbon atom in a hydrocarbon chain.

25.13 A functional group that consists of a carbonyl group with its carbon atom not ending a chain of carbon atoms.

25.14 A functional group which is sometimes written as —COOH.

25.15 A functional group that forms by the condensation of an alcohol with carboxylic acid.

25.16 The reaction of a carboxylic acid with a strong base such as sodium hydroxide.

25.17 A functional group characterized by $-NH_xR_{3-x}$.

25.18 A functional group which is found in peptides and proteins and acts to connect the amino acids.

25.19 A hydrocarbon containing a carbon–carbon double bond.

25.20 The region of our environment that contains the solid earth, natural waters, and the atmosphere.

25.21 A polymer in biochemistry which forms from monomers of amino acids.

25.22 A substance found in biochemical systems which has the formula $C_x(H_2O)_y$.

Terms:

(a) addition	**(l)** carbohydrate
(b) alcohol	**(m)** carbonyl
(c) aldehyde	**(n)** carboxylic acid
(d) alkane	**(o)** cycloalkane
(e) alkene	**(p)** ester
(f) alkyl	**(q)** ether
(g) alkyne	**(r)** functional group
(h) amide	**(s)** geometric
(i) amine	**(t)** ketone
(j) aromatic	**(u)** protein
(k) biosphere	**(v)** saponification

True-False Statements:

25.23 *Nucleotides* condense together to form linear polymers.

25.24 All amino acids exist in nature in a *chiral* form.

25.25 If two molecules are nonsuperimposable mirror images of one another, they are *enantiomers*.

25.26 The sequence of amino acids along a protein chain is the *primary structure* of the protein.

25.27 Hydrogen bonds play an important role in determining the *secondary structure* of a protein.

25.28 The *tertiary structure* of a protein shows the linear arrangements of protein strands.

25.29 *DNA* molecules have lower molecular weights than *RNA* molecules.

25.30 In the *double-helix* structure for DNA, we find that adenine units on the two strands are paired for optimal hydrogen-bond interaction.

25.31 *Nucleic acids* are the primary storehouses of energy in humans.

25.32 *Cellulose* is a polysaccharide of glucose.

25.33 An important storehouse of carbohydrate energy in mammals is *glycogen*.

25.34 The *monosaccharide* fructose is a polyhydroxy aldeyhyde.

25.35 Fructofuranose is an example of a *starch*.

25.36 Biopolymers include proteins, carbohydrates, and nucleic acids.

25.37 A tripeptide consists of four *amino acids* bonded through three peptide bonds in a linear chain.

25.38 A *peptide* bond contains the following unit

$$-\overset{\overset{\displaystyle O}{\|}}{C}-\underset{\underset{\displaystyle H}{|}}{N}-\cdot$$

25.39 Proteins that have molar masses less than 6000 amu are called *polypeptides*.

25.40 The formula of *glucose* is $C_6H_{12}O_6$.

25.41 When *polysaccharides* are reacted in the presence of an acid catalyst, they form glucose and/or fructose.

25.42 A *zwitterion* is an amino acid that has transferred internally a hydrogen ion from the basic amine group to another basic site on a different amino acid.

Problems and Short-Answer Questions

25.43 Which functional group occurs in each of the following organic compounds? Unshaded spheres represent oxygen atoms, lightly shaded spheres represent hydrogen atoms, and dark spheres represent carbon atoms.

(a)

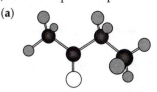

(b)

(c)

25.44 What hybrid atomic orbitals does a carbon atom use when it is in a linear, trigonal planar, and tetrahedral environment?

25.45 Why are aromatic compounds such as benzene more stable than linear alkenes with a similar number of carbon–carbon double bonds?

25.46 Why don't alkanes exhibit *cis-trans* isomerism?

25.47 Why are alkanes relatively unreactive? What type of reaction discussed in the text can they readily undergo?

25.48 What are the characteristic functional groups of an alcohol, a ketone, a carboxylic acid, and an amine?

25.49 Which of the following functional groups contains a carbonyl group: halide, ketone, amine, amide, and ester?

25.50 Would you expect the following compound to be chiral?

$$CH_3CH=CHCH_2CClBrCH_3$$

25.51 What do polypeptides and proteins have in common? How are they different?

25.52 Name the following compounds:

(a) 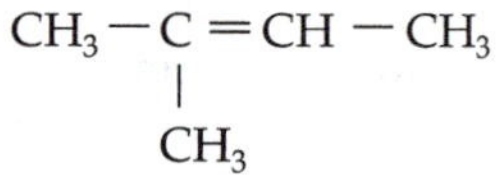 $CH_3-C=CH-CH_3$
                  |
                  CH_3

(b) 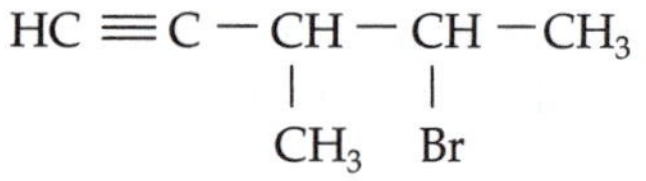 $HC\equiv C-CH-CH-CH_3$
                    | |
                    CH_3 Br

(c) HCOOH

(d) 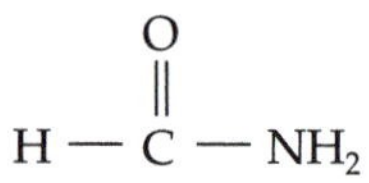
              O
              ‖
       $H-C-NH_2$

(e) 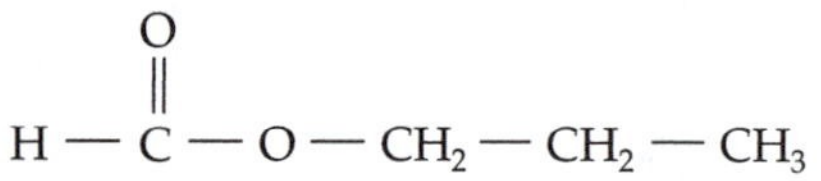
              O
              ‖
       $H-C-O-CH_2-CH_2-CH_3$

25.53 Write and identify all geometrical isomers possible for 2-bromo-2-pentene.

25.54 Write the structural formulas for the products of the following reactions:

(a) one mole each of ethane and bromine in the presence of ultraviolet light;

(b) propene and hydrogen gas under pressure in the presence of platinum;

(c) one mole each of propyne and bromine;

(d) oxidation of CH_3CH_2CHO with potassium permanganate;

(e) one mole each of methanol and propionic acid;

(f) methyl-amine and propionic acid.

25.55 Draw all the structural and geometrical isomers of bromopropene.

25.56 Give the structural formula for an alcohol that is an isomer of dimethylether; for an aldehyde that is an isomer of methyl ethyl ketone.

25.57 Milk from cows contains about 5 percent of the sugar lactose:

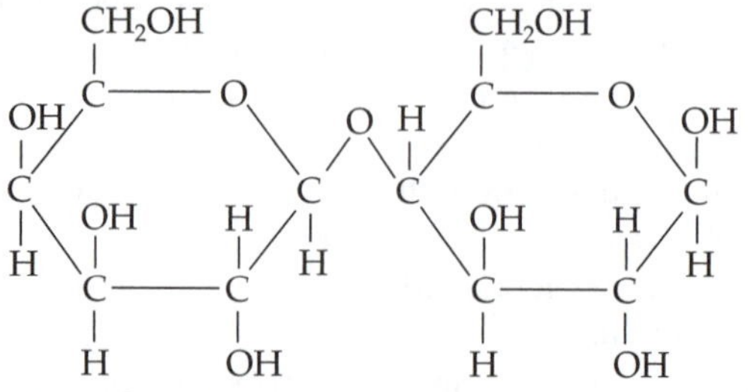

What type of saccharide is lactose?

25.58 Would you expect a compound indicated by the approximate formula $C_{2932}H_{4724}N_{828}S_8Fe_4O_{840}$ to be primarily of a protein, a carbohydrate, or a small peptide?

25.59 Experimentally, the activity range for many enzymes occurs between 10 °C and 50 °C. As the temperature increases significantly above 50 °C, the activity of enzymes decreases rapidly, and eventually all activity is destroyed. Explain why the activity of enzymes is destroyed at temperatures significantly greater than 50 °C.

25.60 What is the principal difference between the structures of the sugar units in DNA and RNA molecules?

25.61 Write the structural formula of the tripeptide val-thr-gly. Refer to Section 25.7 in the text for assistance in doing this problem.

25.62 A student finds that the boiling points for three straight-chain hydrocarbons are 68 °C, 98 °C, and 125 °C. The student is told that the hydrocarbons have the chemical formulas for hexane, heptane, and octane. Is this sufficient information to assign the boiling points to the three hydrocarbons? If not, what additional information is needed and why?

25.63 An unknown alcohol is oxidized to form an aldehyde. Additional oxidation leads to a new substance that has a pH in water of approximately 4. Is this sufficient information to determine the type of organic substance produced by the second oxidation? If not, what additional information is needed and why?

25.64 A student believes s/he has a substance that is a nucleic acid. In this substance s/he finds repeating units that contain phosphoric acid and a five-carbon sugar. Is this sufficient information for the student to conclude that the substance is indeed part of a nucleic acid? If not, what additional information is needed and why?

25.65 A molecule is found to have three amide linkages. The molecule also has amine and carboxylic acid functional groups. Is this sufficient information to conclude that it is a tetrapeptide? If not, what additional information is needed and why?

Integrative Exercises

25.66 Glycine, an amino acid, is a small molecule yet it has a surprisingly high melting point, 262°C. Most of the amino acids are moderately soluble in water. Explain these observations by drawing the Lewis structure of glycine and inspecting its functional groups and their expected properties.

25.67 Polyethylene is formed by a free-radical reaction using ethylene as the monomer.

(a) What property of ethylene enables the free-radical reaction to occur? Explain.

(b) What is the expected sign of ΔS for the polymerization reaction?

(c) Given that the polymerization reaction is spontaneous, what is the expected sign of ΔH for the polymerization reaction?

25.68 A C, H, O-containing compound has 6.71% hydrogen and 40.00% carbon. Its molecular weight is 60.05 amu. If it is reacted with pentanol in an acid environment it forms a new substance that has the odor of banana.

(a) What is the molecular formula of the unknown substance?

(b) What is the structural formula of it?

(c) What is the substance formed after the reaction with pentanol?

Multiple-Choice Questions

25.69 Which alcohol forms a ketone when oxidized?

(a) 1-propanol
(b) methanol
(c) 2-methyl-2-propanol
(d) 2-propanol
(e) all of the above

25.70 Which substance is the missing reactant in the reaction?

$$\underline{\hspace{2cm}} \xrightarrow[140°C]{H_2SO_4} CH_3CH_2CH_2OCH_2CH_2CH_3 + H_2O$$

(a) $CH_3CH_2CH_3$
(b) $CH_3CH_2CH_2Cl$
(c) $CH_3CH_2CH_2OH$
(d) CH_3CH_2CHO
(e) $CH_3CHOHCH_3$

25.71 Which formula represents an ether?

(a) CH_2CH_2
(b) CH_3CHO
(c) $CH_3CH_2OCH_2CH_3$
(d) CH_2OHCH_2OH
(e) CH_3COOCH_3

25.72 Which structure *best* represents a ketone unit? (*X* represents any single element.)

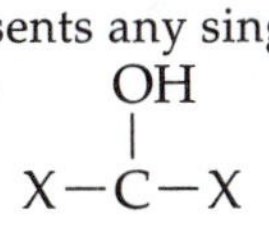

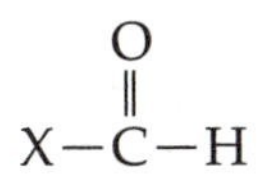

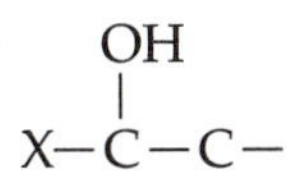

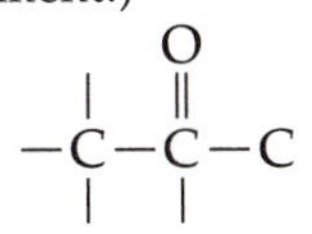

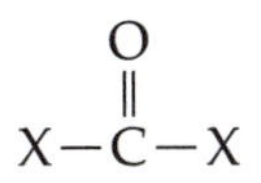

25.73 Which name is associated with this compound?

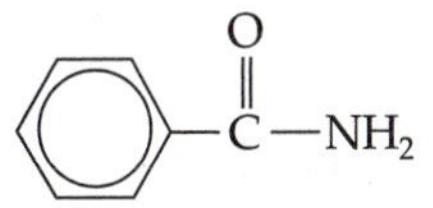

(a) benzoic acid
(b) ammonium benzoate
(c) benzamide
(d) benzoamine
(e) benzene

25.74 Which name best corresponds to the structural formula

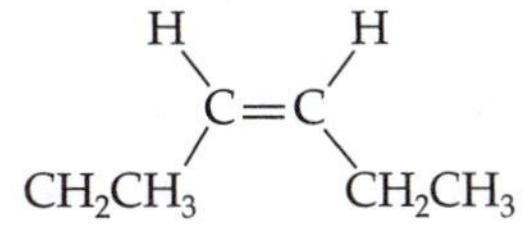

(a) *trans*-3-hexene
(b) *cis*-3-hexene
(c) *trans*-hexane
(d) *cis*-hexene
(e) *cis*-diethylene

25.75 Which figure best represents the structure of *meta*-dimethylbenzene?

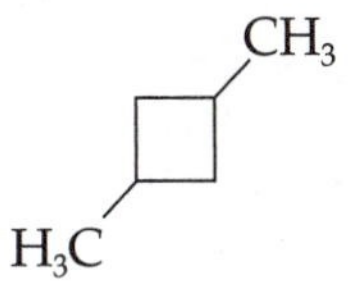

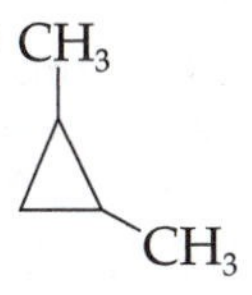

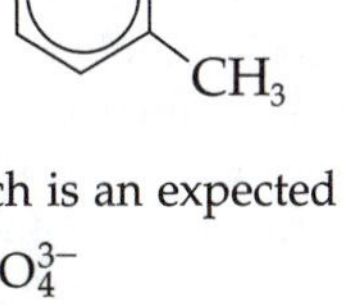

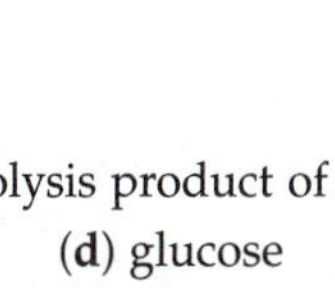

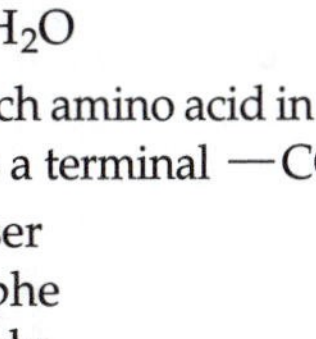

25.76 Which is an expected hydrolysis product of DNA?

(a) PO_4^{3-}
(b) CO_2
(c) H_2O
(d) glucose
(e) ribose

25.77 Which amino acid in the polypeptide ser-phe-gly-ala-gly-lys has a terminal $-CO_2H$ free group?

(a) ser
(b) phe
(c) gly
(d) ala
(e) lys

25.78 For glycine in water, $K_a = 1.6 \times 10^{-10}$ for the ionization of the $-CO_2H$ group and $K_b = 2.5 \times 10^{-12}$ for the reaction of the $-NH_2$ group with water. In water, the expected predominant form of glycine is which of the following?

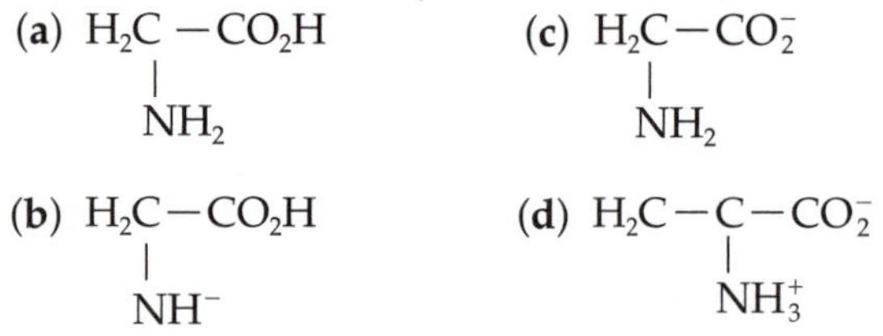

(a) H_2C-CO_2H
      |
      NH_2

(c) $H_2C-CO_2^-$
      |
      NH_2

(b) H_2C-CO_2H
      |
      NH^-

(d) $H_2C-C-CO_2^-$
           |
           NH_3^+

(e) none of the above

25.79 Natural silk from silkworms and spiders contains linear biopolymers formed from simple proteins. The structure of natural silk shows a "linear" zigzag arrangement of protein molecules; the strands of protein are not coiled around one another. From this information, we can conclude that linear silk strands have which of the following structures associated with proteins?

(a) primary
(b) secondary
(c) tertiary
(d) α-helix
(e) double helix

SELF-TEST SOLUTIONS

25.1 (j). **25.2**(d). **25.3** (g). **25.4** (r). **25.5** (b). **25.6** (a). **25.7** (o). **25.8** (m). **25.9** (s). **25.10** (f). **25.11** (q). **25.12** (c). **25.13** (t). **25.14** (n). **25.15** (p). **25.16** (v). **25.17** (i). **25.18** (h). **25.19** (e). **25.20** (k). **25.21** (u). **25.22** (l). **25.23** True. **25.24** False. All but glycine are chiral. **25.25** True. **25.26** True. **25.27** True. **25.28** False. The tertiary structure of a protein refers to the overall folding of protein strands upon themselves; the structure is not linear. **25.29** False. The reverse is true. **25.30** False. Adenine is paired with thymine. **25.31** False. Nucleic acids are store-houses of information that enable regulation of cell reproduction and development. **25.32** True. **25.33** True. **25.34** False. Fructose is a monosaccharide that is a polyhydroxy ketone. **25.35** False. Starch refers to a group of polysaccharides, not to a monosaccharide such as fructofuranose. **25.36** True. **25.37** False. A tripeptide consists of three amino acids linked together. **25.38** True. **25.39** False. Molar masses are in the range of 6000-50 million amu. **25.40** True. **25.41** True. **25.42** False. A *zwitterion* is a doubly charged ion. An example is an amino acid that has a hydrogen ion transferred internally from the carboxyl group (acid site) to the amine group (base site). A protein can also show this behavior.

25.43 (a) Ketone. The carbonyl group is within the carbon chain.

(b) Ether. There is within the carbon chain the $-C-O-C-$ ether linkage.

(c) Carboxylic acid. The $-COOH$ group is at the left end of the molecule.

25.44 Linear: sp; trigonal planar; sp^2; and tetrahedral, sp^3.

25.45 Aromatic compounds contain carbon atoms in a cyclic arrangement bonded together via sigma bonds and alternating pi bonds. The electrons in the pi bonds are delocalized over the entire cyclic structure and this gives extra stability to the $C-C$ bonds.

25.46 *Cis-trans* isomerism in organic substances exists when there are two structures in which the same two groups attached to adjacent carbon atoms are either adjacent to one another or approximately 180° apart across a $C=C$. This requires that they have fixed spatial positions. The adjacent carbon atoms that hold the two attached groups must not be able to rotate about their $C-C$ bond; Alkanes have $C-C$ single bonds and the carbon atoms are able to rotate about the sigma bond. It requires the presence of a pi bond between the two carbon atoms to prevent rotation.

25.47 The primary reason for the unreactivity of alkanes is the strength of the $C-C$ and $C-H$ bonds. An important reaction of alkanes is combustion.

25.48 Alcohol: $R-OH$; ketone: 
$R-C-R'$ where R or R' is not a hydrogen atom;
     ||
     R
Carboxylic acid: $-\overset{\overset{O}{||}}{C}-OH$; Amine: $R-NH_2$

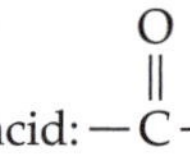

25.49 The carbonyl group is $C=O$; thus any functional group that has this as a component has a carbonyl group. These are ketone, amide, and ester.

25.50 We need to draw a structural formula to determine if any carbon atom has four different groups attached.

```
      H                H  Br  H
      |                |   |   |
H  -  C - C = C - C  -  C  -  C - H
      |                |   |   |
      H   H   H   H   Cl   H
```

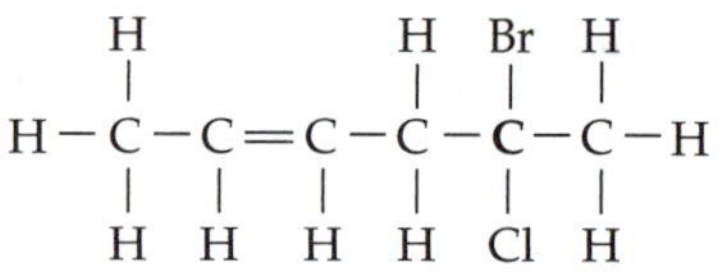

the atom in boldface font has four different groups attached and thus it is a chiral carbon atom. This means that the substance is also chiral.

25.51 Both are formed by amino acids undergoing condensation reactions to form amide bonds. An amine group on one amino acid reacts with the carboxyl group of a different amino acid. A polypeptide has from 2 to approximately 50 amino acids, whereas proteins are larger compounds often with hundreds of amide bonds in their structures.

25.52 (a) 2-methyl-2-butene;
(b) 4-bromo-3-methyl-1-pentyne;
(c) formic acid;
(d) formamide;
(e) propyl formate

25.53

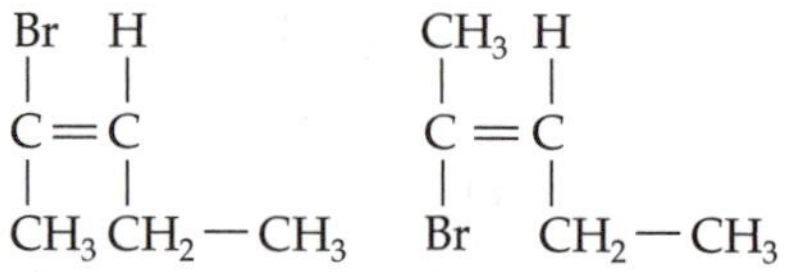

```
Br   H                CH3  H
|    |                |    |
C  = C                C  = C
|    |                |    |
CH3  CH2 - CH3        Br   CH2 - CH3
cis                   trans
```

25.54

(a)

$$CH_2-CH_2 \quad \text{or} \quad Br$$
$$\;|\quad\;\;\; | \qquad\qquad\quad |$$
$$Br\quad Br \qquad\quad H-C-CH_3$$
$$\qquad\qquad\qquad\qquad |$$
$$\qquad\qquad\qquad\qquad Br$$

(b) $CH_3-CH_2-CH_2$

(c) $CH_3-CH=CH$
$$\qquad\qquad\;\; |\quad\;\; |$$
$$\qquad\qquad\;\; Br\;\; Br$$

(d)
$$\qquad\qquad O$$
$$\qquad\qquad \|$$
$$CH_3CH_2C-OH$$

(e) $CH_3-CH_2-C-O-CH_3$
$$\qquad\qquad\qquad\;\; \|$$
$$\qquad\qquad\qquad\;\; O$$

(f) $CH_3-CH_2-C-N-CH_3$
$$\qquad\qquad\qquad\;\; \|\;\; |$$
$$\qquad\qquad\qquad\;\; O\;\; H$$

25.55 The chemical formula of bromopropene is C_3H_5Br. A double bond exists between two carbon atoms as indicated by the *-ene* ending of the chemical name. The structural and geometrical forms are

$$\begin{array}{cccc} H\quad H & Br\quad H & H\quad H & H\quad Br \\ |\qquad | & |\qquad | & |\qquad | & |\qquad | \\ C=C & C=C & C=C & C=C \\ |\qquad | & |\qquad | & |\qquad | & |\qquad | \\ H\quad CH_2Br & H\quad CH_3 & Br\quad CH_3 & H\quad CH_3 \end{array}$$

25.56 Dimethyl ether is $H_3C-O-CH_3$. An alcohol has a $-OH$ group; thus, an isomer is H_3C-CH_2-OH, ethanol. Methyl ethyl ketone is

$$\qquad\qquad O$$
$$\qquad\qquad \|$$
$$CH_3-C-CH_2CH_3$$

An aldehyde has a carbonyl group with a hydrogen atom attached; thus an isomer that is an aldehyde is

$$\qquad\quad O$$
$$\qquad\quad \|$$
$$HC-CH_2CH_2CH_3$$

25.57 Disaccharide; it contains two glucose units.

25.58 The compound is very large, as indicated by the large number of carbon, hydrogen, and nitrogen atoms. Thus small peptides are eliminated because they do not contain that many carbon atoms—nor iron or sulfur atoms. Carbohydrates are eliminated because they do not contain nitrogen, sulfur, or iron atoms. Proteins may contain carbon, hydrogen, nitrogen, oxygen, and sulfur atoms. Experimentally, this is the approximate formula for oxyhemoglobin, an iron-containing protein.

25.59 Proteins are denatured by the application of heat. At temperatures significantly greater than 50 °C, enzymes are denatured; consequently, their activity is lost.

25.60 The sugar unit of RNA contains one $-OH$ group, whereas that of DNA contains no $-OH$ unit (see chemical diagram in Exercise 6).

25.61

$$\qquad\qquad\quad O \qquad\qquad\qquad\quad O$$
$$\qquad\qquad\quad \| \qquad\qquad\qquad\quad \|$$
$$CH_3-CH-CH-C-N-CH-C-N-CH_2-COOH$$
$$\qquad\qquad |\qquad\;\; |\qquad\;\; |\qquad |\qquad\qquad |$$
$$\qquad\qquad CH_3\;\; NH_2\quad H\; HCOH\quad\; H$$
$$\qquad\qquad\qquad\qquad\qquad\qquad |$$
$$\qquad\qquad\qquad\qquad\qquad\qquad CH_3$$

Valine   Threonine   Glycine

25.62 There is sufficient information. Straight-chain hydrocarbons are nonpolar; thus London dispersion forces exist between molecules and permit the formation of a liquid. Their boiling points increase with increasing extent of London dispersion forces within the liquid. This trend is related to increasing molecular volume or the size of the hydrocarbon. Therefore, the assignments should be: 68 °C (hexane), 98 °C (heptane), and 125 °C (octane).

25.63 There is sufficient information. Controlled oxidation of an alcohol leads to the formation of a ketone or aldehyde. Further oxidation can lead to the formation of a carboxylic acid. The fact that the pH of this substance in water is approximately four supports the conclusion that it is a weak acid, a carboxylic acid.

25.64 There is insufficient information. The monomers of nucleic acids are nucleotides and have three components: a phosphoric acid unit, a five-carbon sugar, and a nitrogen-containing organic base. The student has failed to confirm the third unit present in nucleotides. Additional analysis is required.

25.65 There is insufficient information. A peptide linkage is an amide linkage occurring between two amino acids. To determine if the molecule is a tetrapeptide we need to determine if the units making up the molecule are amino acids. It is possible that the units are carboxylic acids with amine groups that are not at the alpha-carbon atom.

25.66 Glycine and other amino acids contain an amine functional group at the alpha-carbon and a carboxylic acid functional group. The first is a moderately weak base and the second is a weak acid. These react to form $R-NH_3^+$ and $R-COO^-$ in the amino acid. Thus the Lewis structure of glycine is

$$\qquad\quad H$$
$$\qquad\quad |$$
$$H-C-COO^-$$
$$\qquad\quad |$$
$$\qquad\quad NH_3^+$$

This structure is similar to those of ionic compounds. The enhanced ion–ion attractive forces between glycine particles in the solid state result in a higher melting point than normally expected. A similar structure occurs for other amino acids. When placed in water significant ion–dipole interactions occur and solubility is enhanced.

25.67 **(a)** Ethylene, C_2H_4, is an alkene with a double bond. Double bonds are reactive functional groups and participate in addition reactions and reactions activated by UV light. The UV light causes the carbon–carbon double bond to "open" and each carbon atom possesses a single unpaired electron, resulting in a free radical. A carbon-free radical is reactive and links to another carbon-free radical to form a carbon–carbon bond between ethylene molecules. This process continues and polyethylene forms.

(b) The polymer is a more organized structure than individual ethylene particles. Also, ethylene is a gas at room temperature and polyethylene is a solid. Thus the entropy change of the polymerization process is negative since the product is more ordered than the reactants.

(c) $\Delta G = \Delta H - T\Delta S$. Substituting the signs for each results in $(-) = ? - T(-) = ? + T$. Thus, the sign of the enthalpy change, ?, is $? = (-) - T$. Since T is a positive number the sign of enthalpy must be a negative number.

25.68 **(a)** First calculate the percent of oxygen in the compound by subtracting the percent hydrogen and percent carbon from 100%; $100\% - 6.71\% - 40.00\% = 53.29\%$. The relative number of moles of each element in a 100.0 g sample of the compound is:

mol H = (6.71 g)(1 mol/1.008 g/mol) = 6.65 mol H

mol C = (40.00 g)(1 mol/12.00 g/mol) = 3.33 mol C

mol O = (53.29 g)(1 mol/16.00 g/mol) = 3.331 mol O

Dividing by the smallest number of moles gives the empirical formula: H_2CO. The empirical formula weight is 30.02 amu. Dividing the molecular weight by 30.02 amu gives: 60.52 amu/30.02 amu = 2.016 or simply 2. The molecular formula consists of two empirical formula units or $H_4C_2O_2$.

(b) The banana smell is one of the keys to helping you determine the nature of the product of the reaction with pentanol. Pentanol is an alcohol and when it reacts with a carboxylic acid it produces an ester. Esters are sweet-smelling substances that are used to make perfumes and constitute the smell of many fruits. If our conclusion is correct, then the structural formula is

$$
\begin{array}{ccc}
\text{H} & \text{O} \\
| & \| \\
\text{H}-\text{C}-\text{C}-\text{OH} \\
| \\
\text{H}
\end{array}
$$

It is acetic acid.

(c) The ester formed is pentyl acetate:

$$
\begin{array}{ccccccccc}
\text{H} & \text{O} & & \text{H} & \text{H} & \text{H} & \text{H} & \text{H} \\
| & \| & & | & | & | & | & | \\
\text{H}-\text{C}-\text{C}-\text{O}-\text{C}-\text{C}-\text{C}-\text{C}-\text{C}-\text{H} \\
| & & & | & | & | & | & | \\
\text{H} & & & \text{H} & \text{H} & \text{H} & \text{H} & \text{H}
\end{array}
$$

25.69 **(d)**. **25.70** **(c)**. **25.71** **(c)**.

25.72 **(e)** Structure **(c)** is not correct because if X = H, we have an aldehyde.

25.73 **(c)**. **25.74** **(b)**. **25.75** **(e)**. **25.76** **(a)**. **25.77** **(e)**. **25.78** **(d)**. **25.79** **(a)**.

MCAT and DAT Practice Questions VIII

Note: The topic of coordination chemistry is not covered in the MCAT and DAT exams. Questions relating to organic chemistry are covered in the biological sciences section of MCAT and in the organic chemistry section of the DAT. However, descriptive passages in the physical sciences section of MCAT may use examples from organic chemistry. The material covering organic chemistry is useful in preparing to work with such examples. This section is provided to help users of the *Student's Guide*.

Passage

Transition metal compounds exhibit a variety of physical properties, some of which depend on the electronic structures of the transition metals. Examples of these properties include their colors, magnetism, and structures.

A transition metal compound exhibits color when a *d* electron is excited from a lower energy *d* orbital to a higher energy *d* orbital and the energy difference between the two is in the visible region of radiation. The energy of the visible radiation absorbed determines the color observed. An eye sees the light reflected or transmitted by the substance. The color of the light transmitted is the complementary color of the color of light absorbed. Table 1 shows limited data for wavelengths of radiation in the visible region of light and their associated colors and complementary colors.

TABLE 1 Wavelengths of radiation in visible region and colors

λ(nm) of absorbed energy	Color	Complementary color
720	Red	Green
680	Red-orange	Blue-green
610	Orange	Blue
580	Yellow	Indigo
530	Green	Purple
500	Blue-green	Red
480	Blue	Orange
430	Indigo	Yellow
410	Violet	Lemon-yellow

Radiant energy is quantized and the energy of a photon is $E = h\nu$, where h is Planck's constant, 6.626×10^{-34} J-s. The relationship between frequency and wavelength of radiation is $\nu\lambda = c$, where c is the speed of light, 3.00×10^8 m s^{-1}.

The crystal field model provides an explanation for energy differences between *d* orbitals when a central metal atom is surrounded by ligands. The energy difference between two sets of d orbitals in an octahedral transition metal complex is called delta, Δ_o. Δ_o in metal complexes of the first row of transition metals is the difference in energy between an orbital in a lower-energy set of three orbitals, $3d_{xy}$, $3d_{xz}$, and $3d_{yz}$, and an orbital in a higher-energy set of two orbitals, $3d_{x^2-y^2}$, and $3d_{z^2}$.

The value of Δ_o depends on the charge and size of the transition metal as well as the properties of the attached ligands. We find for a transition metal in a fixed ion state and with the same six ligands that the magnitude of Δ_o tends to increase in the following order of ligands (the spectrochemical series):

FIGURE 1 Spectrochemical Series

$$I^-<Br^-<Cl^-<F^-<H_2O<NH_3<en<NO_2^-<CN^-<CO$$

Increasing Δ

1. What is the likely observed color of $[Cr(CN)_6]^{3-}$?
   (a) Red　　(b) Yellow　(c) Blue　(d) Purple

2. Which absorbed color corresponds to the lowest value of Δ when a d electron is excited in an octahedral transition metal compound?
   (a) Red　　(b) Green　(c) Indigo　(d) Yellow

3. Which metal complex ion is most likely to have cobalt with completely paired $3d$ electrons?
   (a) $[Co(CN)_6]^{3-}$ 　　　　(b) $[Co(Br)_6]^{3-}$
   (c) $[Co(H_2O)_6]^{3+}$ 　　　(d) $[Co(F)_6]^{3-}$

4. $NH_2CH_2CH_2NH_2$ is a bidentate ligand. Which organic functional group does it contain?
   (a) Carbonyl　(b) Halide　(c) Amine　(d) Ketone

5. Another ligand is the oxalate ion, $C_2O_4^{2-}$. Which statement characterizes this ligand?
   (a) It is a Lewis acid.
   (b) Contains the hydroxo functional group
   (c) Contains a carbon–carbon double bond.
   (d) It is a chelating ligand.

6. Which ion should *not* form colored transition metal complexes?
   (a) Cr^{3+}　　(b) Zn^{2+}　(c) Fe^{3+}　(d) Ni^{2+}

7. Which is *not* likely to act as a ligand towards transition metal ions?
   (a) Methane　　　　　(b) Nitrite ion
   (c) Ammonia　　　　　(d) Carbon monoxide

| Questions 8 through 12 are **not** based on a descriptive passage. |

8. What is the oxidation number of iron in $[Fe(NH_3)_4(H_2O)_2](NO_3)_3$?
   (a) +1　　(b) +2　　(c) +3　　(d) +4

9. Which metal complex is correctly named?
   (a) $[Fe(CN)_4Cl_2]^{3-}$, dichlorotetracyanoironate(III) ion
   (b) $[Co(NH_3)_6]_2(SO_4)_3$, hexamminecobalt(III) sulfato
   (c) $Na[MnO_4]$, sodium tetraoxidemanganate(VII)
   (d) $[HgBr_4]^{2-}$, tetrabromomercuate(II) ion

10. Which is an ester?
    (a) CH_3Br
    (b) $CH_3CH_2CH_2OH$

    (c) $CH_3-\overset{\overset{\displaystyle O}{\|}}{C}-OH$
    (d) $CH_3-\overset{\overset{\displaystyle O}{\|}}{C}-O-CH_3$

11. Which is a monosaccharide?
    (a) Maltose   (b) Starch   (c) Cellulose (d) Glucose
12. All are components of a typical nucleotide except
    (a) a sugar in a furanose form.
    (b) a nitrogen-containing organic base.
    (c) a carboxylic acid.
    (d) phosphoric acid.

ANSWERS

1. (b) The ligand CN^- lies far to the right in the series in Figure 1. This means Δ_o will be relatively large. The energy absorbed will have a small wavelength because $E = \dfrac{hc}{\lambda}$. If we look at the data in Table 1 we see that the observed color, yellow, is the complementary color of indigo, which has the smallest associated wavelength of the colors listed in the question.

2. (a) The lowest Δ_o is associated with an abosorbed color that represents the smallest energy. As shown by the equation $E = \dfrac{hc}{\lambda}$, the smaller the energy, the larger the wavelength. In the list of colors given in the question, the largest wavelength is associated with the color of red.

3. (a) All of the complexes contain Co^{3+}. Cobalt has the valence electron configuration $3d^7 4s^2$; thus, the valence electron configuration for Co^{3+} is $3d^6$. With six valence electrons to be distributed into two sets of $3d$ orbitals, the only way to obtain completely paired electrons is for the following electron distribution to exist:

$$\overline{3d_{x^2-y^2}} \quad \overline{3d_{z^2}}$$
$$\underset{3d_{xy},\, 3d_{xz},\, 3d_{yz}}{\underline{\uparrow\downarrow} \quad \underline{\uparrow\downarrow} \quad \underline{\uparrow\downarrow}}$$

This requires a strong field which is produced by ligands which lie far to the right in Table I. CN^- fits this requirement.

4. (c) An amine contains the $-C-NH_2$ functional group, which occurs in $NH_2CH_2CH_2NH_2$.

5. (d) The Lewis structure for the oxalate ion is

$$
\begin{array}{cc}
:\!O\!: & :\!O\!: \\
\| & \| \\
C & - C \\
| & | \\
:\!\overset{..}{\underset{..}{O}}\!\bar{} & :\!\overset{..}{\underset{..}{O}}\!\bar{}
\end{array}
$$

The two negatively charged oxygen atoms can simultaneously bind to the same central metal atom, which is the characteristic of chelating ligands.

6. (b) A transition metal complex shows a color when a $3d$ electron moves from a lower energy orbital to a higher energy orbital and the energy difference between the two orbitals is in the visible range of light. This requires a vacancy for an electron in the higher energy orbital. Zn^{2+} has a valence electron configuration of $3d^{10}$. This set of $3d$ orbitals is completely filled with electrons. An electron cannot be excited from a lower energy $3d$ orbital to a higher one. The other transition metal ions have valence electron configurations consisting of $3d^y$ where y is less than 10; a higher energy $3d$ orbital with an electron vacancy can exist and electron excitation can occur.

7. (a) A ligand is a Lewis base. It has available electron density (nonbonding electrons) which can bind to a central metal atom in a metal complex. The Lewis structure of methane is

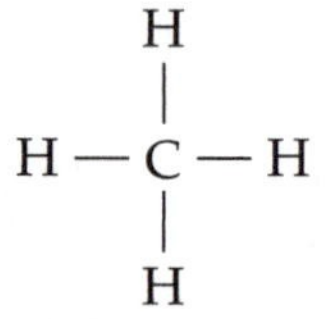

$$
\begin{array}{c}
\text{H} \\
| \\
\text{H} - \text{C} - \text{H} \\
| \\
\text{H}
\end{array}
$$

It does not possess nonbonding pairs of electrons to bind to a metal atom. The other Lewis structures have atoms with nonbonding electrons which are available to bind to a metal atom.

8. (c) The charge of the metal complex, $[Fe(NH_3)_4(H_2O)_2]^{3+}$, is determined by the presence of three NO_3^- ions which are not bonded to iron. Both ligands are neutral molecules; thus the oxidation number of iron is $+3$, the charge of the metal complex.

9. (d) The correct names for the others are: $[Fe(CN)_4Cl_2]^{3-}$, dichlorotetracyanoferrate(III) ion; $[Co(NH_3)_6]_2(SO_4)_3$, hexaamminecobalt(III) sulfate; $Na[MnO_4]$, sodium tetraoxomanganate(VII). These are the named using rules for naming metal complexes in Chapter 24.

10. (d) An ester contains the following functional group

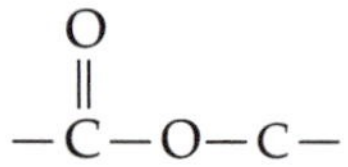

$$
\begin{array}{c}
\text{O} \\
\| \\
- \text{C} - \text{O} - \text{C} -
\end{array}
$$

11. (d) A monosaccharide is a simple sugar that cannot be broken into smaller molecules by acid hydrolysis. Glucose is the only monosaccharide of the sugars that are listed. Maltose is a disaccharide. Starch and cellulose are polysaccharides.

12. (c) A nucleotide contains a phosphoric acid, a five-carbon sugar, and a nitrogen-containing organic base. A carboxylic acid is not part of a nucleotide.